Arguing About Science

Arguing About Science is an outstanding, engaging introduction to the essential topics in philosophy of science, edited by two leading experts in the field. This exciting and innovative anthology contains a selection of classic and contemporary readings that examine a broad range of issues, from classic problems such as scientific reasoning, causation, and scientific realism, to more recent topics such as science and race, forensic science, and the scientific status of medicine.

The editors bring together some of the most influential contributions of famous philosophers in the field, including John Stuart Mill and Karl Popper, as well as more recent extracts from philosophers and scientists such as Ian Hacking, Stephen Jay Gould, Bas van Fraassen, Nancy Cartwright and John Worrall. The anthology is organised into nine clear sections:

- Science, non science and pseudo-science
- Science, race and gender
- Scientific reasoning
- Scientific explanation
- Laws and causation
- Science and medicine
- Probability and forensic science
- Risk, uncertainty and science policy
- Scientific realism and antirealism.

The articles chosen are clear, interesting, and free from unnecessary jargon. The editors provide lucid introductions to each section in which they give an overview of the debate, as well as suggestions for further reading.

Alexander Bird is Professor of Philosophy and Faculty of Arts Research Director at Bristol University, UK. His previous publications include *Nature's Metaphysics: Laws and Properties* (2007) and *Philosophy of Science* (1998).

James Ladyman is Professor of Philosophy at Bristol University, UK. He is the author (with Ross, Spurrett and Collier) of *Every Thing Must Go: Metaphysics Naturalised* (2007) and of the *Choice* awarding-winning book *Understanding Philosophy of Science* (Routledge, 2001).

Arguing About Philosophy

This exciting and lively series introduces key subjects in philosophy with the help of a vibrant set of readings. In contrast to many standard anthologies which often reprint the same technical and remote extracts, each volume in the *Arguing About Philosophy* series is built around essential but fresher philosophical readings, designed to attract the curiosity of students coming to the subject for the first time. A key feature of the series is the inclusion of well-known yet often neglected readings from related fields, such as popular science, film and fiction. Each volume is edited by leading figures in their chosen field and each section carefully introduced and set in context, making the series an exciting starting point for those looking to get to grips with philosophy.

Arguing About Metaethics
Edited by Andrew Fisher and Simon Kirchin

Arguing About the Mind
Edited by Brie Gertler and Lawrence Shapiro

Arguing About Art 3rd Edition
Edited by Alex Neill and Aaron Ridley

Arguing About Knowledge
Edited by Duncan Pritchard and Ram Neta

Arguing About Law
Edited by John Oberdiek and Aileen Kanvanagh

Arguing About Metaphysics
Edited by Michael Rea

Arguing About Religion
Edited by Kevin Timpe

Arguing About Political Philosophy
Edited by Matt Zwolinski

Arguing About Language
Edited by Darragh Byrne and Max Kolbel

Arguing About Bioethics
Edited by Stephen Holland

Forthcoming titles:

Arguing About Political Philosophy 2nd Edition
Edited by Matt Zwolinski

Arguing About Science

Edited by

Alexander Bird and
James Ladyman

Routledge
Taylor & Francis Group

LONDON AND NEW YORK

First published 2013
by Routledge
2 Park Square, Milton Park, Abingdon, Oxon OX14 4RN

Simultaneously published in the USA and Canada
by Routledge
711 Third Avenue, New York, NY 10017

Routledge is an imprint of the Taylor & Francis Group, an informa business

British Library Cataloguing in Publication Data
A catalogue record for this book is available from the British Library

Library of Congress Cataloging in Publication Data
   Arguing about science / edited by Alexander Bird and James Ladyman.
      p. cm. – (Arguing about philosophy)
   Includes bibliographical references and index.
   1. Science–Philosophy.   I. Bird, Alexander, 1964–II. Ladyman,
   James, 1969–
   Q175.3.A72 2012
   501–dc23
                                                   2012009171

ISBN: 978-0-415-49229-4 (hbk)
ISBN: 978-0-415-49230-0 (pbk)

Typeset in Joanna and Bell Gothic
by RefineCatch Limited, Bungay, Suffolk

Printed and bound in the United States of America
by Edwards Brothers, Inc. Lillington, North Carolina

Contents

Acknowledgements

The editors and publishers wish to thank the following for permission to use copyrighted material:

Part One: What is science?

Karl Popper. 'Science: Conjectures and Refutations', Chapter 1 of his *Conjectures and Refutations: The Growth of Scientific Knowledge*. © 2002, The Estate of Karl Popper.

Adolf Grünbaum. 'Is Freudian Psychoanalytic Theory Pseudo-Scientific by Karl Popper's Criterion of Demarcation?', in *American Philosophical Quarterly*, **16**, 1979, 131–41. © 1979, North American Philosophical Publications.

Stephen Jay Gould. Excerpts from *The Mismeasure of Man*. © 1996, Stephen Jay Gould.

Thomas Kuhn. 'Objectivity, Value Judgement, and Theory Choice', from *The Essential Tension: Selected Studies in Scientific Tradition and Change*. © 1977, University of Chicago.

Part Two: Science, race and gender

Sally Haslanger. 'Gender and Race: (What) Are They? (What) Do We Want Them To Be?', in *Noûs*, **34**, 2000, 31–55. © 2000, Blackwell Publishers Inc.

Helena Cronin. 'The Battle of the Sexes Revisited', in Alan Grafen and Mark Ridley (eds) *Richard Dawkins: How a Scientist Changed the Way We Think: Reflections by Scientists, Writers, and Philosophers*. © 2006, Oxford University Press.

Evelyn Fox Keller. 'Beyond the Gene but Beneath the Skin', in Susan Oyama, Paul E. Griffiths, and Russell D. Gray (eds) *Cycles of Contingency: Developmental Systems and Evolution*. © 2001, Massachusetts Institute of Technology.

Lucius T. Outlaw. 'Toward a Critical Theory of "Race" ', in David Theo Goldberg (ed.) *Anatomy of Racism*. © 1990, the Regents of the University of Minnesota.

Robin O. Andreasen. 'Race: Biological Reality or Social Construct?', in *Philosophy of Science*, **67**, Supplementary Volume, 2000, S653–S666. © 2000, Philosophy, of Science Association.

Joshua M. Glasgow. 'On the New Biology of Race', in *The Journal of Philosophy*, **100**, 2003, 456–74. © 2003, Journal of Philosophy Inc.

Clark Glymour. 'What Went Wrong? Reflections on Science by Observation and *The Bell Curve*', in *Philosophy of Science*, **65**, 1988, 1–32. © 1988, Philosophy of Science Association.

Part Three: Scientific reasoning

Peter Lipton. 'Induction', Chapter 1 of his *Inference to the Best Explanation* (Second Edition). © 2004, Routledge.

Hilary Putnam. 'The "Corroboration" of Theories', in Paul Schilpp (ed.) *The Philosophy of Karl Popper, The Library of Living Philosophers*, Volume XIV. © 1974, Open Court.

John Stuart Mill. Excerpts from *A System of Logic*. London, 1875.

William Whewell. Excerpts from *Theory of the Scientific Method*. Indianapolis, 1968.

Peter Achinstein. 'Waves and Scientific Method', in *PSA: Proceedings of the Biennial Meeting of the Philosophy of Science Association*, 1992, Volume Two: Symposia and Invited Papers, 193–204.

Michael Strevens. '*Notes on Bayesian Confirmation Theory*', available at: http://www.strevens.org/bct/.

Part Four: Explanation

Wesley C. Salmon. 'Scientific Explanation', in Merrilee H. Salmon, John Earman, Clark Glymour, James G. Lennox, Peter Machamer, J.E. McGuire, John D. Norton, Wesley C. Salmon and Kenneth F. Schaffner (eds) *Introduction to the Philosophy of Science*. © 1992, Prentice-Hall Inc.

Bas van Fraassen. 'The Pragmatics of Explanation', Chapter 5 of his *The Scientific Image*. © 1980, Bas van Fraassen.

Peter Lipton. 'Explanation', Chapter 2 of his *Inference to the Best Explanation* (Second Edition). © 2004, Peter Lipton.

Part Five: Laws and causation

Fred I. Dretske. 'Laws of Nature', in *Philosophy of Science*, **44**, 1977, 248–68. © 1977, Philosophy of Science Association.

Bas van Fraassen. 'What are Laws of Nature?', Chapter 2 of his *Laws and Symmetry*. © 1989, Bas van Fraassen.

Marc Lange. 'Natural Laws and the Problem of Provisos', in *Erkenntnis*, **38**, 1993, 233–248. © 1993, Kluwer Academic Publishers.

Nancy Cartwright. 'Causal Laws and Effective Strategies', Chapter 1 of her *How the Laws of Physics Lie*. © 1983, Nancy Cartwright.

Part Six: Science and medicine

Ronald Munson. 'Why Medicine Cannot Be a Science', in *The Journal of Medicine and Philosophy*, **6**, 1981, 183–208. © 1981, The Society for Health and Human Values.

Kenneth F. Schaffner. 'Philosophy of Medicine', in Merrilee H. Salmon, John Earman, Clark Glymour, James G. Lennox, Peter Machamer, J.E. McGuire, John D. Norton, Wesley C. Salmon and Kenneth F. Schaffner (eds) *Introduction to the Philosophy of Science*. © 1992, Prentice-Hall Inc.

School of Health and Related Research, University of Sheffield. 'Hierarchy of Evidence', from www.shef.ac.uk/scharr/ir/units/systrev/hierarchy.htm.

John Worrall. '*What* Evidence in Evidence-Based Medicine?', in *Philosophy of Science*, **69**, 2002, S316–S330.

Part Seven: Probability in action: forensic science

William C. Thompson and Edward L. Schumann. 'Interpretation of Statistical Evidence in Criminal Trials: The Prosecutor's Fallacy and the Defense Attorney's Fallacy', in *Law and Human Behaviour*, **11**, 1987, 167–87.

Neven Sesardić. 'Sudden Infant Death or Murder? A Royal Confusion about Probabilities', in *British Journal for the Philosophy of Science*, **58**, 2007, 299–329. © 2007, Neven Sesardić.

Part Eight: Risk, uncertainty and science policy

Neil A. Manson. 'Formulating the Precautionary Principle', in *Environmental Ethics*, **24**, 2002, 263–74.

Per Sandin, Martin Peterson, Sven Ove Hansson, Christina Rudén, and André Juthe. 'Five Charges Against the Precautionary Principle', in *Journal of Risk Research*, **5**, 2002, 287–99.

Sven Ove Hansson. 'Risk and ethics: three approaches', in Tim Lewens (ed.) *Risk: Philosophical Perspectives*. © 2007, Sven Ove Hansson.

Part Nine: Scientific realism and antirealism

Pierre Duhem. Excerpts from '*The Aim and Structure of Physical Theory*.' © 1982, Princeton University Press.

Henri Poincaré. Excerpts from '*Science and Hypothesis*'. Dover, 1952.

Larry Laudan. 'A Confutation of Convergent Realism', in *Philosophy of Science*, **48**, 1981, 19–49. © 1981, Philosophy of Science Association.

Bas van Fraassen. 'Arguments Concerning Scientific Realism', Chapter 2 of his *The Scientific Image*. © 1980, Bas van Fraassen.

Ian Hacking. 'Experimentation and Scientific Realism', in *Philosophical Topics*, **13**, 1, 71–87.

John Worrall. 'Structural Realism: The Best of Both Worlds?', in *Dialectica*, **43**, 1989, 99–124.

GENERAL INTRODUCTION

PHILOSOPHY OF SCIENCE is a subject of profound intellectual interest and great social and political importance. Science itself has produced knowledge that is relevant to answering questions in almost all areas of philosophy. Sometimes it is argued by both scientists and philosophers that philosophy is increasingly redundant as science expands into what was once its exclusive domain. For example, the ancient Greek atomists speculated about the nature of matter, but such speculation has been replaced by detailed experimental knowledge. Similarly, experimental psychology and neuroscience are now rapidly colonizing the territory that was once the sovereign domain of philosophers, poets and artists; aspects of the human mind such as the imagination, language and free will are all now extensively studied by science. On the other hand, science's nature, methods and goals are continually contested and evolving, and philosophical reflection on it has much to contribute to both science and culture in general. Many aspects of the scientific method that contemporary scientists hold dear were first explicitly articulated by philosophers, and philosophers of science have carefully studied aspects of science in general as well as specializing in the study of particular theories and sciences and their history and foundations. Science is not always reliable and there are many examples of scientific orthodoxy being subsequently found to be erroneous. Usually, in such episodes, established knowledge is refined and corrected at the margins. However, there are also revisions in fundamental components of theories, and subjecting scientific theories and methods to scrutiny is essential if errors are to be found and corrected.

In this collection we have included material pertaining to classic issues in academic philosophy of science. We have also included papers on areas that are less familiar and which are not covered in most of the existing anthologies that aim to introduce students to the subject. We begin with the first classic question in the philosophy of science, concerning the nature of science in Part One: What is Science?. Part Two: Science, Race and Gender reveals science's complex interface with politics and ideology. Science is often set up as a source of knowledge of the most basic facts of existence that is an alternative to those on which most cultures base their beliefs. In Part Three we return to the traditional heartland of philosophy of science and consider scientific reasoning, including induction, probability and the confirmation of theories by evidence. Many philosophers think that providing explanations is a key function of science and element of the scientific method, and accordingly we devote Part Four to the nature of scientific explanation. Explanations often appeal to laws of nature and causes of which we have no direct experience; in Part Five we examine laws and causes. For many people their most immediate experience of the scientific method is when they visit the doctor, who administers a scientifically designed and interpreted test and perhaps then drugs that have been subject to scientifically organized clinical trials. However, medicine has distinctive features centrally involving care, which we do not associate with science. The relationship between science and medicine is the subject of Part Six. In Part Seven we introduce the reader to the complexities of probabilistic reasoning, the well-documented susceptibility of people to probabilistic fallacies and how these issues are of central importance in the interpretation of forensic evidence in legal contexts. In Part Eight we turn to the problems the public and policymakers face in applying scientific knowledge to the management of risk and uncertainty, notably in the context of climate change, where the stakes are incredibly high and where the accurate estimation of both risk and uncertainty is of paramount importance. In the last part of the book we return to the question of the ultimate nature of reality. Whilst science undoubtedly embodies ever greater and more precise knowledge of how the world behaves, many philosophers and scientists question whether it will ever answer questions about the ultimate nature of matter and of space and time.

Each section begins with an introductory essay in which we give an overview of the issues and explain how the articles selected address them. The selection of articles was much improved by extensive feedback and suggestions from several referees for Routledge, to whom we are extremely grateful. We also wish to record our thanks to our teachers and students in philosophy of science, with whom we hope to spend many more years exploring this incredibly rich subject, which we both love. We very much hope that the readers of this volume will find studying the subject as interesting and rewarding as we do. A note of thanks is also due to Tony Bruce for conceiving of and encouraging us to pursue this project, and for his patience in waiting for it to be brought to fruition. Finally, we owe a massive debt to Kit Patrick, whose work assembling the articles and checking and correcting the text was outstanding, and who also helped shape the final volume through his suggestions and comments. You probably would not be reading this book if it was not for him, and we are extremely grateful.

PART ONE

What is science?

INTRODUCTION TO PART ONE

WE BEGIN WITH the most fundamental question in the philosophy of science, and it is one that concerns many people other than philosophers. The definition of science matters to education, public policy, medicine, religion and the law. Governments must draw upon genuine science when making decisions, must license only those medicines that have passed scientific tests and standards of evidence, and must ensure that children are taught science not pseudo-science in areas such as biology, cosmology, medicine and physics. Courts must often take the opinion of scientific expert witnesses when considering forensic evidence. Where laws provide for the separation of church and state, the definition of science may be important because there are religiously-inspired bodies of teaching for which scientific status is claimed. When we consider the social sciences, the definition of science becomes ethically and politically important. If social scientists use concepts such as 'race', 'gender', 'sexuality' and so on, it is important that how those concepts are defined is made clear so that it can be established whether or not their use is properly scientific, or in fact just a vehicle for an ideological agenda. Similarly, when policymakers claim to be using 'evidence-based' decision making we need to be able to investigate the dividing line between the scientific content of their deliberations and the influence of vested interests and ideology in its interpretation.

Philosophers of science and scientists investigate these issues and there is no sharp dividing line between theoretical science and philosophical inquiry in it. Disputes about the proper definition of terms, such as 'species', 'gene' or 'fitness', and arguments about the right methodology to use in interpreting data, are an important part of everyday scientific life. When there are profound economic and social consequences at stake, the debate about what is properly scientific becomes of interest to everyone in society. Almost everyone has an opinion about anthropogenic climate change, for example. The famous hockey stick graph of the average temperature in the Northern hemisphere has become the subject of everyday discussion, yet is a fairly sophisticated statistical construction that shows departures from an average plotted against time. Even the fairly elementary science behind it requires a level of mathematical statistics that some people never learn, applied to measurements that we will not perform ourselves. In this, as in many issues, the definition of science matters because we need to defer to scientific experts in arriving at our own beliefs.

In everyday parlance science is associated with logic, with empirical investigation, and with reason and rationality. For some, these connotations are overwhelmingly positive; for others, all harbour dangers and threaten to overlook important things about human existence and to denude the world of meaning. There was a reaction against science in the Romantic movement of the late eighteenth and early nineteenth centuries, and there have been similar such criticisms of science ever since. The original meaning of the term 'science' is from the Latin *scientia* meaning knowledge, where knowledge is not the same thing as true belief. There is much debate about the nature of knowledge among philosophers, but it is always taken to be distinct from true belief that is somehow accidental or the result of completely spurious reasoning. To know is to be possessed of evidence, justification and reason; or at least to be reliable or connected to the world or the truth in the right kind of way.

It is important to note that there is a lot of bona fide knowledge that is not scientific. The science/non-science distinction does not track the difference between rational enquiry and non-rational enquiry either. A person can be rational, making deductions and inferences, and use empirical methods, measuring, gathering and collating data, without being part of science (though it may be argued that science is continuous with, and has its origins in, good everyday reasoning). Philosophers of science are often interested in the difference between science and pseudo-science, and usually regard the former as good and the latter as bad, but this should not be taken to imply the lack of a critical attitude to science itself. There is good and bad science and shades of grey in between, and much contemporary science is hopefully to be corrected and refined in the future. The word 'scientist' was not used in its contemporary sense until the nineteenth century when the emergence of scientific institutions and disciplines had become mature and established after their beginnings in the seventeenth century. Science is an emergent historical phenomenon and, according to a widely held though contested view, came into being, from ancient and medieval roots, in the Scientific Revolution in Europe around the mid-seventeenth century. There are two vital facts to know about this period.

The first is that the larger Scientific Revolution had as its most important component the Copernican Revolution, when the view that the Earth is the centre of the universe (geocentrism) was replaced by the hypothesis that it orbits the Sun along with the other planets (heliocentrism). This was not just a dispute about astronomy, since the hypothesis that the Earth is at the centre of the universe was essential to the medieval orthodoxy of Aristotelian physics, especially to the explanation of gravity. It was also linked to the understanding of heaven, hell, and the place of human beings in the cosmos. The Catholic Church had a very great investment in the established tradition of scholastic Aristotelian natural philosophy, and the former's intellectual authority had been challenged by the thinkers of the Reformation. Hence, when Galileo publicly advocated Copernicanism, this eventually led to his trial and imprisonment. Eventually, even the Catholic Church came to accept Copernicanism. A similarly heated conflict between science and religion would not arise again until Darwin's theory of evolution by natural selection challenged the idea that each species on the Earth was individually created by God.

The second feature of the Scientific Revolution that must be mentioned even in the briefest of histories is that there was an extraordinary growth in knowledge in numerous domains in a relatively short period of time, so that the term 'revolution' is not a misnomer. A standard list of scientific advances of the period would probably include:

- Kepler's laws of planetary motion (1609)
- Newton's mechanics (1687)
- the first law of ideal gases (Robert Boyle, 1662)
- the circulation of the blood by the heart (William Harvey, 1628).

All but the third involved the abandonment of entrenched orthodoxies. Kepler overthrew the ancient idea that all motion in the perfect heavenly realm was composed of circular movements. Newton overthrew the recent orthodoxy that all motion was caused by contact, with his mysterious but mathematically precise gravitational force. Harvey rejected the orthodoxy originating with Galen that blood is created by the liver and that the heart exists solely to produce heat. It is often argued that there is something inherently scientific about scepticism towards received ideas. We will come back to the idea that science proceeds by the continual questioning and overthrowing of established ideas below, since it is central to the first account we will consider of the nature of science.

There are two reasons why scientists and others need an account of the nature of science. The first is that people often want to criticize particular theories or research programmes as poor science by invoking criteria for good science, and the latter are in part derived from and definitive of what they think science is. The second is that people often want to argue that particular theories, research programmes and even whole areas of activity are not scientific at all. This judgement can clearly only be defended by appeal to conditions that must be satisfied for something to be scientific. Such conditions are called *necessary* conditions. For example, a necessary condition for a shape to be a square is that is has four sides. Sometimes a necessary condition is also a *sufficient* condition. A sufficient condition for a shape to be a triangle is that it has three (straight) sides. This is also necessary because any shape lacking three sides is not a triangle. But having four sides is not sufficient for being a square, because squares must have four equal sides and four equal internal angles. We have here two necessary conditions and they are jointly sufficient but not each individually sufficient. The ideal endpoint of our inquiry is that we arrive at a set of necessary conditions for something to be science and that they are also together sufficient. We would have then a definition of 'science' and we could use it to tell science apart from pseudo-science (and bad science). The question of how to separate science from 'pseudo-science' is 'the *demarcation problem*'.

Conditions for science, often known as *demarcation criteria*, may be applied to different things. Roughly speaking there are three plausible candidates, and accounts of the nature of science can be considered for what they say about each and the relative emphasis they place on them. The first candidate is the theories that are the output of science, the second is the methods by which the theories are produced, and the third is the character and disposition of the scientists who produce them.

It will help to have an example in mind, and we shall take Newtonian mechanics. It introduced the contemporary understanding of mass, force and velocity; it consisted of three laws of motion and a law of universal gravitation; it successfully unified Kepler's laws of planetary motion, Galileo's laws of free-fall and the pendulum; it explained the tides in terms of the Moon's gravity; and it predicted the return of Halley's comet, the existence of Neptune and the fact that the Earth is not quite spherical but bulged at the equator.

If we are to put theories at the centre of our account of science, we might consider the above good features of Newtonian mechanics, and generalize arguing that proper scientific theories must be mathematically precise, have a few simple laws that unify diverse phenomena, and be predictive in the sense of implying facts that we have not yet observed. If we focus on the method by which the theories are produced then we will cite the fact that Newton sought to recover by a mathematical analysis the known laws that describe the motions of the planets and so arrived at his inverse square law of gravitation. Those laws themselves were arrived at after Kepler had considered a vast and accurate body of astronomical data about the observed positions of the planets in the night sky at different times. Newton himself claimed to have inferred his laws from the data and not to have made hypotheses, and his example led others to advocate a theory of the scientific method according to which theories must always be arrived at this way. Finally, if we focus on the scientist himself we will emphasize Newton's mathematical genius, his obsessive character and rigorous approach to foundational matters, and the role of the individual genius working largely alone but building on the work of the scientists who had gone before him, especially, in this case, Galileo and Kepler.

The problem with focusing on theories is that science contains very heterogeneous theories covering a vast range of subject matters, and these theories change over time, so it would be rash to associate science in general with the characteristics of particular theories. So, for example, Darwin's theory was not at all mathematical and nor was it, originally, especially precisely predictive. It is often said that there are no strict mathematical laws outside of physics and, although that may be an exaggeration, there is certainly a lot of difference between theories in physics and theories in the other physical and social sciences. When it comes to the characteristics of scientists they too vary greatly: some being highly collaborative and others loners, some being very open-minded while others cling to their preferred theory with the conviction of a zealot, and some being mathematical geniuses while others are meticulous observers and field workers. Nonetheless, many philosophers of science have argued that scientists must have certain qualities, or that the scientific community must be organized in a particular way. Most accounts of the nature of science have something to say about the theories, methods and virtues of good scientists, but it is probably fair to say that most have focused on the scientific method. For many people, it is its methods that fundamentally distinguish science, making it a self-correcting and successful form of inquiry that is unrivalled.

Our readings begin with a classic chapter on the demarcation of science from non-science from Karl Popper. No philosopher has had such an influence on what scientists themselves say about science since, and although his ideas were developed in the mid-twentieth century, his influence remains strong today (for example, a recent book on philosophy of science by the physicist David Deutsch has scant regard for any philosopher other than Popper). The fundamental idea he introduced was a negative one. Science is not about proof and certainty, according to Popper; it is about refutation and conjecture. He called this account of the methodology of science 'critical rationalism'. He argued that scientists must always accept theories in a provisional way, and that we must live without certainty in scientific knowledge (fallibilism). This can be seen as learning the lessons of history, for Popper wrote at a time when it was widely accepted that classical mechanics, despite being extremely accurate, was not in fact true when applied to the novel domains of the very fast, the very small and the very large that are the province of the relativity and quantum theories. More generally, we have learned in all the sciences as they have matured that even very good theories with lots of evidence behind them can be replaced by even more accurate ones.

Popper was originally very interested in Freudianism and Marxism, which were claimed by their adherents to be scientific. Popper came to the conclusion that they were not, and he wanted to work out why. He realized that what he admired about relativity theory is that it made precise predictions about what was not yet observed. On the other hand, he observed that Marxists and Freudians were always able to explain anything that happened with their theories but were unable to predict it in advance. This insight of Popper's is profound. Science proceeds because theories are required to produce predictions of as yet unknown facts, not just explanations of known ones. For Popper, the riskier the prediction the better. A risky prediction is when a theory predicts an event that would otherwise be considered to be of very low probability.

Prior accounts of the scientific method tended to focus on *induction*. Induction is any form of reasoning that generalizes from observed cases to unobserved cases. It is a basic component of human cognition that we use unwittingly all the time, especially when we learn how the world works as children. In everyday life induction may be based on a small data set. For example, people visit a restaurant once or twice, enjoy the food and infer that it will always be a good place to eat. In science, however, it is usually stipulated that induction must be based on a large number and wide range of observations. A classic example is the claim that all metals expand when heated. The evidence for this must include observations of many different kinds of metals, of different shapes and sizes, not just on a couple of cases. As mentioned above, Newton claimed that his own method was inductive, and much of the official self-image of scientists since has emphasized two features of the scientific method, namely the gathering of a lot of empirical data and generalizing from it to laws. This was indeed the account of the scientific method due to the first philosopher of science of the modern era, namely Francis Bacon, who argued for a new method of collective knowledge production based on experiment, observation and induction at the turn of the seventeenth century (see Part Four for more on induction).

Popper noticed that having lots of empirical data and generalizing from it could just as well lead to pseudo-science, since, for example, astrology and Freudianism were based on huge amounts of empirical observations. As mentioned above, Popper thought that generalizations that explain phenomena are much easier to come by than theories that make predictions, especially risky predictions. The apparent strength of Marxism and Freudianism, namely their ability to provide explanations of all the phenomena with which their advocates were presented, was in fact a weakness, according to Popper. In sum, Popper argues, truly scientific theories are *falsifiable* in the sense that they make predictions that may turn out to be false. The more predictions there are, and the more precise and risky these predictions are, the more falsifiable the theory and the better it is.

The falsifiability criterion is applied to theories. Popper then argues that the scientific method consists in attempting to falsify them, not in attempting to confirm them. Accordingly, proper scientists are disposed to believe in their theories only provisionally insofar as they have so far avoided falsification, and will not resort to ad hoc modifications of theories to save them from falsification as when, for example, Marxists explain away the apparent falsification of the prediction of Marx's theory that there would in quick order have been a worldwide proletarian revolution. As Popper explains there are therefore various ways that pseudo-science can arise. There may be unfalsifiable theories, or there may be falsification that is ignored and masked by ad hoc modification. Either way, Popper's solution to the demarcation problem is to posit falsifiability as a necessary criterion for science, though as he stresses he does not propose falsifiability as a criterion for meaningfulness (unlike his rivals the logical positivists, who did propose empirical verifiability as a criterion for meaningfulness).

Popper goes on to argue that Hume's problem of induction, namely the problem of explaining how we can justify generalizing about the unobserved from the observed, cannot be solved, but that this does not matter because while no amount of finite positive instances can ever entail a generalization that covers all possible cases, a single negative instance can falsify such a generalization. It is fair to say that almost all philosophers of science have rejected Popper's claim that the positive notion of a theory being confirmed by evidence is completely dispensable from our understanding of science. Popper does resort to a notion of 'corroboration', and argues that theories are corroborated when they have survived a number of attempts at falsification. Most philosophers have regarded this as confirmation by another name and so argued that Popper never really escapes the problem of induction. Popper's criticism of inductivism includes many important points however, especially the idea that scientific theorizing can start from a hypothesis arrived at by any means at all and not necessarily by generalizing from data. Popper's idea of 'hypothetico-deductivism' remains central to many people's idea of the scientific method. Hypotheses may have any source but empirical consequences must be deduced from them and then tested. This is related to the distinction that is often made between the 'context of discovery' and the 'context of justification': the same rules need not apply in each as inductivists had often insisted, and theories may be arrived at by dreams, divine inspiration or guess-work; their origins are irrelevant to their scientific status, which only depends on their relationship to the evidence.

Our next reading is a detailed assessment of Popper's critique of psychoanalysis by Adolf Grünbaum, who takes issue both with Popper's argument that psychoanalysis is unfalsifiable, and with his claim that inductivism is flawed as an account of the scientific method because psychoanalysis does well according to its criteria for good science. Grünbaum distinguishes between the treatment associated with Freud's theories and the theories themselves, and argues that both are falsifiable. He goes on to argue that Popper's own admission that falsifiability admits of degrees undermines his arguments. There are a number of terms defined by Grünbaum in his paper so that he can make distinctions between key concepts and hence make his arguments precise; readers will be introduced to the rigorous and often intricate style of argument common in analytic philosophy.

Having warmed up with the classic debate about demarcation and the qualitative example of psychoanalysis, we must now enter the baffling world of probability and statistics. Contemporary science is unimaginable without the latter because the amounts of data that are handled routinely are now so large, and because much depends on sophisticated methods of turning raw data into appropriate mathematical representations. Statistics describes distributions of different data points in a population. For example, are there equal numbers of people for all the different values of height, or is the distribution non-uniform? Roughly speaking, statistics is in large part about the different proportions of different values of quantities in a population of some kind. More generally it is the science of data. As a kind of metascience (a science that talks generally about data and hypotheses in general and is drawn upon by all the other sciences), statistics arose relatively recently. It involves the mathematical theory of probability and the application of advanced techniques from the calculus. Once a body of data has been obtained concerning the distribution of various quantities in a population we may ask questions about it, and one of the most important such questions is 'what correlations are there among the quantities?' For example, we know that there is a correlation in the population between those who smoke cigarettes and those who develop lung cancer, in the sense that the proportion of individuals who develop lung cancer is greater among smokers than among the population as a whole. One of the most important tasks of scientists interpreting statistical data is to tell when a correlation is evidence of a causal link, and when it is 'spurious', meaning that it is accidental or non-causal. We live in a world that seems structured with cause and effect, but is this more than mere regularity? Hume famously said not. However, many philosophers and scientists disagree with him, and they pursue methods to extract causal information from statistics assuming that mere regularity is not evidence of causation. (The crucial question of how to infer causation from statistical data is addressed in more detail in Parts Three, Five, and Eight.)

The biologist Stephen Jay Gould combated the misunderstanding and misuse of statistics in the context of the measurement of human beings, and especially their intelligence. Racist theories of intelligence have a long history within the scientific establishment. This establishes, as do other cases, that pseudo-science may be within and not on the margins of mainstream science. The extract from Gould illustrates how the complexity of the mathematics used to infer causal structure in science may be abused and used to produce seemingly technically bona fide-looking material that can masquerade as science. In the end, the only way such examples can be dealt with is by careful study. Gould initiates the reader who is not familiar with statistical methods into some of their intricacies and associated fallacies of reasoning. He explains how *factor analysis* can be used to introduce a measure and thereby a concept of intelligence when it is applied to the correlations between individuals' scores in tests, and also explains its limitations. The second part of the reading from Gould is a critique of grand claims that have been made about explaining human behavioural traits in biological terms by advocates of 'sociobiology'.

The excerpt from Gould shows that debates about pseudo-science are not confined to philosophy departments; eminent scientists have also been at loggerheads over the scientific status of 'IQ'. For example, in a debate in *New Scientist* (Eysenck and Rose, 1979), Hans Eysenck defends the scientific worth of IQ and the claim that differences in average IQ are associated with race, while Steven Rose takes the view that IQ is of little scientific interest and argues that the folk concept of intelligence does not pick out a natural kind. By 'natural kind' he means a classification that is independent of all human beliefs and values. He takes a similar line with the idea of race, so according to him the association of IQ with race is doubly unscientific. (It is notable that many of the issues from the earlier IQ debate are recapitulated in the debate about *The Bell Curve* addressed in Part Two.) The idea that categories applied to human beings in the social sciences might be socially constructed is one important component of the legacy of Thomas Kuhn's work discussed below. Some have argued that even the kind terms in the natural sciences fail to 'carve nature at its joints' and instead reflect our interests and cognition. In Part Two we will return to the issue of whether kind terms like 'race' and 'gender' pick out natural kinds like 'electron' and 'oxygen' presumably do, or socially constructed ones like 'student' and 'pensioner' do.

This part ends with an article by Thomas Kuhn that presents a nuanced perspective on the nature of science written in response to critics of his incredibly influential book *The Structure of Scientific Revolutions*. The latter is responsible for the prevalence of the word 'paradigm' in our culture. Kuhn's book started a dispute about the nature of science. Both Popper and his adversaries, the inductivists, regard science as rational, progressive and governed by methodological principles that make it independent of psychological, social, economic and political factors. Kuhn wrote a history of science that suggested that evidence and reason did not determine which theories were abandoned and which accepted. He cited social forces as factors in causing scientific developments. Whether or not he intended to paint a picture of science as somewhat irrational and arbitrary, he succeeded in dividing his readers into those who did not accept his account, and those who embraced it and learned general lessons about reason and objectivity from it. Hence, his book became one of the most cited works of the twentieth century.

Further reading

Chalmers, A.F. 1999 *What Is This Thing Called Science?*, Buckingham: Open University Press.

Lakatos, I., Feyerabend, P. and Motterlini, M. 1999 *For and Against Method*, Chicago: University of Chicago Press. See especially 'Lecture 1. The Demarcation Problem', pp. 20–30 and 'Lecture VIII. The Methodology of Scientific Research Programmes', pp. 96–108.

Laudan, L. 1983 'The Demise of the Demarcation Problem', in Adolf Grünbaum and Robert Sonné Cohen (eds), *Physics, Philosophy, and Psychoanalysis: Essays in Honor of Adolf Grünbaum*. Dordrecht: Reidel, pp. 111–27.

Schilpp, P.A. (ed.) 1974 *The Philosophy of Karl Popper*, Library of Living Philosophers, La Salle, IL. See especially Imre Lakatos 'Popper on demarcation and induction', pp. 241–73 and Karl Popper's 'Reply to my critics', pp. 961–1197.

Reference

Eysenck, H.J. and Rose, S. 1979 'Race, Intelligence and Education', *New Scientist*, 82: 849–52.

Chapter 1

Karl Popper

SCIENCE: CONJECTURES AND REFUTATIONS

There could be no fairer destiny for any . . . theory than that it should point the way to a more comprehensive theory in which it lives on, as a limiting case.

Albert Einstein

Mr. Turnbull had predicted evil consequences, . . . and was now doing the best in his power to bring about the verification of his own prophecies.

Anthony Trollope

I

When I received the list of participants in this course and realized that I had been asked to speak to philosophical colleagues I thought, after some hesitation and consultation, that you would probably prefer me to speak about those problems which interest me most, and about those developments with which I am most intimately acquainted. I therefore decided to do what I have never done before: to give you a report on my own work in the philosophy of science, since the autumn of 1919 when I first began to grapple with the problem, 'When should a theory be ranked as scientific?' or 'Is there a criterion for the scientific character or status of a theory?'

The problem which troubled me at the time was neither, 'When is a theory true?' nor, 'When

is a theory acceptable?' My problem was different. I wished to distinguish between science and pseudo-science; knowing very well that science often errs, and that pseudo-science may happen to stumble on the truth.

I knew, of course, the most widely accepted answer to my problem: that science is distinguished from pseudo-science – or from 'metaphysics' – by its empirical method, which is essentially inductive, proceeding from observation or experiment. But this did not satisfy me. On the contrary, I often formulated my problem as one of distinguishing between a genuinely empirical method and a non-empirical or even a pseudo-empirical method – that is to say, a method which, although it appeals to observation and experiment, nevertheless does not come up to scientific standards. The latter method may be exemplified by astrology, with its stupendous mass of empirical evidence based on observation – on horoscopes and on biographies.

But as it was not the example of astrology which led me to my problem I should perhaps briefly describe the atmosphere in which my problem arose and the examples by which it was stimulated. After the collapse of the Austrian Empire there had been a revolution in Austria: the air was full of revolutionary slogans and ideas, and new and often wild theories. Among the theories which interested me Einstein's theory of relativity was no doubt by far the most important. Three others were Marx's theory of history, Freud's psychoanalysis, and Alfred Adler's so-called 'individual psychology'.

There was a lot of popular nonsense talked about these theories, and especially about relativity (as still happens even today), but I was fortunate in those who introduced me to the study of this theory. We all – the small circle of students to which I belonged – were thrilled with the result of Eddington's eclipse observations which in 1919 brought the first important confirmation of Einstein's theory of gravitation. It was a great experience for us, and one which had a lasting influence on my intellectual development.

The three other theories I have mentioned were also widely discussed among students at that time. I myself happened to come into personal contact with Alfred Adler, and even to co-operate with him in his social work among the children and young people in the working-class districts of Vienna where he had established social guidance clinics.

It was during the summer of 1919 that I began to feel more and more dissatisfied with these three theories – the Marxist theory of history, psychoanalysis, and individual psychology; and I began to feel dubious about their claims to scientific status. My problem perhaps first took the simple form, 'What is wrong with Marxism, psychoanalysis, and individual psychology? Why are they so different from physical theories, from Newton's theory, and especially from the theory of relativity?'

To make this contrast clear I should explain that few of us at the time would have said that we believed in the *truth* of Einstein's theory of gravitation. This shows that it was not my doubting the *truth* of those other three theories which bothered me, but something else. Yet neither was it that I merely felt mathematical

physics to be more *exact* than the sociological or psychological type of theory. Thus what worried me was neither the problem of truth, at that stage at least, nor the problem of exactness or measurability. It was rather that I felt that these other three theories, though posing as sciences, had in fact more in common with primitive myths than with science; that they resembled astrology rather than astronomy.

I found that those of my friends who were admirers of Marx, Freud, and Adler, were impressed by a number of points common to these theories, and especially by their apparent *explanatory power*. These theories appeared to be able to explain practically everything that happened within the fields to which they referred. The study of any of them seemed to have the effect of an intellectual conversion or revelation, opening your eyes to a new truth hidden from those not yet initiated. Once your eyes were thus opened you saw confirming instances everywhere: the world was full of *verifications* of the theory. Whatever happened always confirmed it. Thus its truth appeared manifest; and unbelievers were clearly people who did not want to see the manifest truth; who refused to see it, either because it was against their class interest, or because of their repressions which were still 'un-analysed' and crying aloud for treatment.

The most characteristic element in this situation seemed to me the incessant stream of confirmations, of observations which 'verified' the theories in question; and this point was constantly emphasized by their adherents. A Marxist could not open a newspaper without finding on every page confirming evidence for his interpretation of history; not only in the news, but also in its presentation – which revealed the class bias of the paper – and especially of course in what the paper did *not* say. The Freudian analysts emphasized that their theories were constantly verified by their 'clinical observations'. As for Adler, I was much impressed by a personal experience. Once, in 1919, I reported to him a case which to me did not seem particularly Adlerian, but which he found no difficulty in analysing in terms of his theory of inferiority feelings, although he had not even seen the child. Slightly shocked, I asked him how he could be so sure. 'Because of my thousandfold experience,' he replied; whereupon I could not help saying: 'And with this new case, I suppose, your experience has become thousand-and-one-fold.'

What I had in mind was that his previous observations may not have been much sounder than this new one; that each in its turn had been interpreted in the light of 'previous experience', and at the same time counted as additional confirmation. What, I asked myself, did it confirm? No more than that a case could be interpreted in the light of the theory. But this meant very little, I reflected, since every conceivable case could be interpreted in the light of Adler's theory, or equally of Freud's. I may illustrate this by two very different examples of human behaviour: that of a man who pushes a child into the water with the intention of drowning it; and that of a man who sacrifices his life in an attempt to save the child. Each of these two cases can be explained with equal ease in Freudian and in Adlerian terms. According to Freud the first man suffered from repression (say, of some component of his Oedipus complex), while the second man had achieved sublimation. According to Adler the first man suffered from feelings of inferiority (producing perhaps the need to prove to himself that he dared to commit some crime), and so did the second man (whose need was to prove to himself that he dared to rescue the child). I could not think of any human behaviour which could not be interpreted in terms of either theory. It was precisely this fact – that they always fitted, that they were always

confirmed – which in the eyes of their admirers constituted the strongest argument in favour of these theories. It began to dawn on me that this apparent strength was in fact their weakness.

With Einstein's theory the situation was strikingly different. Take one typical instance – Einstein's prediction, just then confirmed by the findings of Eddington's expedition. Einstein's gravitational theory had led to the result that light must be attracted by heavy bodies (such as the sun), precisely as material bodies were attracted. As a consequence it could be calculated that light from a distant fixed star whose apparent position was close to the sun would reach the earth from such a direction that the star would seem to be slightly shifted away from the sun; or, in other words, that stars close to the sun would look as if they had moved a little away from the sun, and from one another. This is a thing which cannot normally be observed since such stars are rendered invisible in daytime by the sun's overwhelming brightness; but during an eclipse it is possible to take photographs of them. If the same constellation is photographed at night one can measure the distances on the two photographs, and check the predicted effect.

Now the impressive thing about this case is the *risk* involved in a prediction of this kind. If observation shows that the predicted effect is definitely absent, then the theory is simply refuted. The theory is *incompatible with certain possible results of observation* – in fact with results which everybody before Einstein would have expected.[1] This is quite different from the situation I have previously described, when it turned out that the theories in question were compatible with the most divergent human behaviour, so that it was practically impossible to describe any human behaviour that might not be claimed to be a verification of these theories.

These considerations led me in the winter of 1919–20 to conclusions which I may now reformulate as follows.

1   It is easy to obtain confirmations, or verifications, for nearly every theory – if we look for confirmations.

2   Confirmations should count only if they are the result of *risky predictions*; that is to say, if, unenlightened by the theory in question, we should have expected an event which was incompatible with the theory – an event which would have refuted the theory.

3   Every 'good' scientific theory is a prohibition: it forbids certain things to happen. The more a theory forbids, the better it is.

4   A theory which is not refutable by any conceivable event is non-scientific. Irrefutability is not a virtue of a theory (as people often think) but a vice.

5   Every genuine *test* of a theory is an attempt to falsify it, or to refute it. Testability is falsifiability; but there are degrees of testability: some theories are more testable, more exposed to refutation, than others; they take, as it were, greater risks.

6   Confirming evidence should not count *except when it is the result of a genuine test of the theory*; and this means that it can be presented as a serious but unsuccessful attempt to falsify the theory. (I now speak in such cases of 'corroborating evidence'.)

7   Some genuinely testable theories, when found to be false, are still upheld by their admirers – for example by introducing *ad hoc* some auxiliary assumption, or by re-interpreting the theory *ad hoc* in such a way that it escapes refutation. Such a procedure is always possible, but it rescues the theory from refutation only at the price of destroying, or at least lowering, its scientific status. (I later described such a

rescuing operation as a '*conventionalist twist*' or a '*conventionalist stratagem*'.)

One can sum up all this by saying that *the criterion of the scientific status of a theory is its falsifiability, or refutability, or testability.*

II

I may perhaps exemplify this with the help of the various theories so far mentioned. Einstein's theory of gravitation clearly satisfied the criterion of falsifiability. Even if our measuring instruments at the time did not allow us to pronounce on the results of the tests with complete assurance, there was clearly a possibility of refuting the theory.

Astrology did not pass the test. Astrologers were greatly impressed, and misled, by what they believed to be confirming evidence – so much so that they were quite unimpressed by any unfavourable evidence. Moreover, by making their interpretations and prophecies sufficiently vague they were able to explain away anything that might have been a refutation of the theory had the theory and the prophecies been more precise. In order to escape falsification they destroyed the testability of their theory. It is a typical soothsayer's trick to predict things so vaguely that the predictions can hardly fail: that they become irrefutable.

The Marxist theory of history, in spite of the serious efforts of some of its founders and followers, ultimately adopted this soothsaying practice. In some of its earlier formulations (for example in Marx's analysis of the character of the 'coming social revolution') their predictions were testable, and in fact falsified.[2] Yet instead of accepting the refutations the followers of Marx re-interpreted both the theory and the evidence in order to make them agree. In this way they rescued the theory from refutation; but they did so at the price of adopting a device which made it irrefutable. They thus gave a 'conventionalist twist' to the theory; and by this stratagem they destroyed its much advertised claim to scientific status.

The two psychoanalytic theories were in a different class. They were simply non-testable, irrefutable. There was no conceivable human behaviour which could contradict them. This does not mean that Freud and Adler were not seeing certain things correctly: I personally do not doubt that much of what they say is of considerable importance, and may well play its part one day in a psychological science which is testable. But it does mean that those 'clinical observations' which analysts naïvely believe confirm their theory cannot do this any more than the daily confirmations which astrologers find in their practice.[3] And as for Freud's epic of the Ego, the Super-ego, and the Id, no substantially stronger claim to scientific status can be made for it than for Homer's collected stories from Olympus. These theories describe some facts, but in the manner of myths. They contain most interesting psychological suggestions, but not in a testable form.

At the same time I realized that such myths may be developed, and become testable; that historically speaking all – or very nearly all – scientific theories originate from myths, and that a myth may contain important anticipations of scientific theories. Examples are Empedocles' theory of evolution by trial and error, or Parmenides' myth of the unchanging block universe in which nothing ever happens and which, if we add another dimension, becomes Einstein's block universe (in which, too, nothing ever happens, since everything is, four-dimensionally speaking, determined and laid down from, the beginning). I thus felt that if a theory is found to be non-scientific, or 'metaphysical' (as we might say), it is not thereby found to be unimportant, or insignificant, or 'meaningless', or 'nonsensical'.[4] But it cannot

claim to be backed by empirical evidence in the scientific sense – although it may easily be, in some genetic sense, the 'result of observation'.

(There were a great many other theories of this pre-scientific or pseudo-scientific character, some of them, unfortunately, as influential as the Marxist interpretation of history; for example, the racialist interpretation of history – another of those impressive and all-explanatory theories which act upon weak minds like revelations.)

Thus the problem which I tried to solve by proposing the criterion of falsifiability was neither a problem of meaningfulness or significance, nor a problem of truth or acceptability. It was the problem of drawing a line (as well as this can be done) between the statements, or systems of statements, of the empirical sciences, and all other statements – whether they are of a religious or of a metaphysical character, or simply pseudo-scientific. Years later – it must have been in 1928 or 1929 – I called this first problem of mine the '*problem of demarcation*'. The criterion of falsifiability is a solution to this problem of demarcation, for it says that statements or systems of statements, in order to be ranked as scientific, must be capable of conflicting with possible, or conceivable, observations.

III

Today I know, of course, that this *criterion of demarcation* – the criterion of testability, or falsifiability, or refutability – is far from obvious; for even now its significance is seldom realized. At that time, in 1920, it seemed to me almost trivial, although it solved for me an intellectual problem which had worried me deeply, and one which also had obvious practical consequences (for example, political ones). But I did not yet realize its full implications, or its philosophical significance. When I explained it to a fellow student of the Mathematics Department (now a distinguished mathematician in Great Britain), he suggested that I should publish it. At the time I thought this absurd; for I was convinced that my problem, since it was so important for me, must have agitated many scientists and philosophers who would surely have reached my rather obvious solution. That this was not the case I learnt from Wittgenstein's work, and from its reception; and so I published my results thirteen years later in the form of a criticism of Wittgenstein's *criterion of meaningfulness*.

Wittgenstein, as you all know, tried to show in the *Tractatus* (see for example his propositions 6.53; 6.54; and 5) that all so-called philosophical or metaphysical propositions were actually non-propositions or pseudo-propositions: that they were senseless or meaningless. All genuine (or meaningful) propositions were truth functions of the elementary or atomic propositions which described 'atomic facts', i.e. – facts which can in principle be ascertained by observation. In other words, meaningful propositions were fully reducible to elementary or atomic propositions which were simple statements describing possible states of affairs, and which could in principle be established or rejected by observation. If we call a statement an 'observation statement' not only if it states an actual observation but also if it states anything that *may* be observed, we shall have to say (according to the *Tractatus*, 5 and 4.52) that every genuine proposition must be a truth-function of, and therefore deducible from, observation statements. All other apparent propositions will be meaningless pseudo-propositions; in fact they will be nothing but nonsensical gibberish.

This idea was used by Wittgenstein for a characterization of science, as opposed to philosophy. We read (for example in 4.11, where natural science is taken to stand in opposition to philosophy): 'The totality of true propositions is the total natural science (or the totality of the natural sciences).' This means that the propositions which belong to science are those deducible from *true* observation statements; they are those propositions which can be *verified* by true observation statements. Could we know all true observation statements, we should also know all that may be asserted by natural science.

This amounts to a crude verifiability criterion of demarcation. To make it slightly less crude, it could be amended thus: 'The statements which may possibly fall within the province of science are those which may possibly be verified by observation statements; and these statements, again, coincide with the class of *all* genuine or meaningful statements.' For this approach, then, *verifiability, meaningfulness, and scientific character all coincide*.

I personally was never interested in the so-called problem of meaning; on the contrary, it appeared to me a verbal problem, a typical pseudo-problem. I was interested only in the problem of demarcation, i.e. in finding a criterion of the scientific character of theories. It was just this interest which made me see at once that Wittgenstein's verifiability criterion of meaning was intended to play the part of a criterion of demarcation as well; and which made me see that, as such, it was totally inadequate, even if all misgivings about the dubious concept of meaning were set aside. For Wittgenstein's criterion of demarcation – to use my own terminology in this context – is verifiability, or deducibility from observation statements. But this criterion is too narrow (and too wide): it excludes from science practically everything that is, in fact, characteristic of it (while failing in effect to exclude astrology). No scientific theory can ever be deduced from observation statements, or be described as a truth-function of observation statements.

All this I pointed out on various occasions to Wittgensteinians and members of the Vienna Circle. In 1931–2 I summarized my ideas in a largish book (read by several members of the

Circle but never published; although part of it was incorporated in my *Logic of Scientific Discovery*); and in 1933 I published a letter to the Editor of *Erkenntnis* in which I tried to compress into two pages my ideas on the problems of demarcation and induction.[5] In this letter and elsewhere I described the problem of meaning as a pseudo-problem, in contrast to the problem of demarcation. But my contribution was classified by members of the Circle as a proposal to replace the verifiability criterion of *meaning* by a falsifiability criterion of *meaning* — which effectively made nonsense of my views.[6] My protests that I was trying to solve, not their pseudo-problem of meaning, but the problem of demarcation, were of no avail.

My attacks upon verification had some effect, however. They soon led to complete confusion in the camp of the verificationist philosophers of sense and nonsense. The original proposal of verifiability as the criterion of meaning was at least clear, simple, and forceful. The modifications and shifts which were now introduced were the very opposite.[7] This, I should say, is now seen even by the participants. But since I am usually quoted as one of them I wish to repeat that although I created this confusion I never participated in it. Neither falsifiability nor testability were proposed by me as criteria of meaning; and although I may plead guilty to having introduced both terms into the discussion, it was not I who introduced them into the theory of meaning.

Criticism of my alleged views was widespread and highly successful. I have yet to meet a criticism of my views.[8] Meanwhile, testability is being widely accepted as a criterion of demarcation.

IV

I have discussed the problem of demarcation in some detail because I believe that its solution is the key to most of the fundamental problems of the philosophy of science. I am going to give you later a list of some of these other problems, but only one of them — the *problem of induction* — can be discussed here at any length.

I had become interested in the problem of induction in 1923. Although this problem is very closely connected with the problem of demarcation, I did not fully appreciate the connection for about five years.

I approached the problem of induction through Hume. Hume, I felt, was perfectly right in pointing out that induction cannot be logically justified. He held that there can be no valid logical[9] arguments allowing us to establish 'that those instances, of which we have had no experience, resemble those, of which we have had experience'. Consequently 'even after the observation of the frequent or constant conjunction of objects, we have no reason to draw any inference concerning any object beyond those of which we have had experience'. For 'shou'd it be said that we have experience'[10] — experience teaching us that objects constantly conjoined with certain other objects continue to be so conjoined — then, Hume says, 'I wou'd renew my question, why from this experience we form any conclusion beyond those past instances, of which we have had experience'. In other words, an attempt to justify the practice of induction by an appeal to experience must lead to an *infinite regress*. As a result we can say that theories can never be inferred from observation statements, or rationally justified by them.

I found Hume's refutation of inductive inference clear and conclusive. But I felt completely dissatisfied with his psychological explanation of induction in terms of custom or habit.

It has often been noticed that this explanation of Hume's is philosophically not very satisfactory. It is, however, without doubt intended as a *psychological* rather than a philosophical theory; for it tries to give a causal explanation of a psychological fact — the fact that we believe in laws, in statements asserting regularities or constantly conjoined kinds of events – by asserting that this fact is due to (i.e. constantly conjoined with) custom or habit. But even this reformulation of Hume's theory is still unsatisfactory; for what I have just called a 'psychological fact' may itself be described as a custom or habit – the custom or habit of believing in laws or regularities; and it is neither very surprising nor very enlightening to hear that such a custom or habit must be explained as due to, or conjoined with, a custom or habit (even though a different one). Only when we remember that the words 'custom' and 'habit' are used by Hume, as they are in ordinary language, not merely to *describe* regular behaviour, but rather to *theorize about its origin* (ascribed to frequent repetition), can we reformulate his psychological theory in a more satisfactory way. We can then say that, like other habits, *our habit of believing in laws is the product of frequent repetition* – of the repeated observation that things of a certain kind are constantly conjoined with things of another kind.

This genetico-psychological theory is, as indicated, incorporated in ordinary language, and it is therefore hardly as revolutionary as Hume thought. It is no doubt an extremely popular psychological theory – part of 'common sense', one might say. But in spite of my love of both common sense and Hume, I felt convinced that this psychological theory was mistaken; and that it was in fact refutable on purely logical grounds.

Hume's psychology, which is the popular psychology, was mistaken, I felt, about at least three different things: (a) the typical result of repetition; (b) the genesis of habits; and especially (c) the character of those experiences or modes of behaviour which may be described as 'believing in a law' or 'expecting a law-like succession of events'.

(a)   The typical result of repetition – say, of repeating a difficult passage on the piano – is that movements which at first needed attention are in the end executed without attention. We might say that the process becomes radically abbreviated, and ceases to be conscious: it becomes 'physiological'. Such a process, far from creating a conscious expectation of law-like succession, or a belief in a law, may on the contrary begin with a conscious belief and destroy it by making it superfluous. In learning to ride a bicycle we may start with the belief that we can avoid falling if we steer in the direction in which we threaten to fall, and this belief may be useful for guiding our movements. After sufficient practice we may forget the rule; in any case, we do not need it any longer. On the other hand, even if it is true that repetition may create unconscious expectations, these become conscious only if something goes wrong (we may not have heard the clock tick, but we may hear that it has stopped).

(b)   Habits or customs do not, as a rule, *originate* in repetition. Even the habit of walking, or of speaking, or of feeding at certain hours, *begins* before repetition can play any part whatever. We may say, if we like, that they deserve to be called 'habits' or 'customs' only after repetition has played its typical part; but we must not say that the practices in question originated as the result of many repetitions.

(c)   Belief in a law is not quite the same thing as behaviour which betrays an expectation of a law-like succession of events; but these two are sufficiently closely connected to be treated together. They may, perhaps, in exceptional cases, result from a mere repetition of sense impressions (as in the case of the stopping clock). I was prepared to concede this, but I contended that normally, and in most cases of any interest, they cannot be so explained. As Hume admits, even a single striking observation may be sufficient to create a belief or an expectation – a fact which he tries to explain as due to an inductive habit, formed as the result of a vast number of long repetitive sequences which had been experienced at an earlier period of life.[11] But this, I contended, was merely his attempt to explain away unfavourable facts which threatened his theory; an unsuccessful attempt, since these unfavourable facts could be observed in very young animals and babies – as early, indeed, as we like. 'A lighted cigarette was held near the noses of the young puppies', reports F. Bäge. 'They sniffed at it once, turned tail, and nothing would induce them to come back to the source of the smell and to sniff again. A few days later, they reacted to the mere sight of a cigarette or even of a rolled piece of white paper, by bounding away, and sneezing.'[12] If we try to explain cases like this by postulating a vast number of long repetitive sequences at a still earlier age we are not only romancing, but forgetting that in the clever puppies' short lives there must be room not only for repetition but also for a great deal of novelty, and consequently of non-repetition.

But it is not only that certain empirical facts do not support Hume; there are decisive arguments of a *purely logical* nature against his psychological theory.

The central idea of Hume's theory is that of *repetition, based upon similarity* (or 'resemblance'). This idea is used in a very uncritical way. We are led to think of the water-drop that hollows the stone: of sequences of unquestionably like events slowly forcing themselves upon us, as does the tick of the clock. But we ought to realize that in a psychological theory such as Hume's,

only repetition-for-us, based upon similarity-for-us, can be allowed to have any effect upon us. We must respond to situations as if they were equivalent; *take* them as similar; *interpret* them as repetitions. The clever puppies, we may assume, showed by their response, their way of acting or of reacting, that they recognized or interpreted the second situation as a repetition of the first: that they expected its main element, the objectionable smell, to be present. The situation was a repetition-for-them because they responded to it by *anticipating* its similarity to the previous one.

This apparently psychological criticism has a purely logical basis which may be summed up in the following simple argument. (It happens to be the one from which I originally started my criticism.) The kind of repetition envisaged by Hume can never be perfect; the cases he has in mind cannot be cases of perfect sameness; they can only be cases of similarity. Thus *they are repetitions only from a certain point of view*. (What has the effect upon me of a repetition may not have this effect upon a spider.) But this means that, for logical reasons, there must always be a point of view – such as a system of expectations, anticipations, assumptions, or interests – *before* there can be any repetition; which point of view, consequently, cannot be merely the result of repetition. (See now also appendix *x, (1), to my *L.Sc.D.*)

We must thus replace, for the purposes of a psychological theory of the origin of our beliefs, the naïve idea of events which *are* similar by the idea of events to which we react by *interpreting* them as being similar. But if this is so (and I can see no escape from it) then Hume's psychological theory of induction leads to an infinite regress, precisely analogous to that other infinite regress which was discovered by Hume himself, and used by him to explode the logical theory of induction. For what do we wish to explain? In the example of the puppies we wish to explain behaviour which may be described as *recognizing* or *interpreting* a situation as a repetition of another. Clearly, we cannot hope to explain this by an appeal to earlier repetitions, once we realize that the earlier repetitions must also have been repetitions-for-them, so that precisely the same problem arises again: that of *recognizing or interpreting* a situation as a repetition of another.

To put it more concisely, similarity-for-us is the product of a response involving interpretations (which may be inadequate) and anticipations or expectations (which may never be fulfilled). It is therefore impossible to explain anticipations, or expectations, as resulting from many repetitions, as suggested by Hume. For even the first repetition-for-us must be based upon similarity-for-us, and therefore upon expectations – precisely the kind of thing we wished to explain.

This shows that there is an infinite regress involved in Hume's psychological theory.

Hume, I felt, had never accepted the full force of his own logical analysis. Having refuted the logical idea of induction he was faced with the following problem: how do we actually obtain our knowledge, as a matter of psychological fact, if induction is a procedure which is logically invalid and rationally unjustifiable? There are two possible answers: (1) We obtain our knowledge by a non-inductive procedure. This answer would have allowed Hume to retain a form of rationalism. (2) We obtain our knowledge by repetition and induction, and therefore by a logically invalid and rationally unjustifiable procedure, so that all apparent knowledge is merely a kind of belief – belief based on habit. This answer would imply that even scientific knowledge is irrational, so that rationalism is absurd, and must be given up. (I shall not discuss here the age-old attempts, now again fashionable, to get out of the difficulty by asserting that though induction is of course logically invalid if we mean by 'logic' the same as 'deductive logic', it is not irrational by its own standards, as may

be seen from the fact that every reasonable man applies it *as a matter of fact*: it was Hume's great achievement to break this uncritical identification of the question of fact – *quid facti?* – and the question of justification or validity – *quid juris?*. (See below, point (13) of the appendix to the present chapter.)

It seems that Hume never seriously considered the first alternative. Having cast out the logical theory of induction by repetition he struck a bargain with common sense, meekly allowing the re-entry of induction by repetition, in the guise of a psychological theory. I proposed to turn the tables upon this theory of Hume's. Instead of explaining our propensity to expect regularities as the result of repetition, I proposed to explain repetition-for-us as the result of our propensity to expect regularities and to search for them.

Thus I was led by purely logical considerations to replace the psychological theory of induction by the following view. Without waiting, passively, for repetitions to impress or impose regularities upon us, we actively try to impose regularities upon the world. We try to discover similarities in it, and to interpret it in terms of laws invented by us. Without waiting for premises we jump to conclusions. These may have to be discarded later, should observation show that they are wrong.

This was a theory of trial and error – of *conjectures and refutations*. It made it possible to understand why our attempts to force interpretations upon the world were logically prior to the observation of similarities. Since there were logical reasons behind this procedure, I thought that it would apply in the field of science also; that scientific theories were not the digest of observations, but that they were inventions – conjectures boldly put forward for trial, to be eliminated if they clashed with observations; with observations which were rarely accidental but as a rule undertaken with the definite intention of testing a theory by obtaining, if possible, a decisive refutation.

V

The belief that science proceeds from observation to theory is still so widely and so firmly held that my denial of it is often met with incredulity. I have even been suspected of being insincere – of denying what nobody in his senses can doubt.

But in fact the belief that we can start with pure observations alone, without anything in the nature of a theory, is absurd; as may be illustrated by the story of the man who dedicated his life to natural science, wrote down everything he could observe, and bequeathed his priceless collection of observations to the Royal Society to be used as inductive evidence. This story should show us that though beetles may profitably be collected, observations may not.

Twenty-five years ago I tried to bring home the same point to a group of physics students in Vienna by beginning a lecture with the following instructions: 'Take pencil and paper; carefully observe, and write down what you have observed!' They asked, of course, *what* I wanted them to observe. Clearly the instruction, 'Observe!' is absurd.[13] (It is not even idiomatic, unless the object of the transitive verb can be taken as understood.) Observation is always selective. It needs a chosen object, a definite task, an interest, a point of view, a problem. And its description presupposes a descriptive language, with property words; it presupposes similarity and classification, which in its turn presupposes interests, points of view, and problems. 'A hungry animal', writes Katz,[14] 'divides the environment into edible and inedible things. An animal in flight sees roads to escape and hiding places. . . . Generally speaking, objects change . . . according to the needs of the animal.' We may add that objects can be classified, and can become similar or dissimilar, only in this way – by being related to needs and interests. This rule applies not only to animals but also to scientists. For the animal a point of view is provided by its needs, the task of the moment, and its expectations; for the scientist by his theoretical interests, the special problem under investigation, his conjectures and anticipations, and the theories which he accepts as a kind of background: his frame of reference, his 'horizon of expectations'.

The problem 'Which comes first, the hypothesis (H) or the observation (O),' is soluble; as is the problem, 'Which comes first, the hen (H) or the egg (O)'. The reply to the latter is, 'An earlier kind of egg'; to the former, 'An earlier kind of hypothesis'. It is quite true that any particular hypothesis we choose will have been preceded by observations – the observations, for example, which it is designed to explain. But these observations, in their turn, presupposed the adoption of a frame of reference: a frame of expectations: a frame of theories. If they were significant, if they created a need for explanation and thus gave rise to the invention of a hypothesis, it was because they could not be explained within the old theoretical framework, the old horizon of expectations. There is no danger here of an infinite regress. Going back to more and more primitive theories and myths we shall in the end find unconscious, *inborn* expectations.

The theory of inborn *ideas* is absurd, I think; but every organism has inborn *reactions* or *responses*; and among them, responses adapted to impending events. These responses we may describe as 'expectations' without implying that these 'expectations' are conscious. The new-born baby 'expects', in this sense, to be fed (and, one could even argue, to be protected and loved). In view of the close relation between expectation and knowledge we may even speak in quite a reasonable sense of 'inborn knowledge'. This 'knowledge' is not, however, *valid a priori*; an inborn expectation, no matter how strong and specific, may be mistaken. (The newborn child may be abandoned, and starve.)

Thus we are born with expectations; with 'knowledge' which, although not *valid a priori*, is *psychologically or genetically a priori*, i.e. prior to all observational experience. One of the most important of these expectations is the expectation of finding a regularity. It is connected with an inborn propensity to look out for regularities, or with a *need* to *find* regularities, as we may see from the pleasure of the child who satisfies this need.

This 'instinctive' expectation of finding regularities, which is psychologically a priori, corresponds very closely to the 'law of causality' which Kant believed to be part of our mental outfit and to be a priori valid. One might thus be inclined to say that Kant failed to distinguish between psychologically a priori ways of thinking or responding and a priori valid beliefs. But I do not think that his mistake was quite as crude as that. For the expectation of finding regularities is not only psychologically a priori, but also logically a priori: it is logically prior to all observational experience, for it is prior to any recognition of similarities, as we have seen; and all observation involves the recognition of similarities (or dissimilarities). But in spite of being logically a priori in this sense the expectation is not valid a priori. For it may fail: we can easily construct an environment (it would be a lethal one) which, compared with our ordinary environment, is so chaotic that we completely fail to find regularities. (All natural laws could remain valid: environments of this kind have been used in the animal experiments mentioned in the next section.)

Thus Kant's reply to Hume came near to being right; for the distinction between an a priori valid expectation and one which is both genetically *and* logically prior to observation, but not a priori valid, is really somewhat subtle. But Kant proved too much. In trying to show how

knowledge is possible, he proposed a theory which had the unavoidable consequence that our quest for knowledge must necessarily succeed, which is clearly mistaken. When Kant said, 'Our intellect does not draw its laws from nature but imposes its laws upon nature', he was right. But in thinking that these laws are necessarily true, or that we necessarily succeed in imposing them upon nature, he was wrong.[15] Nature very often resists quite successfully, forcing us to discard our laws as refuted; but if we live we may try again.

To sum up this logical criticism of Hume's psychology of induction we may consider the idea of building an induction machine. Placed in a simplified 'world' (for example, one of sequences of coloured counters) such a machine may through repetition 'learn', or even 'formulate', laws of succession which hold in its 'world'. If such a machine can be constructed (and I have no doubt that it can) then, it might be argued, my theory must be wrong; for if a machine is capable of performing inductions on the basis of repetition, there can be no logical reasons preventing us from doing the same.

The argument sounds convincing, but it is mistaken. In constructing an induction machine we, the architects of the machine, must decide a priori what constitutes its 'world'; what things are to be taken as similar or equal; and what *kind* of 'laws' we wish the machine to be able to 'discover' in its 'world'. In other words we must build into the machine a framework determining what is relevant or interesting in its world: the machine will have its 'inborn' selection principles. The problems of similarity will have been solved for it by its makers who thus have interpreted the 'world' for the machine.

VI

Our propensity to look out for regularities, and to impose laws upon nature, leads to the psychological phenomenon of *dogmatic thinking* or, more generally, dogmatic behaviour: we expect regularities everywhere and attempt to find them even where there are none; events which do not yield to these attempts we are inclined to treat as a kind of 'background noise'; and we stick to our expectations even when they are inadequate and we ought to accept defeat. This dogmatism is to some extent necessary. It is demanded by a situation which can only be dealt with by forcing our conjectures upon the world. Moreover, this dogmatism allows us to approach a good theory in stages, by way of approximations: if we accept defeat too easily, we may prevent ourselves from finding that we were very nearly right.

It is clear that this *dogmatic attitude*, which makes us stick to our first impressions, is indicative of a strong belief; while a *critical attitude*, which is ready to modify its tenets, which admits doubt and demands tests, is indicative of a weaker belief. Now according to Hume's theory, and to the popular theory, the strength of a belief should be a product of repetition; thus it should always grow with experience, and always be greater in less primitive persons. But dogmatic thinking, an uncontrolled wish to impose regularities, a manifest pleasure in rites and in repetition as such, are characteristic of primitives and children; and increasing experience and maturity sometimes create an attitude of caution and criticism rather than of dogmatism.

I may perhaps mention here a point of agreement with psychoanalysis. Psycho-analysts assert that neurotics and others interpret the world in accordance with a personal set pattern which is not easily given up, and which can often be traced back to early childhood. A pattern or scheme which was adopted very early in life is maintained throughout, and every new experience is interpreted in terms of it; verifying it, as it were, and contributing to its rigidity. This is a description of what I have called the dogmatic attitude, as distinct from the critical attitude, which shares with the dogmatic attitude the quick adoption of a schema of expectations – a myth, perhaps, or a conjecture or hypothesis – but which is ready to modify it, to correct it, and even to give it up. I am inclined to suggest that most neuroses may be due to a partially arrested development of the critical attitude; to an arrested rather than a natural dogmatism; to resistance to demands for the modification and adjustment of certain schematic interpretations and responses. This resistance in its turn may perhaps be explained, in some cases, as due to an injury or shock, resulting in fear and in an increased need for assurance or certainty, analogous to the way in which an injury to a limb makes us afraid to move it, so that it becomes stiff. (It might even be argued that the case of the limb is not merely analogous to the dogmatic response, but an instance of it.) The explanation of any concrete case will have to take into account the weight of the difficulties involved in making the necessary adjustments – difficulties which may be considerable, especially in a complex and changing world: we know from experiments on animals that varying degrees of neurotic behaviour may be produced at will by correspondingly varying difficulties.

I found many other links between the psychology of knowledge and psychological fields which are often considered remote from it – for example the psychology of art and music; in fact, my ideas about induction originated in a conjecture about the evolution of Western polyphony. But you will be spared this story.

VII

My logical criticism of Hume's psychological theory, and the considerations connected with it (most of which I elaborated in 1926–7, in a thesis entitled 'On Habit and Belief in Laws'[16]) may seem a little removed from the field of the philosophy of science. But the distinction between dogmatic and critical thinking, or the dogmatic and the critical attitude, brings us right back to our central problem. For the dogmatic attitude is clearly related to the tendency to *verify* our laws and schemata by seeking to apply them and to confirm them, even to the point of neglecting refutations, whereas the critical attitude is one of readiness to change them – to test them; to refute them; to *falsify* them, if possible. This suggests that we may identify the critical attitude with the scientific attitude, and the dogmatic attitude with the one which we have described as pseudo-scientific.

It further suggests that genetically speaking the pseudo-scientific attitude is more primitive than, and prior to, the scientific attitude: that it is a pre-scientific attitude. And this primitivity or priority also has its logical aspect. For the critical attitude is not so much opposed to the dogmatic attitude as super-imposed upon it: criticism must be directed against existing and influential beliefs in need of critical revision – in other words, dogmatic beliefs. A critical attitude needs for its raw material, as it were, theories or beliefs which are held more or less dogmatically.

Thus science must begin with myths, and with the criticism of myths; neither with the collection of observations, nor with the invention of experiments, but with the critical discussion of myths, and of magical techniques and practices. The scientific tradition is distinguished from the pre-scientific tradition in having two layers. Like the latter, it passes on its theories; but it also passes on a critical attitude towards them. The theories are passed on, not as dogmas, but rather with the challenge to discuss them and improve upon them. This tradition is Hellenic: it may be traced back to Thales, founder of the first *school* (I do not mean 'of the first *philosophical* school', but simply 'of the first school') which

was not mainly concerned with the preservation of a dogma.[17]

The critical attitude, the tradition of free discussion of theories with the aim of discovering their weak spots so that they may be improved upon, is the attitude of reasonableness, of rationality. It makes far-reaching use of both verbal argument and observation – of observation in the interest of argument, however. The Greeks' discovery of the critical method gave rise at first to the mistaken hope that it would lead to the solution of all the great old problems; that it would establish certainty; that it would help to *prove* our theories, to *justify* them. But this hope was a residue of the dogmatic way of thinking; in fact nothing can be justified or proved (outside of mathematics and logic). The demand for rational proofs in science indicates a failure to keep distinct the broad realm of rationality and the narrow realm of rational certainty: it is an untenable, an unreasonable demand.

Nevertheless, the role of logical argument, of deductive logical reasoning, remains all-important for the critical approach; not because it allows us to prove our theories, or to infer them from observation statements, but because only by purely deductive reasoning is it possible for us to discover what our theories imply, and thus to criticize them effectively. Criticism, I said, is an attempt to find the weak spots in a theory, and these, as a rule, can be found only in the more remote logical consequences which can be derived from it. It is here that purely logical reasoning plays an important part in science.

Hume was right in stressing that our theories cannot be validly inferred from what we can know to be true – neither from observations nor from anything else. He concluded from this that our belief in them was irrational. If 'belief' means here our inability to doubt our natural laws, and the constancy of natural regularities, then Hume is again right: this kind of dogmatic belief has, one might say, a physiological rather than a rational basis. If, however, the term 'belief' is taken to cover our critical acceptance of scientific theories – a *tentative* acceptance combined with an eagerness to revise the theory if we succeed in designing a test which it cannot pass – then Hume was wrong. In such an acceptance of theories there is nothing irrational. There is not even anything irrational in relying for practical purposes upon well-tested theories, for no more rational course of action is open to us.

Assume that we have deliberately made it our task to live in this unknown world of ours; to adjust ourselves to it as well as we can; to take advantage of the opportunities we can find in it; and to explain it, if possible (we need not assume that it is), and as far as possible, with the help of laws and explanatory theories. *If we have made this our task, then there is no more rational procedure than the method of trial and error – of conjecture and refutation: of* boldly proposing theories; of trying our best to show that these are erroneous; and of accepting them tentatively if our critical efforts are unsuccessful.

From the point of view here developed all laws, all theories, remain essentially tentative, or conjectural, or hypothetical, even when we feel unable to doubt them any longer. Before a theory has been refuted we can never know in what way it may have to be modified. That the sun will always rise and set within twenty-four hours is still proverbial as a law 'established by induction beyond reasonable doubt'. It is odd that this example is still in use, though it may have served well enough in the days of Aristotle and Pytheas of Massalia – the great traveller who for centuries was called a liar because of his tales of Thule, the land of the frozen sea and the *midnight sun*.

The method of trial and error is not, of course, simply identical with the scientific or critical approach – with the method of conjecture and refutation. The method of trial and error is applied not only by Einstein but, in a more dogmatic fashion, by the amoeba also. The difference lies not so much in the trials as in a critical and constructive attitude towards errors; errors which the scientist consciously and cautiously tries to uncover in order to refute his theories with searching arguments, including appeals to the most severe experimental tests which his theories and his ingenuity permit him to design.

The critical attitude may be described as the conscious attempt to make our theories, our conjectures, suffer in our stead in the struggle for the survival of the fittest. It gives us a chance to survive the elimination of an inadequate hypothesis – when a more dogmatic attitude would eliminate it by eliminating us. (There is a touching story of an Indian community which disappeared because of its belief in the holiness of life, including that of tigers.) We thus obtain the fittest theory within our reach by the elimination of those which are less fit. (By 'fitness' I do not mean merely 'usefulness' but truth; see chapters 3 and 10, below.) [sic] I do not think that this procedure is irrational or in need of any further rational justification.

VIII

Let us now turn from our logical criticism of the *psychology of experience* to our real problem – the problem of *the logic of science*. Although some of the things I have said may help us here, in so far as they may have eliminated certain psychological prejudices in favour of induction, my treatment of the *logical problem of induction* is completely independent of this criticism, and of all psychological considerations. Provided you do not dogmatically believe in the alleged psychological fact that we make inductions, you may now forget my whole story with the exception of two logical points: my logical remarks on testability or falsifiability as the criterion of demarcation; and Hume's logical criticism of induction.

From what I have said it is obvious that there was a close link between the two problems which interested me at that time: demarcation, and induction or scientific method. It was easy to see that the method of science is criticism, i.e. attempted falsifications. Yet it took me a few years to notice that the two problems – of demarcation and of induction – were in a sense one.

Why, I asked, do so many scientists believe in induction? I found they did so because they believed natural science to be characterized by the inductive method – by a method starting from, and relying upon, long sequences of observations and experiments. They believed that the difference between genuine science and metaphysical or pseudo-scientific speculation depended solely upon whether or not the inductive method was employed. They believed (to put it in my own terminology) that only the inductive method could provide a satisfactory *criterion of demarcation*.

I recently came across an interesting formulation of this belief in a remarkable philosophical book by a great physicist – Max Born's *Natural Philosophy of Cause and Chance*.[18] He writes: 'Induction allows us to generalize a number of observations into a general rule: that night follows day and day follows night . . . But while everyday life has no definite criterion for the validity of an induction, . . . science has worked out a code, or rule of craft, for its application.' Born nowhere reveals the contents of this inductive code (which, as his wording shows, contains a 'definite criterion for the validity of an induction'); but he stresses that 'there is no logical argument' for its acceptance: 'it is a question of faith'; and he is therefore 'willing to call induction a metaphysical principle'. But why does he believe that such a code of valid inductive rules must exist? This becomes clear when he speaks of the 'vast communities of people ignorant of, or rejecting, the rule of science, among them the

members of anti-vaccination societies and believers in astrology. It is useless to argue with them; I cannot compel them to accept the same criteria of valid induction in which I believe: the code of scientific rules.' This makes it quite clear that *'valid induction' was here meant to serve as a criterion of demarcation between science and pseudo-science.*

But it is obvious that this rule or craft of 'valid induction' is not even metaphysical: it simply does not exist. No rule can ever guarantee that a generalization inferred from true observations, however often repeated, is true. (Born himself does not believe in the truth of Newtonian physics, in spite of its success, although he believes that it is based on induction.) And the success of science is not based upon rules of induction, but depends upon luck, ingenuity, and the purely deductive rules of critical argument.

I may summarize some of my conclusions as follows:

1   Induction, i.e. inference based on many observations, is a myth. It is neither a psychological fact, nor a fact of ordinary life, nor one of scientific procedure.

2   The actual procedure of science is to operate with conjectures: to jump to conclusions – often after one single observation (as noticed for example by Hume and Born).

3   Repeated observations and experiments function in science as *tests* of our conjectures or hypotheses, i.e. as attempted refutations.

4   The mistaken belief in induction is fortified by the need for a criterion of demarcation which, it is traditionally but wrongly believed, only the inductive method can provide.

5   The conception of such an inductive method, like the criterion of verifiability, implies a faulty demarcation.

6   None of this is altered in the least if we say that induction makes theories only probable rather than certain. (See especially chapter 10, below.) [*sic*]

IX

If, as I have suggested, the problem of induction is only an instance or facet of the problem of demarcation, then the solution to the problem of demarcation must provide us with a solution to the problem of induction. This is indeed the case, I believe, although it is perhaps not immediately obvious.

For a brief formulation of the problem of induction we can turn again to Born, who writes: '. . . no observation or experiment, however extended, can give more than a finite number of repetitions'; therefore, 'the statement of a law – B depends on A – always transcends experience. Yet this kind of statement is made everywhere and all the time, and sometimes from scanty material.'[19]

In other words, the logical problem of induction arises from (*a*) Hume's discovery (so well expressed by Born) that it is impossible to justify a law by observation or experiment, since it 'transcends experience'; (*b*) the fact that science proposes and uses laws 'everywhere and all the time'. (Like Hume, Born is struck by the 'scanty material', i.e. the few observed instances upon which the law may be based.) To this we have to add (*c*) *the principle of empiricism* which asserts that in science, only observation and experiment may decide upon the *acceptance or rejection* of scientific statements, including laws and theories.

These three principles, (*a*), (*b*), and (*c*), appear at first sight to clash; and this apparent clash constitutes the *logical problem of induction*.

Faced with this clash, Born gives up (*c*), the principle of empiricism (as Kant and many others, including Bertrand Russell, have done before him), in favour of what he calls a 'metaphysical principle'; a metaphysical principle which he does not even attempt to formulate; which he vaguely describes as a 'code or rule of craft'; and of which I have never seen any formulation which even looked promising and was not clearly untenable.

But in fact the principles (*a*) to (*c*) do not clash. We can see this the moment we realize that the acceptance by science of a law or of a theory is *tentative only*; which is to say that all laws and theories are conjectures, or tentative *hypotheses* (a position which I have sometimes called 'hypotheticism'); and that we may reject a law or theory on the basis of new evidence, without necessarily discarding the old evidence which originally led us to accept it.[20]

The principle of empiricism (*c*) can be fully preserved, since the fate of a theory, its acceptance or rejection, is decided by observation and experiment – by the result of tests. So long as a theory stands up to the severest tests we can design, it is accepted; if it does not, it is rejected. But it is never inferred, in any sense, from the empirical evidence. There is neither a psychological nor a logical induction. *Only the falsity of the theory can be inferred from empirical evidence, and this inference is a purely deductive one.*

Hume showed that it is not possible to infer a theory from observation statements; but this does not affect the possibility of refuting a theory by observation statements. The full appreciation of this possibility makes the relation between theories and observations perfectly clear.

This solves the problem of the alleged clash between the principles (*a*), (*b*), and (*c*), and with it Hume's problem of induction.

X

Thus the problem of induction is solved. But nothing seems less wanted than a simple solution to an age-old philosophical problem. Wittgenstein and his school hold that genuine philosophical problems do not exist;[21] from which it clearly follows that they cannot be solved. Others among my contemporaries do believe that there are philosophical problems, and respect them; but they seem to respect them too much; they seem to believe that they are insoluble, if not taboo; and they are shocked and horrified by the claim that there is a simple, neat, and lucid, solution to any of them. If there is a solution it must be deep, they feel, or at least complicated.

However this may be, I am still waiting for a simple, neat and lucid criticism of the solution which I published first in 1933 in my letter to the Editor of *Erkenntnis*,[22] and later in *The Logic of Scientific Discovery*.

Of course, one can invent new problems of induction, different from the one I have formulated and solved. (Its formulation was half its solution.) But I have yet to see any reformulation of the problem whose solution cannot be easily obtained from my old solution. I am now going to discuss some of these re-formulations.

One question which may be asked is this: how do we really jump from an observation statement to a theory?

Although this question appears to be psychological rather than philosophical, one can say something positive about it without invoking psychology. One can say first that the jump is not from an observation statement, but from a problem-situation, and that the theory must allow us *to explain* the observations which created the problem (that is, *to deduce* them from the theory strengthened by other accepted theories and by other observation statements, the so-called initial conditions). This leaves, of course, an immense number of possible theories, good and bad; and it thus appears that our question has not been answered.

But this makes it fairly clear that when we asked our question we had more in mind than, 'How do we jump from an observation statement to a theory?' The question we had in mind was, it now appears, 'How do we jump from an observation statement to a *good* theory?' But to this the answer is: by jumping first to *any* theory and then testing it, to find whether it is good or not; i.e. by repeatedly applying the critical method, eliminating many bad theories, and inventing many new ones. Not everybody is able to do this; but there is no other way.

Other questions have sometimes been asked. The original problem of induction, it was said, is the problem of *justifying* induction, i.e. of justifying inductive inference. If you answer this problem by saying that what is called an 'inductive inference' is always invalid and therefore clearly not justifiable, the following new problem must arise: how do you justify your method of trial and error? Reply: the method of trial and error is a *method of eliminating false theories* by observation statements; and the justification for this is the purely logical relationship of deducibility which allows us to assert the falsity of universal statements if we accept the truth of singular ones.

Another question sometimes asked is this: why is it reasonable to prefer non-falsified statements to falsified ones? To this question some involved answers have been produced, for example pragmatic answers. But from a pragmatic point of view the question does not arise, since false theories often serve well enough: most formulae used in engineering or navigation are known to be false, although they may be excellent approximations and easy to handle; and they are used with confidence by people who know them to be false.

The only correct answer is the straightforward one: because we search for truth (even though we can never be sure we have found it), and because the falsified theories are known or believed to be false, while the non-falsified theories may still be true. Besides, we do not prefer *every* non-falsified theory – only one which, in the light of criticism, appears to be better than its competitors: which solves our problems, which is well tested, and of which we think, or rather conjecture or hope (considering other provisionally accepted theories), that it will stand up to further tests.

It has also been said that the problem of induction is, 'Why is it *reasonable* to believe that the future will be like the past?', and that a satisfactory answer to this question should make it plain that such a belief is, in fact, reasonable. My reply is that it is reasonable to believe that the future will be very different from the past in many vitally important respects. Admittedly it is perfectly reasonable to *act* on the assumption that it will, in many respects, be like the past, and that well-tested laws will continue to hold (since we can have no better assumption to act upon); but it is also reasonable to believe that such a course of action will lead us at times into severe trouble, since some of the laws upon which we now heavily rely may easily prove unreliable. (Remember the midnight sun!) One might even say that to judge from past experience, and from our general scientific knowledge, the future will *not* be like the past, in perhaps most of the ways which those have in mind who say that it will. Water will sometimes not quench thirst, and air will choke those who breathe it. An apparent way out is to say that the future will be like the past *in the sense that the laws of nature will not change*, but this is begging the question. We speak of a 'law of nature' only if we think that we have before us a regularity which does not change; and if we find that it changes then we shall not continue to call it a 'law of nature'. Of course our search for natural laws indicates that we hope to find them, and that we believe that there are natural laws; but our belief in any particular natural law cannot have a safer

basis than our unsuccessful critical attempts to refute it.

I think that those who put the problem of induction in terms of the *reasonableness* of our beliefs are perfectly right if they are dissatisfied with a Humean, or post-Humean, sceptical despair of reason. We must indeed reject the view that a belief in science is as irrational as a belief in primitive magical practices – that both are a matter of accepting a 'total ideology', a convention or a tradition based on faith. But we must be cautious if we formulate our problem, with Hume, as one of the reasonableness of our *beliefs*. We should split this problem into three – our old problem of demarcation, or of how to *distinguish* between science and primitive magic; the problem of the rationality of the scientific or critical *procedure*, and of the role of observation within it; and lastly the problem of the rationality of our *acceptance* of theories for scientific and for practical purposes. To all these three problems solutions have been offered here.

One should also be careful not to confuse the problem of the reasonableness of the scientific procedure and the (tentative) acceptance of the results of this procedure – i.e. the scientific theories – with the problem of the rationality or otherwise *of the belief that this procedure will succeed*. In practice, in practical scientific research, this belief is no doubt unavoidable and reasonable, there being no better alternative. But the belief is certainly unjustifiable in a theoretical sense, as I have argued (in section v). Moreover, if we could show, on general logical grounds, that the scientific quest is likely to succeed, one could not understand why anything like success has been so rare in the long history of human endeavours to know more about our world.

Yet another way of putting the problem of induction is in terms of probability. Let t be the theory and e the evidence: we can ask for $P(t,e)$, that is to say, the probability of t, given e. The problem of induction, it is often believed, can then be put thus: construct a *calculus of probability* which allows us to work out for any theory t what its probability is, relative to any given empirical evidence *e*; and show that P(t,e) increases with the accumulation of supporting evidence, and reaches high values – at any rate values greater than ½.

In *The Logic of Scientific Discovery* I explained why I think that this approach to the problem is fundamentally mistaken.[23] To make this clear, I introduced there the distinction between *probability* and *degree of corroboration or confirmation*. (The term 'confirmation' has lately been so much used and misused that I have decided to surrender it to the verificationists and to use for my own purposes 'corroboration' only. The term 'probability' is best used in some of the many senses which satisfy the well-known calculus of probability, axiomatized, for example, by Keynes, Jeffreys, and myself; but nothing of course depends on the choice of words, as long as we do not *assume*, uncritically, that degree of corroboration must also be a probability – that is to say, that it must satisfy the calculus of probability.)

I explained in my book why we are interested in theories with a *high degree of corroboration*. And I explained why it is a mistake to conclude from this that we are interested in *highly probable* theories. I pointed out that the probability of a statement (or set of statements) is always the greater the less the statement says: it is inverse to the content or the deductive power of the statement, and thus to its explanatory power. Accordingly every interesting and powerful statement must have a low probability; and *vice versa*: a statement with a high probability will be scientifically uninteresting, because it says little and has no explanatory power. Although we seek theories with a high degree of corroboration, *as scientists we do not seek highly probable theories but explanations; that is to say, powerful and improbable theories.*[24] The opposite view – that science aims at high

probability – is a characteristic development of verificationism: if you find that you cannot verify a theory, or make it certain by induction, you may turn to probability as a kind of '*Ersatz*' for certainty, in the hope that induction may yield at least that much.

I have discussed the two problems of demarcation and induction at some length. Yet since I set out to give you in this lecture a kind of report on the work I have done in this field I shall have to add, in the form of an Appendix, a few words about some other problems on which I have been working, between 1934 and 1953. I was led to most of these problems by trying to think out the consequences of the solutions to the two problems of demarcation and induction. But time does not allow me to continue my narrative, and to tell you how my new problems arose out of my old ones. Since I cannot even start a discussion of these further problems now, I shall have to confine myself to giving you a bare list of them, with a few explanatory words here and there. But even a bare list may be useful, I think. It may serve to give an idea of the fertility of the approach. It may help to illustrate what our problems look like; and it may show how many there are, and so convince you that there is no need whatever to worry over the question whether philosophical problems exist, or what philosophy is really about. So this list contains, by implication, an apology for my unwillingness to break with the old tradition of trying to solve problems with the help of rational argument, and thus for my unwillingness to participate wholeheartedly in the developments, trends, and drifts, of contemporary philosophy.

Appendix: some problems in the philosophy of science

My first three items in this list of additional problems are connected with the calculus of probabilities.

1   The frequency theory of probability. In *The Logic of Scientific Discovery* I was interested in developing a consistent theory of probability as it is used in science; which means, a statistical or frequency theory of probability. But I also operated there with another concept which I called 'logical probability'. I therefore felt the need for a generalization – for a formal theory of probability which allows different *interpretations*: (a) as a theory of the logical probability of a statement relative to any given evidence; including a theory of absolute logical probability, i.e. of the measure of the probability of a statement relative to zero evidence; (b) as a theory of the probability of an event relative to any given *ensemble* (or 'collective') of events. In solving this problem I obtained a simple theory which allows a number of further interpretations: it may be interpreted as a calculus of contents, or of deductive systems, or as a class calculus (Boolean algebra) or as propositional calculus; and also as a calculus of *propensities*.[25]

2   This problem of a *propensity interpretation of probability* arose out of my interest in Quantum Theory. It is usually believed that Quantum Theory has to be interpreted statistically, and no doubt statistics is essential for its empirical tests. But this is a point where, I believe, the dangers of the testability theory of meaning become clear. Although the tests of the theory are statistical, and although the theory (say, Schrödinger's equation) may imply statistical consequences, it need not have a statistical meaning: and one can give examples of objective propensities (which are something like generalized forces) and of fields of propensities, which can be measured by statistical methods without being themselves statistical. (See also the last paragraph of chapter 3, below, with note 35.) [*sic*]

3   The use of statistics in such cases is, in the main, to provide *empirical tests* of theories which need not be purely statistical; and this raises the question of the *refutability of statistical statements* – a problem treated, but not to my full satisfaction, in the 1934 edition of my *The Logic of Scientific Discovery*. I later found, however, that all the elements for constructing a satisfactory solution lay ready for use in that book; certain examples I had given allow a mathematical characterization of a class of infinite chance-like sequences which are, in a certain sense, the *shortest sequences* of their kind.[26] A statistical statement may now be said to be testable by comparison with these 'shortest sequences'; it is refuted if the statistical properties of the tested *ensembles* differ from the statistical properties of the initial sections of these 'shortest sequences'.

4   There are a number of further problems connected with the interpretation of the formalism of a quantum theory. In a chapter of *The Logic of Scientific Discovery* I criticized the 'official' interpretation, and I still think that my criticism is valid in all points but one: one example which I used (in section 77) is mistaken. But since I wrote that section, Einstein, Podolski, and Rosen have published a thought-experiment which can be substituted for my example, although their tendency (which is deterministic) is quite different from mine. Einstein's belief in determinism (which I had occasion to discuss with him) is, I believe, unfounded, and also unfortunate: it robs his criticism of much of its force, and it must be emphasized that much of his criticism is quite independent of his determinism.

5   As to the problem of determinism itself, I have tried to show that even classical physics, which is deterministic in a certain *prima facie* sense, is misinterpreted if used to support a deterministic view of the physical world in Laplace's sense.

6   In this connection, I may also mention the *problem of simplicity* — of the simplicity of a theory, which I have been able to connect with the content of a theory. It can be shown that what is usually called the simplicity of a theory is associated with its logical improbability, and not with its probability, as has often been supposed. This, indeed, allows us to deduce, from the theory of science outlined above, why it is always advantageous to try the simplest theories first. They are those which offer us the best chance to submit them to severe tests: the simpler theory has always a higher degree of testability than the more complicated one.[27] (Yet I do not think that this settles all problems about simplicity. See also chapter 10, section xviii, below.) [*sic*]

7   Closely related to this problem is the problem of the *ad hoc* character of a hypothesis, and of degrees of this *ad hoc* character (of '*ad hocness*', if I may so call it). One can show that the methodology of science (and the history of science also) becomes understandable in its details if we assume that the aim of science is to get explanatory theories which are as little *ad hoc* as possible: a 'good' theory is not *ad hoc*, while a 'bad' theory is. On the other hand one can show that the probability theories of induction imply, inadvertently but necessarily, the unacceptable rule: always use the theory which is the most *ad hoc*, i.e. which transcends the available evidence as little as possible. (See also my paper 'The Aim of Science', mentioned in note 28 below.)

8   An important problem is the problem of the *layers of explanatory hypotheses* which we find in the more developed theoretical sciences, and of the relations between these layers. It is often asserted that Newton's theory can be induced or even deduced from Kepler's and Galileo's laws. But it can be shown that Newton's theory (including his theory of absolute space) strictly speaking contradicts Kepler's (even if we confine ourselves to the two-body problem[28] and neglect the mutual attraction between the planets) and also Galileo's; although approximations to these two theories can, of course, be deduced from Newton's. But it is clear that neither a deductive nor an inductive inference can lead, from consistent premises, to a conclusion which contradicts them. These considerations allow us to analyse the logical relations between 'layers' of theories, and also the idea of an *approximation*, in the two senses of (*a*) The theory *x* is an approximation to the theory *y*; and (*b*) The theory *x* is 'a good approximation to the facts'. (See also chapter 10, below.) [*sic*]

9   A host of interesting problems is raised by *operationalism*, the doctrine that theoretical concepts have to be defined in terms of measuring operations. Against this view, it can be shown that *measurements presuppose theories*. There is no measurement without a theory and no operation which can be satisfactorily described in non-theoretical terms. The attempts to do so are always circular; for example, the description of the measurement of length needs a (rudimentary) theory of heat and temperature-measurement; but these, in turn, involve measurements of length.

The analysis of operationalism shows the need for a *general theory of measurement*; a theory which does not, naïvely, take the practice of measuring as 'given', but explains it by analysing its function in the testing of scientific hypotheses. This can be done with the help of the doctrine of degrees of testability.

Connected with, and closely parallel to, operationalism is the doctrine of *behaviourism*, i.e. the doctrine that, since all test-statements describe behaviour, our theories too must be stated in terms of possible behaviour. But the inference is as invalid as the phenomenalist doctrine which asserts that since all test-statements are observational, theories too must be stated in terms of possible observations. All these doctrines are forms of the verifiability theory of meaning; that is to say, of inductivism.

Closely related to operationalism is *instrumentalism*, i.e. the interpretation of scientific theories as practical instruments or tools for such purposes as the prediction of impending events. That theories may be used in this way cannot be doubted; but instrumentalism asserts that they can be best understood as instruments; and that this is mistaken, I have tried to show by a comparison of the *different functions* of the formulae of applied and pure science. In this context the problem of the *theoretical* (i.e. non-practical) function of predictions can also be solved. (See chapter 3, section 5, below.) [sic]

It is interesting to analyse from the same point of view the function of language – as an instrument. One immediate finding of this analysis is that we use descriptive language in order to talk *about the world*. This provides new arguments in favour of *realism*.

Operationalism and instrumentalism must, I believe, be replaced by 'theoreticism', if I may call it so: by the recognition of the fact that we are always operating within a complex framework of theories, and that we do not aim simply at correlations, but at explanations.

10   The problem of *explanation* itself. It has often been said that scientific explanation is reduction of the unknown to the known. If pure science is meant, nothing could be further from the truth. It can be said without paradox that scientific explanation is, on the contrary, the reduction of the known to the unknown. In pure science, as opposed to an applied science which takes pure science as 'given' or 'known', explanation is always the logical reduction of hypotheses to others which are of a higher level of universality; of 'known' facts and 'known' theories to assumptions of which we know very little as yet, and which have still to be tested. The analysis of degrees of explanatory power, and of the relationship between genuine and sham explanation and between explanation and prediction, are examples of problems which are of great interest in this context.

11   This brings me to the problem of the relationship between explanation in the natural sciences and historical explanation (which, strangely enough, is logically somewhat analogous to the problem of explanation in the pure and applied sciences); and to the vast field of problems in the methodology of the social sciences, especially the problems of *historical prediction; historicism* and *historical determinism*; and *historical relativism*. These problems are linked, again, with the more general problems of determinism and relativism, including the problems of linguistic relativism.[29]

12   A further problem of interest is the analysis of what is called 'scientific objectivity'.

I have treated this problem in several places, especially in connection with a criticism of the so-called 'sociology of knowledge'.[30]

13   One type of solution of the problem of induction should be mentioned here again (see section iv, above), in order to warn against it. (Solutions of this kind are, as a rule, put forth without a clear formulation of the problem which they are supposed to solve.) The view I have in mind may be described as follows. It is first taken for granted that nobody seriously doubts that we do, in fact, make inductions, and successful ones. (My suggestion that this is a myth, and that the apparent cases of induction turn out, if analysed more carefully, to be cases of the method of trial and error, is treated with the contempt which an utterly unreasonable suggestion of this kind deserves.) It is then said that the task of a theory of induction is to describe and classify our inductive policies or procedures, and perhaps to point out which of them are the most successful and reliable ones and which are less successful or reliable; and that any further question of justification is misplaced. Thus the view I have in mind is characterized by the contention that the distinction between the factual problem of describing how we argue inductively (*quid facti?*), and the problem of the justification of our inductive arguments (*quid juris?*) is a misplaced distinction. It is also said that the justification required is unreasonable, since we cannot expect inductive arguments to be 'valid' in the same sense in which deductive ones may be 'valid': induction simply is not deduction, and it is unreasonable to demand from it that it should conform to the standards of logical − that is, deductive − validity. We must therefore judge it by its own standards − by inductive standards − of reasonableness.

I think that this defence of induction is mistaken. It not only takes a myth for a fact, and the alleged fact for a standard of rationality, with the result that a myth becomes a standard of rationality; but it also propagates, in this way, a principle which may be used to defend *any* dogma against *any* criticism. Moreover, it mistakes the status of formal or 'deductive' logic. (It mistakes it just as much as those who saw it as the systematization of our factual, that is, psychological, 'laws of thought'.) For deduction, I contend, is not valid because we choose or decide to adopt its rules as a standard, or decree that they shall be accepted; rather, it is valid because it adopts, and incorporates, the rules by which truth is transmitted from (logically stronger) premises to (logically weaker) conclusions, and by which falsity is re-transmitted from conclusions to premises. (This re-transmission of falsity makes formal logic the *Organon of rational criticism* − that is, of refutation.)

One point that may be conceded to those who hold the view I am criticizing here is this. In arguing from premises to the conclusion (or in what may be called the 'deductive direction'), we argue from the truth or the certainty or the probability of the premises to the corresponding property of the conclusion; while if we argue from the conclusion to the premises (and thus in what we have called the 'inductive direction'), we argue from the falsity or the uncertainty or the impossibility or the improbability of the conclusion to the corresponding property of the premises; accordingly, we must indeed concede that standards such as, more especially, *certainty*, which apply to arguments in the deductive direction, do not also apply to arguments in the inductive direction. Yet even this concession of mine turns in the end against those who hold the view which I am criticizing here; for they assume, wrongly, that we may argue in the inductive direction, though not to the certainty, yet to the *probability of our 'generalizations'*. But this assumption is mistaken, for all the intuitive ideas of probability which have ever been suggested.

This is a list of just a few of the problems of the philosophy of science to which I was led in my pursuit of the two fertile and fundamental problems whose story I have tried to tell you.[31]

Notes

1   This is a slight oversimplification, for about half of the Einstein effect may be derived from the classical theory, provided we assume a ballistic theory of light.

2   See, for example, my *Open Society and Its Enemies*, ch. 15, section iii, and notes 13–14.

3   'Clinical observations', like all other observations, are *interpretations in the light of theories* (see below, sections iv ff.); and for this reason alone they are apt to seem to support those theories in the light of which they were interpreted. But real support can be obtained only from observations undertaken as tests (by 'attempted refutations'); and for this purpose *criteria of refutation* have to be laid down beforehand: it must be agreed which observable situations, if actually observed, mean that the theory is refuted. But what kind of clinical responses would refute to the satisfaction of the analyst not merely a particular analytic diagnosis but psychoanalysis itself? And have such criteria ever been discussed or agreed upon by analysts? Is there not, on the contrary, a whole family of analytic concepts, such as 'ambivalence' (I do not suggest that there is no such thing as

ambivalence), which would make it difficult, if not impossible, to agree upon such criteria? Moreover, how much headway has been made in investigating the question of the extent to which the (conscious or unconscious) expectations and theories held by the analyst influence the 'clinical responses' of the patient? (To say nothing about the conscious attempts to influence the patient by proposing interpretations to him, etc.) Years ago I introduced the term '*Oedipus effect*' to describe the influence of a theory or expectation or prediction *upon the event which it predicts* or describes: it will be remembered that the causal chain leading to Oedipus' parricide was started by the oracle's prediction of this event. This is a characteristic and recurrent theme of such myths, but one which seems to have failed to attract the interest of the analysts, perhaps not accidentally. (The problem of confirmatory dreams suggested by the analyst is discussed by Freud, for example in *Gesammelte Schriften*, III, 1925, where he says on p. 314: 'If anybody asserts that most of the dreams which can be utilized in an analysis . . . owe their origin to [the analyst's] suggestion, then no objection can be made from the point of view of analytic theory. Yet there is nothing in this fact', he surprisingly adds, 'which would detract from the reliability of our results.')

4   The case of astrology, nowadays a typical pseudo-science, may illustrate this point. It was attacked, by Aristotelians and other rationalists, down to Newton's day, for the wrong reason – for its now accepted assertion that the planets had an 'influence' upon terrestrial ('sublunar') events. In fact Newton's theory of gravity, and especially the lunar theory of the tides, was historically speaking an offspring of astrological lore. Newton, it seems, was most reluctant to adopt a theory which came from the same stable as for example the theory that 'influenza' epidemics are due to an astral 'influence'. And Galileo, no doubt for the same reason, actually rejected the lunar theory of the tides; and his misgivings about Kepler may easily be explained by his misgivings about astrology.

5   My *Logic of Scientific Discovery* (1959, 1960, 1961), here usually referred to as *L.Sc.D.*, is the translation

of *Logik der Forschung* (1934), with a number of additional notes and appendices, including (on pp. 312–14) the letter to the Editor of *Erkenntnis* mentioned here in the text which was first published in *Erkenntnis*, **3**, 1933, pp. 426 f.

Concerning my never published book mentioned here in the text, see R. Carnap's paper '*Ueber Protokollstäze*' (On Protocol-Sentences), *Erkenntnis*, **3**, 1932, pp. 215–28 where he gives an outline of my theory on pp. 223–8, and accepts it. He calls my theory 'procedure B', and says (p. 224, top): 'Starting from a point of view different from Neurath's' (who developed what Carnap calls on p. 223 'procedure A'), 'Popper developed procedure B as part of his system.' And after describing in detail my theory of tests, Carnap sums up his views as follows (p. 228): 'After weighing the various arguments here discussed, it appears to me that the second language form with procedure B – that is in the form here described – is the most adequate among the forms of scientific language at present advocated . . . in the . . . theory of knowledge.' This paper of Carnap's contained the first published report of my theory of critical testing. (See also my critical remarks in *L.Sc.D.*, note 1 to section 29, p. 104, where the date '1933' should read '1932'; and ch. 11, below, text to note 39.) [*sic*]

6 Wittgenstein's example of a nonsensical pseudo-proposition is: 'Socrates is identical'. Obviously, 'Socrates is not identical' must also be nonsense. Thus the negation of any nonsense will be nonsense, and that of a meaningful statement will be meaningful. But *the negation of a testable (or falsifiable) statement need not be testable*, as was pointed out, first in my *L.Sc.D.*, (e.g. pp. 38 f.) and later by my critics. The confusion caused by taking testability as a criterion of *meaning* rather than of *demarcation* can easily be imagined.

7 The most recent example of the way in which the history of this problem is misunderstood is A.R. White's 'Note on Meaning and Verification', *Mind*, **63**, 1954, pp. 66 ff. J.L. Evans's article, *Mind*, **62**, 1953, pp. 1 ff.; which Mr. White criticizes, is excellent in my opinion, and unusually perceptive. Understandably enough, neither of the authors

can quite reconstruct the story. (Some hints may be found in my *Open Society*, notes 46, 51 and 52 to ch. 11; and a fuller analysis in ch. 11 of the present volume.) [*sic*]

8 In *L.Sc.D.* I discussed, and replied to, some likely objections which afterwards were indeed raised, without reference to my replies. One of them is the contention that the falsification of a natural law is just as impossible as its verification. The answer is that this objection mixes two entirely different levels of analysis (like the objection that mathematical demonstrations are impossible since checking, no matter how often repeated, can never make it quite certain that we have not overlooked a mistake). On the first level, there is a logical asymmetry: one singular statement – say about the perihelion of Mercury – can formally falsify Kepler's laws; but these cannot be formally verified by any number of singular statements. The attempt to minimize this asymmetry can only lead to confusion. On another level, we may hesitate to accept any statement, even the simplest observation statement; and we may point out that every statement involves *interpretation in the light of theories*, and that it is therefore uncertain. This does not affect the fundamental asymmetry, but it is important: most dissectors of the heart before Harvey observed the wrong things – those, which they expected to see. There can never be anything like a completely safe observation, free from the dangers of misinterpretation. (This is one of the reasons why the theory of induction does not work.) The 'empirical basis' consists largely of a mixture of *theories of lower degree of universality* (of 'reproducible effects'). But the fact remains that, relative to whatever basis the investigator may accept (at his peril), he can test his theory only by trying to refute it.

9 Hume does not say 'logical' but 'demonstrative', a terminology which, I think, is a little misleading. The following two quotations are from the *Treatise of Human Nature*, Book I, Part III sections vi and xii. (The italics are all Hume's.)

10 This and the next quotation are from op. cit., section vi. See also Hume's *Enquiry Concerning Human Understanding*, section IV, Part II, and his *Abstract*,

11 *Treatise*, section xiii; section xv, rule 4.

12 F. Bäge, 'Zur Entwicklung, etc.', *Zeitschrift f. Hundeforschung*, 1933; cp. D. Katz, *Animals and Men*, ch. VI, footnote.

edited 1938 by J.M. Keynes and P. Sraffa, p. 15, and quoted in *L.Sc.D.*, new appendix *VII, text to note 6.

13 See section 30 of *L.Sc.D.*

14 Katz, op. cit.

15 Kant believed that Newton's dynamics was a priori valid. (See his *Metaphysical Foundations of Natural Science*, published between the first and the second editions of the *Critique of Pure Reason*.) But if, as he thought, we can explain the validity of Newton's theory by the fact that our intellect imposes its laws upon nature, it follows, I think, that our intellect *must succeed* in this; which makes it hard to understand why a priori knowledge such as Newton's should be so hard to come by. A somewhat fuller statement of this criticism can be found in ch. 2, especially section x, and chs 7 and 8 of the present volume. [*sic*]

16 A thesis submitted under the title '*Gewohnheit und Gesetzerlebnis*' to the Institute of Education of the City of Vienna in 1927. (Unpublished.)

17 Further comments on these developments may be found in chs 4 and 5, below. [*sic*]

18 Max Born, *Natural Philosophy of Cause and Chance*, Oxford, 1949, p. 7.

19 *Natural Philosophy of Cause and Chance*, p. 6.

20 I do not doubt that Born and many others would agree that theories are accepted only tentatively. But the widespread belief in induction shows that the far-reaching implications of this view are rarely seen.

21 Wittgenstein still held this belief in 1946; see note 8 to ch. 2, below. [*sic*]

22 See note 5 above.

23 *L.Sc.D.* (see note 5 above), ch. x, especially sections 80 to 83, also section 34 ff. See also my note 'A Set of Independent Axioms for Probability', *Mind*, N.S. **47**, 1938, p. 275. (This note has since been reprinted, with corrections, in the new appendix *ii of *L.Sc.D.* See also the next note but one to the present chapter.)

24 A definition, in terms of probabilities (see the next note), of $C(t,e)$, i.e. of the degree of corroboration (of a theory t relative to the evidence e) satisfying the demands indicated in my *L.Sc.D.*, sections 82 to 83, is the following:

$$C(t,e) = E(t,e) \, (1 + P(t)P(t,e)),$$

where $E(t,e) = (P(e,t) - P(e))/(P(e,t) + P(e))$ is a (non-additive) measure of the explanatory power of t with respect to e. Note that $C(t,e)$ is not a probability: it may have values between -1 (refutation of t by e) and $C(t,t) \leq + 1$. Statements t which are lawlike and thus non-verifiable cannot even reach $C(t,e) = C(t,t)$ upon empirical evidence e. $C(t,t)$ is the *degree of corroborability* of t, and is equal to the *degree of testability* of t, or to the *content* of t. Because of the demands implied in point (6) at the end of section I above, I do not think, however, that it is possible to give a complete formalization of the idea of corroboration (or, as I previously used to say, of confirmation).

(Added 1955 to the first proofs of this paper:) See also my note 'Degree of Confirmation', *British Journal for the Philosophy of Science*, **5**, 1954, pp. 143 ff. (See also **5**, pp. 334.) I have since simplified this definition as follows (B.J.P.S., 1955, **5**, pp. 359:)

$$C(t,e) = (P(e,t) - P(e))/(P(e,t) - P(et) + P(e))$$

For a further improvement, see B.J.P.S. **6**, 1955, pp. 56.

25 See my note in *Mind*, op. cit. The axiom system given there for elementary (i.e. non-continuous) probability can be simplified as follows ('$\bar{x}$' denotes the complement of x; 'xy' the intersection or conjunction of x and y):

(A1)	$P(xy) \geq P(yx)$	(Commutation)
(A2)	$P(x(yz)) \geq P((xy)z)$	(Association)
(A3)	$P(xx) \geq P(x)$	(Tautology)
(B1)	$P(x) \geq P(xy)$	(Monotony)
(B2)	$P(xy) + P(x\bar{y}) = P(x)$	(Addition)
(B3)	$(x) \, (Ey) \, (P(y) \neq 0 \; and$	
	$P(xy) = P(x)P(y))$	(Independence)

(C1)   If $P(y) \neq 0$, then

$P(x,y) = P(xy)/P(y)$ (Definition of relative)

(C2)   If $P(y) = 0$, then

$P(x,y) = P(x,x) = P(y,y)$ (Probability)

Axiom (C2) holds, in this form, for the finitist theory only; it may be omitted if we are prepared to put up with a condition such as $P(y) \neq 0$ in most of the theorems on relative probability. For relative probability, (A1) − (B2) and (C1) − (C2), is sufficient; (B3) is not needed. For absolute probability, (Al) − (B3) is necessary and sufficient: without (B3) we cannot, for example, derive the definition of absolute in terms of relative probability,

$$P(x) = P(x, x\bar{x})$$

nor its weakened corollary

$$(x)(Ey)\ (P(y) \neq 0 \text{ and } P(x) = P(x,y))$$

from which (B3) results immediately (by substituting for '$P(x,y)$' its definiens). Thus (B3), like all other axioms with the possible exception of (C2), expresses part of the intended meaning of the concepts involved, and we must not look upon $1 \geq P(x)$ or $1 \geq P(x,y)$, which are derivable from (B1), with (B3) or with (C1) and (C2), as 'inessential conventions' (as Carnap and others have suggested).

(Added 1955 to the first proofs of this paper; see also note 31, below.)

I have since developed an axiom system for *relative probability* which holds for finite *and* infinite systems (and in which absolute probability can be defined as in the penultimate formula above). Its axioms are:

(B1) $P(x,z) \geq P(xy,z)$

(B2)   If $P(y,y) \neq P(u,y)$ then $P(x,y) + P(\bar{x}, y) = P(y,y)$

(B3) $P(xy,z) = P(x,yz)P(y,z)$

(C1) $P(x,x) = P(y,y)$

(D1)   If $((u)P(x,u) = P(y,u))$ then $P(w,x) = P(w,y)$

(E1) $(Ex)\ (Ey)\ (Eu)\ (Ew)\ P(x,y) \neq P(u,w)$

This is a slight improvement on a system which I published in B.J.P.S., **6**, 1955, pp. 56 f.; 'Postulate 3' is here called 'D1'. (See also vol. cit., bottom of p. 176. Moreover, in line 3 of the last paragraph on p. 57, the words 'and that the limit exists' should be inserted, between brackets, before the word 'all'.)

(Added 1961 to the proofs of the present volume.)

A fairly full treatment of all these questions will now be found in the new addenda to L.Sc.D.

I have left this note as in the first publication because I have referred to it in various places. The problems dealt with in this and the preceding note have since been more fully treated in the new appendices to L.Sc.D. (To its 1961 American Edition I have added a system of only three axioms; see also *Addendum* 2 to the present volume.) [sic]

26   See L.Sc.D., p. 163 (section 55); see especially the new appendix *xvi.

27   Ibid., sections 41 to 46. But see now also ch. 10, section xviii. [sic]

28   The contradictions mentioned in this sentence of the text were pointed out, for the case of the many-body problem, by P. Duhem, *The Aim and Structure of Physical Theory* (1905; trans. by P.P. Wiener, 1954). In the case of the two-body problem, the contradictions arise in connection with Kepler's third law, which may be reformulated for the two-body problem as follows. 'Let S be any *set of pairs* of bodies such that *one* body of each pair is of the mass of our sun; then $a^3/T^2 = $ constant, for any set S.' Clearly this contradicts Newton's theory, which yields for appropriately chosen units $a^3/T^2 = m_0 + m_1$ (where $m_0 = $ mass of the sun = constant, and $m_1 = $ mass of the second body, which varies with this body). But '$a^3/T^2 = $ constant' is, of course, an excellent approximation, *provided* the varying masses of the second bodies are all negligible compared with that of our sun. (See also my paper 'The Aim of Science', *Ratio*, **1**, 1957, pp. 24 ff., and section 15 of the *Postscript* to my *Logic of Scientific Discovery*.)

29   See my *Poverty of Historicism*, 1957, section 28 and note, 30 to 32; also the Addendum 1 to vol. ii of my *Open Society* (added to the 4th edition 1962).

30 *Poverty of Historicism*, section 32; *L.Sc.D.*, section 8; *Open Society*, ch. 23 and addendum to vol. ii (Fourth Edition). The passages are complementary.

31 (13) was added in 1961. Since 1953, when this lecture was delivered, and 1955, when I read the proofs, the list given in this appendix has grown considerably, and some more recent contributions which deal with problems not listed here will be found in this volume (see especially ch. 10, below) [sic] and in my other books (see especially the new appendices to my *L.Sc.D.*, and the new *Addendum* to vol. ii of my *Open Society* which I have added to the fourth edition, 1962). See especially also my paper 'Probability Magic, or Knowledge out of Ignorance', *Dialectica*, **11**, 1957, pp. 354–374.

Chapter 2

Adolf Grünbaum

IS FREUDIAN PSYCHOANALYTIC THEORY PSEUDO-SCIENTIFIC BY KARL POPPER'S CRITERION OF DEMARCATION?

I Introduction

Karl Popper has indicted Freud's psycho-analytic theory as a non-falsifiable pseudo-science or myth (Popper 1963, chapter 1). And he has claimed that traditional *inductive* methods of theory-validation do accord good scientific credentials to psychoanalysis. Hence Popper adduced Freudian theory to lend poignancy to his general castigation of inductive confirmation as a criterion for distinguishing science from pseudo-science or non-science. Popper's censure of Freud and his general indictment of induc-tivism have each had considerable influence in our intellectual culture at large. Thus, the biolo-gist Peter Medawar (1975) has endorsed both of Popper's complaints with gusto, while the literary critic Frederick Crews credits Popper with having vindicated his own repudiation of Freudian explanations (Crews 1976; 1965, pp. 125–137). And very recently the cosmolo-gist Hermann Bondi echoed Popper's anti-inductivism amid extolling the falsifiability criterion of demarcation between science and pseudo-science as enunciated by Popper (Bondi 1976). According to this criterion, the hallmark of the scientific status of a theory is that empir-ical findings which would refute it are logically possible: Any such theory is said to be empir-ically "*falsifiable*" in the sense that the *actual*

occurrence of findings contrary to it would be the warrant of its falsity.

I have offered a critique of the major facets of Popper's falsificationist conception of scientific rationality in a sequence of four papers published in 1976 (Grünbaum 1976a, 1976b, 1976c, 1976d). In one of these articles (1976a, pp. 222–229) and in more recent publications (1977a, 1977b and 1978), I took only limited issue with his charge that psychoanalysis is immune to refutation by his standards. There I confined myself to Freud's *psychotherapy* and argued as follows: Ironically by Popper's falsifiability-criterion the claim that psycho-analytic treatment is therapeutic does qualify as scientific, while failing to pass muster as induc-tively well-founded! But I explicitly *doubted* that the Popperian falsifiability of the purported therapeutic efficacy can itself guarantee the like falsifiability of the psychogenetic and psycho-dynamic core of the Freudian theoretical corpus. Thus, I had emphasized that it is logically possible for Freud's account of pathogenesis to be true while his therapy may have no value.

But according to John Watkins, my conten-tion that Freud's therapeutic theses do indeed meet Popper's scientific standards is irrelevant as a criticism of Popper's complaint against Freud.[1] Watkins' retort is that Popper had *not* been concerned with the question "Is psychoanalysis an effective therapy?" Instead, Popper had presumably meant to deal only with the issue "Is psychoanalytical theory a scientific theory and genuinely corroborated by the clinical observations adduced in its support?" This challenge now prompts me to scrutinize Popper's handling of the specific question of *scientific status* which, according to Watkins, Popper was concerned to answer. And I shall offer the following principal contentions:

(i) Popper's charge of pseudo-science against the core of Freudian theory is predicated on a caricature of his target no less than his indictment of the traditional inductivist conception of science, (ii) just as Freud's therapeutic doctrine qualifies as falsifiable and hence as scientific *by Popper's criterion*, so also does the foundation and bulk of the psychoanalytic theoretical edifice, and (iii) Popper's declaration that "the [falsifiability] criterion of demarcation [enunciated by him] cannot be an absolutely sharp one but will itself have degrees" (1963, p. 252) is incompatible with the remainder of his account as to just how he construes his criterion. Indeed the premises he offers for attributing fuzziness to the line of demarcation do not sustain this attribution, although there may, of course, be *other* very good grounds for it. But if the partition of non-tautological theories into falsifiable and non-falsifiable subclasses *is* blurred at the interface, then the *mere* existence of such a penumbra cannot itself gainsay that Freud's theory qualifies for Popperian scientific status.

II Popper's joint misdepiction of Freudian theory and of the inductivist appraisal of its scientific status

Popper tells us (1963, pp. 33–41 and p. 256) that his twentieth century falsifiability criterion of demarcation between science and non-science is not only *new* but constitutes a *successful* advance over its major traditional predecessor, which he disapprovingly labels "inductivism." Popper is referring disparagingly to the time-honored idea that theories about natural or cultural phenomena are scientific to the extent that they are made more or less credible by so-called "induction" from observation or experiment. In his view, the inductive method cannot discriminate pseudo-science from science, despite its appeal to observation and experiment, because it "nevertheless does not come up to scientific standards" (1963, pp. 33–34). And one of Popper's major avowed reasons for his rejection of any inductivist criterion of demarcation is the following (1963, ch. 1): Psychoanalysis has much more in common with *astrology* than with the genuine sciences, and yet inductivism is bound to accord good scientific credentials to psychoanalysis. According to Popper, inductivist canons of validation are simply *helpless* to discount as insufficiently supportive the great wealth of clinical, anthropological and even experimental findings which Freudians adduce as being favorable to psychoanalysis.

Hence he tells us that when he came upon the philosophic scene in 1919, the legacy of inductivism was such that "there clearly was a need for a different criterion of demarcation" (1963, p. 256). And he considers his falsifiability criterion to be far more stringent than the inductivist one. Thus he believes that his own recipe does indict psychoanalysis as pseudo-scientific along with astrology, whereas the canons of inductivism *willy-nilly* authenticate Freud's theory and Karl Marx's theory of history as *bona fide* science.

For brevity, I shall say of a theory T that "T is I-scientific" iff it qualifies as *inductively* scientific or *well*-supported by the neo-Baconian standards of controlled inquiry: For our purposes, the attribution of inductive well-supportedness to T by all the available pertinent evidence E does *not* require that E render T more likely to be true than false, but this attribution requires that *E* confer *greater credibility* on *T* than on any of its available rivals. Vague as this inductive criterion may still be, its vagueness is *not* the target of Popper's indictment, but rather its alleged *permissiveness*. Furthermore, I shall speak of *T* as being "P-scientific" iff it meets *Popper's* standard of conceivable falsifiability. Accordingly, I shall speak interchangeably of a theory as being "P-scientific" and "P-falsifiable."

Furthermore, I intend the prefix "P" to alert us to the fact that Popper (1963, p. 112) appreciates the *fallibility* of falsifications. Thus he recognizes that the observational findings invoked in empirical refutations of theories are themselves dubitable. Furthermore, he concedes to Pierre Duhem that a presumed falsification of a major hypothesis within a wider theory is revocable. Duhem had emphasized that the falsification of the major hypothesis by adverse evidence was predicated on the truth of various collateral hypotheses or initial conditions. And Duhem then pointed out that these latter auxiliary assumptions might in fact be false while the major hypothesis may be true. Popper explicitly countenances the use of such corrigible auxiliary assumptions as part of his schema for falsifications. Hence when I assert below the P-falsifiability of a certain high-level psychoanalytic hypothesis or even of a low-level empirical claim, I wish it to be explicitly remembered that any presumed P-falsification is avowedly fallible in several ways and hence revocable. Yet elsewhere (Grünbaum 1976a, §3, Section V), I have maintained that nonetheless Popper made altogether quite insufficient allowance for the cogency of Duhem's polemic against the feasibility of deductively falsifying any one constituent hypothesis of a wider theoretical system by *modus tollens.*

Suffice it to add here that Popper did not go far enough, when he grudgingly conceded to Duhem that even a presumably refuted hypothesis may subsequently be *reinstated* with evidential warrant. This concession is insufficient, if only for the following reason: Popper's attempt to assimilate all empirical discreditation of hypotheses to the deductive model of *modus tollens* simply does *not* do justice to the important role of merely *probabilistic indicators* in such discreditation.

To avoid the complexities of an illustration from physics – say a particle detection experiment – consider the example of discrediting a tentative diagnosis of cervical cancer of the uterus. The hypothesis that such cancer is present does *not* guarantee deductively via a causal law that cancer cells will be found in the Pap test. Hence a *negative* Pap test outcome does *not* deductively refute a diagnosis of cervical cancer. Indeed, there are *bona fide* cases of such cancer whose actual existence is *belied* by a negative cell test. Moreover, there are diseases that can be *asymptomatic*, so that the absence of an otherwise common manifestation of such a disease does not disprove its presence. Yet there might well be telltale manifestations of the disease that are even pathognomonic for it. These examples are mere illustrations of the following medical commonplace: The discreditation of a given diagnostic hypothesis by a negative test outcome may be only *spurious* because that negative outcome is only a *probabilistic* indicator! In order to call attention to such spurious discreditation of an actually *true* diagnosis, physicians caution against so-called "false negatives." And thus it can only be cold comfort to Popper that doctors also tell us to beware of "false positives."

As I see it, the upshot of such considerations is this: In various *bona fide* sciences, and even in physics – which is Popper's paradigm of a genuine science – his requirement of strictly deductive falsifiability cannot generally be met. Thus, suppose that some branch of psychoanalysis can be discredited by potential findings via merely probabilistic empirical indicators.

In that event, it would boomerang if Popperians were to object that such discreditability of Freudian hypotheses does not measure up to *deductive* falsifiability, and hence does not redound to the P-scientificality of psychoanalysis. This objection would commit the Popperian to throw out the allegedly clean baby of physics with the bathwater of psychoanalysis.

Let us now lay the groundwork for examining Popper's contention that Freud's theory poignantly illustrates the imperativeness of supplanting the inductivist criterion of demarcation by his purportedly more stringent one.

In any given context of inquiry, it can usually be taken for granted what kinds of properties and relations of certain individuals are considered "observable" or "empirical." Indeed, this is done more or less explicitly in some of the literature on the experimental investigation of psychoanalysis (e.g., in, Kline 1972, pp. 2–3; Silverman 1976, pp. 621–622, and Eysenck and Wilson 1973, p. 113). Hence it can also be taken as more or less explicitly understood in the given context, what ascriptions of properties or relations count as empirical propositions. Now suppose that for a certain domain of occurrences thus characterized as "empirical," a theory T is offered as providing explanations and perhaps also predictions of some of these observable events. Popper can hardly deny that *well before him*, inductivists insisted that *the construal of* T *at any given stage of its development* be such as to allow the fulfillment of the following requirement by any empirical statement S which is compatible with T: If, at a particular time, S is *declared* to be a logical consequence of T under the assumption of stated initial conditions, or is declared not to be such a consequence, then *neither* declaration is allowed to *depend* on *knowing* at the time whether S is true. More specifically, suppose that – under the assumption of given initial conditions – T is to warrant that S be declared a deductive *consequence* of T, or as probabilistically expected on the basis of T to a stated degree. Then the declaration that S is such a deductive or probabilistic *consequence* of T must *not* be made to require that S has already been observationally found to be true at the time of declaration. For brevity, I shall refer to this restriction as "the declared consequence restriction." It was imposed by inductivists at least implicitly to preclude "retroactive" tampering with the very entailments of T as follows: The construal of T is so protean that S is only *ex post facto* held to have followed from T after S has been found to be true.

Now assume this venerable inductivist restriction on the construal of a theory *vis-à-vis* its declared commitments. And consider Heinz Hartmann's neo-Freudian version of the *metapsychology* of psychic energy, a version which was cut off from the neurobiological moorings that Freud had originally aspired to furnish for his metapsychology. Then in Hartmann's version, metapsychology may perhaps be devoid of any *determinate* empirical consequences in the domain of phenomena to which it is addressed. Corresponding remarks apply to at least some construals of the doctrine of fatalism. Clearly, any such empirically sterile theory A fails to be *testable* by findings which count as "observational" in the given context. The obvious reason is that *no empirical findings can be evidentially or probatively relevant to A, be it favorably or unfavorably.* Hence no such theory A is inductively supportable at all by conceivable empirical findings, *any more* than it is falsifiable by logically possible observational results. Thus, any such theory A will fail to be I-scientific, *no less* than it fails to be P-scientific. Therefore, at least with respect to theories of kind A, the inductivist demarcation criterion is clearly *not* more permissive than Popper's criterion.

This capability of the inductivist criterion to match Popper's in restrictiveness is illustrated by the *equally negative* verdicts which have been obtained when these respective criteria were employed by Ernest Nagel (1959, pp. 40–43) and Frederick Crews (1976) to assess the scientific status of Heinz Hartmann's version of Freud's metapsychology, which pertains to the distribution and transformation of psychic energy.

Popper points to Freudian and Adlerian psychology as telling illustrations of his thesis that the inductivist demarcation criterion must be supplanted by his own falsifiability requirement (1963, ch. I). He relates (1963, p. 35) that he himself "could not think of any human behavior which could not be interpreted in terms of either theory." And moreover he reports that champions of Freud and Adler made the following rash claim: For any and all actual human behavior, their favorite psychological theory "fits" the behavior in question *and* is indeed "confirmed" by it. Popper recalls his rejection of this extravagant claim of ubiquitous confirmation by saying (1963, p. 35):

> I could not think of any human behavior which could not be interpreted in terms of either theory. It was precisely this fact – that they always fitted, that they were always confirmed [sic!] – which in the eyes of their admirers constituted the strongest argument in favour of these theories. It began to dawn on me that this apparent strength was in fact their weakness.

But unfortunately and astonishingly, Popper took no cognizance at all of the following cardinal fact: Allegations of wholesale confirmation of a largely deterministic psychological theory by every case of actual or conceivable human behavior, no matter what the initial conditions, transgress the declared consequence restriction. And since they transgress this restriction, such allegations of super-abundant confirmation plainly violate the inductivist canons of theory-validation. By blithely ignoring this fact,

Popper paved the way for his gross misportrayal of the inductivist demarcation criterion as effectively sanctioning the following patent blunder: The slide from the mere presumed or actual compatibility of given behavior B with a psychological theory T to the claim that B inductively supports T. Thus, Popper has caricatured post-Baconian inductivism by saddling its demarcation criterion with the pseudo-confirmational sins of those Freudian and Adlerian advocates who "saw confirming instances everywhere." These advocates deemed their pet theory to be such that "Whatever happened always confirmed it" (Popper 1963, p. 35). Moreover, Popper simply takes it for granted without any further ado that these epistemological excesses of some of Freud's disciples plainly derive from a corresponding logical and methodological laxity in Freud's conception of the explanatory import of his theory.

Indeed, as I shall now illustrate, Popper is no less slipshod in his account of Freudian explanations than in his depiction of inductivism. In fact, his caricature of the latter will turn out to be of-a-piece with his misportrayal of the former. He writes (1963, p. 35):

> ... every conceivable case could be interpreted in the light of Adler's theory, or equally of Freud's. I may illustrate this by two very different examples of human behaviour: that of a man who pushes a child into the water with the intention of drowning it; and that of a man who sacrifices his life in an attempt to save the child. Each of these two cases can be explained with equal ease in Freudian and in Adlerian terms. According to Freud the first man suffered from repression (say, of some component of his Oedipus complex), while the second man had achieved sublimation. According to Adler the first man suffered from feelings of inferiority (producing perhaps the need to prove to himself that he dared to

commit some crime), and so did the second man (whose need was to prove to himself that he dared to rescue the child). I could not think of any human behaviour which could not be interpreted in terms of either theory.

I deem Popper's criticism here to be slipshod for several reasons as follows.

Popper claims that each of his two examples of behavior are interpretable in terms of Freud's theory, which presumably means that they are at least logically compatible with this theory. But furthermore he tells us that "each of these cases can be *explained* [deductively or probabilistically] with equal ease in Freudian or Adlerian terms" (italics added). Just for the sake of argument, let us make the following quite questionable assumption for now: Freud's theory as such would countenance the postulation *at will* of the particular initial conditions stated by Popper, regardless of the availability of any independent evidence of their fulfillment. Even then, it is at best very unclear that Freudian theory yields *explanations* of Popper's two cases of behavior by deductive or probabilistic subsumption, as he maintains here without giving the requisite details. Of course, I do not deny that the Freudians whom Popper encountered may well have euphorically asserted easy explainability with abandon in such cases. But at least *prima facie*, such an assertion is gratuitous. And so long as it remains unsubstantiated, it provides warrant only for censuring the intellectual conduct of these enthusiasts, *not* also for Popper's logical indictment of Freud's theory *as such*. Besides, why would it necessarily be a liability of psychoanalysis, if it *actually* could *explain* the two cases of behavior with equal ease? Presumably there actually are such instances of self-sacrificing child-*rescuing* behavior no less than such cases of *infanticidal* conduct. And a fruitful psychological theory might well succeed in *actually* explaining each of them, perhaps even *deductively*.

Moreover, even if we assume that Popper's assumed initial conditions do generate the explanations claimed by him, why does Popper simply take it for granted without careful documentation that the permissiveness of psychoanalytic theory in regard to postulating potentially explanatory initial conditions is *wholly unbridled*? Furthermore, is it clear that the postulation of initial conditions *ad libitum* without any *independent* evidence of their fulfillment is quite generally countenanced by that theory to a far greater extent than in, say, physics, which Popper deems to be a *bona fide* science?

To illustrate this comparison with physics quite simply, let us take Newton's second law of motion, which states that the *net* force on a mass particle is equal to the product of its constant mass and its acceleration. Now suppose it were observed that the acceleration of a particular particle is *non-zero* even though all the *known* kinds of forces that are recognized to be acting on it yield a zero net force. Clearly, if we were to assume the initial condition that the net force is actually zero, then the posited acceleration would falsify Newton's second law. Hence let us ask: Would Newton's theory *interdict* the postulation of an *alternative* initial condition that would enable his second law to explain the supposed acceleration? Surely Newton's mechanics does allow that there might be as yet unknown kinds of forces. And hence it does not preclude that, unbeknownst to us, one of these is acting on the given particle such that the net force on the particle is non-zero and can explain the supposed acceleration. It would seem, therefore, that there is nothing in Newton's theory of motion to rule out the postulation of the latter quite speculative initial condition, although there is no *independent* evidence of its fulfillment. Yet this postulational feasibility within Newton's theory is *not* tantamount to an outright epistemological endorsement of such an *ad hoc* procedure by that theory.

Let us briefly return to Popper's example of the two men, one of whom tries to drown a child, while the other perishes in a vain attempt to save it. Popper gives no indication whatever whether either Freud or Adler actually stated anywhere how they would propose to explain either case of behavior, if at all. Yet this seemingly contrived example is Popper's *centerpiece* illustration of his cardinal thesis that Freud's explanatory hypotheses are not falsifiable, *because* they are inevitably confirmed, *come what may*!

I am simply astonished that Popper did not see fit to try to deal with some of the hypotheses actually invoked by Freud to furnish explanations, such as his famous infantile seduction aetiology of hysteria. And I cannot help wondering, for example, whether Popper knew the following: Freud had hoped to make his reputation by solving the riddle of hysteria, but in 1897 he was painfully driven by unfavorable evidence to abandon his strongly cherished aetiological hypothesis. More generally, upon looking at the actual development of Freud's thought, one finds that, as a rule, his repeated *modifications* of his theories were clearly motivated by evidence and hardly idiosyncratic or capricious. Why, I ask, were Popper and his followers *not* given pause by their obligation to carry out some actual exegesis of Freud?

Whatever the answer to this question, let me briefly consider Freud's avowed 1897 rationale for *abandoning* his erstwhile hypothesis of the aetiology of hysteria. On this hypothesis, actual episodes of traumatic seduction in childhood by close relatives were the *causally necessary* though *not* sufficient pathogens of adult hysteria. My purpose just now is to point out the following: Freud was driven to discount as spurious the original *prima facie confirmations* that the hypothesized seductions of hysterical patients had actually occurred. And far from relying on such confirmations to render his aetiological hypothesis unfalsifiable, he rejected this hypothesis for

several explicit reasons. One of these reasons is furnished by means of the following ancillary facts: The incidence of hysteria was unexpectedly high, but sexual molestation in childhood was causally quite *insufficient* to generate hysteria. Hence if actual childhood seductions were causally *necessary* for hysteria as Freud had hypothesized, then the required incidence of perverted acts against children was *preposterously* high, even in the face of the attempted concealment of these transgressions by the guilty adults. And this *over-taxed* Freud's own belief in his seduction aetiology.

The mentioned *prima facie* confirmations of the postulated seduction episodes had been furnished by the seemingly vivid and presumably repressed memories that Freud had been able to elicit from his hysterical patients in the course of their analysis. But very strong *extra-clinical* evidence had induced him to discount the reliability of the subjective certainty felt by his adult patients in the reality of purported memories going back to childhood (cf. Freud's Letter 69 to Wilhelm Fliess, dated 21.9.1897).

I have indicated why I contend that Popper's parody of inductivism is of-a-piece with his regrettably cursory wholesale dismissive account of all Freudian explanations as inevitably "confirmed" and hence unfalsifiable pseudo-explanations. It will not come as a surprise therefore that, in my forthcoming book *Is Freudian Psychoanalysis a Pseudo-Science?* ("FPPS"), I am able to demonstrate the following result: At least for a large class of important actual theories featuring universal *causal* hypotheses, the *logical possibility* of empirical instances that neo-Baconian inductivist canons would countenance as genuinely supportive *goes hand-in-hand* with the logical possibility of *falsifying* empirical instances. But this very class of theories comprises Freud's psychogenetic, aetiological and psychodynamic hypotheses! As a corollary, it then follows that if Freud's theory were not to qualify as P-scientific, then it also could not qualify as I-scientific. Indeed, for the specified large class of theories, the inductivist criterion of demarcation is *more stringent* than Popper's rather than less so: Clearly, the mere *logical possibility* of an empirical P-falsification can hardly vouchsafe that sufficient inductively supporting evidence will *actually materialize* via appropriate predictive or explanatory success. And it is altogether fair to Popper to assess the comparative stringency of his P-scientificality *vis-à-vis* the stated criterion of I-scientificality. After all, it was *he* who chose these very criteria for just such comparison when adducing Freud's theory as an illustration of the alleged greater restrictiveness of his P-falsifiability! Yet, as we just noted, the rank-ordering with respect to stringency is precisely the opposite, as illustrated by Freud's theory, which might well be P-scientific without necessarily also being I-scientific. Hence, contrary to Popper, the case of Freud's theory can hardly exhibit the imperativeness of supplanting the inductivist criterion of demarcation by his purportedly more stringent one. Moreover, ironically, Freudian developmental and psychodynamic theory will indeed soon turn out to be P-scientific.

Before arguing for this contention, let me ask the following: Are there any *other* kinds of actual interesting theories that might qualify as inductively scientific while turning out to be pseudo-scientific by Popper's criterion? And if so, does the actual judgment of the scientific community vindicate Popper in the case of such theories? If Popper is not vindicated in this way, I shall ask: As between the two presumedly diverging norms of scientificality, can we decide which one ought to be adopted as being more conducive to the attainment of the aims of science?

In order to answer the first of these three questions, take the hypothesis H that "There are black holes." And consider H pretty much in isolation from the bulk of the remaining

physical theory within which it is embedded. As stated by physicists, H says that there are black holes *somewhere* in space-time without telling us just where. Then if our space-time is suitably infinite, a spatio-temporally unrestricted existential hypothesis like our H cannot be refuted deductively even by our *continued* failure to find a black hole in the *limited* regions of this space-time that are accessible to us. The simple reason is that such failure does not rule out the existence of black holes in as yet unexplored regions. Yet if we reliably detect one or more black holes somewhere – hopefully without being sucked into them! – H is inductively well-confirmed and thus inductively scientific. Hence I ask: Does our kind of existential hypothesis actually show that, at least for *such* hypotheses, Popper's criterion is more stringent than the traditional inductive one? And if so, should we then simply indict black hole physics as pseudo-scientific with thanks to Popper?

Let me point out that Popper (1963, p. 258) was careful to speak of "*isolated*" hypotheses when he indicted spatio-temporally unrestricted existential hypotheses as pseudo-scientific in virtue of their *non*-falsifiability. I take his qualification of isolatedness to indicate his awareness that our hypothesis H is not tested in isolation but rather as part of a wider network of physical postulates. And hence I take him to *allow* that in that wider context the following be the case: Even the given kind of existential hypothesis may perhaps *contribute* to the *falsifiable* empirical content of the theoretical network and thus be P-scientific in its larger context.

But now suppose *instead* that Popper construes his criterion more narrowly such that it yields a verdict of "pseudo-scientific" on our H after all. Then I must ask: Why is it not otherwise rational and fruitful to countenance the black hole hypothesis as scientific *once it is inductively confirmed*, even if it is pseudo-scientific *by Popper's criterion*? If the hypothesis is confirmed and fruitful, would

Popper's criterion not be stultifying for scientific activity?

A philosopher can try to legislate norms of scientificality and then let the chips fall where they may. But philosophical legislation that is unencumbered by a concern with such scientific aims as fruitfulness runs a serious risk: The philosopher who presumes to sit on the legislative pedestal may be left to contemplate his own normative navel. This consideration may apply not only to Popper's account of the scientific value of the stated kind of existential hypothesis but also to his treatment of irreducibly *statistical* theories, which do pose a serious problem for Popper's criterion of demarcation.

In any case, the kind of existentially quantified hypothesis typically encountered in psychoanalysis is spatio-temporally restricted or finitist in the respects relevant to its empirical falsifiability. For example, Freud conjectured that *all* cases of depression result from the loss of *some* emotionally important object, *usually* a loved person. And I take it that this universal but existentially finitist statement has already been discredited by the evidence for the existence of the so-called "non-reactive" depressions alongside the "reactive" ones.

III The P-scientificality of Freudian psychodynamic theory

I shall present several sets of examples of Freudian hypotheses that I deem to be P-falsifiable.

1. In Freud's typology of the "oral" character, preoccupation with oral satisfactions is claimed to tend to be associated with dependency, submissiveness, need for approval, and pessimism. And in his typology of the "anal" sort of character, we are told that there is a clustering tendency among the triad of orderliness, parsimony and obstinacy. Beyond the latter claims of correlations in adulthood, Freud conjectured

that these character types originated causally in such unfavorable infantile experiences as premature weaning and fierce toilet training. Yet, Eysenck and Wilson (1973, p. 96) point out that only the postulated anal *aetiology* of the obsessive compulsive trait cluster but not the delineation of that cluster itself was original with Freud. Of course, the *label* "anal" for that presumed constellation of traits derives from the pathogenesis that Freud had posited for it.

Freud's typologies of the oral and anal characters as coupled with his avowals of corresponding aetiologies are at least *prima facie* P-falsifiable hypotheses. Hence one wonders why Popper felt entitled to dismiss psychoanalysis flatly as P-unscientific so cursorily. On what grounds would he discount the *prima facie* P-falsifiability of Freud's aetiologies as being only spurious?

2. Ten years before Popper precociously formulated his demarcation criterion at the age of seventeen, Freud himself had recognized (1909) the following refutation: The best available evidence concerning the actual life history of his "Rat Man" Paul Lorenz had *refuted* his prior hypothesis as to the specifics of the sexual aetiology which he had postulated for adult obsessional neurosis (cf. Glymour 1974, pp. 299–304). Similarly for Freud's abandonment (1905 and 1914, pp. 299–300) of his erstwhile 1896 view of the role of passive infantile sexual experiences (of seduction) in the traumatic aetiology of hysteria. Clearly, the actual P-falsification of a Freudian hypothesis H (under the assumption of suitable auxiliary assertions) assures that H is P-falsifiable and hence P-scientific, though *not* conversely. And the major issue here is, of course, only whether the central tenets of psychoanalytic theory collectively have the weaker property of P-falsifiability. Hence I must now go on to illustrate that the P-falsifiability of Freud's developmental theory, just exemplified from the Rat Man case, is *not*

confined to relatively "peripheral" or "inessential" parts of his psychogenetic edifice.

3. In a recent paper (Holmes 1974), a leading laboratory investigator of memory selectivity and ego-threat has argued in detail that there is indeed observational evidence *adverse* to the Freudian doctrine of repression. And the paramount role of repression in the psychoanalytic corpus is emphasized by Freud's declaration (1914, pp. 297–298) that "the doctrine of repression is the foundation-stone on which the whole structure of psychoanalysis rests, the most essential part of it." For example, repression is held to be a pathogenic defense mechanism, because it induces neurotic anxiety (Kline 1972, p. 152).

4. There is evidence which P-falsifies Freud's theory of dreams. In their monumental recent book, Fisher and Greenberg (1977, p. 394) have summarized the import of this evidence as follows: "His understanding of the nature of dreaming has been contradicted by many scientific observations. There is no empirical backing for his thesis that the dream is a camouflage wrapped around an inner concealed wish. Likewise, the accumulated research grossly contradicts his theory that the dream functions importantly to preserve sleep." Furthermore, as I explain in my book FPPS, Freud's view of dreams as being wish-fulfilling is *not* rendered unfalsifiable by his handling of "counter-wish" dreams and of "masochistic" ones.

By not having come to grips with details of the kind we have illustrated, Popper failed to address adequately the issue of the P-falsifiability of Freud's psychogenetics. And I trust that I have provided enough examples to show that much of the developmental and psychodynamic subtheory of the Freudian corpus can reasonably be held to be P-falsifiable and hence P-scientific. Popper's indictment of psychoanalysis as intrinsically *not* being P-falsifiable has unwarrantedly derived plausibility from his failure to allow for

the following distinction: The (revocable) falsifiability of the theory as such in the context of its semantic anchorage is a logical property of the theory itself, whereas the tenacious unwillingness of the majority of its defenders to accept adverse evidence as refuting is an all-too-human property of these advocates. Thus lack of methodological honesty on the part of the defenders of a theory and even on the part of its originator does *not* necessarily render the theory itself unfalsifiable.

This distinction was unfortunately played down by Michael Martin (1964, p. 90). He emphasizes that the putative logical falsifiability of psychoanalytic theory would *not* vouchsafe the methodological honesty of its proponents. And his requirements for falsifiability seem to be so stringent as to be utopian even for physics.

But Hans Eysenck (Eysenck and Wilson 1973, p. 109) has impugned the P-falsifiability of psychoanalytic theory. As he sees it, Freud's psychodynamic edifice seems to have a quite *specific* kind of *built-in* protection against refutation, which can be harnessed to accommodate any observed behavior that is contrary to its predictions. On Eysenck's view, Freud did *not* restrict the *post hoc* postulation of the following sort of *rescuing* event: There was a conversion of the originally predicted behavior into the actually observed *contrary* conduct by the intervention of the so-called mechanism of "reaction formation." Eysenck believes that there is nothing in Freud's theory to hamper just such a retrospective transmutation of *every* predictive fiasco into either a supportive explanatory triumph or, at worst, into an outcome that does not jeopardize the theory. Let me now give a very brief ancillary sketch of Freud's notion of reaction-formation in order to facilitate my impending espousal of the following conclusion: Popper's complaint of non-falsifiability against Freud cannot be vindicated on the strength of the purported extricating role of reaction-formation within psychoanalysis.

In Freud's 1911, he hypothesized the aetiology of male paranoia along the following lines. Given the social taboo on male homosexuality, the failure to repress homosexual impulses may well issue in feelings of severe anxiety and guilt. And the latter anxiety could then be eliminated by converting the love emotion "I love him" into its opposite "I hate him," a type of transformation which Freud labeled "reaction formation." When the defense of reaction formation proves insufficient to alleviate the anxiety, however, the afflicted party may resort to the further defensive maneuver of "projection" in which "I hate him" is converted into "He hates me." This final stage of the employment of defenses is the full-blown paranoia (cf. Sears 1943, p. 71, and Kline 1972, p. 264). Earlier illustrations of reaction formation had been given by Freud in 1908 (cf. Kline 1972, p. 153).

Interestingly enough, even the pro-Freudian psychologist Paul Kline makes the following comment (1972, p. 153) on the latter illustrations:

> . . . reaction-formations create attitudes opposite to those being defended against. Thus . . . disgust is a reaction-formation against pleasure in handling faeces. It is to be noted that this kind of concept gives a bad reputation to psychoanalysis. Thus even if the observed behaviour is quite opposed to prediction it can always be interpreted as a reaction-formation.

In order to lend specificity to our scrutiny of this claim of non-falsifiability, let us recall the previously mentioned Freudian aetiology of the *oral* personality type. For brevity, we shall refer to that typology *and* aetiology as "the Freudian orality hypothesis." In 1957, the psychologist A. Scodel reported the results of an experiment that he had designed to test one rather strong

construal of the Freudian orality hypothesis (reprinted in Eysenck and Wilson 1973, pp. 102–110). On that construal, males exhibiting the greater-than-average dependency trait of the oral trait cluster should also display a correspondingly greater preference for *large-breasted* females. But Scodel found that men exhibiting greater dependency on the thematic apperception test preferred *small*-breasted women. And he adduced this result as being contrary to the Freudian orality hypothesis (ibid., pp. 109–110).

Paul Kline (1972, pp. 91–92) discussed Scodel's claim. And Kline was presumably referring to how "confirmed Freudians" would react to Scodel's findings, when he declared: "The fact that small rather than large breasts were preferred would have to be attributed to reaction-formation." Kline's discussion of Scodel's results fueled Eysenck's contention that Freudians lend substance to Popper's charge of non-falsifiability by licensing the *post hoc* postulation of reaction-formation. Responding to Kline's comments, Eysenck writes (Eysenck and Wilson 1973, p. 109):

> These comments clearly illustrate the unscientific nature of some Freudian theorizing. True, Scodel's results can be "explained" in terms of the theory if the concept of "reaction-formation" is invoked, but would this defence mechanism have been called upon if the results had not turned out that way? There was no mention of reaction-formation at the time the [orality] hypothesis was formulated.

Eysenck shares Popper's judgment that modern physics and astronomy are P-scientific. Hence his response to Kline here prompts me to offer four critical retorts to Eysenck as follows:

1. Concerning his objection that reaction-formation was being invoked only *post hoc*, I ask:

Would the astronomers have postulated the existence of an extra-Uranian planet Neptune, if the observed orbit of Uranus had not turned out to *disagree* with the one calculated by means of Newtonian theory in conjunction with the perturbations by the intra-Uranian planets? (cf. Grünbaum 1976d). Thus, was Neptune's postulation not patently *post hoc?* And was not the neutrino likewise postulated *post hoc* by Pauli about 1930 to rescue energy conservation in the theory of radioactive nuclear disintegration? Indeed, the neutrino was postulated well before the 1956 neutrino detection experiment of Cowan and Reines was even *envisioned*, and before the neutrino was known to have any other theoretical sanction!

2. Eysenck himself (Eysenck and Wilson 1973, p. 5) characterizes as *ad hoc* Newton's postulation of an auxiliary hypothesis for the sake of rescuing his corpuscular optics from refutation. And Eysenck likens that procedure to the postulation of Freudian reaction-formation. Newton had to explain within the framework of his particle theory of light why, if all light corpuscles are alike, both reflection and refraction occur simultaneously when a light beam hits a water surface. In order to do so, he postulated that upon reaching the interface, some of the corpuscles were in a "fit of easy transmission" while others were in a "fit of easy reflection." For argument's sake, let me grant Eysenck's complaint against this optical auxiliary hypothesis.

Then I ask: Is there any warrant for his claim that just as fits of easy reflection or transmission are a supposedly sterile insular *post hoc* dangler in Newton's optics, so also there is nothing in Freud's theory to indicate just when the *post hoc* invocation of reaction-formation could be *independently* legitimated? I think not. As we recall, in Freud's aetiology of male paranoia, reaction-formation is characterized as a defense needed to cope with the presence of a specified kind of

anxiety. The occurrence of such anxiety as the trigger of a reactive transition from "I love him" to "I hate him" ought to be testable, at least in principle, in a longitudinal study of young homosexual males. More generally, Freud's theory can therefore be held to envision *independent* evidence for the *actuation* of reaction-formation by the pertinent kind of anxiety. Hence suppose now that in the case of Scodel's study, there is *no separate evidence* for assuming that the psychologically dependent males were specifically anxious about their presumed initial preference for large breasts. Why then does Eysenck suppose that Freud's theory nonetheless *licenses* or *endorses* the *ad hoc* postulation of reaction-formation to explain Scodel's opposite preference findings at the cost of courting non-falsifiability? It may well be true that there is nothing in Freud's theory to *preclude* outright such an *ad hoc* invocation of reaction-formation. But if this is to be damaging to psychoanalytic theory, why is it not similarly detrimental to Newton's mechanics that it does not interdict outright the *ad hoc* postulation of as yet unknown dynamical agencies which, as we saw, can serve as a rescuing *deus ex machina* for the second law of motion?

3. The actuation of reaction-formation by anxiety seems to me to be testable in yet another way *as part* of the empirical scrutiny of the Freudian aetiology of male paranoia. For if the latter is right, the decline of the taboo on homosexuality in our society should be accompanied by a decreased incidence of male paranoia under the proviso that no other potential causes of it become operative so as to compensate for that decrease. And, by the same token, there ought to have been relatively less paranoia in those ancient societies in which male homosexuality was condoned or even sanctioned.

4. Let us suppose, merely for the sake of argument, that Freud's theory does explicitly give free rein to the assumption of reaction-formation in

order to deal with otherwise anomalous findings. But, as we saw, reaction-formation is specifically a defense mechanism against anxiety. How then can even its entirely untrammeled *post hoc* postulation disqualify or neutralize all of the *diverse* examples of P-falsifiability of psychoanalysis that I have adduced here and in my FPPS? For example, just how could reaction-formation altogether rescue the theory of dreams?

I conclude that in view of the very subtleties of Freud's theory, the putative unbridled resort to reaction-formation by its unshakable defenders can not cogently be used to impugn the P-falsifiability of that theory itself. Indeed, as I read the literature, the scientific liabilities of Freud's central hypotheses do *not* derive from their failure to be collectively P-scientific. Instead, what troubles me about them is the rather marginal status of their I-scientificality and *P-corroboration* over and above their P-falsifiability.

IV Psychoanalysis *vis-à-vis* Popper's degrees of P-scientificality

Popper tells us that "the [P-falsifiability] criterion of demarcation cannot be an absolutely sharp one but will itself have degrees" (1963, p. 252). It behooves us to examine the reasoning which he offers as his basis for this contention. For, *prima facie*, it may seem that if the line of demarcation is thus blurred, its fuzziness could be a double-edged sword. On the one hand, it might perhaps serve to impugn Popper's own *categorical* charge of pseudo-science against psychoanalysis. On the other hand, a penumbra at Popper's interface between science and pseudo-science possibly might gainsay my opposing thesis that Freud's theory is P-scientific (cf. Farrell 1970, pp. 502–503 and Cioffi 1970). Hence let us see how Popper reasoned that his demarcation criterion *itself* has degrees.

Consider *non*-tautological theories A and B such that A is logically stronger than B *and* also has some *empirical* import not possessed by B. Popper (1963, p. 256) pointed out cogently enough that if there is a class of conceivable empirical findings that would each P-falsify B, this *non*-empty class is only a proper subclass of the class of those potential test-results that would each similarly refute A. This much guarantees the following: *Within* the set of those theories that do have non-empty classes of potential empirical P-falsifiers, we can thus speak of *degrees* of P-falsifiability in the case of any one subset of such P-falsifiable theories that are rank-ordered according to logical strength. But note what Popper makes of this result when he writes:

> There are, moreover (as I found later), degrees of testability: some theories expose themselves to possible refutations more boldly than others (1963, p. 256) . . . This indicates that the criterion of demarcation cannot be an absolutely sharp one but will itself have degrees. There will be well-testable theories, hardly testable theories, and non-testable theories. Those which are non-testable are of no interest to empirical scientists. They may be described as metaphysical (1963, p. 257).

Let us be mindful that for Popper (1963, p. 256) degrees of testability are avowedly interchangeable with his previously articulated degrees of P-falsifiability or refutability. Then I must ask: Why does Popper reason that the mere existence of rankings of comparatively "well-testable" and "hardly testable" non-tautological theories *suffices* to blur the sharpness of the demarcation between testable theories on the one hand, and the "non-testable" theories *of which he himself speaks* on the other? In an analogical vein, let me ask the altogether trivial question: Do rank-order differences between relatively large and very small positive real numbers *within* the totally-ordered class of positive reals blur at all the sharpness of the delineation between the membership of this class on the one hand, and numbers not belonging to it on the other? Plainly the hierarchy *within* the class does not make for any fuzziness of the partition *between* that class and its complement.

Why then should the *mere* existence of the stated comparative degrees of P-falsifiability automatically blur the line of demarcation between Popper's testable and non-testable theories? Even if the partial ordering with respect to P-falsifiability *were* sufficient to guarantee such fuzziness, it would not suffice as well to gainsay the following assertion: Freud's psychogenetic and psychodynamic theory falls squarely into the subclass of unambiguously P-testable theories *outside* the penumbra. Indeed, I fully allow that such a penumbra exists, although I did fault Popper's argument for its existence.

I conclude against Popper that with respect to P-falsifiability, Freud's theory is *not* a pseudo-science, although I believe that psychoanalysis does suffer from other serious foibles.

Cioffi (1970) has proposed to modify Popper's demarcation criterion along the following lines: A theory's refutability is quite insufficient for its scientificality: Even if its claims "are eminently refutable," Cioffi tells us, the theory's pseudo-scientificality can be assured by its advocates through the "habitual and wilful employment" of methodological procedures that are *calculated* to lessen the uncovering of potentially falsifying evidence. Thus, Cioffi argues at length that Freud's theory is pseudo-scientific despite its actual empirical falsifiability, because Freud's attempts to validate it are rife with just such methodological evasions. But I have argued elsewhere (Grünbaum 1979) that Cioffi's indictment greatly exaggerates Freud's culpability on this score.

Notes

The author thanks the Fritz Thyssen Stiftung for the support of research relevant to this paper.

1 Watkins' objection was presented as part of his "London School of Economics Position Paper" at the conference on "Progress and Rationality in Science" held in Kronberg, West Germany in July of 1975 with support from the Fritz Thyssen Stiftung.

References

Bondi, Hermann (1976), "Setting the Scene" in *Cosmology Now*, ed. by L. John (New York).

Cioffi, Frank (1970), "Freud and the Idea of a Pseudo-Science" and "Reply" in *Explanation in the Behavioural Sciences*, ed. by F. Cioffi and R. Borger (Cambridge).

Crews, Frederick (1965), *The Pooh Perplex* (New York).

—— (1976), Letter to the Editor in *New York Review of Books*, vol. xxiii (Feb. 5), p. 34.

Eysenck, H.J. and Wilson, G.D. (1973), *The Experimental Study of Freudian Theories* (London).

Farrell, B.A. (1970), Comment in *Explanation in the Behavioural Sciences*, ed. by F. Cioffi and R. Borger (Cambridge).

Fisher, Seymour and Greenberg, R.P. (1977), *The Scientific Credibility of Freud's Theories and Therapy* (New York).

Freud, Sigmund (1905), "My Views on the Part Played by Sexuality in the Aetiology of the Neuroses" in *Collected Papers*, vol. 1, tr. J. Riviere (New York 1959).

—— (1909), "Notes Upon a Case of Obsessional Neurosis" in *Collected Papers*, vol. 3, tr. A. and J. Strachey (New York, 1959).

—— (1911), "Psychoanalytic Notes Upon an Autobiographical Account of a Case of Paranoia" in *Collected Papers*, vol. 3, tr. A. and J. Strachey (New York, 1959).

—— (1914), "On the History of the Psycho-analytic Movement" in *Collected Papers*, vol. 1, tr. by J. Riviere (New York, 1959).

Glymour, Clark (1974), "Freud, Kepler and the Clinical Evidence" in *Freud*, ed. by R. Wollheim (New York).

Grunbaüm, Adolf (1976a), "Is Falsifiability the Touchstone of Scientific Rationality? Karl Popper versus Inductivism" in *Essays in Memory of Imre Lakatos, Boston Studies in the Philosophy of Science*, vol. 39, ed. by R.S. Cohen, P.K Feyerabend and M.W. Wartofsky (Dordrecht).

—— (1976b), "Can a Theory Answer More Questions Than One of Its Rivals?" *The British Journal for the Philosophy of Science*, vol. 27, pp. 1–23.

—— (1976c), "Is the Method of Bold Conjectures and Attempted Refutations Justifiably the Method of Science?" *The British Journal for the Philosophy of Science*, vol. 27, pp. 105–136.

—— (1976d), "*Ad Hoc* Auxiliary Hypotheses and Falsificationism," *The British Journal for the Philosophy of Science*, vol. 27, pp. 329–362.

—— (1977a), "How Scientific is Psychoanalysis?" in *Science and Psychotherapy*, ed. by R. Stern, L. Horowitz and J. Lynes. (New York).

—— (1977b), "Is Psychoanalysis a Pseudo-Science? Karl Popper versus Sigmund Freud," Part I, *Zeitschrift für philosophische Forschung*, vol. 31, Heft 3, pp. 333–353.

—— (1978), "Is Psychoanalysis a Pseudo-Science? Karl Popper versus Sigmund Freud," Part II, *Zeitschrift für philosophische Forschung*, vol. 32, Heft 1, pp. 49–69.

—— (1979), "The Role of Psychological Explanations of the Rejection or Acceptance of Scientific Theories," *Transactions of the New York Academy of Sciences*, Series II, vol. 39, A Festschrift for Robert Merton.

Holmes, D.S. (1974), "Investigations of Repression: Differential Recall of Material Experimentally or Naturally Associated with Ego Threat," *Psychological Bulletin*, vol. 81, pp. 632–653.

Kline, Paul (1972), *Fact and Fantasy in Freudian Theory* (London).

Martin, Michael (1964), "Mr. Farrell and the Refutability of Psychoanalysis," *Inquiry*, vol. 7, pp. 80–98.

Medawar, P.B. (1975), Review of "The Victim is Always the Same" by I.S. Cooper in *New York Review of Books*, vol. xxi (Jan. 23), p. 17.

Nagel, Ernest (1959), "Methodological Issues in Psychoanalytic Theory" in *Psychoanalysis, Scientific Method and Philosophy*, ed. by S. Hook (New York).

Popper, K.R. (1963), *Conjectures and Refutations* (London).

Sears, R.R. (1943), *Survey of Objective Studies of Psychoanalytic Concepts* (New York).

Silverman, L.H. (1976), "Psychoanalytic Theory, 'The Reports of My Death Are Greatly Exaggerated,' " *American Psychologist*, vol. 31, pp. 621–637.

Chapter 3

Stephen Jay Gould

THE MISMEASURE OF MAN

1 Correlation, cause, and factor analysis

Correlation and cause

The spirit of Plato dies hard. We have been unable to escape the philosophical tradition that what we can see and measure in the world is merely the superficial and imperfect representation of an underlying reality. Much of the fascination of statistics lies embedded in our gut feeling – and never trust a gut feeling – that abstract measures summarizing large tables of data must express something more real and fundamental than the data themselves. (Much professional training in statistics involves a conscious effort to counteract this gut feeling.) The technique of *correlation* has been particularly subject to such misuse because it seems to provide a path for inferences about causality (and indeed it does, sometimes – but only sometimes).

Correlation assesses the tendency of one measure to vary in concert with another. As a child grows, for example, both its arms and legs get longer; this joint tendency to change in the same direction is called a *positive correlation*. Not all parts of the body display such positive correlations during growth. Teeth, for example, do not grow after they erupt. The relationship between first incisor length and leg length from, say, age ten to adulthood would represent *zero correlation* – legs would get longer while teeth changed not at all. Other correlations can be negative – one measure increases while the other decreases. We begin to lose neurons at a distressingly early age, and they are not replaced. Thus, the relationship between leg length and number of neurons after mid-childhood represents *negative correlation* – leg length increases while number of neurons decreases. Notice that I have said nothing about causality. We do not know why these correlations exist or do not exist, only that they are present or not present.

The standard measure of correlation is called Pearson's product moment correlation coefficient or, for short, simply the correlation coefficient, symbolized as r. The correlation coefficient ranges from +1 for perfect positive correlation, to 0 for no correlation, to −1 for perfect negative correlation.[1]

In rough terms, r measures the shape of an ellipse of plotted points (see Fig. 3.1). Very skinny ellipses represent high correlations – the skinniest of all, a straight line, reflects an r of 1.0. Fat ellipses represent lower correlations, and the fattest of all, a circle, reflects zero correlation (increase in one measure permits no prediction about whether the other will increase, decrease, or remain the same).

The correlation coefficient, though easily calculated, has been plagued by errors of

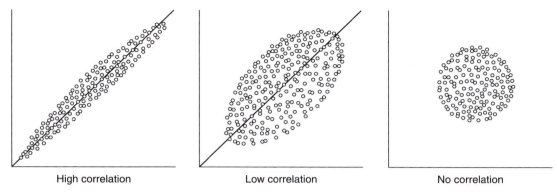

High correlation   Low correlation   No correlation

Figure 3.1 Strength of correlation as a function of the shape of an ellipse of points. The more elongate the ellipse, the higher the correlation.

interpretation. These can be illustrated by example. Suppose that I plot arm length vs. leg length during the growth of a child. I will obtain a high correlation with two interesting implications. First, I have achieved *simplification*. I began with two dimensions (leg and arm length), which I have now, effectively, reduced to one. Since the correlation is so strong, we may say that the line itself (a single dimension) represents nearly all the information originally supplied as two dimensions. Secondly, I can, in this case, make a reasonable inference about the *cause* of this reduction to one dimension. Arm and leg length are tightly correlated because they are both partial measures of an underlying biological phenomenon, namely growth itself.

Yet, lest anyone become too hopeful that correlation represents a magic method for the unambiguous identification of cause, consider the relationship between my age and the price of gasoline during the past ten years. The correlation is nearly perfect, but no one would suggest any assignment of cause. The fact of correlation implies nothing about cause. It is not even true that intense correlations are more likely to represent cause than weak ones, for the correlation of my age with the price of gasoline is nearly 1.0. I spoke of cause for arm and leg lengths not because their correlation was high, but because

I know something about the biology of the situation. The inference of cause must come from somewhere else, not from the simple fact of correlation – though an unexpected correlation may lead us to search for causes so long as we remember that we may not find them. The vast majority of correlations in our world are, without doubt, noncasual. Anything that has been decreasing steadily during the past few years will be strongly correlated with the distance between the earth and Halley's comet (which has also been decreasing of late) – but even the most dedicated astrologer would not discern causality in most of these relationships. The invalid assumption that correlation implies cause is probably among the two or three most serious and common errors of human reasoning.

Few people would be fooled by such a *reductio ad absurdum* as the age–gas correlation. But consider an intermediate case. I am given a table of data showing how far twenty children can hit and throw a baseball. I graph these data and calculate a high r. Most people, I think, would share my intuition that this is not a meaningless correlation; yet in the absence of further information, the correlation itself teaches me nothing about underlying causes. For I can suggest at least three different and reasonable causal interpretations for the correlation (and the true reason is probably some combination of them):

1. The children are simply of different ages, and older children can hit and throw farther.

2. The differences represent variation in practice and training. Some children are Little League stars and can tell you the year that Rogers Hornsby hit .424 (1924 – I was a bratty little kid like that); others know Billy Martin only as a figure in Lite beer commercials.

3. The differences represent disparities in native ability that cannot be erased even by intense training. (The situation would be even more complex if the sample included both boys and girls of conventional upbringing. The correlation might then be attributed primarily to a fourth cause – sexual differences; and we would have to worry, in addition, about the cause of the sexual difference: training, inborn constitution, or some combination of nature and nurture).

In summary, most correlations are noncausal; when correlations are causal, the fact and strength of the correlation rarely specify the nature of the cause.

Correlation in more than two dimensions

These two-dimensional examples are easy to grasp (however difficult they are to interpret). But what of correlations among more than two measures? A body is composed of many parts, not just arms and legs, and we may want to know how several measures interact during growth. Suppose, for simplicity, that we add just one more measure, head length, to make a three-dimensional system. We may now depict the correlation structure among the three measures in two ways:

1. We may gather all correlation coefficients between pairs of measures into a single table, or *matrix* of correlation coefficients (Fig. 3.2). The line from upper left to lower right records the necessarily perfect correlation of each variable with itself. It is called the *principal diagonal*, and

	Arm	Leg	Head
Arm	1.0	0.91	0.72
Leg	0.91	1.0	0.63
Head	0.72	0.63	1.0

Figure 3.2 *A correlation matrix for three measurements.*

all correlations along it are 1.0. The matrix is symmetrical around the principal diagonal, since the correlation of measure 1 with measure 2 is the same as the correlation of 2 with 1. Thus, the three values either above or below the principal diagonal are the correlations we seek: arm with leg, arm with head, and leg with head.

2. We may plot the points for all individuals onto a three-dimensional graph (Fig. 3.3). Since the correlations are all positive, the points are

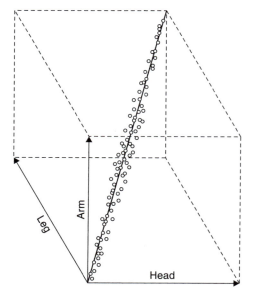

Figure 3.3 *A three-dimensional graph showing the correlations for three measurements.*

oriented as an ellipsoid (or football). (In two dimensions, they formed an ellipse.) A line running along the major axis of the football expresses the strong positive correlations between all measures.

We can grasp the three-dimensional case, both mentally and pictorially. But what about 20 dimensions, or 100? If we measured 100 parts of a growing body, our correlation matrix would contain 10,000 items. To plot this information, we would have to work in a 100-dimensional space, with 100 mutually perpendicular axes representing the original measures. Although these 100 axes present no mathematical problem (they form, in technical terms, a hyperspace), we cannot plot them in our three-dimensional Euclidian world.

These 100 measures of a growing body probably do not represent 100 different biological phenomena. Just as most of the information in our three-dimensional example could be resolved into a single dimension (the long axis of the football), so might our 100 measures be simplified into fewer dimensions. We will lose some information in the process to be sure – as we did when we collapsed the long and skinny football, still a three-dimensional structure, into the single line representing its long axis. But we may be willing to accept this loss in exchange for simplification and for the possibility of interpreting the dimensions that we do retain in biological terms.

Factor analysis and its goals

With this example, we come to the heart of what *factor analysis* attempts to do. Factor analysis is a mathematical technique for reducing a complex system of correlations into fewer dimensions. It works, literally, by factoring a matrix, usually a matrix of correlation coefficients. (Remember the high-school algebra exercise called "factoring," where you simplified horrendous expressions by removing common multipliers of all terms?) Geometrically, the process of factoring amounts to placing axes through a football of points. In the 100-dimensional case, we are not likely to recover enough information on a single line down the hyperfootball's long axis – a line called the *first principal component*. We will need additional axes. By convention, we represent the second dimension by a line *perpendicular* to the first principal component. This second axis, or *second principal component*, is defined as the line that resolves more of the remaining variation than any other line that could be drawn perpendicular to the first principal component. If, for example, the hyperfootball were squashed flat like a flounder, the first principal component would run through the middle, from head to tail, and the second also through the middle, but from side to side. Subsequent lines would be perpendicular to all previous axes, and would resolve a steadily decreasing amount of remaining variation. We might find that five principal components resolve almost all the variation in our hyperfootball – that is, the hyperfootball drawn in five dimensions looks sufficiently like the original to satisfy us, just as a pizza or a flounder drawn in two dimensions may express all the information we need, even though both original objects contain three dimensions. If we elect to stop at five dimensions, we may achieve a considerable simplification at the acceptable price of minimal loss of information. We can grasp the five dimensions conceptually; we may even be able to interpret them biologically.

Since factoring is performed on a correlation matrix, I shall use a geometrical representation of the correlation coefficients themselves in order to explain better how the technique operates. The original measures may be represented as vectors of unit length,[2] radiating from a common point. If two measures are highly correlated, their vectors lie close to each other.

The cosine of the angle between any two vectors records the correlation coefficient between them. If two vectors overlap, their correlation is perfect, or 1.0; the cosine of 0° is 1.0. If two vectors lie at right angles, they are completely independent, with a correlation of zero; the cosine of 90° is zero. If two vectors point in opposite directions, their correlation is perfectly negative, or −1.0; the cosine of 180° is −1.0. A matrix of high positive correlation coefficients will be represented by a cluster of vectors, each separated from each other vector by a small acute angle (Fig. 3.4). When we factor such a cluster into fewer dimensions by computing principal components, we choose as our first component the axis of maximal resolving power, a kind of grand average among all vectors. We assess resolving power by projecting each vector onto the axis. This is done by drawing a line from the tip of the vector to the axis, perpendicular to the axis. The ratio of projected length

on the axis to the actual length of the vector itself measures the percentage of a vector's information resolved by the axis. (This is difficult to express verbally, but I think that Fig. 3.5 will dispel confusion.) If a vector lies near the axis, it is highly resolved and the axis encompasses most of its information. As a vector moves away from the axis toward a maximal separation of 90°, the axis resolves less and less of it.[3]

We position the first principal component (or axis) so that it resolves more information among all the vectors than any other axis could.[4] For our matrix of high positive correlation coefficients, represented by a set of tightly clustered

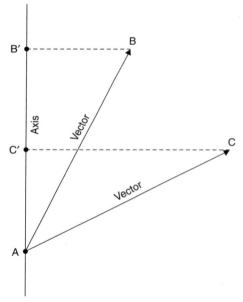

Figure 3.5 *Computing the amount of information in a vector explained by an axis. Draw a line from the tip of the vector to the axis, perpendicular to the axis. The amount of information resolved by the axis is the ratio of the projected length on the axis to the true length of the vector. If a vector lies close to the axis, then this ratio is high and most of the information in the vector is resolved by the axis. Vector **AB** lies close to the axis and the ratio of the projection **AB′** to the vector itself, **AB**, is high. Vector **AC** lies far from the axis and the ratio of its projected length **AC′** to the vector itself, **AC**, is low.*

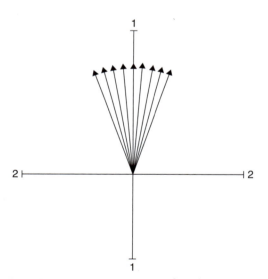

Figure 3.4 *Geometric representation of correlations among eight tests when all correlation coefficients are high and positive. The first principal component, labeled 1, lies close to all the vectors, while the second principal component, labeled 2, lies at right angles to the first and does not explain much information in the vectors.*

vectors, the first principal component runs through the middle of the set (Fig. 3.4). The second principal component lies at right angles to the first and resolves a maximal amount of remaining information. But if the first component has already resolved most of the information in all the vectors, then the second and subsequent principal axes can only deal with the small amount of information that remains (Fig. 3.4).

Such systems of high positive correlation are found frequently in nature. In my own first study in factor analysis, for example, I considered fourteen measurements on the bones of twenty-two species of pelycosaurian reptiles (the fossil beasts with the sails on their backs, often confused with dinosaurs, but actually the ancestors of mammals). My first principal component resolved 97.1 percent of the information in all fourteen vectors, leaving only 2.9 percent for subsequent axes. My fourteen vectors formed an extremely tight swarm (all practically overlapping); the first axis went through the middle of the swarm. My pelycosaurs ranged in body length from less than two to more than eleven feet. They all look pretty much alike, and big animals have larger measures for all fourteen bones. All correlation coefficients of bones with other bones are very high; in fact, the lowest is still a whopping 0.912. Scarcely surprising. After all, large animals have large bones, and small animals small bones. I can interpret my first principal component as an abstracted size factor, thus reducing (with minimal loss of information) my fourteen original measurements into a single dimension interpreted as increasing body size. In this case, factor analysis has achieved both *simplification* by reduction of dimensions (from fourteen to effectively one), and *explanation* by reasonable biological interpretation of the first axis as a size factor.

But – and here comes an enormous but – before we rejoice and extol factor analysis as a panacea for understanding complex systems of correlation, we should recognize that it is subject to the same cautions and objections previously examined for the correlation coefficients themselves. I consider two major problems in the following sections.

The error of reification

The first principal component is a mathematical abstraction that can be calculated for any matrix of correlation coefficients; it is not a "thing" with physical reality. Factorists have often fallen prey to a temptation for *reification* – for awarding *physical meaning* to all strong principal components. Sometimes this is justified; I believe that I can make a good case for interpreting my first pelycosaurian axis as a size factor. But such a claim can never arise from the mathematics alone, only from additional knowledge of the physical nature of the measures themselves. For nonsensical systems of correlation have principal components as well, and they may resolve more information than meaningful components do in other systems. A factor analysis for a five-by-five correlation matrix of my age, the population of Mexico, the price of Swiss cheese, my pet turtle's weight, and the average distance between galaxies during the past ten years will yield a strong first principal component. This component – since all the correlations are so strongly positive – will probably resolve as high a percentage of information as the first axis in my study of pelycosaurs. It will also have no enlightening physical meaning whatever.

In studies of intelligence, factor analysis has been applied to matrices of correlation among mental tests. Ten tests may, for example, be given to each of one hundred people. Each meaningful entry in the ten-by-ten correlation matrix is a correlation coefficient between scores on two tests taken by each of the one hundred persons. We have known since the early days of mental testing – and it should surprise no one – that most of these correlation coefficients are positive: that is, people who score highly on one kind of test tend, on average, to score highly on others as well. Most correlation matrices for mental tests contain a preponderance of positive entries. This basic observation served as the starting point for factor analysis. Charles Spearman virtually invented the technique in 1904 as a device for inferring causes from correlation matrices of mental tests.

Since most correlation coefficients in the matrix are positive, factor analysis must yield a reasonably strong first principal component. Spearman calculated such a component indirectly in 1904 and then made the cardinal invalid inference that has plagued factor analysis ever since. He reified it as an "entity" and tried to give it an unambiguous causal interpretation. He called it g, or general intelligence, and imagined that he had identified a unitary quality underlying all cognitive mental activity – a quality that could be expressed as a single number and used to rank people on a unilinear scale of intellectual worth.

Spearman's g – the first principal component of the correlation matrix of mental tests – never attains the predominant role that a first component plays in many growth studies (as in my pelycosaurs). At best, g resolves 50 to 60 percent of all information in the matrix of tests. Correlations between tests are usually far weaker than correlations between two parts of a growing body. In most cases, the highest correlation in a matrix of tests does not come close to reaching the lowest value in my pelycosaur matrix – 0.912.

Although g never matches the strength of a first principal component of some growth studies, I do not regard its fair resolving power as accidental. Causal reasons lie behind the positive correlations of most mental tests. But what reasons? We cannot infer the reasons from a strong first principal component any more than

we can induce the cause of a single correlation coefficient from its magnitude. We cannot reify g as a "thing" unless we have convincing, independent information beyond the fact of correlation itself.

The situation for mental tests resembles the hypothetical case I presented earlier of correlation between throwing and hitting a baseball. The relationship is strong and we have a right to regard it as non-accidental. But we cannot infer the cause from the correlation, and the cause is certainly complex.

Spearman's g is particularly subject to ambiguity in interpretation, if only because the two most contradictory causal hypotheses are both fully consistent with it: (1) that it reflects an inherited level of mental acuity (some people do well on most tests because they are born smarter); or (2) that it records environmental advantages and deficits (some people do well on most tests because they are well schooled, grew up with enough to eat, books in the home, and loving parents). If the simple existence of g can be theoretically interpreted in either a purely hereditarian or purely environmentalist way, then its mere presence – even its reasonable strength – cannot justly lead to any reification at all. The temptation to reify is powerful. The idea that we have detected something "underlying" the externalities of a large set of correlation coefficients, something perhaps more real than the superficial measurements themselves, can be intoxicating. It is Plato's essence, the abstract, eternal reality underlying superficial appearances. But it is a temptation that we must resist, for it reflects an ancient prejudice of thought, not a truth of nature.

Rotation and the nonnecessity of principal components

Another, more technical, argument clearly demonstrates why principal components cannot be automatically reified as causal entities. If principal components represented the only way to simplify a correlation matrix, then some special status for them might be legitimately sought. But they represent only one method among many for inserting axes into a multidimensional space. Principal components have a definite geometric arrangement, specified by the criterion used to construct them — that the first principal component shall resolve a maximal amount of information in a set of vectors and that subsequent components shall all be mutually perpendicular. But there is nothing sacrosanct about this criterion; vectors may be resolved into any set of axes placed within their space. Principal components provide insight in some cases, but other criteria are often more useful.

Consider the following situation, in which another scheme for placing axes might be preferred. In Fig. 3.6 I show correlations between

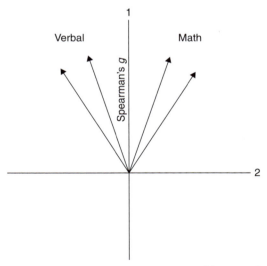

Figure 3.6 *A principal components analysis of four mental tests. All correlations are high and the first principal component, Spearman's g, expresses the overall correlation. But the group factors for verbal and mathematical aptitude are not well resolved in this style of analysis.*

four mental tests, two of verbal and two of arithmetical aptitude. Two "clusters" are evident, even though all tests are positively correlated. Suppose that we wish to identify these clusters by factor analysis. If we use principal components, we may not recognize them at all. The first principal component (Spearman's g) goes right up the middle, between the two clusters. It lies close to no vector and resolves an approximately equal amount of each, thereby masking the existence of verbal and arithmetic clusters. Is this component an entity? Does a "general intelligence" exist? Or is g, in this case, merely a meaningless average based on the invalid amalgamation of two types of information?

We may pick up verbal and arithmetic clusters on the second principal component (called a "bipolar factor" because some projections upon it will be positive and others negative when vectors lie on both sides of the first principal component). In this case, verbal tests project on the negative side of the second component, and arithmetic tests on the positive side. But we may fail to detect these clusters altogether if the first principal component dominates all vectors. For projections on the second component will then be small, and the pattern can easily be lost (see Fig. 3.6).

During the 1930s factorists developed methods to treat this dilemma and to recognize clusters of vectors that principal components often obscured. They did this by rotating factor axes from the principal components orientation to new positions. The rotations, established by several criteria, had as their common aim the positioning of axes near clusters. In Fig. 3.7, for example, we use the criterion: place axes near vectors occupying extreme or outlying positions in the total set. If we now resolve all vectors into these rotated axes, we detect the clusters easily; for arithmetic tests project high on rotated axis 1 and low on rotated axis 2, while verbal tests project high on 2 and low on 1. Moreover, g has

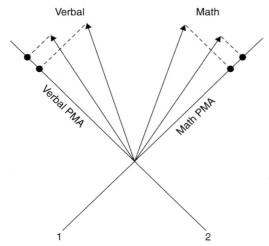

Figure 3.7 *Rotated factor axes for the same four mental tests depicted in Fig. 3.6. Axes are now placed near vectors lying at the periphery of the cluster. The group factors for verbal and mathematical aptitude are now well identified (see high projections on the axes indicated by dots), but g has disappeared.*

disappeared. We no longer find a "general factor" of intelligence, nothing that can be reified as a single number expressing overall ability. Yet we have lost no information. The two rotated axes resolve as much information in the four vectors as did the two principal components. They simply distribute the same information differently upon the resolving axes. How can we argue that g has any claim to reified status as an entity if it represents but one of numerous possible ways to position axes within a set of vectors?

In short, factor analysis simplifies large sets of data by reducing dimensionality and trading some loss of information for the recognition of ordered structure in fewer dimensions. As a tool for simplification, it has proved its great value in many disciplines. But many factorists have gone beyond simplification, and tried to define factors as causal entities. This error of reification has plagued the technique since its inception. It was "present at the creation" since Spearman invented factor analysis to study the correlation matrix of mental tests and then reified his principal component as g or innate, general intelligence. Factor analysis may help us to understand causes by directing us to information beyond the mathematics of correlation. But factors, by themselves, are neither things nor causes; they are mathematical abstractions. Since the same set of vectors (see Fig. 3.6, 3.7) can be partitioned into g and a small residual axis, or into two axes of equal strength that identify verbal and arithmetical clusters and dispense with g entirely, we cannot claim that Spearman's "general intelligence" is an ineluctable entity necessarily underlying and causing the correlations among mental tests. Even if we choose to defend g as a non-accidental result, neither its strength nor its geometric position can specify what it means in causal terms – if only because its features are equally consistent with extreme hereditarian and extreme environmentalist views of intelligence.

2 A positive conclusion

Walt Whitman, that great man of little brain, advised us to "make much of negatives," and this book [sic] has heeded his words, some might say with a vengeance. While most of us can appreciate a cleansing broom, such an object rarely elicits much affection; it certainly produces no integration. But I do not regard this book [sic] as a negative exercise in debunking, offering nothing in return once the errors of biological determinism are exposed as social prejudice. I believe that we have much to learn about ourselves from the undeniable fact that we are evolved animals. This understanding cannot permeate through entrenched habits of thought that lead us to reify and rank – habits that arise within social contexts and support them in return. My message, as I hope to convey it at least, is strongly positive for three major reasons.

Debunking as positive science

The popular impression that disproof represents a negative side of science arises from a common, but erroneous, view of history. The idea of unilinear progress not only lies behind the racial rankings that I have criticized as social prejudice throughout this book; it also suggests a false concept of how science develops. In this view, any science begins in the nothingness of ignorance and moves toward truth by gathering more and more information, constructing theories as facts accumulate. In such a world, debunking would be primarily negative, for it would only shuck some rotten apples from the barrel of accumulating knowledge. But the barrel of theory is always full; sciences work with elaborated contexts for explaining facts from the very outset. Creationist biology was dead wrong about the origin of species, but Cuvier's brand of creationism was not an emptier or less-developed world view than Darwin's. Science advances primarily by replacement, not by addition. If the barrel is always full, then the rotten apples must be discarded before better ones can be added.

Scientists do not debunk only to cleanse and purge. They refute older ideas in the light of a different view about the nature of things.

Learning by debunking

If it is to have any enduring value, sound debunking must do more than replace one social prejudice with another. It must use more adequate biology to drive out fallacious ideas. (Social prejudices themselves may be refractory, but particular biological supports for them can be dislodged.)

We have rejected many specific theories of biological determinism because our knowledge about human biology, evolution, and genetics has increased. For example, Morton's egregious errors could not be repeated in so bald a way by modern scientists constrained to follow canons of statistical procedure. The antidote to Goddard's claim that a single gene causes feeble-mindedness was not primarily a shift in social preferences, but an important advance in genetical theory – the idea of polygenic inheritance. Absurd as it seems today, the early Mendelians did try to attribute even the most subtle and complex traits (of apolitical anatomy as well as character) to the action of single genes. Polygenic inheritance affirms the participation of many genes – and a host of environmental and interactive effects – in such characters as human skin color.

More importantly, and as a plea for the necessity of biological knowledge, the remarkable lack of genetic differentiation among human groups – a major biological basis for debunking determinism – is a contingent fact of evolutionary history, not an a priori or necessary truth. The world might have been ordered differently. Suppose, for example, that one or several species of our ancestral genus *Australopithecus* had survived – a perfectly reasonable scenario in theory, since new species arise by splitting off from old ones (with ancestors usually surviving, at least for a time), not by the wholesale transformation of ancestors to descendants. We – that is, *Homo sapiens* – would then have faced all the moral dilemmas involved in treating a human species of distinctly inferior mental capacity. What would we have done with them – slavery? extirpation? coexistence? menial labor? reservations? zoos?

Similarly, our own species, *Homo sapiens*, might have included a set of subspecies (races) with meaningfully different genetic capacities. If our species were millions of years old (many are), and if its races had been geographically separated for most of this time without significant genetic interchange, then large genetic differences might have slowly accumulated between groups. But *Homo sapiens* is tens of thousands, or at most a few hundred thousand, years old, and all modern human races probably split from a common ancestral stock only tens of thousands of years ago. A few outstanding traits of external appearance lead to our subjective judgment of important differences. But biologists have recently affirmed – as long suspected – that the overall genetic differences among human races are astonishingly small. Although frequencies for different states of a gene differ among races, we have found no "race genes" – that is, states fixed in certain races and absent from all others. Lewontin (1972) studied variation in seventeen genes coding for differences in blood and found that only 6.3 percent of the variation can be attributed to racial membership. Fully 85.4 percent of the variation occurred within local populations (the remaining 8.3 percent records differences among local populations within a race). As Lewontin remarked (personal communication): if the holocaust comes and a small tribe deep in the New Guinea forests are the only survivors, almost all the genetic variation now expressed among the innumerable groups of our four billion people will be preserved.

This information about limited genetic differences among human groups is useful as well as interesting, often in the deepest sense – for saving lives. When American eugenicists attributed diseases of poverty to the inferior genetic construction of poor people, they could propose no systematic remedy other than sterilization. When Joseph Goldberger proved that pellagra was not a genetic disorder, but a result of vitamin deficiency among the poor, he could cure it.

Biology and human nature

If people are so similar genetically, and if previous claims for a direct biological mapping of human affairs have recorded cultural prejudice and not nature, then does biology come up empty as a guide in our search to know ourselves? Are we after all, at birth, the *tabula rasa*, or blank slate, imagined by some eighteenth-century empiricist philosophers? As an evolutionary biologist, I cannot adopt such a nihilistic position without denying the fundamental insight of my profession. The evolutionary unity of humans with all other organisms is the cardinal message of Darwin's revolution for nature's most arrogant species.

We are inextricably part of nature, but human uniqueness is not negated thereby. "Nothing but" an animal is as fallacious a statement as "created in God's own image." It is not mere hubris to argue that *Homo sapiens* is special in some sense – for each species is unique in its own way; shall we judge among the dance of the bees, the song of the humpback whale, and human intelligence?

The impact of human uniqueness upon the world has been enormous because it has established a new kind of evolution to support the transmission across generations of learned knowledge and behavior. Human uniqueness resides primarily in our brains. It is expressed in the culture built upon our intelligence and the power it gives us to manipulate the world. Human societies change by cultural evolution, not as a result of biological alteration. We have no evidence for biological change in brain size or structure since *Homo sapiens* appeared in the fossil record some fifty thousand years ago. (Broca was right in stating that the cranial capacity of Cro Magnon skulls was equal if not superior to ours.) All that we have done since then – the greatest transformation in the shortest time that our planet has experienced since its crust solidified nearly four billion years ago – is the product of cultural evolution. Biological (Darwinian) evolution continues in our species, but its rate, compared with cultural evolution, is so incomparably slow that its impact upon the history of *Homo sapiens* has been small. While the gene for sickle-cell anemia declines in frequency among black Americans, we have invented the railroad, the automobile, radio and television, the atom bomb, the computer, the airplane and spaceship.

Cultural evolution can proceed so quickly because it operates, as biological evolution does not, in the "Lamarckian" mode – by the inheritance of acquired characters. Whatever one generation learns, it can pass to the next by writing, instruction, inculcation, ritual, tradition, and a host of methods that humans have developed to assure continuity in culture. Darwinian evolution, on the other hand, is an indirect process: genetic variation must first be available to construct an advantageous feature, and natural selection must then preserve it. Since genetic variation arises at random, not preferentially directed toward advantageous features, the Darwinian process works slowly. Cultural evolution is not only rapid; it is also readily reversible because its products are not coded in our genes.

The classical arguments of biological determinism fail because the features they invoke to make distinctions among groups are usually the products of cultural evolution. Determinists did seek evidence in anatomical traits built by biological, not cultural, evolution. But, in so doing, they tried to use anatomy for making inferences about capacities and behaviors that they linked to anatomy and we regard as engendered by culture. Cranial capacity per se held as little interest for Morton and Broca as variation in third-toe length; they cared only about the mental characteristics supposedly associated with differences in average brain size among groups. We now believe that different attitudes and styles of thought among human groups are usually the nongenetic products of cultural evolution. In short, the *biological* basis of human uniqueness leads us to reject biological determinism. Our large brain is the biological foundation of intelligence; intelligence is the ground of culture; and cultural transmission builds a new mode of evolution more effective than Darwinian processes in its limited realm – the "inheritance" and modification of learned behavior. As philosopher Stephen Toulmin stated (1977, p. 4): "Culture has the power to impose itself on nature from within."

Yet, if human biology engenders culture, it is also true that culture, once developed, evolved with little or no reference to genetic *variation* among human groups. Does biology, then, play no other valid role in the analysis of human behavior? Is it only a foundation without any insight to offer beyond the unenlightening recognition that complex culture requires a certain level of intelligence?

Most biologists would follow my argument in denying a genetic basis for most behavioral *differences* between groups and for *change* in the complexity of human societies through the recent history of our species. But what about the supposed constancies of personality and behavior, the traits of mind that humans share in all cultures? What, in short, about a general "human nature"? Some biologists would grant Darwinian processes a substantial role not only in establishing long ago, but also in actively maintaining now, a set of specific adaptive behaviors forming a biologically conditioned "human nature." I believe that this old tradition of argument – which has found its most recent expression as "human sociobiology" – is invalid not because biology is irrelevant and human behavior only reflects a disembodied culture, but because human *biology* suggests a different and less constraining role for genetics in the analysis of human nature.

Sociobiology begins with a modern reading of what natural selection is all about – differential reproductive success of individuals. According to the Darwinian imperative, individuals are selected to maximize the contribution of their own genes to future generations, and that is all. (Darwinism is not a theory of progress, increasing complexity, or evolved harmony for the good of species or ecosystems.) Paradoxically (as it seems to many), altruism as well as selfishness can be selected under this criterion – acts of kindness may benefit individuals either because they establish bonds of reciprocal obligation, or because they aid kin who carry copies of the altruist's genes.

Human sociobiologists then survey our behaviors with this criterion in mind. When they identify a behavior that seems to be adaptive in helping an individual's genes along, they develop a story for its origin by natural selection operating upon genetic variation influencing the specific act itself. (These stories are rarely backed by any evidence beyond the inference of adaptation.) Human sociobiology is a theory for the origin and maintenance of *specific, adaptive behaviors* by *natural selection*[5]; these behaviors must therefore have a *genetic basis*, since natural selection cannot operate in the absence of genetic variation. Sociobiologists have tried, for example, to identify an adaptive and genetic foundation for aggression, spite, xenophobia, conformity, homosexuality,[6] and perhaps upward mobility as well (Wilson, 1975).

I believe that modern biology provides a model standing between the despairing claim that biology has nothing to teach us about human behavior and the deterministic theory that specific items of behavior are genetically programmed by the action of natural selection. I see two major areas for biological insight:

1. Fruitful analogies. Much of human behavior is surely adaptive; if it weren't, we wouldn't be around any more. But adaptation, in humans, is neither an adequate, nor even a good argument for genetic influence. For in humans, as I argued above (p. 324) [sic], adaptation may arise by the alternate route of nongenetic, cultural evolution. Since cultural evolution is so much more rapid than Darwinian evolution, its influence should prevail in the behavioral diversity displayed by human groups. But even when an adaptive behavior is nongenetic, biological analogy may be useful in interpreting its meaning. Adaptive constraints are often strong, and some functions may have to proceed in a certain way whether their underlying impetus be learning or genetic programming.

For example, ecologists have developed a powerful quantitative theory, called optimal foraging strategy, for studying patterns of exploitation in nature (herbivores by carnivores, plants by herbivores). Cornell University anthropologist Bruce Winterhalder has shown that a community of Cree-speaking peoples in northern Ontario follow some predictions of the theory in their hunting and trapping behavior. Although Winterhalder used a biological theory to understand some aspects of human hunting, he does not believe that the people he studied were genetically selected to hunt as ecological theory predicts they should. He writes (personal communication, July 1978):

> It should go without saying . . . that the causes of human variability of hunting and gathering behavior lie in the socio-cultural realm. For that reason, the models that I used were adapted, not adopted, and then applied to a very circumscribed realm of analysis. . . . For instance, the models assist in analyzing what species a hunter will seek from those available *once a decision has been made to go hunting* [his italics].

They are, however, useless for analyzing why the Cree still hunt (they don't need to), how they decide on a particular day whether to hunt or join a construction crew, the meaning of hunting to a Cree, or any of a plethora of important questions.

In this area, sociobiologists have often fallen into one of the most common errors of reasoning: discovering an analogy and inferring a genetic similarity (literally, in this case!). Analogies are useful but limited; they may reflect common constraints, but not common causes.

2. Biological potentiality vs. biological determinism. Humans are animals, and everything we do is constrained, in some sense, by our biology. Some constraints are so integral to our being that we rarely even recognize them, for we never imagine that life might proceed in another way. Consider our narrow range of average adult size and the consequences of living in the gravitational world of large organisms, not the world of surface forces inhabited by insects (Went, 1968; Gould, 1977). Or the fact that we are born helpless (many animals are not), that we mature slowly, that we must sleep for a large part of the day, that we do not photosynthesize, that we can digest both meat and plants, that we age and die. These are all results of our genetic construction, and all are important influences upon human nature and society.

These biological boundaries are so evident that they have never engendered controversy. The contentious subjects are specific behaviors that distress us and that we struggle with difficulty to change (or enjoy and fear to abandon): aggression, xenophobia, male dominance, for example. Sociobiologists are not genetic determinists in the old eugenical sense of postulating single genes for such complex behaviors. All biologists know that there is no gene "for" aggression, any more than for your lower-left wisdom tooth. We all recognize that genetic influence can be spread diffusely among many genes and that genes set limits to ranges; they do not provide blueprints for exact replicas. In one sense, the debate between sociobiologists and their critics is an argument about the breadth of ranges. For sociobiologists, ranges are narrow enough to program a specific behavior as the predictable result of possessing certain genes. Critics argue that the ranges permitted by these

genetic factors are wide enough to include all behaviors that sociobiologists atomize into distinct traits coded by separate genes.

Notes

1   Pearson's r is not an appropriate measure for all kinds of correlation, for it assesses only what statisticians call the intensity of linear relationship between two measures – the tendency for all points to fall on a single straight line. Other relationships of strict dependence will not achieve a value of 1.0 for r. If, for example, each increase of 2 units in one variable were matched by an increase in 2^2 units in the other variable, r would be less than 1.0, even though the two variables might be perfectly "correlated" in the vernacular sense. Their plot would be a parabola, not a straight line, and Pearson's r measures the intensity of linear resemblance.

2   (Footnote for aficionados – others may safely skip.) Here, I am technically discussing a procedure called "principal components analysis," not quite the same thing as factor analysis. In principal components analysis, we preserve all information in the original measures and fit new axes to them by the same criterion used in factor analysis in principal components orientation – that is, the first axis explains more data than any other axis could and subsequent axes lie at right angles to all other axes and encompass steadily decreasing amounts of information. In true factor analysis, we decide beforehand (by various procedures) not to include all information on our factor axes. But the two techniques – true factor analysis in principal components orientation and principal components analysis – play the same conceptual role and differ only in mode of calculation. In both, the first axis (Spearman's g for intelligence tests) is a "best fit" dimension that resolves more information in a set of vectors than any other axis could.

3   During the past decade or so, semantic confusion has spread in statistical circles through a tendency to restrict the term "factor analysis" only to the rotations of axes usually performed after the calculation of principal components, and to extend the

term "principal components analysis" both to true principal components analysis (all information retained) and to factor analysis done in principal components orientation (reduced dimensionality and loss of information). This shift in definition is completely out of keeping with the history of the subject and terms. Spearman, Burt, and hosts of other psychometricians worked for decades in this area before Thurstone and others invented axial rotations. They performed all their calculations in the principal components orientation, and they called themselves "factor analysts." I continue, therefore, to use the term "factor analysis" in its original sense to include any orientation of axes – principal components or rotated, orthogonal or oblique.

4 I will also use a common, if somewhat sloppy, shorthand in discussing what factor axes do. Technically, factor axes resolve variance in original measures. I will, as is often done, speak of them as "explaining" or "resolving" information – as they do in the vernacular (though not in the technical) sense of information. That is, when the vector of an original variable projects strongly on a set of factor axes, little of its variance lies unresolved in higher dimensions outside the system of factor axes.

5 The brouhaha over sociobiology during the past few years was engendered by this hard version of the argument – genetic proposals (based on an inference of adaptation) for specific human behaviors. Other evolutionists call themselves "sociobiologists," but reject this style of guesswork about specifics. If a sociobiologist is anyone who believes that biological evolution is not irrelevant to human behavior, then I suppose that everybody (creationists excluded) is a sociobiologist. At this point, however, the term loses its meaning and might as well be dropped. Human sociobiology entered the literature (professional and popular) as a definite theory about the adaptive and genetic basis of specific traits of human behavior. If it has failed in this goal – as I believe it has – then the study of

valid relationships between biology and human behavior should receive another name. In a world awash in jargon, I don't see why "behavioral biology" can't extend its umbrella sufficiently to encompass this legitimate material.

6 Lest homosexuality seem an unlikely candidate for adaptation since exclusive homosexuals have no children, I report the following story, advocated by E. O. Wilson (1975, 1978). Ancestral human society was organized as a large number of competing family units. Some units were exclusively heterosexual; the gene pool of other units included factors for homosexuality. Homosexuals functioned as helpers to raise the offspring of their heterosexual kin. This behavior aided their genes since the large number of kin they helped to raise held more copies of their genes than their own offspring (had they been heterosexual) might have carried. Groups with homosexual helpers raised more offspring, since they could more than balance, by extra care and higher rates of survival, the potential loss by nonfecundity of their homosexual members. Thus, groups with homosexual members ultimately prevailed over exclusively heterosexual groups, and genes for homosexuality have survived.

References

Gould, S.J. (1977). *Ever since Darwin*. New York: W.W. Norton.

Gould, S.J. (1977). *Ontogeny and phylogeny*. Cambridge, MA: Harvard University Press.

Lewontin, R.C. (1972). The apportionment of human diversity. *Evolutionary Biology* 6: 381–398.

Toulmin, S. (1977). Back to nature. *New York Review of Books*, 9 June, pp. 3–6.

Went, F.W. (1968). The size of man. *American Scientist* 56: 400–413.

Wilson, E.O. (1975). *Sociobiology*. Cambridge, MA: Harvard University Press.

Wilson, E.O. (1978). *On human nature*. Cambridge, MA: Harvard University Press.

Chapter 4

Thomas S. Kuhn

OBJECTIVITY, VALUE JUDGMENT, AND THEORY CHOICE

In the penultimate chapter of a controversial book first published fifteen years ago, I considered the ways scientists are brought to abandon one time-honored theory or paradigm in favor of another. Such decision problems, I wrote, "cannot be resolved by proof." To discuss their mechanism is, therefore, to talk "about techniques of persuasion, or about argument and counterargument in a situation in which there can be no proof." Under these circumstances, I continued, "lifelong resistance [to a new theory] . . . is not a violation of scientific standards. . . . Though the historian can always find men – Priestley, for instance – who were unreasonable to resist for as long as they did, he will not find a point at which resistance becomes illogical or unscientific."[1] Statements of that sort obviously raise the question of why, in the absence of binding criteria for scientific choice, both the number of solved scientific problems and the precision of individual problem solutions should increase so markedly with the passage of time. Confronting that issue, I sketched in my closing chapter a number of characteristics that scientists share by virtue of the training which licenses their membership in one or another community of specialists. In the absence of criteria able to dictate the choice of each individual, I argued, we do well to trust the collective judgment of scientists trained in this way. "What better criterion could there be," I asked rhetorically, "than the decision of the scientific group?"[2]

A number of philosophers have greeted remarks like these in a way that continues to surprise me. My views, it is said, make of theory choice "a matter for mob psychology."[3] Kuhn believes, I am told, that "the decision of a scientific group to adopt a new paradigm cannot be based on good reasons of any kind, factual or otherwise."[4] The debates surrounding such choices must, my critics claim, be for me "mere persuasive displays without deliberative substance."[5] Reports of this sort manifest total misunderstanding, and I have occasionally said as much in papers directed primarily to other ends. But those passing protestations have had negligible effect, and the misunderstandings continue to be important. I conclude that it is past time for me to describe, at greater length and with greater precision, what has been on my mind when I have uttered statements like the ones with which I just began. If I have been reluctant to do so in the past, that is largely because I have preferred to devote attention to areas in which my views diverge more sharply from those currently received than they do with respect to theory choice.

* * *

What, I ask to begin with, are the characteristics of a good scientific theory? Among a number of quite usual answers I select five, not because they are exhaustive, but because they are individually important and collectively sufficiently varied to indicate what is at stake. First, a theory should be accurate: within its domain, that is, consequences deducible from a theory should be in demonstrated agreement with the results of existing experiments and observations. Second, a theory should be consistent, not only internally or with itself, but also with other currently accepted theories applicable to related aspects of nature. Third, it should have broad scope: in particular, a theory's consequences should extend far beyond the particular observations, laws, or subtheories it was initially designed to explain. Fourth, and closely related, it should be simple, bringing order to phenomena that in its absence would be individually isolated and, as a set, confused. Fifth — a somewhat less standard item, but one of special importance to actual scientific decisions — a theory should be fruitful of new research findings: it should, that is, disclose new phenomena or previously unnoted relationships among those already known.[6] These five characteristics — accuracy, consistency, scope, simplicity, and fruitfulness — are all standard criteria for evaluating the adequacy of a theory. If they had not been, I would have devoted far more space to them in my book, for I agree entirely with the traditional view that they play a vital role when scientists must choose between an established theory and an upstart competitor. Together with others of much the same sort, they provide *the* shared basis for theory choice.

Nevertheless, two sorts of difficulties are regularly encountered by the men who must use these criteria in choosing, say, between Ptolemy's astronomical theory and Copernicus's, between the oxygen and phlogiston theories of combustion, or between Newtonian mechanics and the quantum theory. Individually the criteria are imprecise: individuals may legitimately differ about their application to concrete cases. In addition, when deployed together, they repeatedly prove to conflict with one another; accuracy may, for example, dictate the choice of one theory, scope the choice of its competitor. Since these difficulties, especially the first, are also relatively familiar, I shall devote little time to their elaboration. Though my argument does demand that I illustrate them briefly, my views will begin to depart from those long current only after I have done so.

Begin with accuracy, which for present purposes I take to include not only quantitative agreement but qualitative as well. Ultimately it proves the most nearly decisive of all the criteria, partly because it is less equivocal than the others but especially because predictive and explanatory powers, which depend on it, are characteristics that scientists are particularly unwilling to give up. Unfortunately, however, theories cannot always be discriminated in terms of accuracy. Copernicus's system, for example, was not more accurate than Ptolemy's until drastically revised by Kepler more than sixty years after Copernicus's death. If Kepler or someone else had not found other reasons to choose heliocentric astronomy, those improvements in accuracy would never have been made, and Copernicus's work might have been forgotten. More typically, of course, accuracy does permit discriminations, but not the sort that lead regularly to unequivocal choice. The oxygen theory, for example, was universally acknowledged to account for observed weight relations in chemical reactions, something the phlogiston theory had previously scarcely attempted to do. But the phlogiston theory, unlike its rival, could account for the metals' being much more alike than the ores from which they were formed. One theory thus matched experience better in one area, the other in another. To choose between them on the basis of accuracy, a scientist would need to decide the area in which accuracy was more significant. About that matter chemists could and did differ without violating any of the criteria outlined above, or any others yet to be suggested.

However important it may be, therefore, accuracy by itself is seldom or never a sufficient criterion for theory choice. Other criteria must function as well, but they do not eliminate problems. To illustrate I select just two – consistency and simplicity – asking how they functioned in the choice between the heliocentric and geocentric systems. As astronomical theories both Ptolemy's and Copernicus's were internally consistent, but their relation to related theories in other fields was very different. The stationary central earth was an essential ingredient of received physical theory, a tight-knit body of doctrine which explained, among other things, how stones fall, how water pumps function, and why the clouds move slowly across the skies. Heliocentric astronomy, which required the earth's motion, was inconsistent with the existing scientific explanation of these and other terrestrial phenomena. The consistency criterion, by itself, therefore, spoke unequivocally for the geocentric tradition.

Simplicity, however, favored Copernicus, but only when evaluated in a quite special way. If, on the one hand, the two systems were compared in terms of the actual computational labor required to predict the position of a planet at a particular time, then they proved substantially equivalent. Such computations were what astronomers did, and Copernicus's system offered them no labor-saving techniques; in that sense it was not simpler than Ptolemy's. If, on the other hand, one asked about the amount of mathematical apparatus required to explain, not the detailed quantitative motions of the planets, but merely their gross qualitative features – limited elongation, retrograde motion, and the like – then, as every schoolchild knows, Copernicus required only one circle per planet, Ptolemy two. In that sense the Copernican theory was the simpler, a fact vitally important to the choices made by both Kepler and Galileo and thus essential to the ultimate triumph of Copernicanism. But that sense of simplicity was not the only one available, nor even the one most natural to professional astronomers, men whose task was the actual computation of planetary position.

Because time is short and I have multiplied examples elsewhere, I shall here simply assert that these difficulties in applying standard criteria of choice are typical and that they arise no less forcefully in twentieth-century situations than in the earlier and better-known examples I have just sketched. When scientists must choose between competing theories, two men fully committed to the same list of criteria for choice may nevertheless reach different conclusions. Perhaps they interpret simplicity differently or have different convictions about the range of fields within which the consistency criterion must be met. Or perhaps they agree about these matters but differ about the relative weights to be accorded to these or to other criteria when several are deployed together. With respect to divergences of this sort, no set of choice criteria yet proposed is of any use. One can explain, as the historian characteristically does, why particular men made particular choices at particular times. But for that purpose one must go beyond the list of shared criteria to characteristics of the individuals who make the choice. One must, that is, deal with characteristics which vary from one scientist to another without thereby in the least jeopardizing their adherence to the canons that make science scientific. Though such canons do exist and should be discoverable (doubtless the criteria of choice with which I began are among them), they are not by themselves sufficient to determine the decisions of individual scientists. For that purpose the shared canons must be fleshed out in ways that differ from one individual to another.

Some of the differences I have in mind result from the individual's previous experience as a scientist. In what part of the field was he at work when confronted by the need to choose? How long had he worked there; how successful had he been; and how much of his work depended on concepts and techniques challenged by the new theory? Other factors relevant to choice lie outside the sciences. Kepler's early election of Copernicanism was due in part to his immersion in the Neoplatonic and Hermetic movements of his day; German Romanticism predisposed those it affected toward both recognition and acceptance of energy conservation; nineteenth-century British social thought had a similar influence on the availability and acceptability of Darwin's concept of the struggle for existence. Still other significant differences are functions of personality. Some scientists place more premium than others on originality and are correspondingly more willing to take risks; some scientists prefer comprehensive, unified theories to precise and detailed problem solutions of apparently narrower scope. Differentiating factors like these are described by my critics as subjective and are contrasted with the shared or objective criteria from which I began. Though I shall later question that use of terms, let me for the moment accept it. My point is, then, that every individual choice between competing theories depends on a mixture of objective and subjective factors, or of shared and individual criteria. Since the latter have not ordinarily figured in the philosophy of science, my emphasis upon them has made my belief in the former hard for my critics to see.

* * *

What I have said so far is primarily simply descriptive of what goes on in the sciences at times of theory choice. As description, furthermore, it has not been challenged by my critics, who reject instead my claim that these facts of scientific life have philosophic import. Taking up that issue, I shall begin to isolate some, though I think not vast, differences of opinion. Let me begin by asking how philosophers of science can for so long have neglected the subjective elements which, they freely grant, enter regularly into the actual theory choices made by individual scientists? Why have these elements seemed to them an index only of human weakness, not at all of the nature of scientific knowledge?

One answer to that question is, of course, that few philosophers, if any, have claimed to possess either a complete or an entirely well-articulated list of criteria. For some time, therefore, they could reasonably expect that further research would eliminate residual imperfections and produce an algorithm able to dictate rational, unanimous choice. Pending that achievement, scientists would have no alternative but to supply subjectively what the best current list of objective criteria still lacked. That some of them might still do so even with a perfected list at hand would then be an index only of the inevitable imperfection of human nature.

That sort of answer may still prove to be correct, but I think no philosopher still expects that it will. The search for algorithmic decision procedures has continued for some time and produced both powerful and illuminating results. But those results all presuppose that individual criteria of choice can be unambiguously stated and also that, if more than one proves relevant, an appropriate weight function is at hand for their joint application. Unfortunately, where the choice at issue is between scientific theories, little progress has been made toward the first of these desiderata and none toward the second. Most philosophers of science would, therefore, I think, now regard the sort of algorithm which has traditionally been sought as a not quite attainable ideal. I entirely agree and shall henceforth take that much for granted.

Even an ideal, however, if it is to remain credible, requires some demonstrated relevance to the situations in which it is supposed to apply. Claiming that such demonstration requires no recourse to subjective factors, my critics seem to appeal, implicitly or explicitly, to the well-known distinction between the contexts of discovery and of justification.[7] They concede, that is, that the subjective factors I invoke play a significant role in the discovery or invention of new theories, but they also insist that that inevitably intuitive process lies outside of the bounds of philosophy of science and is irrelevant to the question of scientific objectivity. Objectivity enters science, they continue, through the processes by which theories are tested, justified, or judged. Those processes do not, or at least need not, involve subjective factors at all. They can be governed by a set of (objective) criteria shared by the entire group competent to judge.

I have already argued that that position does not fit observations of scientific life and shall now assume that that much has been conceded. What is now at issue is a different point: whether or not this invocation of the distinction between contexts of discovery and of justification provides even a plausible and useful idealization. I think it does not and can best make my point by suggesting first a likely source of its apparent cogency. I suspect that my critics have been misled by science pedagogy or what I have elsewhere called textbook science. In science teaching, theories are presented together with exemplary applications, and those applications may be viewed as evidence. But that is not their primary pedagogic function (science students are distressingly willing to receive the word from professors and texts). Doubtless *some* of them were *part* of the evidence at the time actual decisions were being made, but they represent only a fraction of the considerations relevant to the decision process. The context of pedagogy differs almost as much from the context of justification as it does from that of discovery.

Full documentation of that point would require longer argument than is appropriate here, but two aspects of the way in which philosophers ordinarily demonstrate the relevance of choice criteria are worth noting. Like the science textbooks on which they are often modelled, books and articles on the philosophy of science refer again and again to the famous crucial experiments: Foucault's pendulum, which demonstrates the motion of the earth; Cavendish's demonstration of gravitational attraction; or Fizeau's measurement of the relative speed of sound in water and air. These experiments are paradigms of good reason for scientific choice; they illustrate the most effective of all the sorts of argument which could be available to a scientist uncertain which of two theories to follow; they are vehicles for the transmission of criteria of choice. But they also have another characteristic in common. By the time they were performed no scientist still needed to be convinced of the validity of the theory their outcome is now used to demonstrate. Those decisions had long since been made on the basis of significantly more equivocal evidence. The exemplary crucial experiments to which philosophers again and again refer would have been historically relevant to theory choice only if they had yielded unexpected results. Their use as illustrations provides needed economy to science pedagogy, but they scarcely illuminate the character of the choices that scientists are called upon to make.

Standard philosophical illustrations of scientific choice have another troublesome characteristic. The only arguments discussed are, as I have previously indicated, the ones favorable to the theory that, in fact, ultimately triumphed. Oxygen, we read, could explain weight relations, phlogiston could not; but nothing is said about the phlogiston theory's power or about the oxygen theory's limitations. Comparisons of Ptolemy's theory with Copernicus's proceed in the same way. Perhaps these examples should not be given since they contrast a developed theory with one still in its infancy. But philosophers regularly use them nonetheless. If the only result of their doing so were to simplify the decision situation, one could not object. Even historians do not claim to deal with the full factual complexity of the situations they describe. But these simplifications emasculate by making choice totally unproblematic. They eliminate, that is, one essential element of the decision situations that scientists must resolve if their field is to move ahead. In those situations there are always at least some good reasons for each possible choice. Considerations relevant to the context of discovery are then relevant to justification as well; scientists who share the concerns and sensibilities of the individual who discovers a new theory are *ipso facto* likely to appear disproportionately frequently among that theory's first supporters. That is why it has been difficult to construct algorithms for theory choice, and also why such difficulties have seemed so thoroughly worth resolving. Choices that present problems are the ones philosophers of science need to understand. Philosophically interesting decision procedures must function where, in their absence, the decision might still be in doubt.

That much I have said before, if only briefly. Recently, however, I have recognized another, subtler source for the apparent plausibility of my critics' position. To present it, I shall briefly describe a hypothetical dialogue with one of them. Both of us agree that each scientist chooses between competing theories by deploying some Bayesian algorithm which permits him to compute a value for $p(T,E)$, i.e., for the probability of a theory T on the evidence E available both to him and to the other members of his professional group at a particular period of time. "Evidence," furthermore, we both interpret broadly to include such considerations as simplicity and fruitfulness. My critic asserts, however, that there is only one such value of p, that corresponding to objective choice, and he believes that all rational members of the group must arrive at it. I assert, on the other hand, for reasons previously given, that the factors he calls objective are insufficient to determine in full any algorithm at all. For the sake of the discussion I have conceded that each individual has an algorithm and that all their algorithms have much in common. Nevertheless, I continue to hold that the algorithms of individuals are all ultimately different by virtue of the subjective considerations with which each must complete the objective criteria before any computations can be done. If my hypothetical critic is liberal, he may now grant that these subjective differences do play a role in determining the hypothetical algorithm on which each individual relies during the early stages of the competition between rival theories. But he is also likely to claim that, as evidence increases with the passage of time, the algorithms of different individuals converge to the algorithm of objective choice with which his presentation began. For him the increasing unanimity of individual choices is evidence for their increasing objectivity and thus for the elimination of subjective elements from the decision process.

So much for the dialogue, which I have, of course, contrived to disclose the non sequitur underlying an apparently plausible position. What converges as the evidence changes over time need only be the values of p that individuals compute from their individual algorithms. Conceivably those algorithms themselves also become more alike with time, but the ultimate unanimity of theory choice provides no evidence whatsoever that they do so. If subjective factors are required to account for the decisions that initially divide the profession, they may still be present later when the profession agrees. Though I shall not here argue the point, consideration of the occasions on which a scientific community divides suggests that they actually do so.

* * *

My argument has so far been directed to two points. It first provided evidence that the choices scientists make between competing theories depend not only on shared criteria – those my critics call objective – but also on idiosyncratic factors dependent on individual biography and personality. The latter are, in my critics' vocabulary, subjective, and the second part of my argument has attempted to bar some likely ways of denying their philosophic import. Let me now shift to a more positive approach, returning briefly to the list of shared criteria – accuracy, simplicity, and the like – with which I began. The considerable effectiveness of such criteria does not, I now wish to suggest, depend on their being sufficiently articulated to dictate the choice of each individual who subscribes to them. Indeed, if they were articulated to that extent, a behavior mechanism fundamental to scientific advance would cease to function. What the tradition sees as eliminable imperfections in its rules of choice I take to be in part responses to the essential nature of science.

As so often, I begin with the obvious. Criteria that influence decisions without specifying what those decisions must be are familiar in many aspects of human life. Ordinarily, however, they are called, not criteria or rules, but maxims, norms, or values. Consider maxims first. The individual who invokes them when choice is urgent usually finds them frustratingly vague and often also in conflict one with another.

Contrast "He who hesitates is lost" with "Look before you leap," or compare "Many hands make light work" with "Too many cooks spoil the broth." Individually maxims dictate different choices, collectively none at all. Yet no one suggests that supplying children with contradictory tags like these is irrelevant to their education. Opposing maxims alter the nature of the decision to be made, highlight the essential issues it presents, and point to those remaining aspects of the decision for which each individual must take responsibility himself. Once invoked, maxims like these alter the nature of the decision process and can thus change its outcome.

Values and norms provide even clearer examples of effective guidance in the presence of conflict and equivocation. Improving the quality of life is a value, and a car in every garage once followed from it as a norm. But quality of life has other aspects, and the old norm has become problematic. Or again, freedom of speech is a value, but so is preservation of life and property. In application, the two often conflict, so that judicial soul-searching, which still continues, has been required to prohibit such behavior as inciting to riot or shouting fire in a crowded theater. Difficulties like these are an appropriate source for frustration, but they rarely result in charges that values have no function or in calls for their abandonment. That response is barred to most of us by an acute consciousness that there are societies with other values and that these value differences result in other ways of life, other decisions about what may and what may not be done.

I am suggesting, of course, that the criteria of choice with which I began function not as rules, which determine choice, but as values, which influence it. Two men deeply committed to the same values may nevertheless, in particular situations, make different choices as, in fact, they do. But that difference in outcome ought not to suggest that the values scientists share are less than critically important either to their decisions or to the development of the enterprise in which they participate. Values like accuracy, consistency, and scope may prove ambiguous in application, both individually and collectively; they may, that is, be an insufficient basis for a *shared* algorithm of choice. But they do specify a great deal: what each scientist must consider in reaching a decision, what he may and may not consider relevant, and what he can legitimately be required to report as the basis for the choice he has made. Change the list, for example by adding social utility as a criterion, and some particular choices will be different, more like those one expects from an engineer. Subtract accuracy of fit to nature from the list, and the enterprise that results may not resemble science at all, but perhaps philosophy instead. Different creative disciplines are characterized, among other things, by different sets of shared values. If philosophy and engineering lie too close to the sciences, think of literature or the plastic arts. Milton's failure to set *Paradise Lost* in a Copernican universe does not indicate that he agreed with Ptolemy but that he had things other than science to do.

Recognizing that criteria of choice can function as values when incomplete as rules has, I think, a number of striking advantages. First, as I have already argued at length, it accounts in detail for aspects of scientific behavior which the tradition has seen as anomalous or even irrational. More important, it allows the standard criteria to function fully in the earliest stages of theory choice, the period when they are most needed but when, on the traditional view, they function badly or not at all. Copernicus was responding to them during the years required to convert heliocentric astronomy from a global conceptual scheme to mathematical machinery for predicting planetary position. Such predictions were what astronomers valued; in their

absence, Copernicus would scarcely have been heard, something which had happened to the idea of a moving earth before. That his own version convinced very few is less important than his acknowledgment of the basis on which judgments would have to be reached if heliocentricism were to survive. Though idiosyncrasy must be invoked to explain why Kepler and Galileo were early converts to Copernicus's system, the gaps filled by their efforts to perfect it were specified by shared values alone.

That point has a corollary which may be more important still. Most newly suggested theories do not survive. Usually the difficulties that evoked them are accounted for by more traditional means. Even when this does not occur, much work, both theoretical and experimental, is ordinarily required before the new theory can display sufficient accuracy and scope to generate widespread conviction. In short, before the group accepts it, a new theory has been tested over time by the research of a number of men, some working within it, others within its traditional rival. Such a mode of development, however, *requires* a decision process which permits rational men to disagree, and such disagreement would be barred by the shared algorithm which philosophers have generally sought. If it were at hand, all conforming scientists would make the same decision at the same time. With standards for acceptance set too low, they would move from one attractive global viewpoint to another, never giving traditional theory an opportunity to supply equivalent attractions. With standards set higher, no one satisfying the criterion of rationality would be inclined to try out the new theory, to articulate it in ways which showed its fruitfulness or displayed its accuracy and scope. I doubt that science would survive the change. What from one viewpoint may seem the looseness and imperfection of choice criteria conceived as rules may, when the same criteria are seen as values, appear an indispensable means of spreading the risk which the introduction or support of novelty always entails.

Even those who have followed me this far will want to know how a value-based enterprise of the sort I have described can develop as a science does, repeatedly producing powerful new techniques for prediction and control. To that question, unfortunately, I have no answer at all, but that is only another way of saying that I make no claim to have solved the problem of induction. If science did progress by virtue of some shared and binding algorithm of choice, I would be equally at a loss to explain its success. The lacuna is one I feel acutely, but its presence does not differentiate my position from the tradition.

It is, after all, no accident that my list of the values guiding scientific choice is, as nearly as makes any difference, identical with the tradition's list of rules dictating choice. Given any concrete situation to which the philosopher's rules could be applied, my values would function like his rules, producing the same choice. Any justification of induction, any explanation of why the rules worked, would apply equally to my values. Now consider a situation in which choice by shared rules proves impossible, not because the rules are wrong but because they are, as rules, intrinsically incomplete. Individuals must then still choose and be guided by the rules (now values) when they do so. For that purpose, however, each must first flesh out the rules, and each will do so in a somewhat different way even though the decision dictated by the variously completed rules may prove unanimous. If I now assume, in addition, that the group is large enough so that individual differences distribute on some normal curve, then any argument that justifies the philosopher's choice by rule should be immediately adaptable to my choice by value. A group too small, or a distribution excessively skewed by external historical pressures, would, of course,

prevent the argument's transfer.[8] But those are just the circumstances under which scientific progress is itself problematic. The transfer is not then to be expected.

I shall be glad if these references to a normal distribution of individual differences and to the problem of induction make my position appear very close to more traditional views. With respect to theory choice, I have never thought my departures large and have been correspondingly startled by such charges as "mob psychology," quoted at the start. It is worth noting, however, that the positions are not quite identical, and for that purpose an analogy may be helpful. Many properties of liquids and gases can be accounted for on the kinetic theory by supposing that all molecules travel at the same speed. Among such properties are the regularities known as Boyle's and Charles's law. Other characteristics, most obviously evaporation, cannot be explained in so simple a way. To deal with them one must assume that molecular speeds differ, that they are distributed at random, governed by the laws of chance. What I have been suggesting here is that theory choice, too, can be explained only in part by a theory which attributes the same properties to all the scientists who must do the choosing. Essential aspects of the process generally known as verification will be understood only by recourse to the features with respect to which men may differ while still remaining scientists. The tradition takes it for granted that such features are vital to the process of discovery, which it at once and for that reason rules out of philosophical bounds. That they may have significant functions also in the philosophically central problem of justifying theory choice is what philosophers of science have to date categorically denied.

* * *

What remains to be said can be grouped in a somewhat miscellaneous epilogue. For the sake of clarity and to avoid writing a book, I have throughout this paper utilized some traditional concepts and locutions about the viability of which I have elsewhere expressed serious doubts. For those who know the work in which I have done so, I close by indicating three aspects of what I have said which would better represent my views if cast in other terms, simultaneously indicating the main directions in which such recasting should proceed. The areas I have in mind are: value invariance, subjectivity, and partial communication. If my views of scientific development are novel − a matter about which there is legitimate room for doubt − it is in areas such as these, rather than theory choice, that my main departures from tradition should be sought.

Throughout this paper I have implicitly assumed that, whatever their initial source, the criteria or values deployed in theory choice are fixed once and for all, unaffected by their participation in transitions from one theory to another. Roughly speaking, but only very roughly, I take that to be the case. If the list of relevant values is kept short (I have mentioned five, not all independent) and if their specification is left vague, then such values as accuracy, scope, and fruitfulness are permanent attributes of science. But little knowledge of history is required to suggest that both the application of these values and, more obviously, the relative weights attached to them have varied markedly with time and also with the field of application. Furthermore, many of these variations in value have been associated with particular changes in scientific theory. Though the experience of scientists provides no philosophical justification for the values they deploy (such justification would solve the problem of induction), those values are in part learned from that experience, and they evolve with it.

The whole subject needs more study (historians have usually taken scientific values, though not scientific methods, for granted), but a few remarks will illustrate the sort of variations I have in mind. Accuracy, as a value, has with time increasingly denoted quantitative or numerical agreement, sometimes at the expense of qualitative. Before early modern times, however, accuracy in that sense was a criterion only for astronomy, the science of the celestial region. Elsewhere it was neither expected nor sought. During the seventeenth century, however, the criterion of numerical agreement was extended to mechanics, during the late eighteenth and early nineteenth centuries to chemistry and such other subjects as electricity and heat, and in this century to many parts of biology. Or think of utility, an item of value not on my initial list. It too has figured significantly in scientific development, but far more strongly and steadily for chemists than for, say, mathematicians and physicists. Or consider scope. It is still an important scientific value, but important scientific advances have repeatedly been achieved at its expense, and the weight attributed to it at times of choice has diminished correspondingly.

What may seem particularly troublesome about changes like these is, of course, that they ordinarily occur in the aftermath of a theory change. One of the objections to Lavoisier's new chemistry was the roadblocks with which it confronted the achievement of what had previously been one of chemistry's traditional goals: the explanation of qualities, such as color and texture, as well as of their changes. With the acceptance of Lavoisier's theory such explanations ceased for some time to be a value for chemists; the ability to explain qualitative variation was no longer a criterion relevant to the evaluation of chemical theory. Clearly, if such value changes had occurred as rapidly or been as complete as the theory changes to which they related, then theory choice would be value choice, and neither could provide justification for the other. But, historically, value change is ordinarily a belated and largely unconscious concomitant of theory choice, and the former's magnitude is regularly smaller than the latter's. For the functions I have here ascribed to values, such relative stability provides a sufficient basis. The existence of a feedback loop through which theory change affects the values which led to that change does not make the decision process circular in any damaging sense.

About a second respect in which my resort to tradition may be misleading, I must be far more tentative. It demands the skills of an ordinary language philosopher, which I do not possess. Still, no very acute ear for language is required to generate discomfort with the ways in which the terms "objectivity" and, more especially, "subjectivity" have functioned in this paper. Let me briefly suggest the respects in which I believe language has gone astray. "Subjective" is a term with several established uses: in one of these it is opposed to "objective," in another to "judgmental." When my critics describe the idiosyncratic features to which I appeal as subjective, they resort, erroneously I think, to the second of these senses. When they complain that I deprive science of objectivity, they conflate that second sense of subjective with the first.

A standard application of the term "subjective" is to matters of taste, and my critics appear to suppose that that is what I have made of theory choice. But they are missing a distinction standard since Kant when they do so. Like sensation reports, which are also subjective in the sense now at issue, matters of taste are undiscussable. Suppose that, leaving a movie theater with a friend after seeing a Western, I exclaim: "How I liked that terrible potboiler!" My friend, if he disliked the film, may tell me I have low tastes, a matter about which, in these circumstances, I would readily agree. But, short of saying that I lied, he cannot disagree with my report that I liked the film or try to persuade me that what I said about my reaction was wrong. What is discussable in my remark is not my

characterization of my internal state, my exemplification of taste, but rather my *judgment* that the film was a potboiler. Should my friend disagree on that point, we may argue most of the night, each comparing the film with good or great ones we have seen, each revealing, implicitly or explicitly, something about how he *judges* cinematic merit, about his aesthetic. Though one of us may, before retiring, have persuaded the other, he need not have done so to demonstrate that our difference is one of judgment, not taste.

Evaluations or choices of theory have, I think, exactly this character. Not that scientists never say merely, I like such and such a theory, or I do not. After 1926 Einstein said little more than that about his opposition to the quantum theory. But scientists may always be asked to explain their choices, to exhibit the bases for their judgments. Such judgments are eminently discussable, and the man who refuses to discuss his own cannot expect to be taken seriously. Though there are, very occasionally, leaders of scientific taste, their existence tends to prove the rule. Einstein was one of the few, and his increasing isolation from the scientific community in later life shows how very limited a role taste alone can play in theory choice. Bohr, unlike Einstein, did discuss the bases for his judgment, and he carried the day. If my critics introduce the term "subjective" in a sense that opposes it to judgmental – thus suggesting that I make theory choice undiscussable, a matter of taste – they have seriously mistaken my position.

Turn now to the sense in which "subjectivity" is opposed to "objectivity," and note first that it raises issues quite separate from those just discussed. Whether my taste is low or refined, my report that I liked the film is objective unless I have lied. To my judgment that the film was a potboiler, however, the objective-subjective distinction does not apply at all, at least not obviously and directly. When my critics say I deprive theory choice of objectivity, they must, therefore, have recourse to some very different sense of subjective, presumably the one in which bias and personal likes or dislikes function instead of, or in the face of, the actual facts. But that sense of subjective does not fit the process I have been describing any better than the first. Where factors dependent on individual biography or personality must be introduced to make values applicable, no standards of factuality or actuality are being set aside. Conceivably my discussion of theory choice indicates some limitations of objectivity, but not by isolating elements properly called subjective. Nor am I even quite content with the notion that what I have been displaying are limitations. Objectivity ought to be analyzable in terms of criteria like accuracy and consistency. If these criteria do not supply all the guidance that we have customarily expected of them, then it may be the meaning rather than the limits of objectivity that my argument shows.

Turn, in conclusion, to a third respect, or set of respects, in which this paper needs to be recast. I have assumed throughout that the discussions surrounding theory choice are unproblematic, that the facts appealed to in such discussions are independent of theory, and that the discussions' outcome is appropriately called a choice. Elsewhere I have challenged all three of these assumptions, arguing that communication between proponents of different theories is inevitably partial, that what each takes to be facts depends in part on the theory he espouses, and that an individual's transfer of allegiance from theory to theory is often better described as conversion than as choice. Though all these theses are problematic as well as controversial, my commitment to them is undiminished. I shall not now defend them, but must at least attempt to indicate how what I have said here can be adjusted to conform with these more central aspects of my view of scientific development.

For that purpose I resort to an analogy I have developed in other places. Proponents of different theories are, I have claimed, like native speakers of different languages. Communication between them goes on by translation, and it raises all translation's familiar difficulties. That analogy is, of course, incomplete, for the vocabulary of the two theories may be identical, and most words function in the same ways in both. But some words in the basic as well as in the theoretical vocabularies of the two theories — words like "star" and "planet," "mixture" and "compound," or "force" and "matter" — do function differently. Those differences are unexpected and will be discovered and localized, if at all, only by repeated experience of communication breakdown. Without pursuing the matter further, I simply assert the existence of significant limits to what the proponents of different theories can communicate to one another. The same limits make it difficult or, more likely, impossible for an individual to hold both theories in mind together and compare them point by point with each other and with nature. That sort of comparison is, however, the process on which the appropriateness of any word like "choice" depends.

Nevertheless, despite the incompleteness of their communication, proponents of different theories can exhibit to each other, not always easily, the concrete technical results achievable by those who practice within each theory. Little or no translation is required to apply at least some value criteria to those results. (Accuracy and fruitfulness are most immediately applicable, perhaps followed by scope. Consistency and simplicity are far more problematic.) However incomprehensible the new theory may be to the proponents of tradition, the exhibit of impressive concrete results will persuade at least a few of them that they must discover how such results are achieved. For that purpose they must learn to translate, perhaps by treating already published papers as a Rosetta stone or, often more effective, by visiting the innovator, talking with him, watching him and his students at work. Those exposures may not result in the adoption of the theory; some advocates of the tradition may return home and attempt to adjust the old theory to produce equivalent results. But others, if the new theory is to survive, will find that at some point in the language-learning process they have ceased to translate and begun instead to speak the language like a native. No process quite like choice has occurred, but they are practicing the new theory nonetheless. Furthermore, the factors that have led them to risk the conversion they have undergone are just the ones this paper has underscored in discussing a somewhat different process, one which, following the philosophical tradition, it has labelled theory choice.

Notes

1   *The Structure of Scientific Revolutions*, 2d ed. (Chicago, 1970), pp. 148, 151–52, 159. All the passages from which these fragments are taken appeared in the same form in the first edition, published in 1962.

2   Ibid., p. 170.

3   Imre Lakatos, "Falsification and the Methodology of Scientific Research Programmes," in I. Lakatos and A. Musgrave, eds, *Criticism and the Growth of Knowledge* (Cambridge, Cambridge University Press, 1970), pp. 91–195. The quoted phrase, which appears on p. 178, is italicized in the original.

4   Dudley Shapere, "Meaning and Scientific Change," in R.G. Colodny, ed., *Mind and Cosmos: Essays in Contemporary Science and Philosophy*, University of Pittsburgh Series in the Philosophy of Science,

vol. 3 (Pittsburgh, University of Pittsburgh Press, 1966), pp. 41–85. The quotation will be found on p. 67.

5   Israel Scheffler, *Science and Subjectivity* (Indianapolis, Bobbs-Merrill, 1967), p. 81.

6   The last criterion, fruitfulness, deserves more emphasis than it has yet received. A scientist choosing between two theories ordinarily knows that his decision will have a bearing on his subsequent research career. Of course he is especially attracted by a theory that promises the concrete successes for which scientists are ordinarily rewarded.

7   The least equivocal example of this position is probably the one developed in Scheffler, *Science and Subjectivity*, chap. 4.

8   If the group is small, it is more likely that random fluctuations will result in its members' sharing an atypical set of values and therefore making choices different from those that would be made by a larger and more representative group. External environment – intellectual, ideological, or economic – must systematically affect the value system of much larger groups, and the consequences can include difficulties in introducing the scientific enterprise to societies with inimical values or perhaps even the end of that enterprise within societies where it had once flourished. In this area, however, great caution is required. Changes in the environment where science is practiced can also have fruitful effects on research. Historians often resort, for example, to differences between national environments to explain why particular innovations were initiated and at first disproportionately pursued in particular countries, e.g. Darwinism in Britain, energy conservation in Germany. At present we know substantially nothing about the minimum requisites of the social milieux within which a sciencelike enterprise might flourish.

PART TWO

Science, race and gender

INTRODUCTION TO PART TWO

IN PART ONE it was mentioned that in the past poor *scientific* theories were the result of, and produced in support of, racism and sexism. The contemporary scientific view is that variability in race and sex are not correlated with variability in intelligence and aptitude to any great extent, if at all. However, there is still considerable dispute at the margins. For example, some scientists claim to have identified some small but significant differences in aspects of brain function between the sexes. The notion of race is the subject of intense recent debate, with some arguing that it is meaningless as a scientific category, and others argue that it deserves a role as a predictive and explanatory category and/or as a political and legal category. It is important to note that some feminists are concerned to defend the essential difference between the sexes, and also that some people believe the concept of race is important for solidarity and emancipation. Finally, the so-called 'science wars' centrally concern the question of whether science itself can or should aspire to be racially and gender neutral.

As we saw in Part One, Stephen Jay Gould discussed pseudo-science within mainstream science that involved erroneous claims about the intelligence of human beings expressed in terms of racial generalisations. In this part we will look much more closely at the notion of race, and in particular how folk concepts of race relate to theoretical concepts, if any, that are scientifically respectable and useful. (By 'folk concepts' we mean the everyday concepts that people have.) While there is much debate about whether the concept of race has any worth at all in science, it is uncontroversial that folk characterisations of race do not well approximate scientific ones. For example, it is often the case that ethnic groups between which there is so-called interracial tension are in fact biologically closely related, though they may be culturally, religiously and linguistically quite different. We will also look at the concept of gender and how it relates to the biological notions of male and female. In both cases we have folk concepts that are heavily value-laden and scientific ones associated with them. Any association between the everyday and scientific concepts must be very carefully considered, otherwise fallacious reasoning may result from conflating them, inferring false claims about the one from true claims about the other. We ought to be very open-minded about the possible outcomes of our investigations. It may be that concepts of race and gender are sometimes important and useful in both social, political and legal contexts, and/or in science. But it may also be the case that the concepts differ greatly between, and perhaps also within, contexts depending on the purpose to which they are to be put. So, for example, the concept of race that is likely, if at all, to be useful in ensuring representative results in a drug trial or genetic study may not be the same as that employed in anti-discriminatory legislation.

Insofar as distinct groups are reasonably identified, it is of fundamental importance that between-group differences in populations may exist but nonetheless be
small compared to within-group differences. This is the case with whatever differences there are in the brains of men and women. Even if it is the case that there are
small overall differences in the average brains of men and women, this does not imply
that whether a given individual has a given characteristic is well predicted by whether
they are male or female. This is true even when it comes to physical characteristics.
How strong people are is best predicted by what kind of lives they live, and there are
a great many women with greater upper body strength than a great many men, even
though there is a biological propensity for men to have greater upper body strength.
The differences in aptitude for various cognitive tasks that have been found between
men and women are very small, and easily swamped by education and other factors.
While very few women used to pursue science and mathematics to doctoral and post-
doctoral level, now very many do. It has become clear that the barriers to female
achievement in these domains were cultural, not biological. Likewise, the relative
degrees to which different ethnic groups participate and achieve in education and the
various professions have been transformed by cultural, not biological, changes.

The issues of race and gender are not the same, though there are some parallels. So,
turning specifically to race, we must begin by noting that folk racial taxonomies vary
greatly from culture to culture and are often completely arbitrary both with respect to
actual human geographic origins and cultures, and with respect to such biological
differences as exist among groups. The next thing to note is that superficial similarities
and diversity are often a very poor guide to genetic diversity. For example, the genetic
diversity among Africans is greater than the genetic diversity between Africans and
the rest of the human population of the world. One of the most natural features of
people that is associated with different races in people's minds is skin colour; however,
this is a biological superficial characteristic that may disappear completely from view
in a small number of generations, while much genetic material is preserved. We must
also note, of course, that migration has been a pervasive feature of human history and
prehistory and that populations that seem indigenous may in fact be relatively recent
arrivals from the scientific point of view. Furthermore, local historical circumstances
may make small differences seem more than they are. As noted above, people tend to
focus on what may be superficial differences from the biological perspective.

Folk racial categorisations are often heavily laden with values and norms, and
disentangling them from their residue, if any, is hard. The same goes for gender and
its relationship to biological sex. For example, a woman's biological role in bearing
children may be overlaid with norms about the nuclear family that have no biological
grounding whatsoever. Furthermore, evolution is fundamentally about how biological
individuals change over time, and so arguments for the fixity of social roles (for
example, child rearing in the context of an advanced industrialized society) based on
evolutionary function (bearing children) are spurious. Most of our traits were fixed in
an environment that in many ways differs massively from the one in which we must
now survive. The extent to which human individuals should depart from historic roles

depends on what is adaptive, but when it comes to race, biology delivers one clear lesson: genetic diversity is associated with healthier and better adapted populations. Whereas we often find ideologues arguing for racial inbreeding, insofar as we can make sense of the idea of race at all, racial crossbreeding helps avoid congenital diseases and other damaging conditions.

In the first reading Sally Haslanger explicitly argues for a plurality of concepts of race and gender rather than searching for the one true set of categories. This may be sensible. For example, it may be politically and socially important to have a certain racial or ethnic group identified in the laws of a particular society because that group is perceived to be important and identifiable within the culture and has historically been persecuted. In such circumstances it seems the kind of group identifications that should be made need not be the same at all as those made by science. Rather, depending on the purpose, different concepts are appropriate.

Haslanger begins with a helpful distinction between investigating a particular concept and investigating what, if any, kind it picks out in the world. As is well known, at one time the concept 'fish' was applied to whales until the concept of mammals was developed. There is then the further question as to whether a concept is useful. Concepts that track the objective kinds of things in the world would seem to often be more useful, but not always; as we have seen above where a concept of a particular racial subgroup may have little scientific value while being useful in a legal code. Haslanger refers to the 'extension' of a concept; this is the collection of things in the world that fall under it, if any. There is no extension of the concept 'unicorn', or it is the empty set. She is primarily interested in pursuing the analytical project of deciding which concepts we should bother using rather than investigating existing concepts and their extensions. When it comes to the uses to which concepts of race and gender can be put, Haslanger points out that these may go beyond the need to state truths about the world. In the case of gender, Haslanger discusses the 'commonality problem' and the 'normativity problem'. The former concerns what, if anything, can be identified as uniting all women other than biological sex; the latter concerns the danger that any such identification will marginalize some females. Haslanger concludes that both gender and race are real social categories but contingent ones that could be replaced by others. She claims there is no one right way of classifying people in either respect, and that our purposes must dictate the concepts we use. Haslanger is clear that her purpose is that of combating injustice, and much of her discussion concerns feminist conceptions of gender that involve defining 'man' and 'woman' in terms of privilege and subordination, respectively. She then turns to a comparative discussion of race and racial categories.

Whereas Haslanger begins with the social, the political and the ethical, Helena Cronin examines gender from a biological and, specifically, an evolutionary view. She also adopts a particularly reductionist and much-criticised interpretation of Darwin's theory of evolution by natural selection, namely the 'selfish gene' account of Richard Dawkins. According to the latter, the units of natural selection are genes and only genes; it is genes that reproduce themselves via organisms, not the other way round.

Cronin presents a host of examples that show how different organisms solve the problems of allocating resource to reproduction and parenting, and how the form and behaviour of organisms can be explained by their emergence from the competition between the genes of the parents. She points out that the 'battle of the sexes' involves competition at the margins of what is fundamentally a co-operative endeavour for human beings, namely sexual reproduction. The reading from Evelyn Fox Keller shows how complex the issues concerning genetic determination are, and undermines the idea that biology reduces to genetics.

Lucius Outlaw is, like Haslanger, explicitly concerned with theorizing about race in the cause of emancipation. He outlines the history of the concept of 'race' and its complex relationship with science and politics as well as with the practice of slavery, before explaining and advocating a 'critical theory' of race. The following discussion by Robin Andreasen is analytical and based on our best contemporary biology. It poses a simple question: is race biologically real, or is it a social construct, or is it perhaps both? Within biology it has been argued for some time that there is no need for any taxonomic units smaller than species, such as subspecies. Indeed, many contemporary biologists argue that even the species concept is redundant in the light of the accurate and detailed tracking of heredity made possible by genetics. Another line of argument based on the study of human genetics concludes that, whether or not the notion of a subspecies is generally needed, there are no human subspecies. Andreasen argues that there are biologically real races where these are breeding populations, but also that this is compatible with the social construction of racial categories because the former cladistic concept of race that is well-defined in biology cross-cuts the commonsense notion of race that is at issue in social and political contexts. In a direct response to Andreasen, and to related ideas due to Philip Kitcher and others, Joshua Glasgow argues that it is wrong to think that the cladistic idea of a breeding population provides the basis for a biological concept of race, on the grounds that the two notions are simply too different. The reader should appreciate how such debates in the philosophy of science depend on both careful philosophical work analyzing concepts and arguments, and detailed knowledge of the relevant science.

We saw in Part One how factor analysis provides the foundation for the theory of IQ. Clark Glymour's article explains and assesses the use of the former and the statistical technique of regression in a critique of an infamous social science book that purported to show that intelligence is a better predictor of such things as income, job performance, unwanted pregnancy and criminality than education or parental socio-economic status. The book also claimed that there are racial differences in intelligence and suggested that they are genetic. Glymour agrees with Gould that factor analysis can be unreliable but disagrees with him about why. His article contains a wealth of information about the methods by which causal information can be extracted from statistical information and their limitations and dangers. The reader will learn how the issues in the headlines about nature versus nurture explanations of human characteristics depend on extremely subtle and complicated matters of scientific theory and methodology.

In the 1990s there arose intellectual disputes about science, conducted especially in North America, that have become known as the 'science wars', with scientific realists and empiricist instrumentalists alike pitted against those who denied the objectivity of science. We have seen that science has been contaminated by racial and gender prejudice. Some critics of science have gone much further and argued that science is never value-neutral and so always incorporates and reflects the dominant ideology and interests of those who produce it. These critiques of science come in various forms (Koertge, 2000). It is common for appeal to be made to many issues – such as the theory–ladenness of observation and underdetermination – that are also discussed in philosophy of science that does not regard itself as critical. It seems that investigations into the scientific method failed to show how logic and probability theory alone could settle the question of what scientists ought to believe. At the same time, Kuhn and others argued that historians of science had paid insufficient attention to the complexity of theory change and the contribution of social factors. If one combines this with knowledge of appalling cases of pseudo-science that were much to the detriment of already disadvantaged social groups then it is not too hard to be led to the conclusion that science reflects more than the facts.

However, it is one thing to admit a role for social factors in science and to appreciate that sometimes bias enters into scientific findings, and quite another to deny that science is generally reliable, progressive and an objectively good basis for belief. Nonetheless, many critics of science have denied some or all of these claims. The sociology of knowledge programme describes the history of science without according the results of experiments a privileged role – the emphasis being instead on personal and social properties and relations. Some feminist criticism of science argues that various fundamental aspects of scientific thought and methodology are inherently androcentric. Similarly, it has been claimed that science is Eurocentric. Other critics are more moderate and argue that objectivity can be achieved in science if effort is expended to ensure that it is representative and no particular group or social values are dominant. Koertge argues that the best of such critical work is well worth engaging with, but she also argues that there has been a lot of academic commentary about science written by people who know very little about it. We have already seen above in our consideration of the status of the concept of race that the arguments in philosophy of science can depend on the details of our best science.

In 1996 Alan Sokal published a spoof article in the journal *Social Text*. 'Transgressing the boundaries: Towards a transformative hermeneutics of quantum gravity' purported to be a serious discussion of the relationship between the most advanced mathematical physics and various postmodernist and poststructuralist ideas. However, it was submitted to the journal to test whether complete nonsense about science and mathematics would be mistaken for profound insight, if it was surrounded by many citations of the right people, and suitably sprinkled with buzzwords. It seems that Sokal told the editors what they wanted to hear, lent scientific credibility to their favoured theorists, and did so in the kind of language they liked, and that this was sufficient for publication. Yet, because of the erroneous claims about scientific matters upon which the reasoning relied, and because of the passages of pure nonsense, the article was completely intellectually worthless. In 'Some comments on the parody' (Sokal and Bricmont, pp. 241–7) Sokal explains his hoax and why he did it. Sokal's article was the single most important intervention in the 'science wars'. To an extent they were never resolved and constructivist and other critiques still challenge science and philosophy of science, which tends to assume, as we have explicitly in this book, that science has a privileged epistemic status that we are trying to understand.

Further reading

Andreasen, R.O. 2004 'The cladistic race concept: A defense', *Biology and Philosophy* 19:425–42.

Keller, E.F. 1996 *Reflections on Gender and Science*. Yale University Press.

Koertge, N. 2000 '"New Age" Philosophies of Science: Constructivism, Feminism and Postmodernism', *British Journal for the Philosophy of Science* 51(4): 667–83.

Longino, H.E. 1990 *Science as Social Knowledge: Values and Objectivity in Scientific Inquiry*. Princeton University Press.

Sesardić, N. 2000 'Philosophy of science that ignores science: Race, IQ, and heritability', *Philosophy of Science* 67:580–602.

Sokal, A.D. and Bricmont, J. 2003 *Intellectual Impostures: Postmodern Philosophers' Abuse of Science*. Profile Books.

Zack, N. 2002 *Philosophy of Science and Race*. Routledge.

Chapter 5

Sally Haslanger

GENDER AND RACE: (WHAT) ARE THEY? (WHAT) DO WE WANT THEM TO BE?[1]

If her functioning as a female is not enough to define woman, if we decline also to explain her through "the eternal feminine," and if nevertheless we admit, provisionally, that women do exist, then we must face the question: what is a woman?

Simone de Beauvoir, *The Second Sex*

I guess you could chuckle and say that I'm just a woman trapped in a woman's body.

Ellen DeGeneres, *My Point . . . and I Do Have One*

The truth is that there are no races: there is nothing in the world that can do all we ask race to do for us.

Kwame Anthony Appiah, *In My Father's House*

It is always awkward when someone asks me informally what I'm working on and I answer that I'm trying to figure out what gender is. For outside a rather narrow segment of the academic world, the term 'gender' has come to function as the polite way to talk about the sexes. And one thing people feel pretty confident about is their knowledge of the difference between males and females. Males are those human beings with a range of familiar primary and secondary sex characteristics, most important being the penis; females are those with a different set, most important being the vagina or, perhaps, the

uterus. Enough said. Against this background, it isn't clear what could be the point of an inquiry, especially a philosophical inquiry, into "what gender is".

But within that rather narrow segment of the academic world concerned with gender issues, not only is there no simple equation of sex and gender, but the seemingly straightforward anatomical distinction between the sexes has been challenged as well. What began as an effort to note that men and women differ socially as well as anatomically has prompted an explosion of different uses of the term 'gender'. Within these debates, not only is it unclear what gender is and how we should go about understanding it, but whether it is anything at all.

The situation is similar, if not worse, with respect to race. The self-evidence of racial distinctions in everyday American life is at striking odds with the uncertainty about the category of race in law and the academy. Work in the biological sciences has informed us that our practices of racial categorization don't map neatly onto any useful biological classification; but that doesn't settle much, if anything. For what should we make of our tendency to classify individuals according to race, apparently on the basis of physical appearance? And what are we to make of the social and economic consequences of such classifications? Is race real or is it not?

This paper is part of a larger project, the goal of which is to offer accounts of gender and race informed by a feminist epistemology. Here my aim is to sketch some of the central ideas of those accounts. Let me emphasize at the beginning that I do not want to argue that my proposals provide the only acceptable ways to define race or gender; in fact, the epistemological framework I employ is explicitly designed to allow for different definitions responding to different concerns. It is sometimes valuable to consider race or gender alone or to highlight the differences between them; however, here I will begin by exploring some significant parallels. Although there are dangers in drawing close analogies between gender and race, I hope my discussion will show that theorizing them together can provide us valuable resources for thinking about a wide range of issues. Working with a model that demonstrates some of the parallels between race and gender also helps us locate important differences between them.

I The question(s)

It is useful to begin by reflecting on the questions: "What is gender?", "What is race?" and related questions such as: "What is it to be a man or a woman?"[2], "What is it to be White? Latino? or Asian?" There are several different ways to understand, and so respond to, questions of the form, "What is X?" or "What is it to be an X?" For example, the question "What is knowledge?" might be construed in several ways. One might be asking: What is our concept of knowledge? (looking to a priori methods for an answer). On a more naturalistic reading one might be asking: What (natural) kind (if any) does our epistemic vocabulary track? Or one might be undertaking a more revisionary project: What is the point of having a concept of knowledge? What concept (if any) would do that work best?[3] These different sorts of projects

cannot be kept entirely distinct, but draw upon different methodological strategies. Returning to the questions, "What is race?" or "What is gender?" we can distinguish, then, three projects with importantly different priorities: *conceptual*, *descriptive*, and *analytical*.

A *conceptual* inquiry into race or gender would seek an articulation of our *concepts* of race or gender (Riley 1988). To answer the conceptual question, one way to proceed would be to use the method of reflective equilibrium. (Although within the context of analytic philosophy this might be seen as a call for a conceptual *analysis* of the term(s), I want to reserve the term 'analytical' for a different sort of project, described below.)

In contrast to the conceptual project, a *descriptive* project is not concerned with exploring the nuances of our concepts (or anyone else's for that matter); it focuses instead on their extension. Here, the task is to develop potentially more accurate concepts through careful consideration of the phenomena, usually relying on empirical or quasi-empirical methods. Paradigm descriptive projects occur in studying natural phenomena. I offered the example of naturalistic approaches to knowledge above: the goal is to determine the (natural) kind, if any, we are referring to (or are attempting to refer to) with our epistemic talk. However, a descriptive approach need not be confined to a search for *natural* or *physical* kinds; inquiry into what it is to be, e.g., a human right, a citizen, a democracy, might begin by considering the full range of what has counted as such to determine whether there is an underlying (possibly social) kind that explains the temptation to group the cases together. Just as natural science can enrich our "folk" conceptualization of natural phenomena, social sciences (as well as the arts and humanities) can enrich our "folk" conceptualization of social phenomena. So, a descriptive inquiry into race and gender need not presuppose that race and gender are biological kinds; instead it might ask whether our uses of race and gender vocabularies are tracking social kinds, and if so which ones.

The third sort of project takes an *analytical* approach to the question, "What is gender?" or "What is race?" (Scott 1986). On this approach the task is not to explicate our ordinary concepts; nor is it to investigate the kind that we may or may not be tracking with our everyday conceptual apparatus; instead we begin by considering more fully the pragmatics of our talk employing the terms in question. What is the point of having these concepts? What cognitive or practical task do they (or should they) enable us to accomplish? Are they effective tools to accomplish our (legitimate) purposes; if not, what concepts would serve these purposes better? In the limit case of an analytical approach the concept in question is introduced by stipulating the meaning of a new term, and its content is determined entirely by the role it plays in the theory. But if we allow that our everyday vocabularies serve both cognitive and practical purposes, purposes that might also be served by our theorizing, then a theory offering an improved understanding of our (legitimate) purposes and/or improved conceptual resources for the tasks at hand might reasonably represent itself as providing a (possibly revisionary) account of the everyday concepts.[4]

So, on an analytical approach, the questions "What is gender?" or "What is race?" require us to consider what work we want these concepts to do for us; why do we need them at all? The responsibility is ours to define them for our purposes. In doing so we will want to be responsive to some aspects of ordinary usage (and to aspects of both the connotation and extension of the terms). However, neither ordinary usage nor empirical investigation is overriding, for there is a stipulative element to the project: *this* is the phenomenon we need to be thinking about. Let the term in question refer to it. On this approach, the world by itself can't tell us what gender is, or

what race is; it is up to us to decide what in the world, if anything, they are.

This essay pursues an analytical approach to defining race and gender. However, its analytical objectives are linked to the descriptive project of determining whether our gender and race vocabularies in fact track social kinds that are typically obscured by the manifest content of our everyday race and gender concepts.[5] Although the analyses I offer will point to existing social kinds (and this is no accident), I am not prepared to defend the claim that these social kinds are what our race and gender talk is "really" about. My priority in this inquiry is not to capture what we do mean, but how we might usefully revise what we mean for certain theoretical and political purposes.

My characterization of all three approaches remains vague, but there is one reason to be skeptical of the analytical approach that should be addressed at the outset. The different approaches I've sketched differ both in their methods and their subject matter. However, we come to inquiry with a conceptual repertoire in terms of which we frame our questions and search for answers: hence, the subject matter of any inquiry would seem to be set from the start. In asking what *race* is, or what *gender* is, our initial questions are expressed in *everyday* vocabularies of race and gender, so how can we meaningfully answer these questions without owing obedience to the everyday concepts? Or at least to our everyday usage? Revisionary projects are in danger of providing answers to questions that weren't being asked.

But ordinary concepts are notoriously vague; individual conceptions and linguistic usage varies widely. Moreover, inquiry often demonstrates that the ordinary concepts used initially to frame a project are not, as they stand, well-suited to the theoretical task at hand. (This is one reason why we may shift from a *conceptual* project to an *analytical* one.) But precisely because our ordinary concepts are vague (or it is vague which concept we are expressing by our everyday use of terms), there is room to stretch, shrink, or refigure what exactly we are talking about in new and sometimes unexpected directions.

However, in an explicitly revisionary project, it is not at all clear when we are warranted in appropriating existing terminology. Given the difficulty of determining what "our" concept is, it isn't entirely clear when a project crosses over from being explicative to revisionary, or when it is no longer even revisionary but simply changes the subject. If our goal is to offer an analysis of "our" concept of X, then the line between what's explication and what's not matters. But if our goal is to identify a concept that serves our broader purposes, then the question of terminology is primarily a pragmatic and sometimes a political one: should we employ the terms of ordinary discourse to refer to our theoretical categories, or instead make up new terms? The issue of terminological appropriation is especially important, and especially sensitive, when the terms in question designate categories of social identity such as 'race' and 'gender'.

Are there principles that determine when it is legitimate to appropriate the terms of ordinary discourse for theoretical purposes? An answer, it seems to me, should include both a semantic and a political condition (though in some cases the politics of the appropriation will be uncontroversial). The semantic condition is not surprising: the proposed shift in meaning of the term would seem semantically warranted if central functions of the term remain the same, e.g., if it helps organize or explain a core set of phenomena that the ordinary terms are used to identify or describe.[6] Framing a political condition in general terms is much more difficult, however, for the politics of such appropriation will depend on the acceptability of the goals being served, the intended and unintended effects of the change, the politics of the speech context, and whether the underlying values are justified. We will return to some of these issues later in the paper once my analyses have been presented.

II Critical (feminist, anti-racist) theory

In an analytical project we must begin by considering what we want the concept in question for. Someone might argue, however, that the answer is simple: our concepts must do the work of enabling us to articulate truths. But of course an unconstrained search for truth would yield chaos, not theory; truths are too easy to come by, there are too many of them. Given time and inclination, I could tell you many truths – some trivial, some interesting, many boring – about my physical surroundings. But a random collection of facts does not make a theory; they are a disorganized jumble. In the context of theorizing, some truths are more significant than others because they are relevant to answering the question that guides the inquiry. (Anderson 1995.)

Theorizing – even when it is sincerely undertaken as a search for truth – must be guided by more than the goal of achieving justified true belief. Good theories are systematic bodies of knowledge that select from the mass of truths those that address our broader cognitive and practical demands. In many contexts the questions and purposes that frame the project are understood and progress does not require one to investigate them. But in other contexts, e.g., especially when debate has seemed to break down and parties are talking at cross-purposes, an adequate evaluation of an existing theory or success in developing a new one is only possible when it is made clear what the broader goals are.

With this sketch of some of the theoretical options, I want to frame my own project as a *critical analytical* effort to answer the questions: "What is gender?", "What is race?" and the related questions "What is it to be a man?" ". . .

a woman?", ". . . White?" ". . . Latino?" etc. More specifically, the goal of the project is to consider what work the concepts of gender and race might do for us in a critical − specifically feminist and anti-racist − social theory, and to suggest concepts that can accomplish at least important elements of that work (Geuss 1981). So to start: why might feminist anti-racists want or need the concepts of gender and race? What work can they do for us?

At the most general level, the task is to develop accounts of gender and race that will be effective tools in the fight against injustice. The broad project is guided by four concerns:

(i)   The need to identify and explain persistent inequalities between females and males, and between people of different "colors"[7]; this includes the concern to identify how social forces, often under the guise of biological forces, work to perpetuate such inequalities.

(ii) The need for a framework that will be sensitive to both the similarities and differences among males and females, and the similarities and differences among individuals in groups demarcated by "color"; this includes the concern to identify the effects of interlocking oppressions, e.g., the intersectionality of race, class, and gender. (Crenshaw 1993.)

(iii) The need for an account that will track how gender and race are implicated in a broad range of social phenomena extending beyond those that obviously concern sexual or racial difference, e.g., whether art, religion, philosophy, science, or law might be "gendered" and/or "racialized".

(iv) The need for accounts of gender and race that take seriously the agency of women and people of color of both genders, and within which we can develop an understanding of agency that will aid feminist and anti-racist efforts to empower critical social agents.

In this paper I will begin to address the first two concerns, though the fourth will become relevant later in the discussion. Let me emphasize, however, that my goal in this paper is not to provide a thoroughgoing explanation of sexism and racism, if one wants by way of explanation a causal account of why and how females have come to be systematically subordinated throughout history, or why and how "color" has come to be a basis for social stratification. My goal here is in some ways more modest, and in other ways more contentious. Prior to explanation it is valuable to provide clear conceptual categories to identify the phenomenon needing explanation, e.g., categories that identify the kind of injustice at issue and the groups subject to it. In the case of racial and sexual subordination this is not as easy as it may seem. In the first place, the forms of racial and sexual subordination are tremendously heterogeneous and it would help to have accounts that enable us to distinguish *racial* subordination and *sexual* subordination from other sorts. But further, we must be cautious about treating familiar demarcations of "color" and "sex" as purely natural categories, as if the question at hand is simply why one's "color" or sex − where we take for granted our familiar understandings of these terms − has ever seemed to be socially significant. At least at this stage of the inquiry we must allow that the criteria for distinguishing "colors" or "sexes" differ across time and place, and that the boundaries are at least partly political; but in spite of this variation, we are still dealing with an overarching phenomenon of racial and sexual subordination.

III What is gender?

Even a quick survey of the literature reveals that a range of things have counted as "gender" within feminist theorizing. The guiding idea is sometimes expressed with the slogan: "gender is the social meaning of sex". But like any slogan, this one allows for different interpretations. Some theorists use the term 'gender' to refer to the subjective experience of sexed embodiment, or a broad psychological orientation to the world ("gender identity"[8]); others to a set of attributes or ideals that function as norms for males and females ("masculinity" and "femininity"); others to a system of sexual symbolism; and still others to the traditional social roles of men and women. My strategy is to offer a focal analysis that defines gender, in the primary sense, as a social class. A focal analysis undertakes to explain a variety of connected phenomena in terms of their relations to one that is theorized as the central or core phenomenon. As I see it, the core phenomenon to be addressed is the pattern of social relations that constitute the social classes of men as dominant and women as subordinate; norms, symbols, and identities are gendered in relation to the social relations that constitute gender.[9] As will become clearer below, I see my emphasis as falling within, though not following uncritically, the tradition of materialist feminism.[10]

Among feminist theorists there are two problems that have generated pessimism about providing any unified account of women; I'll call them the *commonality problem* and the *normativity problem*. Very briefly, the commonality problem questions whether there is anything social that females have in common that could count as their "gender". If we consider *all* females – females of different times, places, and cultures – there are reasons to doubt that there is anything beyond body type (if even that) that they all share (Spelman 1988). The normativity problem raises the concern that any definition of "what woman is" is value-laden, and will marginalize certain females, privilege others, and reinforce current gender norms (Butler 1990, Ch. 1).

It is important to note, even briefly, that these problems take on a different cast when they arise within a *critical analytical* project. The emphasis of an analytical project is not on discovering commonalities among females: although the empirical similarities and differences between females are relevant, the primary goal is an analysis of gender that will serve as a tool in the quest for sexual justice (see section II). Moreover, a critical project can accept the result that an effort to define "what women is" carries normative implications, for critical projects explicitly embrace normative results; the hope is that the account's implications would not reinforce but would help undermine the structures of sexual oppression. However, we will return to these issues below.

Given the priority I place on concerns with justice and sexual inequality, I take the primary motivation for distinguishing sex from gender to arise in the recognition that males and females do not only differ physically, but also systematically differ in their social positions. What is of concern, to put it simply, is that societies, on the whole, privilege individuals with male bodies. Although the particular forms and mechanisms of oppression vary from culture to culture, societies have found many ways – some ingenious, some crude – to control and exploit the sexual and reproductive capacities of females.

The main strategy of materialist feminist accounts of gender has been to define gender in terms of women's subordinate position in systems of male dominance.[11] Although there are materialist feminist roots in Marxism, contemporary versions resist the thought that all social phenomena can be explained in or reduced to economic terms; and although materialist feminists emphasize the role of language and culture in women's oppression, there is a wariness of extreme forms of linguistic constructivism and a commitment to staying grounded in the material realities of women's lives. In effect, there is a concerted effort to show how gender oppression is jointly sustained by both cultural and material forces.

Critiques of universalizing feminisms have taught us to be attentive to the variety of forms gender takes and the concrete social positions females occupy. However it is compatible with these commitments to treat the category of gender as a genus that is realized in different ways in different contexts; doing so enables us to recognize significant patterns in the ways that gender is instituted and embodied. Working at the most general level, then, the materialist strategy offers us three basic principles to guide us in understanding gender:

(i) Gender categories are defined in terms of how one is socially positioned, where this is a function of, e.g., how one is viewed, how one is treated, and how one's life is structured socially, legally, and economically; gender is not defined in terms of an individual's intrinsic physical or psychological features.

   (This allows that there may be other categories – such as sex – that are defined in terms of intrinsic physical features. Note, however, that once we focus our attention on gender as social position, we must allow that one can be a woman without ever (in the ordinary sense) "acting like a woman", "feeling like a woman", or even having a female body.)

(ii) Gender categories are defined hierarchically within a broader complex of oppressive relations; one group (viz., women) is socially positioned as subordinate to the other (viz., men), typically within the context of other forms of economic and social oppression.

(iii) Sexual difference functions as the physical marker to distinguish the two groups, and is used in the justification of viewing and

treating the members of each group differently.

(Tentatively) we can capture these main points in the following analyses:

S *is a woman* iff$_{df}$ S is systematically subordinated along some dimension (economic, political, legal, social, etc.), and S is "marked" as a target for this treatment by observed or imagined bodily features presumed to be evidence of a female's biological role in reproduction.[12]

S *is a man* iff$_{df}$ S is systematically privileged along some dimension (economic, political, legal, social, etc.), and S is "marked" as a target for this treatment by observed or imagined bodily features presumed to be evidence of a male's biological role in reproduction.

It is a virtue, I believe, of these accounts, that depending on context, one's sex may have a very different meaning and it may position one in very different kinds of hierarchies. The variation will clearly occur from culture to culture (and subculture to sub-culture); so, e.g., to be a Chinese woman of the 1790s, a Brazilian woman of the 1890s, or an American woman of the 1990s may involve very different social relations, and very different kinds of oppression. Yet on the analysis suggested, these groups count as women insofar as their subordinate positions are marked and justified by reference to (female) sex. (Also Hurtado 1994, esp. 142.) Similarly, this account allows that the substantive import of gender varies even from individual to individual within a culture depending on how the meaning of sex interacts with other socially salient characteristics (e.g., race, class, sexuality, etc.). For example, a privileged White woman and a Black woman of the underclass will both be women insofar as their social positions are affected by the social meanings of being female; and yet the social implications of being female vary for each because sexism is intertwined with race and class oppression.

There are points in the proposed analysis that require clarification, however. What does it mean to say that someone is "systematically subordinated" or "privileged", and further, that the subordination occurs "on the basis of" certain features? The background idea is that women are *oppressed*, and that they are oppressed *as women*. But we still need to ask: What does it mean to say that women are oppressed, and what does the qualification "as women" add?

Marilyn Frye's account of oppression with Iris Young's elaborations provides a valuable starting point (Frye 1983; Young 1990). Although these ideas are commonplace within certain intellectual circles, it is useful to summarize them very briefly here. There are of course unresolved difficulties in working out a satisfactory theory of oppression; I'm afraid I can't take on that further task here, so I can only invoke the rough outlines of the background view with the hope that an adequate account can at some point be supplied. Nonetheless, oppression in the intended sense is a structural phenomenon that positions certain groups as disadvantaged and others as advantaged or privileged in relation to them. Oppression consists of, "an enclosing structure of forces and barriers which tends to the immobilization and reduction of a group or category of people." (Frye 1983, p. 11.) Importantly, such structures, at least as we know them, are not designed and policed by those in power, rather,

... oppression refers to the vast and deep injustices some groups suffer as a consequence of often unconscious assumptions and reactions of well-meaning people in ordinary interactions, media and cultural stereotypes, and structural features of bureaucratic hierarchies and market mechanisms – in short, the normal processes of everyday life. (Young 1990, p. 41.)

Developing this concept of oppression, Young specifies five forms it can take: exploitation, marginalization, powerlessness, cultural imperialism, and (systematic) violence. The key point for us is that oppression comes in different forms, and even if one is privileged along some dimension (e.g., in income or respect), one might be oppressed in others.[13] In fact, one might be systematically subordinated along some social axis, and yet still be tremendously privileged in one's *overall* social position.

It is clear that women are oppressed in the sense that women are members of groups that suffer exploitation, marginalization, etc. But how should we understand the claim that women are oppressed *as women*. Frye explains this as follows:

One is marked for application of oppressive pressures by one's membership in some group or category . . . In the case at hand, it is the category, *woman*. . . . If a woman has little or no economic or political power, or achieves little of what she wants to achieve, a major causal factor in this is that she is a woman. For any woman of any race or economic class, being a woman is significantly attached to whatever disadvantages and deprivations she suffers, be they great or small. . . . [In contrast,] being male is something [a man] has going for him, even if race or class or age or disability is going against him. (Frye 1983, pp. 15–16.)

But given the diffusion of power in a model of structural oppression how are we to make sense of one's being "marked" and the "application" of pressures? In the context of oppression, certain properties of individuals are socially meaningful. This is to say that the properties play a role in a broadly accepted (though usually not fully explicit) representation of the world that functions to justify and motivate particular forms of social intercourse. The significant properties in question – in the cases at hand, assumed or actual properties of the body – mark you "for application of oppressive pressures" insofar as the attribution of these properties is interpreted as adequate, in light of this background representation, to explain and/or justify your position in a structure of oppressive social relations. In the case of women, the idea is that societies are guided by representations that link being female with other facts that have implications for how one should be viewed and treated; insofar as we structure our social life to accommodate the cultural meanings of the female (and male) body, females occupy an oppressed social position.

Although I agree with Frye that in sexist societies social institutions are structured in ways that on the whole disadvantage females and advantage males, we must keep in mind that societies are not monolithic and that sexism is not the only source of oppression. For example, in the contemporary US, there are contexts in which being Black *and male* marks one as a target for certain forms of systematic violence (e.g., by the police). In those contexts, contrary to Frye's suggestion, *being male* is not something that a man "has going for him"; though there are other contexts (also in the contemporary US) in which Black males benefit from being male. In examples of this sort, the systematic violence against males *as males* is emasculating (and may be intended as such); but there are important differences between an emasculated man and a woman. On the sort of view we're considering, a woman is someone whose subordinated status is marked by reference to (assumed) *female* anatomy; someone marked for subordination by reference to (assumed) *male* anatomy does not

qualify as a woman, but also, *in the particular context*, is not socially positioned as a man.

These considerations suggests that it may be useful to bring context explicitly into our account. Recent work on gender socialization also supports the idea that although most of us develop a relatively fixed gender identity by the age of three, the degree to which the marked body makes a difference varies from context to context. In her study of elementary school children, Barrie Thorne suggests:

> Gender boundaries are episodic and ambiguous, and the notion of "borderwork" [i.e., the work of contesting and policing gender boundaries] should be coupled with a parallel term – such as "neutralization" – for processes through which girls and boys (and adults . . .) neutralize or undermine a sense of gender as division and opposition. (Thorne 1993, p. 84.)

Thorne's study is motivated by a recognition that gender is a well-entrenched system of oppression. However, her comments here are intended as an antidote to two problematic tendencies in speaking of girls and boys, men and women: first, the tendency to overgeneralize gender differences based on paradigm or stereotyped interactions; second, the tendency to view individuals (specifically children) as passive participants in gender socialization and, more generally, gendered life.

In some respects, Frye's and Thorne's approaches appear to be in tension with one another. Frye is keen to highlight the structural facts of sexist oppression: like it or not, your body positions you within a social hierarchy. Thorne, on the other hand, examines how oppression is lived, enforced, and resisted at the micro-level. There are important advantages to both: without a recognition of oppressive structures and the overall patterns of advantage and disadvantage, individual slights or conflicts can seem harmless. But without a recognition of individual variation and agency, the structures take on a life of their own and come to seem inevitable and insurmountable. But can both perspectives be accommodated in an account of gender? The idea seems simple enough: there are dominant ideologies and dominant social structures that work together to bias the micro-level interactions, however varied and complex they may be, so that for the most part males are privileged and females are disadvantaged.

Although an adequate account of gender must be highly sensitive to contextual variation, if we focus entirely on the narrowly defined contexts in which one's gender is negotiated, we could easily lose sight of the fact that for most of us there is a relatively fixed interpretation of our bodies as sexed either male or female, an inter-pretation that marks us within the dominant ideology as eligible for only certain positions or opportunities in a system of sexist oppression. Given our priority in theorizing systems of inequality, it is important first to locate the social classes men and women in a broad structure of subordination and privilege[14]:

S *is a woman* iff

(i)   S is regularly and for the most part observed or imagined to have certain bodily features presumed to be evidence of a female's biological role in reproduction;

(ii)   that S has these features marks S within the dominant ideology of S's society as someone who ought to occupy certain kinds of social position that are in fact subordinate (and so motivates and justi-fies S's occupying such a position); and

(iii) the fact that S satisfies (i) and (ii) plays a role in S's systematic subordination, i.e., *along some dimension*, S's social posi-tion is oppressive, and S's satisfying (i)

and (ii) plays a role in that dimension of subordination.

S *is a man* iff

(i)   S is regularly and for the most part observed or imagined to have certain bodily features presumed to be evidence of a male's biological role in reproduction;

(ii)   that S has these features marks S within the dominant ideology of S's society as someone who ought to occupy certain kinds of social position that are in fact privileged (and so motivates and justi-fies S's occupying such a position); and

(iii) the fact that S satisfies (i) and (ii) plays a role in S's systematic privilege, i.e., *along some dimension*, S's social position is privi-leged, and S's satisfying (i) and (ii) plays a role in that dimension of privilege.

These accounts are, however, compatible with the idea that (at least for some of us) one's gender may not be entirely stable, and that other systems of oppression may disrupt gender in particular contexts: a woman may not always function socially as a woman; a man may not always func-tion socially as a man.[15] To return to a previous example, when systems of White supremacy and male dominance collide, a Black man's male privilege may be seen as so threatening that it must be violently wrested from him. In an effort to accommodate this variation, we can add:

S *functions as a woman* in context C iff$_{df}$

(i)   S is observed or imagined in C to have certain bodily features presumed to be evidence of a female's biological role in reproduction;

(ii)   that S has these features marks S within the background ideology of C as someone who ought to occupy certain kinds of social position that are in fact subordinate (and so motivates and justifies S's occupying such a position); and

(iii)   the fact that S satisfies (i) and (ii) plays a role in S's systematic subordination in C, i.e., *along some dimension*, S's social position in C is oppressive, and S's satisfying (i) and (ii) plays a role in that dimension of subordination.

And mutatis mutandis for functioning as a man in context C.

It is important to note that the definitions don't require that the background ideology in question must use (assumed) reproductive function as itself the justification for treating men or women in the way deemed "appropriate"; (assumed) reproductive features may instead simply be "markers" of supposedly "deeper" (and morally relevant?) characteristics that the ideology supposes justifies the treatment in question. (Appiah 1992, pp. 13–15.)

Although ultimately I will defend these analyses of *man* and *woman*, I'll argue below that there are reasons to modify the broader materialist strategy in defining *gender*. In short, I believe that gender can be fruitfully understood as a higher-order genus that includes not only the hierarchical social positions of man and woman, but potentially other non-hierarchical social positions defined in part by reference to reproductive function. I believe gender *as we know it* takes hierarchical forms as men and women; but the theoretical move of treating men and women as only two kinds of gender provides resources for thinking about other (actual) genders, and the political possibility of constructing non-hierarchical genders.

IV What is race?

One advantage of this account of gender is the parallel it offers for race. To begin, let me review a couple of points that I take to be matters of established fact: First, there are no racial genes responsible for the complex morphologies and cultural patterns we associate with different races. Second, in different contexts racial distinctions are drawn on the basis of different characteristics, e.g., the Brazilian and US classification schemes for who counts as "Black" differ. For these reasons and others, it appears that race, like gender, could be fruitfully understood as a position within a broad social network.

Although suggestive, this idea is not easy to develop. It is one thing to acknowledge that race is *socially real*, even if a biological fiction; but it is another thing to capture in general terms "the social meaning of color". There seem to be too many different forms race takes. Note, however, that we encountered a similar problem with gender: is there any prospect for a unified analysis of "the social meaning of sex"? The materialist feminist approach offered a helpful strategy: don't look for an analysis that assumes that the meaning is always and everywhere the same; rather, consider how members of the group are *socially positioned*, and what *physical markers* serve as a supposed basis for such treatment.

How might we extend this strategy to race? Transposing the slogan, we might say that race is the social meaning of the geographically marked body, familiar markers being skin color, hair type, eye shape, physique. To develop this, I propose the following account.[16]

First definition:

A group is *racialized* iff$_{df}$ its members are socially positioned as subordinate or privileged along some dimension (economic, political, legal, social, etc.), and the group is "marked" as a target for this treatment by observed or imagined bodily features presumed to be evidence of ancestral links to a certain geographical region.

Or in the more elaborate version:

A group G is *racialized* relative to context C iff$_{df}$ members of G are (all and only) those:

(i)   who are observed or imagined to have certain bodily features presumed in C to be evidence of ancestral links to a certain geographical region (or regions);

(ii)   whose having (or being imagined to have) these features marks them within the context of the background ideology in C as appropriately occupying certain kinds of social position that are in fact either subordinate or privileged (and so motivates and justifies their occupying such a position); and

(iii)   whose satisfying (i) and (ii) plays (or would play) a role in their systematic subordination or privilege in C, i.e., who are *along some dimension* systematically subordinated or privileged when in C, and satisfying (i) and (ii) plays (or would play) a role in that dimension of privilege or subordination.[17]

In other words, races are those groups demarcated by the geographical associations accompanying perceived body type, when those associations take on evaluative significance concerning how members of the group should be viewed and treated. As in the case of gender, the ideology need not use physical morphology or geography as the entire basis for "appropriate" treatment; these features may instead simply be "markers" of other characteristics that the ideology uses to justify the treatment in question.

Given this definition, we can say that S is of the White (Black, Asian . . .) race [in C] iff Whites (Blacks, Asians . . .) are a racialized group [in C], and S is a member.[18] On this view, whether a group is racialized, and so how and whether an individual is raced, is not an absolute fact, but will depend on context. For example, Blacks, Whites, Asians, Native Americans, are currently racialized in the US insofar as these are all groups defined in terms of physical features associated with places of origin, and insofar as membership in the group functions socially as a basis for evaluation. However, some groups are not currently racialized in the US, but have been so in the past and possibly could be again (and in other contexts are), e.g., the Italians, the Germans, the Irish.

It is useful to note a possible contrast between race and ethnicity. I don't have a theory of ethnicity to offer; these are some preliminary comparisons. One's ethnicity concerns one's ancestral links to a certain geographical region (perhaps together with participation in the cultural practices of that region); often ethnicity is associated with characteristic physical features. For our purposes, however, it might be useful to employ the notion of "ethnicity" for those groups that are like races as I've defined them except that they do not experience systematic subordination or privilege in the context in question.[19] Ethnic groups can be (and are) racialized, however, and when they are, one's membership in the group positions one in a social hierarchy; but (on the view I'm sketching) the occurrence of this hierarchical positioning means that the group has gone beyond simply being an ethnic group and functions in that context as a race. In short, we can distinguish between grouping individuals on the basis of their (assumed) origins, and grouping them *hierarchically* on the basis of their (assumed) origins, and the contrast between race and ethnicity might be a useful way to capture this distinction.

V Normativity and commonality

So what, if anything, is achieved by adopting the above analyses? Are they the tools we need? Let's first consider the problems of commonality and normativity, and begin with gender.

Remember, the problem of commonality questions whether there is anything social that all females can plausibly be said to have in common. If we ask whether females share any intrinsic (non-anatomical) features such as psychological makeup, character traits, beliefs, values, experiences or, alternatively, whether there is a particular social role that all females have occupied across culture and history, the answer seems to be "no".

On my analysis women are those who occupy a particular *kind* of social position, viz., one of sexually-marked subordinate. So women have in common that their (assumed) sex has socially disadvantaged them; but this is compatible with the kinds of cultural variation that feminist inquiry has revealed, for the substantive content of women's position and the ways of justifying it can vary enormously. Admittedly, the account accommodates such variation by being very abstract; nonetheless, it provides a schematic account that highlights the interdependence between the material forces that subordinate women, *and* the ideological frameworks that sustain them.

One might complain, however, that there must be *some* women (or rather, females) who aren't oppressed, and in particular, aren't oppressed *as women*. Perhaps there are; e.g., some may "pass" as men, others may be recognizably female but not be subordinated in any way linked to that recognition. I'm not convinced that there are many cases (if any) of the latter, but I'll certainly grant that there *could be* females who did not satisfy the definition that I've offered. In fact, I believe it is part of the project of feminism to bring about a day when there are no more women (though, of course, we should not aim to do away with females!). I'm happy to admit that there could be females who aren't women in the sense I've defined, but these

individuals (or possible individuals) are not counterexamples to the analysis. The analysis is intended to capture a meaningful political category for critical feminist efforts, and non-oppressed females do not fall within that category (though they may be interesting for other reasons).

But this leads us directly from the commonality problem to the normativity problem. The normativity problem raises the challenge that any effort to define *women* will problematically privilege some women and (theoretically) marginalize others, and will itself become normative. One worry is that bias inevitably occurs in deciding which experiences or social roles are definitive; a second worry is that if someone wants to be a "real" woman, she should conform to the definition of women provided, and this will reinforce rather than challenge male dominance.

On the account I've offered, it is true that certain females don't count as "real" women; and it is true that I've privileged certain facts of women's lives as definitive. But given the epistemological framework outlined above, it is both inevitable and important for us to choose what facts are significant on the basis of explicit and considered values. For the purposes of a critical feminist inquiry, oppression is a significant fact around which we should organize our theoretical categories; it may be that non-oppressed females are marginalized within my account, but that is because for the broader purposes at hand – relative to the feminist and anti-racist values guiding our project – they are not the ones who matter. The important issue is not whether a particular account "marginalizes" some individuals, but whether its doing so is in conflict with the feminist values that motivate the inquiry. And as far as I can tell, *not* focusing our theoretical efforts on understanding the position of oppressed females would pose just such a conflict.

The question remains whether my definition of woman helps sustain gender hierarchy by implicitly offering a normative ideal of woman. Given that women on my definition are an oppressed group, I certainly hope not! Instead, the definition is more likely to offer a negative ideal that challenges male dominance.

I won't defend here my account of racialized groups against an extension of the normativity and commonality complaints, for I would simply repeat the strategy just employed. Although there are interesting nuances in adapting the arguments to apply to racialized groups, I don't see anything peculiar to race that would present an obstacle to developing the same sort of response.

VI Negotiating terms

Let me now turn to summarize some of the advantages of the proposed definitions. At this point we could bracket the terminological issues and just consider whether the groups in question are ones that are important to consider given the goals of our inquiry. I hope it is clear from what I've already said how the analyses can help us identify and critique broad patterns of racial and sexual oppression (MacKinnon 1987), and how they accommodate the intersectionality of social categories. But a further and, I think, more interesting question is whether it is useful to think of these groups in *these terms*: Does it serve both the goal of understanding racial and sexual oppression, and of achieving sexual and racial equality to think of ourselves as *men* or *women*, or *raced* in the ways proposed?

By appropriating the everyday terminology of race and gender, the analyses I've offered invite us to acknowledge the force of oppressive systems in framing our personal and political identities. Each of us has some investment in our race and gender: I am a White woman. On my accounts, this claim locates me within social systems that in some respects privilege and in others subordinate me. Because gender and racial inequality are not simply a matter of public policy but implicate each of us at the heart of our self-understandings, the terminological shift calls us to reconsider who we think we are.

This point highlights why the issue of terminological appropriation is especially sensitive when the terms designate categories of social identity. Writing in response to a *NY Times* editorial supporting the terminological shift from "Black" to "African-American," Trey Ellis responded:

> When somebody tries to tell me what to call myself in all its uses just because they come to some decision at a cocktail party to which I wasn't even invited, my mama raised me to tell them to kiss my black ass. In many cases, *African-American* just won't do.[20]

The issue is not just what words we should use, and who gets to say what words to use, but who we take ourselves to be, and so, in some sense, who we are. Terms for social groups can function as descriptive terms: it may be accurate to say that someone is a woman when she satisfies certain conditions. However, terms for social groups serve other rhetorical purposes. Typically the act of classifying someone as a member of a social group invokes a set of "appropriate" (contextually specific) norms and expectations. It positions her in a social framework and makes available certain kinds of evaluation; in short, it carries prescriptive force. Accepting or identifying with the classification typically involves an endorsement of some norms and expectations, however, not always the socially sanctioned ones. The question whether I should be called a "woman" or a "wommon", "White" or "Euro-American", is not just a matter of what words to use, but what norms and expectations are taken to be appropriate; to ask what I should be called is to ask what norms I should be judged by (Haslanger 1993, esp. pp. 89–91).

Although "identifying" someone as a member of a social group invokes a set of "appropriate" norms, what these norms are is not fixed. What it means to be a woman, or to be White, or to be Latino, in this sense, is unstable and always open to contest. The instability across time is necessary to maintain the basic structure of gender and race relations through other social changes: as social roles change – prompted by the economy, immigration, political movements, natural disasters, war, etc. – the contents of normative race and gender identities adjust. The flexibility across contexts accommodates the complexity of social life: what norms are assumed to apply depends on the dominant social structure, the ideological context, and other dimensions of one's identity (such as class, age, ability, sexuality). But this instability and flexibility is exactly what opens the door for groups to redefine themselves in new ways. One strategy is for the group to adopt new names ('African-American', 'womyn'); another is to appropriate old names with a normative twist ('queer'); but in some cases the contest is over the meanings of the standard terms ("Ain't I a woman?"). Because individuals are so deeply invested in gender and, at least in the US, race categories, it remains of crucial importance to be and to be perceived as a 'woman' or a 'man' and as a member of one of the standard races. But even so, (although this is something of an exaggeration) it is possible to view our gender and race vocabulary as, in effect, providing terminological place-holders marking space for the collective negotiation of our social identities.

Given the normative force and political potential of identifying someone (or self-identifying) in racial or gendered terms, how do we evaluate a terminological appropriation of the kind I'm proposing? For example, isn't there something disingenuous about appropriating race and gender terminology *because* it is used to

frame how we think of ourselves and each other, in order to use them for new concepts that are not part of our self-understandings?

This latter question is especially pressing because the appropriation under consideration intentionally invokes what many find to be positive self-understandings – being Latina, being a White man – and offers analyses of them which emphasize the broader context of injustice. Thus there is an invitation not only to revise one's understanding of these categories (given their instability, this happens often enough), but to revise one's relationship to their prescriptive force. By offering these analyses of our ordinary terms, I call upon us to reject what seemed to be positive social identities. I'm suggesting that we should work to undermine those forces that make being a man, a woman, or a member of a racialized group possible; we should refuse to be gendered man or woman, refuse to be raced. This goes beyond denying essentialist claims about one's embodiment and involves an active political commitment to live one's life differently (Stoltenberg 1989). In one sense this appropriation is "just semantics": I'm asking us to use an old term in a new way. But it is also politics: I'm asking us to understand ourselves and those around us as deeply molded by injustice and to draw the appropriate prescriptive inference. This, I hope, will contribute to empowering critical social agents. However, whether the terminological shift I'm suggesting is politically useful will depend on the contexts in which it is employed and the individuals employing it. The point is not to legislate what terms to use in all contexts, but to offer resources that should be used judiciously.

VII Lingering concerns, promising alternatives

There is, nonetheless, a broader concern one might have about the strategy I've employed: Why build hierarchy into the definitions? Why not define gender and race as those social positions motivated and justified by cultural responses to the body, without requiring that the social positions are hierarchical? Wouldn't that provide what we need without implying (implausibly) that women are, by definition, subordinate, men, by definition, privileged, and races, by definition, hierarchically positioned?

If we were to remove hierarchy from the definitions, then there would be two other benefits: first, by providing a place in our model for cultural representations of the body *besides* those that contribute to maintaining subordination and privilege, we could better acknowledge that there are positive aspects to having a gender and a race. And second, the accounts would provide a framework for envisioning the sorts of constructive changes needed to create a more just world. The suggestion that we must eliminate race and gender may be a powerful rallying call to those who identify with radical causes, but it is not at all clear that societies can or should avoid giving meanings to the body, or organizing themselves to take sexual and reproductive differences into account. Don't we at least need a concept of gender that will be useful in the reconstructive effort, not only the destructive one?

Consider gender. I am sympathetic to radical rethinkings of sex and gender. In particular, I believe that we should refuse to use anatomy as a primary basis for classifying individuals and that any distinctions between kinds of sexual and reproductive bodies are importantly political and open to contest. Some authors have argued that we should acknowledge the continuum of anatomical differences and recognize at least five sexes (Fausto-Sterling 1993). And if sexual distinctions become more complex, we would also need to rethink sexuality, given that sexual desire would not fit neatly within existing homosexual/heterosexual paradigms.

However, one can encourage the proliferation of sexual and reproductive options without maintaining that we can or should eliminate *all* social implications of anatomical sex and reproduction. Given that as a species there are substantial differences in what human bodies contribute to reproduction, and what sorts of bodies bear the main physical burdens of reproduction, and given further that reproduction cannot really help but be a socially significant fact (it does, after all, produce children), it can seem difficult to imagine a functioning society, more specifically, a functioning *feminist* society, that doesn't acknowledge in some way the difference between those kinds of bodies that are likely able to bear children, and those that aren't. One could argue that we should work towards a society free of gender in a materialist sense – one in which sex-oppression does not exist – while still allowing that sexual and reproductive differences should be taken into account in a just society. (Frye 1996; Gatens 1996.)

I will not debate here the degree to which a just society must be attentive to sexual and reproductive differences. Whether we, as feminists, ought to recommend the construction of (new) non-hierarchical genders or work to abolish gender entirely is a normative issue I leave for another occasion. Nonetheless, at the very least it would help to have terminology to debate these issues. I propose that we use the definitions of *man* and *woman* offered above: it is clear that these dominant nodes of our current gender structures are hierarchical. But borrowing strategies employed before, we can define gender in generic terms under which the previous definitions of man and women fall,[21] thus allowing the possibility of non-hierarchical genders and breaking the binary opposition between man and woman.

A group G is *a gender* relative to context C iff$_{df}$ members of G are (all and only) those:

(i)   who are regularly observed or imagined to have certain bodily features presumed in C to be evidence of their reproductive capacities[22];

(ii) whose having (or being imagined to have) these features marks them within the context of the ideology in C as motivating and justifying some aspect(s) of their social position; and

(iii) whose satisfying (i) and (ii) plays (or would play) a role in C in their social position's having one or another of these designated aspects.

I offer this analysis as a way of capturing the standard slogan: gender is the social meaning of sex. Note, however, that in imagining "alternative" genders we should be careful not to take for granted that the relevant biological divisions will correspond to what *we* consider "sex".[23] (Alternative groupings could include: "pregnant persons," "lactating persons," "menstruating persons," "infertile persons," (perhaps "homosexuals," depending on the story given about physical causes)). Neither should we assume that membership in a gender will constitute one's personal or psychological identity to any significant degree. Recall that on the accounts of gender and race I am proposing, both are to be understood first and foremost as social groups defined within a structure of social relations; whatever links there might be to identities and norms are highly contingent and would depend on the details of the picture. For example, we might imagine that "after the revolution" gender is a component of one's overall social position because, for example, there are legal protections or medical entitlements granted to individuals classified as having a certain sort of "sexed" body; but this need not have broad implications for psychological identity or everyday social interactions, for the "sex" of bodies might not even be publicly marked.

Turning briefly to race, the parallel issue arises: Do we need a concept of non-hierarchical "races" in order to frame and debate different visions of a "racially" just society? It would seem that we have the terminological resources available without a further definition: let races be, as previously defined, those hierarchically organized groups that are defined (roughly) by physical features and (assumed) geographical origins, and call those that aren't hierarchically organized (in the context in question) "ethnicities". Admittedly, ethnicity as we know it does have implications for social status and power, so my proposal is to employ the term for a somewhat idealized conception.

As in the case of gender, the question arises whether it ought to be part of an anti-racist project to recommend the preservation of existing ethnic groups or the formation of "new" ethnicities. And more generally, we need to ask whether a feminist anti-racism should treat genders and ethno-racial groups in the same way over the long term. Should we seek, e.g., to eliminate all genders and ethno-racial groupings; to preserve and proliferate them; to eliminate gender but not ethnicity (or vice versa)? These questions deserve careful attention but I cannot address them here.

Because the structure of definitions has become quite complex, it may help at this point to provide a diagram:

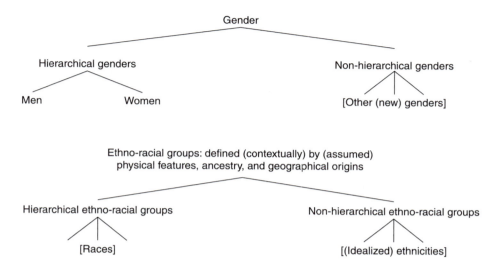

Figure 5.1

VIII Conclusion

On the accounts I've offered, there are striking parallels between race and gender. Both gender and race are real, and both are social categories. Neither gender nor race is chosen, but the forms they take can be resisted or mutated. Both race and gender (as we know it) are hierarchical, but the systems that sustain the hierarchy are contingent. And although the ideologies of race and gender and the hierarchical structures they sustain are substantively very different, they are intertwined.

There are many different types of human bodies; it is not the case that there is a unique "right" way of classifying them, though certain classifications will be more useful for some purposes than others. How we classify bodies can and does matter politically, for our laws, social institutions, and personal identities are profoundly linked to understandings of the body and its possibilities. This is compatible with the idea that what possibilities a human body has is not wholly a function of our understandings of it. Our bodies often outdo us, and undo us, in spite of the meanings we give them.

Within the framework I've sketched, there is room for theoretical categories such as *man*, *woman*,

and *race* (and particular racial groups), that take hierarchy to be a constitutive element, and those such as *gender* and *ethnicity* that do not. As I have suggested before, I am willing to grant that there are other ways to define race or gender, man or woman, that are useful to answer different questions, motivated by different concerns and priorities. I'm sure we need several concepts to do all the work needed in understanding the complex systems of racial and gender subordination.

In short, (speaking of my analyses) I'm less committed to saying that this is what gender is and what race is, than to saying that these are important categories that a feminist anti-racist theory needs. As I've explained above, I think there are rhetorical advantages to using the terms 'gender', 'man' and 'woman,' and 'race' for the concepts I've defined, but if someone else is determined to have those terms, I'll use different ones. To return to the point made much earlier in characterizing analytic projects: it is our responsibility to define gender and race for our theoretical purposes. The world itself can't tell us what gender is. The same is true for race. It may be as Appiah claims that "there is nothing in the world that can do all we ask race to do for us" (Appiah 1992, p. 45), if our project inevitably inherits the concept's complex history; but we might instead ask "race" to do different things than have been asked before. Of course, in defining our terms, we must keep clearly in mind our political aims both in analyzing the past and present, and in envisioning alternative futures. But rather than worrying, "what is gender, really?" or "what is race, really?" I think we should begin by asking (both in the theoretical and political sense) what, if anything, we want them to be.

Notes

1   Special thanks to: Elizabeth Anderson, Larry Blum, Tracy Edwards, Marilyn Frye, Stephen Darwall, Elizabeth Hackett, Elizabeth Harman, Donald Herzog, Will Kymlicka, Ishani Maitra, Mika Lavaque-Manty, Joe Levine, Elisabeth Lloyd, Mary Kate McGowan, Toril Moi, Christine Overall, Gerald Postema, Phyllis Rooney, Debra Satz, Geoff Sayre-McCord, Barry Smith, Jacqueline Stevens, Natalie Stoljar, Martin Stone, Ásta Sveinsdóttir, Paul Taylor, Greg Velazco y Trianosky, Catherine Wearing, Ralph Wedgwood, and Stephen Yablo for helpful comments on earlier versions of this paper. Extra thanks to Louise Antony for her extensive and tremendously insightful comments and editorial advice. Thanks to audiences in the philosophy departments at the University of Kentucky, University of North Carolina, Queens University, Stanford University, Tufts University, and the University of Utah where I presented this material in talks. Research on this project was supported by the National Humanities Center where I was a fellow during 1995–6; thanks to Delta Delta Delta Sorority whose support of the Center underwrote my fellowship there.

2   I use the terms 'man' and 'woman' to distinguish individuals on the basis of gender, the terms 'male' and 'female' to distinguish individuals on the basis of sex.

3   See Stich 1988. Stich uses the term "analytical epistemology" for what I would call a "conceptual" rather than an "analytical" project.

4   Cf. Appiah and Gutmann 1996, pp. 30–105. Appiah doesn't consider an analytical approach to race except rather elliptically on p. 42.

5   On the distinction between manifest and operative concepts, see Haslanger 1995, esp. p. 102.

6   It is important to keep in mind that what's at issue is not a criterion for *sameness* of meaning, but the boundary between what could count as a revisionary project and a new project altogether.

7   We need here a term for those physical features of individuals that mark them as members of a race. One might refer to them as "racial" features, but to avoid any suggestion of racial essences I will use the term 'color' to refer to the (contextually variable) physical "markers" of race, just as I use the term 'sex' to refer to the (contextually variable) physical "markers" of gender. I mean to include in "color" more than just skin tone: common markers also include eye, nose, and lip shape, hair texture, physique, etc. Although the term 'people of color' is used to refer to non-Whites, I want to allow that the markers of "Whiteness" count as "color".

8   There are at least four different uses of the term 'identity' that are relevant in considering the issue of gender or racial "identity"; here my comments about "gender identity" are admittedly superficial.

9   Very roughly, feminine norms are those that enable one to excel in the social position constituting the class *women*; feminine gender identity (at least in one sense of the term) is a psychological orientation to the world that includes the internalization of feminine norms; and feminine symbols are those that encode idealized feminine norms. What counts as a "feminine" norm, a "feminine" gender identity, or a "feminine" symbol is derivative (norms, symbols, and identities are not intrinsically feminine or masculine), and depends on how the social class of women is locally constituted.

10 For a sample of materialist feminist work, see Hennessy and Ingraham 1997.

11 Some theorists (Delphy, Hartmann) focus on the economic exploitation of women in domestic relations of production; others (Wittig) focus on sexual and reproductive exploitation under compulsory heterosexuality; others (MacKinnon) focus on sexual objectification.

12 These analyses allow that there isn't a common understanding of "sex" across time and place. On my account, gendered social positions are those marked by reference to features that are generally assumed *in the context in question* to either explain or provide evidence of reproductive role, whether or not these are features that *we* consider "sex".

13 On the importance of disaggregating power and oppression, see Ortner 1996.

14 This proposal depends on the claim that at least some societies have a "dominant ideology". Others have employed the notions of "background," "hegemony," "habitus," for the same purpose. Rather than debating what is the preferred notion, I'm happy to let the term "dominant ideology" serve as a placeholder for an account to be decided upon later. Given the strategy of my accounts, however, we must be sure to allow for multiple ideological strands in any society. See Geuss 1981, Hoy 1994.

15 We noted before that on a materialist account sex and gender don't always coincide. I'm making here a further claim: one may be gendered man or woman without functioning socially in that gender every moment of one's life.

16 On this I am deeply indebted to Stevens 1999, Ch. 4, and Omi and Winant 1994, esp. pp. 53–61.

17 There are aspects of this definition that need further elaboration or qualification. I will mention four here.

First, on my account, those who actually have the ancestral links to the specified region but who "pass", do not count as members of the racialized group in question. This is parallel to the case of a female functioning socially as a man or a male functioning socially as a woman. Because the goal is to define race and gender as social positions, I endorse this consequence of the definitions.

Second, as it stands the definition does not accommodate contexts such as Brazil in which membership in "racial" groups is partly a function of education and class. It excludes privileged ("Whitened") members from the subordinate races they might seem – considering only "color" – to belong to, and subordinated ("darkened")

members from privileged races, because they don't satisfy the third condition. But it cannot handle the inclusion of the "Whitened" members in the privileged group or the "darkened" members in the subordinated group because they don't satisfy the first condition. However, we could take the definition to capture a *strong* version of racialization, and develop another version on which appropriate "color" is relevant but not necessary by modifying the second condition:

> (ii*) having (or being imagined to have) these features – *possibly in combination with others* – marks them within the context of C's cultural ideology as appropriately occupying the kinds of social position that are in fact either subordinate or privileged (and so motivates and justifies their occupying such a position);

The first condition already allows that the group's members may have supposed origins in more than one region (originally necessary to accommodate the racialization of "mixed-race" groups); modifying the second condition allows that racialized groups may include people of different "colors", and depend on a variety of factors.

Third, need racialized groups be "marked" by actual or assumed body type? What about Jews, Native Americans, and Romanies? (Romanies are also interesting because it isn't entirely clear that there is a supposed place of origin, though I take "no place of origin" to be a factor in their racialization, and to serve as the limit case.) I would suggest that there are *some* (perhaps imagined) physical features that are regarded as salient in contexts where Jews and Native Americans are racialized, though not every member of the group need have those features if there is other evidence of ancestral links. However, ultimately it might be more useful to allow racial membership to be determined by a cluster of features (such as physical appearance, ancestry, and class) weighted differently in different contexts.

Finally, I want the definition to capture the idea that members of racial groups may be scattered across social contexts and may not all actually be (immediately) affected by local structures of privilege and subordination. So, for example, Black Africans and African-Americans are together members of a group currently racialized in the US, even if a certain ideological interpretation of their "color" has not played a role in the subordination of all Black Africans. So I suggest that members of a group racialized in C are those who are *or would be* marked and correspondingly subordinated or privileged when in C. Those who think (plausibly) that all Blacks worldwide have been affected by the structures and ideology of White supremacy do not need this added clause; and those who want a potentially more fine-grained basis for racial membership can drop it.

18 As in the case of gender, I recommend that we view membership in a racial/ethnic group in terms of how one is viewed and treated regularly and for the most part in the context in question; though as before, one could distinguish *being* a member of a given race from *functioning as* one by considering the degree of one's entrenchment in the racialized social position (not on the basis of biology or ancestry).

19 We may want to allow there to be kinds of social stratification between ethnic groups that falls short of the kind of systematic subordination constitutive of race. My account remains vague on this point. Clarification might be gained by plugging in a more sophisticated account of social hierarchies. The body is also relevant: are ethnicities distinguishable from races by the degree to which they are perceived as capable of assimilation?

20 Trey Ellis, *Village Voice*, June 13, 1989; quoted in H.L. Gates 1992, "What's In a Name?", p. 139. Gates quotes the passage differently, leaving out "black" before "ass". Although he adds Ellis's conclusion, he robs the quote of its self-exemplifying power by the alteration.

21 Thanks to Geoff Sayre-McCord for suggesting this approach.

22 It is important here that the "observations" or "imaginings" in question not be idiosyncratic but part of a broader pattern of social perception; however, they need not occur, as in the case of *man* and *woman*, "for the most part". They may even be both regular and rare.

23 I leave it an open question whether groups that have been identified as "third genders" count as genders on my account. Some accounts of gender that purport to include third genders pay inadequate attention to the body, so cannot distinguish, e.g., race from gender. See, e.g., Roscoe 1996.

References

Anderson, Elizabeth. (1995) "Knowledge, Human Interests, and Objectivity in Feminist Epistemology," *Philosophical Topics* **23**:2:27–58.

Appiah, K. Anthony and Amy Gutmann. (1996) *Color Conscious*. Princeton, NJ: Princeton University Press.

Appiah, K. Anthony. (1992) *In My Father's House*. New York: Oxford University Press.

Butler, Judith. (1990) *Gender Trouble*. New York: Routledge.

Crenshaw, Kimberlé. (1993) "Beyond Racism and Misogyny: Black Feminism and 2 Live Crew," *Words that Wound*, ed., M. Matsuda, C. Lawrence, R. Delgado, and K. Crenshaw. Boulder, CO: Westview, 111–132.

Fausto-Sterling, Anne. (1993) "The Five Sexes: Why Male and Female Are Not Enough," *The Sciences* **33**:2: 20–24.

Frye, Marilyn. (1996) "The Necessity of Differences: Constructing a Positive Category of Women," *Signs* **21**:4: 991–1010.

Frye, Marilyn. (1983) *The Politics of Reality*. Freedom, CA: Crossing Press.

Gatens, Moira. (1996) "A Critique of the Sex-Gender Distinction," *Imaginary Bodies*. New York: Routledge, 3–20.

Gates, Jr., Henry Louis. (1992) *Loose Canons*. New York: Oxford University Press.

Geuss, Raymond. (1981) *The Idea of a Critical Theory*. Cambridge: Cambridge University Press.

Haslanger, Sally. (1993) "On Being Objective and Being Objectified," *A Mind of One's Own*, ed., L. Antony and C. Witt. Boulder: Westview, 85–125.

Haslanger, Sally. (1995) "Ontology and Social Construction," *Philosophical Topics* **23**:2: 95–125.

Hennessy, Rosemary and Chrys Ingraham, eds (1997) *Materialist Feminism*. New York: Routledge.

Hoy, David. (1994) "Deconstructing Ideology," *Philosophy and Literature* **18**:1.

Hurtado, Aída. (1994) "Relating to Privilege: Seduction and Rejection in the Subordination of White Women and Women of Color," *Theorizing Feminism*, ed., Anne Hermann and Abigail Stewart. Boulder, CO: Westview Press, 136–154.

MacKinnon, Catharine. (1987) "Difference and Dominance: On Sex Discrimination," *Feminism Unmodified*. Cambridge, MA: Harvard University Press, 32–45.

Omi, M. and H. Winant. (1994) *Racial Formation in the United States*. New York: Routledge.

Ortner, Sherry. (1996) "Gender Hegemonies," *Making Gender*. Boston: Beacon Press, 139–172.

Riley, Denise. (1988) *Am I That Name?* Minneapolis: University of Minnesota Press.

Roscoe, Will. (1996) "How to Become a Berdache: Toward a Unified Analysis of Gender Diversity," *Third Sex, Third Gender*, ed., Gilbert Herdt. New York: Zone Books, 329–372.

Scott, Joan. (1986) "Gender: A Useful Category of Historical Analysis," *American Historical Review* 91:5:1053–75.

Spelman, Elizabeth. (1988) *The Inessential Woman*. Boston: Beacon Press.

Stevens, Jacqueline. (1999) *Reproducing the State*. Princeton: Princeton University Press.

Stich, Stephen. (1988) "Reflective Equilibrium, Analytic Epistemology, and the Problem of Cognitive Diversity," *Synthese* **74**: 391–413.

Stoltenberg, John. (1989) *Refusing To Be a Man*. New York: Meridian Books.

Thorne, Barrie. (1993) *Gender Play*. New Brunswick, NJ: Rutgers University Press.

Young, Iris. (1990) *Justice and the Politics of Difference*. Princeton: Princeton University Press.

Chapter 6

Helena Cronin

THE BATTLE OF THE SEXES REVISITED

A fruit fly delivers his sperm in a toxic cocktail that consigns his consort to an early death. A female dung fly, caught in a scrum of eager suitors, is ignominiously drowned in a cowpat. A spider eats her mate in the very act of copulation. Dramas such as these epitomize the battle of the sexes. Or do they? What is that notorious battle really about? And how can we tell it from life's many other conflicts?

In *The Selfish Gene*, (Dawkins, 1976) Richard Dawkins invites us to take a gene's-eye view of these and other Darwinian questions. Like Einstein's imagined ride on a beam of light, this is an invitation to journey into unreachable worlds for a clearer understanding of reality. It envisages the strategies of genes as they take their paths down the generations, over evolutionary time; the eloquent biographies speak to us of natural selection's design. Unlike most thought-experiments, this is not a solution to one particular problem but a way of perceiving the entire world of living things. And it is a method of great potency. It has immense explanatory power – not surprisingly, for it precisely captures the logic of natural selection's problem-solving; and thus it can generate testable hypotheses. It has remarkable predictive power, prising out telling but otherwise unappreciated evidence. It transforms our view of the familiar, turning into questions what had unthinkingly been regarded as answers. It reveals worlds undreamed of, alerting us to counterintuitive realms. And it dispels confusions, even the tenacious and the wilful.

I started in philosophy, where Darwinism was persistently maligned. Surveying the science, I rapidly concluded that the philosophers were profoundly wrong. *The Selfish Gene* became my staunchest guide. Here was a Darwinian world that was gene-centred, adaptationist; this had to be how natural selection worked. That, and *The Extended Phenotype* (Dawkins, 1982), introduced me to fundamental questions of evolutionary theory. And they taught me how, holding steadily to a gene-centred view, I could find the way through muddle; follow that gene and the rest will fall into place.

So how does this perspective contribute to our understanding of the battle of the sexes? Let's begin by reviewing the gene's-eye view of life. Genes are machines for turning out more genes; they are selfishly engaged in the dedicated pursuit of self-replication. The means by which genes propagate themselves are adaptations, devices that enable them to exploit whatever they can of the world's potential resources to survive, flourish, and replicate. Adaptations manifest themselves as the familiar design features of living things – tails, shells, petals, scents, the ability to glide on the breeze or beguile a mate.

Differences in genes give rise to differences in adaptations. Natural selection acts on these differences and thereby on genes. Thus genes come to be represented in successive generations according to the success of their adaptations.

Among genes all is selfishness, every gene out for its own replication. But from conflict can come forth harmony; the very selfishness of genes can give rise to cooperation. For among the potential resources that genes can exploit is the potential for cooperation with other genes. And, if it pays to cooperate, natural selection will favour genes that do so. Thus selfish genes can come to be accomplished cooperators – selfish cooperators, pragmatic cooperators, but accomplished cooperators nonetheless. Their cooperation arises not in spite of but because of genetic selfishness.

And so, by working together, genes can enjoy the fruits of such magnificent feats of cooperation as the high-tech bio-chemical factory that is a cell; the orchestrated assembly line that is embryonic development; the elaborately equipped vehicles that are bodies. Each of these adaptations is incomparably more intricate, more effective than any gene could build alone. And each cooperative enterprise creates a platform for the next. Thus, from modest beginnings and from a foundation of implacable genetic selfishness, genes have evolved the means to transform the world's resources in ever more ingenious ways, proliferating adaptations of ever greater complexity and sophistication.

Cooperation is in principle possible wherever interests coincide. But interests are of course seldom, if ever, identical. So cooperation gives rise to potential conflict. It should, then, be no surprise if conflict is a persistent accompaniment to a cooperative enterprise – in particular if it occurs over the very resources that cooperation creates, the spoils of joint venture. Each cooperative endeavour generates new resources and thus new arenas of potential conflict.

So when conflict arises within a game of cooperation it arises at the margins, for that is where interests diverge. Even small margins can generate heated disagreement. This is because haggling over the margins is a zero-sum game, a game of 'your loss is my gain'. Think of a buyer and seller arguing over the price of a rug. A stranger to rug-markets would assume that they were arenas where strife runs deep. But such bargaining takes place within a game of cooperation – in this case, mutually beneficial trade.

Thus wherever we see cooperation, at whatever level, we should be prepared also to see conflict. This is crucial to bear in mind when we look at the battle of the sexes. For it is a battle played on the margins of a prodigiously successful cooperative enterprise – sexual reproduction. So the presence of conflict should not be interpreted as lack of cooperation; on the contrary, it will be the result of cooperation.

And so to sexual reproduction. It evolved as a solution to two troubles that cloning organisms are heir to: mutations and parasites. The problem with mutations – copying mistakes – is that they get faithfully copied along with the other genes, whether or not they work well together; the mistakes accumulate down the generations; and eventually they drive the lineage to extinction. The problem with parasites is that, once they have specialized in exploiting a particular host, that host's descendants can't shake off their sitting tenant and so are obliged to provide free board and lodging for as long as the host's lineage persists. Enter sexual reproduction, shuffling genes thoroughly with each generation, creating entirely novel collections of genes with each organism – and thereby both cleansing the gene pool of mutations that undermine the workings of organisms and presenting to parasites not the sitting target of clones but a target forever on the move.

With the advent of full-scale sexual reproduction – some 800 million years ago – came new tasks. A sexual organism must divide its total reproductive investment into two — competing for mates and caring for offspring; and whatever is spent on one is unavailable for the other.

From the very beginning of sexual reproduction there was an asymmetry in investment, a sex difference — one sex specializing slightly more in competing for mates and the other slightly more in caring for offspring. This arose because there were two distinct groups of genes and they got into the next generation in different ways. Genes were housed (as they are today) in two parts of the cell — the vast majority in the nucleus and a few outside it, mainly in the mitochondria, the cell's powerhouse. Nuclear genes went into both kinds of sex cells, half of them in each cell. But, because two sets of mitochondria would lead to disruptive conflict over which should be the powerhouse and which would be surplus to requirements, they were allowed into only one kind of sex cell. So one sex cell started out larger and with more essential resources than the other. And thus began the great divide into fat, resource-laden eggs, already investing in providing for offspring, and slim, streamlined sperm, already competing for that investment.

Once that divergence had opened up, evolutionary logic dictated that it became self-reinforcing. If you specialize in competing, you gain most selective advantage by putting more into competing; and the same for caring. And so the divergence widened over evolutionary time, with natural selection proliferating and amplifying the differences, down the generations, in every sexually reproducing species that has ever existed.

Thus, from such inauspicious beginnings, from this slight initial asymmetry, flow all the characteristic differences between males and females throughout the living world, differences that pervade what constitutes being male or female. Indeed, for evolutionary biologists,

downloading mitochondria into future generations or offloading them into genetic oblivion is the fundamental difference between females and males.

What do those sex differences typically look like? A female possesses a scarce and valuable resource: her eggs and the investment that goes with them. She needs to be judicious about how she ties up this precious commodity. So she goes for quality. Which qualities? Good genes and good resources, particularly food, shelter, and protection for herself and her offspring. Meanwhile, a male's reproductive success is limited only by the number of matings that he can get. So he goes for quantity. Thus males compete with one another to provide what females want. They display their quality with costly, elaborate ornaments; and they strive to get resources and to hold on to them.

So males, far more than females, compete to be the biggest, brightest, brashest, and best; they are larger and stronger; they take more risks and fight more; they care more about status and power; they are more promiscuous and do less child-care; and they expend vast quantities of time, energy, and resources just strutting their stuff — singing, dancing, roaring; flaunting colours and iridescence; displaying tails and horns . . . adaptations in glorious profusion.

Now that we know what the two sexes want, we can see how and why males and females come into conflict. He wants to mate more often than she does; she wants to be choosier than he does. And she wants to extract more parental investment from him than he wants to give; he — vice versa.

Conflicts over mate choice have led males into advertising and deception, stealth and force — and females into counter-adaptations ranging from lie-detectors to anti-clamping devices. The escalation of extravagant male advertisement driven by female scrutiny is famously exemplified by the peacock's tail, a story told in the second edition of *The Selfish Gene*. Male force and a woman's fight to choose are vividly illustrated by the female reproductive tract of a multitude of insect species. She has evolved Byzantine arrangements of chambers and corridors, lumber rooms and labyrinths for storing sperm to be used when she chooses. And he has evolved an armature of scoops and toxins, hooks and horns to oust the sperm of other males, and an array of glues and plugs to prevent her or would-be later lovers from ousting his. As for adaptations for conflicts over parental investment – tactics for monopolizing or increasing the other's investment and minimizing one's own – this story, too, is told in *The Selfish Gene*. And any sample of recent papers is striking testimony to how much we now know about the multitude of inventive arguments to be had over who looks after the children: 'why don't male mammals lactate?'; 'mother/father differences in response to infant crying'; female starlings 'increase their copulation rates when faced with the risk of polygyny'; 'female-coerced monogamy in burying beetles'.

Let's now parse a case involving both mate choice and parental care, disentangling conflict from cooperation and the strategies of genes from the behaviour of the males and females that house them. Picture a pair of titi monkeys, husband and wife in close embrace, their tails entwined, in sleep cuddled together, when awake always close, preferring one another's company above that of all others – and with so little sex difference that they look more like twins than spouses. What is the recipe for their successful marriage? It's that he takes on a huge burden of child-care; he's an authentic New Man. During the baby's first fortnight, apart from its feeding times, he carries it constantly; and his care continues to be so assiduous that the baby is more distressed at his disappearance than at its mother's. And as a wealth of evidence shows – including cross-species comparisons

and behavioural and hormonal measures – their parental division of labour reflects cooperation among their 'parental-investment' genes. But continue to trace the interests of their respective genes and we find that what might appear to be the very epitome of their happy pact – the entwined tails, the constant cuddles – reflects instead genetic conflict. For they are mate-guarding: an adaptation that reflects an evolutionary legacy of less than perfect monogamy (which continues still in titis, albeit rarely). He is protecting himself from lavishing his investment on another male's offspring; she, however, could benefit if her putative lover had superior genes. She is protecting herself from the danger of losing his paternal investment; he, however, could benefit from additional matings elsewhere. So their conflict is over mate choice. And it is engendered by the very resource, parental investment, that their cooperation has created. What joins them together has also – among their genes – put them asunder.

Now to a case of parental conflict more arcane and more wondrous. It takes us to a prediction that, when *The Selfish Gene* was published, was a mere twinkle in the gene's-eye view; although not spelt out explicitly in the book, these tactics of genes brimmed from the logic. To begin, let's remind ourselves of how to enter into the gene's view. It involves following the trajectory of genes, tracing the careers of the strategies by which they replicate themselves. We must think, then, not about a single encounter between a male and a female, nor about their success over a mating season, nor even about all the encounters during their lifetimes but about the average success of a gene over all its numerous instantiations, in different individuals, over many generations, down evolutionary time. And we must bear in mind that, in sexually reproducing species, most genes move back and forth between male and female bodies, spending 50 per cent of their time in each; so genes responsible for sex differences will alter their tactics to fit with their current neighbours, following the rule: 'If in a male body, do this; if in a female body, do that'.

Now to the example. 'Don't argue in front of the children!' goes the standard advice. But what about arguing in the children? Consider this gene-centred thought-experiment. Imagine a gene travelling down the generations in successive offspring and having different interests depending on whether it has just come from the father or the mother. For the interests of paternally- and maternally-derived genes will indeed diverge, particularly if the species cannot boast a history of unalloyed female monogamy (which no species can). And one divergence arises because maternally-derived genes will be in all the future offspring of that mother whereas paternally-derived genes might be in none. Now imagine, too, that the mother provides generous nutrients inside her body for the fertilized egg; so the entire burden of care is on her and the father is absolved. That immediately creates an arena for conflict over maternal provisions. The maternally-derived genes want the mother to provide not only for themselves but also for copies of themselves in her future offspring; but the paternally-derived genes, having no guaranteed interest in future offspring, want to exploit the mother's body beyond that 'fair share'.

Impeccable gene-logic. But thirty years ago the facts looked unsympathetic; orthodox thinking was that genes couldn't possibly know which parent they came from. It turns out, however, that they can – it's called 'genomic imprinting' – and, what's more, that they behave precisely as predicted. Molecular biologists, after initial astonishment, are now accustomed to discovering such genes regularly in both mammals and flowering plants.

Take a mouse species in which the mother is near-monogamous; all the offspring in a litter have the same father and there's an 80 per cent

chance that he'll father the next litter. And take also a closely related species in which there is a plethora of fathers within litters and from litter to litter. Cross a 'multiple-paternity'-species male with a 'monogamous-mother'-species female. Result: offspring larger than normal. This is because the father-derived genes wrest all that they can from the mother and the mother-derived genes are unaccustomed to putting up much resistance. Cross the other way – a 'monogamous-mother' male with a 'multiple-paternity' female. Result: offspring smaller than normal. The mother-derived genes are taking their miserly 'fair share' without the usual top-up from the father-derived genes. Compare the results: dramatic differences in birth weight. The genes' battle is a tug-of-war, each side assuming resistance from the other, the tugs escalating in strength over the generations. The experiment forces one side to let go; and the entire game comes tumbling down. Even without laboratory intervention one side sometimes lets go. This explains, for example, much about the typical pathologies of human pregnancy, otherwise so baffling in a well-honed adaptation. Note that the tug-of-war is a marginal conflict arising in a cooperative game. The parents have a vast area of overlapping interests, for both want the offspring to develop normally. But conflict over the size of the mother's investment has triggered an evolutionary arms race and that has settled into the tug-of-war. When both sides tug normally, the result is a compromise growth rate and a normal baby. But either side letting go spells disaster for all – for genes in the embryo and in the parents.

Why does this particular battle of the sexes occur in mammals and flowering plants? Because, in both, the mother provides nutrients after fertilization. Maternal genes have hit on the same strategies for solving a 'caring' problem; and so they have generated the same arena for conflict, the father trying to get just that bit more than the mother's rationing stipulates. Thus the gene's-eye view has revealed hitherto unexplored commonalities across vast taxonomic divides – genes that have converged on the same strategic solutions and faced the same resultant conflicts.

And the embryo is just one such battleground. The same logic could get to work wherever the interests of relatives on the mother's side and the father's side diverge. So watch out for it in closely related groups; it could shed unexpected light on family rows.

Having seen what the battle of the sexes is, we are now in a position to see what it is not – to identify cases that might look persuasively like such a battle but aren't. For the battle of the sexes, there must be a conflict of interest between 'male' and 'female' genes (genes implementing male and female strategies) – not just a tussle between male and female bodies. And the conflict must be over mate choice or parental investment.

We can begin by clearing away that dead dung fly. Although she was drowned in a scrum of over-eager males, neither she nor her sisters who are often cited as victims of 'male-induced harm' – insect, bird, and reptile, drowned, crushed, and suffocated – are casualties of the battle between the sexes. Such battles are within one sex, males against males; the females are civilians caught in crossfire. There is no adaptation for courting by drowning; indeed, the winner's prize is a Pyrrhic victory.

The corpse of the poisoned fruit fly is, by and large, also collateral damage. The toxins in her suitor's seminal fluids are a cocktail that search-and-destroy the sperm of previous suitors; trigger her egg production; and slip her an anti-aphrodisiac intended to last until her eggs are fertilized. Insofar as his genes attempt to control when and by whom she is fertilized, she is certainly a victim of the battle of the sexes. But the main cause of her early death is a conflict between males who use her body as their battleground – sperm competition so fierce that the drugs take an unintended toll. Indeed, remove sperm competition from their mating and give him a vested interest in her as a long-term baby-machine – as was done in an experiment that imposed monogamy on both of them – and, within a few generations, his seminal fluid becomes increasingly less toxic.

Thus reducing the poison when monogamous is an adaptation; poisoning her when promiscuous is not. Jesuitical as such distinctions might seem to a drowned or poisoned bride, for a gene-centred analysis it is crucial to ask whether the adaptation is a weapon in the battle of the sexes and, if so, whether designed to harm her. Admittedly, such scrupulous partitioning of sexual conflicts into male–female and male–male is not always germane. But, before we attribute a conflict to the battle of the sexes, we should bear in mind that both selective forces could be at work.

Now to an example of harm to one of the partners that is not collateral damage and indeed, contrary to dramatic appearances, is probably not a result of conflict at all. The female redback spider consumes her mate while they copulate. But genetic conflict? No. He is investing in his offspring. Some males feed their young with bodies that they have caught; he cuts out the middleman and offers up his own. In a virtuoso delivery of fast food, he flips a somersault that lands him neatly in her mouth. Such suicidal behaviour usually evolves when males have little chance of finding another mate, which is indeed the case for him. However, he might also have a further agenda: sperm competition. By preoccupying her with food, the disappearing male can prolong copulation – far longer than needed to transfer sperm but just right for adding a poisoned parting shot for rival sperm. If so, he has moved from giving paternal care to circumventing her mate choice: from cooperation to the battlefield of the sexes.

Female cannibalism can be against the lover's will but even so not necessarily a battle of the sexes. The female praying mantis is notorious for her voracious sexual cannibalism. About one-third of her matings are also meals, consumed during or after copulation. The male has evolved elaborate adaptations against being eaten; he approaches stealthily and pounces suddenly on her back while she is distracted, a finely orchestrated suite of moves. So there is certainly conflict in their encounter. But whereas on a poor diet she devours about three-quarters of her mates, on a rich diet she downs less than a quarter. So perhaps her adaptation is not to eat her lover but just to eat, not gratefully to receive paternal investment but just to have lunch. If, to her, he is indeed meat not mate, then their battle is not that of the sexes but that of predator and prey.

Finally, an apparently flagrant case, the very cutting edge of sexual conflict: females castrating males, a practice rife among flowering plants. But it turns out not to be a battle between the sexes although, among organisms, the male parts of the plant wither while the female parts on that very same plant flourish. The battle is between mitochondrial and nuclear genes. Remember that, from the beginning of sexual reproduction, mitochondria have gone only into eggs, never into sperm; so they have travelled solely through the female line. Thus – although they replicate by cloning – they rely on the sexual reproduction of the body that they are in to exit, in fertilized eggs, to the next generation. However, whereas when they exit into a daughter they are again in a vehicle to the next generation, when they exit into a son the vehicle is their hearse, for they cannot get into sperm. And so mitochondria in male parts in those flowering plants, rather than helping in the sperm-making factory as they should do, subvert the assembly line so that the sperm aren't viable. Such unruly behaviour is strongly predicted by

gene-centred theory. Genes that travel as a band down the generations through only one sex would not want to get into the other sex; and so they would want to gang up against it. The ancient arrangement with mitochondria was a recipe for conflict. For plant species, it becomes a conflict about whether to reproduce sexually at all.

We are now ready to survey afresh the battlefield of the sexes. With *The Selfish Gene* as our guide, we can compile a gene-centred inventory of any participants that we come across. There are the principal antagonists: 'male' and 'female' genes in conflict and the resulting multifarious adaptations in male and female creatures. There are the victors and victims that we would otherwise have missed: the agents of genomic imprinting, visiting the feuds of the previous generation upon the children. There are the subversive opportunists: mitochondria waging their own private war against nuclear genes over their mitochondrial mausoleum, the male parts of plants. And there are the innocent bystanders: victims of collateral damage from other's battles, such as drowned and poisoned females.

So the banner of the battle of the sexes should depict not a spider's quietus in his lover's jaws nor the withered anthers of a flower; for they represent other concerns of genes. Nor need it depict struggle or pain, injury or death. It could instead portray the dazzling beauty of the peacock's tail, the deep intimacy of the titis' embrace, the finely poised equilibrium that builds a newborn baby. Such are the ways of genes that even their conflicts – 'male' against 'female' genes – can appear deceptively to us as harmony and beauty in their bearers.

A wider survey of the battlefield reminds us that, however salient the conflicts, however fierce the battles, they are but marginal within the vast and intricate game of cooperation that is sexual reproduction. And this cooperation reminds us – for there are those that need reminding still – that selfish genes do cooperate with one another. Indeed, as *The Ancestor's Tale* emphasizes, even with the gene-shuffling of sexual reproduction, the genes of a species become good cooperators.

> [G]iven sex, . . . genes are continually being tried out against different genetic backgrounds. In every generation, a gene is shuffled into a new team of companions, meaning the other genes with which it shares a body . . . Genes that are habitually good companions . . . tend to be in winning teams – meaning successful individual bodies that pass them on to offspring . . . But in the long term, the set of genes with which it [a gene] has to co-operate are all the genes of the gene pool, for they are the ones that it repeatedly encounters as it hops from body to body down the generations.[1]

And, finally, moving beyond the battlefield, *The Selfish Gene* reminds us that, although the bearers in which genes move down the generations are engaged in enterprises of apparently great pith and moment, a genes'-eye perspective takes our understanding far beyond these transitory players, opening up an immense reach down time, along the deep flow of immortal genes.

Note

1 Richard Dawkins, *The Ancestor's Tale*, London: Weidenfeld & Nicolson, 2004, pp. 359.

References

Richard Dawkins, *The Extended Phenotype*, Oxford: Oxford University Press, 1982.

Richard Dawkins, *The Selfish Gene*, Oxford: Oxford University Press, 1976.

Evelyn Fox Keller

BEYOND THE GENE BUT BENEATH THE SKIN

The past decade has witnessed an efflorescence of critical commentary deploring the excessively genocentric focus of contemporary molecular and evolutionary biology. Among philosophers, the best known work may well be that of developmental systems theorists (DST) Oyama, Griffiths, and Gray (and sometimes including Lewontin and Moss), but related critiques have also emerged from a variety of other quarters. Expressing diverse intellectual and philosophical preoccupations, and motivated by a variety of scientific and political concerns, these analyses have converged on a number of common themes and sometimes even on strikingly similar formulations.[1] Common themes include: conceptual problems with the attribution of causal primacy (or even causal efficacy) to genes;[2] disarray in contemporary uses of the very term *gene*; confusions and misapprehensions generated by use of the particular locution of "genetic program." Much of the impetus behind these critiques issues from long-standing concerns, and indeed, many of the critical observations could have been (and in some cases have been) made long ago. Why then their particular visibility today? An obvious answer lies close at hand: Critiques of genocentrism have found powerful support in many of the recent findings of molecular biologists. Indeed, I would argue that it is from these empirical findings that the major impetus for a reformulation of genetic phenomena now comes.

Three developments (or findings) are of particular importance here: (1) the need for elaborate mechanisms for editing and repair of DNA to ensure sequence stability and fidelity of replication; (2) the importance of complex (and nonlinear) networks of epigenetic interactions in the regulation of transcription; (3) the extent to which the "sense" of the messenger transcript depends on highly regulated mechanisms of editing and splicing. The implication of the first is that the structure of the gene (or sequence of DNA) may be the (or, from the perspective of DST, "a") locus of heredity constancy, but it can no longer be supposed to be its source: Particular genes (or sequences) persist as stable entities only as long as the machinery responsible for that stability persists. The dependence of gene function on complex epigenetic networks challenges (or at least seriously complicates) the attribution of causal agency to individual genes. Finally, the third finding radically undermines the assumption that proteins are simply and directly encoded in the DNA (indeed, it undermines the very notion of the gene as a functional unit residing on the chromosome).

On this much, one finds a certain general agreement. But differences – deriving in part from the different intellectual, scientific, and political perspectives of their authors – can also be found. Sometimes these are matters of emphasis, sometimes of focus, and sometimes of more substantive import. In this chapter, I want to focus on what I believe to be a substantive issue distinguishing my own perspective from that which tends to dominate the DST literature, and that issue can be put in the form of a question: Is there a place on our biological map for the material body of the organism, for that which lies beyond the gene yet beneath the skin? And if so, where is that place?

The body in question

I share with proponents of DST the conviction that the oppositional terms in which the nature/nurture debate has historically been framed are both artificial and counterproductive. But the particular question I pose here reflects an additional source of unease, and that is over the tacit elision of the body implied not only by the framing of the classical controversies, but at least partially continued in the solutions that have been thus far been put forth.

Without question, the most conspicuous roots of my concern are to be found in the history of genetics and neo-Darwinian evolutionary theory. With the emergence of genetics in the early part of the century, debates over the relative force of nature and nurture [sic] (first framed as such by Francis Galton in 1874) were recast, initially, in terms of heredity and environment (see, e.g., Barrington and Pearson 1909; Morgan 1911; Conklin 1915), and soon after, in terms of genes and environment (see, e.g., analyses of the relative importance of heredity and environment in Fisher 1918 and Wright 1920). This second reframing may have been an inevitable consequence of the terminological shift in the biological literature of the 1920s and later, in which the term *heredity* came

to be replaced by the newer term *genetics*,[3] but the consequence of this shift was more than terminological: it amounted to a conceptual reduction of "nature" to "genes," and with that reduction, only one of two possible statuses for the extragenetic body: either its complete elision or its relegation to the category of "nurture" or "environment." Genetics further contributed to this relegation with its recasting of another and far older controversy, namely that concerning the relations between form (generally construed as active) and matter (construed as passive). That ancient discussion could now be (and was) reconceptualized in terms of genes as the agents of "action" (later, as sources of "information"), and of a cellular or extracellular environment that is simultaneously acted upon and informed, serving as passive material substrate for the development (or unfolding) of the organism.[4] But where many of the discussions of heredity and environment among geneticists focused on their relative force in individual development, elsewhere, such debates more commonly focused on their relative force in shaping the course of evolution. Here, the neo-Darwinian synthesis was of particular importance. In identifying genetic continuity and change as the sole fundament of evolution, it contributed powerfully to the polarization of debates over the relative force of genes and environment in such highly charged arenas as eugenics, or the "heritability" of intelligence and other behavioral attributes; it also helped pave the way for the mid-twentieth-century recasting of the nature-nurture debate in one of its crudest forms, that is, as a battle between advocates of "Darwinian" and "Lamarckian" evolution (see, e.g., Jablonka and Lamb 1995).

To the extent that such debates imply a logical disjunction (form *or* matter; nature *or* nurture; genes *or* environment), they are clearly counterproductive. But my particular argument here is that replacing an implied disjunction by an explicit conjunction (nature *and* nurture; genes *and* environment) does little to ameliorate the particular problem of the role of the body that resides beyond the gene yet beneath the skin. To be sure, the disjunctive framing absolutely denies informational function to the material environment whereas the conjunctive framing advocated by DST clearly does permit such a function.[5] But both exhibit a discernible tendency to figure the organismic body qua environment (and qua nature), and accordingly, to leave any distinctive informing role the organism's "internal environment" might play in development and evolution concealed from view.

Which body, which skin? And, anyway, why stop at the skin?

Should the organismic body be singled out as having particular biological significance? And if so, which body, and which skin? Biology recognizes many bodies, corresponding to many skins: in higher organisms, there is the multi-cellular body contained within an outer integument; in all organisms, cellular bodies are contained by cell membranes; and in eukaryotic organisms nuclear bodies are contained by nuclear membranes. To avoid some of this ambiguity, I choose to focus on that moment in the life cycle of higher (metazoan) organisms in which the outer integument *is* the cell membrane, and the organismic body *is* the cellular body – that is, in which the body in question is the fertilized egg or zygote. But there remains the question, why stop at the skin? Certainly, no biological integument provides an absolute divide between interior and exterior, and the cell membrane of a fertilized egg is, of necessity, more porous than most. Furthermore, because it regulates so much of the traffic between inside and out, the cell membrane is itself an active agent in shaping the body it

contains, indeed, in determining the very meaning of interiority. These facts constitute a warning against conceptualizing the organism as an autonomous individual, sealed off from an exterior world by a static or preexisting boundary. Yet even so, the cell membrane, dynamic and permeable though it may be, defines a boundary which evolution has not only crafted into a cornerstone of biological organization but has endowed with absolutely vital significance. And given the dire effect the physical erasure of this boundary would have on the survival of the organism or cell, it scarcely seems necessary to elaborate upon the inappropriateness of its conceptual erasure.

In other words, the immediate and most obvious reason for taking this boundary seriously is grounded in its manifest indispensability for viability. But this said, we are still not any closer to understanding why is it so important. By way of addressing this last question, I would like to suggest that the primary function of the cell membrane (as of any other biological skin) is simply that it holds things together – more specifically, that it keeps in proximity the many large molecules and subcellular structures required for growth and development. Proximity is crucial, for it enables a degree of interconnectivity and interactive parallelism that would otherwise not be possible, but that is required for what I take to be the fundamental feature of the kind of developmental system we find in a fertilized egg, namely, its robustness. Prior to all its other remarkable properties – in fact, a precondition of these – is the capacity of a developmentally competent zygote to maintain its functional specificity in the face of all the vicissitudes it inevitably counters. This paradigmatic body may not be autonomous, but as embryologists have always known, it is far more tolerant of changes in its external environment than in its internal milieu. Indeed, were it not for their robustness, that is, for the tolerance of (at least many) early embryos to being moved from one environment to another, embryonic manipulation would not be possible, and much of what we know of as experimental embryology would never have come into being.

All of this may seem too obvious to need saying, but there are times when the obvious is what most needs saying. We have learned that no elision is innocent. Nor, for that matter, is any reminder of elision. Politics are everywhere. Just as there are important political dimensions to the history of debates over genes and environment,[6] so too, there are political dimensions to the elision of the body in genetical discourse,[7] as there also are, inevitably, to my insistence here on the boundary of the skin. In fact, the title of this essay contains its own elision: it reminds us of another title (Barbara Duden's *The Woman Beneath the Skin*), while at the same time suppressing the subject of that other title. There is, of course, a reason. This is not an essay in feminist theory, nor is it about women. The "woman" in my title is signified only by its absence – intended, by that absence, to evoke nothing more than a recognition of the trace of the woman beneath the skin that still lurks, if not in the body generally, surely, in the reproductive body of the fertilized egg. Because, in sexual reproduction, the cytoplasm derives almost entirely from the unfertilized egg, it is no mere figure of speech to refer to it as the maternal contribution. Furthermore, the representation of that body as "genetic environment," as nothing more than a source of nurture for the developing organism, is a bit too reminiscent of conventional maternal discourse for at least this author's comfort. My title, in short, is deliberate in its allusivity: I want to indicate the possibility that gender politics has been implicated in the historic elision of the body in question, without, at the same time, reinscribing the woman in that or any other body. The primary aim of this essay is, finally, a biological one, by which I mean that

it is to reclaim the possibility of finding biological significance and agency in that no man's land beyond the gene but beneath the skin. Contra Oyama (1992),[8] I want to argue that taking the cell rather than the gene as a unit of development *does* make a difference: not only does it yield a significant conceptual gain in the attempt to understand development, but also, it permits better conformation to the facts of development as we know them.

Is there a program for development? And if so, where is it to be found?

Many authors have taken issue with the concept of a *program* for development, noting its teleological implications, its metaphoric reliance on computer science, its implication of a unidirectional flow of information.[9] But my concern here is not with the concept of *program* per se: rather, it is with the more specific notion of a genetic program, especially in contradistinction to its companion notion, that of a developmental program. In other words, for my particular purposes here, I accept the metaphor of *program*, warts and all, and focus my critical attention instead on the implications of attaching to that metaphor the modifier "genetic." And I ask two questions: First, what is the meaning of a "genetic program"? Second, how did this concept come to be so widely accepted as an "explanation" of biological development?[10]

Taken as a composite, the meaning of the term *genetic program* simultaneously depends upon and underwrites the particular presumption that a "plan of procedure" for development is itself written in the sequence of nucleotide bases. Is this presumption correct? Certainly, it is almost universally taken for granted, but I want to argue that, at best, it must be said to be misleading, and at worst, simply false: To the extent that we may speak at all of a developmental program, or of a set of instructions for development, in contradistinction to the data or resources for such a program, current research obliges us to acknowledge that these "instructions" are not written into the DNA itself (or at least, are not all written in the DNA), but rather are distributed throughout the fertilized egg. Indeed, if the distinction between program and data is to have any meaning in biology, it has become abundantly clear that it does not align (as had earlier been assumed) either with a distinction between "genetic" and "epigenetic," or with the precursor distinction between nucleus and cytoplasm. To be sure, the informational content of the DNA is essential – without it development (life itself) cannot proceed. But for many developmental processes, it is far more appropriate to refer to this informational content as data than as program (Atlan and Koppel 1990). Indeed, I want to suggest that the notion of genetic program both depends upon and sustains a fundamental category error in which two independent distinctions, one between "genetic" and "epigenetic," and the other between program and data, are pulled into mistaken alignment. The net effect of such alignment is to reinforce two outmoded associations: on the one hand, between "genetic" and active, and, on the other, between "epigenetic" and passive.

Development results from the temporally and spatially specific activation of particular genes which in turn depends on a vastly complex network of interacting components including not only the "hereditary codescript" of the DNA, but also a densely interconnected cellular machinery made up of proteins and RNA molecules. Necessarily, each of these systems functions in relation to the others alternatively as data and as a program. If development cannot proceed without the "blueprint" of genetic memory, neither can it proceed without the "machinery" embodied in cellular structures. To be sure, the elements of these structures are fixed by genetic memory, but their assembly is

dictated by cellular memory.[11] Furthermore, one must remember that more than genes are passed from parent to offspring. To forget this is to be guilty of what Richard Lewontin calls an "error of vulgar biology." As he reminds us, "an egg, before fertilization, contains a complete apparatus of production deposited there in the course of its cellular development. We inherit not only genes made of DNA but an intricate structure of cellular machinery made up of proteins" (Lewontin 1992: 33).

Assuming one is not misled by Lewontin's colloquial use of the term *inherit* to refer to transmission over a single generation (as distinct from multigenerational transmission), none of this is either controversial or new, nor does it depend on the extraordinary techniques now available for molecular analysis. Yet, however surprisingly, it is only within the last decade or two that the developmental and evolutionary implications of so called "maternal effects" has begun to be appreciated.[12] Current research now provides us with an understanding of the mechanisms involved in the processing of genetic data that make the errors of what Lewontin calls "vulgar biology" manifest. Yet, even when elaborated by the kind of detail we now have available, such facts are still not sufficient to dislodge the confidence that many distinguished biologists continue to have in both the meaning and explanatory force of the genetic program. The question I want therefore to ask is, how come? What grants the "genetic program" its apparent explanatory force, even in the face of such obvious caveats as those above? To look for answers, I will turn to history, more specifically, to the history of the term itself.

"Programs" in the biological literature of the 1960s

The metaphor of a "program," borrowed directly from computer science, entered the biological literature in the 1960s not once, but several times, and in at least two distinctly different registers. In its first introduction, simultaneously by Mayr (1961) and by Monod and Jacob (1961), the locus of the "program" was explicitly identified as the genome, but, over the course of that decade, another notion of "program," a "developmental program," also surfaced, and repeatedly so. This program was not located in the genome, but instead, distributed throughout the fertilized egg (see, e.g., Apter 1966). By the 1970s, however, the "program" for development had effectively collapsed into a "genetic program," with the alternative, distributed, sense of a "developmental program" all but forgotten.

Francois Jacob, one of the earliest to use the concept "genetic program" contributed crucially to its popularization. In *The Logic of Life*, first published in French in 1970, Jacob describes the organism as "the realization of a programme prescribed by its heredity" (Jacob 1976: 2), claiming that "when heredity is described as a coded programme in a sequence of chemical radicals, the paradox [of development] disappears" (Jacob 1976: 4). For Jacob, the genetic program, written in the alphabet of nucleotides, is what is responsible for the apparent purposiveness of biological development; it and it alone gives rise to "the order of biological order." (Jacob 1976: 8) He refers to the oft-quoted characterization of teleology as a "mistress" whom biologists "could not do without, but did not care to be seen with in public," and writes, "The concept of programme has made an honest woman of teleology" (Jacob 1976: 8–9). Although Jacob does not exactly define the term, he notes that "[t]he programme is a model borrowed from electronic computers. It equates the genetic material of an egg with the magnetic tape of a computer" (Jacob 1976: 9).

However, equating the genetic material of an egg with the magnetic tape of a computer does not imply that that material encodes a "program"; it might just as well be thought of as encoding "data" to be processed by a cellular "program." Or by a program residing in the machinery of transcription and translation complexes. Or by extranucleic chromatin structures in the nucleus. Computers have provided a rich source of metaphors for molecular biology, but they cannot by themselves be held responsible for the notion of "genetic program." Indeed, as already indicated, other, quite different, uses of the program metaphor for biological development were already in use. One such use was in the notion of a "developmental program" – a term that surfaced repeatedly through the 1960s, and that stood in notable contrast to that of a "genetic program."

Let me give an example of this alternative use. In 1965, a young graduate student, Michael Apter, steeped in information theory and cybernetics, teamed up with the developmental biologist Lewis Wolpert to argue for a direct analogy not between computer programs and the genome, but between computer programs and the egg:

> if the genes are analogous with the subroutine, by specifying how particular proteins are to be made ... then the cytoplasm might be analogous to the main programme specifying the nature and sequence of operations, combined with the numbers specifying the particular form in which these events are to manifest themselves. ... In this kind of system, instructions do not exist at particular localized sites, but the system acts as a dynamic whole. (Apter and Wolpert 1965: 257)

Indeed, throughout the 1960s, a number of developmental biologists attempted to employ ideas from cybernetics to illuminate development, and almost all shared Apter's starting assumptions (see Keller 1995, chap. 3 for examples) – that is, they located the program (or "instructions") for development in the cell as a whole.

The difference in where the program is said to be located is crucial, for it bears precisely on the controversy over the adequacy of genes to account for development that had been raging among biologists since the beginning of the century. By the beginning of the 1960s, this debate had subsided, largely as result of the eclipse of embryology as a discipline during the 1940s and 1950s. Genetics had triumphed, and after the identification of DNA as the genetic material, the successes of molecular biology had vastly consolidated that triumph. Yet the problems of development, still unresolved, lay dormant. Molecular biology had revealed a stunningly simple mechanism for the transmission and translation of genetic information, but, at least until 1960, it had been able to offer no account of developmental regulation.

James Bonner, a professor of biology at CalTech, in an early attempt to bring molecular biology to bear on development, put the problem well. Granting that "the picture of life given to us by molecular biology . . . applies to cells of all creatures," he goes on to observe that this picture

> is a description of the manner in which all cells are similar. But higher creatures, such as people and pea plants, possess different kinds of cell. The time has come for us to find out what molecular biology can tell us about why different cells in the same body are different from one another, and how such differences arise (Bonner 1965: v)

Bonner's own work was on the biochemistry and physiology of regulation in plants, in an institution well known for its importance in the birth of molecular biology. Here, in this work, published in 1965, like Apter and a number of others of that period, Bonner too employs the conceptual

apparatus of automata theory to deal with the problem of developmental regulation. But unlike them, he does not locate the "program" in the cell as a whole, but rather, in the chromosomes, and more specifically in the genome. Indeed, he begins with the by then standard credo of molecular biology, asserting that "[w]e know that . . . the directions for all cell life [are] written in the DNA of their chromosomes" (Bonner 1965: v). Why? An obvious answer is suggested by his location. Unlike Apter and unlike other developmental biologists of the time, Bonner was situated at a major thoroughfare for molecular biologists, and it is hard to imagine that he was uninfluenced by the enthusiasm of his colleagues at CalTech. In any case Bonner's struggle to reconcile the conceptual demands posed by the problems of developmental regulation with the received wisdom among molecular biologists is at the very least instructive, especially given its location in time, and I suggest it is worth examining in some detail for the insight it has to offer on our question of how the presumption of a "genetic program" came – in fact, over the course of that very decade – to seem self-evident. In short. I want to take Bonner as representative of a generation of careful thinkers about an extremely difficult problem who opted for this (in retrospect, inadequate) conceptual shortcut.

Explanatory logic of the "genetic program"

From molecular biology, Bonner inherited a language encoding a number of critical if tacit presuppositions. That language shapes his efforts in decisive ways. Summarizing the then current understanding of transcription and translation, he writes:

> Enzyme synthesis is therefore an information-requiring task and . . . the essential

information-containing component is the long punched tape which contains, in coded form, the instructions concerning which amino acid molecule to put next to which in order to produce a particular enzyme. (Bonner 1965: 3)

At the same time, he clearly recognized that only the composition of the protein had thus been accounted for, and not the regulation of its production required for the formation of specialized cells, that is, cell differentiation remained unexplained. As he wrote, "Each kind of specialized cell of the higher organism contains its characteristic enzymes but each produces only a portion of all the enzymes for which its genomal DNA contains information" (Bonner 1965: 6). But, he continues: "Clearly then, the nucleus contains some further mechanism which determines in which cells and at which times during development each gene is to be active and produce its characteristic messenger RNA, and in which cells each gene is to be inactive, to be repressed" (Bonner 1965: 6).

Two important moves have been made here. Bonner argues that something other than the information for protein synthesis encoded in the DNA is required to explain cell differentiation (and this is his main point), but on the way to making this point, he has placed this "further mechanism" in the nucleus, with nothing more by way of argument or evidence than his "Clearly then." Why does such an inference follow? And why does it follow "clearly"? Perhaps the next paragraph will help:

The egg is activated by fertilization. . . . As division proceeds cells begin to differ from one another and to acquire the characteristics of specialized cells of the adult creature. There is then within the nucleus some kind of programme which determines the property [sic] sequenced repression and derepression of genes and which brings about orderly development. (Bonner 1965: 6)

Here, the required "further mechanism" is explicitly called a "program," and, once again, it is located in the nucleus. But this time around, a clue to the reasoning behind the inference has been provided in the first sentence, "The egg is activated by fertilization." This is how I believe the (largely tacit) reasoning goes: If the egg is "activated by fertilization," the implication is that it is entirely inactive prior to fertilization. What does fertilization provide? The entrance of the sperm, of course, and unlike the egg, the sperm has almost no cytoplasm: it can be thought of as pure nucleus. Ergo, the active component must reside in the nucleus and not in the cytoplasm. Today, the supposition of an inactive cytoplasm would be challenged, but in Bonner's time, it would have been taken for granted as a carryover from what I have called "the discourse of gene action" of classical genetics (Keller 1995). And even then, it might have been challenged had it been made explicit, but as an implicit assumption encoded in the language of "activation," it was likely to go unnoticed by Bonner's readers as by Bonner himself.

Bonner then goes on to ask the obvious questions: "What is the mechanism of gene repression and derepression which makes possible development? Of what does the programme consist and where does it live?" (Bonner 1965: 6) And he answers them as best he can: "We can say that the programme which sequences gene activity must itself be a part of the genetic information since the course of development and the final form are heritable. Further than this we cannot go by classical approaches to differentiation" (Bonner 1965: 6).

In these few sentences. Bonner has completed the line of argument leading him to the conclusion that the program must be part of the genetic information, that is, to the "genetic program." And again, we can try to unpack his reasoning. Why does the heritability of the course of development and the final form imply that the program must be part of the genetic information? Because – and only because – of the unspoken assumption that it is only the genetic material that is inherited. The obvious fact – that the reproductive process passes on (or transmits) not only the genes but also the cytoplasm (the latter through the egg for sexually reproducing organisms) – is not mentioned. But even if it were, this act would almost certainly be regarded as irrelevant, simply because of the prior assumption that the cytoplasm contains no active components. The conviction that the cytoplasm could neither carry nor transmit effective traces of intergenerational memory had been a mainstay of genetics for so long that it had become part of the "memory" of that discipline, working silently but effectively to shape the very logic of inference employed by geneticists.

Yet another ellipsis becomes evident (now, even to Bonner himself) as he attempts to integrate his own work on the role of histones in genetic regulation. Not all copies of a gene (or a genome) are in fact the same: Because of the presence of proteins in the nucleus, capable of binding to the DNA, "in the higher creature, if it is to be a proper higher creature, one and the same gene must possess different attributes, different attitudes, in different cells" (Bonner 1965: 102). The difference is a function of the histones. How can we reconcile this fact with the notion of a "genetic program"? There is one simple way, and Bonner takes it – namely, to elide the distinction between genome and chromosome. The "genetic program" is saved (for this discussion) by just a slight shift in reference: now it refers to a program built into the chromosomal structure – that is, into the complex of genes and histones, where that complex is itself here referred to as the "genome."

But the most conspicuous inadequacy of the location of the developmental program in the genetic information becomes evident in the final chapter, in which Bonner attempts to sketch out an actual computer program for development. Here, the author undertakes to reframe what is known about the induction of developmental pathways in terms of a "master program," proposing to "consider the concept of the life cycle as made up of a master programme constituted in turn of a set of subprogrammes or subroutines" (Bonner 1965: 134). Each subroutine specifies a specific task to be performed. For a plant, his list includes: cell life, embryonic development, how to be a seed, bud development, leaf development, stem development, root development, reproductive development. Within each of these subroutines is a list of cellular instructions or commands, such as, "divide tangentially with growth"; "divide transversely with growth", "grow without dividing"; and "test for size or cell number" (Bonner 1965: 137). He then asks the obvious next question: "[H]ow might these subroutines be related to one another? Exactly how are they to be wired together to constitute a whole programme?" (Bonner 1965: 135). Conveniently, this question is never answered. If it had been, the answer would have necessarily undermined Bonner's core assumption. To see this, two points emerging from his discussion need to be understood: First, the list of subroutines, although laid out in a linear sequence (as if following from an initial "master program") actually constitute a circle, as indeed they must if they are to describe a life cycle. Bonner's own "master program" is in fact nothing but this composite set of programs, wired together in a structure exhibiting the characteristic cybernetic logic of "circular causality."

The second point bears on Bonner's earlier question, "Of what does the programme consist and where does it live?" The first physical structures that were built to embody the logic of computer programs were built out of electrical networks[13] (hence the term "switching networks"), and this is Bonner's frame of reference. As he writes, "[t]hat the logic of development is based upon [a developmental switching] network, there can be no doubt" (Bonner 1965: 148). But what would serve as the biological analogue of an electric (or electronic) switching network? How are the instructions specified in the subroutines that comprise the life cycle actually embodied? Given the dependence of development on the regulating activation of particular genes, Bonner reasonably enough calls the developmental switching network a "genetic switching network." But this does not quite answer our question; rather, it obfuscates it. The clear implication is that such a network is constituted of nothing but genes, whereas in fact, many other kinds of entities also figure in this network, all playing critical roles in the control of genetic activity. Bonner himself writes of the roles played by histones, hormones, and RNA molecules; today, the list has expanded considerably to include enzymatic networks, metabolic networks, transcription complexes, signal transduction pathways, and so on, with many of these additional factors embodying their own "switches." We could, of course, still refer to this extraordinarily complex set of interacting controlling factors as a "genetic switching network" – insofar as the regulation of gene activation remains central to development – but only if we avoid the implication (an implication tantamount to a category error) that that network is embodied in and by the genes themselves.

Indeed, it is this "category error" that confounds the very notion of a "genetic program." If we were now to ask Bonner's question, "Of what does the programme consist and where does it live?" we would have to say, just as Apter saw long ago, that it consists not of particular gene entities, and lives not in the genome itself, but of and in the cellular machinery integrated into a dynamic whole. As Garcia-Bellido writes, "Development results from local effects, and there is no brain or mysterious entity governing the whole: there are local computations and they explain the specificity of something that is historically defined" (1998: 113). Thus, if we wish to preserve the computer metaphor, it would seem more reasonable to describe the fertilized egg as a massively parallel processor in which "programs" (or networks) are distributed throughout the cell.[14] The roles of "data" and "program" here are relative, for what counts as "data" for one "program" is often the output of a second "program," and the output of the first is "data" for yet another "program," or even for the very "program" that provided its own initial "data." Thus, for some developmental stages, the DNA might be seen as encoding "programs" or switches which process the data provided by gradients of transcription activators, or alternatively, one might say that DNA sequences provide data for the machinery of transcription activation (some of which is acquired directly from the cytoplasm of the egg). In later developmental stages, the products of transcription serve as data for splicing machines, translation machines, and the like. In turn, the output of these processes make up the very machinery or programs needed to process the data in the first place. Sometimes, this exchange of data and programs can be represented sequentially, sometimes as occurring in simultaneity.

Into the present

In the mid 1960s, when Bonner, Apter, and others were attempting to represent development in the language of computer programs, automata theory was in its infancy, and cybernetics was at the height of its popularity. During the 1970s and 1980s, these efforts lay forgotten: cybernetics had lost its appeal to computer scientists and biologists alike, and molecular biologists found they had no need of such models. The mere notion of a "genetic program" sufficed by itself to guide their research. Today, however, provoked in large part by the construction of hard-wired parallel processors, the project to simulate biological development on the computer has returned in full force, and in some places has become a flourishing industry. It goes by various names – Artificial Life, adaptive complexity, or genetic algorithms. But what is a genetic algorithm? Like Bonner's subroutines, it is "a sequence of computational operations needed to solve a problem" (see, e.g., Emmeche 1994). And once again, we need to ask, why "genetic"? Furthermore, not only are the individual algorithms referred to as "genetic," but also "in the fields of genetic algorithms and artificial evolution, the [full] representation scheme is often called a 'genome' or 'genotype'" (Fleischer 1995: 1). And, in an account of the sciences of *complexity* written for the lay reader, Mitchell Waldrop quotes Chris Langton, the founder of *Artificial Life*, as saying:

> [Y]ou can think of the genotype as a collection of little computer programs executing in parallel, one program per gene. When activated, each of these programs enters into the logical fray by competing and cooperating with all the other active programs. And collectively, these interacting programs carry out an overall computation that is the phenotype: the structure that unfolds during an organism's development. (Waldrop 1992:194)

Like their counterparts in molecular genetics, workers in Artificial Life are not confused. They well understand, and when pressed readily acknowledge, that the biological analogs of these computer programs are not in fact "genes" (at least as the term is used in biology), but complex biochemical structures or networks constituted of proteins, RNA molecules, and metabolites that often, although certainly not always, execute their tasks in interaction with particular stretches of DNA.[15] Artificial Life's "genome" typically consists of instructions such as "reproduce," "edit," "transport," or "metabolize," and the biological instantiation of these algorithms is found not in the nucleotide sequences of DNA, but in specific kinds of cellular machinery such as transcription complexes, spliceosomes, and metabolic networks. Why then are they called "genetic," and why is the full representation called a "genome"? Is it not simply because it so readily follows from the usage the term "genetic program" had already acquired in genetics?

Words have a history, and their usage depends on this history, as does their meaning. History does not fix the meaning of words: rather, it builds into them a kind of memory. In the field of genetic programming, "genes" have come to refer not to particular sequences of DNA, but to the computer programs required to execute particular tasks (as Langton puts it, "one program per gene"); yet, at the same time, the history of the term ensures that the word *gene*, even as adapted by computer scientists, continues to carry its original meaning. And perhaps most importantly, that earlier meaning remains available for deployment whenever it is convenient to do so. Much the same can be said for the use of the terms *gene* and *genetic programs* by geneticists.

A recapitulation

I have taken some time in examining Bonner's argument for "genetic programs," not because his book played a major role in establishing the centrality of this notion in biological discourse, but rather because of the critical moment in time at which it was written and because of the relative accessibility of the kind of slippage on which his argument depends. The very first use of the term *program* that I have been able to find in the molecular biology literature had appeared only four years earlier.[16] In 1961, Jacob and Monod published a review of their immensely influential work on a genetic mechanism for enzymatic adaptation in *E. coli*, that is, the operon model. The introduction of the term *program* appears in their concluding sentence: "The discovery of regulator and operator genes, and of repressive regulation of the activity of structural genes, reveals that the genome contains not only a series of blue-prints, but a coordinated program of protein synthesis and the means of controlling its execution" (Jacob and Monod 1961: 354).

Three decades later, Sydney Brenner refers to the belief "that all development could be reduced to [the operon] paradigm" – that "It was simply a matter of turning on the right genes in the right places at the right times" – in rather scathing terms. As he puts it, "[o]f course, while absolutely true this is also absolutely vacuous. The paradigm does not tell us how to make a mouse but only how to make a switch" (Brenner *et al.* 1990: 485).[17] And even in the first flush of enthusiasm, not everyone was persuaded of the adequacy of this particular regulatory mechanism to explain development.[18] Lewis Wolpert was one of the early skeptics. In the late 1960s, he seemed certain that an understanding of development required a focus not simply on genetic information, but also on cellular mechanisms.[19] But by the mid 1970s, even Wolpert had been converted to the notion of a "genetic program" (see, e.g., Wolpert and Lewis 1975).

What carried the day? Certainly not more information about actual developmental processes. Far more than most histories of scientific terms, the history of *genetic programs* bears the conspicuous marks of a history of discourse and power. Initially founded on a simple category error, in which the role of genes as subjects (or agents) of development was unwittingly conflated with their role as objects of developmental dynamics, the remarkable popularity of this term in molecular genetics over the last three decades cries out for an accounting. Certainly, it provided a convenient gloss, an easy way to talk that rarely if ever trips scientists up in their daily laboratory work. But it does trip them up in their efforts to explain development; indeed, the term has proven remarkably effective in obscuring enduring gaps in our understanding of developmental logic. Arguably, it has also contributed to the endurance of such gaps.

So why did it prevail? If its popularity cannot be accounted for in strictly scientific or cognitive terms, we must look elsewhere. I suggest we look to the consonance of this formulation with the prior history of genetic discourse, particularly with the discourse of "gene action" that earlier prevailed. Fortifying the "genetic program" in the postwar era, with its easy and continuing elision of the cytoplasmic body, were an entirely new set of resources. Primary among these were the new science of computers, the imprimateur of Schrödinger, and the phenomenal success of the new molecular biology. Jacob's *Logic of Life* was of key importance in the popularization of the concept of "genetic program." Invoking the approval of both Schrödinger and Wiener, Jacob endows the transition from past to future metaphors with the stamp of authority.[20]

What's in a word? As I have already suggested, quite a lot. Words shape the ways we think, and how we think shapes the ways we act. In particular, the use of the term *genetic* to describe developmental instructions (or programs) encourages the belief even in the most careful of readers (as well as writers) that it is only the DNA that matters; it helps all of us to lose sight of the fact that, if that term is to have any applicability at all, it is primarily to refer to the *entities upon which instructions directly or indirectly act and not of which these instructions are constituted.* The necessary dependency of genes on their cellular context, not simply as nutrient but as embodying causal agency, is all too easily forgotten. It is forgotten in laboratory practice, in medical counseling, and perhaps above all, seduced by the promise of utopian transformation, in popular culture.

Notes

1 See, e.g., Griffiths and Neumann-Held (1999); Keller (1995, 2000).

2 See, e.g., Lewontin (1992): Moss (1992): Keller (1995): Strohman (1997).

3 Earlier, in the late nineteenth century, the term *heredity* had commonly been used far more inclusively. Encompassing both the study of both genetics and embryology (see, e.g., Sapp 1987). Furthermore, in the 1920s, the term *genetics* was largely understood to refer solely to transmission genetics.

4 For further discussion, see Griesemer (forthcoming): Keller (1995).

5 Indeed, the brunt of much of this literature is to argue for symmetry between the role of genes and other developmental resources. Thus, for example, in arguing against the conventional view that genes code for traits, Griffiths and Gray suggest that "we can talk with equal legitimacy of cytoplasmic or landscape features coding for traits in standard genic backgrounds" (1994: 283).

6 See, e.g., Kevles (1986) and Paul (1995, 1998) for discussions of the politics of eugenics debates, and Sapp (1987) for a discussion of the impact of Lysenko's anti-genetic crusade in the Soviet Union on American genetics just before and during the cold war.

7 See my discussion of the "discourse of gene action" in Keller (1995, chap. 1).

8 I refer in particular to Oyama's discussion of Brenner's abandonment of the concept of a "genetic program" and his emerging conviction that the proper "unit of development is the cell." She writes, "Having given up genetic programs, [Brenner] now speaks of internal representations and descriptions. In doing so he is like many workers who have been faced with the contradictions and inadequacies of traditional notions of genetic forms and have tried to resolve them, not by seriously altering their concepts, but by making the forms in the genome more abstract: not noses in the genes, but instructions for noses, or potential for noses, or symbolic descriptions of them. This solves nothing" (Oyama 1992: 55).

9 See, e.g., Stent (1985), Newman (1988), Oyama (1989), Moss (1992), and de Chardarevian (1994).

10 The remainder of this paper is adapted from Keller (2000).

11 A vivid demonstration of this interdependency was provided in the 1950s and 1960s with the development of techniques for interspecific nuclear transplantation. Such hybrids almost always fail to develop past gastrulation, and in the rare cases when they do, the resultant embryo exhibits characteristics intermediate between the two parental species. This dependency of genomic function on cytoplasmic structure follows as well from the asymmetric outcomes of reciprocal crosses demonstrated in earlier studies of interspecific hybrids (Markert and Ursprung 1971: 135–137).

12 "Maternal (or cytoplasmic) effects" refers only to the effective agency of maternal (or cytoplasmic) contributions (such as, e.g., gradients). Because such effects need not be (and usually are not) associated with the existence of permanent structures that are transmitted through the generations, they should not be confused with "maternal inheritance."

13 In modern computers, such networks are electronic.

14 Supplementing Lenny Moss's observation that a genetic program is "an object nowhere to be found" (Moss 1992: 335), I would propose the developmental program as an entity that is everywhere to be found.

15 Executing a task means processing data provided both by the DNA and by the products of other

programs – that is, by information given in nucleotide sequences, chromosomal structure, gradients of proteins and RNA molecules, the structure of protein complexes, and so on.

16 Simultaneously, and probably independently Ernst Mayr introduced the notion of "program" in his 1961 article on "Cause and Effect in Biology" (adapted from a lecture given at MIT on Feb. 1, 1961). There he wrote, "The complete individualistic and yet also species-specific DNA code of every zygote (fertilized egg cell), which controls the development of the central and peripheral nervous system ... is the *program* for the behavior computer of this individual" (Mayr 1961: 1504).

17 As Soraya de Chadarevian points out (1994), Brenner had taken a critical stance toward the use of the operon model for development as early as 1974 (see his comments in Brenner 1974).

18 Or even of the appropriateness of the nomenclature. Waddington, for example, noted not only that it "seems too early to decide whether all systems controlling gene-action systems have as their last link an influence which impinges on the gene itself," but also redescribed this system as "genotropic" rather than "genetic" in order "to indicate the site of action of the substances they are interested in" (Waddington 1962: 23).

19 For example, Wolpert wrote in 1969: "Dealing as it does with intracellular regulatory phenomena, it is not directly relevant to problems where the cellular bases of the phenomena are far from clear" (Wolpert 1969: 2–3).

20 He writes: "According to Norbert Wiener, there is no obstacle to using a metaphor 'in which the organism as seen as a message'" (Jacob 1976: 251–252). Two pages later, he adds, "According to Schrödinger, the chromosomes 'contain in some kind of code-script the entire pattern of the individual's future development and of its functioning in the mature state.... The chromosome structures are at the same time instrumental in bringing about the development they foreshadow. They are law-code and executive power – or, to use another simile, they are architect's plan and builder's craft all in one'" (Jacob 1976: 254).

References

Apter, M.J. (1966). *Cybernetics and Development*. Oxford: Pergamon Press.

Apter, M.J., and L. Wolpert. (1965). Cybernetics and development. *Journal of Theoretical Biology* **8**: 244–257.

Atlan, H., and M. Koppel. (1990). The cellular computer DNA: Program or data. *Bulletin of Mathematical Biology* **52**(3): 335–348.

Barrington, A., and Pearson, K. (1909). *A First Study of the Inheritance of Vision and of the Relative Influence of Heredity and Environment on Sight.* Cambridge: Cambridge University Press.

Bonner, J. (1965). *The Molecular Biology of Development.* Oxford: Oxford University Press.

Brenner, S. (1974). New directions in molecular biology. *Nature* **248**: 785–787.

Brenner, S., W. Dove, I. Herskowitz, and R. Thomas. (1990). Genes and development: Molecular and logical themes. *Genetics* **126**: 479–486.

Conklin, E.G. (1915). *Heredity and Environment in the Development of Men.* Princeton: Princeton University Press.

de Chardarevian, S. (1994). Development, programs and computers: Work on the worm (1963–1988). Paper presented at the Summer Academy, Berlin, Germany.

Duden, B. (1991). *The woman Beneath the Skin.* Cambridge, MA: Harvard University Press.

Emmeche, C. (1994). *The Garden in the Machine* Princeton, NJ: Princeton University Press.

Fisher, R.A. (1918). The correlation between relatives on the supposition of Mendelian inheritance. *Transactions of the Royal Society of Edinburgh* **52**: 399–433.

Fleischer, K. (1995). *A Multiple-Mechanism Developmental Model for Defining Self-Organizing Geometric Structures.* Doctoral dissertation, California Institute of Technology.

Galton, F. (1874/1970). *English Men of Science: Their Nature and Nurture.* London: Cass.

Garcia-Bellido, A. (1998). "Discussion." In *The limits of reductionism.* Chichester: John Wiley & Sons.

Griesemer, J.R. (forthcoming). The informational gene and the substantial body: On the generalization of evolutionary theory by abstraction. In N. Cartwright and M. Jones (eds), *Varieties of Idealization.* Poznan

Studies (Leszek Nowak, series ed.). Amsterdam: Rodopi Publishers.

Griffiths, P.E., and R. Gray. (1994). Developmental systems and evolutionary explanation. *Journal of Philosophy* **91**: 277–304.

Griffiths, P.E., and E.M. Neumann-Held. (1999). The many faces of the gene. *BioScience* **49**: 656–662.

Hull, D., and M. Ruse. (1998). *The Philosophy of Biology*. Oxford: Oxford University Press.

Jablonka, E., and M. Lamb. (1995). *Epigenetic Inheritance and Evolution*. New York: Oxford University Press.

Jacob, F. (1976). *The Logic of Life*. New York: Vanguard.

Jacob, F., and J. Monod. (1961). Genetic regulatory mechanisms in the synthesis of proteins. *Journal of Molecular Biology* **3**: 318–356.

Keller, E.F. (1995). *Refiguring Life*. New York: Columbia University Press.

Keller, E.F. (2000). Decoding the genetic program. In P.J. Beurton, R. Falk, and H.-J. Rheinberger (eds), *The Concept of the Gene in Development and Evolution: Historical and Epistemological Perspectives*, pp. 159–177. Cambridge: Cambridge University Press.

Kevles, Daniel J. (1986). *In the Name of Eugenics: Genetics and the Uses of Human Heredity*. Berkeley: University of California Press.

Lewontin, R. (1992). The dream of the human genome. *New York Review of Books*, May 28, pp. 31–40.

Markert, C.L. and H. Ursprung. (1971). *Developmental Genetics*. Englewood Cliffs. NJ: Prentice-Hall.

Mayr, E. (1961). Cause and effect in biology. *Science* **134**: 1501–1506.

Monod, J., and F. Jacob. (1961). General conclusions: Teleonomic mechanisms in cellular metabolism, growth, and differentiation. *Cold Spring Harbor Symposia Quantative Biology* **26**: 389–401.

Morgan, T.H. (1911). The influence of heredity and of environment in determining the coat colors in mice. *Annals of the New York Academy of Science XXI*: 87–118.

Moss, L. (1992). A kernel of truth? On the reality of the genetic program. *Philosophy of Science Association* **1**: 335–348.

Newman, S.A. (1988). Idealist biology. *Perspectives in Biology and Medicine* **31**(3): 353–368.

Oyama, S. (1989). Ontogeny and the central dogma: Do we need the concept of genetic programming in order to have an evolutionary perspective? In M. R. Gunnar and E. Thelen (eds), *Systems and Development*, pp. 1–34. Hillside. NJ: Lawrence Erlbaum Associates.

Oyama, S. (1992). Transmission and construction: Levels and the problem of heredity. In G. Greenberg and E. Tobach (eds), *Levels of Social Behavior: Evolutionary and Genetic Aspects*, pp. 51–60. Wichita, KS: T. C. Schneirla Research Fund.

Oyama, S. (2000). *Evolution's Eye: A Systems View of the Biology-Culture Divide*. Durham, NC: Duke University Press.

Paul, D. (1995). *Controlling Human Heredity*. Atlantic Highlands, NJ: Humanities Press.

Paul, D. (1998). *The Politics of Heredity*. New York: New York University Press.

Sapp, J. (1987). *Beyond the Gene*. Oxford: Oxford University Press.

Stent, G.S. (1985). Hermeneutics and the analysis of complex biological systems. In D.J. Depew and B. Weber (eds), *Evolution at a Crossroads*, pp. 209–225. Cambridge, MA: MIT Press.

Strohman, R.C. (1997). The coming Kuhnian revolution in biology. *Nature Biotechnology* **15**: 194–200.

Waddington, C.H. (1962). *New Patterns in Genetics and Development*. New York: Columbia University Press.

Waldrop, J.M. (1992). *Complexity*. New York: Simon and Schuster.

Wolpert, L. (1969). Positional information and the spatial pattern of cellular differentiation. *Journal of Theoretical Biology* **25**(1): 1–48.

Wolpert, L., and J.H. Lewis. (1975). Towards a theory of development. *Federation Proceedings* **34**(1): 14–20.

Wright, S. (1920). The relative importance of heredity and environment in determining the piebald patterns of guinea-pigs. *Proceedings of the National Academy of Sciences* **6**: 320–332.

Chapter 8

Lucius T. Outlaw

TOWARD A CRITICAL THEORY OF "RACE"

I A need for rethinking

For most of us that there are different races of people is one of the most obvious features of our social worlds. The term "race" is a vehicle for notions deployed in the organization of these worlds in our encounters with persons who are significantly different from us particularly in terms of physical features (skin color and other anatomical features), but also, often combined with these, when they are different with respect to language, behavior, ideas, and other "cultural" matters.

In the United States in particular, "race" is a constitutive element of our common sense and thus is a key component of our "taken-for-granted valid reference schema" through which we get on in the world.[1] And, as we are constantly burdened by the need to resolve difficulties, posing varying degrees of danger to the social whole, in which "race" is the focal point of contention (or serves as a shorthand explanation for the source of contentious differences), we are likewise constantly reinforced in our assumption that "race" is self-evident.

Here has entered "critical" thought: as self-appointed mediator for the resolution of such difficulties by the promotion (and practical effort to realize) a given society's "progressive" evolution, that is, its development of new forms of shared self-understanding – and corresponding forms of social practice – void of the conflicts thought to rest on inappropriate valorizations and rationalizations of "race." Such efforts notwithstanding, however, the "emancipatory project"[2] has foundered on the crucible of "race." True to the prediction of W. E. B. Du Bois, the twentieth century has indeed been dominated by "the problem of the color line." It will clearly be so for the remainder of the century, and well into the twenty-first. For on one insightful reading, we are now in a period in which a major political struggle is being waged, led by the administrations of Ronald Reagan and George Bush, to "rearticulate"[3] racial meanings as part of a larger project to consolidate the victory of control of the state by those on the Right, control that allows them to set the historical agenda for America, thus for the Western "free" world.

Of course, it must be said that the persistence of social struggles – in the United States and elsewhere – in which "race" is a key factor is not due simply to a failure to realize emancipatory projects on the part of those who championed them. While there is some truth to such an analysis, the fuller story is much more complex. Nor has the failure been total. It is possible to identify numerous points in history, and various concrete developments, that were significantly influenced – if not inspired entirely – by emancipatory projects informed by traditions of critical theoretical thought: from the New Deal to the modern freedom (i.e., civil rights), Black Power, and antiwar movements; to the modern women's and environmental movements in the United States and elsewhere; to anticolonial, anticapitalist, antidictatorial, anti-racist struggles throughout the so-called Third World and Europe.

Still, the persistence of struggles around matters involving "race" requires that those of us who continue to be informed by leftist traditions of critical thought and practice confront, on the one hand, unresolved problems. On the other, by way of a critical review of our own traditions, we must determine the extent to which those traditions have failed to account appropriately for "race" (i.e., provide an understanding that is sufficiently compelling for self-understanding and enlightening of social reality) in a way that makes practically possible mobilization sufficient to effect social reconstructions that realize emancipatory promises. It may well be that we will need to review what we think will constitute "emancipation" and whether our notions coincide with those of liberation and self-realization indigenous to persons and traditions of various "racial" groups that would be assisted by us, or who wage their own struggles with assistance from leftist traditions.

No more compelling need is required for our undertaking such reviews than that of getting beyond the interminable debate whether "race" or "class" is the proper vehicle for understanding (and mobilizing against) social problems with invidiously racial components. The present essay is another installment in this ongoing rethinking.[4] Here the focus will be less on the limitations of traditions of critical theory and practice with respect to the privileging of "class" over "race" and more on rethinking "race." A primary concern will be to question "race" as an obvious, biologically or metaphysically given, thereby self-evident reality – to challenge the presumptions sedimented in the "reference schemata" that, when socially shared, become common sense, whether through a group's construction of its life world and/or through hegemonic imposition.[5]

This rethinking will involve, first, a review of the career of "race" as a concept: the context of its emergence and reworking, and the changing agendas of its deployment. Second, a brief recounting of approaches to "race" within traditions of critical theory will facilitate responding to the central question of the essay: "Why a critical theory of 'race' today?" This question is generated by the need to face a persistent problem within Western societies but, in the United States and European societies in particular, one that today presents a new historical conjuncture of crisis proportions: the prospects – and the concrete configurations – of democracy in the context of historic shifts in the demographics of "racial" pluralism. The centripetal, possibly balkanizing forces of racial pluralism have been intensified during the past quarter-century by heightened group (and individual) "racial" self-consciousness as the basis for political mobilization and organization without the constraining effects of the once dominant paradigm of "ethnicity," in which differences are seen as a function of sociology and culture rather than biology.[6]

According to the logic of "ethnicity" as the paradigm for conceptualizing group differences and fashioning social policy to deal with them, the socially divisive effects of "ethnic" differences were to disappear in the social-cultural "melting pot" through assimilation, or, according to the pluralists, ethnic identity would be maintained across time but would be mediated by principles of the body politic: all *individuals*, "without regard to race, creed, color, or national origin," were to win their places in society on the basis of demonstrated achievement (i.e., merit). For both assimilationists and pluralists, *group* characteristics (ethnicity) were to have no play in the determination of merit; their legitimacy was restricted to the private sphere of "culture." This has been the officially sanctioned, and widely socially shared, interpretation of the basic principles of the body politic in the United States in the modern period, even though it was, in significant measure, a cover for the otherwise sometimes explicit, but always programmatic, domination of Africans and of other peoples.

For the past twenty years, however, "race" has been the primary vehicle for conceptualizing and organizing precisely around group differences with the demand that social justice be applied to *groups* and that "justice" be measured by *results*, not just by opportunities. With the assimilation project of the ethnic paradigm no longer hegemonic, combined with the rising demographics of the "unmeltable ethnics" in the American population (and the populations of other Western countries, including Great Britain, France, and West Germany) and the preponderance of "race thinking" infecting political life, we have the battleground on which many of the key issues of social development into the twenty-first century will continue to be waged. Will "critical theory" provide assistance in this area in keeping with its traditions – that is, enlightenment leading to emancipation – or will it become more and more marginalized and irrelevant?

II On "race"

There is, of course, nothing more fascinating than the question of the various types of mankind and their intermixture. The whole question of heredity and human gift depends upon such knowledge; but ever since the African slave trade and before the rise of modern biology and sociology, we have been afraid in America that scientific study in this direction might lead to confusions with which we were loath to agree; and this fear was in reality because the economic foundation of the modern world was based on the recognition and preservation of so-called racial distinctions. In accordance with this, not only Negro slavery could be justified, but the Asiatic coolie profitably used and the labor classes in white countries kept in their places by low wage.[7]

Race theory . . . had up until fairly modern times no firm hold on European thought. On the other hand, race theory and race prejudice were by no means unknown at the time when the English colonists came to North America. Undoubtedly, the age of exploration led many to speculate on race difference at a period when neither Europeans nor Englishmen were prepared to make allowances for vast cultural diversities. Even though race theories had not then secured wide acceptance or even sophisticated formulation, the first contacts of the Spanish with the Indians in the Americas can now be recognized as the beginning of a struggle between conceptions of the nature of primitive peoples which has not yet been wholly settled. . . . Although in the seventeenth century race theories had not as yet developed any strong scientific or theological rationale, the contact of the English with Indians, and soon afterward with Negroes, in the New World led to the formation of institutions and relationships which were later justified by appeals to race theories.[8]

The notion of "race" as a fundamental component of "race thinking" – that is, a way of conceptualizing and organizing social worlds composed of persons whose differences allow for arranging them into groups that come to be called "race" – has had a powerful career in Western history (though such thinking has not been limited to the "West") and continues to be a matter of significant social weight. Even a cursory review of this history should do much to dislodge the concept from its place as provider of access to a self-evident, obvious, even ontologically *given* characteristic of humankind. For what comes out of such a review is the recognition that although "race" is continually with us as an organizing, explanatory concept, what the term refers to – that is, the origin and basis of "racial" differences – has not remained constant. When this insight is added to the abundant knowledge that the deployment of "race" has virtually always been in service to political agendas, beyond more "disinterested" endeavors simply to "understand" the basis of perceptually obvious (and otherwise not obvious, but real nonetheless) differences among human groups, we will have firm grounds for a rethinking of "race." Such a rethinking might profitably be situated in a more sociohistorically "constructivist" framework, namely, one in which "race" is viewed, in the words of Michael Omi and Howard Winant, as a social "formation."[9] But first, something of the career of the concept.

III "Race" and science

The career of "race" does not begin in science but predates it and emerges from a general need to account for the unfamiliar or, simply, to classify objects of experience, thus to organize the life world. How – or why – it was that "race" came to play important classifying, organizing roles is not clear:

> The career of the race concept begins in obscurity, for experts dispute whether the word derives from an Arabic, a Latin, or a German source. The first recorded use in

English of the word "race" was in a poem by William Dunbar of 1508. . . . During the next three centuries the word was used with growing frequency in a literary sense as denoting simply a class of persons or even things. . . . In the nineteenth, and increasingly in the twentieth century, this loose usage began to give way and the word came to signify groups that were distinguished biologically.[10]

This nineteenth-century development was preceded by others in earlier centuries that apparently generated a more compelling need for classificatory ordering in the social world and, subsequently, the use of "race" as such a device. First, there were the tensions within Europe arising from encounters between different groups of peoples, particularly "barbarians" – whether defined culturally or, more narrowly, religiously. (And it should be noted that within European thought, and elsewhere, the color black was associated with evil and death, with "sin" in the Christian context. The valorizing power inherent in this was ready-to-hand with Europe's encounter with Africa.) A more basic impetus, intensified by these tensions, came from the need to account for human origins in general, for human diversity in particular. Finally, there were the quite decisive European voyages to America and Africa, and the development of capitalism and the slave trade.[11]

The function of "race" as an ongoing, classificatory device gained new authority and a new stage in the concept's career developed when, in the eighteenth century, "evidence from geology, zoology, anatomy and other fields of scientific enquiry was assembled to support a claim that racial classification would help explain many human differences."[12] The concept provided a form of "typological thinking," a mode of conceptualization that was at the center of the agenda of emerging scientific praxis at the time, that served well in the classification of human groups. Plato and Aristotle, of course, were precursors of such thinking: the former with his theory of Forms; the latter through his classification of things in terms of their "nature". In the modern period the science of "race" began in comparative morphology with stress on pure "types" as classificatory vehicles. A key figure contributing to this unfolding agenda was the botanist Linnaeus.[13]

A number of persons were key contributors to the development of theories of racial types. According to Banton and Harwood, Johann Friedrich Blumenbach provided the first systematic racial classification in his *Generis humani varietate nativa liber* ("On the Natural Variety of Mankind," 1776). This was followed by the work of James Cowles Prichard (*Generis humani varietate*, 1808).[14] Georges Cuvier, a French anatomist, put forth a physical-cause theory of races in 1800 in arguing that physical nature determined culture. He classified humans into three major groups along an implied descending scale: whites, yellows, and blacks. As Banton and Harwood interpreted his work, central to his thinking was the notion of "type" more than that of "race": "Underlying the variety of the natural world was a limited number of pure types and if their nature could be grasped it was possible to interpret the diverse forms which could temporarily appear as a result of hybrid mating."[15]

Other important contributions to the developing science of "race" include S. G. Morton's publication of a volume on the skulls of American Indians (1839) and one on Egyptian skulls (1845). His work was extended and made popular by J. C. Nott and G. R. Gliddon in their *Types of Mankind* (1854). Charles Hamilton Smith (*The Natural History of the Human Species*, 1848) developed Cuvier's line of argument in Britain. By Smith's reckoning, according to Banton and

Harwood, "The Negro's lowly place in the human order was a consequence of the small volume of his brain."[16] Smith's former student, Robert Knox (*The Races of Man*, 1850), argued likewise. Finally, there was Count Joseph Arthur de Gobineau's four-volume *Essay on the Inequality of Human Races* (1854) in which he argued that, in the words of Banton and Harwood, "the major world civilizations . . . were the creations of different races and that race-mixing was leading to the inevitable deterioration of humanity."[17]

Two significant achievements resulted from these efforts. First, drawing on the rising authority of "science" as the realization and guardian of systematic, certain knowledge, there was the legitimation of "race" as a gathering concept for morphological features that were thought to distinguish varieties of *Homo sapiens* supposedly related to one another through the logic of a *natural* hierarchy of groups. Second, there was the legitimation of the view that the behavior of a group and its members was determined by their place in this hierarchy. "*Homo sapiens* was presented as a species divided into a number of races of different capacity and temperament. Human affairs could be understood only if individuals were seen as representatives of races for it was there that the driving forces of human history resided."[18] These science-authorized and -legitimated notions about "race," when combined with social projects involving the distinguishing and, ultimately, the control of "racially different" persons and groups (as in the case of the enslavement of Africans) took root and grew to become part of common sense. "Race" was now "obvious."

For Banton and Harwood, this science of "race" peaked during the middle of the nineteenth century. By the century's end, however, a variety of racial classifications had brought confusion, in part because "no one was quite sure what races were to be classified for. A classification is a tool. The same object may be classified differently for different purposes. No one can tell what is the best classification without knowing what it has to do."[19] The situation was both assisted and complicated by the work of Darwin and Mendel. Social Darwinism emerged as an effort by some (notably Herbert Spencer and Ludwig Gumplowicz) to apply Darwin's principles regarding heredity and natural selection to human groups and endeavors and thereby provide firmer grounding for the science of "race" (something Darwin was reluctant to do). Such moves were particularly useful in justifying the dominance of certain groups over others (British over Irish; Europeans over Africans ...). On the other hand, however, Darwin's *Origins* shifted the terrain of scientific discourse from morphology and the stability of "pure types" to a subsequent genetics-based approach to individual characteristics and the effects on them of processes of change, thus to a focus on the analysis of variety. In the additional work of Mendel, this development proved revolutionary:

> A racial type was defined by a number of features which are supposed to go together.... The racial theorists of the nineteenth century assumed there was a natural law which said that such traits were invariably associated and were transmitted to the next generation as part of a package deal. Gregor Mendel's research showed that this was not necessarily the case.... [It] also showed that trait variation *within* a population was just as significant as trait variations *between* populations ... traits do not form part of a package but can be shuffled like a pack of playing cards.[20]

And, since environmental impacts that condition natural selection, in addition to heredity and the interplay between dominant and recessive traits, are important factors in the "shuffling" of traits, the notion of "pure" racial types with fixed essential characteristics was displaced: biologically (i.e., genetically) one can only speak of "clines."[21]

The biology of "races" thus became more a matter of studying diversities within – as well as among – groups, and, of particular interest, the study of how groups "evolve" across both time and space. To these efforts were joined others from the *social* science of "race": that is, understanding groups as sharing some distinctive biological features – though not constituting pure types – but with respect to which sociocultural factors are of particular importance (but in ways significantly different from the thinking of the nineteenth-century theorists of racial types).

For many scientists the old (nineteenth-century) notion of "race" had become useless as a classificatory concept, hence certainly did not support in any truly scientific way the political agendas of racists. As noted by Livingstone, "Yesterday's science is today's common sense and tomorrow's nonsense."[22] Revolutions within science (natural and social) conditioned transformed approaches to "race" (although the consequences have still not completely supplanted the popular, commonsensical notions of "races" as pure types as the Ku Klux Klan, among others, indicates).

The conceptual terrain for this later, primarily twentieth-century approach to "race" continues to be, in large part, the notion of "evolution" and was significantly conditioned by the precursive work of Mendel and Darwin, social Darwinists notwithstanding. In the space opened by this concept it became possible at least to work at synthesizing insights drawn from both natural science (genetics, biochemistry) and social science (anthropology, sociology, psychology, ethnology) for a fuller understanding of "geographical races":[23] studies of *organic* evolution focus on changes in the gene pool of a group or groups; studies of *super-organic* evolution are concerned with changes in the "behavior repertoire" of a group or groups – that is, with the sociocultural development.[24] And it is a legitimate question – though one difficult to answer – to what extent, if at all, super-organic evolution is a function of organic evolution or, to add even more complexity, to what extent, if at all, the two forms of evolution are mutually influential. The question of the relations between both forms of development continues to be a major challenge.

But what is a "race" in the framework of organic evolution and the global social context of the late twentieth century? Certainly not a group of persons who share genetic homogeneity. That is likely only in the few places where one might find groups that have remained completely isolated from other groups, with no intergroup sexual reproductions. Among other things, the logics of the capitalist world system have drawn virtually all peoples into the "global village" and facilitated much "interbreeding." But capitalism notwithstanding, "raciation" (i.e., the development of the distinctive gene pools of various groups that determine the relative frequencies of characteristics shared by their members, but certainly not by them alone) has also been a function, in part, of chance. Consequently:

> Since populations' genetic compositions vary over time, race classifications can never be permanent; today's classification may be obsolete in 100 generations. More importantly, modern race classifications attempt to avoid being arbitrary by putting populations *of presumed common evolutionary descent* into the same racial group. Common descent, however, is inferred from similarity in gene frequencies, and here the problem lies. For . . . a population's gene frequencies are

determined not only by its ancestry but also by the processes of natural selection and genetic drift. This means that two populations could, in principle, be historically unrelated but genetically quite similar if they had been independently subject to similar evolutionary forces. To place them in the same racial group would, as a step in the study of evolution, be quite misleading. In the absence of historical evidence of descent, therefore, it is difficult to avoid the conclusion that classifying races is merely a convenient but biologically arbitrary way of breaking down the variety of gene frequency data into a manageable number of categories.[25]

When we classify a group as a "race," then, at best we refer to generally shared characteristics derived from a "pool" of genes. Social, cultural, and geographical factors, in addition to those of natural selection, all impact on this pool, thus on raciation: sometimes to sustain the pool's relative configuration (for example, by isolating the group – culturally or physically – from outbreeding); sometimes to modify it (as when "mulattoes" were produced in the United States in significant part through slave masters of European descent appropriating African women for their – the "masters' " – sexual pleasure). It is possible to study, with some success, the evolution of a particular group over time (a case of *specific* evolution). The prospects for success are more limited, however, when the context of concern is *general* evolution – that is, the grouping of all of the world's peoples in ordered categories "with the largest and most heterogeneous societies in the top category and the smallest and most homogeneous in the bottom."[26] In either case – general or specific evolution – the concern is with super-organic evolution: changes in behavior repertoires. And such changes are not tied to the genetic specificities of "races."

But not all persons (or groups) think so. Although evolutionary – as opposed to typological – thinking, in some form, is at present the dominant intellectual framework for systematic reconstructions and explanations of human natural and social history, it, too, has been enlisted in the service of those who would have "science" pass absolution on their political agendas: that is, to legitimate the empowerment of certain groups, certain "races," over others. Even shorn of the more crude outfittings of social Darwinism's "survival of the fittest" (those in power, or seeking power, over others being the "fittest," of course), the field of the science of "race" is still occupied by those offering orderings of human groups along an *ascending* scale with a particular group's placement on the scale being a function of the level of their supposed development (or lack thereof) toward human perfectibility: from "primitive" to "civilized" (circa the nineteenth century); from "undeveloped" or "underdeveloped" to "developed" or "advanced" (circa the twentieth century).

Such arguments find fertile soil for nourishment and growth now that "evolution" (organic and super-organic, often without distinction), frequently conceived as linear development along a single path which *all* "races" have to traverse, is now a basic feature of our "common sense," Creationists excepted, as we still face political problems emerging from conflicts among "racial" groups. "Race" continues to function as a critical yardstick for the rank-ordering of racial groups both "scientifically" and sociopolitically, the latter with support from the former. At bottom, then, "race" – sometimes explicitly, quite often implicitly – continues to be a major fulcrum of struggles over the distribution and exercise of power.

Certainly one of the more prominent contemporary struggles has centered on the validity of measurements of the "intelligence" of persons from different "racial" groups that purport to demonstrate the comparative "intelligence" of the groups. This struggle is propelled by the social weight given to "intelligence" as an important basis for achievement and rewards in a meritocratic social order. At its center is the question of the dominant roles played by either the genes or the environment in determining "intelligence" (and, by extension, in determining raciation).

Whichever way the question is answered is not insignificant for social policy. If the genes predominate, some argue, then social efforts on behalf of particular groups (e.g., blacks, women, Hispanics, etc.) intending to ameliorate the effects of disadvantageous sociohistorical conditions and practices are misguided and should be discontinued. It would be more "rational" to rechannel the resources poured into such efforts into "more socially productive" pursuits. On the other hand, if environmental factors dominate, then in a liberal democracy, for example, where justice prevails disparities of opportunities (and results?) among "racial" groups must be corrected, especially when the disparities are the result of years, even centuries, of invidious discrimination and oppression.

The politics of "race" are played out on other fields besides that of "intelligence." Modern science has also been concerned with whether the genes of a "race" determine its cultural practices and/or social characteristics. The findings?

All the known differences between geographical races in the frequency of genes which affect behavior are . . . quite trivial. Yet in principle it is possible that there may be genetic differences affecting socially, politically or economically significant behaviors and it seems reasonable to expect that the more population geneticists and physical anthropologists look for such genetic differences, the more will they discover. Because, however, of (1) the relative plasticity of human behaviour, (2) the genetic heterogeneity of all human populations, and (3) the mass of data suggesting the importance of situational determinants (e.g., economic and political factors) in explaining race relations, there is at present little reason to expect that a substantial part of intergroup relations will ever be explicable in genetic terms.[27]

But if not the genes, what about "evolution"? Has it produced differences in behavior and biological mechanisms for survival in different "races"? Is it possible to extrapolate from studies of the evolution of animal behavior to the evolution of human behavior? According to Banton and Harwood, such efforts are inconclusive, the conclusions being at best hypothetical and difficult to test on humans. Moreover:

> ... the difficulty with generalising about evolution is that it is a process that has happened just once. With relatively few exceptions it is impossible to compare evolutionary change with anything else, or to say what would have happened had one of the components been absent. Therefore everything has its place in evolution ... If everything has its place then, by implication, everything is justified.[28]

What, then, after this extended review of the science of "race," are we left with by way of understanding? With the decisive conclusion, certainly, that "race" is *not* wholly and completely determined by biology, but is only partially so. Even then biology does not *determine* "race," but in complex interplay with environmental, cultural, and social factors provides certain boundary conditions and possibilities that affect raciation and the development of "geographical" races. In addition, the definition of "race" is partly political, partly cultural. Nor does the

modern conceptual terrain of "evolution" provide scientifically secure access to race-determining biological, cultural, social developmental complexes distributed among various groups that fix a group's rank-ordered place on an ascending "great chain of being." Racial categories are fundamentally *social* in nature and rest on shifting sands of biological heterogeneity.[29] The biological aspects of "race" are conscripted into projects of cultural, political, and social construction. "Race" is a *social* formation.

This being the case, the notion of "evolution" is particularly fruitful for critical-theoretical rethinking of "race." As has been indicated, in the biological sciences it dislodged the nineteenth-century notion of races as being determined by specific, fixed, natural characteristics and made possible better understandings of racial diversities *and* similarities. In addition, as a concept for organizing our thinking about change, "evolution" continues to provide a powerful vehicle for studying human sociohistorical development. It is a notion that is part and parcel of the terrain of critical social thought of the nineteenth and twentieth centuries.

IV On "critical theory" and "race"

There is some ambiguity surrounding the notion of "critical theory" within traditions of *social* theory — beyond the fact that it is a phrase now used in reference to certain contemporary efforts in literary studies. On the one hand, the phrase is used to refer to a tradition of significantly revised and extended Marxism initiated by a group of theorists often referred to as "the Frankfurt School."[30] In this case "critical theory" is the name Max Horkheimer, an early director of the Institute for Social Research (established in Frankfurt, Germany in the late 1920s, hence the name "Frankfurt School"), gave to what he projected as the appropriate character and agenda for theoretical work directed at understanding – and contributing to the transformation of – social formations that, in various ways, blocked concrete realizations of increased human freedom.[31] This characterization of the nature of social theorizing and its agenda was shared by other members of the Institute (Herbert Marcuse, Theodor Adorno) even though still other members (Erich Fromm, Henryk Grossman) approached matters differently and used different methods in doing so. Further, there were theoretical differences between Horkheimer, Adorno, and Marcuse (and in Horkheimer's own thinking) over time that are masked by the label "critical theory."[32] Still, the label stuck and even today is used to identify a mode of social thought in the Frankfurt School tradition that continues in the work of a number of persons, Jürgen Habermas no doubt being one of the most widely known. Particularly through the influences of Marcuse on many in the generation coming of age in the 1960s during the socially transforming years of the great social mobilizations of the civil rights, black power, and antiwar movements, it is a tradition that has been especially influential in the United States, in part because it brought many of us of that generation to Marx, without question the major intellectual precursor to Frankfurt School critical theory (along with Kant, Hegel, Freud, Lukács, and others). And here lies the ambiguity, for, on the other hand, the phrase is often expanded to include Marx's work, and that in the various currents of Marxism as well, the Frankfurt School included. In the words of Erich Fromm: "There is no 'critical theory'; there is only Marxism."[33] Thus, while the various schools of Marxism, of whatever pedigree, all share important "family resemblances," there are, as well, significant differences among them sufficient to demand that each be viewed in its own right.[34] This is particularly the case when we come to the issue of "race" in "critical theory."

For a number of complex reasons, the Frankfurt School, for all of its influence on a generation of "new" leftists of various racial/ethnic groups many of whom were being radicalized in struggles in which "race" was a key factor, was not known initially so much for its theorizing about "racial" problems and their resolution as for its insightful critique of social domination generally. Although members of the Institute, according to Martin Jay, were overwhelmingly of Jewish origins, and the Institute itself was made possible by funds provided by a member of a wealthy Jewish family expressly, in part, to study anti-Semitism, all in the context of Germany of the 1920s and early 1930s, "the Jewish question" was not at the center of the Institute's work.[35]

This changed in the late 1930s and early 1940s. With the rise of Hitler and the Nazis, the Institute was eventually moved to New York in 1935 (and California in 1941) where its work continued until after the war (when it was reestablished in West Germany in the 1950s). The focus of the Institute's work during this time was the battle against fascism with debates centering on the character of the changed nature of the economy in twentieth-century capitalism; that is, the expression of group sentiments were to be understood in the historical context of the society.[36]

In this, notes Jay, the Institute broke significant new ground. No less so in another major contribution they made to the Marxian legacy, through the work of Fromm especially, that made their studies of anti-Semitism so informative: the articulation, later supported by extensive empirical studies, of a social psychology – and of individual psychology and character structure in the context of the social – drawing off the work of Freud (among others), in the context of Marxian social theory. This made possible analyses that linked cultural, political, and economic structural and dynamic features of the social world, and the character structure of the person, which helped to illuminate the de facto conditions of possibility for the emergence and social maintenance of Nazi fascism and anti-Semitism. Here, particularly, is to be found the significance of Frankfurt School critical theory for our discussion of "race."

In the course of the Institute's work during its stay in the United States, the concern with anti-Semitism became less and less the focus as members of the Institute concentrated increasingly on "prejudice" more generally, although still fundamentally as related to authority and authoritarianism. Initiated by the American Jewish Committee in 1944 and conducted through its Department of Scientific Research established for that purpose, with the collaboration of the Berkeley Public Opinion Study, the Institute conducted major empirical studies, with critical philosophical analyses of the findings, of "one or another facet of the phenomenon we call prejudice." The object of the studies, it was noted, was "not merely to describe prejudice but to explain it in order to help in its eradication." The sweep of the project involved studies of the bases and dynamics of prejudice on individual, group, institutional, and community levels all in the context of the social whole.[37]

The Authoritarian Personality, the result of an integrated set of studies and analyses, was one among a number of volumes that grew out of this project. As Horkheimer notes in its preface, it is a book that deals with "social discrimination," and its authors, in the terms of the *credo* of critical theory, were "imbued with the conviction that the sincere and systematic scientific elucidation of a phenomenon of such great historical meaning can contribute directly to an amelioration of the cultural atmosphere in which hatred breeds."[38] It is especially pertinent to this discussion of "race," Daniel Levinson's chapter on ethnocentric ideology in particular.[39]

Here two conceptual moves are to be noted. First, Levinson substitutes "ethnocentrism" for "prejudice":

> Prejudice is commonly regarded as a feeling of dislike against a specific group; ethnocentrism, on the other hand, refers to a relatively consistent frame of mind concerning "aliens" generally. . . . Ethnocentrism refers to group relations generally; it has to do not only with numerous groups toward which the individual has hostile opinions and attitudes but, equally important, with groups toward which he is positively disposed.
>
> A theory of ethnocentrism offers a starting point for the understanding of the psychological aspect of *group* relations. (p. 102, my emphasis)

Equipped with a wider gathering concept, Levinson is able to make yet another move, one he thinks crucial to gaining the understanding being sought: "The term 'ethnocentrism' shifts the emphasis from 'race' to 'ethnic group' " (p. 103). What was gained by this?

> . . . apart from the arbitrariness of the organic basis of classification, the greatest dangers of the race concept lie in its hereditarian psychological implications and in its misapplication to cultures. Psychologically, the race theory implies, whether or not this is always made explicit, that people of a given race (e.g., skin color) are also very similar psychologically because they have a common hereditary family tree . . . Furthermore, the term "race" is often applied to groups which are not races at all in the technical sense . . . There is no adequate term, other than "ethnic," by which to describe cultures (that is, systems of social ways, institutions, traditions, language, and so forth) which are not nations . . . From the point of view of sociology, cultural

anthropology, and social psychology, the important concepts are not race and heredity but social organization (national, regional, subcultural, communal) and the interaction of social forms and individual personalities. To the extent that relative uniformities in psychological characteristics are found within any cultural grouping, these uniformities must be explained primarily in terms of social organization rather than "racial heredity." (p. 103)

As noted in the previous section, the conclusion had been reached in contemporary natural and social science that, at the very least, "something other than racial heredity", understood as biological homogeneity, had to serve as a basis for understanding group characteristics and intergroup dynamics. Frankfurt School critical theory was distinctive as critical philosophical theory and material, social, analysis (*a la* Marx), fortified by Freudian psychology, deployed in cultural analyses of authority and mass culture. In the Institute's American sojourn particularly, there developed an explicit concern to bring critical thought to bear on the problems of invidious group-based and group-directed discrimination and oppression. "Race" was viewed as an adequate vehicle for such a task. Conditioned by a commitment to engage in critical praxis as an interdisciplinary venture that drew on the best science available (including that on "race"), these social theorists, through an approach to prejudice *cum* ethnocentrism fashioned from Hegelian, Marxian, Freudian elements, provided a means for getting at the problems of "race" – more precisely of race-ism – that was both critical and radical: within the context of an emancipatory project, it cut through social thought based on a reified, erroneous, even fraudulent philosophical anthropology that derived the culture, psychology, and social position of various groups from the biologizing of their "racial types."

Herbert Marcuse, among all members of the Frankfurt School, is most responsible for conveying this legacy to the "New Left" generation of the United States and Western Europe. In contrast to other members of the Institute, he became the most integrated into the American scene and chose to remain in the country when other members returned to Germany in the 1950s.[40] Influential as a teacher and colleague in a number of institutions, his *One Dimensional Man* inducted many of us into critical theory.[41] Here was an understanding of the social order in a way the necessity of which had been driven home to many of us as, in the context of concrete struggles, we came up against the limits of the idealism fueled by the thought of liberal democracy. For a significant group of persons involved in struggles over the "color line," the limits – and their attempted transcendence – were indicated in the evolution of the struggle for "civil rights" to one seeking "Black Power."[42]

But the Frankfurt School did *not* introduce Marxism to the United States. Nor, consequently, was it the first group of Marxian radical theorists to confront the problems of "race." There were other, much older legacies, in fact.[43] It is this history of multiple legacies that makes for the ambiguity of "a critical theory of race" when "critical theory" covers both the Frankfurt School *and* Marxian traditions in general. For an obvious, critically important question is, "Why, given other Marxian legacies, did the New Left seek guidance in the work of the Frankfurt School which might be applied to the problems of 'race,' among others"?

With respect to what we might call the black New Left, but with regard to many nonblack New Leftists as well, this question has been insightfully probed by Harold Cruse. For him, a crucial reason had to do with what he termed the "serious disease of 'historical discontinuity' ":

... since World War I a series of world-shaking events, social upheavals and aborted movements have intruded and sharply set succeeding generations of Negroes apart in terms of social experiences. The youngest elements in the Negro movement today are activists, of one quality or another, who enter the arena unfortified with the knowledge or meaning of many of the vital experiences of Negro radicals born in 1900, 1910, 1920, or even 1930. The problem is that too many of the earlier-twentieth-century-vintage Negro radicals have become too conservative for the 1940ers. Worse than that, the oldsters have nothing to hand down to the 1940ers in the way of refined principles of struggle, original social theory, historical analysis of previous Negro social trends or radical philosophy suitable for black people. . . . All the evidence indicates that the roots of the current crisis of the Negro movement are to be found in the period between the end of World War I and the years of the Great Depression. . . . most of the social issues that absorb the attention of all the Negro radical elements today were prominently foreshadowed in these years. Yet the strands between the period called by some the "Fabulous Twenties" and the current Negro movement have been broken.[44]

The disease of discontinuity affected more than black youth. It was further facilitated by the anti-Communist repression led by Senator Joseph McCarthy, which had "a distinctly deleterious effect" not only on the leadership of black movements at the time, as Cruse notes, but on "radical" leadership in general.[45]

This discontinuity, bolstered by McCarthyism, was institutionalized in the curricula of most American colleges and universities, both black and white: virtually none provided systematically mediated learning regarding the history of previous struggles in which "radicals" had

played important roles. Thus, when we remember that the US New Left generation emerged principally on campuses and was forged in the crucibles of the modern civil rights and antiwar movements whose troops and general staff included thousands of students, the availability and attractiveness of Frankfurt School critical theory was in part a function of happy historical conjuncture: it was available when members of a generation were in need – and actively in search – of understandings to guide them in the transformation of a society that, when measured by its own best principles, was found seriously deficient. Those who suffered the deficits were no longer willing to do so, and were moving to secure their "freedom." Many others were moved to share in the struggles committed to the realization of what the principles called for. Marcuse, himself a teacher and scholar, was among others a major contributor to the recovery from discontinuity by providing an important linkage with Marxian (and Freudian) critical social thought that aided the conceptualization and understanding of the social order as a whole, within a global, historical context, in which it was possible to situate particular problems that were the focus of struggle, including, to some extent, those of the color line.

But only in part was this a matter of happy coincidence. The linkages between the old and new Lefts were never *completely* broken. Many young whites, in particular, were supported in their efforts by parents and others who themselves had been – and still were – radical activists of previous generations. There was another crucial factor, particularly as experienced by blacks "on the Left," an experience that has been formed into its own legacy: the felt *inadequacy* of Marxian Communist and Socialist projects with respect to "the Negro question," the ultimate test case of the problem of "race." At the core of this legacy is the *other side* of the science of "race":

not its scientific, critical conceptualization, but the lived experiences of *real* persons whose experiences are forged in life worlds in part constituted by self-understandings that are in large measure "racial," no matter how "scientifically" inadequate.[46] Other Left theoretical and practical activities, advanced by various groups and parties, ran aground on this reality. Frankfurt School critical theory, unconstrained by dogmatic adherence to "the party line," offered a conceptualization of revolutionary social transformation while, at the same time, it took democratic freedom seriously. Since, at the time, on the black side of struggles involving "race," Black Nationalism was an increasingly ascendant force that even those on the white side had to contend with, and since participants from both "sides" had been forged in large part by liberal democracy, the vision of a new society that decidedly antidogmatic Frankfurt School critical theory helped to shape (particularly by not centering on *class* theory) was potentially more promising as a resolution of racism while preserving black integrity. In this regard there was the promise that the legacy of inadequacy of other traditions of Marxist thought might be overcome.

Oversimplified, the inadequacy had to do with the reductionism in the theorizing about "race" in those Marxian traditions that attempted to confront problems of the color line through approaches that rested on close adherence to a particular reading of the *classic* texts of the "mature" Marx and Engels, a reading sanctified after the Russian Revolution of 1917 by the subsequent Communist Internationals: *class* was the central – indeed, the only – vehicle for fully and properly understanding social organization and struggle. Problems of "race" are to be understood, then, as secondary to the "primary contradiction" of class conflict that is indigenous to social relations in capitalist social formations given the relations of the various classes to the means of production, relations that, at the very least, determine classes "in the last instance." The prospects for progressive social transformation and development, within and beyond capitalism, on this view, are dependent on successful organization and struggle by the international working class, racial differences notwithstanding. Such differences were to be transcended in the brotherhood of class solidarity beyond their opportunistic manipulation by the class of owners and managers, who used them as devices to foster divisions among workers, and by supposedly misguided, chauvinistic blacks (e.g., Marcus Garvey).

The history of Marxian Communist and Socialist organizations in the United States and elsewhere, populated, on the whole, by persons of European descent, is littered with errors, tragedies, and farces resulting from the dogmatic application of this approach.[47] A key source of the difficulty is the inadequate philosophical anthropology presumed by the privileging of "relations to the means of production" as the definitive determinant of groups defined by these relations, thus of the persons in those groups.[48] Aside from problems involving the racism of white workers in the class struggle, and, frequently, the paternalism of the white leadership, for many African-Americans "proletarianism internationalism" was not enough of a basis for forging a new Communist or Socialist world; it disregards – or explicitly treats as unimportant – much that they take to be definitive of African-Americans as a *people*. Identifying and nurturing these characteristics, and the institutions and practices that generate, shape, sustain, and mediate them, constitutes a complex tradition of its own, that of "Black Nationalism."[49] It is a tradition that continues to inform approaches to "race" from the black side, within Marxian critical theory as well (though not that of the Frankfurt School). In 1928–29, for example, with impetus from black Communists

(Cyril Briggs, Richard B. Moore, and Harry Haywood), who also had roots in the decidedly nationalist African Black Brotherhood, the Communist International took the position that blacks in the "black belt" of the southern United States were an oppressed "nation." The program for their liberation thus called for "self-determination" and "national independence." This was the official position, on and off, for nearly thirty years (1928–57) and was carried out in this country by the Communist Party of the United States of America (CPUSA).[50]

The house of "critical theory" has thus been divided on the issue of "race," sometimes against itself: the approach of the tradition of the Frankfurt School on one side; those of other Socialist and Communist organizations, of many persuasions, on the other, with numerous schools of thought and practice in between: "race" is without scientific basis as an explanatory notion (Frankfurt School); "race," while real, is a factor of conflict secondary to the primary contradiction of class struggle ("classical," "official" Marxism); "race" is the basis of a nation – a group whose members share common history and culture ("official" Marxism of 1928–57). Certainly the divergences have as much to do with social matters as with matters theoretical: the concrete histories of different groups, their agendas, their locations, the personal histories of their members, and so forth. Still, those of us who continue to be informed by legacies and agendas of "critical social theory" must move past this "Tower of Babel" in our own midst if we are to meet the challenges of the present and near future.[51]

V Why a critical theory of "race" today?

Since the Black Nationalist tradition has continued to stress "race" over class, and classical Marxism class over "race," the "class or race" debates have persisted, at great expenditures of paper and ink, not to mention years of interminable struggle, confusion, and failure to conceive and secure the realization of promised emancipation. As we continue to struggle over matters of "race" in the United States and other societies, with very real possibilities for increased conflict, it is not enough to view today's problems as being brought on by the "heightened contradictions" of late capitalism attendant to the policies of neoconservative administrations conflicting with struggles for national liberation and socialism/communism in the "Third World." More is needed, both theoretically and practically.

"Both race *and* class" has been the response of some participants in the debate; "Left Nationalists" such as Manning Marable, on the one hand; theorists of the role of race in market relations and in social stratification (i.e., the social distribution of resources) such as William J. Wilson and Edna Bonacich, on the other.[52] Still others have proposed notions of "people-class," "eth-class," and "nation-class."[53] Yet all of these approaches, mindful of nationalist traditions from the black side, as well as of previous running-agrounds on "race," still presuppose the reality of "race."

But what is that reality? And "real" for whom? Would it be helpful for contemporary critical theory to recover the insights of twentieth-century science of "race" and those of the Frankfurt School regarding "race," "prejudice," and "ethnocentrism" and join them to recently developed critical-theoretic notions of social evolution to assist us in understanding and contributing to the emancipatory transformation of the "racial state" in its present configuration?[54] For, if Omi and Winant are correct: in the United States, the state is *inherently* racial, every state institution is a *racial* institution, and the entire social order is equilibrated (unstably) by the state to preserve the prevailing racial order (i.e., the dominance of "whites" over blacks and other "racial" groups);[55] during the decades of the 1950s through the 1970s, the civil rights, Black Power, Chicano, and other movements assaulted and attempted the "great transformation" of this racial state; however, the assaults were partial, and thus were not successful (as evidenced by the powerful rearticulation of "race" and reforming of the racial state consolidating power and dominance in the hands of a few "whites" in service to "whites" presently under way), because "all failed to grasp the comprehensive manner by which race is structured into the US social fabric. All *reduced* race: to interest group, class faction, nationality, or cultural identity. Perhaps most importantly, all these approaches lacked adequate conceptions of the racial state"[56] – if they are correct, might this not be case enough (if more is needed) for a new critical theory of "race" cognizant of these realities?

Omi and Winant think so, and propose their notion of "racial formation." It is a notion intended to displace that of "race" as an "essence" ("as something fixed, concrete and objective ..."), or, alternatively, as a "mere illusion, which an ideal social order would eliminate." Thus, race should be understood as

> ... *an unstable and "decentered" complex of social meanings constantly being transformed by political struggle* ... The crucial task ... is to suggest how the widely disparate circumstances of individual and group racial identities, and of the racial institutions and social practices with which these identities are intertwined, are formed and transformed over time. This takes place ... through *political contestation over racial meanings*.[57]

Central to their argument is the idea that "race" is socially and historically constructed and changes as a consequence of social struggle. "Race," in a racial state, is thereby irreducibly political.

The discussions and analyses of Omi and Winant, facilitated by their notion of "racial formation," are insightful and informative, particularly for their reading of the "rearticulation" of "race" by the Reagan administration. What these theorists offer is an important contribution to a revised and much needed critical theory of race for the present and near future. And part of the strength of their theorizing lies in the advance it makes beyond the reductionist thinking of other leftist theorists while preserving the socio-historical constructivist (socially formed) dimensions of "race."

Part of the strength lies, as well, in the resituating of "race" as a "formation." For what this allows is an appreciation of the historical and socially constructive aspects of "race" within the context of a theory of social evolution where *learning* is a central feature.[58] Then we would have at our disposal the prospects of an understanding of "race" in keeping with the original promises of critical theory: enlightenment leading to emancipation. Social learning regarding "race," steered by critical social thought, might help us to move beyond racism, without reductionism, to pluralist socialist democracy.

Lest we move too fast on this, however, there is still to be explored the "other side" of "race": namely, the lived experiences of those within racial groups (e.g., blacks for whom Black Nationalism, in many ways, is fundamental). That "race" is without a scientific basis in biological terms does *not* mean, thereby, that it is without any social value, racism notwithstanding. The exploration of "race" from this "other side" is required before we will have an adequate critical theory, one that truly contributes to enlightenment and emancipation, in part by appreciating the integrity of those who see themselves through the prism of "race." We must not err yet again in thinking that "race thinking" must be completely eliminated on the way to emancipated society.

That elimination I think unlikely – and unnecessary. Certainly, however, the social divisive forms and consequences of "race thinking" ought to be eliminated, to whatever extent possible. For, in the United States in particular, a new historical conjuncture has been reached: the effort to achieve democracy in a multi-"ethnic," multi-"racial" society where "group thinking" is a decisive feature of social and political life. A critical theory of "race" that contributes to the learning and social evolution that secures socialist, democratic emancipation in the context of this diversity would, then, be of no small consequence.

Notes

1 Alfred Schutz and Thomas Luckmann, *The Structures of the Life-World*, trans. Richard M. Zaner and H. Tristram Engelhardt Jr. (Evanston, Ill.: Northwestern University Press, 1973), 8.

2 The anticipation of "a release of emancipatory reflection and a transformed social praxis" that emerges as a result of the restoration, via critical reflection, of "missing parts of the historical self-formation process to man and, in this way, to release a self-positing comprehension which enables him to see through socially unnecessary authority and control systems." Trent Schroyer, *The Critique of Domination: The Origins and Development of Critical Theory* (Boston: Beacon Press, 1973), 31.

3 "Rearticulation is the process of redefinition of political interests and identities, through a process of recombination of familiar ideas and values in hitherto unrecognized ways." Michael Omi and Howard Winant, *Racial Formation in the United States: From the 1960s to the 1980s* (London: Routledge & Kegan Paul, 1986), 146 n. 8.

4 For previous installments in this discussion, see my "Race and Class in the Theory and Practice of Emancipatory Social Transformation," in Leonard Harris (ed.), *Philosophy Born of Struggle: Anthology of Afro-American Philosophy from 1917* (Dubuque: Kendal/ Hunt, 1983), 117–29; "Critical Theory in a Period of Radical Transformation," *Praxis International*, **3** (July 1983), 138–46; and "On Race and Class; or,

On the Prospects of 'Rainbow Socialism,' " in Mike Davis, Manning Marable, Fred Pfeil, and Michael Sprinker (eds), *The Year Left 2: An American Socialist Yearbook* (London: Verso, 1987), 106–21.

5 On life-world construction see Schutz and Luckmann, *The Structures of the Life-World*, and Peter L. Berger and Thomas Luckmann, *The Social Construction of Reality* (Garden City, NY: Doubleday, 1966). "Hegemonic imposition" is a notion much influenced by the ideas of Antonio Gramsci (e.g., *Selections from the Prison Notebooks*, ed. and trans. Quintin Hoare and Geoffrey Nowell Smith [New York: International, 1971]), although a now classic formulation of the basic insight was provided by Marx (and Engels?) in *The German Ideology*: "The ideas of the ruling class are in every epoch the ruling ideas; i.e., the class which is the ruling *material* force of society, is at the same time its ruling *intellectual* force" (in *The Marx–Engels Reader*, 2nd edn, ed. Robert C. Tucker [New York: Norton, 1978], 172; emphasis in original).

6 "In contrast to biologically oriented approaches, the ethnicity-based paradigm was an insurgent theory which suggested that race was a *social* category. Race was but one of a number of determinants of ethnic group identity or ethnicity. Ethnicity itself was understood as the result of a group formation process based on culture and descent." Omi and Winant, *Racial Formation in the United States*, 15.

7 W.E.B. Du Bois, "The Concept of Race," in *Dusk of Dawn: An Essay toward an Autobiography of a Race Concept* (New York: Schocken Books, 1968 [1940]), 103.

8 Thomas F. Gossett, *Race: The History of an Idea in America* (Dallas: Southern Methodist University Press, 1963), 16–17.

9 "Our theory of racial formation emphasizes the social nature of race, the absence of any essential racial characteristics, the historical flexibility of racial meanings and categories, the conflictual character of race at both the 'micro-' and 'macro-social' levels, and the irreducible political aspect of racial dynamics." Omi and Winant, *Racial Formation in the United States*, 4.

10 Michael Banton and Jonathan Harwood, *The Race Concept* (New York: Praeger, 1975), 13.

11 Ibid. 14.

12 Ibid. 13.

13 "The eighteenth-century Swedish botanist Linnaeus achieved fame by producing a classification of all known plants which extracted order from natural diversity. Scientists of his generation believed that by finding the categories to which animals, plants and objects belonged they were uncovering new sections of God's plan for the universe. Nineteenth-century race theorists inherited much of this way of looking at things." Banton and Harwood, *The Race Concept*, 46.

14 Ibid. 24–5. Both works were closely studied in Europe and the United States.

15 Ibid. 27.

16 Ibid. 28.

17 Ibid. 29–30. These authors observe that while Gobineau's volumes were not very influential at the time of their publication, they were later to become so when used by Hitler in support of his claims regarding the supposed superiority of the "Aryan race."

18 Ibid. 30.

19 Ibid. 38.

20 Banton and Harwood, *The Race Concept*, 47–9; emphasis in original.

21 "An article by an anthropologist published in 1962 declared in the sharpest terms that the old racial classifications were worse than useless and that a new approach had established its superiority. This article, entitled 'On the Non-existence of Human Races', by Frank B. Livingstone, did not advance any new findings or concepts, but it brought out more dramatically than previous writers the sort of change that had occurred in scientific thinking . . . The kernel of Livingstone's argument is contained in his phrase 'there are no races, there are only clines'. A cline is a gradient of change in a measurable genetic character. Skin colour provides an easily noticed example." Ibid. 56–7.

22 Ibid. 58, quoted by Banton and Harwood.

23 "When we refer to races we have in mind their geographically defined categories which are sometimes called 'geographical races', to indicate that while they have some distinctive biological characteristics they are not pure types." Ibid. 62.

24 Ibid. 63. "The main mistake of the early racial theorists was their failure to appreciate the indifference between organic and super-organic evolution. They wished to explain all changes in biological terms." Ibid. 66.

25 Ibid 72–3; emphasis in original.

26 Banton and Harwood, *The Race Concept*, 77.

27 Ibid. 127–8.

28 Banton and Harwood, *The Race Concept*, 137.

29 Ibid. 147.

30 For discussions of Frankfurt School critical theory see, for example, Martin Jay, *The Dialectical Imagination: A History of the Frankfurt School and the Institute of Social Research, 1923–1950* (Boston: Little, Brown, 1973); Zoltán Tar, *The Frankfurt School: The Critical Theories of Max Horkheimer and Theodor W. Adorno* (New York: Wiley, 1977); David Held, *Introduction to Critical Theory: Horkheimer to Habermas* (Berkeley: University of California Press, 1980); and Schroyer, *The Critique of Domination*.

31 Among Horkheimer's characterizations of critical theory is his now classic essay "Traditional and Critical Theory," repr. in Max Horkheimer, *Critical Theory*, trans. Matthew J. O'Connell et al. (New York: Herder & Herder, 1972), 188–243.

32 Zoltán Tar, *The Frankfurt School: The Critical Theories of Max Horkheimer and Theodor W. Adorno* (New York: Wiley, 1977), 34.

33 From a personal telephone conversation with Fromm during one of his last visits to the United States in 1976.

34 For a particularly conversant overview of the various currents of Marxism and their philosophical and historical backgrounds, see Leszek Kolakowski, *Main Currents of Marxism*, 3 vols. (Oxford: Oxford University Press, 1978).

35 "If one seeks a common thread running through individual biographies of the inner circle [of the Institute], the one that immediately comes to mind is their birth into families of middle or upper-middle class Jews. . . . If one were to characterize the Institute's general attitude towards the 'Jewish question', it would have to be seen as similar to that expressed by another radical Jew almost a century before, Karl Marx. In both cases the religious or ethnic issue was clearly subordinated to the social. . . . In fact, the members

of the Institute were anxious to deny any significance at all to their ethnic roots." Jay, *The Dialectical
Imagination*, 31–2.

36 Ibid. 143, 152.

37 Max Horkheimer and Samuel H. Flowerman,
"Foreword to Studies in Prejudice," in Theodor W.
Adorno, E. Frenkel-Brunswik, DJ Levinson, and
RN Sanford, *The Authoritarian Personality* (New York:
Norton, 1950), pp. vi, vii.

38 Ibid., p. ix.

39 The following discussion centers on the fourth
chapter in *The Authoritarian Personality*, "The Study of
Ethnocentric Ideology," by Daniel J. Levinson. Page
references will be included in the text.

40 The significance of the Studies in Prejudice
notwithstanding, Martin Jay, for example, has
noted the strategic moves adopted by Institute
members on their movement to New York (e.g.,
continuing to publish their works in German,
rather than English) that limited their integration
into the mainstream of American social science.
See *The Dialectical Imagination*, 113–14; and Held,
Introduction to Critical Theory, 36.

41 Subtitled *Studies in the Ideology of Advanced Industrial
Society*, the book was published by Beacon Press
(Boston) in 1964. His *An Essay on Liberation* (Beacon
Press, 1969) was an important – though problematic – sequel that attempted to come to terms with
the massive mobilizations of the late 1960s in the
United States and Western Europe, Paris (1968) in
particular. In the latter case (Paris), during a
student-initiated national strike, Marcuse was cele-
brated as one of the "three 'M's" of revolutionary
heroes: "Marx, Mao, Marcuse." For pertinent writings in regard to Marcuse, see, for example: *The
Critical Spirit: Essays in Honor of Herbert Marcuse* (eds)
Kurt H. Wolff and Barrington Moore Jr. (Boston:
Beacon Press, 1967); Paul Breines (ed.), *Critical
Interruptions: New Left Perspectives on Herbert Marcuse* (New
York: Herder & Herder, 1970). *The Critical Spirit*
includes a helpful Marcuse bibiliography.

42 See Clayborne Carson, *In Struggle: SNCC and the Black
Awakening of the 1960s* (Cambridge, Mass.: Harvard
University Press, 1981); and Robert Allen, *Black
Awakening in Capitalist America* (Garden City, NY:
Doubleday, 1969).

43 For important discussions, see T.H. Kennedy and
T.F. Leary, "Communist Thought on the Negro,"
Phylon, **8** (1947), 116–23; Wilson Record, "The
Development of the Communist Position on the
Negro Question in the United States," *Phylon*, **19**
(Fall 1958), 306–26; and Philip Foner, *American
Socialism and Black Americans* (Westport, Conn.:
Greenwood Press, 1977). For discussions by black
thinkers, see, among others, Cedric J. Robinson,
Black Marxism: The Making of the Black Radical Tradition
(London: Zed Press, 1983); Henry Winston, *Class,
Race and Black Liberation* (New York: International
Publishers, 1977); Harry Haywood, *Black Bolshevik:
Autobiography of an Afro-American Communist* (Chicago:
Liberator Press, 1978); James Boggs, *Racism and the
Class Struggle: Further Pages from a Black Worker's Notebook*
(New York: Monthly Review Press, 1970);
Manning Marable, *Blackwater: Historical Studies in Race,
Class Consciousness and Revolution* (Dayton: Black Praxis
Press, 1981); Oliver Cox, *Caste, Class and Race* (New
York: Modern Reader, 1970); and Harold Cruse,
The Crisis of the Negro Intellectual (New York: Morrow,
1967).

44 Harold Cruse, *Rebellion or Revolution?* (New York:
William Morrow, 1968), 127, 130.

45 "The hysteria of the time (which was labeled as
McCarthyism, but which ranged far beyond the
man) had shaken many persons, cowed others,
silenced large numbers, and broken the radical
impetus that might have been expected to follow
the ferment and agitation of the 1930s and 1940s."
Vincent Harding, *The Other American Revolution*, Center
for Afro-American Studies Monograph Series, iv,
Center for Afro-American Studies (Los Angeles,
Calif.) and Institute of the Black World (Atlanta,
Ga. 1980), 148.

46 A full exploration of "race" in the context of critical theory from the black side, if you will, requires
a separate writing. For some of my thinking, see
the previous installments cited in n. 4, as well as in
the writings of persons listed in n. 43.

47 In the African context, for example, note Aimé
Césaire's protest in his resignation from the
Communist Party in 1956: "What I demand of
Marxism and Communism. Philosophies and
movements must serve the people, not the people

the doctrine and the movement. . . . A doctrine is of value only if it is conceived by us and for us, and revised through us. . . . We consider it our duty to make common cause with all who cherish truth and justice, in order to form organizations able to support effectively the black peoples in their present and future struggle – their struggle for justice, for culture, for dignity, for liberty." Cedric Robinson, *Black Marxism*, 260, as cited by David Caute, *Communism and the French Intellectuals, 1914–1960* (New York: Macmillan, 1964), 211.

48 For a characterization and critique of this philosophical anthropology and its relation to class theory in Marx *et al.*, see my "Race and Class in the Theory and Practice of Emancipatory Social Transformation."

49 Literature on this tradition is abundant. See, for example, John Bracey Jr., August Meier, and Elliott Rudwick, (eds), *Black Nationalism in American* (New York: Bobbs-Merrill, 1970); Sterling Stuckey, *The Ideological Origins of Black Nationalism* (Boston: Beacon Press, 1970); Alphonso Pinkney, *Red, Black and Green: Black Nationalism in the United States* (New York: Cambridge University Press, 1976); and M. Ron Karenga, "Afro-American Nationalism: Beyond Mystification and Misconception," in *Black Books Bulletin* (Spring 1978), 7–12. In addition, each of these includes a substantial bibliography.

50 See Cedric J. Robinson, *Black Marxism: The Making of the Black Radical Tradition*, 300, and Kennedy and Leary, "Communist Thought on the Negro."

51 "There is a kind of progressive Tower of Babel, where we are engaged in building an edifice for social transformation, but none of us are speaking the same language. None understands where the rest are going." Manning Marable, "Common Program: Transitional Strategies for Black and Progressive Politics in America," in *Blackwater: Historical Studies in Race, Class Consciousness and Revolution*, 177.

52 See Marable's *Blackwater: Historical Studies in Race, Class Consciousness and Revolution* and "Through the Prism of Race and Class: Modern Black Nationalism in the US," *Socialist Review* (May–June 1980); William J. Wilson's *The Declining Significance of Race: Blacks and Changing American Institutions* (Chicago: University of Chicago Press, 1978); and Edna Bonacich's "Class Approaches to Ethnicity and Race," *Insurgent Sociologist*, **10** (Fall 1980), 9–23. For a fuller discussion of approaches to "race" through the prism of the paradigm of class theory, see Omi and Winant, *Racial Formation in the United States*, 25–37.

53 On "nation-class," see James A. Geschwender, *Racial Stratification in America* (Dubuque, Ia: Brown, 1978).

54 The recent notions of social evolution I have in mind are those of Jürgen Habermas. See, in particular, his "Historical Materialism and the Development of Normative Structures" and "Toward a Reconstruction of Historical Materialism," in Jürgen Habermas, *Communication and the Evolution of Society*, trans. Thomas McCarthy (Boston: Beacon Press, 1979), 95–177.

55 See Omi and Winant, *Racial Formation in the United States*, 76–9.

56 Ibid. 107.

57 Ibid. 68–9; emphasis in original.

58 See Habermas, *Communication and the Evolution of Society*.

Chapter 9

Robin O. Andreasen

RACE: BIOLOGICAL REALITY OR SOCIAL CONSTRUCT?

1 Introduction

The history of the race debate can be summarized by considering the attitudes that theorists have taken towards three incompatible propositions.

BR: Races are biologically real.

SC: Races are social constructs.

I: Biological realism and social constructivism are incompatible views about race.

Many theorists assume, either implicitly or explicitly, that I is uncontroversial (Banton and Harwood 1975; Goldberg 1993; Omi and Winant 1994; Appiah 1996, 1992; Root 2000). Perhaps this is because I is a special case of the widely-held presumption that social constructivism is always an antirealist thesis (Stove 1991, Fine 1996). The disagreement is over whether races are biologically real or social constructs.

In the nineteenth and early twentieth centuries, biological realism was the dominant view. Races were assumed to be biologically objective categories that exist independently of human classifying activities, and scientists worked towards substantiating this belief. They reasoned that races, if they exist objectively, must be some sort of subspecific taxa; that is, human races must be subspecies of *Homo sapiens*.

Two definitions of 'subspecies' were offered. First there was the *typological subspecies concept*, which treats subspecies as natural kinds defined in terms of essential properties possessed by all and only the members of a subspecies (Mayr and Ashlock 1991). Later came the *geographical subspecies concept*. A 'geographical subspecies' is an aggregate of phenotypically and genetically similar breeding populations that inhabit their own geographic range and differ significantly from other such populations (Mayr and Ashlock 1991). Although both concepts eventually were rejected, each enjoyed a long reign as the accepted definition of human race.

The middle of the twentieth century marked a change in point of view. Biologists began to question the biological reality of subspecies, and broad attacks were launched against the biological reality of race. In one argument, which I call the *no subspecies argument*, theorists maintain that there is no need to posit subspecific taxa for any organisms, including humans (Wilson and Brown 1953, Livingstone 1964). After the typological and geographical concepts were rejected, systematists began to express doubts about the possibility of providing a better definition. They argued that designating subspecies requires identifying *distinct units* and giving them formal names, but the boundaries between "subspecies" are often blurry. Furthermore, since systematists have other methods for studying intra-specific variation, the subspecies concept is dispensable.

In a second argument against the biological reality of race, theorists claim that even if non-human subspecies exist, there are no *human* ones; hence there are no races (Lewontin, Rose, Kamin 1984). I call this the *no human subspecies argument*. Support for this argument comes from detailed work in human genetics which reveals that there is almost as much genetic variation within racial groups (Africans, Asians, Caucasians) as there is between them (Lewontin, Rose, Kamin 1984, Nei and Roychoudhury 1993). Humans are supposedly too genetically similar to each other to justify dividing them into races.

Today, most theorists favor the view that races are social constructs. Although there are many types of social constructivism, as a view about race, constructivism is often formulated as a three-part thesis. The first part is a negative thesis, claiming that BR is false. This is a *local* claim. Race constructivism (RC) allows that some biological categories might be objective; it merely denies the biological reality of *race*. The second part is an explanatory thesis; it aims to explain the origins and persistence of beliefs in the biological reality of race. Constructivists often explain these beliefs by appeal to ideological factors, such as the goal of reinforcing a social order that treats racial inequality as legitimate and inevitable. The third part is a positive thesis about the remaining ontological status of race. What is race if it is not a biologically objective category? Some constructivists argue that "race" is a social fiction; it is entirely a product of the ways that people think about human differences (Appiah 1992, Goldberg 1993). Others argue that race plays a prominent role in human social practices; hence, the social reality of race cannot be denied (Zack 1993, Omi and Winant 1994, Outlaw 1995, Appiah 1996, Root 2000).

In what follows I will set the latter two parts of RC to one side. Instead, I will focus on the assumptions that BR is false and that I is true. I will reject both assumptions. Contrary to popular belief, there is a biologically objective way to define race. Races can be defined in the way that cladistics determines its taxa, as sets of lineages that share a common origin. Moreover, as we will see later, the cladistic concept can coexist with a certain formulation of RC; in fact, there is a sense in which these theories are complementary.

2 Races are biologically real

The philosophical debate concerning the status of systematic categories forms the basis of my examination of race. What is the foundation of an objective classification scheme within systematic biology? Pre-Darwinian naturalists often gave an Aristotelian answer to this question: a biologically objective classification scheme treats taxa as natural kinds defined by appeal to kind-specific essences (Mayr and Ashlock 1991). However, when essentialism fell into disrepute (because it was discovered to be at odds with contemporary evolutionary theory), systematists began debating two new possibilities: *phenetic* and *phylogenetic* classification. 'Pheneticism' defines taxa in terms of overall similarity. Populations of organisms are grouped into subspecies by a criterion of resemblance; subspecies are grouped into species by the same process, etc. 'Phylogenetic classification', on the other hand, defines taxa, specifically higher taxa, in terms of common ancestry.[1] Species that share a recent common ancestor belong to the same genus; species that share a more distant common ancestor belong to the same family, etc.

There are three important differences between these schools of classification. First, phenetic classifications are ahistorical; they define taxa in terms of the similarity of their members, with no reference to the genealogical relations among organisms. In contrast, phylogenetic classifications are historical, since they define taxa in terms of evolutionary history. Second, pheneticism uses similarity to *define* its taxa. The phylogenetic concept, on the other hand, uses similarity as *evidence* for group membership, but taxa are not defined in this way. Finally, phenetic classifications are supposed to be theory neutral. Because pheneticists want an all-purpose classification scheme, no theoretical considerations are supposed to enter into phenetic classification. Phylogenetic classifications, on the other hand, are theory-dependent. Because they aim to represent the patterns and processes of evolution, evolutionary theory plays an important role in phylogenetic classification.

Although both schools claim to be objective, only phylogenetic classifications are in fact objective (Ridley 1986). Phylogenetic classifications aim to represent the evolutionary branching process. Since this process exists independently of human classifying activities, phylogenetic classifications are themselves objective. The problem with pheneticism is that there is no reason to suppose that overall similarity represents an objective feature of reality. As Goodman has argued, "comparative judgments of similarity often require not merely selection of relevant properties but a weighting of their importance" (1970, 21). Because pheneticism defines taxa based upon overall similarity, it purports to take account of *all* of the properties that each individual possesses. However, it is difficult to understand what this totality is supposed to be (Sober 1993). First, there is a problem with understanding what counts as a character. For every organism, there is an infinite number of possible traits. Not only are there many different types of traits (e.g., phenotypic, genotypic, and behavioral) there are countless numbers of traits within each type. Furthermore, each trait can be described in many

different ways (e.g., length, color, or hardness), can be divided in different ways (top half of the leg, top quarter, etc.), and can be combined in different ways. Because pheneticism aims for an all-purpose classification scheme, it offers no non-arbitrary standard for choosing one method of character choice as the right one. The second problem is one of weighting. Once systematists have decided what counts as a trait, they need a way to weight their importance. Pheneticists suggest equal weighting, but again they offer no non-arbitrary reason for preferring equal weighting over another method. Finally, similarity itself can be spelled out in many different ways, which further augments pheneticism's embarrassment of riches.

We are now in a position to see where many theorists have gone wrong in their reasoning about race. They presume that similarity ought to be the foundation of an objective classification scheme without considering the possibility that race can be defined historically. Not only is this assumption implicit in theoretical definitions of race (the typological concept requires that the members of a race share a common essence, and the geographical concept requires overall similarity); it is also inherent in the standard arguments against BR. Defenders of the no subspecies argument advocate abandoning the subspecies concept in part because intraspecific variation is not discrete. Likewise, defenders of the no human subspecies argument reject human subspecific classification by claiming that humans are too genetically similar for races to be biologically real. Yet, as we have just seen, in other areas of systematic biology, shared history has largely replaced similarity as the foundation of a objective classification scheme. It follows that races, if they exist objectively, ought to be defined historically.[2]

Elsewhere, I have argued that race ought to be defined in the way that cladistics determines its taxa (Andreasen 1998). Cladism is a school of classification that defines taxa (traditionally, higher taxa) solely in terms of common ancestry (Hennig 1966). For example, by organizing sets of well-defined species into a branching structure, such as a phylogenetic tree, one can define higher taxa as monophyletic groups, as groups composed of an ancestor and all of its descendants. Let us consider Figure 9.1.

In this tree, the branches represent speciation events and the nodes represent well-defined species. Ancestral species A gives rise to two daughter species B and C, which in turn give rise to species D-L as depicted. Sober (1993) uses what he calls the "cut method" to illustrate the concept of monophyly. If we draw a cut across any branch, the nodes above that cut comprise a monophyletic group. For example, E is a monophyletic group, so is DHIJ, and so are many other groupings. In our tree, then, the terminal nodes (E, F, H–L) represent current species; the next largest monophyletic units (DHIJ and GKL) might represent genera, and so on up the taxonomic hierarchy.[3]

An important feature of the concept of monophyly is that the complement of a monophyletic group is not itself monophyletic. By applying the cut method to Figure 9.1, we can see that DHIJ is monophyletic, but the remaining species are not.

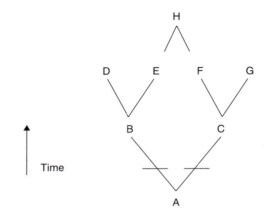

Figure 9.2

Another important fact about monophyly is that it is rarely applied to reticulate structures, such as the one depicted in Figure 9.2. When the concept of monophyly is applied to a reticulate structure, the result is a partial overlap between monophyletic groups. Most systematists choose not to develop classifications in such cases. The reason is that when reticulation is extensive, partial overlap will also be extensive, resulting in a non-hierarchical classification scheme. Again the cut method is useful for seeing why this is so. If we draw two cuts as depicted, the result is two groups (BDEH and CFGH), which overlap partially.

Although the principles of cladistic classification were developed for defining higher taxa, they can be adapted for defining race. A cladistic view of race would require constructing a phylogenetic tree out of human breeding populations; the nodes would represent breeding populations and the branches would represent the births of new breeding populations. A 'breeding population' is a set of local populations that exchange genetic material through reproduction and are reasonably reproductively isolated from other such sets. For example, a tribe of bushmen might constitute a local population. When there is interbreeding among tribes, but no out-breeding, these local populations form a breeding population. A breeding population is 'born' when a

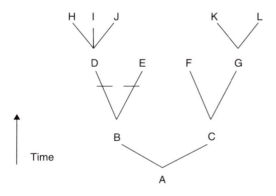

Figure 9.1

local subpopulation becomes separated from its parent population and there is limited gene flow between "parent" and "offspring." Separation often results from the introduction of geographic barriers; however, in the case of humans it can also be due to socio-cultural differences. Referring again to Figure 9.1, races can be defined as follows. The terminal nodes represent current breeding populations, the whole tree represents the human species, and the nested hierarchy of monophyletic units represents a nested hierarchy of races.[4]

Support for this view comes from current work in human evolution. For some time now, anthropologists have been gathering data on the genetics of contemporary populations. Using these data, they can estimate degrees of relatedness among human breeding populations and can reconstruct human evolutionary history (Cavalli-Sforza 1991, Wilson and Cann 1992, Nei and Roychoudhury 1993). For example, Cavalli-Sforza gathered data on 120 different gene states within 42 aboriginal populations (populations that have remained largely reproductively isolated since the late fifteenth

century).[5] Next, he calculated the gene frequency differences (or genetic distances) between populations and used these data to estimate ancestral relations. He reasoned that when two populations are reasonably reproductively isolated over long periods of time, mutations occur and gene frequency differences accumulate. Thus, other things being equal, the larger the genetic distance between two populations, the more distant their ancestral relation. Finally, he confirmed the accuracy of his measures by comparing them with several widely accepted dates suggested by the geological record. For example, the largest genetic distance was between Africans and non-Africans; this distance was approximately twice that between Australians and Asians and was roughly four times the distance between Europeans and Asians. Paleoanthropological research indicates that the dates of separation between breeding populations are in similar ratios: "~100,000 years for the separation between Africans and Asians, about 50,000 years for that between Asians and Australians, and ~35,000 years for that between Asians and Europeans" (Cavalli-Sforza 1991, 106). The

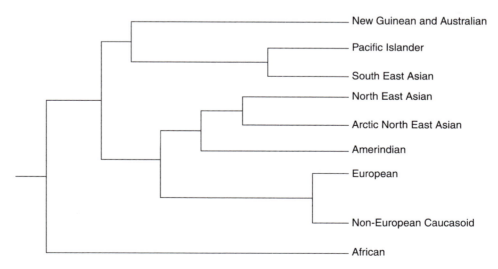

Figure 9.3

result of this research is the phylogenetic tree shown in Fig. 9.3.

This tree represents a racially undifferentiated stock of modern humans evolving in Africa ~200,000 years ago. The first split divides Africans from all other populations. The second split represents a division between Pacific-Southeast Asians and the rest of the world. After that division, the Australopapuans diverged from the rest of Pacific-Southeast Asia, and the fourth split separates northeast Asians and Amerindians from European and non-European Caucasoids.

Some of the above conclusions are controversial. Not only is there controversy surrounding Cavalli-Sforza's method of tree reconstruction; there is some controversy over the specifics of his tree.[6] However, these difficulties need not concern us here. What is important for our purposes is the following conceptual point. Cavalli-Sforza's research illustrates that it is possible to reconstruct human evolutionary history, and this means that it is possible to provide a <u>cladistic definition of race</u>. Even if the empirical details change, this conceptual point will remain in place.

3 Reconstructing race constructivism

We have just seen that the standard arguments against the biological reality of race are unacceptable because they overlook the possibility that race can be defined cladistically. What does this tell us about the thesis of RC? Since most versions assume that races are biologically unreal, one might think that races cannot be social constructs. However, this is not the case. The cladistic view may be incompatible with the standard formulation of RC, but there is a weaker formulation that can coexist with cladism.

When talking about the objective nature of race, there are two kinds of questions that one might ask. First, there are sociological questions concerning how races are conceptualized in different societies. How do people in different societies think about race? What influence do popular conceptions have over a person's self-identity or the identification and treatment of others? Does biology lend support to common sense (CS) conceptions of race? Second, there are biological questions about the extent and nature of human racial variation. Is there a biologically objective way to define subspecies? Can this concept be applied to humans? How much variation is there within each race as compared with that between races?

Although there is some overlap between these sets of questions, they are largely autonomous. Sociologists are mostly interested in CS conceptions of race. Their primary aim is to understand the role that the race concept has played, and continues to play, in human social organization, and this understanding is gained by examining popular beliefs. Answers to the biological questions are useful to the extent that they provide information about whether CS conceptions are biologically real. Yet, if systematists were to discover a biologically objective definition that departs substantially from the CS view, the sociological questions would still be worth asking.

Biologists, on the other hand, want to know whether there is *any* biologically objective way to define race (CS or otherwise). Their aim is to determine whether race represents an objective feature of reality. They might begin by testing the empirical foundations of CS, but if popular conceptions prove to be biologically unjustified, objective biological races might still exist. The objectivity of a kind, biological or otherwise, is not called into question by the fact that ordinary people have mistaken beliefs about the nature of that kind. Those familiar with the causal theory of reference for natural kind terms will be aware of this possibility. According to this theory, natural kind terms have their reference fixed by a baptismal procedure. A speaker indicates what she means by a term either by ostension or by appeal to definite descriptions. The kind is then defined as the class of things that bears the appropriate "sameness relation" to typical samples of that kind (Kripke 1972, Putnam 1975). Ordinary people need not know the conditions for kind membership. The descriptions associated with a kind term do not form part of its meaning, thus even if scientists were to later discover that some of these descriptions are false of the objects originally referred to, the term is still taken to refer.

Yet, my point does not depend on the specifics of the causal theory. One can find in the history of science many instances to support the idea that the objectivity of a kind is not undermined by the fact that ordinary people have mistaken beliefs about its nature. For example, CS has told us that glass is a solid, whales are fish, bats are birds, species have essences, and the heavenly bodies are immutable. Science, however, treats glass as a liquid, whales and bats as mammals, species as lineages without essences, and the heavenly bodies as changeable. In instances such as these, people need not, and often do not, conclude that the kinds in question do not exist. These kinds *do* exist; it is just that ordinary people have (or have had) mistaken beliefs about the natures of these kinds.

It follows that the statement 'races are biologically unreal' is ambiguous. It could mean that biology fails to vindicate CS conceptions of race, or it could mean that there are no biologically objective ways to define race (CS or otherwise). Although constructivists have traditionally tried to defend the stronger claim, they have not succeeded. However, as I am about to argue, they *can* defend a weaker claim – namely, that most CS beliefs about the biological reality of race are empirically unjustified.[7] This would make RC a three-part thesis about CS conceptions only. As already mentioned, the first part would be the claim that biology lends no support to CS beliefs

about the biological reality of race. The second part would explain these false beliefs by appeal to ideological factors. The third part would focus on the social reality, or lack thereof, of CS conceptions of race.

To see that this revision is in keeping with the spirit of RC, let us consider the constructivist project. Constructivists are interested in the sociology of race and race relations. They start with the observation that race often plays a prominent role in human social organization. In the United States, for example, race is a central component of many social policies, many people's identities, and the identification and treatment of others. Constructivists want to understand the concepts of race that are at work in these cases. Of particular interest are invidious conceptions of race, since they often play a role in racist social practices and institutions. Constructivists want to expose these CS conceptions as myths in the hope that our society can begin to move beyond racism.

Although it is ultimately an empirical question what people mean by 'race', very little empirical research has in fact been done on this issue. Even so, there are still some things that can be said about CS beliefs about race. According to some historians, most people in the nineteenth and early twentieth centuries believed that all humans can be sorted into three or more races based upon shared inherited characteristics (Banton and Harwood 1975). Informally, races are demarcated by appeal to observable properties (e.g., skin color, hair type, and eye shape). Yet, many people also assume that these properties are good predictors of more significant inherited differences (e.g., behavioral, intellectual, or physiological differences). I will assume that this concept forms the core of CS beliefs about the biological reality of race. This is not to suggest that all people hold these beliefs. Perhaps many people believe that races are biologically arbitrary. Nor am I suggesting that these beliefs are exhaustive of CS. For example, some people probably assume that a person's race is partly determined by her ancestry, others might believe in shared racial essences, and, unfortunately, some people believe in racial superiority. I am merely claiming that this concept has played, and still plays, an important role in many Western societies. As such, it is the type of concept that constructivists ought to be concerned with when they reject BR.

The central problem with this concept is that it defines races solely in terms of the similarities of their members. However, there is no reason to suppose that these similarities (e.g., skin color, hair type, etc.) represent biologically interesting features of reality. Moreover, although the members of different races differ with respect to their gross morphology, there are few other statistically significant inherited differences among the races. As noted earlier, genetic studies reveal that the genetic variation within CS racial groups is almost as great as that between groups (Lewontin, Rose, and Kamin 1984; Nei and Roychoudhury 1993). Thus, apart from a small handful of arbitrarily selected visible characteristics, the members of different races are not all that different.

4 Proposition I is false

I have just argued that RC ought to be reformulated as a thesis about CS conceptions only. I will now show that the cladistic concept poses no threat to this reformulation of RC. The reason is that it deviates from CS in several important ways. It will follow from this that I is false: Biological realism and social constructivism can sometimes be compatible views about race. Defenders of I often presume that there is a single (CS) meaning of the term 'race'. I argue, however, that the term is ambiguous.

It is often part of CS that the members of a race share many traits with each other that they do not share with members of other races.

According to the cladistic view, however, similarity is neither necessary nor sufficient for race membership. Individuals are members of a cladistic race iff they belong to breeding populations that share a common origin. This will be true regardless of how closely they resemble each other. If two individuals, A and B, are very similar and both differ greatly from a third, C, it still may be true that A and C are in the same race but B is not, if A and C (but not B) are closely related genealogically. As already mentioned, similarities and differences among individuals provide *evidence* for race membership, but cladistic races are not *defined* in this way.

A second feature of CS is the assumption that biological races are *static* categories. Many people probably assume that there will always be the same number of racial groups. The members of a race may change, but the categories themselves never change. However, according to the cladistic view, races can be *dynamic* categories: Not only can races go extinct, new races can come into existence. As we saw above, cladistic classification, if it is to be hierarchical, requires that evolution take the form of a branching process. Subspecific evolution will take this form whenever two breeding populations experience different evolutionary forces under a significant degree of reproductive isolation. Thus, if there is significant outbreeding between two previously isolated breeding populations, these races will go extinct. Similarly, if a subpopulation splits from its parent population, and the two populations are reasonably reproductively isolated over a long period of time, a new race will be born.

Finally, most people divide humans into at least three racial groups (Caucasians, Africans, and Asians). However, if we apply the concept of monophyly to the tree depicted in Figure 9.3, the result is a nested hierarchy of races that cross-classify these standard groupings. Caucasians and Africans are cladistic races, but Asians are not. The reason is that "Asians" do not form a monophyletic group. Pacific-Southeast Asians are more closely related to Australopapuans than they are to Northeast Asians. Moreover, Northeast Asians are more closely related to Amerindians and Caucasians than they are to Southeast Asians. Because there is no group that includes both Southeast and Northeast Asians that does not also include Caucasians, "Asians" do not form a cladistic race.

At this point, one might object that my use of the term 'race' is misleading. The worry is that cladistic races deviate *too far* from CS. If two individuals can be similar in nearly all respects, but be members of different "races" (or if two individuals can differ significantly in their gross morphology and be members of the same "race") then cladistic "races" are not really races. Moreover, the fact that the cladistic concept cross-classifies our standard racial categories might seem to be a further reason for thinking that I am not really talking about race.

Although I agree that the above results are somewhat counterintuitive, they reflect two reasonably common patterns within systematic biology. The possibility that two individuals can differ a great deal and be members of the same race (or that they can be quite similar and be members of different races) is merely the result of defining taxa historically. The same possibilities arise with species and higher taxa, which are also defined historically (Hull 1978). Since it would be a mistake to use this point to deny that species or higher taxa exist, we should not use it to reject the cladistic view of race. Even in the face of cross-classification, it is unacceptable to deny the existence of cladistic races. There are many cases of cross-classification between scientific and CS categories. As I mentioned above, CS once told us that whales are fish, bats are birds, and glass is a solid. Science, on the other hand, says otherwise. In these cases, people did not conclude that the kinds in question do not exist; again, we should not do so in the case of race.

Finally, there might be cause for concern if the cladistic concept were to retain *no* elements of CS. However, there are at least two important elements of CS that the cladistic concept retains. First, many people believe that races are subspecies; they are biologically objective categorical subdivisions of *Homo sapiens*. Second, shared ancestry has played, and probably continues to play, an important role in the ways that ordinary people think about race. This was especially true prior to the nineteenth century, before essentialism was the dominant view about race (Banton and Harwood 1975); however, I suspect it is largely true even today. These two elements of CS are also central to the cladistic concept, hence there is little or no reason to conclude that cladistic races are not really races.

5 Conclusion

In this paper I opposed the trend to reject BR by arguing that cladism, in conjunction with current work in human evolution, provides a new way to define race biologically. I also rejected the widely held assumption that biological realism and social constructivism are incompatible. The reason is that the cladistic concept falls outside the race constructivist's appropriate domain of inquiry.

Notes

1 There are two forms of phylogenetic classification – *evolutionary taxonomy* and *cladism*. 'Evolutionary taxonomy' uses common ancestry and adaptive similarity for defining taxa; 'cladism' relies solely on common ancestry. My definition of race relies on cladism, since evolutionary taxonomy fails to offer a non-arbitrary standard for when similarity matters more than descent (Sober 1993).

2 A somewhat different version of this argument can be found in Andreasen 1998.

3 Cladistic classifications have both a conventional and an objective aspect. The monophyletic groups

are objective, but the way that monophyletic groups get assigned to a taxonomic level is conventional. For example, there is no fact of the matter about whether DHIJ and GKL comprise genera or families. Nonetheless, cladistic classifications are objective because they reflect the evolutionary branching process which is itself objective.

4 Kitcher (1998) has independently proposed a similar definition of race. Like me, he defines races as reasonably reproductively isolated breeding populations. I, however, add that races ought to be monophyletic groups. We also provide different kinds of support for our views. Kitcher uses contemporary data on the rates of interbreeding between major racial groups; I use current work in human evolution. Finally, Kitcher is more optimistic than I am about the existence of races today; I wish to remain agnostic on this issue (see Andreasen 1998).

5 I chose Cavalli-Sforza's research over the alternatives because it is the most comprehensive and uses the largest amount of data.

6 I have discussed these difficulties in Andreasen 1998.

7 Zack (1993) is a constructivist who limits her constructivism to this weaker claim.

References

Andreasen, Robin (1998), "A New Perspective on the Race Debate", *British Journal for the Philosophy of Science* **49**: 199–225.

Appiah, Kwame (1992), *In My Father's House*. New York: Oxford University Press.

— — (1996), "Race, Culture, Identity", in Kwame Appiah and Amy Gutmann (eds), *Color Conscious*. Princeton: Princeton University Press, 30–105.

Banton, Michael and John Harwood (1975), *The Race Concept*. London: David and Charles.

Cavalli-Sforza, Luigi (1991), "Genes, Peoples, Languages", *Scientific American* **265**: 104–110.

Fine, Arthur (1996), "Science Made Up", in David Stump and Peter Galison (eds), *The Disunity of Science*. Stanford: Stanford University Press, 231–254.

Goldberg, David (1993), *Racist Culture*. Cambridge: Blackwell.

Goodman, Nelson (1970), "Seven Strictures on Similarity", in Mary Douglas and David Hull (eds) *How Classification Works*. Edinburgh: Edinburgh University Press, 13–23.

Hennig, Willi (1966), *Phylogenetic Systematics*. Urbana: University of Illinois Press.

Hull, David (1978), "A Matter of Individuality", *Philosophy of Science* **45**: 335–360.

Kitcher, Philip (1998), "Race, Ethnicity, Biology, Culture", unpublished manuscript.

Kripke, Saul (1972), *Naming and Necessity*. Cambridge, MA: Harvard University Press.

Lewontin, Richard, Steven Rose, and Leon Kamin (1984), *Not in Our Genes*. New York: Pantheon.

Livingstone, Frank (1964), "On the Nonexistence of Human Races", in Ashley Montagu (ed.), *The Concept of Race*. New York: Collier Books, 46–60.

Mayr, Ernst and Peter Ashlock (1991), *Principles of Systematic Zoology*, 2nd ed. New York: McGraw-Hill.

Nei, Masatoshi and Arun Roychoudhury (1993), "Evolutionary Relationships of Human Populations on a Global Scale", *Molecular Biology and Evolution* **10**(5): 927–943.

Omi, Michael and Howard Winant (1994), *Racial Formation in the United States*, 2nd ed. New York: Routledge.

Outlaw, Lucius (1995), "On W.E.B. Du Bois's 'The Conservation of Races' ", in Linda Bell and David Blumenfeld (eds), *Overcoming Racism and Sexism*. Lanham, MD: Rowman and Littlefield, 79–102.

Putnam, Hilary (1975), "The Meaning of Meaning", in Hilary Putnam (ed.), *Mind, Language, and Reality*. Cambridge: Cambridge University Press, 215–271.

Ridley, Mark (1986), *Evolution and Classification*. London: Longman.

Root, Michael (2000), "How We Divide the World", *Philosophy of Science* **67**: S628–S639.

Sober, Elliott (1993), *Philosophy of Biology*. Boulder: Westview Press.

Stove, David (1991), *The Plato Cult and Other Philosophical Follies*. Cambridge: Basil Blackwell.

Wilson, Allan and Rebecca Cann (1992), "The Recent Genesis of Humans", *Scientific American* **266**: 68–73.

Wilson, Edward and William Brown (1953), "The Subspecies Concept and Its Taxonomic Application", *Systematic Zoology* **2**: 97–111.

Zack, Naomi (1993), *Race and Mixed Race*. Philadelphia: Temple University Press.

Chapter 10

Joshua M. Glasgow

ON THE NEW BIOLOGY OF RACE

The notion that race has some biological basis has been widely criticized, by both biologists and philosophers. Indeed, the view that race is no more scientifically real than witchcraft is so influential that many who want to argue that race is real divert to understandings of race and reality according to which race is real as a social, rather than natural, kind.[1] Against this trend, however, Robin Andreasen and Philip Kitcher have recently argued for an improved biology of race.[2] The improvements over past biological accounts of race are two-fold. First, the new biology of race avoids the racism of prior biological accounts of race, which often attributed intrinsic significance to racial phenotypic traits or tied intellectual, aesthetic, cultural, and moral potential to those traits. Indeed, both Andreasen and Kitcher, while trying to make biological sense of race, reject the conflict and social division that has surrounded race for so long. Second, the new biology of race actually includes sound scientific research. Briefly, the key idea to this new biology of race is that while perhaps there is no "race gene" or set of necessary and jointly sufficient phenotypic features that can be attributed to each race, races can be understood as *breeding populations*. Here I want to question the viability of this approach.

I

On Andreasen's "cladistic approach," we need not categorize humans according to inherent essences, geography, or conventionally established similarities, for there is another option for a viable biological understanding of race: we can classify via genealogy. A cladistic classification can be represented in a phylogenetic tree, as in Figure 10.1.[3]

Each letter in Figure 10.1 represents a breeding population, that is, a population that is largely reproductively isolated from other populations. Using this model, Andreasen constructs a phylogenetic tree, and hence identifies distinct breeding populations in terms of the relative closeness of different human populations through genetic distance, "a measure of the difference in gene frequencies between two

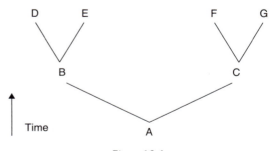

Figure 10.1

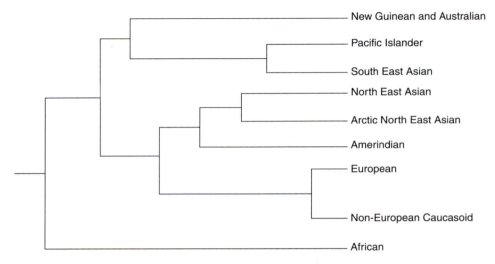

New Guinean and Australian

Pacific Islander

South East Asian

North East Asian

Arctic North East Asian

Amerindian

European

Non-European Caucasoid

African

Figure 10.2

breeding populations" (ibid., p. 210).[4] Drawing from biological data, Andreasen informs us that Africans and non-Africans have the furthest genetic distance. The second split separates Pacific and South East Asians from other non-Africans, and eventually splits occur between Pacific Islanders and South East Asians, North Eurasians and caucasians, and finally within North Eurasians. On the basis of these data, Andreasen borrows the tree in Figure 10.2 from Luigi Luca Cavalli-Sforza.[5]

Since biologists can construct such a "family tree," Andreasen concludes the following: "It means that it is possible to give a biologically objective definition of race. Races are mono-phyletic groups; they are ancestor-descendant sequences of breeding populations, or groups of such sequences, that share a common origin."[6]

This account has several virtues.[7] First, and most obviously, it seems to give a biological backing to our race talk. Second, since it hinges on reproductive isolation, the cladistic approach reveals the dynamic nature of race. That is, insofar as reproductive isolation has increasingly eroded since the European "discovery" and colonization period, races have been slowly

burning out of existence for the last 500 or so years. Thus, we have racial ancestries, even if there are no current biological races. (As we will see, the view that races are disappearing marks a significant point of distinction between Andreasen and Kitcher.) Finally, the cladistic model carries no racist baggage, unlike so many preceding biological notions of race.

II

I have no quarrel with the second and third virtues just mentioned,[8] but I think we ought to take a closer look at the first point – that the cladistic approach affords race some biological reality. Consider first that the folk notion of race does not normally contain the nine races identified in Figure 10.2. As others have noted, it is difficult to determine exactly what races the folk concept of race includes: some speak of "three major races" – African, Asian, and Caucasian; others consider Latinos, or Native Americans, to constitute a race. (Just in terms of numbers, I think it is safe to say that currently it is rare to hear a folk notion of race that involves more than four or five races).[9]

Accordingly, it seems that the nine races in Figure 10.2 do not correspond extensionally with folk notions of race, and Andreasen is quick to agree, with particular mention of the Asian varieties in her schema: "the folk category 'Asian' is not a cladistic race. . . . North East Asians are more closely related to Amerindians and to Caucasians than they are to South East Asians. Similarly, South East Asians are more closely related to Australians than to North East Asians." Andreasen takes this discontinuity with the folk notion of 'race' to be nonproblematic, when she continues, "This conclusion is interesting because it illustrates that the existence of biological races does not depend upon our folk taxonomy being right."[10]

But this exposes what I take to be a central flaw of the cladistic approach. That is, Andreasen has found a way of carving our ancestors into breeding populations, but these populations are not what we call 'races'. In addition to the extensional differences already noted, consider the intension of 'race'. Intensionally, of course, 'race' can mean (and has meant, over the years) a number of different things.[11] It might mean something as putatively benign as groupings based on pigmentation, for instance. Andreasen, however, holds that "Individuals are members of a cladistic race if and only if they belong to breeding populations that share a common origin. This will be true regardless of how closely they resemble each other."[12] Accordingly, her cladistic classification does not match up with the folk concept of race that centers on pigmentation. Presumably, for example, we would at least struggle to reconcile this folk concept with the idea that "South East Asians are more closely related to Australians than to North East Asians." None of this means that the ordinary notion of 'race' is coherent; the point is simply that the cladistic approach does not provide biological backing for it.

Or 'race' might mean something else. It might refer to a cluster of phenotypic features (in addition to skin color), but, since physical resemblance is irrelevant on Andreasen's cladistic approach, that approach seems incapable of matching a phenotypic-cluster folk classification. Or, 'race' might be taken in the way (some) racists mean it, when they attach intellectual, moral, or aesthetic characteristics to phenotypic features in a hierarchical fashion. Or, it might be a concept that attaches some such characteristics to phenotypic "markers," but in a nonhierarchical way, as we find in W. E. B. Du Bois.[13] The point here is not that one of these understandings is better than the others (and, to be sure, each has its problems); again, the point is that Andreasen's cladistic approach does not map onto *any* of the more dominant folk conceptions of 'race', insofar as those conceptions are about more than genealogy.

Surely, however, ancestry – which is at the heart of the cladistic approach – does play a large role in many conventional understandings of race. (This is one horn of Andreasen's two-pronged response to the type of objection offered here.[14]) In the United States, at any rate, the "one-drop rule" has had a crucial role in our system of racial classification, so that persons of mixed black-white ancestry often get labeled as (and identify as) black, morphological indicators notwithstanding. As has been pointed out, of course, the one-drop rule regarding blackness is problematic on several levels, including being inconsistent with US policy regarding Native Americans.[15] The cladistic approach need not rely on the one-drop rule, however, for there is a more general ancestral component to the common-sense notion of race: a person is of race R if and only if her parents are both members of R.

Yet even this general genealogical element in common-sense notions of race does not match the cladistic approach. First, while ancestry is often *part* of the folk meaning of race, people frequently mean more than ancestry when they use racial discourse; skin color, for example, seems like a central – and inextricable – part of the folk meaning of 'race'. Indeed, race is sometimes thought to include even more than skin color. Again, there are overt racists, who adhere to hierarchical racial essences, as well as those, like Du Bois, who seem to hold that the races, while not hierarchically ordered, still have essential characteristics beyond the phenotypic ones.

Second, whatever else conventional notions of race are intended to mean, they all seem to include the idea that races still exist. This, in turn, entails that the folk concept of 'race' at least means more than isolated reproductive groups that are vanishing (or have already vanished). Andreasen, by contrast, holds that because interracial reproduction eliminates isolated breeding populations, races are ceasing to exist. Thus, it seems that the viability of Andreasen's cladistic approach, with its emphasis on ancestral isolated reproductive groups and dismissal of the centrality of phenotypic traits, does not entail that race – in its common-sense meanings – is real.[16]

III

This brings us to the methodological question that is at the heart of Andreasen's theory, for she agrees that she has not shown that race *in its usual meanings* is real: "Questions about biological classification can be about ordinary language classifications, or they can be about scientific classifications. For example, the question 'is there a biologically objective way to define race' could be asking whether biology vindicates our common-sense notions of race. Alternatively, it could be asking whether there are *any* biologically objective ways to divide humans into races."[17] As Andreasen sees it, the theory

proposed by those who think that race is socially constructed rather than real seeks to answer the first question – about common-sense notions of race. The cladistic approach, not inconsistent with the constructionist view, seeks to find a biological notion of race, whether or not it matches up with common sense.

This agenda reveals the core question: How revisionist can one be about the meaning of 'race' and still call it 'race'?[18] For instance, one might argue that because biologists can (to a large extent, though not entirely) divide the human species into two groups, namely those with XX chromosomes and those with XY, there is a biological notion of race, with two races, female and male. The right response to such an argument, I think, is that while it is true that this is one way of dividing up people biologically, it does not converge with what either intellectuals in the race debate or those who employ common sense mean by 'race'. Therefore, the argument has not established that races are biologically real. On the other hand, consider an approach that gave biological backing to race, but only required comparatively minor revisions to the common-sense notion of 'race' (for example, it required putting a group under the racial category R that previously was not thought of as R, but which contained members who were phenotypically similar). Perhaps we would concede that, indeed, races are biologically real and that we should revise our notion of 'race' in this minimal way.

Thus, while some minimal revision to the meaning of 'race' (as for all definitions, of course) is allowable in the search for biological backing for race, we must stay fairly close to the vest, or we risk not talking about race at all. The question, again, is: How much revision is allowable? I cannot offer a good answer to that question here, but I do not think it is necessary. (Although, in section IV, I will suggest a limiting condition on concept revision.) For it seems to me that the burden is on the revisionist to show that her revisions are warranted. We need an argument from Andreasen that we should still call her breeding populations 'races', even though, first, her nine populations do not correspond extensionally to what we usually identify as races, and, second, her concept of 'race' as breeding population does not agree intensionally with the folk concept, insofar as we normally mean something beyond mere isolated reproductive groups, such as groups demarcated by skin color.

Andreasen does offer a response to this concern (this is the other horn of the two-pronged response): "One can find in the history of science many instances to support the idea that the objectivity of a [natural] kind is not undermined by the fact that ordinary people have mistaken beliefs about its nature."[19] Here she cites examples such as whales. Common-sense belief tells us (or once told us, anyway) that whales are fish. However, science classifies whales as mammals. This disagreement between science and common sense does not mean that whales do not exist; rather, it merely means that common sense is wrong. The upshot is that we can replace common-sense concepts with scientific ones when common sense is mistaken.

I think, however, that the analogy between whales and race is tenuous. In the disagreement over the status of whales, the scientist and the layperson can point to a thing they mutually agree is called 'whale', and the scientist can explain why it is more naturally lumped together with mammals than with fish, in terms of common properties like warm-bloodedness. This is a disagreement over how to classify one anomalous species in an otherwise fixed classification schema.

In Andreasen's account of race, however, there is a wholesale reshuffling of the classification schema itself. If, for instance, Andreasen and a layperson were to pick out a person of a certain

genealogy and phenotype, where the layperson would classify that person as (say) 'Asian', and Andreasen would classify her instead as 'North East Asian', there would be a much different ensuing dialogue than that in the case of whales. Since Andreasen's classification schema puts the person in question closer to Caucasians than to South East Asians (while South East Asians are closer to Australians than to that person), it would soon become clear that the disagreement is not primarily about where to put this person in a fixed schema of classification, as it was in the case of labeling whales either fish or mammals. Rather, the disagreement is over the classification schema itself, and, consequently, over each schema's underlying definition of race – one focuses on descent and breeding group, while the other focuses on descent and phenotype. This disagreement accounts for each party using a different label to refer to the same person, in contrast to the case of whales, where both parties can agree on one term – 'whale' – to refer to the object picked out by that term.

At this point the layperson could reasonably assert that now the disagreement is not about how to classify the person in question; rather, it is about what classification schema and definition of race to adopt. And, if our layperson were informed that the common-sense understanding of race, centered on phenotype, is biologically unfounded, it would not be unreasonable for the layperson to reply that perhaps this simply means that there are no races. In this respect, the disagreement is very different from the whale case, where the layperson would be unreasonable to claim that there are no whales. In short, there is a stand-off here that was not present in the whale case, regarding which classification schema and definition of 'race' to choose. Reclassifying anomalous cases of misclassification (like whales) is not analogous to making wholesale changes in the classificatory system.

Accordingly, we need a further argument for revising our concept of race; saying that if we revised our concept of 'race', then it would be more similar to the case of whales than to some nonexistent kind like witches, is not itself a ground for revising our concept of race. Rather, one must argue for revising the concept of race as part of showing that race is more like whales than witchcraft. Only then, if we accept that prior argument, can we explore other conceptions of race.

As a final note, the following claims cannot fill in the missing argument: common-sense races are not biologically real, whereas cladistic races are; therefore, we should replace the folk notion with the cladistic notion. Such an argument begs the question of whether a biologically real notion of 'race' that is minimally related to the common-sense notion is preferable to a conventional, but biologically nonexistent, notion of 'race'. This, I take it, is one of the more crucial questions in the race debate, and any answer must argue for, rather than simply stipulate, one side over the other. Below (section VI), I will further examine the question of replacing or significantly revising common-sense racial discourse.

IV

The above argument against Andreasen's account of race relies on four key premises: (1) that account's nine races present a system of racial classification that is substantially different from folk racial classifications; this extensional difference is based on the facts that (2) the cladistic model's reproductively isolated groups are disappearing (and so its eroding ancestral element does not match up with the folk notion's persisting ancestral element) and (3) reproductively isolated groups do not map onto the phenotypical groupings that seem essential to the folk concept of 'race'; finally, (4) scientists

are not themselves the arbiters of the meaning of 'race'. The first three points are about the meaning of race; the fourth generates a meta-question about how meaning gets settled in the first place. What, then, if there were a viable classification of human races that could disrupt one or more of these premises? Indeed, Kitcher's theory of race seems to bypass the objections in (1)–(3). In this section, I want to examine premise (4); I take up Kitcher's alternative in the next section.

The foregoing suggests that treating races as reproductively isolated breeding populations is too different from the common-sense notion of 'race' to provide an adequate biological account of race. To then conclude, however, that there is no biological basis of race might seem to presuppose a blanket premise that everyday folk, rather than professional scientists, have the authority to determine the meanings of purported natural kind terms. As a general principle, however, this presupposition is not so easily defended, particularly if we adopt the causal theory of reference as found in the work of Saul Kripke and Hilary Putnam.

In Appiah's analysis of the meaning of 'race', for instance, he adopts a method of "semantic deference" that follows Putnam's "linguistic division of labor," a justification for nonspecialists' engaging in discourse that uses terms with meanings that those nonspecialists cannot identify. The only way that such discourse is legitimate is if specialists can identify the meanings of those terms so that they may be freely used in folk discourse.[20] And, if we follow Putnam's Twin Earth thought experiment, we find that water, necessarily, means H_2O, even if nonspecialists point to a watery substance with a chemical composition of XYZ and call that 'water', in part because of the authority vested in specialists. Why, then, should it be a strike against Andreasen's model of biological racial realism if nonspecialists point to a set of entities (for example, the "three major races") and call those 'races' in a way that is different from what the cladistic approach would suggest? Thus one rejoinder to my objection to Andreasen might be that since the experts, rather than common sense, determine the meaning of natural kind terms, it is irrelevant whether the biological data can provide a referent for the folk concept of race.

As I see it, there are two problems with this rejoinder. The first is that semantic deference requires an expert definition of the term in question, and this requires both a defined set of experts and a (near) consensus on what the expert definition of the term is. Chemists, for example, can provide a unified expert judgment that water is composed of H_2O. In the case of race, presumably the experts include biologists and philosophers (among other parties, such as physical anthropologists). However, there is severe disagreement among those experts when it comes to race, and that disagreement exists on two levels. On one level, there is disagreement over which entities are supposed to be identified as races in the first place. As we have seen, Andreasen identifies the nine populations in Figure 10.2. At the same time, however, Appiah's own examination of the concept of 'race' identifies the "three major races" of African, Asian, and Caucasian. Kitcher also identifies those as the three major races.[21]

On another level, even if that first question could be settled, there is widespread disagreement as to whether there is any biological reality to those racial classifications, and, if so, what the underlying biological referent is. For example, while Kitcher argues for the biological reality of the three major races, Appiah disagrees. Also consider Andreasen's use of her main source of biological data, Cavalli-Sforza: while Andreasen takes those data to generate a plausible story about the biological reality of race, her source disagrees.[22] In this respect, then, the expert

consensus that water is H_2O is very different from the ability of experts to fix the meaning (both extensional and intensional) and ontological status of race – unlike water, there simply is no decisive expert opinion on the nature of race.

A second, related, point from above is pertinent here as well. We give preference to the views of experts over common sense only under certain limiting conditions. For instance, we prefer the specialist definition of water as H_2O, even when nonspecialists identify both H_2O and XYZ as 'water', only when there is reasonable overlap between the gross physical substances identified by specialists and nonspecialists concerning the object to be defined. To change the example, suppose chemists pointed to a substance S, with the chemical compound NaCl, and told nonspecialists that chemical analysis reveals that water is NaCl, despite the mass of nonspecialists calling substance W (with the chemical composition H_2O) 'water'. Nonspecialists would rightly respond that NaCl is called 'salt', not 'water'. This is the limiting condition on semantic deference: meanings of folk terms are determined from the ground up by folk usage, rather than from above by specialists, to the extent that technical categories have to overlap reasonably with the folk categories themselves (like the relationship between water and watery substances, but unlike that between salt and water). If there is no scientific backing to some given folk category, and if there is no reasonably overlapping technical category that does have scientific backing, then that is when we determine that there simply are no things of that kind (for example, witches). This point about "reasonable overlap" is not particularly radical: the claim is simply that at some point (anomalous cases aside) what terms designate becomes *rigidified*.

The problem with the chemists' approach in the water-salt example is not that they have incorrectly analyzed the chemical composition of S. Instead, they have made a prespecialist categorization mistake by identifying the wrong substance in need of chemical analysis: they chose to analyze gross macrophysical substance S, which is rigidly designated as 'salt', rather than substance W, which is rigidly designated as 'water', on the incorrect presupposition that anything with the chemical composition of S is designated 'water'. And this is much like the proposed problem with Andreasen's account of race. The objection presented above is not that there is faulty biology, or even faulty analysis of that biology, but, rather, that there is a pre-specialist categorization mistake. The groups offered up as races (for which we can grant that scientists have identified real biological properties) simply are not races, just as salt is not water.

So the causal theory of reference will not bolster Andreasen's cladistic approach.[23] To put it roughly, the causal theory holds that natural kind terms get "baptized" in an initial naming process, and then those terms rigidly designate the objects that they so name. If, later, it is discovered that the baptizers (or other competent language users) identified properties with the object that are not, in fact, constitutive of the object, so much the worse for the folk definitions. On this theory, science, rather than folk usage, tells us what properties are associated with the term in question. The reason this does not aid Andreasen's approach is that her theory of race requires de-rigidifying the term 'race'. On that approach, 'race' no longer picks out the same macrophysical objects (say, the three major races), nor does it pick out a reasonably overlapping object, which subsequently could be determined by scientists to have a different underlying structure than competent language users previously thought.[24] Rather, Andreasen's approach picks out different objects entirely (the nine populations in Figure 10.2).

In short, then, it does not help Andreasen's cladistic approach to privilege specialist over nonspecialist understandings of race. For, first, the specialist understanding of race (as presented by Andreasen) does not reasonably overlap with the folk category; that is, the relationship between cladistic 'races' and folk races seems closer to the relationship between salt and water than that between H_2O and XYZ.[25] Second, in any case, unlike water, there is no scientific consensus that race is a real biological kind; there is no expert consensus about (a) what entities we should identify as races (the three major races, Andreasen's nine, and so forth), or (b) whichever entities we choose, whether they have a biological basis.

V

All of the above does nothing to dispel the idea that if we could formulate a cladistic model that matches the common-sense notion of 'race', race would be biologically real. As it turns out, Kitcher presents a biological picture of race that ends up being strikingly similar to everyday usage. For Kitcher, a pure race (where purity merely connotes reproductive isolation, rather than any kind of racial superiority) is just a subset of *Homo sapiens*, where offspring are of race R when their parents are of race R and parents are of race R when their offspring are of race R. Like Andreasen's account, this identifies races not "on the basis of traits," but, rather, on "patterns of descent." Finally, two further conditions are necessary for this classification to have any "biological significance." First, the members of the pure races must "have some distinctive phenotypic or genetic properties." Second, the mixed-race population cannot be so large that once-existing pure races are no longer reasonably substantial parts of the general population (op. cit., pp. 92–94).

So far, this account is much like Andreasen's. Indeed, Kitcher identifies inbreeding among populations – reproductive isolation – as the factor that ensures the required difference between interracial and intraracial genetic properties. A key point of distinction from Andreasen's picture, however, is that for Kitcher, reproductive isolation persists through the present. On the basis of what he admits to be limited data, Kitcher finds that there are comparatively low rates of sexual union among blacks and whites, though the same is not true of whites and (at least some populations of) Asian Americans (op. cit., pp. 99–100). On this account, human reproductive isolation is not a matter of geographic isolation (though it once was); rather, it is a function of what can be extremely subtle isolating mechanisms, such as cultural barriers to interracial relationships and breeding (op. cit., pp. 105–10). As for the claim that there have been periods of high rates of interracial reproduction in the US (in particular, the widespread rape of black women by white slaveowners), Kitcher notes those high rates were nevertheless much lower than intraracial breeding; and, importantly, we can conceive of this gene pool modification as a "coercive restructuring of the minority race" (op. cit., p. 102). Indeed, since the mechanisms that isolate breeding populations are in this way socio-cultural, for Kitcher race is both socially constructed and biologically real.

This picture, then, is much more faithful to the folk category 'race' than Andreasen's: unlike Andreasen's model, Kitcher (a) finds reproductive isolation, and so race, persisting through the present; and (b) identifies as the races (isolated breeding groups) traditional racial groupings, such as black, white, and so forth.

Nevertheless, I do not think that this picture is sufficiently faithful to the folk category. Kitcher acknowledges that he has only accounted for a difference between Africans and Caucasians in the US, while he has not provided evidence for a distinction between Asians and Africans and little for a division between Asians and Caucasians.[26] Moreover, since this notion of race hinges only on reproductive isolation, different social classes that are reproductively isolated (such as landowners and peasants in England after the Norman conquest) would end up being classified as different races (op. cit., p. 103).

Accordingly, this presents a classification of 'race' that is substantially different from the folk meaning of the term. Kitcher's treatment of the folk category 'Asian' is particularly inconsistent with folk usage. If there is no division between Asians and Caucasians or Asians and Africans, but there is a division between Africans and Caucasians, what happens to Asians in locations where the three meet (op. cit., p. 100)?[27] Are Asians raceless? Or perhaps Asians become Caucasian, or African? Or do Asians combine with either Africans or Caucasians to form a new, fourth race? It is difficult to see how any of these options would be sustained, or how dramatic political implications could be mitigated. More to the point, however, none of them seems to match the folk categorization of race.

One might argue, however, that this problem arises only because of insufficient data on reproductive patterns. That is, if we had more complete data, we might find significantly higher rates of intraracial reproduction than interracial reproduction; and, since these reproductive behaviors are influenced by people's *perceptions* of what race is (rather than any biological facts), the resulting breeding populations are going to end up matching folk racial groupings.

This question leaves us waiting for more complete data. In the meantime, however, Kitcher's account faces two further problems. First, Kitcher himself admits that to be "a workable biological conception of race," there must be mating patterns between the races that are sufficient "to sustain the distinctive traits that mark the races (which must, presumably, lie, at least in part, in terms of phenotypes, since organisms have no direct access to one another's genes)" (op. cit., p. 97). That is, while Kitcher's model identifies races as reproductively isolated breeding populations, and so phenotype alone is not the *basis* for race, this reproductive isolation is *significant* only so long as it maintains phenotypic differences between the races. But this exposes his model to old worries about the possibility of making sense of racial divisions based on phenotype. Naomi Zack puts it this way: "The visual and cultural markers for membership in the black race differ too greatly for there to be any physical traits shared by all black individuals, and likewise for whites."[28] By Kitcher's own standards, then, in order for race to be a significant category, we at least need an additional story about which phenotypic traits are supposed to go with which races. If Zack is right, such a story cannot be told.

Second, the races-as-breeding-populations model seems too broad, as evidenced by the counterintuitive result that peasants and land-owners would have to be considered races. Folk usage, of course, distinguishes between socio-economic classes (for example, peasants) and races (for example, "white people"). And this reflects an intensional difference in the two meanings of race: the folk concept of race seems to include a phenotypical component, including traits such as skin color, that does not correlate one-to-one with class status; class status is orthogonal to membership in any given race.

The more general point here is that potentially there are many breeding populations (based not only on class, but small regions, professions, cultures, and so forth), which are not accurately labeled 'races'. Indeed, it is possible that nonracial breeding populations even could generate distinct phenotypic features, such that one population has, say, "hitchhiker's thumb." On Kitcher's model, such a population

would have to be called a 'race', which seems to stretch the meaning of 'race'. Thus, while Kitcher's account of race might seem to avoid significant divergence from the folk category of race, in the end there do seem to be substantial intensional and extensional differences between the two.

VI

Genetic findings recently published in *Science* by Noah Rosenberg and others might seem capable of plugging the holes that we have so far seen in the races-as-breeding-populations theory.[29] They report that while 93–95% of genetic variation occurs within geographic populations, a further 3–5% of genetic variation distinguishes five populations that correspond to five "major geographic regions": Africa, Eurasia, East Asia, Oceania, and America (where Eurasia includes Europe, the Mid-East, and Central/South Asia). In essence, without prior identification of one's geographic ancestry, genetic information can be used to identify that ancestry. Given that some disease risks are higher for different populations, this is particularly significant for health care treatment and epidemiological research.

It is notable that the authors of this study never use the word 'race' to describe the geographical populations they identify. Yet *The New York Times* reports that they were willing to say in interviews that "[the five major] regions broadly correspond with popular notions of race."[30] Can we therefore say that this is the biological basis for race?

While these data certainly get us closer to the folk concept of 'race', it still seems too distant to say that race is biologically real. Consider, again, both the intension and the extension of the folk concept. Throughout, I have emphasized the significant degree to which the intension of 'race' is tied to gross morphological features, such as skin color. The data reported by Rosenberg and his co-authors provide no indication that those features can be mapped onto population-based properties, and, again, there is the point from Zack that even if such data becomes available, it is difficult to see how we would identify even vague criteria for assigning certain phenotypic properties to one race and not to another.

Relatedly, the geographic populations identified by Rosenberg and his co-authors seem extensionally different from the folk notion of race. For example, consider that the population of Adygei, from the Caucasus (from which the term 'Caucasian' originates), is lumped into the same major geographical population as the French, Palestinians, and Pathan/Pushtuns of Afghanistan. Perhaps we should say that these groups all compose one race, but I think that further argument is required for doing so. This point becomes particularly compelling when one considers the political implications of these categories. Might, for instance, a race-conscious Palestinian categorize herself within a group that does not include the French?

Political questions are important here not just because possible answers reveal something about the way we categorize ourselves. In section III, I noted that the new racial biological realism needs a further argument showing either that we should replace race with the distinct (and more biologically defensible) concept of breeding population, or that we should modify our conception of race in this substantial way. Such a position bears the burden of explaining how it would be practically possible to revise so significantly entrenched racial discourse.[31]

But in addition to the practical problems that would arise, there are also significant political hurdles facing any argument for either replacing race with another discourse entirely or substantially modifying the concept of race. For example, the classifications provided by both Kitcher and

Andreasen struggle to make sense of the folk category 'Asian'. If our political practices ought to contain no descriptive falsehoods, then their accounts of race would disallow political tools focused on Asians, such as those that might fight uniquely anti-Asian discrimination. This is only one of the problems with the "black/white binary" model of race, which others have discussed more extensively.[32] Furthermore, if the races-as-breeding-populations account ends up conflating different political axes, such as race and class (for example, English peasants), we might be left with impoverished political resources for dealing with social problems that are unique to each (which, again, others have discussed more extensively). The point can be stated briefly: a substantive revision to, or replacement of, the concept 'race' must show either that it can offer the requisite conceptual resources to justified political causes, or that those causes must be abandoned for the sake of conceptual coherence.

Yet one might take a different argumentative tack and argue that this substantive revision is already taking place, rendering any political or practical questions moot. After all, since scientific discourse is not wholly isolated from folk discourse, it is conceivable that the biologists' identification of breeding populations is itself changing the meaning of the folk concept of race, particularly in light of significant media coverage of these new biological data. While this change is conceivable, however, it seems doubtful that it has already been effected, given the divergence between the folk concept and the concept of races as breeding populations that has been discussed here. The source of this divergence is evident: the folk meanings of socially charged concepts like race are influenced by social practices (which, in the case of race, have significant normative dimensions) that often are not themselves restricted to being biologically accurate, as much as we might like them to be. Thus, given the divergence between the folk concept of 'race' and the biological facts, racial discourse seems to be at a crossroads: either we must acknowledge that we have no biological basis for that discourse, or the meaning of 'race' must be changed to reflect the biology. In this case, the pressing question is one of social policy: Should such a change in racial discourse be effected? The suggestion here is that any argument in favor of this substantive revision requires a substantive defense against the potential political problems just raised.

VII

All of the above leaves open the possibility that there might be some as yet undiscovered biological basis for race (though any such account of race would need to respond to Zack's challenge that there simply are no phenotypic traits shared by all blacks or all whites). It also leaves intact the idea that we might profitably make distinctions between humans based on reproductively isolated breeding groups, even if we do not cash out race in these terms. Finally, the arguments made here leave open the possibility that race might be real as a social, if not biological, kind. The foregoing, however, does suggest that we have not yet been given an adequate argument for holding that breeding populations are the biological basis of race.

Notes

1 See, for example, Michael Root, "How We Divide the World," *Philosophy of Science*, LXVII, Supplementary Volume (2000): S628–39; Ronald R. Sundstrom, "Race as a Human Kind," *Philosophy and Social Criticism*, XXVIII (2002): 91–115; Sundstrom, "'Racial Nominalism'," *Journal of Social Philosophy*, XXXIII (2002): 193–210; and Paul C. Taylor, "Appiah's Uncompleted Argument: W.E.B. Du Bois

and the Reality of Race," *Social Theory and Practice*, XXVI (2000): 103–28.

2 Andreasen, "A New Perspective on the Race Debate," *British Journal for the Philosophy of Science*, XLIX (1998): 199–225; Andreasen, "Race: Biological Reality or Social Construct?" *Philosophy of Science*, LXVII Supplementary Volume (2000): S653–66; P. Kitcher, "Race, Ethnicity, Biology, Culture," in Leonard Harris, ed., *Racism* (Amherst, NY: Humanity, 1999), pp. 87–120.

3 Andreasen, "A New Perspective," p. 207.

4 It is worth noting that reproductive isolation does not exclusively mean geographic isolation; other mechanisms can foster isolation, as we will see in our discussion of Kitcher. In addition, the isolation does not have to be so strong that there is zero interpopulation reproduction. It only entails that there is enough difference between intrapopulation reproduction and interpopulation reproduction to limit gene flow between populations, thereby resulting in genetic distance.

5 Andreasen, "A New Perspective," p. 212, and "Race," p. S660; L. Cavalli-Sforza, "Genes, Peoples, and Languages," *Scientific American*, CCLXV, **5** (1991): 104–10.

6 Andreasen, "A New Perspective," p. 214.

7 Andreasen, "A New Perspective," pp. 215–17, and "Race," p. S664.

8 Nor with the science behind Andreasen's approach – she considers some possible objections on that front in "A New Perspective."

9 As is often pointed out, such folk categorizations often inconsistently offer groupings that overlap race, ethnicity, and national origin. The history of intellectuals theorizing about race – as opposed to folk categorization – offers an extremely varied set of lists of the races, which differ not only on how many races there are, but also on which races there are (none of which seems to match Andreasen's list). To mention just a few examples, Bernier lists four or five (he is noncommittal about whether Native Americans constitute a distinct race); Voltaire offers seven; Kant offers four or five, depending on the essay; and Du Bois eight. As Robert Bernasconi and Tommy Lott note, by the end of the nineteenth century, the number of races

"grew from four or five to fifty or even eighty," except in the US, which sought to condense everyone of European descent into one race, to the exclusion of blacks, Asians, and Native Americans in particular – "Introduction," in Bernasconi and Lott, ed., *The Idea of Race* (Indianapolis: Hackett, 2000), p. vii. As the question posed here is whether there is any biological referent to what we, especially in the US, currently identify as races, I will be concerned with whether Andreasen and Kitcher can – or even need to – account for a biological basis of the current folk concept of race.

10 Andreasen, "A New Perspective," p. 212–13; cf. Andreasen, "Race," p. S664.

11 For two detailed analyses of what 'race' means and has meant, see K. Anthony Appiah, "Race, Culture, Identity: Misunderstood Connections," in Appiah and Amy Gutmann, *Color Conscious: The Political Morality of Race* (Princeton: University Press, 1996); and David Theo Goldberg, *Racist Culture: Philosophy and the Politics of Meaning* (Cambridge: Blackwell, 1993), chapter 4. Here, I only make some intuitive observations about what people mean or have meant by 'race.' I take it, though these observations are made from the armchair, so to speak, it is more or less obvious that these have been, at one time or another, conventional notions of 'race'. Andreasen offers some overlapping characterizations of common-sense ideas of race in "Race," p. S663. See Luther Wright, Jr., "Who's Black, Who's White, and Who Cares: Reconceptualizing the United States's Definition of Race and Racial Classifications," *Vanderbilt Law Review*, XLVIII (1995): 513–69, for summary and analysis of legal definitions of race in the US. Two relevant results can be found there: when not conflated with ethnicity and national origin, the legal definition of 'race' usually boils down to either overt physical traits or descent (And, since on the second criterion one is of race R when both of one's parents are of R, presumably parental racial classification at some point must be defined by some non-genealogical criterion, most likely physical traits.) Finally, see Charles Hirschman, Richard Alba, and Reynolds Farley, "The Meaning and Measurement of Race in the US Census: Glimpses into the Future,"

Demography, XXXVII (2000): 381–93, for how US citizens self-identify in census reporting.

12 Andreasen, "Race," p. S664.

13 "The Conservation of Races," reprinted in Du Bois, *The Souls of Black Folk*, David W. Blight and Robert Gooding-Williams, (eds) (Boston: Bedford, 1997).

14 Andreasen, "Race," p. S665.

15 See, for instance, Naomi Zack, *Race and Mixed Race* (Philadelphia: Temple, 1993) and M. Annette Jaimes, "Some Kind of Indian," in Zack, ed., *American Mixed Race: The Culture of Microdiversity* (Lanham, MD: Rowman and Littlefield, 1995) pp. 133–53.

16 In their analysis of census-style self reporting on the 1996 Racial and Ethnic Targeted Test, Hirschman *et al.*, "Meaning and Measurement," note that American folk classifications might be more productively captured in terms of origin, rather than race, but these origins are importantly different from Andreasen's ancestral breeding populations. The origins identified in Hirschman *et al.* include more recent origins like 'Hispanic', and national origins like 'Ecuadoran', which do not map on to Andreasen's ancestral breeding populations (intensionally or extensionally).

17 Andreasen, "A New Perspective," p. 218.

18 While Andreasen is working independently of common-sense notions of race, she is also engaging in dialogue with those in "the race debate," as indicated by the title of her paper, "A New Perspective on the Race Debate." As such, it seems all parties ought to be in the same neighborhood, more or less.

19 Andreasen, "Race," p. S662; cf. S665.

20 Appiah, "Race, Culture, Identity," p. 41.

21 Appiah, "Race, Culture, Identity," p. 77; Kitcher, p. 87. Cf. notes 9 and 11 above.

22 Cavalli-Sforza, P. Menozzi, and A. Piazza, *The History and Geography of Human Genes* (Princeton: University Press, 1994). For Andreasen's take on this disagreement, see "A New Perspective," p. 213.

23 For her part, Andreasen holds that her approach "does not depend on" the causal theory – see "Race," p. S662.

24 Nor, for that matter, is it like Putnam's case of lemons that have changed from yellow to blue, since the proposed theory of race is not that the original objects picked out as races have themselves undergone a constitutional change, like lemons that have changed from yellow to blue – see Putnam, "Is Semantics Possible?" reprinted in his *Mind, Language, and Reality: Philosophical Papers*, Volume 2 (New York: Cambridge, 1975), pp. 139–52.

25 This is worded too strongly: perhaps there is an argument for seeing the relationship between Andreasen's race and the folk concept as closer to that between H_2O and XYZ, rather than salt and water. But, as I argued above regarding the analogy with whales, this is not evident, and it requires a non-question begging argument, rather than mere stipulation. Andreasen does offer one analogy in response: like the difference between the folk concept of race and Andreasen's concept, a similar difference is found with species and higher taxa, insofar as they are defined historically by specialists, rather than according to shared phenotypic traits; and, since we would not therefore decide that species or higher taxa do not exist, so we should not decide that races do not exist – "Race," p. S665. The case of species, however, is different from the case of race: 'species' is not de-rigidified in the way that 'race' has been by Andreasen's account.

26 Kitcher does not mention Native Americans, and only briefly mentions Hispanics, in remarking that there are low rates of sexual union between Hispanics and Asians.

27 For Kitcher, since interracial sexual unions may occur at different frequencies at different locations due to different isolating mechanisms at those locations (say, differences between Oakland and

Memphis in the cultural barriers to interracial dating), racial distinctions may shift depending on the locale.

28 N. Zack, "Life after Race," in Zack, ed., *American Mixed Race: The Culture of Microdiversity* (Lanham, MD: Rowman and Littlefield, 1995), pp. 297–307, here p. 303.

29 Rosenberg, Jonathan K. Pritchard, James L. Weber, Howard M. Cann, Kenneth K. Kidd, Lev A. Zhivotovsky, and Marcus W. Feldman, "Genetic Structure of Human Populations," *Science*, CCXCVIII (December 20, 2002): 2381–85.

30 "Gene Study Identifies 5 Main Human Populations," *The New York Times* (December 20, 2002), Late Edition, Section A, p. 37. It should be noted that geographic populations are not identical to breeding populations. In defense of the breeding population model, however, one might argue that reproductive isolation results from geographic barriers. A claim of this sort is made in the *Times'* report on Rosenberg *et al.*, "Genetic Structure."

31 One potential practical problem with the revisionist program is that such a revised concept may retain some hidden references to the previous, inadequate concept. Appiah expresses concern about this in "Social Forces, 'Natural' Kinds," in Abebe Zegeye, Leonard Harris, and Julia Maxted (eds), *Exploitation and Exclusion: Race and Class in Contemporary US Society* (London: Hans Zell, 1991), pp. 1–13, in favor of abandoning race-talk: "if you want to talk about morphology, talk about morphology, if you want to talk about populations, talk about populations" (p. 12, n. 9).

32 See Linda Martin Alcoff, "Latino/as, Asian Americans, and the Black-White Binary," *The Journal of Ethics*, **VII**(2003): 5–27.

Chapter 11

Clark Glymour

WHAT WENT WRONG? REFLECTIONS ON SCIENCE BY OBSERVATION AND *THE BELL CURVE*

1 The thesis of this essay

Philosophical skepticism trades on two maneuvers: a focus on the worst case, and a demand that any method of forming belief find the truth in all logically possible circumstances. When action must be taken, skepticism is in league with obscurantism, with know-nothingism, and in opposition to forces that are more optimistic about the information that inquiry can provide to judgment. In this century, the principal tool of scientific optimism – although not always of social optimism – has been social statistics. Social statistics promised something less than a method of inquiry that is reliable in every possible circumstance, but something more than sheer ignorance; it promised methods that, under explicit and often plausible assumptions, but not in every logically possible circumstance, converge to the truth, whatever that may be, methods whose liability to error in the short run can be quantified and measured.

That promise was kept for two statistical enterprises, hypothesis testing and parameter estimation, which for decades were the cynosure of professional statistical study, but it failed in the more important parts of social inquiry that decide which parameters to estimate and which hypotheses to test. To make

those decisions with the same guarantee of conditional reliability requires methods of search and requires theoretical inquiries into the reliabilities of those methods. Social statistics produced and used a variety of procedures – factor analysis and regression are the principal examples – for searching for appropriate hypotheses, but no analysis of the conditions for their reliability. The reasons their reliabilities were insufficiently analyzed, and alternative methods not sought, are complex. They have to do with a positivism that, to this day, grips much of social statistics, and holds that causal hypotheses are intrinsically unscientific. Since almost all of the hypotheses of social inquiry are causal, this opinion requires a certain mental flexibility that inquiry into the reliabilities of methods of search for causal hypotheses would surely complicate. Perhaps an equally important reason is the view that causal hypotheses are theories, and theories are the special prerogative of experts, not of algorithms. These prejudices combined with a number of more technical disciplinary issues. For example, search methods are difficult to associate with any uniform measure of uncertainty, analogous to the standard error function for a parameter estimator, and social scientists and social statisticians have come to demand such measures without reflection. Again, disciplines are usually blind to their history, and although causal questions motivated much of the development of statistics, the paradigmatic tool for mathematical analysis in statistics is the theory of probability – there is no formal language in the subject for causal analysis (Pearl 1995). In some measure this state of affairs has been abetted by philosophy of science, which for generations taught that there could be no principles, no "logic," to scientific discovery.

The incoherence between practice and methodological theory would do little harm were the methods of searching for causal hypotheses that have developed in social statistics, and that are widely taught to social scientists, and widely used to justify conclusions, reliable under any set of conditions that might reasonably be assumed in the various domains to which the search methods are applied. They are not. We are left with enterprises that use the most rigorous possible methods to estimate parameters in causal models that are often produced by whimsy, prejudice, demonstrably unreliable search procedures, or, often without admission, by *ad hoc* search methods – sometimes reliable, sometimes not.

There is a remedy. Clear representations by directed graphs of causal hypotheses and their statistical implications, in train with rigorous investigation of search procedures, have been developed in the last decade in a thinly populated intersection of computer science, statistics, and philosophy. The empirical results obtained with these methods, including a number of cases in which the causal predictions were independently confirmed, have been good, perhaps surprisingly good. With exceptions, the reception by statisticians and philosophers has chiefly been hostile muddle or worse.

This paper illustrates some of the methodological difficulties of social statistics through a notorious example, *The Bell Curve*. I will use the directed graphical framework to explain the causal claims of the book and the unreliability of the search procedures on which the book relies. I will explain what search procedures that are asymptotically correct under plausible, general assumptions would say about the causal interpretation of the data the book uses. I will briefly describe some of the empirical research that has been carried out with new search procedures. I conclude with some remarks about social policy and about scientific strategy.

2 The Bell Curve

The Bell Curve is distinguished from a thousand and more efforts at social science chiefly by its length, popular style, ambition, and conclusions. The statistical methods of the book are multiple regression, logistic regression and factor analysis, techniques routinely taught to psychology and social science students in almost every graduate program in these subjects and routinely applied to make causal inferences from data of every kind. The methods and kinds of data of *The Bell Curve* are not very different in character from those in celebrated works of social statistics, for example the regression analyses in Peter Blau and Otis Dudley Duncan's *The American Occupational Structure* (1967), or the factor analyses in Melvin Kohn's *Class and Conformity* (1967); many statistical consultants use the same methods to guide business, military, and government policy on endless issues. One of the authors of *The Bell Curve*, Charles Murray, is a well-trained political scientist, and Herrnstein was a prominent psychologist; these are not *naïfs* or incompetents.

When Herrnstein and Murray write "cause" I take them to mean cause – something that varies in the population and whose variation produces variation in other variables, something that, if we could intervene and alter, would alter something else we did not directly wiggle. When they say genes cause IQ scores, I take them to mean that if somehow we could alter the relevant distribution of genes in the population, without altering directly anything else – the "environment" – then a different distribution of IQ scores would result if measured. That is how Sir Ronald Fisher (1958) thought of the causal role of genes in producing phenotypes, and it is how we think of causation in most other contexts.

Some statisticians, notably Paul Holland (1986), have claimed contrary to Fisher that it is nonsensical to talk of genes as causes. The thought seems to be that causation is a relation between individuals or between attributes of an individual, and I, for one, and you, for another, could not be who we are if our respective genetic structures were altered. The objection is wonderfully philosophical, Leibnizian even, but hardly persuasive in an age in which we can stick bits of DNA in chromosomes and re-identify the chromosome before and after the insertion.

There are two parts to the causal argument of *The Bell Curve*. One part argues that there is a feature of people, general intelligence, that is principally responsible for how people perform on IQ tests. The other part argues that this feature, as measured by IQ tests, causes a lot of other things. The first part is argued by appeal to factor analysis, the second part by regression. Because the hypotheses are causal, there is no substitute for making the causal claims explicit, and for that I will use graphical causal models. They explicitly represent important distinctions that are often lost when the discussion is couched in more typical formalisms.

3 Factor analysis

Herrnstein and Murray rely on factor analytic studies to justify the claim that there is a single

unobserved cause – which they, following Charles Spearman, call g for general intelligence – whose variation in the human population is responsible for most of the variation in scores on IQ tests. I want instead to consider the very idea of factor analysis as a reliable method for discovering the unobserved.

The issue is one of those delicate cases where it is important to say the right thing for the right reason. Stephen Jay Gould (in Fraser 1995) says the right thing about factor analysis – it is unreliable – but partly for the wrong reasons: that there exist alternative, distinct causal structures that are "statistically equivalent" and that entities and processes postulated because they explained observed correlations should not be "reified," that is, should not be taken seriously and literally. In the generality they are given, if not intended, Gould's reasons would be the end of science, including his own. Atoms, molecules, gravitational fields, the orbits of the planets, even the reality of the past, are all beyond the eye- and earshot that led our scientific ancestors, and lead us still, to believe in them. Physicists and philosophers of science have known for much of this century that standard physical theories – Newtonian gravitational theory, for example – admit alternative theories with different entities that equally save the phenomena. An objection that, when applied even-handedly, indicts factor analysis along with the best of our science leaves factor analysis in excellent company. The problems of factor analysis are more particular: they are the kinds of alternatives factor analytic procedures allow, the kind of restrictions the factor analytic tradition employs to eliminate alternatives, and, in consequence, the want of correspondence between factor analytic results and actual structures from which data are generated.

Herrnstein and Murray's history of factor analysis requires a correction. They say that Spearman introduced the concept of general intelligence upon noticing that scores on his "mental" tests were all strongly positively correlated. Not exactly. Spearman developed his argument in various roughly equivalent forms over half a century,[1] but it came to this: The correlations of any four mental test scores i, j, k, l satisfy three equations:

$$\rho_{ij}\,\rho_{kl} = \rho_{il}\,\rho_{jk} = \rho_{ik}\,\rho_{jl}$$

Spearman observed that these "tetrad" equations are implied by any linear structure in which scores on tests are all influenced by a single common cause, and otherwise sources of variation in test scores are uncorrelated. Graphically:

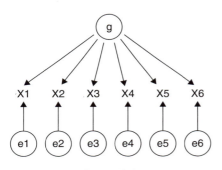

Figure 11.1

Spearman realized that certain alternative structures would also generate the tetrad equations, for example

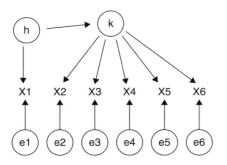

Figure 11.2

but he thought of such structures as simply finer hypotheses about the structure of general intelligence, g.

Spearman must have known that structures with still more latent variables can account for the data. The tetrads, for example, can be made to vanish by suitable choice of the linear coefficients when there are two or more common latent factors affecting the measured variables. Such models might be rejected on the grounds that models that postulate fewer unobserved causes are more likely to be true than those that save the same phenomena by postulating more unobserved causes, but that is a very strong assumption. A much weaker one would serve the purpose: Factor models assume that observed variables that do not influence one another are independent conditional on all of their common causes, an assumption that is a special case of what Terry Speed has called the "Markov condition" for directed graphical models. The rank constraints – of which vanishing tetrads are a special case – used in factor analysis are implied by conditional independencies in factor models, conditional independencies guaranteed by the topological structure of the graph of the model, no matter what values the linear coefficients or "factor loadings" may have. To exclude more latent variables when fewer will do, Spearman needed only to assume that vanishing tetrads do not depend on the constraints on the numerical values of the linear coefficients or "factor loadings," but are implied by the underlying causal structure. A general version of this second assumption has been called "faithfulness." It is known that the set of values of linear parameters (coefficients and variances) that generate probability distributions unfaithful to a directed graph is measure zero in the natural measure on parameter space.

To illustrate the point compare the structures:

188   Clark Glymour

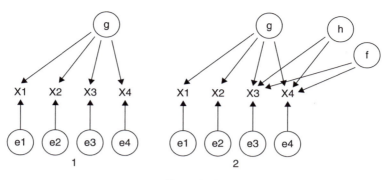

Figure 11.3

Let the factor loadings of g in graphs 1 and 2 be a_i, the factor loadings of h in graph 2 be b_i and the factor loadings of f be c_i where the index is over the measured variable connected to the factor. Then in graph 1 the vanishing tetrad differences follow from the commutativity of multiplication, that is, that $a_i a_j a_k a_l = a_i a_k a_j a_l$. In graph 2, however the tetrad equation $\rho_{12}\rho_{34} = \rho_{13}\rho_{24}$ requires that $a_1 a_2 (a_3 a_4 + b_3 b_4 + c_3 c_4) = a_1 a_3 a_2 a_4$, that is, $b_3 b_4 = -c_3 c_4$.

In the absence of further substantive assumptions, however, neither faithfulness nor the much stronger simplicity assumption would lead from tetrad constraints to Spearman's latent common cause models. Quite different structures also imply his tetrad equations, for example:

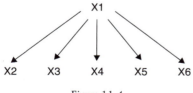

Figure 11.4

where I have omitted the error terms. The vanishing of all tetrads guarantees that a single common cause suffices; it does not guarantee that the common cause is unmeasured. Figures 11.1 and 11.4 are, however, distinguished by the vanishing partial correlations they require among measured variables; Figure 11.1 requires none, Figure 11.4 requires

that all partials on X1 vanish. But Figure 11.5 is statistically indistinguishable from Figure 11.1:

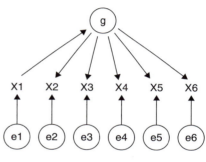

Figure 11.5

So far as I know, Spearman and his followers never considered these matters.

Spearman's original mental tests did not prove well correlated with teachers' and others' judgments of intelligence, and they were replaced by tests in Binet and Simon's mode. These tests in turn had more complicated correlation structures, and typically all tetrads did not vanish. Spearman's followers, notably Karl Holzinger, began the practice of assuming a single common cause, g, and then introducing additional common causes as they were needed to account for residual correlation and prevent the implication of tetrad equations not approximated in the data. Their procedure guaranteed that if most of the correlation among measures could be attributed to one common cause, it would be, even if alternative structures and factor loadings were consistent with the data. Reliability was never an issue.[2]

Thurstone (1947) said he discovered factor analysis when he realized the tetrads were merely the determinant of a second order minor. The mathematical idea in factor analysis was that the rank of the correlation matrix gives information about the minimum number of latent common causes needed to reproduce the matrix. The procedural idea was a method – the centroid method – of forming from the covariances a particular linear causal model, in which all of the correlations of measured variables are due to latent common causes. Thurstone realized that the models his procedure produced were not the only possible linear, latent variable explanations of the data from which they started, and that in fact any non-singular linear combination of latent factors obtained by his centroid method would do as well.

Thurstone's problem is fairly compared to John Dalton's. Thurstone had no means of uniquely determining the latent factor loadings and relations, and Dalton had no means of determining relative atomic weights. Both sought to remove or at least reduce underdetermination with a simplicity principle.[3] In graphical terms, Thurstone's proposal is to find the linear combination of latents that produces the fewest total number of directed edges from latent factors to measured variables. Thurstone thought such a "simple structure" is unique for each correlation matrix, but it is not. More important, why should we think actual mental structures obey Thurstone's rule of simplicity any more than atoms obey Dalton's? Unlike faithfulness, simple structure has no special measure theoretic virtue and no special stability properties.

Thurstone's factor analysis rapidly displaced Spearman's methods. Reliability does not seem to have been one of the reasons. Guilford, who discusses both in his *Psychometric Methods*, recommends factor analysis over tetrad analysis on grounds of computational tractability.

There is no proof of the correctness of any factor analytic procedure in identifying causal structure in the large sample limit. In general, factor analytic procedures are not correct even with perfect data from the marginals of distributions faithful to the actual structure. There is not even a proof that the procedures find a correct simple structure when one exists. I know of no serious study of the reliability of any exploratory factor analysis procedures with simulated data. A serious study would generate random graphs with latent variables, randomly parameterize them, generate random samples (of various sizes) from each model so produced, run the sample data through factor analytic procedures, and measure the average errors in model specification produced by the factor methods.[4]

Consider the more fundamental problem of the existence of asymptotically correct, computable procedures that will, under appropriate distribution assumptions give information about latent structure when the Markov and faithfulness assumptions, or similar structural and stability assumptions appropriate to causal systems,[5] are met. There is nowhere in the statistical literature, at least to my knowledge, any informative mathematical study characterizing the bounds on reliably extracting information about latent structure from measured covariances under general distribution assumptions such as linearity and normality and measure one assumptions such as faithfulness. We do not have a classification of the structures for which faithful distributions will generate any given set of constraints on the covariance matrix of measured variables. Worse, we do not even have a classification of the constraints on the covariance of measured variables that can be implied (via the Markov condition) by partial correlation constraints in latent variable models. Analogous lacunae exist for other factor models, those with discrete observed variables and discrete or continuous latent variables. So far as Reliable Search is concerned, psychometrics is a century of sleep.

There is a little work on the fundamental questions, but so little that it is almost anomalous. Junker and Ellis (1995) have recently provided necessary and sufficient conditions for the existence of a unidimensional latent variable model of any real valued measures – representing the most recent of a sequence of papers by several authors focused on when there exists a single common cause of observed measures. It has been shown (Spirtes *et al.* 1993) that if the investigator provides a correct, initial division of variables into disjoint clusters such that the members of each cluster share at least a distinct latent common cause, then under certain assumptions, including linearity, unidimensional measurement models may be found for each latent, and from such models and the data, some causal relations among latents may reliably be found.

Stephen Jay Gould (in Fraser 1995) claims that one of the essential premises of *The Bell Curve* is that there is a single common factor, g, responsible for performance on intelligence tests. No doubt Herrnstein and Murray make that assumption, but it is largely inessential to their argument. If IQ scores measured a pastiche of substantially heritable features that doom people to misery, the argument of *The Bell Curve* would be much the same. So the more important questions concern causal relations between whatever it is IQ measures and various social outcomes. Which brings us to regression.

4 Regression and discovery

Herrnstein and Murray begin the second part of their book with a description of some of their methods and what the methods are used to do. I ask the reader to keep in mind their account from pages 72–75. I have numbered their paragraphs for subsequent reference:

(1) The basic tool for multivariate analysis in the social sciences is known as regression analysis. The many forms of regression analysis have a common structure. There is a result to explain, the dependent variable. There are some things that might be the causes, the independent variables. Regression analysis tells how much each cause actually affects the result, taking the role of all the other hypothesized causes into account – an enormously useful thing for a statistical procedure to do, hence its widespread use.

(2) In most of the chapters of Part II, we will be looking at a variety of social behaviors, ranging from crime to childbearing to unemployment to citizenship. In each instance, we will look first at the direct relationship of cognitive ability to that behavior. After observing a statistical connection, the next question to come to mind is, What else might be another source of the relationship?

(3) In the case of IQ the obvious answer is socioeconomic status. . . . Our measure of SES is an index combining indicators of parental education, income, and occupational prestige . . . Our basic procedure has been to run regression analyses in which the independent variables include IQ and parental SES. The result is a statement of the form "Here is the relationship of IQ to social behavior X after the effects of socioeconomic background have been extracted," or vice versa. . . .

The causal picture Herrnstein and Murray seem to have in mind is this:

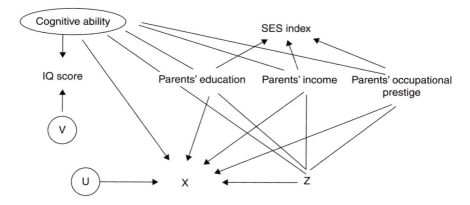

Figure 11.6

where the features in circles or ovals are unobserved, and the lines without arrows indicate statistical associations that may be due to influences in one direction or the other, or to unobserved common causes, or both. Z varies from case to case; often it is age.

If this were the correct causal story, then provided that very little of the variation in IQ scores between individuals were due to V, one could estimate the influence of cognitive ability on X by the two methods Herrnstein and Murray use: multiple regression of X on IQ and SES index when the dependencies are all linear, and by logistic regression on those variables under other distribution assumptions. By "could estimate" I mean the expected values of estimates of parameters would equal their true values.

I will sometimes simplify the diagram in Figure 11.6 as Herrnstein and Murray simplify their discussion:

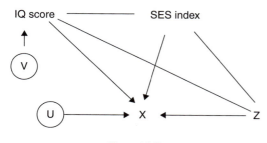

Figure 11.7

Under the assumptions just mentioned, if estimates of the influence of IQ score based on the causal model of Figure 11.6 are correct, so are estimates of IQ based on the simpler surrogate structure of Figure 11.7.

Now the standard objection to assuming something like the structure of Figure 11.6 or Figure 11.7 is put in terms of "correlated error." The objection is that in the corresponding regression equation, the error term, U, for X may be correlated with any of IQ, SES, and Z, that such correlation cannot be detected from the data, and that when it exists the regression estimates of the influence of cognitive ability on X will be incorrect. "Correlated error" is jargon – euphemism, really – for those who want to avoid saying what they are talking about, namely causal relations. Unless correlations arise by sheer chance, the correlation of U and IQ, say, must be due to some common causal pathway connecting IQ scores with whatever features are disguised by the variable U. A "correlated error" between a regressor such as IQ and an outcome variable, X, is the manifestation of some unknown cause or causes influencing both variables.[6]

Suppose something else, denoted by W – mother's character, attention to small children,

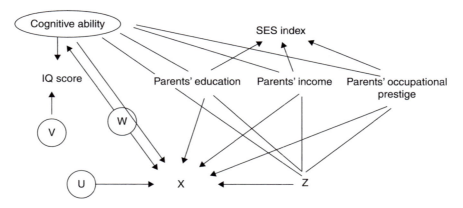

Figure 11.8

the number of siblings, the place in birth order, the presence of two parents, a scholarly tradition, a strong parental positive attitude towards learning, where rather than how long parents went to school, whatever, influences both cognitive ability and X. Then the regression estimates of the influence of cognitive ability on X based on the model in Figure 11.6 will compound that influence with the association between cognitive ability and X produced by W. Or more briefly:

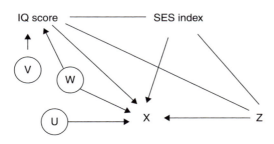

Figure 11.9

Here is how Herrnstein and Murray respond:

(4) We can already hear critics saying, "If only they had added this other variable to the analysis, they would have seen that intelligence has nothing to do with X." A major part of our analysis accordingly has been to anticipate what other variables might be invoked

and seeing if they do in fact attenuate the relationship of IQ to any given social behavior.

This sounds quite sensible, until one notes that none of the possible confounding variables suggested above, nor many others that can easily be imagined, are considered in *The Bell Curve*, and until one reads the following:

(5) At this point, however, statistical analysis can become a bottomless pit ... Our principle was to explore additional dynamics where there was another factor that was not only conceivably important but for clear logical reasons might be important because of dynamics having little or nothing to do with IQ. This last proviso is crucial, for one of the common misuses of regression analysis is to introduce an additional variable that in reality is mostly another expression of variables that are already in the equation.

There is a legitimate concern in this remark, which does not, however, excuse the neglect: If W is an *effect* of cognitive ability, then including W among the regressors will omit the mechanism that involves W and will lead to an incorrect estimate of the overall influence of cognitive ability on X:

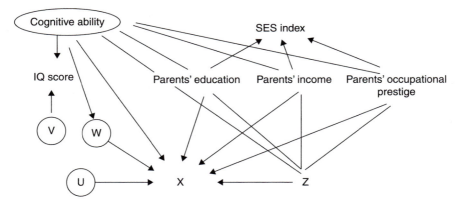

Figure 11.10

Against Herrnstein and Murray's remark in paragraph (5), however, it is exactly the presence of other variables that are common causes of X and of cognitive ability, and therefore "having to do" with cognitive ability, that lead to the "correlated errors" problem in estimating the influence of cognitive ability on X. Omitting such variables, if they exist, ensures that the regression estimates of effects are wrong. *The surprising fact is that the regression estimates may very well be wrong even if such variables are included in the regression.* That requires some explanation.

The authors of *The Bell Curve* have been criticized for omitting the subjects' educations from their set of regressors, an omission about which I will have more to say later. But their analysis would have been no better for including education. Suppose the true causal structure were as in Figure 11.10, with W representing years of education. Then multiple regression with education included would mistake the influence of cognitive ability on X, because it would leave out all pathways from cognitive ability to X that pass through W. At least, one might say, a regression that includes education would tell us how much cognitive ability influences X *other than through* mechanisms involving education, SES and Z. But even that is not so. Depending on whether there are additional unmeasured common causes of

education and X, the error in the estimate of the separate effect of cognitive ability on X might be positive or negative, and arbitrarily large. There are circumstances, arguably quite common circumstances, in which assumptions about distribution families (normal, etc.) are satisfied, and there is no "correlated error" between an outcome variable X and a regressor such as cognitive ability – that is, there is no unmeasured common cause of X and the regressor – but regression nonetheless mistakes the influence of the regressor on the outcome. Suppose the actual structure is as in Figure 11.11:

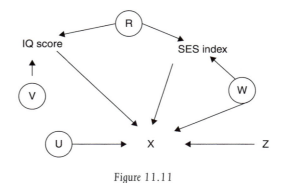

Figure 11.11

Notice that there is no unmeasured common cause of IQ and X, no correlation of the error term with IQ in the regression equation for X, but the error term in the regression equation is correlated with another regressor, SES. In that case, multiple regression of X on IQ, SES and Z will give an incorrect estimate of the influence of IQ on X. The error of the estimate can be arbitrarily large and either positive or negative, depending on the values of the parameters associated with the unmeasured R and W variables. For all we know, the subjects in the data Herrnstein and Murray study are rich in such R's and W's.

Critics have noted that the SES index Murray and Herrnstein use is rather lame, but the criticism is largely beside the point. Suppose they had used a better index, compounded of more measured features of the subjects and their families. The variables in SES indices may be strongly correlated, but they typically have no single common cause – those Murray and Herrnstein use demonstrably do not.[7] So a better index would add a lot of causally disparate measures together. Wouldn't that make it all the more likely that there are unmeasured variables, structurally like W in Figure 11.10, influencing X and also influencing one or more of the components of SES? I think so.

Adding extra variables to their study would not necessarily improve the accuracy of their estimates and might make them much worse, but leaving extra variables out may result in terribly inaccurate estimates.

Herrnstein and Murray remark that an obvious additional variable to control for is education, but they do not, first because years of education are caused by both SES and IQ, second because the effect of education on other variables is not linear and depends on whether certain milestones – graduations – have been passed, third because the correlation of education with SES and IQ makes for unstable estimates of regression coefficients, and fourth

(6) to take education's regression coefficient seriously tacitly assumes that education and intelligence could vary independently and produce similar results. No one can believe this to be true in general: indisputably giving nineteen years of education to a person with an IQ of 75 is not going to have the same impact on life as it would for a person with an IQ of 125.

(7) Our solution to this situation is to report the role of cognitive ability for two sub populations of the NLSY that each have the same level of education: a high school diploma, no more and no less in one group; a bachelor's degree, no more and no less, in the other. This is a simple, but we believe reasonable, way of bounding the degree to which cognitive ability makes a difference independent of education.

The third reason is unconvincing, since SES and IQ are already strongly correlated. The last reason, in paragraph (6), is unconvincing as a ground for omitting education from the analysis, but correct in supposing there is an interaction. The interesting thing, however, is the alternative procedure suggested in paragraph (7), since it reveals a problem about causal inference that may trouble a great deal of work in social science and psychology.

Herrnstein and Murray make it plain – they even draw the graph – that they have in mind a particular causal picture:

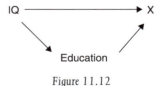

Figure 11.12

If this is the correct structure, then if there is no interaction between IQ and education in their influence on X, one way to estimate the direct effect of IQ on X is to condition on any value of education. The point of conditioning on two values of education, I take it, is to give us some idea of how much the interaction makes this estimate unstable.

Here is the problem: What if Figure 11.12 is *not* the correct causal structure? What if, instead, the correct causal structure is

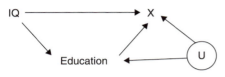

Figure 11.13

whatever U may be. In that case, the association between IQ and X conditional on a value of education will not be a measure of the direct influence of IQ on X, and the error can be as large as you please, positive or negative, depending on U and the parameters associated with it.

This sort of problem, *sample selection bias*, can occur whenever membership in a sample is influenced by variables whose influence on one another is under investigation. It may happen, for example, when using a sample of hospitalized patients, or when using college students as subjects in psychological experiments, or when subjects in a longitudinal study are lost, simply by using a subsample determined by values of a variable with complex causal relations, as Herrnstein and Murray do.

5 The problems of causal inference

Herrnstein and Murray use the tools their professions, and social statistics generally, gave to them. The tools are incompetent for the use Herrnstein and Murray put them to, but what else were they to do? What else can anyone do who is trying to understand the causal structure at work in processes that cannot be controlled experimentally?

Consider for a moment some of the difficulties in the problem of trying to infer causation from observed correlations:

0. Little may actually be known beforehand about the causal relations, or absence of causal relations among variables. In typical social studies, time order often provides the only reliable information – negative information, at that – about cause and effect.

1. Observed associations may be due to unmeasured or unrecorded common causes.

2. A vast number of alternative possible hypotheses – the larger the number of measured variables, the more astronomical the set of possible causal structures. When latent variables are allowed the number of possible causal structures is literally infinite.

3. Several or even a great many hypothetical structures may equally account for the same correlations, no matter how large the sample, and in finite samples a great many models may fit the data quite well.

4. The sample may be unrepresentative of a larger population because membership in the sample is influenced by some of the very features whose causal relations are the object of study.

5. The sample may be unrepresentative by chance.

6. Values for sundry variables may be unrecorded for some units in the sample.

7. The joint distribution of variables may not be well approximated by any of the familiar distributions of statistics. In particular, there may be combinations of continuous variables and variables that take only a finite set of values.

8. Relations among variables may be complicated by feedback, as between education and IQ.

Many of the same difficulties beset causal inference in experimental contexts, even though experimental design aims to remove the possibility of confounding common causes of treatment and to maximize prior knowledge of the causal structure of the experimental system. Psychological experiments often concern unobserved and uncontrolled features; clinical experiments sometimes try to investigate multiple treatments and multiple outcomes simultaneously, with entirely parallel problems about confounding and feedback, especially in longitudinal studies. Sample selection and attrition in experiments, especially experiments with humans, can create selection bias as in (4) and can result in missing values. The distribution of treatments in experiments is controlled by the experimenter but the distribution of outcomes, which may conform to no familiar pattern, is not.

We can imagine a black-box that addresses these problems. Data and relevant beliefs are put in, causal information comes out, and inside the box the problems just listed are taken account of. The box is imaginary, of course. There are no methods available that more or less automatically address all of these problems. There is no computer program that will take the data and prior knowledge, automatically take account of missing values, distributions, possible selection bias, possible feedback, and possible latent variables, and reliably and informatively give back the possible causal explanations that produce good approximations to the data, information about error bounds, or posterior probabilities. But we can think of the box as an ideal, not only for inference but also for forcing practitioners to cleanly separate the claims they make before examining the data from the claims they believe are warranted by the data. How close do the methods used by Herrnstein and Murray and other social scientists come to the ideal box? And how close could they come were they to use available, if nonstandard, methods?

Let us leave aside the problems 4 through 8, and suppose that our samples are nice,

distributed nicely – normally, say – there are no missing values and no feedback, and no sample selection bias. Consider for a moment in this context using regression to decide a simpler question than estimating the influence of cognitive ability on X from ideal data on X, cognitive ability, and a definite set of other regressors: Does cognitive ability have any influence *at all* on X? Multiple regression will lead to a negative answer when the partial regression coefficient for cognitive ability is not significantly different from zero. That is, under a normal distribution, essentially an assumption connecting the absence of causal influence with a conditional independence fact, namely that *cognitive ability does not (directly) influence X if and only if cognitive ability and X are independent conditional on the set of all of the other regressors*.

We have in effect observed by example in the previous section that the italicized principle is false, in fact intensely false. Indeed, without a priori causal knowledge, there is no way to get reliable causal information of any sort from multiple regression. If one should be so fortunate as to know independently of the data analysis that there are no common causes of any of the regressors and the outcome variable, and the outcome variable is not a cause of any of the regressors, then under appropriate distribution assumptions, regression gives the right answer. Otherwise, not.

Regression does a funny thing: to evaluate the influence of one regressor on X, it conditions on *all other* regressors, but not on any proper subsets of other regressors. Stepwise regression procedures typically do investigate the dependence of a regressor and X conditional on various subsets of other regressors, but they do so in completely *ad hoc* ways, with no demonstrable connection between the procedures and getting to the truth about causal structure. Regression and stepwise regression reflect intuitions from experimental design and elsewhere that absence of causation has something to do with conditional independence. They simply do not get the something right. The correct relationship is far more complicated.

Fifteen years ago, Terry Speed and his student, Harry Kiiveri (1982), introduced a correct relation, which, with some historical inaccuracy, they called a Markov condition. Speed has since testified to the correctness of the principle in the most infamous trial of our time. We considered the condition informally in discussing factor analysis; now consider it a little more exactly. Understanding the condition requires the notion of one variable Y, say, being a *direct cause* of another, X, relative to a set of variables **D** to which X and Y both belong. Y is a direct cause of X relative to **D** if there is a causal pathway from Y to X that does not contain any other variable in **D** – in other words, there is no set of variables in **D** such that if we were to intervene to fix values for variables in that set, variations in Y would no longer influence X. We need one further preliminary definition: I will say that any set, **D**, of variables is *causally sufficient*, provided that for every pair, X, Y of variables in **D**, if the directed graph representing all causal connections among variables in **D** contains a variable Z which is the source of two non-intersecting (except at Z) directed paths, respectively to X and to Y, then Z is in **D**.

Causal Markov Condition: For any variable X, and any set of variables **Z** that are not effects of X (and does not have X as a member), and any causally sufficient set **D** of variables including **Z** and having X as a member, X is independent of **Z** conditional on the set of members of **D** that are direct causes of X – the set of parents of X in the directed graph of causal relations.

When true, the Markov condition gives a sufficient causal condition for conditional independence. A converse condition gives necessity:

Faithfulness Condition:[8] All conditional independencies in a causal system result from the Causal Markov condition.

The scope of the Markov condition is occasionally misunderstood by philosophical commentations: *As a formal principle about directed graphs and probability distributions, the Markov Condition is necessarily true if exogenous variables (including errors or noises) are independent, and each variable is a deterministic function of its parents (including among parents, any errors or noises). The form of the functional dependence is irrelevant.*

In systems whose causal structure is represented by a direct acyclic graph and the system generates a probability distributions meeting the Markov condition for that graph, the faithfulness condition can only fail if two variables are connected by two or more causal pathways (either from one variable to another or from a third variable to both) that exactly, perfectly, cancel one another, or if some of the relations among variables (excluding error terms) are deterministic. In practice both the Markov and faithfulness conditions are consistent with almost every causal model in the social scientific literature, non-linear models included, that does not purport to represent feedback or reciprocal influence.

We can use these two conditions to discover what the conditional independencies implied by the structures that Herrnstein and Murray postulate could *possibly* tell us, by any method whatsoever, about those structures. That is, we will suppose their causal story is correct and ask whether they could reasonably infer it is correct from data that nicely agrees with it. To do so we need some simple representations of ignorance about causal structures.[9] Here is a convenient code:

X○───○Y X is a cause of Y, or Y is a cause of X, or there is a common unmeasured cause of X and of Y, or one of X, Y causes the other and there is also an unmeasured common cause.

X○───▶Y X is a cause of Y, or there is a common unmeasured cause of X and of Y, or both.

X◄──►Y There is a common unmeasured cause of X and of Y, but neither X nor Y influences the other.

With these conventions, here is what the conditional independencies implied by the causal hypothesis of Figure 11.7 tell us about causal relations:

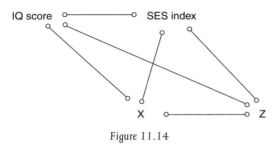

Figure 11.14

Nothing at all about whether IQ score is a cause of X. Suppose that common sense tells us that X is not a cause of the other variables. That does not help much. The result is:

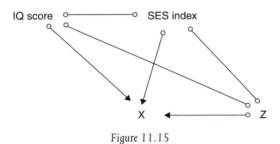

Figure 11.15

We still cannot tell whether IQ has any influence at all on X. For all we know, from the data and the prior knowledge, the association between IQ scores and X is produced entirely by the variation of unmeasured factors that influence both IQ score and X. The sizes and signs of the observed covariances in this case would give no other extra information about the actual causal structure.

The Markov and faithfulness conditions also entail that there are possible causal relations we can determine from observed associations, provided we have none of the problems (4)–(8)

listed above. For example, suppose we have measures of A, B, C and D, and their causal relations are actually as in Figure 11.16:

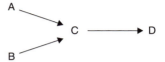

Figure 11.16

Then according to the two conditions, we can determine from independence facts that:

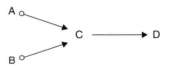

Figure 11.17

and so that C is actually a cause of D. Moreover, there is a certain robustness to the determination, for if we were to decide that the independencies corresponding to Figure 11.16 obtain when in fact they do not quite because of a small common cause of C and D, if the association of A and C or B and C is large, then (in the linear case at any rate) the estimate of the influence of C on D obtained using the result in Figure 11.17 will be a good approximation to the truth. There is here a general moral – almost never observed – about the kind of data one should seek if causal relations are to be inferred from observed data.

The Markov and faithfulness conditions can just as well give us information about the presence of unmeasured common causes (see Spirtes *et al.* 1993 for details).

These remarks would be of little practical use if in any application one were required to prove some intricate theorem, distinct for almost every case, characterizing the structures consistent with prior knowledge and the patterns of independence and conditional independence found in the data. No such effort is necessary. There is a general algorithm,[10] commercially available in the TETRAD II program, that does the computations for any case. The finite sample reliabilities of the program have been pretty thoroughly investigated with simulated data; it is known, for example, that they tend to give too many false arrowheads, and that for reasonable sample sizes they rarely produce false adjacencies. The procedures are rarely used, certainly not by Herrnstein and Murray or their critics. Were they used, social scientists would at least be forced to be entirely explicit about the causal assumptions that they have forced on their data analysis.

In keeping with social scientific tradition, Herrnstein and Murray give endless pages of statistical conclusions but their data are all but hidden; one has to go to the original sources and know the sample they selected from it. Although Herrnstein and Murray report any number of linear regressions with results determined entirely by a simple covariance matrix, they give only one such matrix in the entire book, and no count data. Even so, we have excellent reason to think that scientific searches applied to the data they use would turn up structures such as those in Figure 11.14, permitting no causal inferences of the kind Murray and Herrnstein wish to draw.

6 Some empirical cases

Of course it is easy to avoid making erroneous causal inferences: make none at all. The challenge is to find methods that make correct causal inferences where possible while avoiding, insofar as possible, incorrect causal inferences. The positive value of the search methods can be illustrated by considering a problem posed by the intelligence test Herrnstein and Murray use, the Armed Forces Qualification Test.

Before 1989 the AFQT score was a weighted sum of several component tests, and was itself a part of a larger battery that included tests that were not included in AFQT. Consider data for AFQT and the following seven tests from a sample of 6224 subjects in 1987:

- Arithmetical Reasoning (*AR*)
- Numerical Operations (*NO*)
- Word Knowledge (*WK*)
- Mathematical Knowledge (*MK*)
- Electronics Information (*EI*)
- General Science (*GS*)
- Mechanical Comprehension (*MC*)

Here is the problem. Given the information above, and the correlations, and nothing more, determine which of the seven tests are components of AFQT.[11] The problem is at least partly linear, because we know AFQT depends linearly on some of these tests, perhaps not others, and also on some tests whose scores are not in the data. Of course we do not know that relations among the scores on various tests are produced by linear processes.

If you run the correlations through a linear regression package you will find that two of the seven tests have insignificant regression coefficients, two have significant negative coefficients, and three (NO, WK and AR) have significant positive coefficients. The tests with negative coefficients might be components of AFQT that are negatively correlated with test components not included in the data set, or for some reasons these tests may be weighted with a minus sign in the AFQT score, as they would be if higher scores on the tests indicate less ability.

The TETRAD II program permits the user to determine the structures compatible with background knowledge and a data set under various assumptions, including the assumption that no latent variables are acting, or alternatively without that restriction. Assuming there are no latent variables, the procedure finds that only four of the seven tests are components of AFQT; one of the tests with a significant but negative regression coefficient is eliminated in addition to the two tests with insignificant regression coefficients. The assumption that there are no latent variables is not very plausible, and removing it, the TETRAD II program finds that only NO, WK, and AR are adjacent to AFQT, although the adjacency is a double-headed arrow in all three cases. Recognizing that procedure rarely makes false positive adjacencies, and it is liable to false positive arrowheads, one concludes that the components of AFQT among the seven tests are probably NO, WK and AR, which is the correct answer, one we found with an experimental version of the TETRAD program using only the information given here, not only initially in ignorance of the actual components of AFQT, but burdened with the false information that all seven tests are components.

Several other applications have been made of the techniques, for example:

1   Spirtes *et al.* (1993) used published data on a small observational sample of Spartina grass from the Cape Fear estuary to correctly predict – contrary both to regression results and expert opinion – the outcome of an unpublished greenhouse experiment on the influence of salinity, pH and aeration on growth.
2   Druzdzel and Glymour (1994) used data from the US News and World Report survey of American colleges and universities to predict the effect on dropout rates of manipulating average SAT scores of freshman classes. The prediction was confirmed at Carnegie Mellon University.
3   Waldemark used the techniques to recalibrate a mass spectrometer aboard a Swedish satellite, reducing errors by half.
4   Shipley (1995, 1997) used the techniques to model a variety of biological problems, and developed adaptations of them for small sample problems.

5   Akleman *et al.* (1997) have found that the graphical model search techniques do as well or better than standard time series regression techniques based on statistical loss functions at out of sample predictions for data on exchange rates and corn prices.

7 Projects and attitudes

The TETRAD II program represents only one of several approaches to automated, reliable model specification, steps toward the ideal inference box. Bayesian methods are under investigation, as are methods that combine Bayesian methods with the constraint-based procedures used in the TETRAD programs. Procedures have been described, and under explicit assumptions proved correct, that make a more limited set of causal inferences when sample selection bias may be present. The same is true when feedback may be present in linear systems.[12] Research is under way on understanding constraints other than conditional independence relations.

There may never be an inference box that addresses all of the problems of causal inference from observational data, but there can certainly be boxes that can help social and behavioral scientists do better than they will armed only with their preconceptions, factor analysis, and regression. The more model specification is automated and data driven, the more substantive prior assumptions are separated mechanically from inferences made from the data, the more algorithms in the box give out only information justified by explicit assumptions, the less likely is the kind of work *The Bell Curve* represents.

The statistical community, by and large and no doubt with important exceptions, thinks otherwise. Terry Speed once told me that although he agreed that regression is demonstrably unreliable in general, no other methods should be developed or used, even if they are demonstrably more reliable. When a large simulation study showed that the search procedures in their commercial program, LISREL, are wildly unreliable in cases in which there are latent causes of measured variables, and also influences of measured variables on one another, two prominent statisticians, Karl Joreskog and Dag Sorbom (1990), claimed that no such cases are possible: God has arranged that the world can only be as LISREL finds it. These two disparate attitudes have nearly banished from social statistics serious inquiry into methods of search that reliably get to the truth about the questions we most urgently care about from any data we could possibly have. The questions will not go away, and in the absence of answers psychologists and social scientists, whatever their motives, will make use of the methods in which they have been educated. The Bell Curve, and innumerable less famous works no more worthy of credence, are the result.

8 Policy

It would be too academic and bloodless to pass by the example The Bell Curve offers without saying something about the policy issues at stake. Perhaps surprisingly, Herrnstein and Murray's conclusions about what is happening to American society are substantially the same as those drawn from less evidence by celebrated social democrats (the former Secretary of Labor, Robert Reich, for example), and the same as those of uncountable reports by liberal-minded institutes concerning the state of education and social relations in this country. With liberals, Herrnstein and Murray say the economy increasingly rewards high-level intellectual skills and increasingly penalizes those without such skills; with liberals, they say that unimpeded, the effect is to the disadvantage of nations whose populace is comparatively unskilled in these ways. Within nations those who are talented and skilled at "symbol manipulation" are increasingly segregated from those who are not; public education

achieves a level of knowledge and skill that makes Americans stupid compared with citizens of other industrialized nations; America is still socially segregated by race. The principal difference between liberal writers such as Reich and the authors of The Bell Curve is that Herrnstein and Murray offer a lot more empirical arguments, some relevant, some not, and conclude against the popular social institutions and interventions – public education, affirmative action, Head Start – that Reich would endorse.

The Bell Curve comes apart exactly when it moves from formulating our social problems to recommending solutions. The book reads like the broadsides of intellectual flacks in the Kennedy and Johnson administrations, the Richard Goodwins of the 1960s and 1970s, who reported the evils of the conflict in Vietnam but always ended with an illogical plea for continuing that war. Sensibly read, much of the data of The Bell Curve, as well as other data the book does not report, demands a revived and rational liberal welfare state, but instead the book ends with an incoherent, anti-egalitarian plea for the program of right-wing Republicans.

Consider the decline of the two-parent family. Illegitimacy and divorce come just after atheism in the right wing catalog of the causes of the moral decay of our nation. According to The Bell Curve, the dissolution of the family is caused by stupidity, and for all I know that may indeed be a cause. According to more liberal sociologists, William Julius Wilson for example, a more important cause of single parenthood among urban African Americans is the absence of good jobs for unskilled workers, and that may be a cause, too, and a more important one. But two-parent families are in decline throughout industrialized nations, and past social experimentation has shown that one of the first things poor families do when given a guaranteed income is to divorce (see Murray 1984 for a review of the data).

A household is a business given over to caring for small, temporarily insane people, a business subject to cash-flow problems, endless legal harassments, run by people who expect to have sex with each other, who occupy the same space, and who go nuts when either party has sex with anyone else. Once in marriage, a lot of people try to get out as fast as religious tradition, poverty, or devotion to children permits.

The evil social effects, if any, of illegitimacy and broken homes are best addressed by finding a social structure that will replace whatever benefits to children two parents are supposed to provide, rather than by forcing people who cannot stand each other to live together. We can make some reasonable guesses about what children require: security, discipline, stimulation, care, affection, ideals. Sounds like what a school ought to provide, which brings us to education.

Having described a nation Balkanized by race, gender, cognitive ability, income, and occupation, a nation whose only unifying forces are public schools and MTV, *The Bell Curve* nevertheless concludes that education cannot much help to solve our social problems, and we should privatize education in the way Milton Friedman incessantly urges. The consequences are predictable: Ku Klux Klan Schools, Aryan Nation Schools, The Nation of Northern Idaho Schools, Farrakhan Schools, Pure Creation Schools, Scientology Schools, David Koresh Junior High, and a thousand more schools of ignorance, separation, and hatred will bloom like some evil garden, subsidized by taxes. If ever there was a plan to make America into Yugoslavia, that is it.

American schools are run on a complicated, decentralized system in which state and then local school boards have enormous powers, which they mostly use badly. Herrnstein and Murray document some of the results – low standards, an expectation of mediocrity – every intellectually anxious parent knows the phenomena first hand. (Murray and Herrnstein do not document some of the more imaginative practices of local schools. In parts of Utah, religion is not required of high school students. They have elective options, for example, either Mormon Theology or Advanced Quantum Field Theory, choose one.)

The obvious solution is not privatization or more decentralization, but national educational standards of accomplishment with uniform, national tests, and national school funding. More bureaucrats the Republicans say, as though we are better off with real estate brokers running school boards and fighting, as Herrnstein and Murray point out, against more homework. I say a public school that worked as well as the post office would be a godsend.

Schools are the natural place to provide security, stability, and stimulation to children of parents who, for whatever reasons, cannot provide them. Our present schools will not do, not schools that are closed 16 hours a day nine months a year and 24 hours a day 3 months a year and always closed to people under five. What would do is schools that are always open to children from one month to 17 years, always welcoming, always safe, offering meals and fun and learning and, if need be, a place to sleep. Those schools are the sane and comparatively economical way to create and sustain a civil society.

Herrnstein and Murray review data that all say the same thing: early, intensive pre-schooling, as in good Head Start programs, on average improves children's performance on IQ tests by half a standard deviation or more, and the first three years of public schools eliminates the benefit. With genuine perversity, determined to make three of two plus two, they blame the early educational interventions, not the public schools they blame everywhere else. Herrnstein and Murray say that even if intensive early intervention and schooling works, it is too expensive, requiring too many adult teachers per toddler. It may be too expensive if teachers are paid at some urban and suburban rates, often the equivalent of about seventy dollars an hour. But real, round the clock, round the calendar, throughout infancy and childhood schooling is affordable if teachers are paid reasonably. And the opportunity costs of failing to do so are greater than the 100 billion or so such a program would cost each year.

Herrnstein and Murray just hate affirmative action programs that fade from broadening searches to secretly imposing quotas, and especially they hate racial, ethnic, and gender quotas in university admissions. I dislike them as well, chiefly for a reason that Herrnstein and Murray are unwilling to take seriously even when their data compel it: compensatory efforts come at the wrong time in life, too late to make a difference to most, and too concerned with credentials to make a lot of difference other than to position and income. (Herrnstein and Murray rightly complain that we are an overcredentialed society, and here, at least, I agree: The Commonwealth of Pennsylvania requires a Master's Degree to open a spot in a shopping mall to guard and entertain children while parents shop, and a school nurse, who is prohibited from doing anything except calling home and giving emergency first aid, requires a special college degree, and where, no kidding, a superintendent of schools is paid extra for a mail order doctoral degree.) Herrnstein and Murray lay no clear blame and propose no clear remedies, but the remedy is obvious. The blame is with universities and college professors, who profit from every legal restriction that requires or rewards formal education.

Disappointed and disapproving, Herrnstein and Murray predict we will become a custodial society in which the rich and competent support the many more who are poor and incompetent. They entertain no alternatives except vagaries about living together unequally with "valued places" for all. But if a society wants another way to be, there are ready alternatives. Here is an alternative vision, one I claim better warranted by the phenomena Herrnstein and Murray report: nationalized, serious, educational standards, tax-supported day and night care, minimal universal health care, a living minimum wage, capital invested in systems that enable almost anyone with reasonable training to do a job well.[13]

A third alternative, urged by Newt Gingrich, Phil Gramm, and company, is to remold this country into a nation in which the state does not promise children safety, or nutrition, or education, and does not guarantee adults a living wage, minimal health services, or security against the hazards of industry, a nation pretty much like, well, *Honduras*?

Notes

1 Beginning with Spearman 1904 and ending (so far as I know) with Jones and Spearman 1950.
2 For a more detailed history, see Glymour *et al.* 1987.
3 Glymour 1980 gives an account of Dalton's simplicity principle and its empirical difficulties
4 It might be said that such a study would be otiose: we know the factor analytic procedures are not reliable. But we do not know how unreliable, and in what circumstances, or whether there are heuristics that can be used with factor analysis to improve reliability.
5 For example, the Markov condition fails for feedback systems, although a condition equivalent to it for acyclic graphs holds for linear cyclic systems representing feedback or "non-recursive" models. See Spirtes 1995.
6 There is an open technical issue here. There are cases in which a covariance matrix generated by a model with correlated error cannot be reproduced by that model but with each correlated error replaced by a distinct latent variable and the latent variables are uncorrelated – the question is whether such matrices can always be reproduced from an appropriate latent variable structure.

7 Herrnstein and Murray give a correlation matrix for their four SES variables. The TETRAD II program will automatically test for vanishing tetrad differences not implied by vanishing partial correlations in the matrix. If there is a single common cause there should be three such differences. There are none.

8 The formulation of the condition is due to Pearl (1988).

9 The discussion here applies results in Spirtes *et al.* 1993.

10 Given in Spirtes *et al.* 1993.

11 The covariance matrix is given in Spirtes *et al.* 1993.

12 See Cooper 1995 and Spirtes, Meek, and Richardson 1995.

13 Although there is a lot of anxiety that computerization increases wage segregation, the evidence is inconclusive; economic returns to computer use are about the same as economic returns to pencil use. Arguably, where computer systems are widespread and well-designed, we should expect the reverse. The computer is a cognitive prosthesis: it enables people without special gifts, given a little training, to do many tasks – accounting, inventory control, arithmetic, scheduling, sending messages, etc. – as well or better than those with special talents. It is the great equalizer, provided there is adequate software and hardware capital, and adequate education.

References

Akleman, Derya G., David A. Bessler, and Diana M. Burton (preprint), "Modeling Corn Experts and Exchange Rates with Directed Graphs", Texas A&M University, Department of Economics, February 1997.

Blau, Peter M. and Otis Dudley Duncan (1967), *The American Occupational Structure*. New York: Wiley.

Cooper, Gregory F. (1995), "Causal Discovery from Data in the Presence of Selection Bias", Preliminary Papers of the Fifth International Workshop on Artificial Intelligence and Statistics, Fort Lauderdale, FL, pp. 140–150.

Druzdzel, Marek J., and Clark Glymour (1994), "Application of the TETRAD II Program to the Study of Student Retention in US Colleges", Technical report, American Association for Artificial Intelligence. Menlo Park, CA: AAAI Press, pp. 419–430.

Fisher, Ronald A. (1958), *The Genetical Theory of Natural Selection*. New York: Dover.

Fraser, Steven (ed.) (1995), *The Bell Curve Wars*. New York: Basic Books.

Glymour, Clark (1980), *Theory and Evidence*. Princeton: Princeton University Press.

Glymour, Clark, Richard Scheines, Peter Spirtes, and Kevin Kelly (1987), *Discovering Causal Structure: Artificial Intelligence, Philosophy of Science, and Statistical Modeling*. Orlando: Academic Press.

Hernstein, Richard J. and Charles Murray (1994), *The Bell Curve: Intelligence and Class Structure in American Life*. New York: Free Press.

Holland, Paul (1986), "Statistics and Causal Inference", *Journal of the American Statistical Association* 81: 945–960.

Jones, Ll. Wynn and Charles Spearman (1950), *Human Ability, A Continuation of the "Abilities of Man"*. London: Macmillan.

Joreskog, Karl and Dag Sorbom (1990), "Model Search with Tetrad II and LISREL", *Sociological Methods and Research* **19**: 93–106.

Junker, Brian W. and Seymour Ellis (preprint), "A Characterization of Monotone Uni-dimensional Latent Variables", Department of Statistics, Carnegie Mellon University, 1995.

Kiiveri, Harry and Terry Speed (1982), "Structural Analysis of Multivariate Data: A Review", in Samuel Leinhardt (ed.), *Sociological Methodology*. San Francisco: Jossey-Boss.

Kohn, Melvin L. (1967), *Class and Conformity; a Study of Values*. Homewood, Ill.: Dorsey Press.

Murray, Charles (1984), *Losing Ground: American Social Policy 1950–1980*. New York: Basic Books.

Pearl, Judea (1988), *Probabilistic Reasoning Systems: Networks of Plausible Inference*. San Mateo: Morgan Kaufman.

Pearl, Judea (1995), "Causal Diagrams for Empirical Research", *Biometrika* **82**: 669–688.

Shipley, Bill (1995), "Structured Interspecific Determinants of Specific Leaf Area in 34 Species of Herbaceous Angiosperms", *Functional Ecology* **9**: 312–319.

—. (1997), "Exploratory Path Analysis with Applications in Ecology and Evolution", *The American Naturalist* **149**: 1113–1138.

Spearman, Charles (1904), "General Intelligence Objectively Determined and Measured", *American Journal of Psychology* **10**: 151–293.

Spirtes, Peter (1995), "Directed Cyclic Graphical Representations of Feedback Models", Proceedings of the 1995 Conference on Uncertainty in Artificial Intelligence. San Mateo: Morgan Kaufman, 491–498.

Spirtes, Peter, Clark Glymour, and Richard Schemes (1993), *Causation, Prediction and Search*. New York: Springer-Verlag.

Spirtes, Peter, Christopher Meek, and Thomas Richardson (1995), "Causal Inference in the Presence of Latent Variables and Selection Bias", Proceedings of the 1995 Conference on Uncertainty and Artificial Intelligence. San Mareo: Morgan Kaufman, 499–506.

Thurstone, Louis L. (1947), *Multiple-factor Analysis; A Development and Expansion of the Vectors of the Mind* Chicago: University of Chicago Press.

PART THREE

Scientific reasoning

INTRODUCTION TO PART THREE

TWO CHARACTERISTICS OF science stand out. First, thanks to science we know a vast array of facts that we would not, and probably could not, know without it. Secondly, we come to know these facts by reasoning based on evidence, often large amounts of detailed and precise evidence. Engaged with her environment principally through direct perception, the solitary individual knows a certain number of things. But the information provided by science tells her many things that she cannot perceive, such as the inner structure of matter, the ancestry of living beings over hundreds and thousands of millennia, the age and extent of the universe and the laws of nature. However, the history of science starts with thinkers no better equipped than our solitary individual. Science has developed instruments for extending our perception, but these alone cannot account for the growth in scientific knowledge. Our ability to reason plays a central role in explaining how science generates new knowledge. This ability has itself developed over time; the evolution of mathematical reasoning in general, and more recently statistical reasoning in particular, have played central roles in the progress of science. Mathematics, statistics and scientific methods are technologies that the solitary individual would need many lifetimes to reconstruct. Language too is important, though easily taken for granted. Science draws upon and augments the conceptual resources of natural languages and continually incorporates new words or gives special and precise meaning to familiar words such as 'force' or 'potential'.

Many philosophers of science hold that the reasoning processes of science are *logically ampliative*, which is to say that they can generate conclusions that cannot be obtained from the evidence by the deductive methods of logic and mathematics alone. The methods of logic and mathematics are now well, if not completely, understood. According to Peter Lipton, to understand the ampliative methods of science, there are two issues that face the philosopher of science: the first is the *description* problem, articulating as precisely as possible the methods that scientists do in fact use; the second is the *justification* problem, explaining why these methods are good ones to use (or explaining why they are not, if they are not good methods).

We should not assume without argument that there is a *single* inferential method employed by science. Nor should we assume that the task of justifying whatever methods there are is a purely philosophical one. Talk of *the* scientific method might give that impression. But of course there are a plethora of methods in science and many of these are generated by, and make no sense without, science. For example, tests such as the litmus test for acidity or the fasting plasma glucose test for diabetes permit very specific ampliative inferences about highly theoretical matters. Hund's rules are even more sophisticated inference rules for determining angular momentum quantum numbers in a multi-electron atom. These tests and rules are the products of scientific knowledge, and justifying their use will refer to that knowledge. However, although there are many specific methods and inferential rules of this sort, there might also be some particularly important and general patterns of inference in science; and these patterns – or a single such inference pattern – might be argued to be fundamental (in some sense to be explained). If there are such fundamental methods it might be a philosophical rather than a solely scientific task to explain the utility of these methods (or of this single method).

To take a particularly important theory of scientific inference of this kind, the positivists – a movement of philosophers and scientists working in the late nineteenth century and first half of the twentieth century – held that the aim of science is to articulate general laws under which we subsume all the particular, specific observations that we make. For example, we may record the position in the night sky of the various planets at various times. The great contributions to science of Claudius Ptolemy and later of Nicolaus Copernicus were to articulate general theories concerning the motions of the planets that not only accounted for the actual observations but also allowed for predictions of observations that would be made in the future. Their theories could be used to tell you the positions of the planets at *any* time, not just at times of the actual observations. About sixty years after the publication of Copernicus's work, Johannes Kepler articulated what are now known as Kepler's laws of planetary motion. These are not only more simple than Copernicus's account of the planets, but also more general, telling us about the shape, speed and period of the motion of *any* planet orbiting the Sun. Eighty years later, Newton generalized yet further, providing a single law of gravitation and three laws of motion that encompassed the gravitational attraction and motions of any bodies whatsoever, whether planets, moons, comets in space or apples and other projectiles on the surface of the Earth. According to the positivists this exemplifies the aim and progress of science. Our observations can be encapsulated in general statements, and science aims to articulate such statements in the most general and precise manner possible.

Generalizing from specific cases to a general conclusion is known as *enumerative induction*. There are philosophical debates over exactly how enumerative induction should be understood. Sometimes enumerative induction is expressed as the inference from 'all observed Fs are Gs' to 'all Fs are Gs', where 'F' and 'G' denote properties of things ('F' might be 'planet' and 'G' might be 'an object that travels in an ellipse'). (An alternative with a weaker conclusion is the inference from 'all observed Fs are Gs' to 'the next F to be observed will be a G', but this is of limited interest in the scientific context.) Hans Reichenbach generalizes this form of induction as follows: if m Fs have been observed and n of them are G, then the proportion of all Fs that are G is n/m. Although Reichenbach (1971, p.45) talks of 'observation' here, it should be noted that important inductions can be made from a premise that doesn't involve observation – after all, surely planets are not *observed* to travel in ellipses but inferred to do so. Perhaps the best way to think of enumerative induction is as inference from properties of a *sample* to the corresponding properties of the whole of the population from which the sample is drawn. In this way, we can link enumerative induction with the broader class of statistical inferences.

Another kind of descriptive worry concerns the range of hypotheses to which any form of enumerative induction may be applied. As Nelson Goodman showed (see Lipton's discussion in Chapter 12) not just any F and G will do in the above inference patterns – otherwise any facts will do as premises in an induction supporting any conclusion consistent with the premises. What then constrains the permissible F and G? A final point concerns a common way of conceiving of induction, as saying that the future will resemble the past (in certain respects). But this cannot be all there is to induction, since the conclusions we draw might also be about the past. For example, the planetary models discussed allow us to know not just where the planets will be in the future but where they were in the past (even when no observations were being made of them). Indeed, the interesting thing about Newton's laws is not just that they tell us about past, present and future observations, but that they also tell us about entities we never actually or even could never actually observe (because they are in the very remote past or future, or because they are in the present but very distant, and so forth).

Peter Lipton suggests 'More of the Same' as the slogan for induction. If that is right then there must be more to the processes of reasoning in science according to the scientific realist. That is because many important scientific discoveries are not discoveries of laws or general truths; they concern things that are quite *unlike* things we already know about. So those discoveries cannot be accounted for by an inference pattern that amounts to 'More of the Same'. For example, many discoveries are the discoveries of new entities, processes and causes. When Koch discovered that cholera was caused by a bacterium, later named *Vibrio cholerae*, this was not achieved by applying any principle of enumerative induction. Rather, Koch showed that a certain bacillus, identified under the microscope, was always present in cases of cholera, and never in healthy animals. This bacillus could be isolated and cultured, and when a susceptible animal is inoculated with the cultured bacillus, it develops cholera. Furthermore, the very same bacillus can be retrieved from the sick, inoculated animal. This, argued Koch, establishes a causal relationship between the bacillus and the disease.

While some enumerative induction may be part of Koch's argument – for example the claim that *Vibrio cholerae* is present in *every* case of cholera is the conclusion of an enumerative induction – the important move is from such conclusions to the causal claim. Why does the argument establish causation? Consider the alternatives. Perhaps the presence of the bacillus in the sick animals is a coincidence. But that's a very poor explanation, given the large number of cases studied. In particular, if it is a coincidence, one would expect to find the bacillus in healthy animals. But one never does. Perhaps the bacillus is not the cause of the disease but is a side effect – this would explain why it is never found in healthy animals. However, if it were merely a side-effect, then inoculation with the pure bacillus should be harmless. But susceptible animals that are inoculated with cultured *Vibrio cholerae* do always develop the disease, which the side-effect hypothesis cannot explain.

How should we understand this case? John Stuart Mill articulated his view of induction as central to the production of scientific knowledge. In addition he drew up methods for inferring causal relations between known quantities (Mill's method of difference can be seen as the basic principle behind the controlled clinical trial). But neither seems especially helpful as regards this case. Mill's contemporary, William Whewell, took a rather different view. Whewell emphasized the importance of imaginative hypotheses that would explain phenomena by bringing them under a novel concept or 'conception'. The latter he called 'colligation'. A hypothesis gains credibility when it does two things: it makes novel predictions that are confirmed, and it explains a diversity of phenomena, including phenomena it was not intended to explain. (Whewell also thinks that a good theory will become simpler and more coherent over time.) For example, in Pasteur's hand, the germ theory of disease made novel predictions that were confirmed by experiment and explained otherwise unconnected phenomena, such as putrefaction and the souring of beer, wine and milk. The rise of the germ theory came only at the end of Whewell's life; in his own lifetime the salient scientific debate concerned the revived wave theory of light developed by Fresnel and Young in opposition to Newton's corpuscular theory. Peter Achinstein dissects the structure of the arguments for the wave theory, relating them to the Whewell–Mill debate.

When Whewell emphasizes the diversity of the phenomena being explained by a theory and the importance of simplicity and coherence, he contrasts this with theories that explain only a single phenomenon and whose articulation requires implausible additional ('auxiliary') hypotheses that have no independent confirmation or motivation. One might say that Whewell is spelling out the characteristics of a good explanation in contrast to those of a poor explanation. By these criteria we should prefer good explanations to poor explanations. We should believe Koch's hypothesis because it is able to explain the evidence more effectively than obvious rival hypotheses. This is an example of *inference to the best explanation*.

The key idea of inference to the best explanation (IBE; also known as *abduction*) is straightforward: we have evidence E; various *potential* explanations could explain E; hypothesis *h* is the *best potential* explanation of E; we infer that *h* is the *actual* explanation of E; since *h* is the actual explanation of E, the claims made in *h* are *true*. Spelling out the detail of this idea requires further work. What makes an explanation a *good* explanation? How much better than its rivals does an explanation have to be for us to infer that it is the actual explanation? Does it need to meet a threshold of goodness?

As this example shows, unlike enumerative induction, IBE allows us to have knowledge of processes and entities that are not directly observable. According to the positivists, we cannot have knowledge of the unobservable, and so they reject IBE. All we need is enumerative induction. Some philosophers argue the opposite, that enumerative induction is a special case of IBE. That is because, they argue, we can only legitimately infer the truth of a generalization from a set of instances if something, such as a law of nature or a causal relationship, *explains* the co-occurrence of those instances.

Enumerative induction and inference to the best explanation are two leading answers to the description problem of scientific inference (we will come to a third below). What of the justification problem? Should we use these patterns, and if so why? Can we show that with appropriate evidence, using these inference patterns will lead to justified beliefs or even knowledge? IBE, for example, faces specific questions in relation to justification. Why should the goodness of an explanation be a guide to its truth? Can we be confident that we have thought of the actual explanation among those we consider to be potential explanations? In addition to such specific issues is a very general one, raised by Hume. Since the inference patterns are ampliative, we cannot use deductive logic to show how they justify their conclusions. Their effective- ness will therefore depend on the world being a particular way. How do we know that the world is indeed that way? We cannot use logic to show that the world is that way – if we could, then the inference pattern would not be ampliative after all. Since we need to show that the world is a certain way, but cannot use logic, we must use some ampliative inference procedure. What that shows is that in order to justify an amplia- tive inference procedure we have to *use* an ampliative inference procedure, and that looks to be circular or to lead to an infinite regress. For example, take enumerative induction. To infer a general truth on the basis of a limited number of instances is to assume that nature tends to be uniform in certain respects. But how should we estab- lish that nature is uniform – itself a general proposition – if not by enumerative induc- tion? Suppose we regard enumerative induction as justifying its conclusions because it has a good track record of success; note that to infer from its track record of success to its success in future applications is to argue inductively. So we see that these attempts to justify induction themselves depend on the use of induction.

Attempts to solve, dissolve or reject the problem of induction are a major part of the history of philosophy since David Hume. Some philosophers think that the circu- larity Hume identifies need not arise. Hume says that we would need to justify the claim that nature is uniform in order to use enumerative induction to get to the truth. But these philosophers deny this: so long as it is in fact the case that nature is uniform, then induction will be reliable and that is enough for knowledge and justification.

Other philosophers accept that the problem does indeed show that an ampliative inference does not justify its conclusions. Conclusions drawn using enumerative induc- tion cannot be justified; likewise any beliefs arrived at by IBE are also unjustified. Let us imagine that the descriptive problem of induction has been answered as above: scientists often use enumerative induction and inference as the best explanation. Then, this acceptance of the lesson of Hume's problem has a dramatic consequence for our view of science: the claims of scientists are typically unjustified; science does not lead to knowledge. This is a strongly sceptical conclusion. Some might embrace such scepticism, but since science seems to be rational if anything is, it would be a result one should adopt only as a last resort. If one accepts Hume's argument but regards science as rational, then one is forced to deny that the answers to the description problem made above are correct. Both *falsificationism* and *Bayesianism* do indeed deny this – they both reject the idea that science uses either enumerative induction or IBE.

According to Karl Popper, as we saw in Part One, it is just false that scientists use anything like induction. Rather they make bold conjectures (new theories and hypotheses) and then test these conjectures rigorously. If a conjecture is refuted by a test, then the scientist must exercise her imagination to come up with a new conjecture. If the conjecture survives a battery of tests we should not conclude that the conjecture is true – to do that would be to engage in induction. All we can do is to continue to subject the conjecture to further tests. According to Popper, the relationship between a conjecture and a falsifying test outcome is a purely logical one: if the conjecture is that any susceptible animal inoculated with *Vibrio cholerae* will develop cholera, then a case of an inoculated, susceptible animal that does not develop cholera is one that refutes the hypothesis as a matter of deductive logic. So, according to Popper, the reasoning employed by scientists is solely deductive and not ampliative. Falsificationism has been popular among scientists, not least because it portrays them as creative (generating conjectures is an imaginative process) and honestly self-critical (they set out to refute their own conjectures). On the other hand, falsificationism denies that we are *ever* positively entitled to believe any of our conjectures, however well they have succeeded in the process of rigorous testing. While scientists may hold that their theories are always open to revision, they nonetheless do seem to believe many important scientific claims (that the Sun's energy comes from fusion reactions, that blood circulates pumped by the heart, that humans and chimpanzees have a common ancestor, etc.). Indeed, Thomas Kuhn argues, in his very different description of science, that science in its normal state must depend upon certain ideas (theories, instruments, procedures) as being fully accepted and not up for debate. As Hilary Putnam argues, this has to be the case for the very process of rigorous testing to take place. Without certain assumptions, there is no possibility of falsifying a conjecture.

The subject of the reading by Michael Strevens is Bayesianism, which in its standard, subjectivist form, takes a very different approach. It denies that scientists engage in any process of inference such that the outcome is a belief in some theory or proposition. What scientists do in the face of new evidence is that they change their *credences*. A credence is a subjective degree of belief – how strongly you are committed to a certain proposition. You might have a high credence that the Sun will rise tomorrow and a low credence that it will snow tomorrow. On hearing that there is a large comet heading towards the Earth or on hearing that there is a cold front approaching, you might revise your credences down or up accordingly. How you make this revision is governed by the combination of two claims. The first is Bayes's theorem, which relates your credence in the hypothesis, given the evidence $P(h|e)$ to your antecedent (or *prior*) credence in the hypothesis independently of the evidence $P(h)$, your prior credence in the truth of the evidence $P(e)$, and your prior credence on the truth of the evidence given that the hypothesis is true $P(e|h)$:

$$P(h|e) = \frac{P(e|h)P(h)}{P(e)}$$

Bayes's theorem tells us that if you think that the occurrence of some piece of evidence is inherently unlikely, but if the hypothesis were true, it would be very likely, then your credence in the hypotheses given the evidence should be rather greater than your credence in the hypothesis independent of the evidence. Let's say that, in accordance with Bayes's theorem, your credences are related in just that way. Now we find out that e has indeed occurred. The Bayesian says that you should now change ('update') your credence in h to be equal to $P(h|e)$. This is known as *conditionalization*:

$$P_{new}(h) = P_{old}(h|e)$$

where 'old' and 'new' refer to your credences before and after learning that e has occurred.

The first thing to note about Bayesian conditionalization is that a scientist's new credence in h, $P_{new}(h)$, is determined by her old credences $P_{old}(e)$, $P_{old}(h)$ and $P_{old}(e|h)$. That means that two scientists faced with the same evidence might differ significantly in their assessment of h, because of their prior subjective credences. For the same reason Bayesian conditionalization does not produce an objective probability, but an updated credence. These features may be considered advantages. After all, descriptively, it is true that scientists do disagree about which hypotheses one should believe in the light of the evidence. Perhaps this is explained by their having different prior credences. It also means that conditionalization does not suffer from Hume's problem: since it does not deliver an objective probability there is no issue of whether it delivers the objectively-correct probability. On the other hand, these features might be considered weaknesses. Should not a good account of scientific methodology tell us what one ought to believe (and how strongly one should believe it) in the light of the evidence in a manner that correlates with what is actually true and not just with what we already believe? Bayesian confirmation theory is very popular among many philosophers of science because it seems to provide a way of understanding confirmation that is plausible, quantitative, and justifiable a priori. Its supporters also say that it solves a number of important problems in the philosophy of science. It is, however, highly contentious and cannot help with some key problems such as the problem(s) of induction as classically understood.

Further reading

Bird, A. 2000 *The Philosophy of Science*. London: UCL Press. See especially Part II, pp. 165–287.

Glymour, C. 1981 'Why I am not a Bayesian' in his *Theory and Evidence*. Chicago University Press, pp 63–93. Reprinted in Papineau, D. (ed.) *The Philosophy of Science*. Oxford University Press, 1996, pp 290–313.

Goodman, N. 1983 *Fact, Fiction, and Forecast*. Harvard University Press.

Hempel, C. 1965 'Studies in the Logic of Confirmation' in *Aspects of Scientific Explanation: and other Essays in the Philosophy of Science*. New York: Free Press.

Howson, C. and Urbach, P. 2006 *Scientific Reasoning: the Bayesian Approach*. Open Court.

Reichenbach, H. 1971 *The Theory of Probability*. Berkeley: University of California Press. Translation by Ernest R. Hutton and Maria Reichenbach of Wahrscheinlichkeitslehre. Eine Untersuchung uber die logischen und mathematischen Grundlagen der Wahrscheinlichskeitrechnung, Leiden, 1935.

Chapter 12

Peter Lipton

INDUCTION

Underdetermination

Inductive inference is a matter of weighing evidence and judging likelihood, not of proof. How do we go about making these non-demonstrative judgments, and why should we believe they are reliable? Both the question of description and the question of justification arise from underdetermination. To say that an outcome is underdetermined is to say that some information about initial conditions and rules or principles does not guarantee a unique solution. The information that Tom spent five dollars on apples and oranges and that apples are fifty cents a pound and oranges a dollar a pound underdetermines how much fruit Tom bought, given only the rules of deduction. Similarly, those rules and a finite number of points on a curve underdetermine the curve, since there are many curves that would pass through those points.

Underdetermination may also arise in our description of the way a person learns or makes inferences. A description of the evidence, along with a certain set of rules, not necessarily just those of deduction, may underdetermine what is learned or inferred. Insofar as we have described all the evidence and the person is not behaving erratically, this shows that there are hidden rules. We can then study the patterns of learning or inference to try to discover them.

Noam Chomsky's argument from 'the poverty of the stimulus' is a good example of how under-determination can be used to disclose the existence of additional rules (1965, ch. 1, sec. 8, esp. pp. 58–9). Children learn the language of their elders, an ability that enables them to understand an indefinite number of sentences on first acquaintance. The talk young children hear, however, along with rules of deduction and any plausible general rules of induction, grossly underdetermine the language they learn. What they hear is limited and often includes many ungrammatical sentences, and the little they hear that is well formed is compatible with many possible languages other than the one they learn. Therefore, Chomsky argues, in addition to any general principles of deduction and induction, children must be born with strong linguistic rules or principles that further restrict the class of languages they will learn, so that the actual words they hear are now sufficient to determine a unique language. Moreover, since a child will learn whatever language he is brought up in, these principles cannot be peculiar to a particular human language; instead, they must specify something that is common to all of them. For Chomsky, determining the structure of these universal principles and the way they work is the central task of modern linguistics.

Thomas Kuhn provides another well-known example of using underdetermination as a tool to investigate cognitive principles. He begins from an argument about scientific research strikingly similar to Chomsky's argument about language acquisition (1970; 1977, esp. ch. 12). In most periods in the history of a developed scientific specialty, scientists are in broad agreement about which problems to work on, how to attack them, and what counts as solving them. But the explicit beliefs and rules scientists share, especially their theories, data, general rules of deduction and induction, and any explicit methodological rules, underdetermine these shared judgments. Many possible judgments are compatible with these beliefs and rules other than the ones the scientists make. So Kuhn argues that there must be additional field-specific principles that determine the actual judgments. Unlike Chomsky, Kuhn does not argue for principles that are either innate or in the form of rules, narrowly construed. Instead, scientists acquire through their education a stock of exemplars – concrete problem solutions in their specialty – and use them to guide their research. They pick new problems that look similar to an exemplar problem, they try techniques that are similar to those that worked in that exemplar, and they assess their success by reference to the standards of solution that the exemplars illustrate. Thus the exemplars set up a web of 'perceived similarity relations' that guide future research, and the shared judgments are explained by the shared exemplars. These similarities are not created or governed by rules, but they result in a pattern of research that roughly mimics one that is rule-governed. Just how exemplars do this work, and what happens when they stop working, provide the focus of Kuhn's account of science.

Chomsky and Kuhn are both arguing for unacknowledged principles of induction. In these cases, however, underdetermination is taken to be a symptom of the existence of highly specialized principles, whether of language acquisition or of scientific method in a particular field, since the underdetermination is claimed to remain even if we include general principles of induction among our rules. But the same pattern of argument applies to the general case. If an inference is inductive, then by definition it is underdetermined by the evidence and the rules of deduction. Insofar as our inductive practices are methodical, we must use additional rules or principles of inference, and we may study the patterns of our inferences in an attempt to discover them.

Justification

The two central questions about our general principles of induction concern description and justification. What principles do we use, and how can they be shown to be good principles? The question of description seems at first to take priority. How can we even attempt to justify our principles until we know what they are? Historically, however, the justification question came first. One reason for this is that this question gets its grip from skeptical arguments that seem to apply to any principles that could account for the way we fill the gap between evidence and inference.

The problem of justification is to show that our inferential methods are good methods. The natural way to understand this is in terms of truth. We want our methods of inference to be 'truth-tropic', to take us towards the truth. For deduction, a good argument is one that is valid, a perfect truth conduit, where it is impossible for there to be true premises and a false conclusion. The problem of justification here would be to show that arguments we judge valid are in fact so. For induction, perfect reliability is out of the question. By definition, even a good inductive argument is one where it is possible for there to be true premises but a false conclusion. Moreover, it is clear that the reasonable inductive inferences we make are not 100 per cent reliable even in this world, since they sometimes sadly take us from truth to falsehood. Nevertheless, it remains natural to construe the task of justification as that of showing truth-tropism. We would like to show that those inductive inferences we judge worth making are ones that usually take us from true premises to true conclusions.

A skeptical argument that makes the problem of justification pressing has two components, underdetermination and circularity. The first is an argument that the inferences in question are underdetermined, given only our premises and the rules of deduction; that the premises and those rules are compatible not just with the inferences we make, but also with other, incompatible inferences. This shows that the inferences we make really are inductive and, by showing that there are possible worlds where the principles we use take us from true premises to false conclusions, it also shows that there are worlds where our principles would fail us. Revealing this underdetermination, however, does not yet generate a skeptical argument, since we might have good reason to believe that the actual world is one where our principles are reliable. So the skeptical argument requires a second component, an argument for circularity, which attempts to show that we cannot rule out the possibility of unreliability that underdetermination raises without employing the very principles that are under investigation, and so begging the question.

Although it is not traditionally seen as raising the problem of induction, Descartes' 'First Meditation' is a classic illustration of this technique. Descartes' goal is to cast doubt on the 'testimony of the senses', which leads us to infer that there is, say, a mountain in the distance because that is what it looks like. He begins by arguing that we ought not to trust the senses completely, since we know that they do sometimes mislead us, 'when it is a question of very small and distant things'. This argument relies on the underdetermination component, since it capitalizes on the fact that the way things appear does not entail the way they are, but it does not yet have the circle component. We can corroborate our inferences about small and distant things without circularity by taking a closer look (cf. Williams, 1978, pp. 51–2). But Descartes immediately moves on from the small and the distant to the large and near. No matter how clearly we seem to see something, it may only be a dream, or a misleading experience induced by an evil demon. These arguments describe possible situations where even the most obvious testimony of the senses is misleading. Moreover unlike the worry about small and distant things, these arguments also have a circle component. There is apparently no way to test whether a demon is misleading us with a particular experience since any test would itself rely on experiences that he might have induced. The senses may be liars and give us false testimony, and we should not find any comfort if they also report that they are telling us the truth.

The demon argument begins with the underdetermination of observational belief by observational experience, construes the missing principle of inference on the model of inference from testimony, and then suggests that the reliability of this principle could only be shown by assuming it. Perhaps one of the reasons Descartes' arguments are not traditionally seen as raising the problem of justifying induction is that his response to his own skepticism is to reject the underdetermination upon which it rests. Descartes argues that, since inferences from the senses must be inductive and so raise a skeptical problem, our knowledge must instead have a different sort of foundation for which the problem of underdetermination does not arise.

The *cogito* and the principles of clearness and distinctness that it exemplifies are supposed to provide the non-inductive alternative. In other words, the moral he draws from underdetermination and circularity is not that our principles of induction require some different sort of defense or must be accepted without justification, but that we must use different premises and principles, for which the skeptical problem does not arise. For a skeptical argument about induction that does not lead to the rejection of induction, we must turn to its traditional home, in the arguments of David Hume.

Hume also begins with underdetermination, in this case that our observations do not entail our predictions (1748, sec. IV). He then suggests that the governing principle of all our inductive inferences is that nature is uniform, that the unobserved (but observable) world is much like what we have observed. The question of justification is then the question of showing that nature is indeed uniform. This cannot be deduced from what we have observed, since the claim of uniformity itself incorporates a massive prediction. But the only other way to argue for uniformity is to use an inductive argument, which would rely on the principle of uniformity, leaving the question begged. According to Hume, we are addicted to the practice of induction, but it is a practice that cannot be justified.

To illustrate the problem, suppose our fundamental principle of inductive inference is 'More of the Same'. We believe that strong inductive arguments are those whose conclusions predict the continuation of a pattern described in the premises. Applying this principle of conservative induction, we would infer that the sun will rise tomorrow, since it has always risen in the past; and we would judge worthless the argument that the sun will not rise tomorrow since it has always risen in the past. It is, however, easy to come up with a factitious principle to underwrite the latter argument. According to the

principle of revolutionary induction, 'It's Time for a Change', and this sanctions the dark inference. Hume's argument is that we have no way to show that conservative induction, the principle he claims we actually use for our inferences, will do any better than intuitively wild principles like the principle of revolutionary induction. Of course conservative induction has had the more impressive track record. Most of the inferences from true premises that it has sanctioned have also had true conclusions. Revolutionary induction, by contrast, has been conspicuous in failure, or would have been, had anyone relied on it. The question of justification, however, does not ask which method of inference has been successful; it asks which one will be successful.

Still, the track record of conservative induction appears to be a reason to trust it. That record is imperfect (we are not aspiring to deduction), but very impressive, particularly as compared with revolutionary induction and its ilk. In short, induction will work because it has worked. This seems the only justification our inductive ways could ever have or require. Hume's disturbing observation was that this justification appears circular, no better than trying to convince someone that you are honest by saying that you are. Much as Descartes argued that we should not be moved if the senses give testimony on their own behalf, so Hume argued that we cannot appeal to the history of induction to certify induction. The trouble is that the argument that conservative inductions will work because they have worked is itself an induction. The past success is not supposed to prove future success, only make it very likely. But then we must decide which standards to use to evaluate this argument. It has the form 'More of the Same', so conservatives will give it high marks, but since its conclusion is just to underwrite conservatism, this begs the question. If we apply the revolutionary principle, it counts as a very weak argument. Worse still, by revolutionary standards, conservative induction is likely to fail precisely because it has succeeded in the past, and the past failures of revolutionary induction augur well for its future success (cf. Skyrms, 1986, ch. 2). The justification of revolutionary induction seems no worse than the justification of conservative induction, which is to say that the justification of conservative induction looks very bad indeed.

The problem of justifying induction does not show that there are other inductive standards better than our own. Instead it argues for a deep symmetry: many sets of standards, most of them wildly different from our own and incompatible with each other, are yet completely on a par from a justificatory point of view. This is why the problem of justification can be posed before we have solved the problem of description. Whatever inductive principles we use, the fact that they are inductive seems enough for the skeptic to show that they defy justification. We fill the gap of underdetermination between observation and prediction in one way, but it could be filled in many other ways that would have led to entirely different predictions. We have no way of showing that our way is any better than any of these others. It is not merely that the revolutionaries will not be convinced by the justificatory arguments of the conservatives: the conservatives should not accept their own defense either, since among their standards is one which says that a circular argument is a bad argument, even if it is in one's own aid. Even if I am honest, I ought to admit that the fact that I say so ought not carry any weight. We have a psychological compulsion to favor our own inductive principles but, if Hume is right, we should see that we cannot even provide a cogent rationalization of our behavior.

It seems to me that we do not yet have a satisfying solution to Hume's challenge and that the prospects for one are bleak, but there are other problems of justification that are more tractable. The peculiar difficulty of meeting Hume's skeptical argument against induction is that he casts doubt on our inductive principles as a whole, and so any recourse to induction to justify induction seems hopeless. But one can also ask for the justification of particular inductive principles and, as Descartes' example of small and distant things suggests, this leaves open the possibility of appeal to other principles without begging the question. For example, among our principles of inference is one that makes us more likely to infer a theory if it is supported by a variety of evidence than if it is supported by a similar amount of homogeneous data. This is the sort of principle that might be justified in terms of a more basic inductive principle, say that we have better reason to infer a theory when all the reasonable competitors have been refuted, or that a theory is only worth inferring when each of its major components has been separately tested. Another, more controversial, example of a special principle that might be justified without circularity is that, all else being equal, a theory deserves more credit from its successful predictions than it does from data that the theory was constructed to fit. This appears to be an inductive preference most of us have, but the case is controversial because it is not at all obvious that it is rational. On the one hand, many people feel that only a prediction can be a real test, since a theory cannot possibly be refuted by data it is built to accommodate; on the other, that logical relations between theory and data upon which inductive support exclusively depends cannot be affected by the merely historical fact that the data were available before or only after the theory was proposed. In any event, this is an issue of inductive principle that is susceptible to non-circular evaluation, as we will see in chapter eight [sic]. Finally, though a really satisfying solution to Hume's problem would have to be an argument for the reliability of our principles that had force against the inductive skeptic, there may be arguments for reliability that do not meet this condition yet still have probative value for those of us who already accept some forms of induction. We will consider some candidates in chapter nine [sic].

Description

We can now see why the problem of justification, the problem of showing that our inductive principles are reliable, did not have to wait for a detailed description of those principles. The problem of justifying our principles gets its bite from skeptical arguments, and these appear to depend only on the fact that these principles are principles of induction, not on the particular form they take. The crucial argument is that the only way to justify our principles would be to reason with the very same principles, which is illegitimate; an argument that seems to work whatever the details of our inferences. The irrelevance of the details comes out in the symmetry of Hume's argument: just as the future success of conservative induction gains no plausibility from its past success, so the future success of revolutionary induction gains nothing from its past failures. As the practice varies, so does the justificatory argument, preserving the pernicious circularity. Thus the question of justification has had a life of its own: it has not waited for a detailed description of the practice whose warrant it calls into doubt.

By the same token, the question of description has fortunately not waited for an answer to the skeptical arguments. Even if our inferences were unjustifiable, one still might be interested in saying how they work. The problem of description is not to show that our inferential practices are reliable; it is simply to describe them as they stand. One might have thought that this would be a relatively trivial problem. First of all, there are no powerful reasons for thinking

that the problem is impossible, as there are for the problem of justification. There is no great skeptical argument against the possibility of description. It is true that any account of our principles will require inductive support, since we must see whether it jibes with our observed inductive practice. This, however, raises no general problem of circularity now that justification is not the issue. Using induction to investigate induction is no more a problem here than using observation to study the structure and function of the eye. Second, it is not just that a solution to the problem of describing our inductive principles should be possible, but that it should be easy. After all, they are our principles, and we use them constantly. It thus comes as something of a shock to discover how extraordinarily difficult the problem of description has turned out to be. It is not merely that ordinary reasoners are unable to describe what they are doing: at least one hundred and fifty years of focused effort by epistemologists and philosophers of science has yielded little better. Again, it is not merely that we have yet to capture all the details, but that the most popular accounts of the gross structure of induction are wildly at variance with our actual practice.

Why is description so hard? One reason is a quite general gap between what we can do and what we can describe. You may know how to do something without knowing how you do it; indeed, this is the usual situation. It is one thing to know how to tie one's shoes or to ride a bike; it is quite another thing to be able to give a principled description of what it is that one knows. Chomsky's work on principles of language acquisition and Kuhn's work on scientific method are good cognitive examples. Their investigations would not be so important and controversial if the ordinary speaker of a language knew how she distinguished grammatical from ungrammatical sentences or the normal scientist knew how he made his methodological judgments. The speaker and the scientist employ various principles, but they are not conscious of them. The situation is similar in the case of inductive inference generally. Although we may partially articulate some of our inferences if, for example, we are called upon to defend them, we are not conscious of the fundamental principles of inductive inference we constantly use.

Since our principles of induction are neither available to introspection, nor observable in any other way, the evidence for their structure must be highly indirect. The project of description is one of black box inference, where we try to reconstruct the underlying mechanism on the basis of the superficial patterns of evidence and inference we observe in ourselves. This is no trivial problem. Part of the difficulty is simply the fact of under-determination. As the examples of Chomsky and Kuhn show, underdetermination can be a symptom of missing principles and a clue to their nature, but it is one that does not itself determine a unique answer. In other words, where the evidence and the rules of deduction underdetermine inference, that information also underdetermines the missing principles. There will always be many different possible mechanisms that would produce the same patterns, so how can one decide which one is actually operating? In practice, however, we usually have the opposite problem: we can not come up with even one description that would yield the patterns we observe. The situation is the same in scientific theorizing generally. There is always more than one account of the unobserved and often unobservable world that would account for what we observe, but scientists' actual difficulty is often to come up with even one theory that fits the observed facts. On reflection, then, it should not surprise us that the problem of description has turned out to be so difficult. Why should we suppose that the project of describing our inductive

principles is going to be easier than it would be, say, to give a detailed account of the working of a computer on the basis of the correlations between keys pressed and images on the screen?

Now that we are prepared for the worst, we may turn to some of the popular attempts at description. In my discussion of the problem of justification, I suggested that, following Hume's idea of induction as habit formation, we describe our pattern of inference as 'More of the Same'. This is pleasingly simple, but the conservative principle is at best a caricature of our actual practice. We sometimes do not infer that things will remain the same and we sometimes infer that things are going to change. When my mechanic tells me that my brakes are about to fail, I do not suppose that he is therefore a revolutionary inductivist. Again, we often make inductive inferences from something we observe to something invisible, such as from people's behavior to their beliefs or from the scientific evidence to unobservable entities and processes, and this does not fit into the conservative mold. 'More of the Same' might enable me to predict what you will do on the basis of what you have done (if you are a creature of habit), but it will not tell me what you are or will be thinking.

Faced with the difficulty of providing a general description, a reasonable strategy is to begin by trying to describe one part of our inductive practice. This is a risky procedure, since the part one picks may not really be describable in isolation, but there are sometimes reasons to believe that a particular part is independent enough to permit a useful separation. Chomsky must believe this about our principles of linguistic inference. Similarly, one might plausibly hold that, while simple habit formation cannot be the whole of our inductive practice, it is a core mechanism that can be treated in isolation. Thus one might try to salvage the intuition behind the conservative principle by giving a more precise account of the cases where we are willing to project a pattern into the future, leaving to one side the apparently more difficult problems of accounting for predictions of change and inferences to the unobservable. What we may call the instantial model of inductive confirmation may be seen in this spirit. According to it, a hypothesis of the form 'All A's are B' is supported by its positive instances, by observed A's that are also B (cf. Hempel, 1965, ch. 1). This is not, strictly speaking, an account of inductive *inference*, since it does not say either how we come up with the hypothesis in the first place or how many supporting instances are required before we actually infer it, but this switching of the problem from inference to support may also be taken as a strategic simplification. In any event, the underlying idea is that, if enough positive instances and no refuting instances (A's that are not B) are observed, we will infer the hypothesis, from which we may then deduce the prediction that the next A we observe will be B.

This model could only be a very partial description of our inductive principles but, within its restricted range, it strikes many people initially as a truism, and one that captures Hume's point about our propensity to extend observed patterns. Observed positive instances are not necessary for inductive support, as inferences to the unobserved and to change show, but they might seem at least sufficient. The instantial model, however, has been shown to be wildly over-permissive. Some hypotheses are supported by their positive instances, but many are not. Observing only black ravens may lead one to believe that all ravens are black, but observing only bearded philosophers would probably not lead one to infer that all philosophers are bearded. Nelson Goodman has generalized this problem, by showing how the instantial model sanctions any prediction at all if there is no restriction on the hypotheses to which it can be applied (Goodman, 1983,

ch. 3). His technique is to construct hypotheses with factitious predicates. Black ravens provide no reason to believe that the next swan we see will be white, but they do provide positive instances of the artificial hypothesis that 'All raveswans are blight', where something is a raveswan just in case it is either observed before today and a raven, or not so observed and a swan, and where something is blight just in case it is either observed before today and black, or not so observed and white. But the hypothesis that all raveswans are blight entails that the next observed raveswan will be blight which, given the definitions, is just to say that the next swan will be white.

The other famous difficulty facing the instantial model arises for hypotheses that do seem to be supported by their instances. Black ravens support the hypothesis that all ravens are black. This hypothesis is logically equivalent to the contrapositive hypothesis that all non-black things are non-ravens: there is no possible situation where one hypothesis would be true but the other false. According to the instantial model, the contrapositive hypothesis is supported by non-black non-ravens, such as green leaves. The rub comes with the observation that whatever supports a hypothesis also supports anything logically equivalent to it. This is very plausible, since support provides a reason to believe true, and we know that if a hypothesis is true, then so must be anything logically equivalent to it. But then the instantial model once again makes inductive support far too easy, counting green leaves as evidence that all ravens are black (Hempel, 1965, ch. 1). We will discuss this paradox of the ravens in chapter six [*sic*].

Another account of inductive support is the hypothetico-deductive model (Hempel, 1966, chs 2, 3). On this view, a hypothesis or theory is supported when it, along with various other statements, deductively entails a datum. Thus a theory is supported by its successful predictions.

This account has a number of attractions. First, although it leaves to one side the important question of the source of hypotheses, it has much wider scope than the instantial model, since it allows for the support of hypotheses that appeal to unobservable entities and processes. The big bang theory of the origin of the universe obviously cannot be directly supported but, along with other statements, it entails that we ought to find ourselves today traveling through a uniform background radiation, like the ripples left by a rock falling into a pond. The fact that we do now observe this radiation (or effects of it) provides some reason to believe the big bang theory. Thus, even if a hypothesis cannot be supported by its instances, because its instances are not observable, it can be supported by its observable logical consequences. Second, the model enables us to co-opt our accounts of deduction for an account of induction, an attractive possibility since our understanding of deductive principles is so much better than our understanding of inductive principles. Lastly, the hypothetico-deductive model seems genuinely to reflect scientific practice, which is perhaps why it has become the scientists' philosophy of science.

In spite of all its attractions, our criticism of the hypothetico-deductive model here can be brief, since it inherits all the over-permissiveness of the instantial model. Any case of support by positive instances will also be a case of support by consequences. The hypothesis that all A's are B, along with the premise that an individual is A, entails that it will also be B, so the thing observed to be B supports the hypothesis, according to the hypothetico-deductive model. That is, any case of instantial support is also a case of hypothetico-deductive support, so the model has to face the problem of insupportable hypotheses and the raven paradox. Moreover, the hypothetico-deductive model is similarly over-permissive in the case of vertical inferences to hypotheses about unobservables, a problem that the instantial model avoided by ignoring such inferences altogether. The difficulty is structurally similar to Goodman's problem of factitious predicates. Consider the conjunction of the hypotheses that all ravens are black and that all swans are white. This conjunction, along with premises concerning the identity of various ravens, entails that they will be black. According to the model, the conjunction is supported by black ravens, and it entails its own conjunct about swans. The model thus appears to sanction the inference from black ravens to white swans (cf. Goodman, 1983, pp. 67–8). Similarly, the hypothesis that all swans are white taken alone entails the inclusive disjunction that either all swans are white or there is a black raven, a disjunction we could establish by seeing a black raven, thus giving illicit hypothetico-deductive support to the swan hypothesis. These maneuvers are obviously artificial, but nobody has managed to show how the model can be modified to avoid them without also eliminating most genuine cases of inductive support (cf. Glymour, 1980, ch. 2). Finally, in addition to being too permissive, finding support where none exists, the model is also too strict, since data may support a hypothesis which does not, along with reasonable auxiliary premises, entail them. We will investigate this problem in chapter five and return to the problem of over-permissiveness in chapter six [sic].

We have now canvassed three attempts to tackle the descriptive problem: 'More of the Same', the instantial model, and the hypothetico-deductive model. The first two could, at best, account for a special class of particularly simple inferences, and all three are massively over-permissive, finding inductive support where there is none to be had. They do not give enough structure to the black box of our inductive principles to determine the inferences we actually make. This is not to say that these accounts describe mechanisms that would yield too many inferences: they would probably yield too few. A 'hypothetico-deductive box', for example, would probably have little or no inferential output, given the plausible additional principle that we will not make inferences we know to be contradictory. For every hypothesis that we would be inclined to infer on the basis of the deductive support it enjoys, there will be an incompatible hypothesis that is similarly supported, and the result is no inference at all, so long as both hypotheses are considered and their incompatibility recognized.

A fourth account of induction, the last I will consider in this section, focuses on causal inference. It is a striking fact about our inductive practice, both lay and scientific, that so many of our inferences depend on inferring from effects to their probable causes. This is something that Hume himself emphasized (Hume, 1748, sec. IV). The cases of direct causal inference are legion, such as the doctor's inference from symptom to disease, the detective's inference from evidence to perpetrator, the mechanic's inference from the engine noises to what is broken, and many scientific inferences from data to theoretical explanation. Moreover, it is striking that we often make a causal inference even when our main interest is in prediction. Indeed, the detour through causal theory on the route from data to prediction seems to be at the center of many of the dramatic successes of scientific prediction. All this suggests that we might do well to consider an account of the way causal inference works as a central component of a description of our inductive practice.

The best known account of causal inference is John Stuart Mill's discussion of the 'methods of experimental inquiry' (Mill, 1904, book III, ch. VIII). The two central methods are the Method of Agreement and especially the Method of Difference. According to the Method of Agreement, in idealized form, when we find

that there is only one antecedent that is shared by all the observed instances of an effect, we infer that it is a cause (book III, ch. VIII, sec. 1; references to Mill hereafter as, e.g., 'III. VIII.1'). Thus we come to believe that hangovers are caused by heavy drinking. According to the Method of Difference, when we find that there is only one prior difference between a situation where the effect occurs and an otherwise similar situation where it does not, we infer that the antecedent that is only present in the case of the effect is a cause (III.VIII.2). If we add sodium to a blue flame, and the flame turns yellow, we infer that the presence of sodium is a cause of the new color, since that is the only difference between the flame before and after the sodium was added. If we once successfully follow a recipe for baking bread, but fail another time because we have left out the yeast and the bread does not rise, we would infer that the yeast is a cause of the rising in the first case. Both methods work by a combination of retention and variation. When we apply the Method of Agreement, we hold the effect constant and try to vary the background as much as we can, and see what stays the same; when we apply the Method of Difference, we vary the effect, and try to hold as much of the background constant as we can, and see what changes.

Mill's methods have a number of attractive features. Many of our inferences are causal inferences, and Mill's methods give a natural account of these. In science, for example, the controlled experiment is a particularly common and self-conscious application of the Method of Difference. The Millian structure of causal inference is often particularly clear in cases of inferential dispute. When you dispute my claim that C is the cause of E, you will often make your case by pointing out that the conditions for Mill's methods are not met; that is, by pointing out C is not the only antecedent common to all cases of E, or that the presence of C is not the only salient difference between a case where E occurs and a similar case where it does not. Mill's methods may also avoid some of the over-permissiveness of other accounts, because of the strong constraints that the requirements of varied or shared backgrounds place on their application. These requirements suggest how our background beliefs influence our inferences, something a good account of inference must do. The methods also help to bring out the roles in inference of competing hypotheses and negative evidence, as we will see in chapter five, and the role of background knowledge, as we will see in chapter seven [sic].

Of course, Mill's methods have their share of liabilities, of which I will mention just two. First, they do not themselves apply to unobservable causes or to any causal inferences where the cause's existence, and not just its causal status, is inferred. Second, if the methods are to apply at all, the requirement that there be only a single agreement or difference in antecedents must be seen as an idealization, since this condition is never met in real life. We need principles for selecting from among multiple agreements or similarities those that are likely to be causes, but these are principles Mill does not himself supply. As we will see in chapters five through seven [sic], however, Mill's method can be modified and expanded in a way that may avoid these and other liabilities it faces in its simple form.

This chapter has set part of the stage for an investigation of our inductive practices. I have suggested that many of the problems those practices raise can be set out in a natural way in terms of the underdetermination that is characteristic of inductive inference. The underdetermination of our inferences by our evidence provides the skeptic with his lever, and so poses the problem of justification. It also elucidates the structure of the descriptive problem, and the black box inferences it will take to solve it.

I have canvassed several solutions to the problem of description, partly to give a sense of some of our options and partly to suggest just how difficult the problem is. But at least one solution to the descriptive problem was conspicuous by its absence, the solution that gives this book its title and which will be at the center of attention from chapter four onwards [sic]. According to Inference to the Best Explanation, we infer what would, if true, be the best explanation of our evidence. On this view, explanatory considerations are our guide to inference. So to develop and assess this view, we need first to look at another sector of our cognitive economy, our explanatory practices. This is the subject of the next two chapters [sic].

References

Chomsky, N. (1965) *Aspects of the Theory of Syntax*, Cambridge, MA: MIT Press.

Descartes, R. (1641) *Meditations on First Philosophy*, Donald Cress (trans.), Indianapolis: Hackett (1979).

Glymour, C. (1980) *Theory and Evidence*, Princeton: Princeton University Press.

Goodman, N. (1983) *Fact, Fiction and Forecast*, 4th ed., Indianapolis: Bobbs-Merrill.

Hempel, C. (1965) *Aspects of Scientific Explanation*, New York: Free Press.

Hempel, C. (1966) *The Philosophy of Natural Science*, Englewood Cliffs: Prentice-Hall.

Hume, D. (1748) *An Enquiry Concerning Human Understanding*, T. Beauchamp (ed.), Oxford: Oxford University Press (1999).

Kuhn, T. (1970) *The Structure of Scientific Revolutions*, 2nd ed., Chicago: University of Chicago Press.

Kuhn, T. (1977) *The Essential Tension*, Chicago: University of Chicago Press.

Mill, J.S. (1904) *A System of Logic*, 8th ed., London: Longmans, Green & Co.

Skyrms, B. (1986) *Choice and Chance*, 3rd ed., Belmont: Wadsworth.

Williams, B. (1978) *Descartes*, Harmondsworth: Penguin.

Chapter 13

Hilary Putnam

THE "CORROBORATION" OF THEORIES

Sir Karl Popper is a philosopher whose work has influenced and stimulated that of virtually every student in the philosophy of science. In part this influence is explainable on the basis of the healthy-mindedness of some of Sir Karl's fundamental attitudes: "There is no method peculiar to philosophy." "The growth of knowledge can be studied best by studying the growth of scientific knowledge."

> Philosophers should not be specialists. For myself, I am interested in science and in philosophy only because I want to learn something about the riddle of the world in which we live, and the riddle of man's knowledge of that world. And I believe that only a revival of interest in these riddles can save the sciences and philosophy from an obscurantist faith in the expert's special skill and in his personal knowledge and authority.

These attitudes are perhaps a little narrow (can the growth of knowledge be studied without also studying nonscientific knowledge? Are the problems Popper mentions of merely theoretical interest – just "riddles"?), but much less narrow than those of many philosophers; and the "obscurantist faith" Popper warns against is a real danger. In part this influence stems from Popper's realism, his refusal to accept the peculiar meaning theories of the positivists, and his separation of the problems of scientific methodology from the various problems about the "interpretation of scientific theories" which are internal to the meaning theories of the positivists and which positivistic philosophers of science have continued to wrangle about.[1]

In this paper I want to examine his views about scientific methodology – about what is generally called "induction", although Popper rejects the concept – and, in particular, to criticize assumptions that Popper has in common with received philosophy of science, rather than assumptions that are peculiar to Popper. For I think that there are a number of such common assumptions, and that they represent a mistaken way of looking at science.

1 Popper's view of "induction"

Popper himself uses the term "induction" to refer to any method for verifying or showing to be true (or even probable) general laws on the basis of observational or experimental data (what he calls "basic statements"). His views are radically Humean: no such method exists or can exist. A principle of induction would have to be either synthetic a priori (a possibility that Popper rejects) or justified by a higher level principle. But the latter course necessarily leads to an infinite regress.

What is novel is that Popper concludes neither that empirical science is impossible nor that empirical science rests upon principles that are themselves incapable of justification. Rather, his position is that empirical science does not really rely upon a principle of induction!

Popper does not deny that scientists state general laws, nor that they test these general laws against observational data. What he says is that when a scientist "corroborates" a general law, that scientist does not thereby assert that law to be true or even probable. "I have corroborated this law to a high degree" only means "I have subjected this law to severe tests and it has withstood them". Scientific laws are *falsifiable*, not verifiable. Since scientists are not even trying to *verify* laws, but only to falsify them, Hume's problem does not arise for empirical scientists.

2 A brief criticism of Popper's view

It is a remarkable fact about Popper's book, *The Logic of Scientific Discovery* that it contains but a half-dozen brief references to the *application* of scientific theories and laws; and then all that is said is that application is yet another *test* of the laws. "My view is that . . . the theorist is interested in explanations as such, that is to say, in testable explanatory theories: applications and predictions interest him only for theoretical reasons – because they may be used as *tests* of theories" (*L.Sc.D.*, p. 59).

When a scientist accepts a law, he is recommending to other men that they rely on it – rely on it, often, in practical contexts. Only by wrenching science altogether out of the context in which it really arises – the context of men trying to change and control the world – can Popper even put forward his peculiar view on induction. Ideas are not just ideas; they are guides to action. Our notions of "knowledge", "probability", "certainty", etc., are all linked to and frequently used in contexts in which action is at issue: may I confidently rely upon a certain idea? Shall I rely upon it tentatively, with a certain caution? Is it necessary to check on it?

If "this law is highly corroborated", "this law is scientifically accepted", and like locutions merely meant "this law has withstood severe tests" – and there were no suggestion at all that a law which has withstood severe tests is likely to withstand further tests, such as the tests involved in an application or attempted application, then Popper would be right; but then science would be a wholly unimportant activity. It would be practically unimportant, because scientists would never tell us that any law or theory is safe to rely upon for practical purposes; and it would be unimportant for the purpose of understanding, since on Popper's view, scientists never tell us that any law or theory is true or even probable. Knowing that certain "conjectures" (according to Popper all scientific laws are "provisional conjectures") have not yet been refuted is *not understanding anything*.

Since the application of scientific laws does involve the anticipation of future successes, Popper is not right in maintaining that induction is unnecessary. Even if scientists do not inductively anticipate the future (and, of course, they do), men who apply scientific laws and theories do so. And "don't make inductions" is hardly reasonable advice to give these men.

The advice to regard all knowledge as "provisional conjectures" is also not reasonable. Consider men striking against sweatshop conditions. Should they say "it is only a provisional conjecture that the boss is a bastard. Let us call off our strike and try appealing to his better nature." The distinction between *knowledge* and *conjecture* does real work in our lives; Popper can maintain his extreme scepticism only because of his extreme tendency to regard theory as an end for itself.

3 Popper's view of corroboration

Logic of Scientific Discovery

Although scientists, on Popper's view, do not make inductions, they do "corroborate" scientific theories. And although the statement that a theory is highly corroborated does not mean, according to Popper, that the theory may be accepted as true, or even as approximately true,[2] or even as probably approximately true, still, there is no doubt that most readers of Popper read his account of corroboration as an account of something like the verification of theories, in spite of his protests. In this sense, Popper has, *contre lui* a theory of induction. And it is this theory, or certain presuppositions of this theory, that I shall criticize in the body of this paper.

Popper's reaction to this way of reading him is as follows:

> My reaction to this reply would be regret at my continued failure to explain my main point with sufficient clarity. For the sole purpose of the elimination advocated by all these inductivists was to *establish as firmly as possible the surviving theory* which, they thought, must be the *true* one (or, perhaps, only a *highly probable* one, in so far as we may not have fully succeeded in eliminating every theory except the true one).
>
> As against this, I do not think that we can ever seriously reduce by elimination, the number of the competing theories, since this number remains always infinite. What we do — or should do — is to *hold on, for the time being, to the most improbable of the surviving theories* or, more precisely, to the one that can be most severely tested. We tentatively *accept* this theory — but only in the sense that we select it as worthy to be subjected to further criticism, and to the severest tests we can design.
>
> On the positive side, we may be entitled to add that the surviving theory is the best

theory — and the best tested theory — of which we know. [*Logic of Scientific Discovery*, p. 419]

If we leave out the last sentence, we have the doctrine we have been criticizing in pure form: when a scientist "accepts" a theory, he does not assert that it is probable. In fact, he "selects" it as most improbable! In the last sentence, however, am I mistaken, or do I detect an inductivist quaver? What does "best theory" mean? Surely Popper cannot mean "most likely"?

4 The scientific method – the received schema

Standard "inductivist" accounts of the confirmation[3] of scientific theories go somewhat like this: Theory implies prediction (basic sentence, or observation sentence); if prediction is false, theory is falsified; if sufficiently many predictions are true, theory is confirmed. For all his attack on inductivism, Popper's schema is not *so* different: Theory implies prediction (basic sentence); if prediction is false, theory is falsified; if sufficiently many predictions are true, and certain further conditions are fulfilled, theory is highly corroborated.

Moreover, this reading of Popper does have certain support. Popper does say that the "surviving theory" is *accepted* – his account is, therefore, an account of the logic of accepting theories. We must separate two questions: is Popper right about what the scientist means – or should mean – when he speaks of a theory as "accepted"; and is Popper right about the methodology involved in according a theory that status? What I am urging is that his account of that methodology fits the received schema, even if his interpretation of the status is very different.

To be sure there are some important conditions that Popper adds. Predictions that one could have made on the basis of background knowledge do not test a theory; it is only predictions that are *improbable* relative to background knowledge that test a theory. And a theory is not corroborated, according to Popper, unless we make sincere attempts to derive false predictions from it. Popper regards these further conditions as anti-Bayesian;[4] but this seems to me to be a confusion, at least in part. A theory which implies an improbable prediction is improbable, that is true, but it may be the most probable of all theories which imply that prediction. If so, and the prediction turns out true, then Bayes's theorem itself explains why the theory receives a high probability. Popper says that we select the most improbable of the *surviving* theories – i.e., the accepted theory is most improbable even *after* the prediction has turned out true; but, of course, this depends on using "probable" in a way no other philosopher of science would accept. And a Bayesian is not committed to the view that *any* true prediction significantly confirms a theory. I share Popper's view that quantitative measures of the probability of theories are not a hopeful venture in the philosophy of science[5]; but that does not mean that Bayes's theorem does not have a certain *qualitative* rightness, at least in many situations.

Be all this as it may, the heart of Popper's schema is the theory-prediction link. It is because theories imply basic sentences in the sense of "imply" associated with deductive logic – because basic sentences are *deducible* from theories – that, according to Popper, theories and general laws can be falsifiable by basic sentences. And this same link is the heart of the "inductivist" schema. Both schemes say: *look at the predictions that a theory implies; see if those predictions are true.*

My criticism is going to be a criticism of this link, of this one point on which Popper and the "inductivists" agree. I claim: in a great many important cases, scientific theories do not imply predictions at all. In the remainder of this paper I want to elaborate this point, and show its significance for the philosophy of science.

5 The theory of universal gravitation

The theory that I will use to illustrate my points is one that the reader will be familiar with: it is Newton's theory of universal gravitation. The theory consists of the law that every body a exerts on every other body b a force F_{ab} whose direction is towards a and whose magnitude is a universal constant g times $M_a M_b / d^2$, together with Newton's three laws. The choice of this particular theory is not essential to my case: Maxwell's theory, or Mendel's, or Darwin's would have done just as well. But this one has the advantage of familiarity.

Note that this theory does not imply a single basic sentence! Indeed, any motions whatsoever are compatible with this theory, since the theory says nothing about what forces other than gravitation may be present. The forces F_{ab} are not themselves directly measurable; consequently not a single *prediction* can be deduced from the theory.

What do we do, then, when we apply this theory to an astronomical situation? Typically we make certain simplifying assumptions. For example, if we are deducing the orbit of the earth we might assume as a first approximation:

(I)   No bodies exist except the sun and the earth.

(II) The sun and the earth exist in a hard vacuum.

(III) The sun and the earth are subject to no forces except mutually induced gravitational forces.

From the conjunction of the theory of universal gravitation (UG) and these auxiliary statements (AS) we can, indeed, deduce certain predictions – e.g., Kepler's laws. By making (I), (II), (III) more "realistic" – i.e., incorporating further bodies in our model solar system – we can obtain better predictions. But it is important to note that these predictions do not come from the theory alone, but from the conjunction of the theory with AS. As scientists actually use the term "theory", the statements AS are hardly part of the "theory" of gravitation.

6 Is the point terminological?

I am not interested in making a merely terminological point, however. The point is not just that scientists don't use the term "theory" to refer to the conjunction of UG with AS, but that such a usage would obscure profound methodological issues. A theory, as the term is actually used, is a set of laws. Laws are statements that we hope to be true; they are supposed to be true by the nature of things, and not just by accident. None of the statements (I), (II), (III) has this character. We do not really believe that no bodies except the sun and the earth exist, for example, but only that all other bodies exert forces small enough to be neglected. This statement is not supposed to be a law of nature: it is a statement about the "boundary conditions" which obtain as a matter of fact in a particular system. To blur the difference between AS and UG is to blur the difference between laws and accidental statements, between statements the scientist wishes to establish as true (the laws), and statements he already knows to be false (the oversimplifications (I), (II), (III)).

7 Uranus, Mercury, "dark companions"

Although the statements AS could be more carefully worded to avoid the objection that they are known to be false, it is striking that they are not in practice. In fact, they are not "worded" at all. Newton's calculation of Kepler's laws makes the assumptions (I), (II), (III) without more than a casual indication that this is what is done. One of the most striking indications of the difference between a theory (such as UG) and a set of AS is the great care which scientists use in stating the theory, as contrasted with the careless way in which they introduce the various assumptions which make up AS.

The AS are also far more subject to revision than the theory. For over two hundred years the law of universal gravitation was accepted as unquestionably true, and used as a premise in countless scientific arguments. If the standard kind of AS had not led to successful prediction in that period, they would have been modified, not the theory. In fact, we have an example of this. When the predictions about the orbit of Uranus that were made on the basis of the theory of universal gravitation and the assumption that the known planets were all there were turned out to be wrong, Leverrier in France and Adams in England simultaneously predicted that there must be another planet. In fact, this planet was discovered – it was Neptune. Had this modification of the AS not been successful, still others might have been tried – e.g., postulating a medium through which the planets are moving, instead of a hard vacuum, or postulating significant nongravitational forces.

It may be argued that it was crucial that the new planet should itself be observable. But this is not so. Certain stars, for example, exhibit irregular behavior. This has been explained by postulating companions. When those companions are not visible through a telescope, this is handled by suggesting that the stars have dark companions – companions which cannot be seen through a telescope. The fact is that many of the assumptions made in the sciences cannot be directly tested – there are many "dark companions" in scientific theory.

Lastly, of course, there is the case of Mercury. The orbit of this planet can almost but not quite be successfully explained by Newton's theory. Does this show that Newton's theory is wrong?

In the light of an alternative theory, say the General Theory of Relativity, one answers "yes". But, in the absence of such a theory, the orbit of Mercury is just a slight anomaly, cause: unknown.

What I am urging is that all this is perfectly good scientific practice. The fact that any one of the statements AS may be false – indeed, they are false, as stated, and even more careful and guarded statements might well be false – is important. We do not know for sure all the bodies in the solar system; we do not know for sure that the medium through which they move is (to a sufficiently high degree of approximation in all cases) a hard vacuum; we do not know that nongravitational forces can be neglected in all cases. Given the overwhelming success of the Law of Universal Gravitation in almost all cases, one or two anomalies are not reason to reject it. It is more likely that the AS are false than that the theory is false, at least when no alternative theory has seriously been put forward.

8 The effect on Popper's doctrine

The effect of this fact on Popper's doctrine is immediate. The Law of Universal Gravitation is not strongly falsifiable at all; yet it is surely a paradigm of a scientific theory. Scientists for over two hundred years did not derive predictions from UG in order to falsify UG; they derived predictions from UG in order to explain various astronomical facts. If a fact proved recalcitrant to this sort of explanation it was put aside as an anomaly (the case of Mercury). Popper's doctrine gives a correct account of neither the nature of the scientific theory nor of the practice of the scientific community in this case.

Popper might reply that he is not describing what scientists do, but what they should do. Should scientists then not have put forward UG? Was Newton a bad scientist? Scientists did not try to falsify UG because they could not try to

falsify it; laboratory tests were excluded by the technology of the time and the weakness of the gravitational interactions. Scientists were thus limited to astronomical data for a long time. And, even in the astronomical cases, the problem arises that one cannot be absolutely sure that no non-gravitational force is relevant in a given situation (or that one has summed all the gravitational forces). It is for this reason that astronomical data can support UG, but they can hardly falsify it. It would have been incorrect to reject UG because of the deviancy of the orbit of Mercury; given that UG predicted the other orbits, to the limits of measurement error, the possibility could not be excluded that the deviancy in this one case was due to an unknown force, gravitational or nongravitational, and in putting the case aside as one they could neither explain nor attach systematic significance to, scientists were acting as they "should".[6]

So far we have said that (1) theories do not imply predictions; it is only the conjunction of a theory with certain "auxiliary statements" (AS) that, in general, implies a prediction. (2) The AS are frequently suppositions about boundary conditions (including initial conditions as a special case of "boundary conditions"), and highly risky suppositions at that. (3) Since we are very unsure of the AS, we cannot regard a false prediction as definitively falsifying a theory; theories are not strongly falsifiable.

All this is not to deny that scientists do sometimes derive predictions from theories and AS in order to test the theories. If Newton had not been able to derive Kepler's laws, for example, he would not have even put forward UG. But even if the predictions Newton had obtained from UG had been wildly wrong, UG might still have been true: the AS might have been wrong. Thus, even if a theory is "knocked out" by an experimental test, the theory may still be right, and the theory may come back in at a later stage when it is discovered the AS were not useful approximations to the true situation. As has previously been pointed out,[7] falsification in science is no more conclusive than verification.

All this refutes Popper's view that what the scientist does is to put forward "highly falsifiable" theories, derive predictions from them, and then attempt to falsify the theories by falsifying the predictions. But it does not refute the standard view (what Popper calls the "inductivist" view) that scientists try to *confirm* theories *and* AS by deriving predictions from them and verifying the predictions. There is the objection that (in the case of UG) the AS were known to be false, so scientists could hardly have been trying to confirm them; but this could be met by saying that the AS could, in principle, have been formulated in a more guarded way, and would not have been false if sufficiently guarded.[8] I think that, in fact, there is some truth in the "inductivist" view: scientific theories are shown to be correct by their successes, just as all human ideas are shown to be correct, to the extent that they are, by their successes in practice. But the inductivist schema is still inadequate, except as a picture of one aspect of scientific procedure. In the next sections, I shall try to show that scientific activity cannot, in general, be thought of as a matter of deriving predictions from the conjunction of theories and AS, whether for the purpose of confirmation or for the purpose of falsification.

9 Kuhn's view of science

Recently a number of philosophers have begun to put forward a rather new view of scientific activity. I believe that I anticipated this view about ten years ago when I urged that some scientific theories cannot be overthrown by experiments and observations *alone*, but only by alternative theories.[9] The view is also anticipated by Hanson,[10] but it reaches its sharpest

expression in the writings of Thomas Kuhn[11] and Louis Althusser.[12] I believe that both of these philosophers commit errors; but I also believe that the tendency they represent (and that I also represent, for that matter) is a needed corrective to the deductivism we have been examining. In this section, I shall present some of Kuhn's views, and then try to advance on them in the direction of a sharper formulation.

The heart of Kuhn's account is the notion of a *paradigm*. Kuhn has been legitimately criticized for some inconsistencies and unclarities in the use of this notion; but at least one of his explanations of the notion seems to me to be quite clear and suitable for his purposes. On this explanation, a paradigm is simply a scientific theory together with an example of a successful and striking application. It is important that the application – say, a successful explanation of some fact, or a successful and novel prediction – be *striking*; what this means is that the success is sufficiently impressive that scientists – especially young scientists choosing a career – are led to try to emulate that success by seeking further explanations, predictions, or whatever on the same model. For example, once UG had been put forward and one had the example of Newton's derivation of Kepler's laws together with the example of the derivation of, say, a planetary orbit or two, then one had a paradigm. The most important paradigms are the ones that generate scientific fields; the field generated by the Newtonian paradigm was, in the first instance, the entire field of Celestial Mechanics. (Of course, this field was only a part of the larger field of Newtonian mechanics, and the paradigm on which Celestial Mechanics is based is only one of a number of paradigms which collectively structure Newtonian mechanics.)

Kuhn maintains that the paradigm that structures a field is highly immune to falsification – in particular, it can only be overthrown by a new paradigm. In one sense, this is an exaggeration: Newtonian physics would probably have been abandoned, even in the absence of a new paradigm, if the world had started to act in a markedly non-Newtonian way. (Although even then – would we have concluded that Newtonian physics was false, or just that we didn't know what the devil was going on?) But then even the old successes, the successes which were paradigmatic for Newtonian physics, would have ceased to be available. What is true, I believe, is that in the absence of such a drastic and unprecedented change in the world, and in the absence of its turning out that the paradigmatic successes had something "phony" about them (e.g., the data were faked, or there was a mistake in the deductions), a theory which is paradigmatic is not given up because of observational and experimental results by themselves, but only because and when a better theory is available.

Once a paradigm has been set up, and a scientific field has grown up around that paradigm, we get an interval of what Kuhn calls "normal science". The activity of scientists during such an interval is described by Kuhn as "puzzle solving" – a notion I shall return to.

In general, the interval of normal science continues even though not all the puzzles of the field can be successfully solved (after all, it is only human experience that some problems are too hard to solve), and even though some of the solutions may look *ad hoc*. What finally terminates the interval is the introduction of a new paradigm which manages to supersede the old.

Kuhn's most controversial assertions have to do with the process whereby a new paradigm supplants an older paradigm. Here he tends to be radically subjectivistic (overly so, in my opinion): data, in the usual sense, cannot establish the superiority of one paradigm over another because data themselves are perceived through the spectacles of one paradigm or another. Changing from one paradigm to another requires a "Gestalt switch". The history and methodology of science get rewritten when there are major paradigm changes; so there are no "neutral" historical and methodological canons to which to appeal. Kuhn also holds views on meaning and truth which are relativistic and, on my view, incorrect; but I do not wish to discuss these here.

What I want to explore is the interval which Kuhn calls "normal science". The term "puzzle solving" is unfortunately trivializing; searching for explanations of phenomena and for ways to harness nature is too important a part of human life to be demeaned (here Kuhn shows the same tendency that leads Popper to call the problem of the nature of knowledge a "riddle"). But the term is also striking: clearly, Kuhn sees normal science as neither an activity of trying to falsify one's paradigm nor as an activity of trying to confirm it, but as something else. I want to try to advance on Kuhn by presenting a schema for normal science, or rather for one aspect of normal science; a schema which may indicate why a major philosopher and historian of science would use the metaphor of solving puzzles in the way Kuhn does.

10 Schemata for scientific problems

Consider the following two schemata:

SCHEMA I
THEORY
AUXILIARY STATEMENTS
_____
PREDICTION – TRUE OR FALSE?

SCHEMA II
THEORY
???????????????
_____
FACT TO BE EXPLAINED

These are both schemata for scientific problems. In the first type of problem we have a theory, we have some AS, we have derived a prediction, and our problem is to see if the prediction is true or false: the situation emphasized by standard philosophy of science. The second type of problem is quite different. In this type of problem we have a theory, we have a fact to be explained, but the AS are missing: the problem is to find AS, if we can, which are true, or approximately true (i.e., useful oversimplifications of the truth), and which have to be conjoined to the theory to get an explanation of the fact.

We might, in passing, mention also a third schema which is neglected by standard philosophy of science:

SCHEMA III
    THEORY
    AUXILIARY STATEMENTS
    ————————————
    ????????????????????

This represents the type of problem in which we have a theory, we have some AS, and we want to know what consequences we can derive. This type of problem is neglected because the problem is "purely mathematical". But knowing whether a set of statements has testable consequences at all depends upon the solution to this type of problem, and the problem is frequently of great difficulty – e.g., little is known to this day concerning just what the physical consequences of Einstein's "unified field theory" are, precisely because the mathematical problem of deriving those consequences is too difficult. Philosophers of science frequently write as if it is *clear*, given a set of statements, just what consequences those statements do and do not have.

Let us, however, return to Schema II. Given the known facts concerning the orbit of Uranus, and given the known facts (prior to 1846) concerning what bodies make up the solar system, and the standard AS that those bodies are moving in a hard vacuum, subject only to mutual gravitational forces, etc., it was clear that there was a problem: the orbit of Uranus could not be successfully calculated if we assumed that Mercury, Venus, Earth, Mars, Saturn, Jupiter, and Uranus were all the planets there are, and that these planets together with the sun make up the whole solar system. Let S_1 be the conjunction of the various AS we just mentioned, including the statement that the solar system consists of at least, but not necessarily of only, the bodies mentioned. Then we have the following problem:

Theory: UG
AS: S_1
Further AS: ????????
————————————
Explanandum: The orbit of Uranus

Note that the problem is not to find further explanatory laws (although sometimes it may be, in a problem of the form of Schema II); it is to find further assumptions about the initial and boundary conditions governing the solar system which, together with the Law of Universal Gravitation and the other laws which make up UG (i.e., the laws of Newtonian mechanics) will enable one to explain the orbit of Uranus. If one does not require that the missing statements be true, or approximately true, then there are an infinite number of solutions, mathematically speaking. Even if one includes in S_1 that no nongravitational forces are acting on the planets or the sun, there are still an infinite number of solutions. But one tries first the simplest assumption, namely:

(S_2) There is one and only one planet in the solar system in addition to the planets mentioned in S_1.

Now one considers the following problem:

Theory: UG

AS: S_1, S_2

Consequence ??? – turns out to be that the unknown planet must have a certain orbit O.

This problem is a mathematical problem – the one Leverrier and Adams both solved (an instance of Schema III). Now one considers the following empirical problem:

Theory: UG

AS: S_1, S_2

Prediction: A planet exists moving in orbit O – TRUE OR FALSE?

This problem is an instance of Schema I – an instance one would not normally consider, because one of the AS, namely the statement S_2, is not at all known to be true. S_2 is, in fact, functioning as a low-level hypothesis which we wish to test. But the test is not an inductive one in the usual sense, because a verification of the prediction is also a verification of S_2 – or rather, of the approximate truth of S_2 (which is all that is of interest in this context) – Neptune was not the only planet unknown in 1846; there was also Pluto to be later discovered. The fact is that we are interested in the above problem in 1846, because we know that if the prediction turns out to be true, then that prediction is precisely the statement S_3 that we need for the following deduction;

Theory: UG

AS: S_1, S_2, S_3

Explanandum: the orbit of Uranus

– i.e., the statement S_3 (that the planet mentioned in S_2 has precisely the orbit O)[13] is the solution to the problem with which we started. In this case we started with a problem of the Schema II type: we introduced the assumption S_2 as a simplifying assumption in the hope of solving the original problem thereby more easily; and we had the good luck to be able to deduce S_3 – the solution to the original problem – from UG together with S_1, S_2, and the more important good luck that S_3 turned out to be true when the Berlin Observatory looked. Problems of the Schema II-type are sometimes mentioned by philosophers of science when the missing AS are *laws*; but the case just examined, in which the missing AS was just a further contingent fact about the particular system is almost never discussed. I want to suggest that Schema II exhibits the logical form of what Kuhn calls a "puzzle".

If we examine Schema II, we can see why the term "puzzle" is so appropriate. When one has a problem of this sort one is looking for something to fill a "hole" – often a thing of rather under-specified sort – and that *is* a sort of *puzzle*. Moreover, this sort of problem is extremely widespread in science. Suppose one wants to explain the fact that water is a liquid (under the standard conditions), and one is given the laws of physics; the fact is that the problem is extremely hard. In fact, quantum mechanical laws are needed. But that does not mean that from classical physics one can deduce that water is *not* a liquid; rather the classical physicist would give up this problem at a certain point as "too hard" – i.e., he would conclude that he could not find the right AS.

The fact that Schema II is the logical form of the "puzzles" of normal science explains a number of facts. When one is tackling a Schema II-type problem there is no question of deriving a prediction from UG plus given AS, the whole problem is to find the AS. The theory – UG, or whichever – is *unfalsifiable in the context*. It is also not up for "confirmation" any more than for "falsification"; *it is not functioning in a hypothetical role*.

Failures do not falsify a theory, because the failure is not a false prediction from a theory together with known and trusted facts, but a failure to find something – in fact, a failure to find an AS. Theories, during their tenure of office, are highly immune to falsification; that tenure of office is ended by the appearance on the scene of a better theory (or a whole new explanatory technique), not by a basic sentence. And successes do not "confirm" a theory, once it has become paradigmatic, because the theory is not a "hypothesis" in need of confirmation, but the basis of a whole explanatory and predictive technique, and possibly of a technology as well.

To sum up: I have suggested that standard philosophy of science, both "Popperian" and non-Popperian, has fixated on the situation in which we derive predictions from a theory, and test those predictions in order to falsify or confirm the theory – i.e., on the situation represented by Schema I. I have suggested that, by way of contrast, we see the "puzzles" of "normal science" as exhibiting the pattern represented by Schema II – the pattern in which we take a theory as fixed, take the fact to be explained as fixed, and seek further facts – frequently contingent[14] facts about the particular system – which will enable us to fill out the explanation of the particular fact on the basis of the theory. I suggest that adopting this point of view will enable us better to appreciate both the relative unfalsifiability of theories which have attained paradigm status, and the fact that the "predictions" of physical theory are frequently facts which were known beforehand, and not things which are surprising relative to background knowledge.

To take Schema II as describing everything that goes on between the introduction of a paradigm and its eventual replacement by a better paradigm would be a gross error in the opposite direction, however. The fact is that normal science exhibits a dialectic between two conflicting (at any rate, potentially conflicting) but interdependent tendencies, and that it is the conflict of these tendencies that drives normal science forward. The desire to solve a Schema II-type problem – explain the orbit of Uranus – led to a new hypothesis (albeit a very low-level one): namely, S_2. Testing S_2 involved deriving S_3 from it, and testing S_3 – a Schema I-type situation. S_3 in turn served as the solution to the original problem. This illustrates the two tendencies, and also the way in which they are interdependent and the way in which their interaction drives science forward.

The tendency represented by Schema I is the *critical* tendency. Popper is right to emphasize the importance of this tendency, and doing this is certainly a contribution on his part – one that has influenced many philosophers. Scientists do want to know if their ideas are wrong, and they try to find out if their ideas are wrong by deriving predictions from them, and testing those predictions – that is, they do this *when they can*. The tendency represented by Schema II is the *explanatory* tendency. The element of conflict arises because in a Schema II-type situation one tends to regard the given theory as something *known*, whereas in a Schema I-type situation one tends to regard it as *problematic*. The interdependence is obvious: the theory which serves as the major premise in Schema II *may* itself have been the survivor of a Popperian test (although it need not have been – UG was accepted on the basis of its explanatory successes, not on the basis of its surviving attempted falsifications). And the solution to a Schema II-type problem must itself be confirmed, frequently by a Schema I-type test. If the solution is a general law, rather than a singular statement, that law may itself become a paradigm, leading to new Schema II-type problems. In short, attempted falsifications do "corroborate" theories – not just in Popper's sense, in which this is a tautology, but in the sense he denies, of showing that they are true, or partly true – and explanations on the basis of laws which are regarded as *known* frequently require the introduction of *hypotheses*. In this way, the tension between the attitudes of explanation and criticism drives science to progress.

11 Kuhn versus Popper

As might be expected, there are substantial differences between Kuhn and Popper on the issue of the falsifiability of scientific theories. Kuhn stresses the way in which a scientific theory may be immune from falsification, whereas Popper stresses falsifiability as the *sine qua non* of a scientific theory. Popper's answers to Kuhn depend upon two notions which must now be examined: the notion of an auxiliary hypothesis and the notion of a *conventionalist stratagem*.

Popper recognizes that the derivation of a prediction from a theory may require the use of auxiliary hypotheses (though the term "hypothesis" is perhaps misleading, in suggesting something like putative laws, rather than assumptions about, say, boundary conditions). But he regards these as part of the total "system" under test. A "conventionalist stratagem" is to save a theory from a contrary experimental result by making an *ad hoc* change in the auxiliary hypotheses. And Popper takes it as a fundamental methodological rule of the empirical method to avoid conventionalist stratagems.

Does this do as a reply to Kuhn's objections? Does it contravene our own objections, in the first part of this paper? It does not. In the first place, the "auxiliary hypotheses" AS are not fixed, in the case of UG, but depend upon the context. One simply cannot think of UG as part of a fixed "system" whose other part is a fixed set of auxiliary hypotheses whose function is to render UG "highly testable".

In the second place, an alteration in one's beliefs may be *ad hoc* without being unreasonable. "*Ad hoc*" merely means "to this specific purpose". Of course, "*ad hoc*" has acquired the connotation of "unreasonable" – but that is a different thing. The assumption that certain stars have dark companions is *ad hoc* in the literal sense: the assumption is made for the specific purpose of accounting for the fact that no companion is visible. It is also highly reasonable.

It has already been pointed out that the AS are not only context-dependent but highly uncertain, in the case of UG and in many other cases. So, changing the AS, or even saying in a particular context "we don't know what the right AS are" may be *ad hoc* in the literal sense just noted, but is not "*ad hoc*" in the extended sense of "unreasonable".

12 Paradigm change

How does a paradigm come to be accepted in the first place? Popper's view is that a theory becomes corroborated by passing severe tests: a prediction (whose truth value is not antecedently known) must be derived from the theory and the truth or falsity of that prediction must be ascertained. The severity of the test depends upon the set of basic sentences excluded by the theory, and also upon the improbability of the prediction relative to background knowledge. The ideal case is one in which a theory which rules out a great many basic sentences implies a prediction which is very improbable relative to background knowledge.

Popper points out that the notion of the number of basic sentences ruled out by a theory cannot be understood in the sense of cardinality; he proposes rather to measure it by means of concepts of *improbability* or *content*. It does not appear true to me that improbability (in the sense of logical [im]probability)[15] measures falsifiability, in Popper's sense: UG excludes *no* basic sentences, for example, but has logical probability *zero*, on any standard metric. And it certainly is not true that the scientist always selects "the most improbable of the surviving hypotheses" on *any* measure of probability, except in the trivial sense that all strictly universal laws have probability zero. But my concern here is not with the technical details of Popper's scheme, but with the leading idea.

To appraise this idea, let us see how UG came to be accepted. Newton first derived Kepler's Laws from UG and the AS we mentioned at the outset: this was not a "test", in Popper's sense, because Kepler's Laws were already known to be true. Then he showed that UG would account for the tides on the basis of the gravitational pull of the moon: this also was not a "test", in Popper's sense, because the tides were already known. Then he spent many years showing that small perturbations (which were already known) in the orbits of the planets could be accounted for by UG. By this time the whole civilized world had accepted – and, indeed, acclaimed – UG; but it had not been "corroborated" at all in Popper's sense!

If we look for a Popperian "test" of UG – a derivation of a new prediction, one risky relative to background knowledge – we do not get one until the Cavendish experiment of 1781 – roughly a hundred years after the theory had been introduced! The prediction of S_3 (the orbit of Neptune) from UG and the auxiliary statements S_1 and S_2 can also be regarded as a confirmation of UG (in 1846!); although it is difficult to regard it as a severe test of UG in view of the fact that the assumption S_2 had a more tentative status than UG.

It is easy to see what has gone wrong. A theory is not accepted unless it has real explanatory successes. Although a theory may legitimately be preserved by changes in the AS which are, in a sense, "*ad hoc*" (although not *unreasonable*), its *successes* must not be ad hoc. Popper requires that the predictions of a theory must not be antecedently known to be true in order to

rule out *ad hoc* "successes"; but the condition is too strong.

Popper is right in thinking that a theory runs a risk during the period of its establishment. In the case of UG, the risk was not a risk of definite falsification; it was the risk that Newton would not find reasonable AS with the aid of which he could obtain real (non-*ad hoc*) explanatory successes for UG. A failure to explain the tides by the gravitational pull of the moon alone would not, for example, have falsified UG; but the success did strongly support UG.

In sum, a theory is only accepted if the theory has substantial, non-*ad hoc*, explanatory successes. This is in accordance with Popper; unfortunately, it is in even better accordance with the "inductivist" accounts that Popper rejects, since these stress *support* rather than *falsification*.

13 On practice

Popper's mistake here is no small isolated failing. What Popper consistently fails to see is that *practice is primary*: ideas are not just an end in themselves (although they are *partly* an end in themselves), nor is the selection of ideas to "criticize" just an end in itself. The primary importance of ideas is that they guide practice, that they structure whole forms of life. Scientific ideas guide practice in science, in technology, and sometimes in public and private life. We are concerned in science with trying to discover correct ideas: Popper to the contrary, this is not *obscurantism* but *responsibility*. We obtain our ideas – our correct ones, and many of our incorrect ones – by close study of the world. Popper denies that the accumulation of perceptual experience leads to theories: he is right that it does not lead to theories in a mechanical or algorithmic sense; but it does lead to theories in the sense that it is a regularity of methodological significance that (1) lack of experience with phenomena and with previous knowledge about phenomena decreases the probability of correct ideas in a marked fashion; and (2) extensive experience increases the probability of correct, or partially correct, ideas in a marked fashion. "There is no logic of discovery" – in that sense, there is no logic of *testing*, either; all the formal algorithms proposed for testing, by Carnap, by Popper, by Chomsky, etc., are, to speak impolitely, *ridiculous*: if you don't believe this, program a computer to employ one of these algorithms and see how well it does at testing theories! There are *maxims* for discovery and maxims for testing: the idea that correct ideas just come from the sky, while the methods for testing them are highly rigid and predetermined, is one of the worst legacies of the Vienna Circle.

But the correctness of an idea is not certified by the fact that it came from close and concrete study of the relevant aspects of the world; in this sense, Popper is right. We judge the correctness of our ideas by applying them and seeing if they succeed; in general, and in the long run, correct ideas lead to success, and ideas lead to failures where and insofar as they are incorrect. Failure to see the importance of practice leads directly to failure to see the importance of success.

Failure to see the primacy of practice also leads Popper to the idea of a sharp "demarcation" between science, on the one hand, and political, philosophical, and ethical ideas, on the other. This "demarcation" is pernicious, on my view; fundamentally, it corresponds to Popper's separation of theory from practice, and his related separation of the critical tendency in science from the explanatory tendency in science. Finally, the failure to see the primacy of practice leads Popper to some rather reactionary political conclusions. Marxists believe that there are laws of society; that these laws can be known; and that men can and should act on this knowledge. It is not my purpose here to argue that this Marxist view is correct; but surely any view that rules this out a priori is reactionary. Yet this

is precisely what Popper does – and in the name of an anti-a priori philosophy of knowledge!

In general, and in the long run, true ideas are the ones that succeed – how do we know this? This statement too is a statement about the world; a statement we have come to from experience of the world; and we believe in the practice to which this idea corresponds, and in the idea as informing that kind of practice, on the basis that we believe in any good idea – it has proved successful! In this sense "induction is circular". But of course it is! Induction has no deductive justification; induction is not deduction. Circular justifications need not be totally self-protecting nor need they be totally uninformative[16]: the past success of "induction" increases our confidence in it, and its past failure tempers that confidence. The fact that a justification is circular only means that that justification has no power to serve as a *reason*, unless the person to whom it is given as a reason already has some propensity to accept the conclusion. We do have a propensity – an a priori propensity, if you like – to reason "inductively", and the past success of "induction" increases that propensity.

The method of testing ideas in practice and relying on the ones that prove successful (for that is what "induction" is) is not unjustified. That is an *empirical* statement. The method does not have a "justification" – if by a justification is meant a proof from eternal and formal principles that justifies reliance on the method. But then, nothing does – not even, in my opinion, pure mathematics and formal logic. Practice is primary.

Notes

1 I have discussed positivistic meaning theory in "What Theories Are Not", published in *Logic, Methodology, and Philosophy of Science*, ed. by A. Tarski, E. Nagel, and P. Suppes (Stanford: Stanford University Press, 1962), pp. 24–51, and also in "How Not to

Talk about Meaning", published in *Boston Studies in the Philosophy of Science*, Vol. II, ed. by R.S. Cohen and M.W. Wartofsky (New York: Humanities Press, 1965), pp. 205–22.

2   For a discussion of "approximate truth", see the second of the papers mentioned in the preceding note.

3   "Confirmation" is the term in standard use for support a positive experimental or observational result gives to a hypothesis; Popper uses the term "corroboration" instead, as a rule, because he objects to the connotations of "showing to be true" (or at least probable) which he sees as attaching to the former term.

4   *Bayes's theorem* asserts, roughly, that the probability of a hypothesis H on given evidence E is directly proportional to the probability of E on the hypothesis H, and also directly proportional to the antecedent probability of H – i.e., the probability of H if one doesn't know that E. The theorem also asserts that the probability of H on the evidence E is less, other things being equal, if the probability of E on the assumption $\bar{H}$ (not-H) is greater. Today probability theorists are divided between those who accept the notion of "antecedent probability of a hypothesis", which is crucial to the theorem, and those who reject this notion, and therefore the notion of the probability of a hypothesis on given evidence. The former school are called "Bayesians"; the latter "anti-Bayesians".

5   Cf. my paper "'Degree of Confirmation' and Inductive Logic", in *The Philosophy of Rudolf Carnap* (The Library of Living Philosophers, Vol. 11), ed. by Paul A. Schilpp (La Salle, Ill.: Open Court Publishing Co., 1963), pp. 761–84.

6   Popper's reply to this sort of criticism is discussed below in the section titled "Kuhn versus Popper".

7   This point is made by many authors. The point that is often missed is that, in cases such as the one discussed, the auxiliary statements are much less certain than the theory under test; without this remark, the criticism that one *might* preserve a theory by revising the AS looks like a bit of formal logic, without real relation to scientific practice. (See below, "Kuhn versus Popper".)

8   I have in mind saying "the planets exert forces on each other which are more than .999 (or whatever) gravitational", rather than "the planets exert *no* non-gravitational forces on each other". Similar changes in the other AS could presumably turn them into true statements – though it is not methodologically unimportant that no scientist, to my knowledge, has bothered to calculate exactly what changes in the AS would render them true while preserving their usefulness.

9   Hilary Putnam, "The Analytic and the Synthetic", in *Minnesota Studies in the Philosophy of Science*, Vol. III, ed. by H. Feigl and G. Maxwell (Minneapolis: University of Minnesota Press, 1962), pp. 358–97.

10   N.R. Hanson, *In Patterns of Discovery* (Cambridge, England: Cambridge University Press, 1958).

11   Thomas S. Kuhn, *The Structure of Scientific Revolutions*, Vol. II, No. 2 of International Encyclopedia of Unified Science (Chicago: University of Chicago Press, 1962).

12   Louis Althusser, *Pour Marx* and *Lire le Capital* (Paris, 1965).

13   I use "orbit" in the sense of space-time trajectory, not just spacial path.

14   By "contingent" I mean *not physically necessary*.

15   "Logical probability" is probability assigning equal weight (in some sense) to logically possible worlds.

16   This has been emphasized by Professor Max Black in a number of papers.

Chapter 14

John Stuart Mill

A SYSTEM OF LOGIC

3 Of the ground of induction [sic]

Axiom of the uniformity of the course of nature

Induction properly so called, as distinguished from those mental operations, sometimes, though improperly, designated by the name, which I have attempted in the preceding chapter to characterize [sic], may, then, be summarily defined as generalization from experience. It consists in inferring from some individual instances in which a phenomenon is observed to occur that it occurs in all instances of a certain class, namely, in all which *resemble* the former in what are regarded as the material circumstances.

In what way the material circumstances are to be distinguished from those which are immaterial, or why some of the circumstances are material and others not so, we are not yet ready to point out. We must first observe that there is a principle implied in the very statement of what induction is; an assumption with regard to the course of nature and the order of the universe, namely, that there are such things in nature as parallel cases; that what happens once will, under a sufficient degree of similarity of circumstances, happen again, and not only again, but as often as the same circumstances recur. This, I say, is an assumption involved in every case of induction. And, if we consult the actual course of nature, we find that the assumption is warranted.

The universe, so far as known to us, is so constituted that whatever is true in any one case is true in all cases of a certain description; the only difficulty is to find what description.

This universal fact, which is our warrant for all inferences from experience, has been described by different philosophers in different forms of language: that the course of nature is uniform; that the universe is governed by general laws; and the like. . . .

Whatever be the most proper mode of expressing it, the proposition that the course of nature is uniform is the fundamental principle or general axiom of induction. It would yet be a great error to offer this large generalization as any explanation of the inductive process. On the contrary, I hold it to be itself an instance of induction, and induction by no means of the most obvious kind. Far from being the first induction we make, it is one of the last or, at all events, one of those which are latest in attaining strict philosophical accuracy. As a general maxim, indeed, it has scarcely entered into the minds of any but philosophers; nor even by them, as we shall have many opportunities of remarking, have its extent and limits been always very justly conceived. The truth is that this great generalization is itself founded on prior generalizations. The obscurer laws of nature were discovered by means of it, but the more obvious ones must have been understood and assented to as general truths before it was ever heard of. We should never have thought of affirming that all phenomena take place according to general laws if we had not first arrived, in the case of a great multitude of phenomena, at some knowledge of the laws themselves, which could be done no otherwise than by induction. In what sense, then, can a principle which is so far from being our earliest induction be regarded as our warrant for all the others? In the only sense in which (as we have already seen) the general propositions which we place at the head of our reasonings when we throw them into syllogisms ever really contribute to their validity. As Archbishop Whately remarks, every induction is a syllogism with the major premise suppressed; or (as I prefer expressing it) every induction may be thrown into the form of a syllogism by supplying a major premise. If this be actually done, the principle which we are now considering, that of the uniformity of the course of nature, will appear as the ultimate major premise of all inductions and will, therefore, stand to all inductions in the relation in which, as has been shown at so much length, the major proposition of a syllogism always stands to the conclusion, not contributing at all to prove it, but being a necessary condition of its being proved; since no conclusion is proved for which there cannot be found a true major premise.

The statement that the uniformity of the course of nature is the ultimate major premise in all cases of induction may be thought to require some explanation. The immediate major premise in every inductive argument it certainly is not. Of that, Archbishop Whately's must be held to be the correct account. The induction, "John, Peter, etc., are mortal, therefore all mankind are mortal," may, as he justly says, be thrown into a syllogism by prefixing as a major premise (what is at any rate a necessary condition of the validity of the argument), namely, that what is true of John, Peter, etc., is true of all mankind. But how

came we by this major premise? It is not self-evident; nay, in all cases of unwarranted generalization, it is not true. How, then, is it arrived at? Necessarily either by induction or ratiocination; and if by induction, the process, like all other inductive arguments, may be thrown into the form of a syllogism. This previous syllogism it is, therefore, necessary to construct. There is, in the long run, only one possible construction. The real proof that what is true of John, Peter, etc., is true of all mankind can only be that a different supposition would be inconsistent with the uniformity which we know to exist in the course of nature. Whether there would be this inconsistency or not may be a matter of long and delicate inquiry, but unless there would, we have no sufficient ground for the major of the inductive syllogism. It hence appears that, if we throw the whole course of any inductive argument into a series of syllogisms, we shall arrive by more or fewer steps at an ultimate syllogism which will have for its major premise the principle or axiom of the uniformity of the course of nature.[1] ...

The question of inductive logic stated

In order to a better understanding of the problem which the logician must solve if he would establish a scientific theory of induction, let us compare a few cases of incorrect inductions with others which are acknowledged to be legitimate. Some, we know, which were believed for centuries to be correct were nevertheless incorrect. That all swans are white cannot have been a good induction, since the conclusion has turned out erroneous. The experience, however, on which the conclusion rested was genuine. From the earliest records, the testimony of the inhabitants of the known world was unanimous on the point. The uniform experience, therefore, of the inhabitants of the known world, agreeing in a common result, without one known instance of deviation from that result, is not always sufficient to establish a general conclusion.

But let us now turn to an instance apparently not very dissimilar to this. Mankind were wrong, it seems, in concluding that all swans were white; are we also wrong when we conclude that all men's heads grow above their shoulders and never below, in spite of the conflicting testimony of the naturalist Pliny? As there were black swans, though civilized people had existed for three thousand years on the earth without meeting with them, may there not also be "men whose heads do grow beneath their shoulders," notwithstanding a rather less perfect unanimity of negative testimony from observers? Most persons would answer, No; it was more credible that a bird should vary in its color than that men should vary in the relative position of their principal organs. And there is no doubt that in so saying they would be right; but to say why they are right would be impossible without entering more deeply than is usually done into the true theory of induction.

Again, there are cases in which we reckon with the most unfailing confidence upon uniformity, and other cases in which we do not count upon it at all. In some we feel complete assurance that the future will resemble the past, the unknown be precisely similar to the known. In others, however invariable may be the result obtained from the instances which have been observed, we draw from them no more than a very feeble presumption that the like result will hold in all other cases. That a straight line is the shortest distance between two points we do not doubt to be true even in the region of the fixed stars.[2] When a chemist announces the existence and properties of a newly-discovered substance, if we confide in his accuracy, we feel assured that the conclusions he has arrived at will hold universally, though the induction be founded but on a single instance. We do not withhold our assent, waiting for a repetition of the experiment; or if we do, it is from a doubt whether the one experiment was properly made, not whether if properly made it would be conclusive. Here, then, is a general law of nature inferred without hesitation from a single instance, a universal proposition from a singular one. Now mark another case, and contrast it with this. Not all the instances which have been observed since the beginning of the world in support of the general proposition that all crows are black, would be deemed a sufficient presumption of the truth of the proposition, to outweigh the testimony of one unexceptionable witness who should affirm that in some region of the earth not fully explored, he had caught and examined a crow, and had found it to be grey.

Why is a single instance, in some cases, sufficient for a complete induction, while in others, myriads of concurring instances, without a single exception known or presumed, go such a very little way towards establishing an universal proposition? Whoever can answer this question knows more of the philosophy of logic than the wisest of the ancients, and has solved the great problem of induction. . . .

6 Of the composition of causes [sic]

Two modes of the conjunct action of causes, the mechanical and the chemical

. . . The preceding discussions have rendered us familiar with the case in which several agents, or causes, concur as conditions to the production of an effect; a case, in truth, almost universal, there being very few effects to the production of which no more than one agent contributes. Suppose, then, that two different agents, operating jointly, are followed, under a certain set of collateral conditions, by a given effect. If either of these agents, instead of being joined with the other, had operated alone, under the same set of conditions in all other respects, some effect would probably have followed which would have been different from the joint effect of the two and more or less dissimilar to it. Now, if we happen to know what would be the effect of each cause when acting separately from the other, we are often able to arrive deductively, or a priori, at a correct prediction of what will arise from their conjunct agency. To render this possible, it is only necessary that the same law which expresses the effect of each cause acting by itself shall also correctly express the part due to that cause of the effect which follows from the two together. This condition is realized in the extensive and important class of phenomena commonly called mechanical, namely, the phenomena of the communication of motion (or of pressure, which is tendency to motion) from one body to another. In this important class of cases of causation, one cause never, properly speaking, defeats or frustrates another; both have their full effect. If a body is propelled in two directions by two forces, one tending to drive it to the north and the other to the east, it is caused to move in a given time exactly as far in both directions as the two forces would separately have carried it, and is left precisely where it would have arrived if it had been acted upon first by one of the two forces and afterward by the other. This law of nature is called, in dynamics, the principle of the composition of forces, and, in imitation of that well-chosen expression, I shall give the name of the composition of causes to the principle which is exemplified in all cases in which the joint effect of several causes is identical with the sum of their separate effects.

This principle, however, by no means prevails in all departments of the field of nature. The chemical combination of two substances produces, as is well known, a third substance, with properties different from those of either of the two substances separately or of both of them taken together. Not a trace of the properties of hydrogen or of oxygen is observable in those of their compound, water. The taste of sugar of lead is not the sum of the tastes of its component elements, acetic acid and lead or its oxide, nor is the color of blue vitriol a mixture of the colors of sulphuric acid and copper. This explains why mechanics is a deductive or demonstrative science, and chemistry not. In the one, we can compute the effects of combinations of causes, whether real or hypothetical, from the laws which we know to govern those causes when acting separately, because they continue to observe the same laws when in combination which they observe when separate; whatever would have happened in consequence of each cause taken by itself, happens when they are together, and we have only to cast up the results. Not so in the phenomena which are the peculiar subject of the science of chemistry. There most of the uniformities to which the causes conform when separate cease altogether when they are conjoined, and we are not, at least in the present state of our knowledge, able to foresee what result will follow from any new combination until we have tried the specific experiment.

If this be true of chemical combinations, it is still more true of those far more complex combinations of elements which constitute organized bodies, and in which those extraordinary new uniformities arise which are called the laws of life. All organized bodies are composed of parts similar to those composing inorganic nature, and which have even themselves existed in an inorganic state, but the phenomena of life, which result from the juxtaposition of those parts in a certain manner, bear no analogy to any of the effects which would be produced by the action of the component substances considered as mere physical agents. To whatever degree we might imagine our knowledge of the properties of the several ingredients of a living body to be extended and perfected, it is certain that no mere summing up of the separate actions of those elements will ever amount to the action of the living body itself. The tongue, for instance, is, like all other parts of the animal frame, composed of gelatine, fibrine, and other products of the chemistry of digestion, but from no knowledge of the properties of those substances could we ever predict that it could taste, unless gelatine or fibrine could themselves taste; for no elementary fact can be in the conclusion which was not in the premises.

There are thus two different modes of the conjunct action of causes, from which arise two modes of conflict, or mutual interference, between laws of nature. Suppose, at a given point of time and space, two or more causes, which, if they acted separately, would produce effects contrary, or at least conflicting with each other, one of them tending to undo, wholly or partially, what the other tends to do. Thus the expansive force of the gases generated by the ignition of gunpowder tends to project a bullet toward the sky, while its gravity tends to make it fall to the ground. A stream running into a reservoir at one end tends to fill it higher and higher, while a drain at the other extremity tends to empty it. Now, in such cases as these, even if the two causes which are in joint action exactly annul one another, still the laws of both are fulfilled; the effect is the same as if the drain had been open for half an hour first, and the stream had flowed in for as long afterward. Each agent produces the same amount of effect as if it had acted separately, though the contrary effect which was taking place during the same time obliterated it as fast as it was produced. Here, then, are two causes, producing by their joint operations an effect which at first seems quite dissimilar to those which they produce separately, but which on examination proves to be really the sum of those separate effects. It will be noticed that we here enlarge the idea of the sum of two effects so as to include what is commonly called their difference but which is in reality the result of the addition of opposites, a conception to which mankind are indebted for that admirable extension of the algebraical calculus, which has so vastly increased its powers as an instrument of discovery by introducing into its reasonings (with the sign of subtraction prefixed, and under the name of negative quantities) every description whatever of positive phenomena, provided they are of such a quality in reference to those previously introduced that to add the one is equivalent to subtracting an equal quantity of the other.

There is, then, one mode of the mutual interference of laws of nature in which, even when the concurrent causes annihilate each other's effects, each exerts its full efficacy according to its own law – its law as a separate agent. But in the other description of cases, the agencies which are brought together cease entirely, and a totally different set of phenomena arise, as in the experiment of two liquids which, when mixed in certain proportions, instantly become not a larger amount of liquid, but a solid mass.

The composition of causes the general rule; the other case exceptional

This difference between the case in which the joint effect of causes is the sum of their separate effects and the case in which it is heterogeneous to them – between laws which work together without alteration, and laws which, when called upon to work together, cease and give place to others – is one of the fundamental distinctions in nature. The former case, that of the composition of causes, is the general one; the other is always special and exceptional. There are no objects which do not, as to some of their phenomena, obey the principle of the composition of causes, none that have not some laws which are rigidly fulfilled in every combination into which the objects enter. . . .

Again, laws which were themselves generated in the second mode may generate others in the first. Though there are laws which, like those of chemistry and physiology, owe their existence to a breach of the principle of composition of causes, it does not follow that these peculiar, or, as they might be termed, *heteropathic*, laws are not capable of composition with one another. The causes which by one combination have had their laws altered may carry their new laws with them unaltered into their ulterior combinations. And hence there is no reason to despair of ultimately raising chemistry and physiology to the condition of deductive sciences, for, though it is impossible to deduce all chemical and physiological truths from the laws or properties of simple substances or elementary agents, they may possibly be deducible from laws which commence when these elementary agents are brought together into some moderate number of not very complex combinations. The laws of life will never be deducible from the mere laws of the ingredients, but the prodigiously complex facts of life may all be deducible from comparatively simple laws of life, which laws (depending indeed on combinations, but on comparatively simple combinations, of antecedents) may, in more complex circumstances, be strictly compounded with one another and with the physical and chemical laws of the ingredients. The details of the vital phenomena, even now, afford innumerable exemplifications of the composition of causes, and, in proportion as these phenomena are more accurately studied, there appears more reason to believe that the same laws which operate in the simpler combinations of circumstances do, in fact, continue to be observed in the more complex. This will be found equally true in the phenomena of mind and even in social and political phenomena, the results of the laws of mind. . . .

7 Of observation and experiment [sic]

The first step of inductive inquiry is a mental analysis of complex phenomena into their elements

It results from the preceding exposition that the process of ascertaining what consequents, in nature, are invariably connected with what antecedents, or, in other words, what phenomena are related to each other as causes and effects, is in some sort a process of analysis. . . . If the whole prior state of the entire universe could again recur, it would again be followed by the present state. The question is how to resolve this complex uniformity into the simpler uniformities which compose it and assign to each portion of the vast antecedent the portion of the consequent which is attendant on it.

This operation, which we have called analytical inasmuch as it is the resolution of a complex whole into the component elements, is more than a merely mental analysis. No mere contemplation of the phenomena and partition of them by the intellect alone will of itself accomplish the end we have now in view. Nevertheless, such a mental partition is an indispensable first step. The order of nature, as perceived at a first glance, presents at every instant a chaos followed by another chaos. We must decompose each chaos into single facts. We must learn to see in the chaotic antecedent a multitude of distinct antecedents, in the chaotic consequent a multitude of distinct consequents. This, supposing it done, will not of itself tell us on which of the antecedents each consequent is invariably attendant. To determine that point, we must endeavor to effect a separation of the facts from one another not in our minds only, but in nature. The mental analysis, however, must take place first. And everyone knows that in the mode of performing it one intellect differs immensely from another. It is the essence of the act of observing, for the observer is not he who merely sees the thing which is before his eyes, but he who sees what parts that thing is composed of. . . .

The extent and minuteness of observation which may be requisite and the degree of decomposition to which it may be necessary to carry the mental analysis depend on the particular purpose in view. To ascertain the state of the whole universe at any particular moment is impossible, but would also be useless. In making chemical experiments, we do not think it necessary to note the position of the planets, because experience has shown, as a very superficial experience is sufficient to show, that in such cases that circumstance is not material to the result; and accordingly, in the ages when men believed in the occult influences of the heavenly bodies, it might have been unphilosophical to omit ascertaining the precise condition of those bodies at the moment of the experiment. As to the degree of minuteness of the mental subdivision, if we were obliged to break down what we observe into its very simplest elements, that is, literally into single facts, it would be difficult to say where we should find them; we can hardly ever affirm that our divisions of any kind have reached the ultimate unit. But this, too, is fortunately unnecessary. The only object of the mental separation is to suggest the requisite physical separation, so that we may either accomplish it ourselves or seek for it in nature, and we have done enough when we have carried the subdivision as far as the point at which we are able to see what observations or experiments we require. It is only essential, at whatever point our mental decomposition of facts may for the present have stopped, that we should hold ourselves ready and able to carry it further as occasion requires and should not allow the freedom of our discriminating faculty to be imprisoned by the swathes and bands of ordinary classification, as was the case with all early speculative inquirers, not excepting the Greeks, to whom it seldom occurred that what was called by one abstract name might, in reality, be several phenomena, or that there was a possibility of decomposing the facts of the universe into any elements but those which ordinary language already recognized.

The next is an actual separation of those elements

The different antecedents and consequents being, then, supposed to be, so far as the case requires, ascertained and discriminated from one another, we are to inquire which is connected with which. In every instance which comes under our observation, there are many antecedents and many consequents. If those antecedents could not be severed from one another except in thought or if those consequents never were found apart, it would be impossible for us to distinguish (a *posteriori*, at least) the real laws, or to assign to any cause its effect, or to any effect its cause. To do so, we must be able to meet with some of the antecedents apart from the rest and observe what follows from them, or some of the consequents and observe by what they are preceded. We must, in short, follow the Baconian rule of *varying the circumstances*. This is, indeed, only the first rule of physical inquiry and not, as some have thought, the sole rule, but it is the foundation of all the rest.

For the purpose of varying the circumstances, we may have recourse (according to a distinction commonly made) either to observation or to experiment; we may either find an instance in nature suited to our purposes or, by an artificial arrangement of circumstances, *make* one. The value of the instance depends on what it is in itself, not on the mode in which it is obtained; its employment for the purposes of induction depends on the same principles in the one case and in the other, as the uses of money are the same whether it is inherited or acquired. There is, in short, no difference in kind, no real logical distinction, between the two processes of investigation. There are, however, practical distinctions to which it is of considerable importance to advert. . . .

8 Of the four methods of experimental inquiry [sic]

Method of agreement

The simplest and most obvious modes of singling out from among the circumstances which precede or follow a phenomenon those with which it is really connected by an invariable law are two in number. One is by comparing together different instances in which the phenomenon occurs. The other is by comparing instances in which the phenomenon does occur with instances in other respects similar in which it does not. These two methods may be respectively denominated the Method of Agreement and the Method of Difference.

In illustrating these methods, it will be necessary to bear in mind the twofold character of inquiries into the laws of phenomena, which may be either inquiries into the cause of a given effect or into the effects or properties of a given cause. We shall consider the methods in their application to either order of investigation and shall draw our examples equally from both.

We shall denote antecedents by the large letters of the alphabet and the consequents corresponding to them by the small. Let A, then, be an agent or cause, and let the object of our inquiry be to ascertain what are the effects of this cause. If we can either find or produce the agent A in such varieties of circumstances that the different cases have no circumstance in common except A, then whatever effect we find to be produced in all our trials is indicated as the effect of A. Suppose, for example, that A is tried along with B and C and that the effect is $a\,b\,c$; and suppose that A is next tried with D and E, but without B and C, and that the effect is $a\,d\,e$. Then we may reason thus: b and c are not effects of A, for they were not produced by it in the second experiment; nor are d and e, for they were not produced in the first. Whatever is really the effect of A must have

been produced in both instances; now this condition is fulfilled by no circumstance except a. The phenomenon a cannot have been the effect of B or C, since it was produced where they were not; nor of D or E, since it was produced where they were not. Therefore, it is the effect of A.

For example, let the antecedent A be the contact of an alkaline substance and an oil. This combination being tried under several varieties of circumstances, resembling each other in nothing else, the results agree in the production of a greasy and detersive or saponaceous substance; it is, therefore, concluded that the combination of an oil and an alkali causes the production of a soap. It is thus we inquire by the Method of Agreement into the effect of a given cause.

In a similar manner we may inquire into the cause of a given effect. Let a be the effect. Here, as shown in the last chapter [sic], we have only the resource of observation without experiment; we cannot take a phenomenon of which we know not the origin and try to find its mode of production by producing it; if we succeeded in such a random trial, it could only be by accident. But if we can observe a in two different combinations, $a\,b\,c$ and $a\,d\,e$, and if we know or can discover that the antecedent circumstances in these cases respectively were A B C and A D E, we may conclude, by a reasoning similar to that in the preceding example, that A is the antecedent connected with the consequent a by a law of causation. B and C, we may say, cannot be causes of a, since on its second occurrence they were not present; nor are D and E, for they were not present on its first occurrence. A, alone of the five circumstances, was found among the antecedents of a in both instances.

For example, let the effect a be crystallization. We compare instances in which bodies are known to assume crystalline structure but which have no other point of agreement, and we find them to have one and, as far as we can observe, only one, antecedent in common: the deposition of a solid matter from a liquid state, either a state of fusion or of solution. We conclude, therefore, that the solidification of a substance from a liquid state is an invariable antecedent of its crystallization.

In this example we may go further and say it is not only the invariable antecedent but the cause, or, at least, the proximate event which completes the cause. For in this case we are able, after detecting the antecedent A, to produce it artificially and, by finding that *a* follows it, verify the result of our induction. The importance of thus reversing the proof was strikingly manifested when, by keeping a phial of water charged with siliceous particles undisturbed for years, a chemist (I believe Dr. Wollaston) succeeded in obtaining crystals of quartz, and in the equally interesting experiment in which Sir James Hall produced artificial marble by the cooling of its materials from fusion under immense pressure; two admirable examples of the light which may be thrown upon the most secret processes of Nature by well-contrived interrogation of her.

But if we cannot artificially produce the phenomenon A, the conclusion that it is the cause of *a* remains subject to very considerable doubt. Though an invariable, it may not be the unconditional antecedent of *a*, but may precede it as day precedes night or night day. This uncertainty arises from the impossibility of assuring ourselves that A is the *only* immediate antecedent common to both the instances. If we could be certain of having ascertained all the invariable antecedents, we might be sure that the unconditional invariable antecedent, or cause, must be found somewhere among them. Unfortunately, it is hardly ever possible to ascertain all the antecedents unless the phenomenon is one which we can produce artificially. Even then, the difficulty is merely lightened, not removed; men knew how to raise water in pumps long before they adverted to what was really the operating circumstance in the means they employed, namely, the pressure of the atmosphere on the open surface of the water. It is, however, much easier to analyze completely a set of arrangements made by ourselves than the whole complex mass of the agencies which nature happens to be exerting at the moment of the production of a given phenomenon. We may overlook some of the material circumstances in an experiment with an electrical machine, but we shall, at the worst, be better acquainted with them than with those of a thunderstorm.

The mode of discovering and proving laws of nature which we have now examined proceeds on the following axiom: whatever circumstances can be excluded without prejudice to the phenomenon, or can be absent notwithstanding its presence, is not connected with it in the way of causation. The casual circumstances being thus eliminated, if only one remains, that one is the cause which we are in search of; if more than one, they either are, or contain among them, the cause; and so, *mutatis mutandis*, of the effect. As this method proceeds by comparing different instances to ascertain in what they agree, I have termed it the Method of Agreement, and we may adopt as its regulating principle the following canon:

> **First Canon**
> If two or more instances of the phenomenon under investigation have only one circumstance in common, the circumstance in which alone all the instances agree is the cause (or effect) of the given phenomenon.

Quitting for the present the Method of Agreement, to which we shall almost immediately return, we proceed to a still more potent instrument of the investigation of nature, the Method of Difference.

Method of Difference

In the Method of Agreement, we endeavored to obtain instances which agreed in the given circumstance but differed in every other; in the present method we require, on the contrary, two instances resembling one another in every other respect, but differing in the presence or absence of the phenomenon we wish to study. If our object be to discover the effects of an agent A, we must procure A in some set of ascertained circumstances, as A B C, and having noted the effects produced, compare them with the effect of the remaining circumstances B C, when A is absent. If the effect of A B C is *a b c*, and the effect of B C *b c*, it is evident that the effect of A is *a*. So again, if we begin at the other end and desire to investigate the cause of an effect *a*, we must select an instance, as *a b c*, in which the effect occurs, and in which the antecedents were A B C, and we must look out for another instance in which the remaining circumstances, *b c*, occur without *a*. If the antecedents, in that instance, are B C, we know that the cause of *a* must be A – either A alone, or A in conjunction with some of the other circumstances present.

It is scarcely necessary to give examples of a logical process to which we owe almost all the inductive conclusions we draw in daily life. When a man is shot through the heart, it is by this method we know that it was the gunshot which killed him, for he was in the fullness of life immediately before, all circumstances being the same except the wound.

The axioms implied in this method are evidently the following: whatever antecedent cannot be excluded without preventing the phenomenon is the cause, or a condition, of that phenomenon; whatever consequent can be excluded, with no other difference in the antecedents than the absence of a particular one, is the effect of that one. Instead of comparing different instances of a phenomenon to discover in what they agree, this method compares an instance of its occurrence with an instance of its non-occurrence to discover in what they differ. The canon which is the regulating principle of

the Method of Difference may be expressed as follows:

Second Canon

If an instance in which the phenomenon under investigation occurs and an instance in which it does not occur have every circumstance in common save one, that one occurring only in the former, the circumstance in which alone the two instances differ, is the effect, or cause, or a necessary part of the cause, of the phenomenon.

Mutual relation of these two methods

The two methods which we have now stated have many features of resemblance, but there are also many distinctions between them. Both are methods of *elimination*. This term (which is employed in the theory of equations to denote the process by which one after another of the elements of a question is excluded, and the solution made to depend upon the relation between the remaining elements only,) is well suited to express the operation, analogous to this, which has been understood since the time of Bacon to be the foundation of experimental inquiry: namely, the successive exclusion of the various circumstances which are found to accompany a phenomenon in a given instance, in order to ascertain what are those among them which can be absent consistently with the existence of the phenomenon. The Method of Agreement stands on the ground that whatever can be eliminated, is not connected with the phenomenon by any law. The Method of Difference has for its foundation, that whatever can *not* be eliminated, is connected with the phenomenon by a law.

Of these methods, that of Difference is more particularly a method of artificial experiment; while that of Agreement is more especially the resource we employ where experimentation is impossible. A few reflections will prove the fact, and point out the reason of it.

It is inherent in the peculiar character of the Method of Difference, that the nature of the combinations which it requires is much more strictly defined than in the Method of Agreement. The two instances which are to be compared with one another must be in the relation of A B C and B C, or of *a b c* and *b c*. It is true that this similarity of circumstances needs not extend to such as are already known to be immaterial to the result. And in the case of most phenomena we learn at once, from the most ordinary experience, that most of the coexistent phenomena of the universe may be either present or absent without affecting the given phenomenon; or, it [sic] present, are present indifferently when the phenomenon does not happen, and when it does. Still, even limiting the identity which is required between the two instances, A B C and B C, to such circumstances as are not already known to be indifferent; it is very seldom that nature affords two instances, of which we can be assured that they stand in this precise relation to one another. In the spontaneous operations of nature there is generally such complication and such obscurity, they are mostly either on so overwhelmingly large or on so inaccessible minute a scale, we are so ignorant of a great part of the facts which really take place, and even those of which we are not ignorant are so multitudinous, and therefore so seldom exactly alike in any two cases, that a spontaneous experiment, of the kind required by the Method of Difference, is commonly not to be found. When, on the contrary, we obtain a phenomenon by an artificial experiment, a pair of instances such as the method requires is obtained almost as a matter of course, provided the process does not last a long time. A certain state of surrounding circumstances existed before we commenced the experiment: this is B C. We then introduce A; say, for instance, by merely bringing an object from another part of the room, before there has been time for any change in the other elements. It is, in short (as M. Comte observes) the very nature of an experiment, to introduce into the pre-existing state of circumstances a change perfectly definite. We choose a previous state of things with which we are well acquainted, so that no unforeseen alternation in that state is likely to pass unobserved; and into this we introduce, as rapidly as possible, the phenomenon which we wish to study; so that we in general are entitled to feel complete assurance, that the pre-existing state, and the state which we have produced, differ in nothing except in the presence or absence of that phenomenon. If a bird is taken from a cage, and instantly plunged into carbonic acid gas, the experimentalist may be fully assured (at all events after one or two repetitions) that no circumstance capable of causing suffocation had supervened in the interim, except the change from immersion in the atmosphere to immersion in carbonic acid gas. There is one doubt, indeed, which may remain in some cases of this description; the effect may have been produced not by the change, but by the means we employed to produce the change. The possibility, however, of this last supposition generally admits of being conclusively tested by other experiments. It thus appears that in the study of the various kinds of phenomena which we can, by our voluntary agency, modify or control, we can, in general, satisfy the requisitions of the Method of Difference, but that by the spontaneous operations of nature those requisitions are seldom fulfilled.

The reverse of this is the case with the Method of Agreement. We do not here require instances of so special and determinate a kind. Any instances whatever in which nature presents us with a phenomenon may be examined for the purposes of this method, and, if all such instances agree in anything, a conclusion of considerable value is already attained. We can seldom, indeed, be sure that the one point of agreement is the only one; but this ignorance does not, as in the Method of Difference, vitiate the conclusion; the certainty of the result, as far as it goes, is not affected. We have ascertained one invariable antecedent or consequent, however many other invariable antecedents or consequents may still remain unascertained. If A B C, A D E, A F G, are all equally followed by a, then a is an invariable consequent of A. If $a\ b\ c, a\ d\ e, a\ f\ g$, all number A among their antecedents, then A is connected as an antecedent, by some invariable law, with a. But to determine whether this invariable antecedent is a cause or this invariable consequent an effect, we must be able, in addition, to produce the one by means of the other, or, at least, to obtain that which alone constitutes our assurance of having produced anything, namely, an instance in which the effect, a, has come into existence with no other change in the pre-existing circumstances than the addition of A. And this, if we can do it, is an application of the Method of Difference, not of the Method of Agreement.

It thus appears to be by the Method of Difference alone that we can ever, in the way of direct experience, arrive with certainty at causes. The Method of Agreement leads only to laws of phenomena (as some writers call them, but improperly, since laws of causation are also laws of phenomena), that is, to uniformities which either are not laws of causation or in which the question of causation must for the present remain undecided. The Method of Agreement is chiefly to be resorted to as a means of suggesting applications of the Method of Difference (as in the last example the comparison of A B C, A D E, A F G, suggested that A was the antecedent on which to try the experiment whether it could produce a), or as an inferior resource, in case the Method of Difference is impracticable, which, as we before showed, generally arises from the impossibility of artificially producing the

phenomena. And hence it is that the Method of Agreement, though applicable in principle to either case, is more emphatically the method of investigation on those subjects where artificial experimentation is impossible, because on those it is, generally, our only resource of a directly inductive nature, while, in the phenomena which we can produce at pleasure, the Method of Difference generally affords a more efficacious process which will ascertain causes as well as mere laws.

Joint Method of Agreement and Difference

There are, however, many cases in which, though our power of producing the phenomenon is complete, the Method of Difference either cannot be made available at all, or not without a previous employment of the Method of Agreement. This occurs when the agency by which we can produce the phenomenon is not that of one single antecedent, but a combination of antecedents which we have no power of separating from each other and exhibiting apart. For instance, suppose the subject of inquiry to be the cause of the double refraction of light. We can produce this phenomenon at pleasure by employing any one of the many substances which are known to refract light in that peculiar manner. But if, taking one of those substances, as Iceland spar, for example, we wish to determine on which of the properties of Iceland spar this remarkable phenomenon depends, we can make no use, for that purpose, of the Method of Difference, for we cannot find another substance precisely resembling Iceland spar except in some one property. The only mode, therefore, of prosecuting this inquiry is that afforded by the Method of Agreement, by which, in fact, through a comparison of all the known substances which have the property of doubly refracting light, it was ascertained that they agree in the circumstance of being crystalline substances, and though the converse does not hold, though all crystalline substances have not the property of double refraction, it was concluded, with reason, that there is a real connection between these two properties, that either crystalline structure or the cause which gives rise to that structure is one of the conditions of double refraction.

Out of this employment of the Method of Agreement arises a peculiar modification of that method which is sometimes of great avail in the investigation of nature. In cases similar to the above, in which it is not possible to obtain the precise pair of instances which our second canon requires – instances agreeing in every antecedent except A or in every consequent except *a* – we may yet be able, by a double employment of the Method of Agreement, to discover in what the instances which contain A or *a* differ from those which do not.

If we compare various instances in which *a* occurs and find that they all have in common the circumstance A, and (as far as can be observed) no other circumstance, the Method of Agreement, so far, bears testimony to a connection between A and *a*. In order to convert this evidence of connection into proof of causation by the direct Method of Difference, we ought to be able, in some one of these instances, as for example, A B C, to leave out A, and observe whether by doing so, *a* is prevented. Now supposing (what is often the case) that we are not able to try this decisive experiment; yet, provided we can by any means discover what would be its result if we could try it, the advantage will be the same. Suppose, then, that as we previously examined a variety of instances in which *a* occurred and found them to agree in containing A, so we now observe a variety of instances in which *a* does not occur and find them agree in not containing A, which establishes, by the Method of Agreement, the same connection between the absence of A and the absence of *a* which was before established between their presence. As, then, it had been shown that whenever A is present *a* is present, so, it being now shown that when A is taken away *a* is removed along with it, we have by the one proposition A B C, *a b c*, by the other B C, *b c*, the positive and negative instances which the Method of Difference requires.

This method may be called the Indirect Method of Difference, or the Joint Method of Agreement and Difference, and consists in a double employment of the Method of Agreement, each proof being independent of the other and corroborating it. But it is not equivalent to a proof by the direct Method of Difference. For the requisitions of the Method of Difference are not satisfied unless we can be quite sure either that the instances affirmative of *a* agree in no antecedent whatever but A, or that the instances negative of *a* agree in nothing but the negation of A. Now, if it were possible, which it never is, to have this assurance, we should not need the joint method, for either of the two sets of instances separately would then be sufficient to prove causation. This indirect method, therefore, can only be regarded as a great extension and improvement of the Method of Agreement, but not as participating in the more cogent nature of the Method of Difference. The following may be stated as its canon:

Third Canon

If two or more instances in which the phenomenon occurs have only one circumstance in common, while two or more instances in which it does not occur have nothing in common save the absence of that circumstance, the circumstance in which alone the two sets of instances differ is the effect, or the cause, or an indispensable part of the cause, of the phenomenon.

We shall presently see that the Joint Method of Agreement and Difference constitutes, in another respect not yet adverted to, an improvement upon the common Method of Agreement, namely, in being unaffected by a characteristic imperfection of that method, the nature of which still remains to be pointed out. But as we cannot enter into this exposition without introducing a new element of complexity into this long and intricate discussion, I shall postpone it to a subsequent chapter [sic] and shall at once proceed to a statement of two other methods, which will complete the enumeration of the means which mankind possess for exploring the laws of nature by specific observation and experience.

Method of Residues

The first of these has been aptly denominated the Method of Residues. Its principle is very simple. Subducting from any given phenomenon all the portions which, by virtue of preceding inductions, can be assigned to known causes, the remainder will be the effect of the antecedents which had been overlooked, or of which the effect was as yet an unknown quantity.

Suppose, as before, that we have the antecedents A B C, followed by the consequents $a\ b\ c$, and that by previous inductions (founded, we will suppose, on the Method of Difference) we have ascertained the causes of some of these effects, or the effects of some of these causes; and are thence apprised that the effect of A is a, and that the effect of B is b. Subtracting the sum of these effects from the total phenomenon, there remains c, which now, without any fresh experiments, we may know to be the effect of C. This Method of Residues is in truth a peculiar modification of the Method of Difference. If the instance A B C, $a\ b\ c$, could have been compared with a single instance A B, $a\ b$, we should have proved C to be the cause of c, by the common process of the Method of Difference. In the present case, however, instead of a single instance A B, we have had to study separately the causes A and B, and to infer from the effects which they produce separately, what effect they must produce in the case A B C where they act together. Of the two instances, therefore, which the Method of Difference requires, – the one positive, the other negative, – the negative one, or that in which the given phenomenon is absent, is not the direct result of observation and experiment, but has been arrived at by deduction. As one of the forms of the Method of Difference, the Method of Residues partakes of its rigorous certainty, provided the previous inductions, those which gave the effects of A and B, were obtained by the same infallible methods, and provided we are certain that C is the only antecedent to which the residual phenomenon c can be referred; the only agent of which we had not already calculated and subducted the effect. But as we can never be quite certain of this, the evidence derived from the Method of Residues is not complete unless we can obtain C artificially and try it separately, or unless its agency when once suggested, can be accounted for, and proved deductively, from known laws.

Even with these reservations, the Method of residues is one of the most important amoung [sic] our instruments of discovery. Of all the methods of investigating laws of nature, this is the most fertile in unexpected results: often informing us of sequences in which neither the cause nor the effect were sufficiently conspicuous to attract of themselves the attention of observers. The agent C may be an obscure circumstance, not likely to have been perceived unless sought for, nor likely to have been sought for until attention had been awakened by the insufficiency of the obvious causes to account for the whole of the effect. And c may be so disguised by its intermixture with a and b, that it would scarcely have presented itself spontaneously as a subject of separate study. Of these uses of the method, we shall presently cite some remarkable examples. The canon of the Method of Residues is as follows:–

Fourth Canon

Subduct from any phenomenon such part as is known by previous inductions to be the effect of certain antecedents, and the residue of the phenomenon is the effect of the remaining antecedents.

[. . .]

Method of Concomitant Variations

The case in which this method admits of the most extensive employment, is that in which the variations of the cause are variations of quantity. Of such variations we may in general affirm with safety, that they will be attended not only with variations, but with similar variations, of the effect: the proposition, that more of the cause is followed by more of the effect, being a corollary from the principle of the Composition of Causes, which, as we have seen, is the general rule of causation; cases of the opposite description, in which causes change their properties on being conjoined with one another, being, on the contrary, special and exceptional. Suppose, then, that when A changes in quantity, *a* also changes in quantity, and in such a manner that we can trace the numerical relation which the changes of the one bear to such changes of the other as take place within our limits of observation. We may then, with certain precautions, safely conclude that the same numerical relation will hold beyond those limits. If, for instance, we find that when A is double, *a* is double; that when A is treble or quadruple, *a* is treble or quadruple; we may conclude that if A were a half or a third, *a* would be a half or a third, and finally, that if A were annihilated, *a* would be annihilated, and that *a* is wholly

the effect of A, or wholly the effect of the same cause with A. And so with any other numerical relation according to which A and *a* would vanish simultaneously; as for instance if *a* were proportional to the square of A. If, on the other hand, *a* is not wholly the effect of A, but yet varies when A varies, it is probably a mathematical function not of A alone but of A and something else: its changes, for example, may be such as would occur if part of it remained constant, or varied on some other principle, and the remainder varied in some numerical relation to the variations of A. In that case, when A diminishes, *a* will seem to approach not towards zero, but towards some other limit: and when the series of variations is such as to indicate what that limit is, if constant, or the law of its variation if variable, the limit will exactly measure how much of *a* is the effect of some other and independent cause, and the remainder will be the effect of A (or of the cause of A).

These conclusions, however, must not be drawn without certain precautions. In the first place, the possibility of drawing them at all, manifestly supposes that we are acquainted not only with the variations, but with the absolute quantities, both of A and *a*. If we do not know the total quantities, we cannot, of course, determine the real numerical relation according to which those quantities vary. It is therefore an error to conclude, as some have concluded, that because increase of heat expands bodies, that is, increases the distance between their particles, therefore the distance is wholly the effect of heat, and that if we could entirely exhaust the body of its heat, the particles would be in complete contact. This is no more than a guess, and of the most hazardous sort, not a legitimate induction: for since we neither know how much heat there is in any body, nor what is the real distance between any two of its particles, we cannot judge whether the contraction of the distance does or does not follow the diminution of the quantity of heat according to such a numerical relation that the two quantities would vanish simultaneously.

In contrast with this, let us consider a case in which the absolute quantities are known; the case contemplated in the first law of motion; viz. that all bodies in motion continue to move in a straight line with uniform velocity until acted upon by some new force. This assertion is in open opposition to first appearances; all terrestrial objects, when in motion, gradually abate their velocity and at last stop; which accordingly the ancients, with their *inductio per enumerationem simplicem*, imagined to be the law. Every moving body, however, encounters various obstacles, as friction, the resistance of the atmosphere, &c., which we know by daily experience to be causes capable of destroying motion. It was suggested that the whole of the retardation might be owing to these causes. How was this inquired into? If the obstacles could have been entirely removed, the case would have been amenable to the Method of Difference. They could not be removed, they could only be diminished, and the case, therefore, admitted only of the Method of Concomitant Variations. This accordingly being employed, it was found that every diminution of the obstacles diminished the retardation of the motion: and inasmuch as in this case (unlike the case of heat) the total quantities both of the antecedent and of the consequent were known; it was practicable to estimate, with an approach to accuracy, both the amount of the retardation and the amount of the retarding causes, or resistances, and to judge how near they both were to being exhausted; and it appeared that the effect dwindled as rapidly, and at each step was as far on the road towards annihilation, as the cause was. The simple oscillation of a weight suspended from a fixed point, and moved a little out of the perpendicular, which in ordinary circumstances lasts but a few minutes, was prolonged in Borda's

experiments to more than thirty hours, by diminishing as much as possible the friction at the point of suspension, and by making the body oscillate in a space exhausted as nearly as possible of its air. There could therefore be no hesitation in assigning the whole of the retardation of motion to the influence of the obstacles: and since, after subducting this retardation from the total phenomenon, the remainder was an uniform velocity, the result was the proposition known as the first law of motion.

There is also another characteristic uncertainty affecting the inference that the law of variation which the quantities observe within our limits of observation, will hold beyond those limits. There is of course, in the first instance, the possibility that beyond the limits, and in circumstances therefore of which we have no direct experience, some counteracting cause might develop itself; either a new agent, or a new property of the agents concerned, which lies dormant in the circumstances we are able to observe. This is an element of uncertainty which enters largely into all our predictions of effects; but it is not peculiarly applicable to the Method of Concomitant Variations. The uncertainty, however, of which I am about to speak, is characteristic of that method; especially in the cases in which the extreme limits of our observation are very narrow, in comparison with the possible variations in the quantities of the phenomena. Any one who has the slightest acquaintance with mathematics, is aware that very different laws of variation may produce numerical results which differ but slightly from one another within narrow limits; and it is often only when the absolute amounts of variation are considerable, that the difference between the results given by one law and by another becomes appreciable. When, therefore, such variations in the quantity of the antecedents as we have the means of observing, are small in comparison with the total quantities, there is much danger lest we should mistake the numerical law, and be led to miscalculate the variations which would take place beyond the limits; a miscalculation which would vitiate any conclusion respecting the dependence of the effect upon the cause, that could be founded on those variations. Examples are not wanting of such mistakes. "The formulæ," says Sir John Herschel,[3] "which have been empirically deduced for the elasticity of steam, (till very recently,) and those for the resistance of fluids, and other similar subjects," when relied on beyond the limits of the observations from which they were deduced, "have almost invariably failed to support the theoretical structures which have been erected on them."

In this uncertainty, the conclusion we may draw from the concomitant variations of *a* and A to the existence of an invariable and exclusive connection between them, or to the permanency of the same numerical relation between their variations when the quantities are much greater or smaller than those which we have had the means of observing, cannot be considered to rest on a complete induction. All that in such a case can be regarded as proved on the subject of causation is that there is some connection between the two phenomena: that A, or something which can influence A, must be *one* of the causes which collectively determine *a*. We may, however, feel assured that the relation which we have observed to exist between the variations of A and *a* will hold true in all cases which fall between the same extreme limits; that is, wherever the utmost increase or diminution in which the result has been found by observation to coincide with the law is not exceeded.

The four methods which it has now been attempted to describe are the only possible modes of experimental inquiry – of direct induction a posteriori, as distinguished from deduction; at least, I know not, nor am able to imagine any others. And even of these, the

Method of Residues, as we have seen, is not independent of deduction, though, as it also requires specific experience, it may, without impropriety, be included among methods of direct observation and experiment.

These, then, with such assistance as can be obtained from deduction, compose the available resources of the human mind for ascertaining the laws of the succession of phenomena. Before proceeding to point out certain circumstances by which the employment of these methods is subjected to an immense increase of complication and of difficulty, it is expedient to illustrate the use of the methods by suitable examples drawn from actual physical investigations. These, accordingly, will form the subject of the succeeding chapter [*sic*].

[. . .]

Dr. Whewell's objections to the four methods

Dr. Whewell has expressed a very unfavorable opinion of the utility of the four methods, as well as of the aptness of the examples by which I have attempted to illustrate them. His words are these:[4]

> Upon these methods, the obvious thing to remark is, that they take for granted the very thing which is most difficult to discover, the reduction of the phenomena to formulae such as are here presented to us. When we have any set of complex facts offered to us; for instance, those which were offered in the cases of discovery which I have mentioned – the facts of the planetary paths, of falling bodies, of refracted rays, of cosmical motions, of chemical analysis; and when, in any of these cases, we would discover the law of

nature which governs them, or, if anyone chooses so to term it, the feature in which all the cases agree, where are we to look for our A, B, C, and *a, b, c?* Nature does not present to us the cases in this form; and how are we to reduce them to this form? You say *when* we find the combination of A B C with *a b c* and A B D with *a b d*, then we may draw our inference. Granted; but when and where are we to find such combinations? Even now that the discoveries are made, who will point out to us what are the A, B, C, and *a, b, c,* elements of the cases which have just been enumerated? Who will tell us which of the methods of inquiry those historically real and successful inquiries exemplify? Who will carry these formulae through the history of the sciences, as they have really grown up, and show us that these four methods have been operative in their formation; or that any light is thrown upon the steps of their progress by reference to these formulae?

He adds that, in this work, the methods have not been applied

"to a large body of conspicuous and undoubted examples of discovery, extending along the whole history of science," which ought to have been done in order that the methods might be shown to possess the "advantage" (which he claims as belonging to his own) of being those "by which all great discoveries in science have really been made" (p. 277).

There is a striking similarity between the objections here made against canons of induction and what was alleged, in the last century, by as able men as Dr. Whewell, against the acknowledged canon of ratiocination. Those who protested against the Aristotelian logic said of the syllogism what Dr. Whewell says of the inductive methods, that it "takes for granted the very thing which is most difficult to discover, the reduction of the argument to formulae such as are here presented to us." The grand difficulty, they said, is to obtain your syllogism, not to judge of its correctness when obtained. On the matter of fact, both they and Dr. Whewell are right. The greatest difficulty in both cases is, first, that of obtaining the evidence and, next, of reducing it to the form which tests its conclusiveness. But if we try to reduce it without knowing what it is to be reduced to, we are not likely to make much progress. It is a more difficult thing to solve a geometrical problem than to judge whether a proposed solution is correct, but if people were not able to judge of the solution when found, they would have little chance of finding it. And it cannot be pretended that to judge of an induction when found is perfectly easy, is a thing for which aids and instruments are superfluous, for erroneous inductions, false inferences from experience, are quite as common, on some subjects much commoner than true ones. The business of inductive logic is to provide rules and models (such as the syllogism and its rules are for ratiocination) to which if inductive arguments conform, those arguments are conclusive, and not otherwise. This is what the four methods profess to be, and what I believe they are universally considered to be by experimental philosophers, who had practiced all of them long before anyone sought to reduce the practice to theory.

The assailants of the syllogism had also anticipated Dr. Whewell in the other branch of his argument. They said that no discoveries were ever made by syllogism, and Dr. Whewell says, or seems to say, that none were ever made by the four methods of induction. To the former objectors, Archbishop Whately very pertinently answered that their argument, if good at all, was good against the reasoning process altogether, for whatever cannot be reduced to syllogism is not reasoning. And Dr. Whewell's argument, if good at all, is good against all inferences from experience. In saying that no discoveries were ever made by the four methods, he affirms that none were ever made by observation and experiment, for, assuredly, if any were, it was by processes reducible to one or other of those methods.

This difference between us accounts for the dissatisfaction which my examples give him, for I did not select them with a view to satisfy anyone who required to be convinced that observation and experiment are modes of acquiring knowledge; I confess that in the choice of them I thought only of illustration and of facilitating the *conception* of the methods by concrete instances. If it had been my object to justify the processes themselves as means of investigation, there would have been no need to look far off or make use of recondite or complicated instances. As a specimen of a truth ascertained by the Method of Agreement, I might have chosen the proposition, "Dogs bark." This dog, and that dog, and the other dog, answer to A B C, A D E, A F G. The circumstance of being a dog answers to A. Barking answers to *a*. As a truth made known by the Method of Difference, "Fire burns" might have sufficed. Before

I touch the fire I am not burned; this is B C; I touch it, and am burned; this is A B C, *a b c*.

Such familiar experimental processes are not regarded as inductions by Dr. Whewell, but they are perfectly homogeneous with those by which, even on his own showing, the pyramid of science is supplied with its base. In vain he attempts to escape from this conclusion by laying the most arbitrary restrictions on the choice of examples admissible as instances of induction; they must neither be such as are still matter of discussion (p. 265), nor must any of them be drawn from mental and social subjects (p. 269), nor from ordinary observation and practical life (pp. 241–247) [sic]. They must be taken exclusively from the generalizations by which scientific thinkers have ascended to great and comprehensive laws of natural phenomena. Now it is seldom possible, in these complicated inquiries, to go much beyond the initial steps without calling in the instrument of deduction and the temporary aid of hypothesis, as I myself, in common with Dr. Whewell, have maintained against the purely empirical school. Since, therefore, such cases could not conveniently be selected to illustrate the principles of mere observation and experiment, Dr. Whewell is misled by their absence into representing the experimental methods as serving no purpose in scientific investigation, forgetting that if those methods had not supplied the first generalizations, there would have been no materials for his own conception of induction to work upon.

His challenge, however, to point out which of the four methods are exemplified in certain important cases of scientific inquiry, is easily answered. "The planetary paths," as far as they are a case of induction at all,[5] fall under the Method of Agreement. The law of "falling bodies," namely, that they describe spaces proportional to the squares of the times, was historically a deduction from the first law of motion, but the experiments by which it was verified and by which it might have been discovered were examples of the Method of Agreement, and the apparent variation from the true law caused by the resistance of the air was cleared up by experiments *in vacuo*, constituting an application of the Method of Difference. The law of "refracted rays" (the constancy of the ratio between the sines of incidence and of refraction for each refracting substance) was ascertained by direct measurement and, therefore, by the Method of Agreement. The "cosmical motions" were determined by highly complex processes of thought in which deduction was predominant, but the Methods of Agreement and of concomitant variations had a large part in establishing the empirical laws. Every case without exception of "chemical analysis" constitutes a well-marked example of the Method of Difference. To anyone acquainted with the subjects – to Dr. Whewell himself – there would not be the smallest difficulty in setting out "the A B C and *a b c* elements" of these cases.

If discoveries are ever made by observation and experiment without deduction, the four methods are methods of discovery; but even if they were not methods of discovery, it would not be the less true that they are the sole methods of proof, and, in that character, even the results of deduction are amenable to them. The great generalizations which begin as hypotheses must end by being proved and are, in

reality (as will be shown hereafter), proved by the four methods. Now it is with proof, as such, that logic is principally concerned. This distinction has indeed no chance of finding favour with Dr. Whewell, for it is the peculiarity of his system not to recognize, in cases of induction, any necessity for proof. If, after assuming an hypothesis and carefully collating it with facts, nothing is brought to light inconsistent with it, that is, if experience does not disprove it, he is content; at least until a simpler hypothesis, equally consistent with experience, presents itself. If this be induction, doubtless there is no necessity for the four methods. But to suppose that it is so appears to me a radical misconception of the nature of the evidence of physical truths.

So real and practical is the need of a test for induction similar to the syllogistic test of ratiocination, that inferences which bid defiance to the most elementary notions of inductive logic are put forth without misgiving by persons eminent in physical science as soon as they are off the ground on which they are conversant with the facts and not reduced to judge only by the arguments; and, as for educated persons in general, it may be doubted if they are better judges of a good or a bad induction than they were before Bacon wrote. The improvement in the results of thinking has seldom extended to the processes, or has reached, if any process, that of investigation only, not that of proof. A knowledge of many laws of nature has doubtless been arrived at by framing hypotheses and finding that the facts corresponded to them, and many errors have been got rid of by coming to a knowledge of facts which were inconsistent with them, but not by discovering that the mode of thought which led to the errors was itself faulty and might have been known to be such independently of the facts which disproved the specific conclusion. Hence it is that while the thoughts of mankind have on many subjects worked themselves practically right, the thinking power remains as weak as ever; and on all subjects on which the facts which would check the result are not accessible, as in what relates to the invisible world, and even, as has been seen lately, to the visible world of the planetary regions, men of the greatest scientific acquirements argue as pitiably as the merest ignoramus. For though they have made many sound inductions, they have not learned from them (and Dr. Whewell thinks there is no necessity that they should learn) the principles of inductive *evidence*.

10 Of plurality of causes and of the intermixture of effects

One effect may have several causes

In the preceding exposition of the four methods of observation and experiment by which we contrive to distinguish among a mass of co-existent phenomena the particular effect due to a given cause or the particular cause which gave birth to a given effect, it has been necessary to suppose, in the first instance, for the sake of simplification, that this analytical operation is encumbered by no other difficulties than what are essentially inherent in its nature, and to represent to ourselves, therefore, every effect, on the one hand as connected exclusively with a single cause, and on the other hand as incapable of being mixed and confounded with any other co-existent effect. We have regarded $a\ b\ c\ d\ e$, the

aggregate of the phenomena existing at any moment, as consisting of dissimilar facts, a, b, c, d, and e, for each of which one, and only one, cause needs be sought, the difficulty being only that of singling out this one cause from the multitude of antecedent circumstances, A, B, C, D, and E. The cause, indeed, may not be simple; it may consist of an assemblage of conditions; but we have supposed that there was only one possible assemblage of conditions from which the given effect could result.

If such were the fact, it would be comparatively an easy task to investigate the laws of nature. But the supposition does not hold in either of its parts. In the first place, it is not true that the same phenomenon is always produced by the same cause; the effect a may sometimes arise from A, sometimes from B. And, secondly, the effects of different causes are often not dissimilar but homogeneous, and marked out by no assignable boundaries from one another; A and B may produce not a and b but different portions of an effect a. The obscurity and difficulty of the investigation of the laws of phenomena is singularly increased by the necessity of adverting to these two circumstances, intermixture of effects and plurality of causes. To the latter, being the simpler of the two considerations, we shall first direct our attention.

It is not true, then, that one effect must be connected with only one cause or assemblage of conditions, that each phenomenon can be produced only in one way. There are often several independent modes in which the same phenomenon could have originated. One fact may be the consequent in several invariable sequences; it may follow, with equal uniformity, any one of several antecedents or collections of antecedents. Many causes may produce mechanical motion; many causes may produce some kinds of sensation; many causes may produce death. A given effect may really be produced by a certain cause and yet be perfectly capable of being produced without it.

– which is the source of a characteristic imperfection of the Method of Agreement

One of the principal consequences of this fact of Plurality of Causes is, to render the first of the inductive methods, that of Agreement, uncertain. To illustrate that method, we supposed two instances, A B C followed by *a b c*, and A D E followed by *a d e*. From these instances it might be concluded that A is an invariable antecedent of *a*, and even that it is the unconditional invariable antecedent, or cause, if we could be sure that there is no other antecedent common to the two cases. That this difficulty may not stand in the way, let us suppose the two cases positively ascertained to have no antecedent in common except A. The moment, however, that we let in the possibility of a plurality of causes, the conclusion fails. For it involves a tacit supposition, that *a* must have been produced in both instances by the same cause. If there can possibly have been two causes, those two may, for example, be C and E: the one may have been the cause of *a* in the former of the instances, the other in the latter, A having no influence in either case.

Suppose, for example, that two great artists, or great philosophers, that two extremely selfish, or extremely generous characters, were compared together as to the circumstances of their education and history, and the two cases were found to agree only in one circumstance: would it follow that this one circumstance was the cause of the quality which characterized both those individuals? Not at all; for the causes which may produce any type of character are innumerable; and the two persons might equally have agreed in their character, though there had been no manner of resemblance in their previous history.

This, therefore, is a characteristic imperfection of the Method of Agreement; from which imperfection the Method of Difference is free. For if we have two instances, A B C and B C, of which B C gives *b c*, and A being added converts it into *a b c*, it is certain that in this instance at least, A was either the cause of *a*, or an indispensable portion of its cause, even though the cause which produces it in other instances may be altogether different. Plurality of Causes, therefore, not only does not diminish the reliance due to the Method of Difference, but does not even render a greater number of observations or experiments necessary: two instances, the one positive and the other negative, are still sufficient for the most complete and rigorous induction. Not so, however, with the Method of Agreement. The conclusions which that yields, when the number of instances compared is small, are of no real value, except as, in the character of suggestions, they may lead either to experiments bringing them to the test of the Method of Difference, or to reasonings which may explain and verify them deductively.

It is only when the instances, being indefinitely multiplied and varied, continue to suggest the same result, that this result acquires any high degree of independent value. If there are but two instances, A B C and A D E, although these instances have no antecedent in common except A, yet as the effect may possibly have been produced in the two cases by different causes, the result is at most only a slight probability in favour of A; there may be causation, but it is almost equally probable that there was only a coincidence. But the oftener we repeat the observation, varying the circumstances, the more we advance towards a solution of this doubt. For if we try A F G, A H K, &c., all unlike one another except in containing the circumstance A, and if we find the effect *a* entering into the result in all these cases, we must suppose one of two things, either that it is caused by A, or that it has as many different causes as there are instances. With each addition, therefore, to the number of instances, the presumption is strengthened in favour of A. The inquirer, of course, will not neglect, if an opportunity present itself, to exclude A from some one of these combinations, from A H K for instance, and by trying H K separately, appeal to the Method of Difference in aid of the Method of Agreement. By the Method of Difference alone can it be ascertained that A is the cause of *a*; but that it is either the cause or another effect of the same cause, may be placed beyond any reasonable doubt by the Method of Agreement, provided the instances are very numerous, as well as sufficiently various.

After how great a multiplication, then, of varied instances, all agreeing in no other antecedent except A, is the supposition of a plurality of causes sufficiently rebutted, and the conclusion that *a* is the effect of A divested of the characteristic imperfection and reduced to a virtual certainty? This is a question which we cannot be exempted from answering; but the consideration of it belongs to what is called the Theory of Probability, which will form the subject of a chapter hereafter [sic]. It is seen, however, at once, that the conclusion does amount to a practical certainty after a sufficient number of instances, and that the method, therefore, is not radically vitiated by the characteristic imperfection. The result of these considerations is only, in the first place, to point out a new source of inferiority in the Method of Agreement as compared with other modes of investigation, and new reasons for never resting contented with the results obtained by it, without attempting to confirm them either by the Method of Difference, or by connecting them deductively with some law or laws already ascertained by that superior method. And, in the second place, we learn from this the true theory of the value of mere *number* of instances in inductive inquiry. The Plurality of

Causes is the only reason why mere number is of any importance. The tendency of unscientific inquirers is to rely too much on number, without analysing the instances; without looking closely enough into their nature, to ascertain what circumstances are or are not eliminated by means of them. Most people hold their conclusions with a degree of assurance proportioned to the mere *mass* of the experience on which they appear to rest; not considering that by the addition of instances to instances, all of the same kind, that is, differing from one another only in points already recognised as immaterial, nothing whatever is added to the evidence of the conclusion. A single instance eliminating some antecedent which existed in all the other cases, is of more value than the greatest multitude of instances which are reckoned by their number alone. It is necessary, no doubt, to assure ourselves, by a repetition of the observation or experiment, that no error has been committed concerning the individual facts observed; and until we have assured ourselves of this, instead of varying the circumstances, we cannot too scrupulously repeat the same experiment or observation without any change. But when once this assurance has been obtained, the multiplication of instances which do not exclude any more circumstances would be entirely useless, were it not for the Plurality of Causes.

It is of importance to remark, that the peculiar modification of the Method of Agreement which, as partaking in some degree of the nature of the Method of Difference, I have called the Joint Method of Agreement and Difference, is not affected by the characteristic imperfection now pointed out. For, in the joint method, it is supposed not only that the instances in which *a* is, agree only in containing A, but also that the instances in which *a* is not, agree only in not containing A. Now, if this be so, A must be not only the cause of *a*, but the only possible cause: for if there were another, as for example B, then in the instances in which *a* is not, B must have been absent as well as A, and it would not be true that these instances agree *only* in not containing A. This, therefore, constitutes an immense advantage of the joint method over the simple Method of Agreement. It may seem, indeed, that the advantage does not belong so much to the joint method, as to one of its two premisses, (if they may be so called,) the negative premiss. The Method of Agreement, when applied to negative instances, or those in which a phenomenon does *not* take place, is certainly free from the characteristic imperfection which affects it in the affirmative case. The negative premiss, it might therefore be supposed, could be worked as a simple case of the Method of Agreement, without requiring an affirmative premiss to be joined with it. But although this is true in principle, it is generally altogether impossible to work the Method of Agreement by negative instances without positive ones: it is so much more difficult to exhaust the field of negation than that of affirmation. For instance, let the question be, what is the cause of the transparency of bodies; with what prospect of success could we set ourselves to inquire directly in what the multifarious substances which are *not* transparent, agree? But we might hope much sooner to seize some point of resemblance among the comparatively few and definite species of objects which *are* transparent; and this being attained, we should quite naturally be put upon examining whether the *absence* of this one circumstance be not precisely the point in which all opaque substances will be found to resemble.

The Joint Method of Agreement and Difference, therefore, or, as I have otherwise called it, the Indirect Method of Difference (because, like the Method of Difference properly so called, it proceeds by ascertaining how and in what the cases where the phenomenon is present, differ from those in which it is absent) is, after the direct Method of Difference, the most powerful of the remaining instruments of inductive investigation; and in the sciences which depend on pure observation, with little or no aid from experiment, this method, so well exemplified in the speculation on the cause of dew, is the primary resource, so far as direct appeals to experience are concerned.

Plurality of Causes, how ascertained

We have thus far treated Plurality of Causes only as a possible supposition, which, until removed, renders our inductions uncertain, and have only considered by what means, where the plurality does not really exist, we may be enabled to disprove it. But we must also consider it as a case actually occurring in nature, and which, as often as it does occur, our methods of induction ought to be capable of ascertaining and establishing. For this, however, there is required no peculiar method. When an effect is really producible by two or more causes, the process for detecting them is in no way different from that by which we discover single causes. They may (first) be discovered as separate sequences, by separate sets of instances. One set of observations or experiments shows that the sun is a cause of heat, another that friction is a source of it, another that percussion, another that electricity, another that chemical action is such a source. Or (secondly) the plurality may come to light in the course of collating a number of instances, when we attempt to find some circumstance in which they all agree, and fail in doing so. We find it impossible to trace, in all the cases in which the effect is met with, any common circumstance. We find that we can eliminate *all* the antecedents; that no one of them is present in all the instances, no one of them indispensable to the effect. On closer scrutiny, however, it appears that though no one is always present, one or other of several always is. If, on further analysis, we can detect in these any common element, we may be able to ascend from them to some one cause which is the really operative circumstance in them all. Thus it might, and perhaps will, be discovered, that in the production of heat by friction, percussion, chemical action, &c., the ultimate source is one and the same. But if (as continually happens) we cannot take this ulterior step, the different antecedents must be set down provisionally as distinct causes, each sufficient of itself to produce the effect.

We here close our remarks on the Plurality of Causes . . .

Notes

1   But though it is a condition of the validity of every induction that there be uniformity in the course of nature, it is not a necessary condition that the uniformity should pervade all nature. It is enough that it pervades the particular class of phenomena to which the induction relates. An induction concerning the motions of the planets or the properties of the magnet would not be vitiated though we were to suppose that wind and weather are the sport of chance, provided it be assumed that astronomical and magnetic phenomena are under the dominion of general laws. Otherwise the early experience of mankind would have rested on a very weak foundation, for in the infancy of science it could not be known that all phenomena are regular in their course.

Neither would it be correct to say that every induction by which we infer any truth implies the general fact of uniformity as foreknown, even in reference to the kind of phenomena concerned. It implies either that this general fact is already known, or that we may now know it; as the conclusion, the Duke of Wellington is mortal, drawn from the instances A, B, and C, implies either that we have already concluded all men to be mortal, or that we are now entitled to do so from the same evidence. A vast amount of confusion and paralogism respecting the grounds of induction would be dispelled by keeping in view these simple considerations.

2   In strictness, wherever the present constitution of space exists, which we have ample reason to believe that it does in the region of the fixed stars.

3   *Discourse on the Study of Natural Philosophy*, p. 179.

4   *Philosophy of Discovery*, pp. 263, 264.

5   See, on this point, the second chapter of the present book [*sic*].

Chapter 15

William Whewell

OF CERTAIN CHARACTERISTICS OF SCIENTIFIC INDUCTION

I Invention a part of induction

1

The two operations spoken of in the preceding chapters [sic], – the Explication of the Conceptions of our own minds, and the Colligation of observed Facts by the aid of such Conceptions, – are, as we have just said, inseparably connected with each other. When united, and employed in collecting knowledge from the phenomena which the world presents to us, they constitute the mental process of *Induction*; which is usually and justly spoken of as the genuine source of all our *real general knowledge* respecting the external world. And we see, from the preceding analysis of this process into its two constituents, from what origin it derives each of its characters. It is *real*, because it arises from the combination of Real Facts, but it is *general*, because it implies the possession of General Ideas. Without the former, it would not be knowledge of the External World, without the latter, it would not be Knowledge at all. When Ideas and Facts are separated from each other, the neglect of Facts gives rise to empty speculations, idle subtleties, visionary inventions, false opinions concerning the laws of phenomena, disregard of the true aspect of nature: while the want of Ideas leaves the mind overwhelmed, bewildered, and stupified by particular sensations, with no means of connecting the past with the future, the absent with the present, the example with the rule; open to the impression of all appearances, but capable of appropriating none. Ideas are the *Form*, facts the *Material* of our structure. Knowledge does not consist in the empty mould, or in the brute mass of matter, but in the rightly-moulded substance. Induction gathers general truths from particular facts; – and in her harvest, the corn and the reaper, the solid ears and the binding band, are alike requisite. All our knowledge of nature is obtained by Induction; the term being understood according to the explanation we have now given. And our knowledge is then most complete, then most truly deserves the name of Science, when both its elements are most perfect; – when the Ideas which have been concerned in its formation have, at every step, been clear and consistent; and when they have, at every step also, been employed in binding together real and certain Facts. Of such Induction, I have already given so many examples and illustrations in the two preceding chapters [sic], that I need not now dwell further upon the subject.

2

Induction is familiarly spoken of as the process by which we collect a *General Proposition* from a number of *Particular Cases*: and it appears to be frequently imagined that the general proposition results from a mere juxta-position of the cases, or at most, from merely conjoining and extending them. But if we consider the process more closely, as exhibited in the cases lately spoken of, we shall perceive that this is an inadequate account of the matter. The particular facts are not merely brought together, but there is a New Element added to the combination by the very act of thought by which they are combined. There is a Conception of the mind introduced in the general proposition, which did not exist in any of the observed facts. When the Greeks, after long observing the motions of the planets, saw that these motions might be rightly considered as produced by the motion of one wheel revolving in the inside of another wheel, these Wheels were Creations of their minds, added to the Facts which they perceived by sense. And even if the wheels were no longer supposed to be material, but were reduced to mere geometrical spheres or circles, they were not the less products of the mind alone, − something additional to the facts observed. The same is the case in all other discoveries. The facts are known, but they are insulated and unconnected, till the discoverer supplies from his own stores a Principle of Connexion. The pearls are there, but they will not hang together till some one provides the String. The distances and periods of the planets were all so many separate facts; by Kepler's Third Law they are connected into a single truth: but the Conceptions which this law involves were supplied by Kepler's mind, and without these, the facts were of no avail. The planets described ellipses round the sun, in the contemplation of others as well as of Newton; but Newton conceived the deflection from the tangent in these elliptical motions in a new light, − as the effect of a Central Force following a certain law; and then it was, that such a force was discovered truly to exist.

Thus[1] in each inference made by Induction, there is introduced some General Conception, which is given, not by the phenomena, but by the mind. The conclusion is not contained in the premises, but includes them by the introduction of a New Generality. In order to obtain our inference, we travel beyond the cases which we have before us; we consider them as mere exemplifications of some Ideal Case in which the relations are complete and intelligible. We take a Standard, and measure the facts by it; and this Standard is constructed by us, not offered by Nature. We assert, for example, that a body left to itself will move on with unaltered velocity; not because our senses ever disclosed to us a body doing this, but because (taking this as our Ideal Case) we find that all actual cases are intelligible and explicable by means of the Conception of *Forces*, causing change and motion, and exerted by surrounding bodies. In like manner, we see bodies striking each other, and thus moving and stopping, accelerating and retarding each other: but in all this, we do not perceive by our senses that abstract quantity, *Momentum*, which is always lost by one body as it is gained by another. This Momentum is a creation of the mind, brought in among the facts, in order to convert their apparent confusion into order, their seeming chance into certainty, their perplexing variety into simplicity. This the Conception of *Momentum gained and lost* does: and in like manner, in any other case in which a truth is established by Induction, some Conception is introduced, some Idea is applied, as the means of binding together the facts, and thus producing the truth.

3

Hence in every inference by Induction, there is some Conception *superinduced* upon the Facts: and we may henceforth conceive this to be the peculiar import of the term *Induction*. I am not to be understood as asserting that the term was originally or anciently employed with this notion of its meaning; for the peculiar feature just pointed out in Induction has generally been overlooked. This appears by the accounts generally given of Induction. 'Induction,' says Aristotle,[2] 'is when by means of one extreme term[3] we infer the other extreme term to be true of the middle term.' Thus, (to take such exemplifications as belong to our subject,) from knowing that Mercury, Venus, Mars, describe ellipses about the Sun, we infer that all Planets describe ellipses about the Sun. In making this inference syllogistically, we assume that the evident proposition, 'Mercury, Venus, Mars, do what all Planets do,' may be taken *conversely*, 'All Planets do what Mercury, Venus, Mars, do.' But we may remark that, in this passage, Aristotle (as was natural in his line of discussion) turns his attention entirely to the *evidence* of the inference; and overlooks a step which is of far more importance to our knowledge, namely, the *invention* of the second extreme term. In the above instance, the particular luminaries, Mercury, Venus, Mars, are one logical *Extreme*; the general designation Planets is the *Middle Term*; but having these before us, how do we come to think of *description of ellipses*, which is the other Extreme of the syllogism? When we have once invented this 'second Extreme Term,' we may, or may not, be satisfied with the evidence of the syllogism; we may, or may not, be convinced that, so far as this property goes, the extremes are co-extensive with the middle term;[4] but the *statement* of the syllogism is the important step in science. We know how long Kepler laboured, how hard he fought, how many devices he tried, before he hit upon this *Term*, the Elliptical Motion. He rejected, as we know, many other 'second extreme Terms,' for example, various combinations of epicyclical constructions, because they did not represent with sufficient accuracy the special facts of observation. When he had established his premiss, that 'Mars does describe an Ellipse

about the Sun,' he does not hesitate to *guess* at least that, in this respect, he might *convert* the other premiss, and assert that 'All the Planets do what Mars does.' But the main business was, the inventing and verifying the proposition respecting the Ellipse. The Invention of the Conception was the great step in the *discovery*; the Verification of the Proposition was the great step in the *proof* of the discovery. If Logic consists in pointing out the conditions of proof, the Logic of Induction must consist in showing what are the conditions of proof, in such inferences as this: but this subject must be pursued in the next chapter [sic]; I now speak principally of the act of *Invention*, which is requisite in every inductive inference.

4

Although in every inductive inference, an act of invention is requisite, the act soon slips out of notice. Although we bind together facts by superinducing upon them a new Conception, this Conception, once introduced and applied, is looked upon as inseparably connected with the facts, and necessarily implied in them. Having once had the phenomena bound together in their minds in virtue of the Conception, men can no longer easily restore them back to the detached and incoherent condition in which they were before they were thus combined. The pearls once strung, they seem to form a chain by their nature. Induction has given them a unity which it is so far from costing us an effort to preserve, that it requires an effort to imagine it dissolved. For instance, we usually represent to ourselves the Earth as *round*, the Earth and the Planets as *revolving* about the Sun, and as *drawn* to the Sun by a Central Force; we can hardly understand how it could cost the Greeks, and Copernicus, and Newton, so much pains and trouble to arrive at a view which to us is so familiar. These are no longer to us Conceptions caught hold of and kept hold of by a severe struggle; they are the simplest modes of conceiving the facts: they are really Facts. We are willing to *own* our obligation to those discoverers, but we hardly *feel* it: for in what other manner (we ask in our thoughts) could we represent the facts to ourselves?

Thus we see why it is that this step of which we now speak, the Invention of a new Conception in every inductive inference, is so generally overlooked that it has hardly been noticed by preceding philosophers. When once performed by the discoverer, it takes a fixed and permanent place in the understanding of every one. It is a thought which, once breathed forth, permeates all men's minds. All fancy they nearly or quite knew it before. It oft was thought, or almost thought, though never till now expressed. Men accept it and retain it, and know it cannot be taken from them, and look upon it as their own. They will not and cannot part with it, even though they may deem it trivial and obvious. It is a secret, which once uttered, cannot be recalled, even though it be despised by those to whom it is imparted. As soon as the leading term of a new theory has been pronounced and understood, all the phenomena change their aspect. There is a standard to which we cannot help referring them. We cannot fall back into the helpless and bewildered state in which we gazed at them when we possessed no principle which gave them unity. Eclipses arrive in mysterious confusion: the notion of a *Cycle* dispels the mystery. The Planets perform a tangled and mazy dance; but *Epicycles* reduce the maze to order. The Epicycles themselves run into confusion; the conception of an *Ellipse* makes all clear and simple. And thus from stage to stage, new elements of intelligible order are introduced. But this intelligible order is so completely adopted by the human understanding, as to seem part of its texture. Men ask Whether Eclipses follow a Cycle; Whether the Planets

describe Ellipses; and they imagine that so long as they do not *answer* such questions rashly, they take nothing for granted. They do not recollect how much they assume in *asking* the question: – how far the conceptions of Cycles and of Ellipses are beyond the visible surface of the celestial phenomena: – how many ages elapsed, how much thought, how much observation, were needed, before men's thoughts were fashioned into the words which they now so familiarly use. And thus they treat the subject, as we have seen Aristotle treating it; as if it were a question, not of invention, but of proof; not of substance, but of form: as if the main thing were not *what* we assert, but *how* we assert it. But for our purpose, it is requisite to bear in mind the feature we have thus attempted to mark; and to recollect that, in every inference by induction, there is a Conception supplied by the mind and superinduced upon the Facts.

5

In collecting scientific truths by Induction, we often find (as has already been observed) a Definition and a Proposition established at the same time, – introduced together, and mutually dependent on each other. The combination of the two constitutes the Inductive act; and we may consider the Definition as representing the superinduced Conception, and the Proposition as exhibiting the Colligation of Facts.

II Use of hypotheses

6

To discover a Conception of the mind which will justly represent a train of observed facts is, in some measure, a process of conjecture, as I have stated already; and as I then observed, the business of conjecture is commonly conducted by calling up before our minds several suppositions, and selecting that one which most agrees with what we know of the observed facts. Hence he who has to discover the laws of nature may have to invent many suppositions before he hits upon the right one; and among the endowments which lead to his success, we must reckon that fertility of invention which ministers to him such imaginary schemes, till at last he finds the one which conforms to the true order of nature. A facility in devising hypotheses, therefore, is so far from being a fault in the intellectual character of a discoverer, that it is, in truth, a faculty indispensable to his task. It is, for his purpose, much better that he should be too ready in contriving, too eager in pursuing systems which promise to introduce law and order among a mass of unarranged facts, than that he should be barren of such inventions and hopeless of such success. Accordingly, as we have already noticed, great discoverers have often invented hypotheses which would not answer to all the facts, as well as those which would; and have fancied themselves to have discovered laws, which a more careful examination of the facts overturned.

The tendencies of our speculative nature,[5] carrying us onwards in pursuit of symmetry and rule, and thus producing all true theories, perpetually show their vigour by overshooting the mark. They obtain something, by aiming at much more. They detect the order and connexion which exist, by conceiving imaginary relations of order and connexion which have no existence. Real discoveries are thus mixed with baseless assumptions; profound sagacity is combined with fanciful conjecture [sic]; not rarely, or in peculiar instances, but commonly, and in most cases; probably in all, if we could read the thoughts of discoverers as we read the books of Kepler. To try wrong guesses is, with most persons, the only way to hit upon right ones. The character of the true philosopher is, not that he never conjectures hazardously, but that his conjectures are clearly conceived, and brought into rigid contact with facts. He sees and compares distinctly the Ideas and the Things; – the relations of his notions to each other and to phenomena. Under these conditions, it is not only excusable, but necessary for him, to snatch at every semblance of general rule, – to try all promising forms of simplicity and symmetry.

Hence advances in knowledge[6] are not commonly made without the previous exercise of some boldness and license in guessing. The discovery of new truths requires, undoubtedly, minds careful and scrupulous in examing what is suggested; but it requires, no less, such as are quick and fertile in suggesting. What is Invention, except the talent of rapidly calling before us the many possibilities, and selecting the appropriate one? It is true, that when we have rejected all the inadmissible suppositions, they are often quickly forgotten; and few think it necessary to dwell on these discarded hypotheses, and on the process by which they were condemned. But all who discover truths, must have reasoned upon many errours [sic] to obtain each truth; every accepted doctrine must have been one chosen out of many candidates. If many of the guesses of philosophers of bygone times now appear fanciful and absurd, because time and observation have refuted them, others, which were at the time equally gratuitous, have been confirmed in a manner which makes them appear marvellously sagacious. To form hypotheses, and then to employ much labour and skill in refuting them, if they do not succeed in establishing them, is a part of the usual process of inventive minds. Such a proceeding belongs to the *rule* of the genius of discovery, rather than (as has often been taught in modern times) to the *exception*.

7

But if it be an advantage for the discoverer of truth that he be ingenious and fertile in inventing hypotheses which may connect the phenomena of nature, it is indispensably requisite that he be diligent and careful in comparing his hypotheses with the facts, and ready to abandon his invention as soon as it appears that it does not agree with the course of actual occurrences. This constant comparison of his own conceptions and supposition with observed facts under all aspects, forms the leading employment of the discoverer: this candid and simple love of truth, which makes him willing to suppress the most favourite production of his own ingenuity as soon as it appears to be at variance with realities, constitutes the first characteristic of his temper. He must have neither the blindness which cannot, nor the obstinacy which will not, perceive the discrepancy of his fancies and his facts. He must allow no indolence, or partial views, or self-complacency, or delight in seeming demonstration, to make him tenacious of the schemes which he devises, any further than they are confirmed by their accordance with nature. The framing of hypotheses is, for the inquirer after truth, not the end, but the beginning of his work. Each of his systems is invented, not that he may admire it and follow it into all its consistent consequences, but that he may make it the occasion of a course of active experiment and observation. And if the results of this process contradict his fundamental assumptions, however ingenious, however symmetrical, however elegant his system may be, he rejects it without hesitation. He allows no natural yearning for the offspring of his own mind to draw him aside from the higher duty of loyalty to his sovereign, Truth: to her he not only gives his affections and his wishes, but strenuous labour and scrupulous minuteness of attention.

We may refer to what we have said of Kepler, Newton, and other eminent philosophers, for illustrations of this character. In Kepler we have remarked[7] the courage and perseverance with which he undertook and executed the task of computing his own hypotheses: and, as a still more admirable characteristic, that he never allowed the labour he had spent upon any conjecture to produce any reluctance in abandoning the hypothesis, as soon as he had evidence of its inaccuracy. And in the history of Newton's discovery that the moon is retained in her orbit by the force of gravity, we have noticed the same moderation in maintaining the hypothesis, after it had once occurred to the author's mind. The hypothesis required that the moon should fall from the tangent of her orbit every second through a space of sixteen feet; but according to his first calculations it appeared that in fact she only fell through a space of thirteen feet in that time. The difference seems small, the approximation encouraging, the theory plausible; a man in love with his own fancies would readily have discovered or invented some probable cause of the difference. But Newton acquiesced in it as a disproof of his conjecture, and 'laid aside at that time any further thoughts of this matter.'[8]

8

It has often happened that those who have undertaken to instruct mankind have not possessed this pure love of truth and comparative indifference to the maintenance of their own inventions. Men have frequently adhered with great tenacity and vehemence to the hypotheses which they have once framed; and in their affection for these, have been prone to overlook, to distort, and to misinterpret facts. In this manner, *Hypotheses* have so often been prejudicial to the genuine pursuit of truth, that they have fallen into a kind of obloquy; and have been considered as dangerous temptations and fallacious guides. Many warnings have been uttered against the fabrication of hypotheses, by those who profess to teach philosophy; many disclaimers of such a course by those who cultivate science.

Thus we shall find Bacon frequently discommending this habit, under the name of 'anticipation of the mind,' and Newton thinks it necessary to say emphatically '*hypotheses non fingo.*' It has been constantly urged that the inductions by which sciences are formed must be *cautious* and *rigorous*; and the various imaginations which passed through Kepler's brain, and to which he has given utterance have been blamed or pitied, as lamentable instances of an unphilosophical frame of mind. Yet it has appeared in the preceding remarks that hypotheses rightly used are among the helps, far more than the dangers, of science; – that scientific induction is not a 'cautious' or a 'rigorous' process in the sense of *abstaining from* such suppositions, but in *not adhering* to them till they are confirmed by fact, and in carefully seeking from facts confirmation or refutation. Kepler's distinctive character was, not that he was peculiarly given to the construction of hypotheses, but that he narrated with extraordinary copiousness and candour the course of his thoughts, his labours, and his feelings. In the minds of most persons, as we have said, the inadmissible suppositions, when rejected, are soon forgotten: and thus the trace of them vanishes from the thoughts, and the successful hypothesis alone holds its place in our memory. But in reality, many other transient suppositions must have been made by all discoverers; – hypotheses which are not afterwards asserted as true systems, but entertained for an instant; – 'tentative hypotheses,' as they have been called. Each of these hypotheses is followed by its corresponding train of observations, from which it derives its power of leading to truth. The hypothesis is like the captain, and the observations like the soldiers of an army: while he appears to command them, and in this way to work his own will, he does in fact derive all his power of conquest from their

obedience, and becomes helpless and useless if they mutiny.

Since the discoverer has thus constantly to work his way onwards by means of hypotheses, false and true, it is highly important for him to possess talents and means for rapidly *testing* each supposition as it offers itself. In this as in other parts of the work of discovery, success has in general been mainly owing to the native ingenuity and sagacity of the discoverer's mind. Yet some Rules tending to further this object have been delivered by eminent philosophers, and some others may perhaps be suggested. Of these we shall here notice only some of the most general, leaving for a future chapter [*sic*] the consideration of some more limited and detailed processes by which, in certain cases, the discovery of the laws of nature may be materially assisted.

III Tests of hypotheses

9

A maxim which it may be useful to recollect is this, – that *hypotheses may often be of service to science, when they involve a certain portion of incompleteness, and even of errour* [*sic*]. The object of such inventions is to bind together facts which without them are loose and detached; and if they do this, they may lead the way to a perception of the true rule by which the phenomena are associated together, even if they themselves somewhat misstate the matter. The imagined arrangement enables us to contemplate, as a whole, a collection of special cases which perplex and overload our minds when they are considered in succession; and if our scheme has so much or truth in it as to conjoin what is really connected, we may afterwards duly correct or limit the mechanism of this connexion. If our hypothesis renders a reason for the agreement of cases really similar, we may afterwards find this reason to be false, but we shall be able to translate it into the language of truth.

A conspicuous example of such an hypothesis, – one which was of the highest value to science, though very incomplete, and as a representation of nature altogether false, – is seen in the *Doctrine of epicycles* by which the ancient astronomers explained the motions of the sun, moon, and planets. This doctrine connected the places and velocities of these bodies at particular times in a manner which was, in its general features, agreeable to nature. Yet this doctrine was erroneous in its assertion of the *circular* nature of all the celestial motions, and in making the heavenly bodies revolve *round the earth*. It was, however, of immense value to the progress of astronomical science; for it enabled men to express and reason upon many important truths which they discovered respecting the motion of the stars, up to the time of Kepler. Indeed we can hardly imagine that astronomy could, in its outset, have made so great a progress under any other form, as it did in consequence of being cultivated in this shape of the incomplete and false *epicyclical hypothesis*.

We may notice another instance of an exploded hypothesis, which is generally mentioned only to be ridiculed, and which undoubtedly is both false in the extent of its assertion, and unphilosophical in its expression; but which still, in its day, was not without merit. I mean the doctrine of *Nature's horrour* [sic] *of a vacuum* (*fuga vacui*), by which the action of siphons and pumps and many other phenomena were explained, till Mersenne and Pascal taught a truer doctrine. This hypothesis was of real service; for it brought together many facts which really belong to the same class, although they are very different in their first aspect. A scientific writer of modern times[9] appears to wonder that men did not at once divine the weight of the air, from which the phenomena formerly ascribed to the *fuga vacui* really result.

'Loaded, compressed by the atmosphere,' he says, 'they did not recognize its action. In vain all nature testified that air was elastic and heavy; they shut their eyes to her testimony. The water rose in pumps and flowed in siphons at that time, as it does at this day. They could not separate the boards of a pair of bellows of which the holes were stopped; and they could not bring together the same boards without difficulty, if they were at first separated. Infants sucked the milk of their mothers; air entered rapidly into the lungs of animals at every inspiration; cupping-glasses produced tumours on the skin; and in spite of all these striking proofs of the weight and elasticity of the air, the ancient philosophers maintained resolutely that air was light, and explained all these phenomena by the horrour which they said nature had for a vacuum.' It is curious that it should not have occurred to the author while writing this, that if these facts, so numerous and various, can all be accounted for by *one* principle, there is a strong presumption that the principle is not altogether baseless. And in reality is it not true that nature *does* abhor a vacuum, and does all she can to avoid it? No doubt this power is not unlimited; and moreover we can trace it to a mechanical cause, the pressure of the circumambient air. But the tendency, arising from this pressure, which the bodies surrounding a space void of air have to rush into it, may be expressed, in no extravagant or unintelligible manner, by saying that nature has a repugnance to a vacuum.

That imperfect and false hypotheses, though they may thus explain *some* phenomena, and may be useful in the progress of science, cannot explain *all* phenomena; – and that we are never to rest in our labours or acquiesce in our results, till we have found some view of the subject which is consistent with *all* the observed facts; – will of course be understood. We shall afterwards have to speak of the other steps of such a progress.

10

Thus the hypotheses which we accept ought to explain phenomena which we have observed. But they ought to do more than this: our hypotheses ought to *foretel* [sic] phenomena which have not yet been observed; at least all phenomena of the same kind as those which the hypothesis was invented to explain. For our assent to the hypothesis implies that it is held to be true of all particular instances. That these cases belong to past or to future times, that they have or have not already occurred, makes no difference in the applicability of the rule to them. Because the rule prevails, it includes all cases; and will determine them all, if we can only calculate its real consequences. Hence it will predict the results of new combinations, as well as explain the appearances which have occurred in old ones. And that it does this with certainty and correctness, is one mode in which the hypothesis is to be verified as right and useful.

The scientific doctrines which have at various periods been established have been verified in this manner. For example, the *Epicyclical Theory* of the heavens was confirmed by its *predicting* truly eclipses of the sun and moon, configurations of the planets, and other celestial phenomena; and by its leading to the construction of Tables by which the places of the heavenly bodies were given at every moment of time. The truth and accuracy of these predictions were a proof that the hypothesis was valuable, and, at least to a great extent, true; although, as was afterwards found, it involved a false representation of the structure of the heavens. In like manner, the discovery of the *Laws of Refraction* enabled mathematicians to *predict*, by calculation, what would be the effect of any new form or combination of transparent lenses. Newton's hypothesis of *Fits of Easy Transmission and Easy Reflection* in the particles of light, although not confirmed by other kinds of facts, involved a true statement of the law of the phenomena which it was framed to include, and served to *predict* the forms and colours of thin plates for a wide range of given cases. The hypothesis that Light operates by *Undulations* and *Interferences*, afforded the means of *predicting* results under a still larger extent of conditions. In like manner in the progress of chemical knowledge, the doctrine of *Phlogiston* supplied the means of *foreseeing* the consequence of many combinations of elements, even before they were tried; but the *Oxygen Theory*, besides affording predictions, at least equally exact, with regard to the general results of chemical operations, included all the facts concerning the relations of weight of the elements and their compounds, and enabled chemists to *foresee* such facts in untried cases. And the Theory of *Electromagnetic Forces*, as soon as it was rightly understood, enabled those who had mastered it to *predict* motions such as had not been before observed, which were accordingly found to take place.

Men cannot help believing that the laws laid down by discoverers must be in a great measure identical with the real laws of nature, when the discoverers thus determine effects beforehand in the same manner in which nature herself determines them when the occasion occurs. Those who can do this, must, to a considerable extent, have detected nature's secret; – must have fixed upon the conditions to which she attends, and must have seized the rules by which she applies them. Such a coincidence of untried facts with speculative assertions cannot be the work of chance, but implies some large portion of truth in the principles on which the reasoning is founded. To trace order and law in that which has been observed, may be considered as interpreting what nature has written down for us, and will commonly prove that we understand her alphabet. But to predict what has not been observed, is to attempt ourselves to use the legislative phrases of nature; and when she responds plainly and precisely to that which we thus utter, we cannot but suppose that we have in a great measure made ourselves masters of the meaning and structure of her language. The prediction of results, even of the same kind as those which have been observed, in new cases, is a proof of real success in our inductive processes.

11

We have here spoken of the prediction of facts of *the same kind* as those from which our rule was collected. But the evidence in favour of our induction is of a much higher and more forcible character when it enables us to explain and determine cases of a *kind different* from those which were contemplated in the formation of our hypothesis. The instances in which this has occurred, indeed, impress us with a conviction that the truth of our hypothesis is certain. No accident could give rise to such an extraordinary coincidence. No false supposition could, after being adjusted to one class of phenomena, exactly represent a different class, where the agreement was unforeseen and uncontemplated. That rules springing from remote and unconnected quarters should thus leap to the same point, can only arise from *that* being the point where truth resides.

Accordingly the cases in which inductions from classes of facts altogether different have thus *jumped together*, belong only to the best established theories which the history of science contains. And as I shall have occasion to refer to this peculiar feature in their evidence, I will take the liberty of describing it by a particular phrase; and will term it the *Consilience of Inductions*.

It is exemplified principally in some of the greatest discoveries. Thus it was found by Newton that the doctrine of the Attraction of the Sun varying according to the Inverse Square of this distance, which explained Kepler's *Third Law*, of the proportionality of the cubes of the distance to the squares of the periodic times of the planets, explained also his *First* and *Second Laws*, of the elliptical motion of each planet; although no connexion of these laws had been visible before. Again, it appeared that the force of Universal Gravitation, which had been inferred from the *Perturbations* of the moon and planets by the sun and by each other, also accounted for the fact, apparently altogether dissimilar and remote, of the *Precession of the equinoxes*. Here was a most striking and surprising coincidence, which gave to the theory a stamp of truth beyond the power of ingenuity to counterfeit. In like manner in Optics; the hypothesis of alternate Fits of Easy Transmission and Reflection would explain the colours of thin plates, and indeed was devised and adjusted for that very purpose; but it could give no account of the phenomena of the fringes of shadows. But the doctrine of Interferences, constructed at first with reference to phenomena of the nature of the *Fringes*, explained also the *Colours of thin plates* better than the supposition of the Fits invented for that very purpose. And we have in Physical Optics another example of the same kind, which is quite as striking as the explanation of Precession by inferences from the facts of Perturbation. The doctrine of Undulations propagated in a Spheroidal Form was contrived at first by Huyghens, with a view to explain the laws of *Double Refraction* in calc-spar; and was pursued with the same view by Fresnel. But in the course of the investigation it appeared, in a most unexpected and wonderful manner, that this same doctrine of spheroidal undulations, when it was so modified as to account for the directions of the two refracted rays, accounted also for the positions of their *Planes of Polarization*,[10] a phenomenon which, taken by itself, it had perplexed previous mathematicians, even to represent.

The Theory of Universal Gravitation, and of the Undulatory Theory of Light, are, indeed, full of examples of this Consilience of Inductions. With regard to the latter, it has been justly asserted by Herschel, that the history of the undulatory theory was a succession of *felicities*.[11] And it is precisely the unexpected coincidences of results drawn from distant parts of the subject which are properly thus described. Thus the Laws of the *Modification of polarization* to which Fresnel was led by his general views, accounted for the Rule respecting the *Angle at which light is polarized*, discovered by Sir D. Brewster.[12] The conceptions of the theory pointed out peculiar *Modifications* of the phenomena when *Newton's rings* were produced by polarised light, which modifications were ascertained to take place in fact, by Arago and Airy.[13] When the beautiful phenomena of *Dipolarized light* were discovered by Arago and Biot, Young was able to declare that they were reducible to the general laws of Interference which he had already established.[14] And what was no less striking a confirmation of the truth of the theory, *Measures* of the same element deduced from various classes of facts were found to coincide. Thus the Length of a luminiferous undulation, calculated by Young from the measurement of *Fringes* of shadows, was found to agree very nearly with the previous calculation from the colours of *Thin plates*.[15]

No example can be pointed out, in the whole history of science, so far as I am aware, in which this Consilience of Inductions has given testimony in favour of an hypothesis afterwards discovered to be false. If we take one class of facts only, knowing the law which they follow, we may construct an hypothesis, or perhaps several, which may represent them: and as new circumstances are discovered, we may often adjust the hypothesis so as to correspond to these also. But when the hypothesis, of itself and without adjustment for the purpose, gives us the rule and reason of a class of facts not

contemplated in its construction, we have a criterion of its reality, which has never yet been produced in favour of falsehood.

12

In the preceding Article I have spoken of the hypothesis with which we compare our facts as being framed *all at once*, each of its parts being included in the original scheme. In reality, however, it often happens that the various suppositions which our system contains are *added* upon occasion of different researches. Thus in the Ptolemaic doctrine of the heavens, new epicycles and eccentrics were added as new inqualities of the motions of the heavenly bodies were discovered; and in the Newtonian doctrine of material rays of light, the supposition that these rays had 'fits,' was added to explain the colours of thin plates; and the supposition that they had 'sides' was introduced on occasion of the phenomena of polarization. In like manner other theories have been built up of parts devised at different times.

This being the mode in which theories are often framed, we have to notice a distinction which is found to prevail in the progress of true and false theories. In the former class all the additional suppositions *tend to simplicity* and harmony; the new suppositions resolve themselves into the old ones, or at least require only some easy modification of the hypothesis first assumed: the system becomes more coherent as it is further extended. The elements which we require for explaining a new class of facts are already contained in our system. Different members of the theory run together, and we have thus a constant convergence to unity. In false theories, the contrary is the case. The new suppositions are something altogether additional; — not suggested by the original scheme; perhaps difficult to reconcile with it. Every such addition adds to the complexity of the hypothetical system, which at last becomes unmanageable, and is compelled to surrender its place to some simpler explanation.

Such a false theory for example was the ancient doctrine of eccentrics and epicycles. It explained the general succession of the Places of the Sun, Moon, and Planets; it would not have explained the proportion of their Magnitudes at different times, if these could have been accurately observed; but this the ancient astronomers were unable to do. When, however, Tycho and other astronomers came to be able to observe the planets accurately in all positions, it was found that *no* combination of *equable* circular motions would exactly represent all the observations. We may see, in Kepler's works, the many new modifications of the epicyclical hypothesis which offered themselves to him; some of which would have agreed with the phenomena with a certain degree of accuracy, but not with so great a degree as Kepler, fortunately for the progress of science, insisted upon obtaining. After these epicycles had been thus accumulated, they all disappeared and gave way to the simpler conception of an *elliptical* motion. In like manner, the discovery of new inequalities in the Moon's motions encumbered her system more and more with new machinery, which was at last rejected all at once in favour of the *elliptical* theory. Astronomers could not but suppose themselves in a wrong path, when the prospect grew darker and more entangled at every step.

Again; the Cartesian system of Vortices might be said to explain the primary phenomena of the revolutions of planets about the sun, and satellites about the planets. But the elliptical form of the orbits required new suppositions. Bernoulli ascribed this curve to the shape of the planet, operating on the stream of the vortex in a manner similar to the rudder of a boat. But then the motions of the aphelia, and of the nodes, – perturbations, – even the action of gravity towards the earth, – could not be accounted for without new and independent suppositions.

Here was none of the simplicity of truth. The theory of Gravitation, on the other hand, became more simple as the facts to be explained became more numerous. The attraction of the sun accounted for the motions of the planets; the attraction of the planets was the cause of the motion of the satellites. But this being assumed, the perturbations, and the motions of the nodes and aphelia, only made it requisite to extend the attraction of the sun to the satellites, and that of the planets to each other: – the tides, the spheroidal form of the earth, the precession, still required nothing more than that the moon and sun should attract the parts of the earth, and that these should attract each other; – so that all the suppositions resolved themselves into the single one, of the universal gravitation of all matter. It is difficult to imagine a more convincing manifestation of simplicity and unity.

Again, to take an example from another science; – the doctrine of Phlogiston brought together many facts in a very plausible manner, – combustion, acidification, and others, – and very naturally prevailed for a while. But the balance came to be used in chemical operations, and the facts of weight as well as of combination were to be accounted for. On the phlogistic theory, it appeared that this could not be done without a new supposition, and *that*, a very strange one; – that phlogiston was an element not only not heavy, but absolutely light, so that it diminished the weight of the compounds into which it entered. Some chemists for a time adopted this extravagant view; but the wiser of them saw, in the necessity of such a supposition to the defence of the theory, an evidence that the hypothesis of an element *phlogiston* was erroneous. And the opposite hypothesis, which taught that oxygen was subtracted, and not phlogiston added, was accepted because it required no such novel and inadmissible assumption.

Again, we find the same evidence of truth in the progress of the Undulatory Theory of light, in the course of its application from one class of facts to another. Thus we explain Reflection and Refraction by undulations; when we come to Thin Plates, the requisite 'fits' are already involved in our fundamental hypothesis, for they are the length of an undulation: the phenomena of Diffraction also require such intervals; and the intervals thus required agree exactly with the others in magnitude, so that no new property is needed. Polarization for a moment appears to require some new hypothesis; yet this is hardly the case; for the direction of our vibrations is hitherto arbitrary: − we allow polarization to decide it, and we suppose the undulations to be transverse. Having done this for the sake of Polarization, we turn to the phenomena of Double Refraction, and inquire what new hypothesis they require. But the answer is, that they require none: the supposition of transverse vibrations, which we have made in order to explain Polarization, gives us also the law of Double Refraction. Truth may give rise to such a coincidence; falsehood cannot. Again, the facts of Dipolarization come into view. But they hardly require any new assumption; for the difference of optical elasticity of crystals in different directions, which is already assumed in uniaxal [sic] crystals,[16] is extended to biaxal [sic] exactly according to the law of symmetry; and this being done, the laws of the phenomena, curious and complex as they are, are fully explained. The phenomena of Circular Polarization by internal reflection, instead of requiring a new hypothesis, are found to be given by an interpretation of an apparently inexplicable result of an old hypothesis. The Circular Polarization of Quartz and its Double Refraction does indeed appear to require a new assumption, but still not one which at all disturbs the form of the theory; and in short, the whole history of this theory is a progress, constant and steady, often striking and startling, from one degree of evidence and consistence to another of higher order.

In the Emission Theory, on the other hand, as in the theory of solid epicycles, we see what we may consider as the natural course of things in the career of a false theory. Such a theory may, to a certain extent, explain the phenomena which it was at first contrived to meet; but every new class of facts requires a new supposition − an addition to the machinery: and as observation goes on, these incoherent appendages accumulate, till they overwhelm and upset the original frame-work. Such has been the hypothesis of the Material Emission of light. In its original form, it explained Reflection and Refraction: but the colours of Thin Plates added to it the Fits of easy Transmission and Reflection; the phenomena of Diffraction further invested the emitted particles with complex laws of Attraction and Repulsion; Polarization gave them Sides: Double Refraction subjected them to peculiar Forces emanating from the axes of the crystal: finally, Dipolarization loaded them with the complex and unconnected contrivance of Moveable Polarization: and even when all this had been done, additional mechanism was wanting. There is here no unexpected success, no happy coincidence, no convergence of principles from remote quarters. The philosopher builds the machine, but its parts do not fit. They hold together only while he presses them. This is not the character of truth.

As another example of the application of the Maxim now under consideration, I may perhaps be allowed to refer to the judgment which, in the History of Thermotics, I have ventured to give respecting Laplace's Theory of Gases. I have stated,[17] that we cannot help forming an unfavourable judgment of this theory, by looking for that great characteristic of true theory; namely, that the hypotheses which were assumed to account for *one class* of facts are found to explain *another class* of a different nature. Thus Laplace's first suppositions explain the connexion of Compression with Density, (the law of Boyle and Mariotte,) and the connexion of Elasticity with Heat, (the law of Dalton and Gay Lussac). But the theory requires other assumptions when we come to Latent Heat; and yet these new assumptions produce no effect upon the calculations in any application of the theory. When the hypothesis, constructed with reference to the Elasticity and Temperature, is applied to another class of facts, those of Latent Heat, we have no Simplication of the Hypothesis, and therefore no evidence of the truth of the theory.

13

The last two sections of this chapter direct our attention to two circumstances, which tend to prove, in a manner which we may term irresistible, the truth of the theories which they characterize: − the *Consilience of Inductions* from different and separate classes of facts; − and the progressive *Simplification of the Theory* as it is extended to new cases. These two Characters are, in fact, hardly different; they are exemplified by the same cases. For if these Inductions, collected from one class of facts, supply an unexpected explanation of a new class, which is the case first spoken of, there will be no need for new machinery in the hypothesis to apply it to the newly-contemplated facts; and thus, we have a case in which the system does not become more complex when its application is extended to a wider field, which was the character of true theory in its second aspect. The Consiliences of our Inductions give rise to a constant Convergence of our Theory towards Simplicity and Unity.

But, moreover, both these cases of the extension of the theory, without difficulty or new suppositions, to a wider range and to new classes of phenomena, may be conveniently considered in yet another point of view; namely, as successive steps by which we gradually ascend in our speculative views to a higher and higher point of generality. For when the theory, either by the concurrence of two indications, or by an extension without complication, has included a new range of phenomena, we have, in fact, a new induction of a more general kind, to which the inductions formerly obtained are subordinate, as particular cases to a general proposition. We have in such examples, in short, an instance of *successive generalization*. This is a subject of great importance, and deserving of being well illustrated; it will come under our notice in the next chapter [*sic*].

Notes

1 I repeat here remarks made at the end of the *Mechanical Euclid*, p. 178.
2 *Analyt. Prior. Lib. Il. C. Xxiii.*
3 The syllogism here alluded to would be this:−

   Mercury, Venus, Mars, describe ellipses about the Sun;
   All Planets do what Mercury, Venus, Mars do;
   Therefore all Planets describe ellipses about the sun.

4 Aristot. Ibid.
5 I here take the liberty of characterizing inventive minds in general in the same phraseology which, in the History of Science, I have employed in reference to particular examples. These expressions are what I have used in speaking of the discoveries of Copernicus − *History of the Inductive Sciences* b. v. c. ii.
6 These observations are made on occasion of Kepler's speculations, and are illustrated by reference to his discoveries − *History of the Inductive Sciences* b. v. c. iv. Sect 1.
7 *History of the Inductive Sciences* b. v. c. iv. Sect 1.
8 *History of the Inductive Sciences* b. v. c. iv. Sect 3
9 Deluc, *Modifications de l'Atmosphère*, Partie i.
10 *History of the Inductive Sciences* b. ix. c. xi. Sect 4.
11 See *History of the Inductive Sciences* b. ix. c. xii.
12 Ibid. c. xi. Sect. 4.
13 See *History of the Inductive Sciences* b. ix. c. xii. Sect. 6
14 Ibid. c. xi. Sect. 5.
15 Ibid. c. xi. Sect. 2.
16 *History of the Inductive Sciences* b. ix. c. xi. Sect. 5.
17 *History of the Inductive Sciences* b. x. c. iv.

Chapter 16

Peter Achinstein

WAVES AND SCIENTIFIC METHOD

1 Introduction

In 1802 a youthful Thomas Young, British physician and scientist, had the audacity to resuscitate the wave theory of light (Young 1802). For this he was excoriated by Henry Brougham (1803) in the *Edinburgh Review*. Brougham, a defender of the Newtonian particle theory, asserted that Young's paper was "destitute of every species of merit" because it was not based on inductions from observations but involved simply the formulation of hypotheses to explain various optical phenomena. And, Brougham continued:

> A discovery in mathematics, or a successful induction of facts, when once completed, cannot be too soon given to the world. But . . . an hypothesis is a work of fancy, useless in science, and fit only for the amusement of a vacant hour. (1803, p. 451)

This dramatic confrontation between Young and Brougham, it has been claimed, is but one example of a general methodological gulf between 19th century wave theorists and 18th and 19th century particle theorists. The wave theorists, it has been urged by Larry Laudan (1981) and Geoffrey Cantor (1975), employed a method of hypothesis in defending their theory. This method was firmly rejected by particle theorists, who insisted, with Brougham, that the only way to proceed in physics is to make inductions from observations and experiments.

In a recent work (Achinstein 1991), I argue, contra Laudan and Cantor, that 19th century wave theorists, both in their practice and in their philosophical reflections on that practice, employed a method that is different from the method of hypothesis in important respects; moreover, there are strong similarities between the method the wave theorists practiced and preached and that of 19th century particle theorists such as Brougham and David Brewster. In the present paper I will focus just on the wave theorists. My aims are these: to review my claims about how in fact wave theorists typically argued for their theory; to see whether, or to what extent, this form of reasoning corresponds to the method of hypothesis or to inductivism in sophisticated versions of these doctrines offered by William Whewell and John Stuart Mill; and finally to deal with a problem of anomalies which I did not develop in *Particles and Waves* and might be said to pose a difficulty for my account.

2 The method of hypothesis and inductivism

According to a simple version of the method of hypothesis, if the observed phenomena are explained by, or derived from, an hypothesis, then one may infer the truth or probability of that hypothesis. Laudan maintains that by the 1830s an important shift occurred in the use of this method. An hypothesis was inferable not simply if it explained known phenomena that prompted it in the first place, but only if it also explained and/or predicted phenomena of a kind different from those it was invented to explain. This version received its most sophisticated formulation in the works of William Whewell, a defender of the wave theory. In what follows I will employ Whewell's version of the method of hypothesis as a foil for my discussion of the wave theorist's argument.

Whewell (1967, pp. 60–74) offered four conditions which, if satisfied, will make an hypothesis inferable with virtual certainty. First, it should explain all the phenomena which initially prompted it. Second, it should predict new phenomena. Third, it should explain and/or predict phenomena of a "kind different from those which were contemplated in the formation of . . . [the] hypothesis" (p. 65). If this third condition is satisfied Whewell says that there is a "consilience of inductions." Whewell's fourth condition derives from the idea that hypotheses are part of a theoretical system the components of which are not framed all at once, but are developed over time. The condition is that as the theoretical system evolves it becomes simpler and more coherent.

Since both Laudan and Cantor claim that the wave theorists followed the method of hypothesis while the particle theorists rejected this method in favor of inductivism, it will be useful to contrast Whewell's version of the former with Mill's account of the latter. This contrast should be of special interest for two reasons. Both Whewell and Mill discuss the wave theory, which Whewell supports and Mill rejects; and each criticizes the other's methodology.

One of the best places to note the contrast in Mill is in his discussion of the "deductive

method" (which he distinguishes from the "hypothetical method" or method of hypothesis) (Mill 1959, pp. 299–305). Mill asserts that the deductive method is to be used in situations where causes subject to various laws operate, in other words, in solving typical problems in physics as well as other sciences. It consists of three steps. First, there is a direct induction from observed phenomena to the various causes and laws governing them. Mill defines induction as "the process by which we conclude that what is true of certain individuals of a class is true of the whole class, or that what is true at certain times will be true in similar circumstances at all times" (p. 188). This concept of inductive generalization is used together with his four famous canons of causal inquiry to infer the causes operating and the laws that govern them. The second part of the deductive method Mill calls "ratiocination." It is a process of calculation, deduction, or explanation: from the causes and laws we calculate what effects will follow. Third, and finally, there is "verification": "the conclusions [derived by ratiocination] must be found, on careful comparison, to accord with the result of direct observation wherever it can be had" (p. 303).

Now, in rejecting the method of hypothesis, Mill writes:

> The Hypothetical Method suppresses the first of the three steps, the induction to ascertain the law, and contents itself with the other two operations, ratiocination and verification, the law which is reasoned from being assumed instead of proved (p. 323).

Mill's major objection to the method of hypothesis is that various conflicting hypotheses are possible from which the phenomena can be derived and verified. In his discussion of the wave theory of light, Mill rejects the hypothesis of the luminiferous ether on these grounds. He writes:

This supposition cannot be looked upon as more than a conjecture; the existence of the ether still rests on the possibility of deducing from its assumed laws a considerable number of actual phenomena . . . most thinkers of any degree of sobriety allow, that an hypothesis of this kind is not to be received as probably true because it accounts for all the known phenomena, since this is a condition some-times fulfilled tolerably well by two conflicting hypotheses; while there are prob-ably many others which are equally possible, but which, for want of anything analogous in our experience, our minds are unfitted to conceive (p. 328).

With Whewell's ideas about prediction and consilience in mind, Mill continues:

But it seems to be thought that an hypothesis of the sort in question is entitled to a more favourable reception if, besides accounting for all the facts previously known it has led to the anticipation and prediction of others which experience afterwards verified. . . . Such predictions and their fulfillment are, indeed, well calculated to impress the unin-formed, whose faith in science rests solely on similar coincidences between its prophecies and what comes to pass. . . . Though twenty such coincidences should occur they would not prove the reality of the undulatory ether. . . . (pp. 328–9)

Although in these passages Mill does not discuss Whewell's ideas about coherence and the evolu-tion of theories, it is clear that Mill would not regard Whewell's four conditions as sufficient to infer an hypothesis with virtual certainty or even high probability. The reason is that Whewell's conditions omit the first crucial step of the deductive method, the induction to the causes and laws.

If Laudan and Cantor are correct in saying that 19th century wave theorists followed the method of hypothesis and rejected inductivism, then, as these opposing methodologies are formulated by Whewell and Mill, this would mean the following: 19th century wave theorists argued for the virtual certainty or high prob-ability of their theory by first assuming, with-out argument, various hypotheses of the wave theory; then showing how these will not only explain the known optical phenomena but will explain and/or predict ones of a kind different from those prompting the wave hypotheses in the first place; and finally arguing that as the theory has evolved it has become simpler and more coherent. Is this an adequate picture? Or, in addition, did wave theorists employ a crucial inductive step to their hypotheses at the outset? Or do neither of these methodologies adequately reflect the wave theorists' argument?

3 The wave theorists' argument

Nineteenth century wave theorists frequently employed the following strategy in defense of their theory.

1   Start with the assumption that light consists either in a wave motion trans-mitted through a rare, elastic medium pervading the universe, or in a stream of particles emanating from luminous bodies. Thomas Young (1845) in his 1807 Lectures, Fresnel (1816) in his prize essay on diffraction, John Herschel (1845) in an 1827 review article of 246 pages, and Humphrey Lloyd (1834) in a 119 page review article,[1] all begin with this assumption in presentations of the wave theory.

2   Show how each theory explains various optical phenomena, including the recti-linear propagation of light, reflection,

refraction, diffraction, Newton's rings, polarization, etc.

3 Argue that in explaining one or more of these phenomena the particle theory introduces improbable auxiliary hypotheses but the wave theory does not. For example, light is diffracted by small obstacles and forms bands both inside and outside the shadow. To explain diffraction particle theorists postulate both attractive and repulsive forces emanating from the obstacle and acting at a distance on the particles of light so as to turn some of them away from the shadow and others into it. Wave theorists such as Young and Fresnel argue that the existence of such forces is very improbable. By contrast, diffraction is explainable from the wave theory (on the basis of Huygens' principle that each point in a wave front can be considered a source of waves), without the introduction of any new improbable assumptions. Similar arguments are given for several other optical phenomena, including interference and the constant velocity of light.

4 Conclude from steps 1 through 3 that the wave theory is true, or very probably true.

This represents, albeit sketchily, the overall structure of the argument. More details are needed before seeing whether, or to what extent, it conforms to Whewell's conditions or Mill's. But even before supplying such details we can see that the strategy is not simply to present a positive argument for the wave theory via an induction to its hypotheses and/or by showing that it can explain various optical phenomena. Whether it does these things or not, the argument depends crucially on showing that the rival particle theory has serious problems.

To be sure, neither Whewell's methodology nor Mill's precludes comparative judgments. For example, Whewell explicitly claims that the wave theory is more consilient and coherent than the particle theory. And Mill (who believed that neither theory satisfied his crucial inductive step) could in principle allow the possibility that new phenomena could be discovered permitting an induction to one theory but not the other. I simply want to stress at the outset that the argument strategy of the wave theorists, as I have outlined it so far, is essentially comparative. The aim is to show at least that the wave theory is better, or more probable, than the rival particle theory.

Is the wave theorist's argument intended to be stronger than that? I believe that it is. Thomas Young, both in his 1802 and 1803 Bakerian lectures (reprinted in Crew 1900), makes it clear that he is attempting to show that hitherto performed experiments, and analogies with sound, and passages in Newton, provide strong support for the wave theory, not merely that the wave theory is better supported than its rival. A similar attitude is taken by Fresnel, whose aim is not simply to show that the wave theory is better in certain respects than the particle theory, but that it is acceptable because it can explain various phenomena, including diffraction, without introducing improbable assumptions; by contrast, the particle theory is not acceptable, since it cannot. Even review articles are not simply comparative. Although he does compare the merits of the wave and particle theories in his 1834 report, Humphrey Lloyd makes it clear that this comparison leads him to assert the truth of the wave theory. In that theory, he claims:

there is thus established that connexion and harmony in its parts which is the never failing attribute of truth. . . . It may be confidently said that it possesses characters which no false theory ever possessed before (1877, p. 79).[2]

Let us now look more closely at the three steps of the argument leading to the conclusion. Wave theorists who make the assumption that light consists either of waves or particles do not do so simply in order to see what follows. They offer reasons, which are generally of two sorts. First, there is an argument from authority: "Leading physicists support one or the other assumption." Second, there is an argument from some observed property of light. For example, Lloyd notes that light travels in space from one point to another with a finite velocity, and that in nature one observes motion from one point to another occurring by the motion of a body or by vibrations of a medium.

Whatever one might think of the validity of these arguments, I suggest that they were being offered in support of the assumption that light consists either of waves or of particles. This is not a mere supposition. Argument from authority was no stranger to optical theorists of this period. Young in his 1802 paper explicitly appeals to passages in Newton in defense of three of his four basic assumptions. And Brougham, a particle theorist, defends his theory in part also by appeal to the authority and success of Newton. Moreover, the second argument, if not the first, can reasonably be interpreted as an induction in Mill's sense, i.e., as claiming that all observed cases of finite motion are due to particles or waves, so in all probability this one is too.[3]

I suggest, then, that wave theorists offered grounds for supposing it to be very probable that light consists either of waves or particles. I will write their claim as

(1) $p(W \text{ or } P/O\&b) \approx 1$,

where W is the wave theory, P is the particle theory, O includes certain observed facts about light including its finite motion, and b is background information including facts about modes of travel in other cases. ($\approx$ means "is close to.")

This is the first step in the earlier argument. I will postpone discussion of the second step for a moment, and turn to the third. Here the wave theorists assert that in order to explain various optical phenomena the rival particle theorists introduce improbable auxiliary hypotheses. By contrast, the wave theorists can explain these phenomena without introducing auxiliary hypotheses, or at least any that are improbable. Why are the particle theorists' auxiliary hypotheses improbable? And even if they are, how does this cast doubt on the central assumptions of the particle theory?

Let us return to diffraction, which particle theorists explained by the auxiliary hypothesis that attractive and repulsive forces emanate from the diffracting obstacle and act at a distance on the light particles bending some into the shadow and others away from it. By experiment Fresnel showed that the observed diffraction patterns do not vary with the mass or shape of the diffracting body. But known attractive and repulsive forces exerted by bodies do vary with the mass and shape of the body. So Fresnel concludes that the existence of such forces of diffraction is highly improbable. Again it seems plausible to construe this argument as an inductive one, making an inference from properties of known forces to what should be (but is not) a property of the newly postulated ones. Fresnel's experiments together with observations of other known forces provide inductive reasons for concluding that the particle theorists' auxiliary assumption about attractive and repulsive forces is highly improbable.

Even if this is so, how would it show that other assumptions of the particle theory are improbable? It would if the probability of the auxiliary force assumption given the other assumptions of the particle theory is much,

much greater than the probability of this auxiliary assumption not given the rest of the particle theory, i.e., if

(2) $p(A/P\&O\&b) \gg p(A/O\&b)$,

where A is the auxiliary assumption, O includes information about diffraction patterns and Fresnel's experimental result that these do not vary with the mass or shape of the diffractor, b includes information about other known forces, and $\gg$ means "is much, much greater than." If this condition is satisfied, it is provable that the other assumptions of the particle theory have a probability close to zero,[4] i.e.,

(3) $p(P/O\&b) \approx 0$.

Although wave theorists did not explicitly argue for (2) above, they clearly had grounds for doing so. If by the particle theory P light consists of particles subject to Newton's laws, and if by observational results O light is diffracted from its rectilinear path, then by Newton's first law a force or set of forces must be acting on the light particles. Since the light is being diffracted in the vicinity of the obstacle, it is highly probable that this obstacle is exerting a force or forces on the light particles. That is, with the assumptions of the particle theory, auxiliary hypothesis A is very probable. However, without these assumptions the situation is very different. Without them the fact that other known forces vary with the mass and shape of the body exerting the force, but diffraction patterns do not, makes it unlikely that such forces exist in the case of diffraction. Or at least their existence is much, much more likely on the assumption that light consists of particles obeying Newton's laws than without such an assumption, i.e., (2) above. An important part of the argument here is inductive, based as it is on information about other mechanical forces.

From (1) and (3) we infer:

(4) $p(W/O\&b) \approx 1$,

that is, the probability of the wave theory is close to 1, given the background information and certain optical phenomena, including diffraction.

Now we can return to the second step of the original argument, the one in which the wave theorist shows that his theory can explain a range of optical phenomena, not just the finite velocity of light and diffraction. What inferential value does this have? The wave theorist wants to show that his theory is probable not just given some limited selection of optical phenomena but given all known optical phenomena. This he can do if he can explain these phenomena by deriving them from his theory. Where $O_1, \ldots, O_n$ represent known optical phenomena other than diffraction and the constant velocity of light – including rectilinear propagation, reflection, refraction, and interference – if the wave theorist can derive these from his theory, then the probability of that theory will be at least sustained if not increased. This is a simple fact about probabilities.

Accordingly, the explanatory step in which the wave theorist derives various optical phenomena $O_1, \ldots, O_n$ from his theory permits an inference from (4) above to:

(5) $p(W/O_1, \ldots, O_n\&O\&b) \approx 1$,

i.e., the high probability of the wave theory given a wide range of observed optical phenomena. This is the conclusion of the wave theorist's argument.

If the explanation of known optical phenomena sustains the high probability of the wave theory without increasing it, does this mean that such phenomena fail to constitute evidence for the wave theory? Not at all.

According to a theory of evidence I have developed (Achinstein 1983, chs 10–11), optical phenomena can count as evidence for the wave theory even if they do not increase its probability. I reject the usual increase-in-probability account of evidence in favor of conditions that require the high probability of the theory T given the putative evidence O_i, and the high probability of an explanatory connection between T and O_i, given T and O_i. Both conditions are satisfied in the case of the wave theory.

In formulating the steps of the argument in the probabilistic manner above, I have clearly gone beyond what wave theorists say. For one thing, they do not appeal to probability in the way I have done. More importantly perhaps, while they argue that auxiliary hypotheses of the particle theorists are very improbable, they do not say that these assumptions are much more probable given the rest of the particle theory than without it. The following points are, I think, reasonably clear. (i) Wave theorists suppose that it is very likely that the wave or the particle theory is true, an assumption for which they have arguments. (ii) They argue against the particle theory by criticizing auxiliary assumptions of that theory, which introduce forces (or whatever) that violate inductively supported principles. (iii) Wave theorists argue that their theory can explain various optical phenomena without introducing any such questionable assumptions. (iv) Their reasoning, although eliminative, is different from typical eliminative reasoning; their first step is not to canvass all possible theories of light, but only two, for which they give arguments; their reasoning is not of the typical eliminative form "these are the only possible explanations of optical phenomena, all of which but one lead to difficulties." Reconstructing the wave theorists' argument in the probabilistic way I have done captures these four points. Whether it introduces too many fanciful ideas is a question I leave for my critics.

Is the argument Whewellian or Millian? It does satisfy the first three of Whewell's conditions. It invokes the fact that various optical phenomena are derived from the wave theory. These include ones that prompted the theory in the first place (rectilinear propagation, reflection, and refraction), hitherto unobserved phenomena that were predicted (e.g., the Poisson spot in diffraction), and phenomena of a kind different from those that prompted it (e.g., diffraction, interference, polarization). The argument does not, however, satisfy Whewell's fourth condition. It does not appeal to the historical tendency of the theory over time to become simpler and more coherent. But the latter is not what divides Whewell from Mill. Nor is it Whewell's first three conditions, each of which Mill allows for in the ratiocinative part of his deductive method. Mill's claim is only that Whewell's conditions are not sufficient to establish the truth or high probability of an hypothesis. They omit the crucial first step, the inductive one to the hypothesized causes and laws.

As I have reconstructed the wave theorists' argument, an appeal to the explanatory power of the theory is a part, but not the whole, of the reasoning. There is also reasoning of a type that Mill would call inductive. It enters at two points. It is used to argue that light is most probably composed either of waves or of particles (e.g., the "finite motion" argument of Lloyd). And it is used to show that light is probably not composed of particles, since auxiliary hypotheses introduced to explain various optical phenomena are very improbable. This improbability is established by inductive generalization (e.g., in the case of diffraction, by inductively generalizing from what observations and experiments show about diffraction effects, and from what they show about forces). My claims are that wave theorists did in fact employ such inductive reasoning; that with it the argument that I have constructed is valid; and that without it the argument is invalid, or at least an appeal to Whewell's explanatory conditions is not sufficient to establish the high probability of the theory (though this last claim requires much more than I say here; see Achinstein 1991, Essay 4).

4 Explanatory anomalies

One objection critics of my account may raise is that it does not do justice to explanatory anomalies in the wave theory. That theory was not able to explain all known optical phenomena. Herschel (1845), e.g., notes dispersion as one such phenomenon – the fact that different colors are refracted at different angles. Now the wave theorist wants to show that his theory is probable given all known optical phenomena, not just some favorable subset. But if dispersion is not derivable from the theory, and if there is no inductive argument from dispersion to that theory, then on the account I offer, the wave theorist cannot reach his desired conclusion. He can say only that his theory is probable given other optical phenomena. And he can take a wait-and-see attitude with respect to the unexplained ones. This is essentially what Herschel himself does in the case of dispersion.[5]

Let me now say how wave theorists could in principle deal with such anomalies that relates to the probabilistic reconstruction I offer. The suggestion I will make is, I think, implicit in their writings, if not explicit. And, interestingly, it is a response that combines certain Whewellian and Millian ideas. In what follows, I restrict the anomalies to phenomena which have not yet been derived from the wave theory by itself or from that theory together with auxiliary assumptions whose probability is very much greater given the wave theory than without it.

As Cantor notes in his very informative book *Optics after Newton*:

Probably the central, and certainly the most repeated, claim [by the 1830s] was that in comparison with its rival the wave theory was more successful in explaining optical phenomena (Cantor 1983, p. 192).

Cantor goes on to cite a table constructed in 1833 by Baden Powell, a wave theorist, listing 23 optical phenomena and evaluating the explanations proposed by wave and particle theories as "perfect," "imperfect," or "none." In the no-explanation category there are 12 entries for the particle theory and only two for the wave theory; while there are 18 "perfects" for the wave theory and only five for the particle theory.

Appealing, then, to the explanatory success of the wave theory, a very simple argument is this:

(6)   Optical phenomena $O_1, \ldots, O_n$ can be coherently explained by the wave theory.
O is another optical phenomenon.
So probably
O can be coherently explained by the wave theory.

By a "coherent" explanation I follow what I take to be Whewell's idea: either the phenomenon is explained from the theory without introducing any additional assumptions, or if they are introduced they cohere both with the theory and with other known phenomena. In particular, no auxiliary assumption is introduced whose probability given the theory is very high but whose probability on the phenomena alone is low. Or, more generally, no such assumption is employed whose probability on the theory is very much greater than its probability without it.

Commenting on argument (6), the particle theorist might offer a similar argument to the conclusion that the particle theory can also explain O. But this does not vitiate the previous argument. For one thing, by the 1830s, even though Powell's table was not constructed by a neutral observer, it was generally agreed that the number of optical phenomena known to be coherently explainable by the wave theory was considerably greater than the number explainable by the particle theory. So the wave theorist's argument for his conclusion would be stronger than the particle theorist's for his. But even more importantly, the conclusion of the argument is only that O can be coherently explained by the wave theory, not that it cannot be coherently explained by the particle theory. This is not eliminative reasoning.

Argument (6) above might be construed in Millian terms as inductive: concluding "that what is true of certain individuals of a class is true of the whole class," and hence of any other particular individual in that class (Mill 1959, p. 188; see note 3 above). Mill's definition is quite general and seems to permit an inference from the explanatory success of a theory to its continued explanatory success. Indeed, in his discussion of the wave theory he notes that "if the laws of propagation of light accord with those of the vibrations of an elastic fluid in as many respects as is necessary to make the hypothesis afford a correct expression of all or most of the phenomena known at the time, it is nothing strange that they should accord with each other in one respect more" (Mill 1959, p. 329). Mill seems to endorse this reasoning. What he objects to is concluding from it that the explanation is true or probable.

Argument (6) might also be construed as exhibiting certain Whewellian features. Whewell stresses the idea that a theory is an historical entity which changes over time and can "tend to simplicity and harmony." One of the important aspects of this tendency is that "the elements which we require for explaining a new class of facts are already contained in our system." He explicitly cites the wave theory, by contrast to the particle theory, as exhibiting this tendency. Accordingly, it seems reasonable to suppose that

it will be able to coherently explain some hitherto unexplained optical phenomenon. The important difference between Whewell and Mill in this connection is not over whether the previous explanatory argument (6) is valid, but over whether from the continued explanatory success of the wave theory one can infer its truth. For Whewell one can, for Mill one cannot.

Let me assume, then, that some such argument as (6) was at least implicit in the wave theorists' thinking; and that it would have been endorsed by both Mill and Whewell. How, if at all, can it be used to supplement the probabilistic reconstruction of the wave theorists' argument that I offer earlier in the paper? More specifically, how does it relate to the question of determining the probability of the wave theory given all the known optical phenomena, not just some subset?

The conclusion of the explanatory success argument (6) is that the wave theory coherently explains optical phenomenon O. This conclusion is made probable by the fact that the wave theory coherently explains optical phenomena $O_1, \ldots, O_n$.

Accordingly, we have:

(7) $p(W$ coherently explains optical phenomenon O/W coherently explains optical phenomena $O_1, \ldots, O_n) > k$

where k is some threshold of high probability, and W is the wave theory. If we construe such explanations as deductive, then

(8) "W coherently explains O" entails that
$$p(W/O \& O_1, \ldots, O_n) \geq p(W/O_1, \ldots, O_n)$$

So from (7) and (8) we get the second-order probability statement

(9) $p(p(W/O\&O_1, \ldots, O_n) \geq p(W/O_1, \ldots, O_n)/W$ coherently explains $O_1, \ldots, O_n) > k$

But the conclusion of the wave theorists' argument is

(10) $p(W/O_1, \ldots, O_n) \approx 1,$

where $O_1, \ldots, O_n$ includes all those phenomena for which the wave theorist supplies a coherent explanation (I suppress reference to background information here). If we add (10) to the conditional side of (9), then from (9) we get

(11) $p(p(W/O\&O_1, \ldots, O_n) \approx 1/W$ coherently explains $O_1, \ldots, O_n$ and $p(W/O_1, \ldots, O_n) \approx 1) > k.$

This says that, given that the wave theory coherently explains optical phenomena $O_1, \ldots, O_n$, and that the probability of the wave theory is close to 1 on these phenomena, the probability is high that the wave theory's probability is close to 1 given O – the hitherto unexplained optical phenomenon – together with the other explained phenomena. If we put all the known but hitherto unexplained optical phenomena into O, then we can conclude that the probability is high that the wave theory's probability is close to 1 given all the known optical phenomena.

How is this to be understood? Suppose we construe the probabilities here as representing reasonable degrees of belief. Then the first-order probability can be understood as representing how much belief it is reasonable to have in W; while the second-order probability is interpreted as representing how reasonable it is to have that much belief. Accordingly, conclusion (11) says this:

Given that the wave theory coherently explains optical phenomena $O_1, \ldots, O_n$, and

that it is reasonable to have a degree of belief in the wave theory, on these explained phenomena, that is close to 1, there is a high degree of reasonableness (greater than k) in having a degree of belief in the wave theory W, on both the explained and the unexplained optical phenomena, that is close to 1.

This, of course, does not permit the wave theorist to conclude that $p(W/O\&O_1, \ldots, O_n) \approx 1$, i.e., that the probability of the wave theory on all known optical phenomena – explained and unexplained – is close to 1. But it does permit him to say something stronger than simply that his theory is probable given a partial set of known optical phenomena. It goes beyond a wait-and-see attitude with respect to the unexplained phenomena.

Notes

1 Reprinted in Lloyd (1877). In what follows page references will be to this.

2 Herschel in his (1845) does not take as strong a position as Lloyd, although there are passages in which he says that the wave theory is confirmed by experiments (e.g., pp. 473, 486). In his later 1830 work he is even more positive. For example:

> It may happen (and it has happened in the case of the undulatory doctrine of light) that such a weight of analogy and probability may become accumulated on the side of an hypothesis that we are compelled to admit one of two things; either that it is an actual statement of what really passes in nature, or that the reality, whatever it be, must run so close a parallel with it, as to admit of expression common to both, at least as far as the phenomena actually known are concerned (Herschel 1987, pp. 196–7).

3 Although Mill defines induction as involving an inference from observed members of a class to the whole class, he clearly includes inferences to other unobserved members of the class. He writes: "It is

true that (as already shown) the process of indirectly ascertaining individual facts is as truly inductive as that by which we establish general truths. But it is not a different kind of induction; it is a form of the same process. . . . (Mill 1959, p. 186).

4   For a proof see Achinstein 1991, pp. 85–6. It might be noted that the introduction of an auxiliary assumption with very low probability does not by itself suffice to show that the other assumptions of the theory are highly improbable.

5   Herschel writes "We hold it better to state it [the difficulty in explaining dispersion] in its broadest terms, and call on the reader to suspend his condemnation of the doctrine for what it apparently will not explain, till he has become acquainted with the immense variety and complication of the phenomena which it will. The fact is, that neither the corpuscular nor the undulatory, nor any other system which has yet been devised, will furnish that complete and satisfactory explanation of all the phenomena of light which is desirable (Herschel 1845, p. 450).

References

Achinstein, P. (1983), *The Nature of Explanation*. New York: Oxford University Press.

——. (1991), *Particles and Waves*. New York: Oxford University Press.

Brougham, H. (1803), *Edinburgh Review* **1**:451ff.

Cantor, G. (1975), "The Reception of the Wave Theory of Light in Britain," *Historical Studies in the Physical Sciences* **6**:109–132.

——. (1983), *Optics after Newton*. Manchester: Manchester University Press.

Crew, H. (ed.) (1900), *The Wave Theory of Light*. New York: American Book Co.

Fresnel, A. (1816), "Memoir on the Diffraction of Light," reprinted in Crew (1900), 79–144.

Herschel, J. (1845), "Light," *Encyclopedia Metropolitana*.

——. (1987), *A Preliminary Discourse on the Study of Natural Philosophy*. Chicago: University of Chicago Press.

Laudan, L. (1981), "The Medium and its Message," in *Conceptions of the Ether*, G. Cantor and M. Hodge (eds). Cambridge: Cambridge University Press.

Lloyd, H. (1834), "Report on the Progress and Present State of Physical Optics," *Reports of the British Association for the Advancement of Science*, 297ff.

——. (1877), *Miscellaneous Papers*. London: Longmans, Green, and Co.

Mill, J. (1959), *A System of Logic*. London: Longmans, Green, and Co.

Whewell, W. (1967), *The Philosophy of the Inductive Sciences*, vol. 2. New York: Johnson Reprint Corporation.

Young, T. (1802), "On the Theory of Light and Colors," reprinted in Crew (1900), 45–61.

——. (1845), *A Course of Lectures on Natural Philosophy and the Mechanical Arts*. London: Taylor and Walton.

Chapter 17

Michael Strevens

NOTES ON BAYESIAN CONFIRMATION THEORY

There were three ravens sat on a tree,
Downe a downe, hay downe, hay downe
There were three ravens sat on a tree,
With a downe
There were three ravens sat on a tree,
They were as blacke as they might be.
With a downe derrie, derrie, derrie, downe,
    downe.

Anonymous, sixteenth century

1 Introduction

Bayesian confirmation theory – abbreviated to BCT in these notes – is the predominant approach to confirmation in late twentieth century philosophy of science. It has many critics, but no rival theory can claim anything like the same following. The popularity of the Bayesian approach is due to its flexibility, its apparently effortless handling of various technical problems, the existence of various a priori arguments for its validity, and its injection of subjective and contextual elements into the process of confirmation in just the places where critics of earlier approaches had come to think that subjectivity and sensitivity to context were necessary.

There are three basic elements to BCT. First, it is assumed that the scientist assigns what we will call *credences* or *subjective probabilities* to different competing hypotheses. These credences are numbers between zero and one reflecting something like the scientist's level of expectation that a particular hypothesis will turn out to be true, with a credence of one corresponding to absolute certainty.

Second, the credences are assumed to behave mathematically like probabilities. Thus they can be legitimately called *subjective probabilities* (subjective because they reflect one particular person's views, however rational).

Third, scientists are assumed to learn from the evidence by what is called the Bayesian conditionalization rule. Under suitable assumptions the conditionalization rule directs you to update your credences in the light of new evidence in a quantitatively exact way – that is, it provides precise new credences to replace the old credences that existed before the evidence came in – provided only that you had precise credences for the competing hypotheses before the evidence arrived. That is, as long as you have some particular opinion about how plausible each of a set of competing hypotheses is before you observe any evidence, the conditionalization rule will tell you *exactly* how to update your opinions as more and more evidence arrives.

My approach to BCT is more pragmatic than a priori, and more in the mode of the philosophy of science than that of epistemology or inductive logic. There is not much emphasis, then, on the considerations, such as the Dutch book argument (see section 3.4), that purport to show that we must all become Bayesians. Bayesianism is offered to the reader as a superior (though far from perfect) choice, rather than as the only alternative to gross stupidity.

This is, I think, the way that most philosophers of science see things – you will find the same tone in Horwich (1982) and Earman (1992) – but you should be warned that it is not the approach of the most prominent Bayesian proselytizers. These latter tend to be strict apriorists, concerned to prove above all that there is no rational alternative to Bayesianism. They would not, on the whole, approve of my methods.

A note to aficionados: Perhaps the most distinctive feature of my approach overall is an emphasis on the need to set subjective likelihoods according to the physical likelihoods, using what is often called Miller's Principle. While Miller's Principle is not itself especially controversial, I depart from the usual Bayesian strategy in assuming that, wherever inductive scientific inference is to proceed, a physical likelihood *must be found*, using auxiliary hypotheses if necessary, to constrain the subjective likelihood.

A note to all readers: some more technical or incidental material is separated from the main text in lovely little boxes. I refer to these as *tech boxes*. Other advanced material, occurring at the end of sections, is separated from what precedes it by a horizontal line, like so.

On a first reading, you should skip this material. Unlike the material in tech boxes, however, it will eventually become relevant.

2 Credence or subjective probability

Bayesianism is built on the notion of credence or subjective probability. We will use the term *credence* until we are able to conclude that credences have the mathematical properties of probability; thereafter, we will call credences *subjective probabilities*.

A credence is something like a person's level of expectation for a hypothesis or event: your credence that it will rain tomorrow, for example, is a measure of the degree to which you expect rain. If your credence for rain is very low, you will be surprised if it rains; if it is very high, you will be surprised if it does not rain. Credence, then, is psychological property. Everyone has their own credences for various events.

The Bayesian's first major assumption is that scientists, and other rational creatures, have credences not only for mundane occurrences like rain, but concerning the truth of various scientific hypotheses. If I am very confident about a hypothesis, my credence for that hypothesis is very high. If I am not at all confident, it is low.

The Bayesian's model of a scientist's mind is much richer, then, than the model typically assumed in classical confirmation theory. In the classical model, the scientist can have one of three attitudes towards a theory:[1] they accept the theory, they reject the theory, or they neither accept nor reject it. A theory is accepted once the evidence in its favor is sufficiently strong, and it is rejected once the evidence against it is sufficiently strong; if the evidence is strong in neither way, it is neither accepted nor rejected.

On the Bayesian model, by contrast, a scientist's attitude to a hypothesis is encapsulated in a level of confidence, or credence, that may take any of a range of different values from total disbelief to total belief. Rather than laying down, as does classical confirmation theory, a set of rules dictating when the evidence is sufficient to accept or reject a theory, BCT lays down a set of rules dictating how an individual's credences should change in response to the evidence.

In order to establish credence as a solid foundation on which to build a theory of confirmation, the Bayesian must, first, provide a formal mathematical apparatus for manipulating credences, and second, provide a material basis for the notion of credence, that is, an argument that credences are psychologically real.

2.1 Bayesian theories of acceptance

Some Bayesian fellow travellers (for example, Levi 1967) add to the Bayesian infrastructure a set of rules for accepting or rejecting hypotheses. The idea is that, once you have decided on your credences over the range of available hypotheses, you then have another decision to make, namely, which of those hypotheses, if any, to accept or reject based on your credences. The conventional Bayesian reaction to this sort of theory is that the second decision is unnecessary: your credences express all the relevant facts about your epistemic commitments.

The formal apparatus comes very easily. Credences are asserted to be, as the term *subjective probability* suggests, a kind of probability. That is, they are real numbers between zero and one, with a credence of one for a theory meaning that the scientist is practically certain that the theory is true, and a credence of zero meaning that the scientist is practically certain that the theory is false. (The difference between practical certainty and absolute certainty is explained in section 6.3.) Declaring credences to be probabilities gives BCT much of its power: the mathematical properties of probabilities turn out to be very apt for representing the relation between theory and evidence.

The psychological reality of credences presents more serious problems for the Bayesian.

While no one denies the existence of levels of expectation for events such as tomorrow's rain, what can reasonably be denied is the existence of a complete set of precisely specified numbers characterizing a level of expectation for all the various events and theories that play a role in the scientific process.

2.2 What do credences range over?

In probability mathematics, probabilities may be attached either to events, such as the event of its raining tomorrow, or to propositions, such as the proposition "It will rain tomorrow". It is more natural to think of a theory as a set of propositions than as an "event", for which reason BCT is usually presented as a theory in which probabilities range over propositions. My formal presentation respects this custom, but the commentary uses the notion of event wherever it seems natural.

The original response to this skeptical attitude was developed by Frank Ramsey (1931), who suggested that credences are closely connected to dispositions to make or accept certain bets. For example, if my credence for rain tomorrow is 0.5, I will accept anything up to an even money bet on rain tomorrow. Suppose we decide that, if it rains, you pay me $10, while if it does not rain, I pay you $5. I will eagerly accept this bet. If I have to pay you $10, so that we are both putting up the same amount of money, I will be indifferent to the bet; I may accept it or I may not. If I have to pay you $15, I will certainly not make the bet. (Some of the formal principles behind Ramsey's definition will be laid out more precisely in section 3.4.)

Ramsey argued that betting patterns are sufficiently widespread – since humans can bet on anything – and sufficiently consistent, to underwrite the existence of credences for all important propositions; one of his major contributions to the topic was to show that only very weak assumptions need be made to achieve the desired level of consistency.

What, exactly, is the nature of the connection between credences and betting behavior? The simplest and cleanest answer is to define credences in terms of betting behavior, so that, for example, your having a credence of one half for a proposition is no more or less than your being prepared to accept anything up to even odds on the proposition's turning out to be true.

Many philosophers, however, resist such a definition. They worry, for example, about the possibility that an aversion to gambling may distort the relation between a person's credences and their betting behavior. The idea underlying this and other such concerns is that credences are not dispositions to bet, but are rather psychological properties in their own right that are intimately, but not indefeasibly, connected to betting behavior (and, one might add, to felt levels of expectation). Ramsey himself held such a view.

This picture strikes me as being a satisfactory basis for modern Bayesian confirmation theory (though some Bayesian apriorists – see below – would likely disagree). Psychologists may one day tell us that there are no credences, or at least not enough for the Bayesian; for the sake of these notes on BCT, though, let me assume that we have all the credences that BCT requires.

3 Axioms of probability

3.1 *The axioms*

The branch of mathematics that deals with the properties of probabilities is called the *probability calculus*. The calculus posits certain *axioms* that state properties asserted to be both necessary and sufficient for a set of quantities to count as probabilities. (Mathematicians normally think of the axioms as constituting a kind of definition of the notion of probability.)

It is very important to the workings of BCT that credences count as probabilities in this mathematical sense, that is, that they satisfy all the axioms of the probability calculus. This section will spell out the content of the axioms; section 3.4 asks why it is reasonable to think that the psychological entities we are calling credences have the necessary mathematical properties.

Begin with an example, a typical statement about a probability, the claim that the probability of obtaining heads on a tossed coin is one half. You may think of this as a credence if you like; for the purposes of this section it does not matter what sort of probability it is.

The claim about the coin involves two elements: an outcome, heads, and a corresponding number, 0.5. It is natural to think of the probabilistic facts about the coin toss as mapping possible outcomes of the toss to probabilities. These facts, then, would be expressed by a simple function $P(\cdot)$ defined for just two outcomes, *heads* and *tails*:

$$P(heads) = 0.5;$$

$$P(tails) = 0.5.$$

This is indeed just how mathematicians think of probabilistic information: they see it as encoded in a function mapping outcomes to numbers that are the probabilities of those outcomes. The mathematics of probability takes as basic, then, two entities: a set of outcomes, and a function mapping the elements of that set to probabilities. The set of outcomes is sometimes called the *outcome space*; the whole thing the *probability space*.

Given such a structure, the axioms of the probability calculus can then be expressed as constraints on the probability function. There are just three axioms, which I will first state rather informally.

1   The probability function must map every outcome to a real number between zero and one.

2   The probability function must map an inevitable outcome (e.g., getting a number less than seven on a toss of a single die) to one, and an impossible outcome (e.g., getting a seven on a toss of single die) to zero.

3   If two outcomes are mutually exclusive, meaning that they cannot both occur at the same time, then the probability of obtaining either one or the other is equal to the sum of the probabilities for the two outcomes occurring separately. For example, since heads and tails are mutually exclusive – you cannot get both heads *and* tails on a single coin toss – the probability of getting either heads or tails is the sum of the probability of heads and the probability of tails, that is, $0.5 + 0.5 = 1$, as you would expect.

On the most conservative versions of the probability calculus, these are the only constraints placed on the probability function. As we will see, a surprising number of properties can be derived from these three simple axioms.

Note that axiom 3 assumes that the probability function ranges over combinations of outcomes as well as individual outcomes. For example, it is assumed that there is a probability not just for heads and for tails, but for the outcome *heads or tails*. A formal statement of the axioms makes explicit just what combinations of outcomes must have probabilities assigned to them. For our purposes, it is enough to know

that any simple combination of outcomes is allowed. For example, if the basic outcomes for a die throw are the first six integers, then a probability must be assigned to outcomes such as *either an even number other than six, or a five* (an outcome that occurs if the die shows two, four, or five). Note that we are allowed to refer to general properties of the outcomes (e.g., being even), and to use the usual logical connectives.

It is useful to have a shorthand for these complex outcomes. Let e and d be two possible outcomes of a die throw. Say that e is the event of getting an odd number, and d is the event of getting a number less than three. Then by ed, I mean the event of both e and d occurring (i.e., getting a one), by $e \vee d$ I mean the event of either e or d occurring (i.e., getting one of 1, 2, 3, or 5), and by $\neg e$ I mean the event of e's not occurring (i.e., getting an even number).

Using this new formalism, let me write out the axioms of the probability calculus more formally. Note that in this new version, the axioms appear to contain less information than in the version above. For example, the axioms do not require explicitly that probabilities are less than one. It is easy to use the new axioms to derive the old ones, however; in other words, the extra information is there, but it is implicit. Here are the axioms.

1   For every outcome e, $P(e) \geq 0$.
2   For any inevitable outcome e, $P(e) = 1$.
3   For mutually exclusive outcomes e and d, $P(e \vee d) = P(e) + P(d)$.

3.1 Sigma algebra

The main text's characterization of the domain of outcomes over which the probability function must be defined is far too vague for the formal needs of mathematics. Given a set of basic outcomes (which are themselves subsets of an even more basic set, though in simple cases such as a die throw, you may think of them as atomic elements), mathematicians require that the probability function be defined over what they call a sigma algebra formed from the basic set. The sigma algebra is composed by taking the closure of the basic set under the set operations of union (infinitely many are allowed), intersection (only finitely many), and complement. The outcome corresponding to the union of two other outcomes is deemed to occur if either outcome occurs; the outcome corresponding to the intersection is deemed to occur if both outcomes occur; and the outcome corresponding to the complement of another outcome is deemed to occur if the latter outcome fails to occur.

The notions of inevitability and mutual exclusivity are typically given a formal interpretation: an outcome is inevitable if it is logically necessary that it occur, and two outcomes are mutually exclusive if it is logically impossible that they both occur.

Now you should use the axioms to prove the following simple theorems of the probability calculus:

1   For every outcome e, $P(e) + P(\neg e) = 1$.
2   For every outcome e, $P(e) \leq 1$.
3   For any two logically equivalent propositions e and d, $P(e) = P(d)$. (You might skip this proof the first time through; you will, however, need to use the theorem in the remaining proofs.)
4   For any two outcomes e and d, $P(e) = P(ed) + P(e\neg d)$.
5   For any two outcomes e and d such that e entails d, $P(e) \leq P(d)$.

6   For any two outcomes e and d such that $P(e \supset d) = 1$ (where $\supset$ is material implication), $P(e) \leq P(d)$. (Remember that $e \supset d \equiv d \vee \neg e \equiv \neg(e\neg d)$.)

3.2 The axiom of countable additivity

Most mathematicians stipulate that axiom 3 should apply to combinations of denumerably many mutually exclusive outcomes. (A set is denumerable if it is infinite but countable.) This additional stipulation is called the *axiom of countable additivity*. Some other mathematicians, and philosophers, concerned to pare the axioms of the probability calculus to the weakest possible set, do their best to argue that the axiom of countable additivity is not necessary for proving any important results.

Having trouble? The main step in all of these proofs is the invocation of axiom 3, the only axiom that relates the probabilities for two different outcomes. In order to invoke axiom 3 in the more complicated proofs, you will need to break down the possible outcomes into mutually exclusive parts. For example, when you are dealing with two events e and d, take a look at the probabilities of the four mutually exclusive events ed, $e\neg d$, $d\neg e$, and $\neg e\neg d$, one of which must occur. When you are done, compare your proofs with those at the end (Strevens 2008).

3.2 *Conditional probability*

We now make two important additions to the probability calculus. These additions are conceptual rather than substantive: it is not new axioms that are introduced, but new definitions.

The first definition is an attempt to capture the notion of a *conditional probability*, that is, a probability of some outcome conditional on some other outcome's occurrence. For example, I may ask: what is the probability of obtaining a two on a die roll, given that the number shown on the die is even? What is the probability of *Mariner* winning tomorrow's race, given that it rains tonight? What is the probability that the third of three coin tosses landed heads, given that two of the three were tails?

The conditional probability of an outcome e given another outcome d is written $P(e|d)$. Conditional probabilities are introduced into the probability calculus by way of the following definition:

$$P(e|d) = \frac{P(ed)}{P(d)}.$$

(If $P(d)$ is zero, then $P(e|d)$ is undefined.) In the case of the die throw above, for example, the probability of a two given that the outcome is even is, according to the definition, the probability of obtaining a two *and* an even number (i.e., the probability of obtaining a two) divided by the probability of obtaining an even number, that is, $1/6$ divided by $1/2$, or $1/3$, as you might expect.

The definition can be given the following informal justification. To determine $P(e|d)$, you ought, intuitively, to reason as follows. Restrict your view to the possible worlds in which the outcome d occurs. Imagine that these are the only possibilities. Then the probability of e conditional on d is the probability of e in this imaginary, restricted universe. What you are calculating, if you think about it, is the proportion, probabilistically weighted, of the probability space corresponding to d that also corresponds to e.

3.3 Conditional probability introduced axiomatically

There are some very good arguments for introducing the notion of conditional probability as a primitive of the probability calculus rather than by way of a definition. On this view, the erstwhile definition, or something like it, is to be interpreted as a fourth axiom of the calculus that acts as a constraint on conditional probabilities $P(e|d)$ in those cases where $P(d)$ is non-zero. When $P(d)$ is zero, the constraint does not apply. The aim of this view is to allow the mathematical treatment of probabilities conditional on events that have either zero probability or an undefined probability.

Conditional probabilities play an essential role in BCT, due to their appearance in two important theorems of which BCT makes extensive use. The first of these theorems is Bayes' theorem:

$$P(e|d) = \frac{P(d|e)}{P(d)} P(e).$$

You do not need to understand the philosophical significance of the theorem yet, but you should be able to prove it. Notice that it follows from the definition of conditional probability alone; you do not need *any* of the axioms to prove it. In this sense, it is hardly correct to call it a theorem at all. All the more reason to marvel at the magic it will work . . .

The second important theorem states that, for an outcome e and a set of mutually exclusive, exhaustive outcomes $d_1, d_2, \ldots$ that

$$P(e) = P(e|d_1)P(d_1) + P(e|d_2)P(d_2) + \ldots$$

This is a version of what is called the *total probability theorem*. A set of outcomes is exhaustive if at least one of the outcomes must occur. It is mutually exclusive if at most one can occur. Thus, if a set of outcomes is mutually exclusive and exhaustive, it is guaranteed that exactly one outcome in the set will occur.

To prove the total probability theorem, you will need the axioms, and also the theorem that, if $P(k) = 1$, then $P(ek) = P(e)$. First show that $P(e) = P(ed_1) + P(ed_2) + \ldots$ (this result is itself sometimes called the theorem of total probability). Then use the definition of conditional probability to obtain the theorem. You can make life a bit easier for yourself if you first notice that axiom 3, which on the surface applies to disjunctions of just two propositions, in fact entails an analogous result for any finite number of propositions. That is, if propositions $e_1, \ldots, e_n$ are mutually exclusive, then axiom 3 implies that $P(e_1 \vee \ldots \vee e_n) = P(e_1) + \ldots + P(e_n)$.

3.3 Probabilistic independence

Two outcomes e and d are said to be *probabilistically independent* if

$$P(ed) = P(e)P(d).$$

The outcomes of distinct coin tosses are independent, for example, because the probability of getting, say, two heads in a row, is equal to the probability for heads squared.

Independence may also be characterized using the notion of conditional probability: outcomes e and d are independent if $P(e|d) = P(e)$. This characterization, while useful, has two defects. First, it does not apply when the probability of d is zero. Thus it is strictly speaking only a sufficient condition for independence; however, it is necessary and sufficient in all the interesting cases, that is, the cases in which neither probability is zero or one, which is why it is useful all the same. Second, it does not make transparent the symmetry of the independence

relation: e is probabilistically independent of d just in case d is probabilistically independent of e. (Of course, if you happen to notice that, for non-zero $P(e)$ and $P(d)$, $P(e|d) = P(e)$ just in case $P(d|e) = P(d)$, then the symmetry can be divined just below the surface.)

In probability mathematics, independence normally appears as an assumption. It is assumed that some set of outcomes is independent, and some other result is shown to follow. For example, you might show (go ahead) that, if e and d are independent, then

$$P(e) + P(d) = P(e \vee d) + P(e)P(d).$$

(Hint: start out by showing, without invoking the independence assumption, that $P(e) + P(d) = P(e \vee d) + P(ed)$.)

In applying these results to real world problems, it becomes very important to know when a pair of outcomes can be safely assumed to be independent. An often used rule of thumb assures us that outcomes produced by causally independent processes are probabilistically independent. (Note that the word *independent* appears twice in the statement of the rule, meaning two rather different things: probabilistic independence is a mathematical relation, relative to a probability function, whereas causal independence is a physical or metaphysical relation.) The rule is very useful; however, in many sciences, for example, kinetic theory and population genetics, outcomes are assumed to be independent even though they are produced by processes that are not causally independent. For an explanation of why these outcomes nevertheless tend to be probabilistically independent, run, don't walk, to the nearest bookstore to get yourself a copy of Strevens (2003).

In problems concerning confirmation, the probabilistic independence relation almost never holds between outcomes of interest, for reasons that I will explain later. Thus, the notion of independence is not so important to BCT, though we have certainly not heard the last of it.

3.4 Justifying the axioms

We have seen that calling credence a species of mathematical probability is not just a matter of naming: it imputes to credences certain mathematical properties that are crucial to the functioning of the Bayesian machinery. We have, so far, identified credences as psychological properties. We have not shown that they have any particular mathematical properties. Or rather – since the aim of confirmation theory is more to prescribe than to describe – we have not shown that credences *ought to have* any particular mathematical properties, that is, that people ought to ensure that their credences conform to the axioms of the probability calculus.

To put things more formally, what we want to do is to show that the credence function – the function $C(\cdot)$ giving a person's credence for any particular hypothesis or event – has all the properties specified for a generic probability function $P(\cdot)$ above. If we succeed, we have shown that $C(\cdot)$ is, or rather ought to be, a probability function; it will follow that everything we have proved for $P(\cdot)$ will be true for $C(\cdot)$.

This issue is especially important to those Bayesians who wish to establish a priori the validity of the Bayesian method. They would like to *prove* that credences should obey the axioms of the probability calculus. For this reason, a prominent strand in the Bayesian literature revolves around attempts to argue that it is irrational to allow your credences to violate the axioms.

The best known argument for this conclusion is known as the *Dutch book* argument. (The relevant mathematical results were motivated and proved independently by Ramsey and de Finetti, neither a Dutchman.) Recall that there is a strong relation, on the Bayesian view, between your credence for an event and your willingness to bet for or against the occurrence of the event in various circumstances. A Dutch book argument establishes that, if your credences do not conform to the axioms, it is possible to concoct a series of gambles that you will accept, yet which is sure to lead to a net loss, however things turn out. (Such a series is called a Dutch book.) To put yourself into a state of mind in which you are disposed to make a series of bets that must lose money is irrational; therefore, to fail to follow the axioms of probability is irrational.

The Dutch book argument assumes the strong connection between credence and betting behavior mentioned in section 2. Let me now specify exactly what the connection is supposed to be.

If your credence for an outcome e is p, then you should accept odds of up to $p : (1 - p)$ to bet on e, and odds of up to $(1 - p) : p$ to bet against e. To accept odds of $a : b$ on e is to accept a bet in which you put an amount proportional to a into the pot, and your opponent puts an amount proportional to b into the pot, on the understanding that, if e occurs, you take the entire pot, while if e does not occur, your opponent takes the pot. The important fact about the odds, note, is the ratio of a to b: the odds $1 : 1.5$, the odds $2 : 3$ and the odds $4 : 6$ are exactly the same odds. Consider some examples of the credence/betting relation.

1   Suppose that your credence for an event e is 0.5, as it might be if, say, e is the event of a tossed coin's landing heads. Then you will accept odds of up to $1 : 1$ (the same as $0.5 : 0.5$) to bet on e. If your opponent puts, say, $10 into the pot, you will accept a bet that involves your putting any amount of money up to $10 in the pot yourself, but not more than $10. (This is the example I used in section 2.)

2   Suppose that your credence for *e* is 0.8. Then you will accept odds of up to 4: 1 (the same as 0.8: 0.2) to bet on *e*. If your opponent puts $10 into the pot, you will accept a bet that involves your putting any amount of money up to $40 in the pot yourself.

3   Suppose that your credence for *e* is 1. Then you will accept any odds on *e*. Even if you have to put a million dollars in the pot and your opponent puts in only one dollar, you will take the bet. Why not? You are sure that you will win a dollar. If your credence for *e* is 0, by contrast, you will never bet on *e*, no matter how favorable the odds.

I will not present the complete Dutch book argument here, but to give you the flavor of the thing, here is the portion of the argument that shows how to make a Dutch book against someone whose credences violate axiom 2. Such a person has a credence for an inevitable event *e* that is less than 1, say, 0.9. They are therefore prepared to bet against *e* at odds of 1: 9 or better. But they are sure to lose such a bet. Moral: assign probability one to inevitable events at all times.

It is worth noting that the Dutch book argument says nothing about conditional probabilities. This is because conditional probabilities do not appear in the axioms; they were introduced by definition. Consequently, any step in mathematical reasoning about credences that involves only the definition of conditional probabilities need not be justified; not to take the step would be to reject the definition. Interestingly enough, the mathematical result about probability that has the greatest significance for BCT – Bayes' theorem – invokes only the definition. Thus the Dutch book argument is not needed to justify Bayes' theorem!

The Dutch book argument has been subjected to a number of criticisms, of which I will mention two. The first objection questions the very strong connection between credence and betting behavior required by the argument. As I noted in section 2, the tactic of defining credence so as to establish the connection as a matter of definition has fallen out of favor, but a connection that is any weaker seems to result in a conclusion, not that the violator of the axioms is guaranteed to accept a Dutch book, but that they have a tendency, all other things being equal, in the right circumstances, to accept a Dutch book. That is good enough for me, but it is not good enough for many aprioristic Bayesians.

The second objection to the Dutch book argument is that it seeks to establish too much. No one can be blamed for failing in some ways to arrange their credences in accordance with the axioms. Consider, for example, the second axiom. In order to follow the axiom, you would have to know which outcomes are inevitable. The axiom is normally interpreted fairly narrowly, so that an outcome is regarded as inevitable only if its non-occurrence is a *conceptual* impossibility (as opposed to, say, a *physical* impossibility). But even so, conforming to the axiom would involve your being aware of all the conceptual possibilities, which means, among other things, being aware of all the theorems of logic. If only we could have such knowledge! The implausibility of Bayesianism's assumption that we are aware of all the conceptual possibilities, or as is sometimes said, that we are *logically omniscient*, will be a theme of the discussion of the problem of old evidence in section 11 [sic].

Bayesians have offered a number of modifications of and alternatives to the Dutch book argument. All are attempts to establish the irrationality of violating the probability axioms. All, then, are affected by the second, logical omniscience objection; but each hopes in its own way to accommodate a weaker link between credence and subjective probability, and so to avoid at least the first objection.

Enough. Let us from this point on assume that a scientist's credences tend to, or ought to, behave in accordance with the axioms of probability. For this reason, I will now call credences, as promised, *subjective probabilities*.

4 Bayesian conditionalization

4.1 Bayes' rule

We have now gone as far in the direction of BCT as the axioms of probability can take us. The final step is to introduce the Bayesian conditionalization rule, a rule that, however intuitive, does not follow from any purely mathematical precepts about the nature of probability.

Suppose that your probability for rain tomorrow, conditional on a sudden drop in temperature tonight, is 0.8, whereas your probability for rain given no temperature drop is 0.3. The temperature drops. What should be your new subjective probability for rain? It seems intuitively obvious that it ought to be 0.8.

The Bayesian conditionalization rule simply formalizes this intuition. It dictates that, if your subjective probability for some outcome d conditional on another outcome e is p, and if you learn that e has in fact occurred (and you do not learn anything else), you should set your unconditional subjective probability for d, that is, $C(d)$, equal to p.

Bayes' rule, then, relates subjective probabilities at two different times, an earlier time when either e has not occurred or you do not know that e has occurred, and a later time when you learn that e has indeed occurred. To write down the rule formally, we need a notation that distinguishes a person's subjective probability distribution at two different times. I write a subjective probability at the earlier time as $C(\cdot)$, and a subjective probability at the later time as $C^+(\cdot)$.

Then Bayes' rule for conditionalization can be written:

$$C^+(d) = C(d \,|\, e),$$

on the understanding that the sole piece of information learned in the interval between the two times is that e has occurred. More generally, if $e_1 \ldots e_n$ are all the pieces of information learned between the two times, then Bayes' rule takes the form

$$C^+(d) = C(d \,|\, e_1 \ldots e_n).$$

If you think of what is learned, that is $e_1 \ldots e_n$, as the *evidence*, then Bayes' rule tells you how to update your beliefs in the light of the evidence, and thus constitutes a theory of confirmation. Before I move on to the application of Bayes' rule to confirmation theory in section 5, however, I have a number of important observations to make about conditionalization in itself.

4.2 Observation

Let me begin by saying some more about what a Bayesian considers to be the kind of event that prompts the application of Bayes' rule. I have said that a Bayesian conditionalizes on e – that is, applies Bayes' rule to e – just when they "learn that e has occurred". In classical Bayesianism, to learn e is to have one's subjective probability for e go to one as the result of some kind of observation. This observation-driven change of e's probability is not, note, due to an application of Bayes' rule. It is, as it were, prompted by a perception, not an inference.

The Bayesian, then, postulates two mechanisms by means of which subjective probabilities may justifiably change:

1   An observation process, which has the effect of sending the subjective probability

for some observable state of affairs (or if you prefer, of some observation sentence) to one. The process is not itself a reasoning process, and affects only individual subjective probabilities.

2   A reasoning process, governed by Bayes' rule. The reasoning process is reactive, in that it must be triggered by a probability change due to some other process; normally, the only such process envisaged by the Bayesian is observation.

That the Bayesian relies on observation to provide the impetus to Bayesian conditionalization prompts two questions. First, what if observation raises the probability of some e, but not all the way to one? Second, what kind of justification can be given for our relying on the probability changes induced by observation?

To the first question, there is a standard answer. If observation changes the credence for some e to a value x not equal to one, use the following rule instead of Bayes' rule:

$$C^+(d) = C(d\,|\,e)x + C(d\,|\,\neg e)(1 - x).$$

You will see that this rule is equivalent to Bayes' rule in the case where x is one. The more general rule is called *Jeffrey conditionalization*, after Jeffrey (1983).

The second question, concerning the justification of our reliance on observation, is not provided with any special answer by BCT. Indeed, philosophers of science typically leave this question to the epistemologists, and take the epistemic status of observation as given.

4.3 Background knowledge

When I observe e, I am, according to Bayes' rule, to set my new probability for d equal to $C(d\,|\,e)$. But $C(d\,|\,e)$, it seems, only expresses the relation between e and d in isolation. What if e is, on its own, irrelevant to d, but is highly relevant when other information is taken into account? It seems that I ought to set $C^+(d)$ equal not to $C(d\,|\,e)$, but to $C(d\,|\,ek)$, where k is all my background knowledge.

Let me give an example. Suppose that d is the proposition "The room contains at least two philosophy professors" and e is the proposition "Professor Wittgenstein is in the room". Then $C(d\,|\,e)$ should be, it seems, moderately large, or at least, greater than $C(d)$. But suppose that I know independently that Professor Wittgenstein despises other philosophers and will leave the room immediately if another philosopher enters. The conditional probability that takes into account this background knowledge, $C(d\,|\,ek)$, will then be close to zero. Clearly, upon seeing Professor Wittgenstein in the room, I should take my background knowledge into account, setting $C(d)$ equal to this latter probability. Thus Bayes' rule must incorporate k.

In fact, although there is no harm in incorporating background knowledge explicitly into Bayes' rule, it is not necessary. The reason is that any relevant background knowledge is already figured into the subjective probability $C(d\,|\,e)$; in other words, at all times, $C(d\,|\,e) = C(d\,|\,ek)$.[2] This follows from the assumption that we assign our background knowledge subjective probability one and the following theorem of the probability calculus:

If $P(k) = 1$, then $P(d\,|\,ek) = P(d\,|\,e)$.

which follows in turn from another theorem: if $P(k) = 1$, then $P(ek) = P(e)$.

I misled you in the example above by suggesting that $C(d\,|\,e)$ is moderately large. In fact, it is equal to $C(d\,|\,ek)$ and therefore close to zero. Precisely because it is easy to be misled in this way, however, it is in some circumstances worth putting the background knowledge explicitly into Bayes' rule, just to remind yourself and others that it is always there regardless.

4.4 Justifying Bayes' rule

Bayes' rule does not follow from the axioms of the probability calculus. You can see this at a glance by noting that the rule relates two different probability functions, $C(\cdot)$ and $C^+(\cdot)$, whereas the axioms concern only a single function. Less formally, the axioms put a constraint on the form of the assignment of subjective probabilities at a particular time, whereas Bayes' rule dictates a relation between subjective probability assignments at two different times.

To get a better feel for this claim, imagine that we have a number of cognitively diverse people whose subjective probabilities obey, at all times, the axioms of the probability calculus, and who conditionalize according to Bayes' rule. Construct a kind of mental Frankenstein's monster, by cutting each person's stream of thoughts into pieces, and stitching them together haphazardly. At any particular moment in the hybrid stream, the subjective probability assignments will obey the calculus, because they belong to the mind of some individual, who by assumption obeys the calculus at all times. But the stream will not obey Bayes' rule, since the value of $C(d\,|\,e)$ at one time may belong to a different person than the value of $C(d)$ at a later time. Indeed, there is no coherence at all to the hybrid stream; the moral is that a stream of thoughts that at all times satisfies the probability calculus can be as messed up as you like when examined for consistency through time.

To justify Bayes' rule, then, you need to posit some kind of connection between a person's thoughts at different times. A number of Bayesians have tried to find a connection secure enough to participate as a premise in an a priori argument for Bayes' rule. One suggestion, due to David Lewis, is to postulate a connection between subjective probabilities at one time and

betting behavior at a later time, and to run a kind of Dutch book argument.

4.1 What the apriorist must do

Let me summarize the things that the Bayesian apriorist must prove in order to establish the apparatus of BCT as compulsory for any rational being. It must be shown that:

1. Everyone has, or ought to have, subjective probabilities for all the elements that play a part in scientific confirmation, in particular, hypotheses and outcomes that would count as evidence.
2. The subjective probabilities ought to conform to the axioms of the probability calculus.
3. The subjective probabilities ought to change in accordance with Bayes' rule. (The exceptions are probability changes due to observation; see section 4.2.)

The "old-fashioned" apriorist tries to get (1) and (2) together by defining subjective probability in terms of betting behavior. In note 3 I suggested trying to get (2) and (3) with a stricter definition of subjective probability; the drawback to this is that the definition, being much more stringent, would no longer be satisfied by a person's betting behavior under the rather weak conditions shown by Ramsey to be sufficient for the existence of subjective probabilities on the "old-fashioned" definition. Thus a definition in terms of extended betting behavior would weaken the argument for the existence of subjective probabilities, to some extent undermining (1).

Two other conclusions that we have not encountered yet may also be the object of the apriorist's desire:

4. Subjective probabilities ought to conform to the probability coordination principle (see section 5.2). To show this is not compulsory, but it is highly desirable.
5. Initial probabilities over hypotheses ought to conform to some kind of symmetry principle (see section 9.4 [sic]). To show this is not compulsory; many would regard it as completely unnecessary.

The connection in question is just what you are thinking: if a person has subjective probability p for d conditional on e, and if e (and nothing else) is observed at some later time, then at that later time, the person should accept odds of up to $p : 1 - p$ on d. Note that this is not the connection between subjective probability and betting behavior used in section 3.4 to run the conventional Dutch book argument. That connection relates subjective probabilities at a time and betting behavior at the same point in time; the new principle relates subjective probabilities at a time and betting behavior at a strictly later time, once new evidence has come in.

This raises a problem for an old-fashioned apriorist. The old-fashioned apriorist justifies the old connection between subjective probability and betting behavior by defining one in terms of the other. But the definitional maneuver cannot be used twice. Once necessary and sufficient conditions are given for having a subjective probability at time t in terms of betting behavior at time t, additional necessary conditions cannot be given in terms of betting behavior at time t'.[3]

There are other approaches to justifying Bayes' rule, but let me move on.

5 The machinery of modern Bayesianism

5.1 From conditionalization to confirmation

In two short steps we will turn Bayesianism into a theory of confirmation. The first step — really just a change of attitude — is to focus on the use of Bayes' rule to update subjective probabilities concerning scientific hypotheses. To mark this newfound focus I will from this point on write Bayes' rule as follows:

$$C^+(h) = C(h \,|\, e).$$

The rule tells you what your new subjective probability for a hypothesis h should be, upon observation of the evidence e.

In this context, it is no longer natural to call the argument h of the probability function an "outcome" (though if you insist, you can think of the outcome as *the hypothesis's being true*). It is far more natural to think of it as a proposition. At the same time, it is more natural to think of most kinds of evidence as events; consider, for example, the event of the litmus paper's turning blue. For this reason, most expositors of BCT talk apparently inconsistently as if subjective probability functions range over both propositions and events. There is no harm in this custom (see tech box 2.2), and I will follow it with relish.

The second step in the transformation of Bayesianism into a theory of confirmation is, I think, the maximally revelatory moment in all of BCT. So far, Bayes' rule does not appear to be especially useful. It says that you should, upon observing e, set your new probability for h equal to your old probability for h conditional on e. But what ought that old probability to be? It seems that there are very few constraints on a probability such as $C(h|e)$, and so that Bayes' rule is not giving you terribly helpful advice. A skeptic might even suggest reading the rule backwards: $C(h|e)$ is just the probability that you would assign to h, if you were to learn that e. (But this would be a mistake: see tech box 5.1.)

The power of BCT consists in its ability to tell you, these appearances to the contrary, what your value for $C(h|e)$ should be, given only information about your probabilities for h and its competitors, that is, given only values of the form $C(h)$. It will take some time (the remainder of this section) to explain exactly how this is done.

5.1 Conditional probability characterized dispositionally

There are some interesting counter-examples to the view that $C(h|e)$ is the probability that I would assign to h, if I were to learn that e (due to Richmond Thomason). Suppose that h is the proposition that I am philosophically without talent, and e is some piece of (purely hypothetical!) evidence that incontrovertibly shows that I am philosophically untalented. Then my $C(h|e)$ is very high. But I may be vain enough that, were I to learn that e, I would resist the conclusion that h. Of course, I would be violating Bayes' rule, and (Bayesians would say "therefore") I would be reasoning irrationally. But the scenario is quite possible – plausible, even – and shows that human psychology is such that a dispositional interpretation of conditional probability in these terms is not realistic. The example does not, of course, rule out the possibility, mentioned above, of denying $C(h|e)$ as the probability I *ought to have* for h on learning that e.

I have promised you a revelatory moment, and now it is time to deliver. Bayes' rule sets $C^+(h)$ equal to $C(h|e)$. We encountered, in section 3.2, a result that I called Bayes' theorem:

$$C(d|e) = \frac{C(e|d)}{C(e)} C(d).$$

Writing h instead of d, substitute this into Bayes' rule, obtaining

$$C^+(h) = \frac{C(e|h)}{C(e)} C(h).$$

In this formulation, Bayes' rule is suddenly full of possibilities. I had no idea what value to give to $C(h|e)$, but I know exactly what value to give to $C(e|h)$: it is just the probability that h itself ascribes to a phenomenon such as e. The value of $C(e)$ is perhaps less certain, but for an observable event e, it seems that I am likely to have some opinion or other. (We will see shortly that I have a more definite opinion than I may suppose.) Now, given values for $C(e|h)$ and $C(e)$, I have determined a value for what you might call the *Bayesian multiplier*, the value by which I multiply my old probability for h to arrive at my new probability for h after observing e. What more could you ask from a theory of confirmation?

To better appreciate the virtues of BCT, before I tax you with the details of the same, let me show how BCT deals with a special case, namely, the case in which a hypothesis h entails an observed phenomenon e. I assume that $C(e)$ is less than one, that is, that it was an open question, in the circumstances, whether e would be observed. Because h entails e, $C(e|h)$ is equal to one. (You proved this back in section 3.1.) On the observation of e, then, your old subjective probability for h is multiplied by the Bayesian multiplier

$$\frac{C(e\,|\,h)}{C(e)}$$

which is, because $C(e)$ is less than one, greater than one. Thus the observation of e will increase your subjective probability for h, confirming the hypothesis, as you would expect.

This simple calculation, then, has reproduced the central principle of hypothetico-deductivism, and – what HD itself never does – *given an argument* for the principle. (We will later see how Bayesianism avoids some of the pitfalls of HD.) What's more, the argument does not turn on any of the subjective aspects of your assignment of subjective probabilities. All that matters is that $C(e\,|\,h)$ is equal to one, an assignment which is forced upon all reasoners by the axioms of probability. You should now be starting to appreciate the miracle of Bayesianism.

5.2 Constraining the likelihood

You have begun to see, and will later see far more clearly, that the power of BCT lies to a great extent in the fact that it can appeal to certain powerful constraints on the way that we set the subjective probabilities of the form $C(e\,|\,h)$.

I will call these probabilities the *subjective likelihoods*. (A likelihood is any probability of the form $P(e\,|\,h)$, where h is some hypothesis and e a piece of potential evidence.)

5.2 Logical probability

Properly constrained subjective probability, or credence, is now the almost universal choice for providing a probabilistic formulation of inductive reasoning. But it was not always so. The first half of the twentieth century saw the ascendancy of what is often called *logical probability* (Keynes 1921; Carnap 1950).

A logical probability relates two propositions, which might be called the hypothesis and the evidence. Like any probability, it is a number between zero and one. Logical probabilists hold that their probabilities quantify the evidential relation between the evidence and the hypothesis, meaning roughly that a logical probability quantifies the degree to which the evidence supports or undermines the hypothesis. When the probability of the hypothesis on the evidence is equal to one, the evidence establishes the truth of the hypothesis for sure. When it is equal to zero, the evidence establishes the falsehood of the hypothesis for sure. When it is between zero and one, the evidence has some lesser effect, positive or negative. (For one particular value, the evidence is irrelevant to the hypothesis. This value is of necessity equal to the probability of the hypothesis on the empty set of evidence. It will differ for different hypotheses.) The degree of support quantified by a logical probability is supposed to be an objective matter – the objective logic in question being, naturally, inductive logic. (The sense of *objective* varies: Carnap's logical probabilities are relative to a "linguistic framework", and so may differ from framework to framework.)

Logical probability has fallen out of favor for the most part because its assumption that there is always a fully determined, objectively correct degree of support between a given body of evidence and a given hypothesis has come to be seen as unrealistic. Yet it should be noted that when BCT is combined with PCP and an objectivist approach to the priors (section 9.4) [*sic*], we are back in a world that is not too different from that of the logical probabilists.

At the end of the last section, I appealed to a very simple constraint on the subjective likelihood $C(e|h)$: if h entails e, then the subjective likelihood must be equal to one. This is a theorem of the probability calculus, and so all who are playing the Bayesian game must conform to the constraint. (Apriorist Bayesians would threaten to inflict Dutch books on non-conformists.)

If the hypothesis in question does not entail the evidence e (and does not entail $\neg e$), however, this constraint does not apply. There are two reasons that h might not make a definitive pronouncement on the occurrence of e. Either

1   The hypothesis concerns physical probabilities of events such as e; it assigns a probability greater than zero but less than one to e, or
2   The hypothesis has nothing definite to say about e at all.

I will discuss the first possibility in this section, and the second in section 6.4.

If h assigns a definite probability to e, then it seems obviously correct to set the subjective likelihood equal to the probability assigned by h, which I will call the *physical likelihood*. For example, if e is the event of a tossed coin's landing heads and h is the hypothesis that tossed coins land heads with a probability of one half, then $C(e|h)$ should also be set to one half. Writing $P_h(e)$ for the probability that h ascribes to e, the rule that we seem tacitly to be following is:

$$C(e|h) = P_h(e).$$

That is, your subjective probability for e conditional on h's turning out to be correct should be the physical probability that h assigns to e. Call this rule the *probability coordination principle*, or PCP.

The principle has an important implication for BCT: if PCP is true, then BCT always favors, relatively speaking, hypotheses that ascribe higher physical probabilities to the observed evidence.

5.3 The probability coordination principle

The rule that subjective likelihoods ought to be set equal to the corresponding physical probabilities is sometimes called Miller's Principle, after David Miller. The name is not particularly apt, since Miller thought that the rule was incoherent (his naming rights are due to his having emphasized its importance in BCT, which he wished to refute). Later, David Lewis (1980) proposed a very influential version of the principle that he called the *Principal Principle*. Lewis later decided (due to issues involving admissibility; see tech box 5.5) that the Principal Principle was false, and ought to be replaced with a new rule of probability coordination called the *New Principle* (Lewis 1994). It seems useful to have a way to talk about all of these principles without favoring any particular one; for this reason, I have introduced the term *probability coordination principle*. In these notes, what I mean by PCP is *whatever probability coordination principle turns out to be correct*.

I cannot emphasize strongly enough that $C(e|h)$ and $P_h(e)$ are two quite different kinds of things. The first is a subjective probability, a psychological fact about a scientist. The second is a physical probability of the sort that might be prescribed by the laws of nature; intuitively, it is a feature of the world that might be present even if there were no sentient beings in the world and so no psychological facts at all. We have beliefs about the values of different physical probabilities; the strength of these beliefs is given by our subjective probabilities.

Because subjective probability and physical probability are such different things, it is an open question why we ought to constrain our subjective probabilities in accordance with PCP. Though Bayesians have offered many proofs that one ought to conform to the constraints imposed by the axioms of probability and Bayes' rule, there has been much less work on PCP.

Before I continue, I ought to tell you two important things about PCP. First, not all Bayesians insist that subjective probabilities conform to PCP: the *radical personalists* are quite happy for the subjective likelihoods to be, well, subjective. Radical personalism, which has strong ties to the subjectivist interpretation of probability (see tech box 5.4), will be discussed in section 9.3 [*sic*].

5.4 Subjectivism about physical probability

An important view about the nature of physical probability holds that the probabilities apparently imputed to the physical world by scientific theories are nothing but projections of scientists' own subjective probabilities. This thesis is usually called *subjectivism*, though confusingly, sometimes when people say *subjectivism* they mean *Bayesianism*. The idea that drives subjectivism, due originally to

Bruno de Finetti, is that certain subjective probabilities are especially robust, in the sense that conditionalizing on most information does not affect the value of the probabilities. An example might be our subjective probability that a tossed coin lands heads: conditionalizing on almost anything we know about the world (except for some fairly specific information about the initial conditions of the coin toss) will not alter the probability of one half. (Actually, the robustness is more subtle than this; Strevens (2006) provides a more complete picture.) According to the subjectivists, this robustness gives the probabilities an objective aspect that is usually interpreted literally, but mistakenly so. Almost all subjectivists are Bayesians, but Bayesians certainly need not be subjectivists.

Subjectivists do not have to worry about justifying PCP; since subjectivism more or less identifies physical probabilities with certain subjective likelihoods, PCP is trivially true.

Second, the formulation of PCP stated above, though adequate for the everyday workings of Bayesian confirmation theory, cannot be entirely right. In some circumstances, it is irrational for me to set my subjective likelihood equal to the corresponding physical probability. The physical probability of obtaining heads on a coin toss is one half. But suppose I know the exact velocity and trajectory of a certain tossed coin. Then I can use this information to calculate whether or not it will land heads. Let's say that I figure it will land heads. Then I should set my subjective probability for heads to one, not one half. (Remember that this subjective probability takes into account tacitly all my background knowledge, including my knowledge of the toss's initial conditions and their consequences, as explained in section 4.2.) David Lewis's now classic paper on the probability coordination principle is, in part, an attempt to frame the principle so that it recuses itself when we have information of the sort just described, which Lewis calls *inadmissible information*. You will find more on this problem in tech box 5.5.

The probability coordination principle is a powerful constraint on the subjective likelihoods, but, as I noted above, it seems that it can only brought to bear when the hypotheses under consideration assign specific physical probabilities to the evidence. Some ways around this limitation will be explored in section 6.4.

5.3 Constraining the probability of the evidence

How should you set a value for $C(e)$, the probability of observing a particular piece of evidence? A very helpful answer is to be found in a theorem of the probability calculus presented in section 3.2, the theorem of total probability, which states that for mutually exclusive and exhaustive $d_1, d_2, \ldots$,

$$C(e) = C(e|d_1)C(d_1) + C(e|d_2)C(d_2) + \ldots.$$

Suppose that the d_is are a set of competing hypotheses. Let us change their name to reflect this supposition: henceforth, they shall be the h_i. This small change in notation gives the theorem the following suggestive formulation:

$$C(e) = C(e|h_1)C(h_1) + C(e|h_2)C(h_2) + \ldots.$$

If you use PCP to set the likelihoods $C(e|h_i)$, then the total probability theorem prescribes a technique for setting $C(e)$ that depends only on your probabilities over the hypotheses h_i, that is, only on the $C(h_i)$.

5.5 Inadmissible information

David Lewis's version of PCP says that you should set $C(e|h)$ equal to $P_h(e)$ provided that your background knowledge includes no inadmissible information about e. Lewis does not provide a definition of inadmissibility, but suggests the following heuristic: normally, all information about the world up until the time that the process producing e begins is admissible. (My talk of the process producing e is a fairly crude paraphrase of Lewis's actual view.)

The heuristic is supposed to be foolproof: Lewis mentions, as an exception, the fact that a reading from a crystal ball predicting whether or not e occurs is inadmissible. There are less outré examples of inadmissible information about the past than this, however: the example in the main text, of information about the initial conditions of a coin toss, is a case in point. Cases such as these make it very difficult to provide a good definition of inadmissibility, except to say: evidence is inadmissible relative to some $P_h(e)$ if it contains information relevant to e that is not contained in the physical probability h ascribes to e. But then how to decide what information is relevant to e? Perhaps you look to confirmation theory . . . oh.

By the way, Lewis would not sanction the application of PCP to the probability of heads, because he would not count it as a "real" probability, due to the fact that the process that determines the outcome of a coin toss is, at root, deterministic, or nearly so. (This means that he does not need to worry about the inadmissibility of information about initial conditions.) There may be some metaphysical justification for this view, but it is not very helpful to students of confirmation theory. Many scientific theories concern probabilities that Lewis would regard as unreal, statistical mechanics and population genetics (more generally, modern evolutionary biology) being perhaps the two most notable examples. If we want to use BCT to make judgments about theories of this sort, we will want to constrain the subjective likelihoods using the physical probabilities and we will need a probability coordination principle to do so.

In order to use the total probability theorem in this way, your set of competing hypotheses must satisfy three requirements:

1   The hypotheses each assign an explicit physical probability to events of *e*'s kind,
2   The hypotheses are mutually exclusive, and
3   The hypotheses are exhaustive.

Of these, assumption (1) has already been made (section 5.2) in order to set a value for the subjective likelihood that constitutes the numerator of the Bayesian multiplier. Assumption (2), that no two of the competing hypotheses could both be true, may seem obviously satisfied in virtue of the meaning of "competing". A little thought, however, shows that often hypotheses compete in science although they are strictly speaking consistent, as in the case of competing Darwinian explanations of a trait that pick out different properties of the trait as selectively advantageous. (Perhaps the competitors can be reformulated so that they are mutually exclusive, however.)

Assumption (3) is perhaps the shakiest of the three. The hypotheses are exhaustive only if there is not some further theory that is incompatible with the h_is. To be sure that there

is no such theory, it seems, I would have to have at least some passing acquaintance with all the possible theories that claim to predict events like *e*. But this is unrealistic. The most striking examples of scientific progress are those where an entirely new explanation of the evidence comes to light.

This problem, it turns out, not only undermines a certain application of the theorem of total probability, but goes to the very root of Bayesianism, for the reason that the whole apparatus of subjective probability assumes that the individual puts a probability distribution over all possibilities. The issues arising will be examined in section 11.4 [*sic*]. Until then, put these worries aside.

5.4 Modern Bayesianism: a summary

We have come a long way. Back in section 5.1, Bayes' rule seemed to be telling you, rather unhelpfully, to set your probabilities after the observation of *e* equal to your earlier probabilities conditional on *e*. With the help of Bayes' theorem, this was seen to amount to multiplying your earlier probability for *h* by the Bayesian multiplier:

$$\frac{C(e|h)}{C(e)}$$

Given the constraints imposed by PCP and the theorem of total probability, the Bayesian multiplier turns out to depend entirely on your earlier probabilities for *h* and its rivals. It is equal to

$$\frac{P_h(e)}{P_{h1}(e)C(h_1)+P_{h2}(e)C(h_2)+\cdots}$$

(where *h* is, by assumption, one of the h_is).

Once you have set subjective probabilities for all of a set of competing hypotheses $h_1,h_2,\ldots$, then, the Bayesian apparatus tells you, given the assumptions stated in the previous two sections, how to update these probabilities upon observation of any piece of evidence *e*. In other words, provided that you have some initial view about the relative plausibility of a set of competing hypotheses, BCT prescribes the *exact, quantitative* changes you must make to these on the observation of any set of evidence. The Bayesian apparatus, it seems, is a complete guide to how your beliefs must change in response to the evidence.

The probabilities that are assigned to various hypotheses before any evidence comes in are referred to as *prior probabilities* (see tech box 5.6 for an important terminological note). If you manage to set your prior probabilities to some values before any evidence comes in, then, BCT will take care of the rest. This should be enough to make you acquire a set of prior probabilities even if you do not already have them!

5.6 Prior probabilities

The term *prior probability* is used in three distinct ways in the Bayesian literature. First, it can mean, as in the main text, your subjective probability for a hypothesis before you have received any evidence. Second, it can mean your subjective probability for a hypothesis before some particular piece of evidence *e* comes in. The probability after receipt of *e* is then called your *posterior probability*. In this sense, the term is entirely relative: your posterior probability for a hypothesis relative to one piece of evidence will be your prior probability for the hypothesis relative to the next. In the third sense, a prior probability is any subjective probability that is unconditioned by evidence. Your prior probabilities for the hypotheses are priors in this sense, but they are not the only priors.

Your probabilities that you will see various kinds of evidence, for example, also count as priors, as do subjective likelihoods (probabilities of the form $C(e|h)$). The achievement of modern Bayesianism might then be stated as follows: of all the prior probabilities in the third sense, only the priors in the first sense really matter.

It is as well to enumerate the more important assumptions that were made in order to reach the dramatic conclusion that the prior probabilities for a set of hypotheses determine the pattern of all subsequent inductive reasoning concerning those hypotheses:

1   You ought to set your subjective probabilities in accordance with PCP.

2   Each of the hypotheses ascribes an exact physical probability to any relevant evidence.

3   The hypotheses are mutually exclusive and exhaustive.

As noted in the previous two sections, ways to relax assumptions (2) and (3) will be explored later in these notes.

6 Modern Bayesianism in action

6.1 *A worked example*

Although there is much to say about the abstract properties of BCT, let us first look at an example of BCT in action.

Suppose that we have on the table just three hypotheses, concerning the physical probability that any given raven is black.

h_1 The probability that any given raven is black is one (thus, all ravens are black).

h_2 The probability that any given raven is black is one half (thus, it is overwhelmingly likely that about one half of all ravens are black).

h_3 The probability that any given raven is black is zero (thus, no ravens are black).

We have yet to observe any ravens. Let us therefore assign equal prior probabilities to these three hypotheses of $1/3$ each. (There is nothing in BCT, as presented so far, to constrain this choice, but it does seem rather reasonable; more on this in section 9.4 [sic].)

Now we observe a black raven. We apply the Bayesian conditionalization formula to see how our subjective probabilities change upon receiving this evidence. The formula is:

$$C^+(h) = \frac{P_h(e)}{C(e)}C(h)$$

where $P_{h1}(e)$ is 1, $P_{h2}(e)$ is ½, $P_{h3}(e)$ is 0, and

$$C(e)=P_{h1}(e)C(h_1)+P_{h2}(e)C(h_2)+P_{h3}(e)C(h_3)$$
$$= 1/3 + 1/6 + 0$$
$$= 1/2$$

(Why can we apply the theorem of total evidence, when the three hypotheses clearly fail to exhaust the possibilities? It is enough that the probabilities of the hypotheses sum to one, that is, that I, the amateur ornithologist, think that the probabilities exhaust the possibilities. This ornithological belief is rather artificial, but the simplicity is welcome.)

Applying the conditionalization rule, we find that the probability of h_1 is doubled, going to $2/3$, that of h_2 remains the same, at $1/3$, and that of h_3 is multiplied by zero, going to zero. There are now only two hypotheses in the running, h_1 and h_2. Of the two, h_1 is ahead, due to its having assigned a higher physical probability to the observed evidence, the fact of a certain raven's being black.

Now suppose that another black raven is observed. Is the probability of h_1 doubled again? One would hope not, or it would be greater than one. It is not, because $C(e)$ has taken on a new value closer to one, reflecting our increased confidence in h_1. Let us do the calculation:

$$C(e)=P_{h1}(e)C(h_1)+P_{h2}(e)C(h_2)$$
$$= 2/3 + 1/6$$
$$= 5/6$$

where $C(\cdot)$, note, now represents our subjective probability distribution *after* observing the first black raven but before observing the second.

On observing the second raven, then, the probability of h_1 is multiplied by $6/5$, and that of h_2 by $3/5$, yielding probabilities of $4/5$ and $1/5$ respectively. Our subjective probability of a third raven's being black is now $9/10$ (since we think that h_1 is very likely correct). If the third raven is indeed black, our probabilities for h_1 and h_2 will go to $8/9$ and $1/9$. To be sure you understand what is going on, calculate the new value for $C(e)$, the probability that a fourth raven will be black, and the probabilities for h_1 and h_2 if the fourth raven is as black as we expect. The evolution of the subjective probabilities as ravens are observed is shown in Table 17.1.

Table 17.1 Change in the probabilities for *e*, the event of the next raven's being black, and the three raven hypotheses h_1, h_2 and h_3 (see main text for meanings) as more and more black ravens are observed, where n is the number of ravens observed so far

n	$C(h_1)$	$C(h_2)$	$C(h_3)$	$C(e)$
0	1/3	1/3	1/3	1/2
1	2/3	1/3	0	5/6
2	4/5	1/5	0	9/10
3	8/9	1/9	0	?

This simple example illustrates a number of facts about BCT:

1   If a hypothesis is logically inconsistent with the evidence, upon conditionalization its probability goes to zero.

2   Once a hypothesis's probability goes to zero, it can never come back. The hypothesis is eliminated.

3   The hypothesis that assigns the highest probability to the observed evidence (h_1 in the example) receives the biggest probability boost from the observation of the evidence. A hypothesis that assigns probability one to the evidence will receive the largest boost possible in the circumstances.

4   If a hypothesis is consistent with the evidence, its probability can never go to zero, though it can go as near zero as you like (as would h_2's probability if nothing but black ravens were observed).

5   After conditionalization, your subjective probabilities for a set of mutually exclusive, exhaustive hypotheses (such as h_1, h_2, and h_3) will always sum to one.

6   As a certain hypothesis becomes dominant, in the sense that its probability approaches one, its probability boost from further successful predictions declines (though there is always a boost).

6.2 *General properties of Bayesian confirmation*

Now to generalize from the last section's example. First, a little terminology. When the observation of *e* causes the probability of *h* to increase, say that *h* is *confirmed*. When the observation of *e* causes the probability of *h* to decrease, say that *h* is *disconfirmed*. When the observation of *e* causes the probability of h_1 to increase by a greater proportion than does the probability of

h_2, say that h_1 is *more strongly confirmed* than h_2. This is not the only measure of strength of confirmation that may be used within a Bayesian context – some other possibilities are mentioned in tech box 6.1 – but it is the most useful for my present expository purposes.

6.1 Weight of evidence

The degree to which a piece of evidence confirms a hypothesis can be quantified in various ways using the Bayesian framework of subjective probabilities. The *difference measure* of relevance equates the degree of confirmation with the difference between the relevant posterior and prior probabilities, so that the degree to which *e* confirms *h* is equal to $C(h|e) - C(h)$. The *ratio measure* equates it with the ratio of the same two probabilities, that is, with the Bayesian multiplier; this is the measure used in the main text. The *likelihood ratio measure* sets the degree of relevance to $C(e|h)/C(e|\neg h)$, and the *log likelihood ratio measure* to the logarithm of this quantity (which has the effect of making degree of confirmation additive). Each of these measures has its uses.

Some writers argue that there is a single, correct way to measure degree of confirmation that gives sense to our everyday talk about the weight of evidence. Most Bayesians take a more pluralistic and pragmatic line, noting that unlike some systems of inductive logic, BCT does not give a notion of weight of evidence any crucial role to play.

Let me define the same terms more compactly. As above, let the Bayesian multiplier be $P_h(e)/C(e)$, the amount by which the probability of *h* is multiplied when conditionalizing on *e*. Then, first, *e* confirms *h* if the Bayesian multiplier is greater than one and disconfirms *h* if the multiplier is less than one. Second, *e* confirms h_1 more strongly than it confirms h_2 if the Bayesian multiplier for h_1 is greater than the multiplier for h_2.

I will state some generalizations about BCT in the form of five "principles", followed by two "remarks".

The Hypothetico-Deductive Principle

If h entails e, then the observation of e confirms h.

We have already seen that this is true. If *h* entails *e*, then $P_h(e)$ is equal to one, so the Bayesian multiplier must be greater than one, on the assumption that $C(e)$ is less than one.

The smaller the value of $C(e)$, the greater will be the value of the multiplier. Thus surprising predictions confirm a hypothesis more strongly than unsurprising predictions. The exception to the hypothetico-deductive principle that occurs when $C(e) = 1$ is an extreme case of this observation: evidence that is expected with certainty has no confirmatory (or disconfirmatory) weight whatsoever. You might think that in practice, this could never happen; but in a certain sense it often does, as we will see in section 11 [*sic*].

The Likelihood Lover's Principle

The higher the physical probability that h assigns to e, the more strongly h is confirmed by the observation of e.[4]

The denominator of the Bayesian multiplier is the same, at any given time, for all competing hypotheses, equal to $C(e)$. The numerator is $P_h(e)$. Therefore the Bayesian multiplier varies among competing hypotheses in proportion to $P_h(e)$, the physical probability that they assign to the evidence.

Some more specific claims along the lines of the likelihood lover's principle:

1 If a hypothesis assigns a physical probability to *e* that is higher than $C(e)$, it is confirmed by the observation of *e* (obviously!). To put it more qualitatively, if *h* predicts *e* with a higher probability than (the probabilistically weighted) average, it is confirmed by the observation of *e*.

2 If there are only two hypotheses in the running, and one ascribes *e* a higher probability than the other, then *e* confirms the one and disconfirms the other. This is because, as a consequence of the theorem of total probability, $C(e)$ is always somewhere between the physical probability assigned by one hypothesis and that assigned by the other.

(The likelihood lover's principle should not be confused with what some philosophers call the likelihood principle, according to which subjective likelihoods (and only subjective likelihoods) are responsible for a piece of evidence's differential impact on two or more different hypotheses. The likelihood principle differs from the likelihood lover's principle in two ways: it concerns subjective, not physical, likelihoods (and thus does not depend on PCP), and it says not only that strength of confirmation varies with the likelihoods, but that it is entirely determined, relatively speaking, by the likelihoods.)

The equal probability principle

Hypotheses that assign equal physical probabilities to a body of evidence are equally strongly confirmed by that evidence.

It is easy to see that such hypotheses have equal Bayesian multipliers. Though it is trivially true, the equal probability principle is essential for an understanding of what BCT can and cannot do; see, in particular, section 7.

The good housekeeping principle

No matter what evidence is observed, the probabilities of a set of mutually exclusive, exhaustive hypotheses will always sum to one.

Bayesianism's ability to keep all of your subjective probabilities in good order as you conditionalize your way through the evidence can seem quite miraculous, not least when you are in the middle of a complex Bayesian calculation. The reason that the good housekeeping principle holds true, though, is easy to see. Consider the case in which there are just two competing hypotheses h_1 and h_2. The probabilities of h_1 and h_2 after conditionalization on some e are

$$C^+(h_1) = \frac{P_{h1}(e)C(h_1)}{P_{h1}(e)C(h_1) + P_{h2}(e)C(h_2)} \text{ and}$$

$$C^+(h_2) = \frac{P_{h2}(e)C(h_2)}{P_{h1}(e)C(h_1) + P_{h2}(e)C(h_2)}.$$

These probabilities have the form

$$\frac{a}{a+b} \text{ and } \frac{b}{a+b}$$

and so they must sum to one.[5] You should have no trouble seeing, first, that the result generalizes to the case where there are any number of competing hypotheses, and second, that the fact that the subjective likelihoods are set in accordance with PCP makes no difference to this result.

The commutativity principle

The order in which the evidence is observed does not alter the cumulative effect on the probability of a hypothesis.

For example, conditionalizing on e and then on d will leave you with the same probability for h as conditionalizing on d and then on e, or as conditionalizing on e and d at the same time.

This result is a straightforward consequence of the definition of conditional probability, but it can be difficult to represent multiple conditionalizations in just the right way to see how. Here is one approach. Define $C_d(\cdot)$ to mean $C(\cdot | d)$. Then upon conditionalizing on d, one's new probability for h ought to be equal to the old $C_d(h)$. Further conditionalizing on e will result in a probability for h equal to the original $C_d(h | e)$. Now

$$C_d(h|e) = \frac{C_d(he)}{C_d(e)}$$

$$= \frac{C(hed)/C(d)}{C(ed)/C(d)}$$

$$= \frac{C(hed)}{C(ed)}$$

$$= C(h|ed)$$

as desired. By symmetrical reasoning, $C_e(h|d) = C(h|ed)$.[6]

Remark: the contrastive aspect of Bayesian confirmation theory

Observe that a hypothesis's degree of confirmation depends not only on its own predictions, but on the predictions of its competitors. One notable consequence of this property of BCT is that a hypothesis that is quite inaccurate can be assigned a subjective probability as near one as you like if its competitors are even more inaccurate. To take a very simple example, if the only hypotheses about raven blackness that have non-zero subjective probability are *One half of all ravens are black* and *No ravens are black*, then if all observed ravens are black, the first hypothesis will come to have subjective probability one. Similarly, a hypothesis that is very accurate can have its subjective probability greatly decrease (though slowly) if its competitors are even more accurate. There is only a limited amount of subjective probability; it flows towards those hypotheses with the physical likelihoods that are relatively the highest, regardless of their absolute size.

Remark: The Contextual Aspect of Bayesian Confirmation Theory

In the Bayesian scheme, any piece of knowledge can, potentially, affect the impact of a piece of evidence on a hypothesis. Relative to one set of background knowledge, a piece of evidence may confirm a hypothesis a great deal, while relative to another, it may confirm it not at all. This is in part because the background can make a difference to whether or not the hypothesis entails the evidence, or to what physical probability it assigns the hypothesis (some examples are given in section 8.1 [*sic*]). But only in part: in principle, almost any kind of information might affect your subjective likelihood $C(e\,|\,h)$.

This contextual sensitivity of BCT has been generally regarded as a good thing, since it has long been accepted that inductive reasoning has a certain holistic aspect. But it also makes it more difficult to establish that BCT will be well behaved and that it will lead to some kind of scientific consensus (see section 9 [*sic*]).

6.3 *Working with infinitely many hypotheses*

In the example of the ravens in section 6.1, I assumed that there were just three hypotheses under serious consideration, hypotheses assigning probabilities of zero, one half, and one, to the event of the next observed raven's being black. In reality, I would want to consider all the possible probabilities for blackness, which is to say that I would seriously entertain all hypotheses of the form *The probability of the next raven's being black is p.*

There are infinitely many such hypotheses. This creates some rather tricky problems for the Bayesian. Most worrying, the only way to assign probabilities to an infinite set of mutually exclusive hypotheses so that the probabilities sum to one (as the axioms insist they must) is to assign almost every hypothesis a probability of zero. But a hypothesis that is assigned a prior probability of zero can never have its probability increase. So it seems that I must rule almost every hypothesis out of the running before I even begin to collect evidence. Nevermore indeed!

There is a purely technical solution to these problems that will be familiar to anyone who has done a little work on probability or statistics. It is to introduce the notion of a *probability density*. Even those of you who know no statistics have very likely seen at least one example of a probability density: the normal curve.

The fundamental idea behind a probability density is that, in a case where there are infinitely many possibilities, I should assign subjective probabilities not to individual possibilities, but to sets of possibilities. For example, in the case of the ravens, I may not have an interesting (that is, non-zero) subjective probability for the hypothesis that the probability of black raven is exactly 0.875, but I may have a subjective probability for the hypothesis that the probability of a black raven is somewhere between 0.8 and 0.9. That is, I may have a subjective probability for the set of hypotheses of the form

{The probability of the next raven's being black is $x : 0.8 < x < 0.9$}

Provided that I have enough subjective probabilities of this sort (in a well-defined sense of *enough*), I can take advantage of all that the Bayesian has to offer by applying conditionalization to the probabilities over sets.

It turns out that there is a very simple way for me to assign subjective probabilities to all the relevant sets. What I can do is to adopt a *probability density* over the competing hypotheses. The properties of a probability density are easiest to appreciate in graphical form. Consider a two-dimensional graph on which the points along the x-axis between zero and one each represent the corresponding raven hypothesis, that is, on which the point x = p represents the hypothesis *The probability of the next raven's being black is p.*

A probability density can then be represented by a well-behaved function f(x) defined for arguments between zero and one, such as the function shown in figure 17.1. The assignment of subjective probabilities inherent in f(x) is to be interpreted as follows: the subjective probability that the physical probability of raven blackness is between *a* and *b* is equal to the area under f(x) between *a* and *b*. (The area under f(x) between zero and one, then, had better be equal to one.)

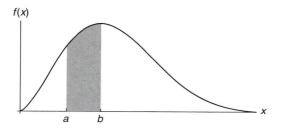

f(x)

Figure 17.1 A probability density over the raven hypotheses. The area of the shaded region is by definition equal to the subjective probability that the physical probability of raven blackness is somewhere between a and b.

Before I observe any ravens, then, I can set my prior probabilities over the infinitely many hypotheses about the physical probability of blackness by adopting a subjective probability density. If I think that all hypotheses are equally likely, I will adopt a flat density like that shown in figure 17.2a. If I think that middling probabilities for blackness are more likely, I will adopt a humped density like that shown in figure 17.2b. If I think that higher probabilities are more likely, my density might look like the density in figure 17.2c. And so on.

How to conditionalize? Very conveniently, I can conditionalize by pretending that f(x) for any x is a probability. That is, when I observe *e*, my new probability density over the hypotheses $f^+(x)$ should be related to my old density f(x) by the following familiar-looking relation:

$$f^+(x) = \frac{P_{h_x}(e)}{C(e)} f(x).$$

The probability $P_{h_x}(e)$ is the probability assigned to the evidence by the hypothesis that the probability of a black raven is x. If *e* is the observation of a black raven, then $P_{h_x}(e) = x$.

The theorem of total probability can be applied by using the integral calculus. You do

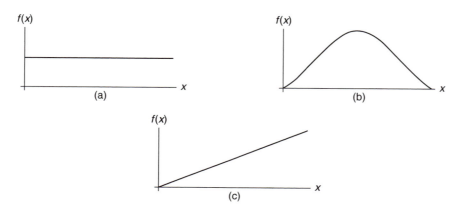

Figure 17.2 Some choices of prior probability density for the raven blackness hypotheses.

not need to understand this. For those of you who do,

$$C(e) = \int_0^1 P_{h_x}(e) f(x)\,dx.$$

In the case where e is the observation of a black raven, then,

$$C(e) = \int_0^1 x f(x)\,dx.$$

Now we can crunch some numbers. Suppose that I begin with a flat prior over the raven probability hypotheses (figure 17.3a). After the observation of a black raven, my new density is that shown in figure 17.3b. After two black ravens, my density is as in figure 17.3c. After three and four black ravens, it is as in figures 17.3d and 17.3e. After ten black ravens, it is as in figure 17.3f.

By way of contrast, figure 17.4 shows the change in my prior if only some of the observed ravens are black. The prior is shown in (a), then (b)–(f) show the density after observing 2, 6, 10, 20, and 36 ravens respectively, in each case assuming that only one half of the observed ravens are black.

As you can see, my density becomes heaped fairly quickly around those hypotheses that ascribe a physical probability to blackness that is close to the observed frequency of blackness. As more and more evidence comes in, my subjective probability density will (with very high physical probability) become more and more heaped around the true hypothesis. (For more on the mathematics of convergence, see tech box 6.2. For more on the use of convergence results to undergird BCT, see section 9.2 [sic].) After observing 100 ravens, exactly 50 of which are black, for example, my

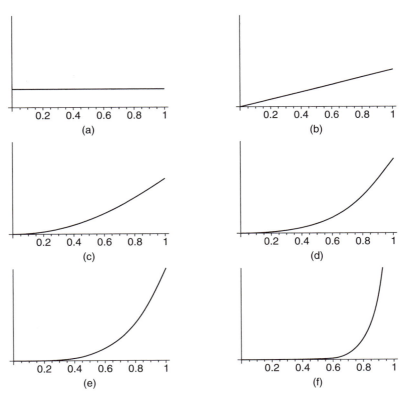

Figure 17.3 *The change in my subjective probability density as more and more black ravens are observed.*

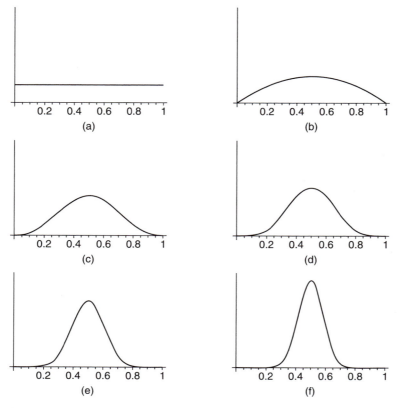

Figure 17.4 *The change in my subjective probability density as more and more ravens are observed, if half of all observed ravens at any stage are black.*

density will be close to zero for any hypothesis that assigns a probability to blackness that differs from one half by more than about 0.1.

6.2 The law of large numbers

The heaping, over time, of subjective probability around the correct hypothesis concerning raven blackness, depends on a mathematical result called the law of large numbers. The law concerns a series of experiments each of which produces an outcome *e* with a probability *x*. It states that, as more and more experiments are performed, the proportion of experiments that produce *e* will, with high probability, converge on *x*. For example, if the probability of observing a black raven is 0.7, then, as more and more ravens are observed, the proportion of observed ravens that are black will converge to 70% with very high probability. Note that there is always a small probability that the proportion of black ravens is quite different from 70%; this probability, though, becomes smaller and smaller as more and more ravens are observed, tending to zero in the long run.

There is more than one version of the law of large numbers. The result that I stated informally in the last paragraph is called the *weak law of large numbers*. It may be stated more formally as follows. Let b be the proportion of observed ravens that are black, and x the probability that any given raven is black. Assume that the observations are probabilistically independent (section 3.3). Then, as the number of ravens observed tends to infinity, for any small quantity ε, the probability that b differs from x by more than ε tends to zero.

What, in this case, is my subjective probability for the hypothesis that the probability of blackness is exactly one half? It is zero! It is zero even after I observe two million ravens, exactly one million of which are black. This is because, although almost all my subjective probability is piled in the area very, very close to $x = 0.5$, there are infinitely many hypotheses even in this tiny area, and the large but finite quantity of probability piled in the area must be shared between these infinitely many hypotheses.

As a consequence, no matter how much evidence I accumulate, I am never in a position to say: I am very confident that the probability of blackness is one half. The best I can do is to say: I am very confident that the probability of blackness is very close to one half. Quite reasonably, most Bayesians would say.

You might think that if I have a subjective probability of zero for a hypothesis, I am certain that it is false. But as the raven case shows, this is not correct. When I have the sort of probability density over the rival raven hypotheses shown in figures 17.2, 17.3, and 17.4, I assign each of them a zero probability. But at the same time, I am sure that one of them is correct. Unless I contradict myself, my assignment of zero probability cannot entail certainty of falsehood. It rather represents something that you might call

practical certainty of falsehood. Given one of the raven hypotheses in particular, I will accept any odds you like that it is false. Yet I acknowledge that it might turn out to be true.

Assigning probability zero to a hypothesis, then, can mean one of two things. It can mean that I am absolutely sure that the hypothesis is false, as when I assign zero probability to an antitautology (a proposition that is false of logical necessity). Or it can mean mere practical certainty of falsehood, that is, a willingness to bet against the hypothesis at any odds, coupled with the belief that it might nevertheless be true. The same is true for probability one.

6.4 *When explicit physical probabilities are not available*

We have proceeded so far on the assumption that our competing hypotheses each ascribe a definite physical probability to the evidence e. But in many cases, this assumption is false.

Consider a very simple case. Suppose that our competing hypotheses concerning raven color specify not a physical probability of blackness, but a frequency of black ravens. For example, $h_{0.5}$ says not *The next observed raven will be black with probability 0.5*, but rather *Half of all ravens are black*. This hypothesis does not entail that the next observed raven will be black with probability 0.5. It is consistent with the possibility that, for example, all of the ravens around here are black, but enough ravens elsewhere are white to compensate for the local predominance of blackness. In such a scenario, the probability of the next raven's being black is more or less one. Then again, it might be that all the local ravens are white, in which case the probability of the next raven's being black is about zero. How to apply BCT to hypotheses such as these?[7]

The usual solution to the problem is to introduce one or more *auxiliary hypotheses*. Traditionally, an auxiliary hypothesis is a posit used to extract from a theory a definite verdict as to whether some piece of evidence e will be observed or not – to get the theory to entail either e or $\neg e$ – but in the present context, it is a hypothesis that, in conjunction with a hypothesis h, assigns a definite physical probability to e.

A simple example of an auxiliary hypothesis suitable for the ravens case is the hypothesis that raven color is uncorrelated with raven habitat, so that roughly the same percentage of black ravens live in any locale. Conjoining this hypothesis with a hypothesis that the frequency of black ravens is x yields the conclusion that the probability of the next observed raven's being black is x.[8]

Call the auxiliary hypothesis a. Let h be the thesis that the proportion of black ravens is x. Then h and a together ascribe a physical probability x to e, or in symbols, $P_{ha}(e) = x$. The subjective likelihood $C(e\,|\,ha)$ can therefore be set, in accordance with PCP, to x.

This is not quite what we wanted. It allows us to use PCP to calculate a shift in our subjective probability for ha, but not in our subjective probability for h. It turns out that, in certain circumstances, a piece of evidence can have a qualitatively different effect on h and ha, raising the subjective probability of one while lowering the subjective probability of the other. Although e's impact on the subjective probability for ha is regimented by PCP, then, its impact on h is, in general, not. The problem, and some solutions, are investigated further in section 10 [sic].

It is worth noting that, in some circumstances, you can make your auxiliary hypotheses true. This is, in effect, what is done in statistical sampling. By choosing ravens from a population at random, you ensure that the probability of the next observed raven's being black is equal to the proportion of black ravens in the population as a whole. In this way, you can learn about the population even if unplanned encounters with population members are far from random.

More generally, when scientists conduct experiments in carefully controlled circumstances, they are ensuring, as best they can, the truth of auxiliary hypotheses that, in combination with the hypotheses being tested, assign definite physical probabilities to the various possible experimental outcomes.

The prospect for a Bayesian philosophy of science that is mediated entirely by physical likelihoods is for these reasons far better than some writers (e.g., Earman 1992, §6.3) would have you believe.

7   Does Bayesianism solve the problem of induction?

7.1   Subjective and Objective In Bayesian Confirmation Theory

The old-fashioned problem of induction is the problem of finding a priori, objective grounds for preferring some hypotheses to others on the basis of what we observe (where the hypotheses have a scope that goes beyond our observations). Bayesian confirmation theory tells us how to change our degrees of belief for the hypotheses given the evidence. Does BCT therefore solve the problem of induction?

Although it is true that BCT tells us how to update our subjective probabilities in the light of the evidence, that is not enough in itself to solve the problem of induction. The recommended changes must be founded in a priori, objective constraints. Modern Bayesian confirmation theory claims three such constraints: subjective probabilities should be set in accordance with

1   The axioms of probability,
2   Bayes' conditionalization rule, and
3   The probability coordination principle.

Let us grant, for the sake of the argument, that these constraints are indeed a priori and objective. Now does BCT solve the problem of induction?

Not quite. The reason is that BCT's recommendations are determined not only by the three objective, a priori constraints, but also by your prior probabilities, that is, the subjective probabilities you assign before any evidence comes in. Assuming, as we have been so far, that the priors are subjectively determined, it follows from the fact that the priors make an essential contribution to BCT's recommendations that these recommendations are partly subjectively determined. Thus, the fact that BCT makes such recommendations does not qualify it as a solution to the old-fashioned problem of induction.

In order to use BCT to solve the problem of induction we must do one of two things. First, we might find some further objective, a priori rule that constrains the prior probabilities so effectively that BCT's recommendations are completely determined by the constraints. The subjective element of Bayesianism would be eliminated entirely. This possibility is investigated further in section 7.4.

The second option is to find some aspect of BCT's recommendations that is entirely determined by the constraints. Such an aspect is enunciated by LLP, the likelihood lover's principle (section 7.2):

> The higher the physical probability that a hypothesis h assigns to the evidence e, the more strongly h is confirmed by the observation of e.[9]

Although the quantitative details of BCT's recommendations depend on the subjectively determined priors, no matter how I set my priors, I will respect LLP. That I should follow LLP, then, is determined by the axioms of probability, Bayes' rule, and PCP. On our working assumption that these constraints are all objective and a priori, LLP is also, for the Bayesian, objective and a priori.

It would seem, then, we have made some progress on induction. Bayesian confirmation theory puts a real constraint on inductive reasoning. But how much, exactly, follows from LLP? And how much depends, by contrast, on the priors?

In what follows, I ask, concerning three features of our inductive reasoning, whether they follow from LLP or some other objective aspect of the Bayesian machinery, or whether they depend on the way that one's prior probabilities are assigned. The three features are:

1   Our expectation that the future will resemble the past, that is, our belief in the uniformity of nature,
2   Our preference for hypotheses that are not framed using "grue-like" predicates, and
3   Our preference for simpler over more complex hypotheses, when the competing hypotheses are otherwise equally successful in accounting for the data.

In each case, sad to say, it is the assignment of the priors, and not anything more objective, that accounts, on the Bayesian story, for these elements of our inductive practice.

7.2 The uniformity of nature

Can BCT justify a belief − or a high degree of belief − in the uniformity of nature? At first blush, it seems that if I am committed to the likelihood lover's principle, I ought to expect the future to be like the past, as illustrated by the raven case. Suppose that every raven I see is black. Then, if I follow LLP, my subjective probabilities will converge on the hypotheses that assign the highest probabilities to these observations, that is, the hypotheses that assign the highest physical probabilities to blackness in ravens. As the convergence occurs, the probability I assign to any future raven's being black grows higher and higher, thanks to the theorem of total probability – as shown by the examples in section 6.1 (see especially Table 17.1).

More generally, if all my hypotheses have the form *The probability of an r being b is x*, and I observe a large number of rs of which a proportion p are b, then my subjective probability for the next r's being b will be roughly equal to p, the more so as more rs are observed. Thus, I will expect the future proportion of rs that are b to be equal, at least roughly, the past proportion of the same – which is to say, I will set my expectations in accordance with the principle of the uniformity of nature.

Did I make any additional assumptions in reaching this conclusion? Unfortunately, yes. I made the least acceptable assumption possible: I assumed that I started out with a very high subjective probability for the future's resembling the past.

The assumption came by way of the assumption that all my hypotheses have the form *The probability of an r being b is x*, which implies the existence of a single, unchanging physical probability for raven blackness. To see how powerful this assumption is, consider the following alternative form for a hypothesis covering the same phenomena:

> The probability of an r first observed before the year 3000's being b is x; the probability of an r first observed after the year 3000's being b is 1 − x.

For the value x = 1, a hypothesis of this form will be strongly confirmed by all black ravens observed before 3000. Indeed, it will receive precisely the same Bayesian multiplier as the raven hypothesis, *All ravens are black*, thanks to the equal probability principle (section 6.2, and explained further in section 7.3). If it is assigned the same prior as the raven hypothesis, then it will have the same posterior probability in the

year 3000, and so will count as heavily *against* the next observed raven's being black as the raven hypothesis counts for it.

Now you see that the argument from LLP to an expectation that the future will resemble the past depends on my discounting – that is, assigning very low or zero probability to – hypotheses that explicitly state that the future will (very likely) not resemble the past.

The point, that inductive expectations were assumed, not justified, can be made even more dramatically. Suppose that my hypotheses concerning the blackness of ravens are $h_1, \ldots, h_n$. Let u be the proposition that the future, with high probability, resembles the past, at least with respect to the color of ravens. Then

$$C(u) = C(u \mid h_1)C(h_1) + \ldots + C(u \mid h_n)C(h_n).$$

But any individual h_i having the form assumed above – *The probability of an r being b is x* – entails u. Thus

$$C(u) = C(h_1) + \ldots + C(h_n) = 1.$$

So by assuming that you consider only hypotheses of the given form to be viable candidates for laws about raven blackness, I assume that you already assign subjective probability one to the uniformity of nature. It is not BCT that compels your allegiance to uniformity; the passion for order has all along been burning in your heart.

7.3 Goodman's new riddle

If you followed the argument in the last section, then you already strongly suspect that BCT will fail to solve Goodman's new riddle of induction, where a solution to the riddle is an a priori argument for preferring hypotheses couched using standard predicates such as *black* to hypotheses couched using non-standard predicates such as *blite*. (An object is *blite* if it is black and first observed before the year 3000, or white and first observed after 3000.)

In order to solve the new riddle, we need a reason currently to prefer, for example, the hypothesis *All ravens are black* to the hypothesis *All ravens are blite*, such that the preference does not depend in any way on the prior probabilities for the hypotheses. But because both hypotheses assign exactly the same physical probability to the evidence observed so far (some large number of ravens, all black), BCT cannot discriminate between them.

Let me explain the point in greater detail. Call the observed evidence *e*. Your current subjective probability for the first hypothesis, call it h, is

$$C^+(h) = \frac{C(e|h)}{C(e)} C(h);$$

that for the second hypothesis, call it h', is

$$C^+(h') = \frac{C(e|h')}{C(e)} C(h').$$

Since both hypotheses predict that all ravens observed until now will be black, their likelihoods are both equal to one, that is, $C(e|h) = C(e|h') = 1$. So we have

$$C^+(h) = \frac{1}{C(e)} C(h); \; C^+(h') = \frac{1}{C(e)} C(h').$$

The only difference between these two expressions is in the prior probabilities assigned to each, thus, any preference a Bayesian has for one over the other must be expressed in the prior probabilities, that is, the probabilities assigned *before* any evidence comes in. If you are predisposed against hypotheses involving the predicate *blite*, and you thus assigned a low prior to h', then BCT will preserve your bias. Equally, if you are predisposed against hypotheses involving *black* and in favor of *blite*, then BCT will preserve that bias, too. If Bayesianism is correct, then, no considerations from confirmation

theory alone militate for or against Goodman predicates.

You will see that the source of BCT's even-handedness is the equal probability principle. This principle is just the flip side of the LLP: whereas LLP directs us to change our probabilities in accordance with the likelihoods, the equal probability principle directs us *only* to change our probabilities in accordance with the likelihoods.

7.4 Simplicity

All things being equal, we prefer the simpler to the more complex of two competing theories. What is the meaning of the ceteris paribus clause, *all things being equal*? Something like: *if the two theories, in conjunction with the same auxiliary hypotheses, assign the same physical probabilities to all the phenomena in their domain.* Say that two such hypotheses are *empirically equivalent*. Of course, our preference for simplicity is not confined to cases of empirical equivalence; we consider theoretical simplicity a virtue wherever we find it. But confining our attention to the case of empirically equivalent theories is like conducting an experiment in a clean room: the influence of confounding variables is minimized.

You should see straight away that BCT does not impose a preference for simplicity in these circumstances. The argument is, of course, identical to the argument of the last section. When two theories are empirically equivalent, their likelihoods relative to any given body of evidence are equal. Thus the difference in anyone's subjective probabilities for the theories must be due entirely to the difference in the prior probabilities that were assigned to the theories before any evidence came in. Bayesian confirmation theory preserves a prior bias towards simplicity, but it implements no additional bias of its own.

You will find an additional comment about the confirmatory virtues of simplicity in tech box 11.1 [sic].

7.5 Conclusion

Let us take stock. Bayesian confirmation theory does impose an objective constraint on inductive inference, in the form of the likelihood lover's principle, but this is not sufficient to commit the Bayesian to assuming the uniformity of nature, or the superiority of "non-grueish" vocabulary or simple theories. The first of these failures, in particular, implies that BCT does not solve the problem of induction in its old-fashioned sense.

If the old-fashioned problem of induction cannot be solved, what can we nevertheless say about BCT's contribution to the justification of induction? There are two kinds of comments that can be made. First, we can identify unconditional, though relatively weak, constraints that BCT puts on induction, most notably the likelihood lover's principle. Second, we can identify conditional constraints on induction, that is, constraints that hold given other, reasonable, or at least psychologically compelling, assumptions. We can say, for example, that if we assign low priors to grueish hypotheses, BCT directs us to expect a future that resembles the past. This is, remember, considerably more than we had before we began.

knowledge includes no inadmissible evidence (see tech box 5.5 for inadmissibility).

5   It is not necessary, note, to assume that $C(h_1)$ and $C(h_2)$ sum to one. Conditionalization actually repairs subjective probabilities that are out of whack! (But of course, you would have no reason to apply the theorem of total probability in a case such as this.)

6   It is an open question whether commutativity holds for Jeffrey conditionalization (Field 1978). (Jeffrey conditionalization is described in section 4.2.) The difficulty is more philosophical than mathematical: it is unclear what would count as making the same observations in two different orders. On one interpretation, Jeffrey conditionalization is commutative, but on others, not.

7   The Bayesian apparatus does not strictly speaking require a physical probability, only a subjective likelihood $C(e|h)$. By insisting on physical probabilities now, I make possible a much stronger response to later worries about the subjectivity of BCT (section 9 [*sic*]).

8   Strictly speaking, of course, you need in addition the assumption that you are sampling randomly from the local habitat.

9   The principle requires that the hypotheses have non-zero priors, and that the background knowledge includes no inadmissible evidence (see tech box 5.5 for inadmissibility).

Notes

1   The classical theorist does not necessarily deny the existence of a richer psychology in individual scientists; what it denies is the relevance of this psychology to questions concerning confirmation.

2   This is only true of subjective probability, not of other varieties of probability you may come across, such as physical probability and logical probability.

3   Exercise to the reader: what are the pitfalls of defining subjective probability so that having a certain subjective probability entails both present *and* future betting behavior? The answer is in tech box 4.1.

4   The principle requires that the hypotheses have non-zero priors, and that the background

References

Carnap, R. (1950). *Logical Foundations of Probability*. University of Chicago Press, Chicago.

Earman, J. (1992). *Bayes or Bust?* MIT Press, Cambridge, MA.

Field, H. (1978). A note on Jeffrey conditionalization. *Philosophy of Science* **45**:361–367.

Horwich, P. (1982). *Probability and Evidence*. Cambridge University Press, Cambridge.

Jeffrey, R. C. (1983). *The Logic of Decision*. Second edition. University of Chicago Press, Chicago.

Keynes, J. M. (1921). *A Treatise on Probability*. Macmillan, London.

Levi, I. (1967). *Gambling with the Truth*. MIT Press, Cambridge, MA.

Lewis, D. (1980). A subjectivist's guide to objective chance. In R. C. Jeffrey (ed.), *Studies in Inductive Logic and Probability*, volume 2. University of California Press, Berkeley, CA.

Lewis, D. (1994). Humean supervenience debugged. Mind **103**:473–490.

Ramsey, F. (1931). Truth and probability. In *The Foundations of Mathematics and Other Logical Essays*. Edited by R. B. Braithwaite. Routledge & Kegan Paul, London.

Strevens, M. (2003). *Bigger than Chaos: Understanding Complexity through Probability*. Harvard University Press, Cambridge, MA.

Strevens, M. (2006). Probability and chance. In D. M. Borchert (ed.), *Encyclopedia of Philosophy*, second edition. Macmillan Reference USA, Detroit.

Strevens, M. (2008). Notes on Bayesian Confirmation Theory, http://www.nyu.edu/classes/strevens/BCT/BCT.pdf.

PART FOUR

Explanation

INTRODUCTION TO PART FOUR

S CIENTIFIC REALISTS THINK that explanation is important, for two reasons. First, they think part of the aim of science is to explain things. Scientists want to understand the world – they have the same curiosity as children and ask *why* things are thus and so. Why is the sky blue? Why does the giraffe have a long neck? Why does potassium react with water? What we are seeking with such questions is explanations. Second, according to most realists, explanation is important in generating knowledge. If, as realists typically accept, inference to the best explanation (see Part Three) is a correct description of an important inferential practice, then it is largely, if not wholly, by seeking explanations and comparing potential explanations that we generate theoretical knowledge.

We should bear in mind that the term 'explain' can be used in a number of different ways:

1 In one use it is *people* who explain things, for example when we might say, 'Einstein explained the photoelectric effect by postulating the quantization of energy'.
2 In a second use, we also say that *theories* explain things, 'the theory of plate tectonics explains why the ocean floor is spreading'.
3 And in a third use it is features of the *world* that explain things, 'the rise in interest rates explains the housing market slump'.

One approach to relating these uses takes the third as primary. Explanation is a relation between things in the world. So an actual, real or correct explanation is some sort of worldly relation. Potential explanations are possible worldly relations. When some person or some theory explains something, the person or the theory asserts or proposes that a worldly explanatory relationship holds.

On the other hand, empiricists tend to deny that explanations are worldly relations of any significant kind. They can take one of two approaches. The first denies that explanation is a worldly affair at all. Rejecting the proposal in the preceding paragraph, the empiricist might say that the focus of explanation is the person who is doing the explanation. Explanation is a subjective affair and the correctness of explanations reflects the interests and context of the explainer. This is the approach taken by Bas van Fraassen. Not surprisingly then, van Fraassen denies that inference to the best explanation can be an inference procedure that leads to the truth (see Part Nine).

On the other hand, an empiricist might accept that explanation is an objective affair, but a very weak one with features acceptable to the empiricist. This is how it is with the *covering-law model* of explanation. The central idea here is simple: to explain a phenomenon is to show how it is a logical consequence of some law plus certain background conditions. The fact to be explained is known as the *explanandum*; what does the explanation is known as the *explanans*. Why does this pendulum have a period of 1.9s? This is our explanandum; we explain this by citing the law of the pendulum due to Galileo: $T = 2\pi\sqrt{(l/g)}$ where T is the period of the pendulum and l is its length and g is the acceleration due to gravity at the Earth's surface $= 9.8\text{ms}^{-2}$; and we also cite the length of the pendulum, which is 90cm. From these facts, the explanans, we can deduce that $T = 1.9$s. Schematically, this can be represented:

$$\begin{array}{ll} \textit{from} & \text{Law} \qquad\quad T = 2\pi\sqrt{(l/g)} \\[4pt] \textit{deduce} & \dfrac{\text{Conditions} \quad l = 0.9\,\text{m}, g = 9.8\text{ms}^{-2}}{\text{Explanandum} \qquad\quad T = 1.9\text{s}} \end{array}$$

In this form, the covering-law approach to explanation is known as the deductive-nomological (or D-N) model of explanation.

The D-N model is acceptable to empiricists because the only relation involved in explanation is a logical one – deduction – and because the notion of *law* is understood in empiricist terms. Typically that will mean that laws are understood as regularities in nature; truths that can be stated as generalizations, for example: all metal objects conduct electricity. Hence, the D-N explanation of why this ring conducts electricity will be: all metal objects conduct electricity; this ring is a metal object; *therefore* this ring conducts electricity. This might strike some readers as not really explanatory at all. After all, the explanation is equivalent to saying: this ring, which is metal, conducts electricity, and all other metal objects conduct electricity *therefore* this ring conducts electricity. That is, it seems the very thing we want to explain, the explanandum, is cited in the facts that do the explaining, the explanans.

There are further problems with the D-N model. For example, looking at the pendulum case, not only can we deduce the period of the pendulum from the law plus the length of the pendulum. We can also deduce the length of the pendulum from the law plus its period (the law can be re-arranged so that: $l = gT^2/4\pi^2$). But typically the period of a pendulum does not explain its length. Wesley Salmon spells out this and related problems for the D-N model (he also explains its statistical sibling). This kind of problem shows that deduction from a law and conditions is not *sufficient* for explanation. While that can allow us to predict an event, being able to predict something does not amount to being able to explain it. For example, one can predict that someone's blood is circulating from the fact that the sound of beating can be heard in their chest. But the sound does not explain the circulation. Rather, both are explained by the pumping of the heart. Is being deducible from a law and conditions *necessary* for explanation? Simple causal explanations seem to be an obvious counterexample. Why does Kate have chickenpox? She caught it from another child at school who had chickenpox. Why did the K-T extinction occur? Because a huge meteor hit the Earth and the resulting dust led to the death of much plant life and hence to the death of many animals dependent upon it. These answers appear to provide satisfactory explanations, but in neither case are we able to deduce the explanandum from the information given in the explanations. In particular, those explanations cite no laws but are best thought of as causal explanations.

We say that these cases *seem* to show that one can have explanations that do not fit the D-N model (all suggest that the D-N model is not *necessary*). But the defender of the latter might wish to maintain that we should distinguish between partial, imperfect explanations and ideal explanations. The latter do conform to the D-N model (or a related version of the covering-law approach). An ideal explanation (or 'explanatory text') of Kate's acquiring chickenpox would provide all the details (e.g. her level of susceptibility, the infectiousness of the child at school, the fact that they held hands, etc.) and laws that would allow the deduction of her being infected (or at least of the high probability of that occurring). However we do not know all these details (nor all the relevant laws), and not all those that we know are of interest to us. So the information we provide in the explanation we give is only a subset of the information in an ideal explanation.

According to Salmon, the D–N model exemplifies one tendency in the theory of explanation, which sees explanations as devices for unifying phenomena. Laws, for example, being a sort of general fact, unify their various instances. Hence, in effect, the D–N model unifies the instance we wish to explain with other instances. Perhaps, then, to generalise, what makes something an explanation is that it unifies the explanandum with other fact – as Philip Kitcher and Michael Friedman have suggested. Newton's unification of the motions of the planets and of objects on and near the Earth's surface under a single theory of gravity is an example of this. Is unification what makes something an explanation? Or is it simply that when a potential explanation unifies phenomena that is a reason for believing the potential explanation to be a true, actual explanation? The latter might be correct without obliging one to think that an explanation ipso facto involves unification or that explanations explain because they unify.

There are further arguments that suggest that the D-N model is not sufficient for explanation. A different approach to explanation accepts that causal explanations (such as the chickenpox and K-T extinction explanations mentioned above) can be perfectly good explanations, and should not be regarded as merely partial. Peter Lipton develops this thought in a direction that in some respects is opposite to the unification idea. For Lipton explanations are explicitly or implicitly *contrastive*. Often the contrasting possibility is one where some cited event does not occur, such as Kate's not getting chickenpox. But they need not be. For example, had another of Kate's school friends had measles, then we might ask why Kate got chickenpox but did not get measles. In such circumstances the answer might be that Kate had been vaccinated against measles but not against chickenpox. Here we are looking not for unification, but for differentiation – what are the *differences* between Kate's relationship to chickenpox and to measles, such that she got the first disease but not the second; the answer being that she was vaccinated against one but not against the other.

Given that there might be both covering-law explanations and causal explanations, which takes precedence? We have seen that the covering-law theorist might attempt to encompass causal explanations; more generally one might suppose that causal relations exist because of deeper relations between things that are explained by the laws of nature. The stone caused the breaking of the window, but underlying this are complex relations and processes involving the atoms of the stone and of the glass governed by the laws of electromagnetism. By contrast, Nancy Cartwright, as we see in Part Five, holds that causes are primary and that what we call the laws are not strictly true, but are exception-ridden generalizations produced by particular experimental or other structured set-ups.

So far we have ignored a class of explanations that is in common use: explanations appealing to the intentions and purposes of people. Jane, for example, threw the stone at the window in order to attract David's attention. On the face of it such explanations are not like causal explanations, since they refer to some goal that is in the future and might not even be achieved at all. Such goal-citing explanations are called *teleological*. Some philosophers think that teleological explanations can be reduced to causal explanations: it was Jane's *desire* to alert David that caused her to throw the stone. The desire is cause that precedes its effect in the normal way. Other philosophers have held that there is something distinctive about teleological explanations: that we can see how the goal gave the subject a reason to act; we explain their actions by putting ourselves in their shoes. Such a claim is often supplemented by the view that this distinction between causal and teleological explanation is what draws the boundary between the natural sciences and the social sciences.

We do find, however, that there are teleological explanations in natural science, principally in biology. Why does the male tragopan bird inflate its throat? In order to attract a mate. Why do hearts beat? In order to pump blood around the body. Why do pupils dilate in the dark? In order to expose the retina to more light. These functional explanations are goal-citing, and so teleological. How should we understand such explanations, given that they obviously do not work by allowing us to understand the goal and action from the subject's point of view? One might reject such explanations as merely metaphorical. But their usefulness to biologists would suggest that a better approach would be to try to accommodate such explanations. How should this be done without attributing desires and intentions to hearts and pupils? Several philosophers hope to articulate a notion of function in terms of the reason why some feature is present (e.g. Millikan 1989). Think first of artefacts. Jane's office telephone has the function of enabling her to make and receive calls. It also happens to be sitting on a pile of papers, weighing them down like a paperweight. But the latter is not the phone's function, even though it may be doing that more of the time than being used in calls. According to the approach of Ruth Millikan and others, the reason is that the existence of the phone is brought about by the fact that it can be used to make calls; its existence is not related in this way to its being able to weigh down papers. Now consider the biological case. What is the function of the tragopan's throat inflation? The answer that it functions to attract a mate is correct because it is this tendency that explains the existence of this trait. The fact that ancestor male tragopans that inflated their throats significantly attracted more and better mates and so had more offspring than those that did not explains why this trait persists and is present in today's tragopans. So we can attribute function in biology in a non-metaphorical way thanks to the fact that function can be understood as a component in the theory of evolution.

Further reading

Achinstein, P. 1985 *The Nature of Explanation*. Oxford University Press.

Friedman M. 1974 'Explanation and Scientific Understanding'. *Journal of Philosophy*, 71: 5–19.

Godfrey-Smith, P. 1993 'Functions: Consensus Without Unity'. *Pacific Philosophical Quarterly* 74: 196–208.

Kitcher, P. 1981 'Explanatory Unification'. *Philosophy of Science*, 48: 507–531.

Millikan, R.G. 1989 'In Defence of Proper Functions'. *Philosophy of Science*, 56: 288–302.

Ruben, D.H. 2004 *Explaining Explanation*. London, Taylor and Francis.

Ruben, D.H. (ed.) 1993 *Explanation*. Oxford University Press.

Salmon, W.C. 2006 *Four Decades of Scientific Explanation*. University of Pittsburgh Press.

Chapter 18

Wesley C. Salmon

SCIENTIFIC EXPLANATION

The fruits of science are many and various. When science is first mentioned, many people think immediately of high technology. Such items as computers, nuclear energy, genetic engineering, and high-temperature superconductors are likely to be included. These are the fruits of applied science. Evaluating the benefits, hazards, and costs of such technological developments often leads to lively and impassioned debate.

In this chapter, however, we are going to focus on a different aspect of science, namely, the intellectual understanding it gives us of the world we live in. This is the fruit of pure science, and it is one we highly prize. All of us frequently ask the question "Why?" in order to achieve some degree of understanding with regard to various phenomena. This seems to be an expression of a natural human curiosity. Why, during a total lunar eclipse, does the moon take on a coppery color instead of just becoming dark when the earth passes between it and the sun? Because the earth's atmosphere acts like a prism, diffracting the sunlight passing through it in such a way that the light in the red region of the spectrum falls upon the lunar surface. This is a rough sketch of a scientific explanation of that phenomenon, and it imparts at least some degree of scientific understanding.

Our task in this chapter is to try to say with some precision just what scientific explanation consists in. Before we embark on that enterprise, however, some preliminary points of clarification are in order.

1 Explanation vs. confirmation

The first step in clarifying the notion of scientific explanation is to draw a sharp distinction between explaining *why* a particular phenomenon occurs and giving reasons for believing *that* it occurs. My reason for believing *that* the moon turns coppery during total eclipse is that I have observed it with my own eyes. I can also appeal to the testimony of other observers. That is how the proposition that the moon turns coppery during a total eclipse is confirmed,[1] and it is entirely different from explaining *why* it happens. Consider another example. According to contemporary cosmology all of the distant galaxies are receding from us at high velocities. The evidence for this is the fact that the light from them is shifted toward the red end of the spectrum; such evidence confirms the statement that the other galaxies are moving away from our galaxy (the Milky Way). The fact that there is such a red shift does not explain *why* the galaxies are moving in that way; instead, the fact that they are receding explains — in terms of the Doppler effect[2] — why the light is shifted toward the red end of the spectrum. The explanation of the recession lies in the "big bang" with which our universe began several billion years ago; this is what makes all of the galaxies recede from one another and, consequently, makes all of the others move away from us.

2 Other kinds of explanation

Another preliminary step in clarifying the notion of *scientific* explanation is to recognize that there are many different kinds of explanation in addition to those we classify as scientific. For example, we often encounter explanations of *how* to do something – how to use a new kitchen gadget, or how to find a certain address in a strange city. There are, in addition, explanations of *what* – what an unfamiliar word means, or what is wrong with an automobile. While many, if not all, scientific explanations can be requested by means of why-questions, requests for explanations of these other sorts would not normally be phrased in why-questions; instead, *how-to*-questions and *what*-questions would be natural.

Still other types of explanation exist. Someone might ask for an explanation of the meaning of a painting or a poem; such a request calls for an artistic interpretation. Or, someone might ask for an explanation of a mathematical proof; an appropriate response would be to fill in additional steps to show how one gets from one step to another in the original demonstration. Neither of these qualifies as scientific explanation. Also excluded from our domain of scientific explanation are explanations of formal facts of pure mathematics, such as the infinitude of the set of prime numbers. We are concerned only with explanation in the empirical sciences.

As we understand the concept of *scientific* explanation, such an explanation is an attempt to render understandable or intelligible some particular event (such as the 1986 accident at the Chernobyl nuclear facility) or some general fact (such as the copper color of the moon during total eclipse) by appealing to other particular and/or general facts drawn from one or more branches of empirical science. This formulation is *not* meant as a definition, because such terms as "understandable" and "intelligible" are as much in need of clarification as is the term "explanation." But it should serve as a rough indication of what we are driving at.

In pointing out the distinction between scientific explanations and explanations of other types we do not mean to disparage the others. The aim is only to emphasize the fact that the word "explanation" is extremely broad – it applies to a great many different things. We simply want to be clear on the type of explanation with which our discussion is concerned.

3 Scientific explanations and why-questions

Many scientific explanations are requested by means of why-questions, and even when the request is not actually formulated in that way, it can often be translated into a why-question. For example, "What caused the Chernobyl accident?" or "For what reason did the Chernobyl accident occur?" are equivalent to "Why did the Chernobyl accident occur?" However, not all why-questions are requests for scientific explanations. A woman employee might ask why she received a smaller raise in salary than a male colleague when her job-performance is just as good as his. Such a why-question might be construed as a request for a justification, or, perhaps, simply a request for more pay. A bereaved widow might ask why her husband died even though she fully understands the medical explanation. Such a why-question is a request for consolation, not explanation. Some why-questions are requests for evidence. To the question, "Why should we believe that the distant galaxies are traveling away from us at high velocities?" the answer, briefly, is the red shift. Recall, as we noted in Section 1, that the red shift does not explain the recession. The recession explains the red shift; the red shift is evidence for the recession. For the sake of clarity we distinguish *explanation-seeking why-questions* from why-questions that seek such other things as justification, consolation, or evidence.

Can all types of scientific explanation be requested by why-questions? Some authors say "yes" and others say "no." It has been suggested, for example, that some scientific explanations are answers to *how-possibly-questions*. There is an old saying that a cat will always land on its feet (paws), no matter what position it falls from. But remembering the law of conservation of angular momentum we might well ask "How is it possible for a cat, released (without imparting any angular momentum) from a height of several feet above the ground with its legs pointing upward, to turn over so that its paws are beneath it when it lands? Is this just an unfounded belief with no basis in fact?" The answer is that the cat can (and does) twist its body in ways that enable it to turn over without ever having a total angular momentum other than zero (see Frohlich 1980).

Other requests for explanation may take a *how-actually* form. A simple commonsense example illustrates the point. "How did the prisoner escape?" calls for an explanation of how he did it, not why he did it. The answer to this question might be that he sawed through some metal bars with a hacksaw blade smuggled in by his wife. If we were to ask why, the answer might be his intense desire to be with his wife outside of the prison. For a somewhat more scientific example, consider the question, "How did large mammals get to New Zealand?" The answer is that they came in boats – the first were humans, and humans brought other large mammals. Or, consider the question, "How is genetic information transmitted from parents to offspring?" The answer to this question involves the structure of the DNA molecule and the genetic code.

In this chapter we will not try to argue one way or other on the issue of whether all scientific explanations can appropriately be requested by means of why-questions. We will leave open the possibility that some explanations cannot suitably be requested by why-questions.

4 Some matters of terminology

As a further step in preliminary clarification we must establish some matters of terminology. In the first place, any explanation consists of two parts, the *explanandum* and the *explanans*. The explanandum is the fact that is to be explained. This fact may be a *particular* fact, such as the explosion of the Challenger space-shuttle vehicle, or a *general* fact, such as the law of conservation of linear momentum. A statement to the effect that the explanandum obtained is called the *explanandum-statement*. Sometimes, when it is important to contrast the fact-to-be-explained with the statement of the explanandum, we may refer to the explanandum itself as the *explanandum-fact*. When the explanandum is a particular fact we often speak of it as an event or occurrence, and there is no harm in this terminology, provided we are clear on one basic point. By and large, the events that happen in our world are highly complex, and we hardly ever try to explain every aspect of such an occurrence. For example, in explaining the explosion of the Challenger vehicle, we are not concerned to explain the fact that a woman was aboard, the fact that she was a teacher, or the fact that her life had been insured for a million dollars. When we speak of a particular fact, it is to be understood that this term refers to certain limited aspects of the event in question, not to the event in its full richness and complexity.

The other part of an explanation is the explanans. The explanans is that which does the explaining. It consists of whatever facts, particular or general, are summoned to explain the explanandum. When we want to refer to the statements of these facts we may speak of the *explanans-statements*; to contrast the facts with the statements of them we may also speak of the *explanans-facts*.

In the philosophical literature on scientific explanation, the term "explanation" is used ambiguously. Most authors use it to apply to a linguistic entity composed of the explanans-statements and the explanandum-statement. Others use it to refer to the collection of facts consisting of the explanans-facts and the explanandum-fact. In most contexts this ambiguity is harmless and does not lead to any confusion. But we should be aware that it exists.

5 Deduction and induction

As we will see, one influential philosophical account of explanation regards all bona fide scientific explanations as arguments. An argument is simply a set of statements, one of which is singled out as the conclusion of the argument. The remaining members of the set are premises. There may be one or more premises; no fixed number of premises is required.[3] The premises provide support for the conclusion.

All logically correct arguments fall into two types, deductive and inductive, and these types differ fundamentally from one another. For purposes of this chapter (and later chapters as well [sic]) we need a reasonably precise characterization of them. Four characteristics are important for our discussion.

DEDUCTION	INDUCTION
1. In a valid deductive argument, all of the content of the conclusion is present, at least implicitly, in the premises. Deduction is *nonampliative*.	1. Induction is *ampliative*. The conclusion of an inductive argument has content that goes beyond the content of its premises.
2. If the premises are true, the conclusion must be true. Valid deduction is *necessarily truth-preserving*.	2. A correct inductive argument may have true premises and a false conclusion. Induction is not *necessarily truth-preserving*.
3. If new premises are added to a valid deductive argument (and none of the original premises is changed or deleted) the argument remains valid. Deduction is *erosion-proof*.	3. New premises may completely undermine a strong inductive argument. Induction is *not erosion-proof*.
4. Deductive validity is an *all-or-nothing* matter; validity does not come in degrees. An argument is totally valid or it is invalid.	4. Inductive arguments come in different *degrees of strength*. In some inductions the premises support the conclusions more strongly than in others.

These characteristics can be illustrated by means of simple time-honored examples.

(1)   All humans are mortal.
Socrates is human.

Socrates is mortal.

Argument (1) is obviously a valid deduction. When we have said that all humans are mortal,

we have already said that Socrates is mortal, given that Socrates is human. Thus, it is nonampliative. Because it is nonampliative, it is necessarily truth-preserving. Since nothing is said by the conclusion that is not already stated by the premises, what the conclusion says *must* be true if what the premises assert is true. Moreover, the argument remains nonampliative, and hence, necessarily truth-preserving, if new premises

– for example, "Xantippe is human" – are added. You can not make a valid deduction invalid just by adding premises. Finally, the premises support the conclusion totally, not just to some degree; to accept the premises and reject the conclusion would be outright self-contradiction.

(2)   <u>All observed ravens have been black.</u>

    All ravens are black.

This argument is obviously ampliative; the premise refers only to ravens that have been observed, while the conclusion makes a statement about all ravens, observed or unobserved. It is not necessarily truth-preserving. Quite possibly there is, was, or will be – at some place or time – a white raven, or one of a different color. It is not erosion-proof; the observation of one non-black raven would undermine it completely. And its strength is a matter of degree. If only a few ravens in one limited environment had been observed, the premise would not support the conclusion very strongly; if vast numbers of ravens have been observed under a wide variety of circumstances, the support would be much stronger. But in neither case would the conclusion be necessitated by the premise.

Deductive validity and inductive correctness do not hinge on the truth of the premises or the conclusion of the argument. A valid deduction may have true premises and a true conclusion, one or more false premises and a false conclusion, and one or more false premises and a true conclusion.[4] When we say that valid deduction is necessarily truth-preserving, we mean that the conclusion would have to be true *if the premises were true*. Thus there cannot be a valid deduction with true premises and a false conclusion. Where correct inductive arguments are concerned, since they are not necessarily truth-preserving, any combination of truth values of premises and conclusion is possible. What we would like to say

is that, if the premises are true (and embody all relevant knowledge), the conclusion is probable. As we will see in Chapter 2 [*sic*], however, many profound difficulties arise in attempting to support this claim about inductive arguments.

We have chosen very simple – indeed, apparently trivial – examples in order to illustrate the basic concepts. In actual science, of course, the arguments are much more complex. Most of the deductive arguments found in serious scientific contexts are mathematical derivations, and these can be extremely complicated. Nevertheless, the basic fact remains that all of them fulfill the four characteristics listed above. Although deep and interesting problems arise in the philosophy of mathematics, they are not our primary concern in this book. Our attention is focused on the empirical sciences, which, as we argue in Chapter 2, necessarily involve induction. In that chapter we encounter much more complex and interesting inductive arguments [*sic*].

6   Is there any such thing as scientific explanation?

The idea that science can furnish explanations of various phenomena goes back to Aristotle (fourth century BC), and it has been reaffirmed by many philosophers and scientists since then. Nevertheless, many other philosophers and scientists have maintained that science must "stick to the facts," and consequently can answer only questions about *what* but not about *why*. To understand "the *why* of things," they felt, it is necessary to appeal to theology or metaphysics. Science can describe natural phenomena and predict future occurrences, but it cannot furnish explanations. This attitude was particularly prevalent in the early decades of the twentieth century. Since it is based upon certain misconceptions regarding scientific explanation, we need to say a bit about it.

It is natural enough, when attempting to find out why a person did something, to seek a conscious (or perhaps unconscious) motive. For example, to the question, "Why did you buy that book?" a satisfactory answer might run, "Because I wanted to read an amusing novel, and I have read several other novels by the same author, all of which I found amusing." This type of explanation is satisfying because we can put ourselves in the place of the subject and *understand* how such motivation works. The concept of *understanding* is critical in this context, for it signifies empathy. If we yearn for that kind of empathetic understanding of nonhuman phenomena, we have to look elsewhere for motivation or purpose. One immediate suggestion is to make the source of purpose supernatural. Thus, prior to Darwin, the variety of species of living things was explained by *special creation* – that is, God's will. Another manifestation of the same viewpoint – held by some, but not all, *vitalists* – was the notion that behind all living phenomena there is a vital force or *entelechy* directing what goes on. These entities – entelechies and vital forces – are not open to empirical investigation.

The insistence that all aspects of nature be explained in human terms is known as *anthropomorphism*. The supposition – common before the rise of modern science – that the universe is a cozy little place, created for our benefit, with humans at its center, is an anthropomorphic conception. The doctrines of special creation and some forms of vitalism are anthropomorphic. So-called "creation science" is anthropomorphic. Teleological explanation of nonhuman phenomena in terms of human-like purposes is anthropomorphic.[5]

Many philosophers and scientists rejected the appeal to anthropomorphic and teleological explanations as an appeal to hypotheses that could not, even in principle, be investigated by empirical science. If this is what is needed for explanation, they said, we want no part of it.

Science is simply not concerned with explaining natural phenomena; anyone who wants explanations will have to look for them outside of science. Such scientists and philosophers were eager to make clear that scientific knowledge does not rest on nonempirical metaphysical principles.

Not all philosophers were willing to forgo the claim that science provides explanations of natural phenomena. Karl R. Popper (1935), Carl G. Hempel (1948), R. B. Braithwaite (1953), and Ernest Nagel (1961) published important works in which they maintained that there are such things as legitimate scientific explanations, and that such explanations can be provided without going beyond the bounds of empirical science. They attempted to provide precise characterizations of scientific explanation, and they were, to a very large degree, in agreement with respect to the core of the account. The line of thought they pursued grew into a theory that enjoyed a great deal of acceptance among philosophers of science. We will discuss it at length in later sections of this chapter.

7 Does explanation involve reduction to the familiar?

It has sometimes been asserted that explanation consists in reducing the mysterious or unfamiliar to that which is familiar. Before Newton, for example, comets were regarded as mysterious and fearsome objects. Even among educated people, the appearance of a comet signified impending disaster, for example, earthquakes, floods, famines, or epidemic diseases. Newton showed that comets could be understood as planet-like objects that travel around the sun in highly eccentric orbits. For that reason, any given comet spends most of its time far from the sun and well beyond the range of human observation. When one appeared it was a surprise. But when we learned that they behave very much as the familiar planets do, their behavior was explained, and they were no longer objects of dread.

Appealing as the notion of reduction of the unfamiliar to the familiar may be, it is not a satisfactory characterization of scientific explanation. The point can best be made in terms of a famous puzzle known as *Olbers's paradox* – which is named after a nineteenth-century astronomer but was actually formulated by Edmund Halley in 1720 – why is the sky dark at night? Nothing could be more familiar than the darkness of the night sky. But Halley and later astronomers realized that, if Newton's conception of the universe were correct, then the whole night sky should shine as brightly as the noonday sun. The question of how to explain the darkness of the sky at night is extremely difficult, and there may be no answer generally accepted by the experts. Among the serious explanations that have been offered, however, appeal is made to such esoteric facts as the non-Euclidean character of space or the mean free path of photons in space. In this case, and in many others as well, a familiar phenomenon is explained by reference to facts that are very unfamiliar indeed.

I suspect that a deep connection exists between the anthropomorphic conception of explanation and the thesis that explanation consists in reduction of the unfamiliar to the familiar. The type of explanation with which we are best acquainted is that in which human action is explained in terms of conscious purposes. If it is possible to explain the phenomena of physics or biology in terms of attempting to realize a goal, that is a striking case of reduction to the familiar. A problem with this approach is, of course, that a great deal of the progress in scientific understanding has resulted in the elimination, not the injection, of purposes.

8 The deductive-nomological pattern of scientific explanation

In a classic 1948 paper, Carl G. Hempel and Paul Oppenheim formulated, with great precision, one pattern of scientific explanation that is central to all discussions of the subject. It is known as the *deductive-nomological* (D-N) model of scientific explanation. Stated very simply, an explanation of this type explains by subsuming its explanandum-fact under a general law. This can best be appreciated by looking at an example.

A figure skater with arms outstretched stands balanced on one skate. Propelling herself with her other skate she begins to rotate slowly. She stops propelling herself, but she continues to rotate slowly for a few moments. Suddenly – without propelling herself again and without being propelled by any external object, such as another skater – she begins spinning very rapidly. Why? Because she drew her arms in close to her body, thus concentrating her total body mass closer to the axis of rotation. Because of the law of conservation of angular momentum, her rate of rotation had to increase to compensate for her more compact body configuration.

More technically, the angular momentum of an object is the product of its angular velocity (rate of rotation) and its moment of inertia. The moment of inertia depends upon the mass of the object and the average distance of the mass from the axis of rotation; for a fixed mass, the moment of inertia is smaller the more compactly the mass is distributed about the axis of rotation. The law of conservation of angular momentum says that the angular momentum of a body that is not being propelled or retarded by external forces does not change; hence, since the moment of inertia is decreased, the rate of rotation must increase to keep the value of the product constant.[6]

According to Hempel and Oppenheim, an explanation of the foregoing sort is to be viewed as a deductive argument. It can be set out more formally as follows:

(3) The angular momentum of any body (whose rate of rotation is not being increased or decreased by external forces) remains constant.

The skater is not interacting with any external object in such a way as to alter her angular velocity.

The skater is rotating (her angular momentum is not zero).

The skater reduces her moment of inertia by drawing her arms in close to her body.

The skater's rate of rotation increases.

The explanandum − the increase in the skater's rate of rotation − is the conclusion of the argument. The premises of the argument constitute the explanans. The first premise states a law of nature − the law of conservation of angular momentum. The remaining three premises state the antecedent conditions. The argument is logically correct; the conclusion follows validly from the premises. For purposes of our discussion, we may take the statements of antecedent conditions as true; expert figure skaters do this maneuver frequently. The law of conservation of angular momentum can also be regarded as true since it is a fundamental law of physics which has been confirmed by a vast quantity of empirical data.

Hempel and Oppenheim set forth four conditions of adequacy for D-N explanations:

1   The explanandum must be a logical consequence of the explanans; that is, the explanation must be a valid deductive argument.
2   The explanans must contain at least one general law, and it must actually be required for the derivation of the explanandum; in other words, if the law or laws were deleted, without adding any new premises, the argument would no longer be valid.
3   The explanans must have empirical content; it must be capable, at least in principle, of test by experiment or observation.
4   The sentences constituting the explanans must be true.

These conditions are evidently fulfilled by our example. The first three are classified as *logical* conditions of adequacy; the fourth is *empirical*. An argument that fulfills all four conditions is an explanation (for emphasis we sometimes say "true explanation"). An argument that fulfills the first three conditions, without necessarily fulfilling the fourth, is called a *potential explanation*. It is an argument that would be an explanation if its premises were true.[7]

According to Hempel and Oppenheim, it is possible to have D-N explanations, not only of particular occurrences as in argument (3), but also of general laws. For example, in the context of Newtonian mechanics, it is possible to set up the following argument:

(4) $F = ma$ (Newton's second law).
For every action there is an equal and opposite reaction (Newton's third law).

In every interaction, the total linear momentum of the system of interacting bodies remains constant (law of conservation of linear momentum).

This argument is valid, and among its premises are statements of general laws. There are no statements of antecedent conditions, but that is not a problem since the conditions of adequacy do not require them. Because we are not concerned to explain any particular facts, no premises regarding particular facts are needed.

Both premises in the explanans are obviously testable, for they have been tested countless times. Thus, argument (4) fulfills the logical conditions of adequacy, and consequently, it qualifies as a potential explanation. Strictly speaking, it does not qualify as a true explanation, for we do not consider Newton's laws of motion literally true, but in many contexts they can be taken as correct because they provide extremely accurate approximations to the truth.

Although Hempel and Oppenheim discussed both deductive explanations of particular facts and deductive explanations of general laws, they offered a precise characterization only of the former, but not of the latter. They declined to attempt to provide a characterization of explanations of general laws because of a problem they recognized but did not know how to solve. Consider Kepler's laws of planetary motion K and Boyle's law of gases B. If, on the one hand, we conjoin the two to form a law $K \cdot B$, we can obviously deduce K from it. But this could not be regarded as an explanation of K, for it is only a pointless derivation of K from itself. On the other hand, the derivation of K from Newton's laws of motion and gravitation constitutes an extremely illuminating explanation of Kepler's laws. Hempel and Oppenheim themselves confessed that they were unable to provide any criterion to distinguish the pointless pseudo-explanations from the genuine explanations of laws (see Hempel and Oppenheim 1948 as reprinted in Hempel 1965b, 273, f.n. 33).

Hempel and Oppenheim envisioned two types of D-N explanation, though they were able to provide an account of only one of them. In addition, they remarked that other types of explanation are to be found in the sciences, namely, explanations that appeal, not to universal generalizations, but to statistical laws instead (ibid., 250–251).

Table 18.1 shows the four kinds of explanations to which Hempel and Oppenheim called

Table 18.1

Laws \ Explananda	Particular facts	General regularities
Universal laws	D-N Deductive-Nomological	D-N Deductive-Nomological
Statistical laws	I-S Inductive-Statistical	I-S Inductive-Statistical

attention; they furnished an account only for the type found in the upper left-hand box. Some years later Hempel (1962) offered an account of the I-S pattern in the lower left-hand box. In Hempel (1965b) he treated both the I-S and the D-S patterns. In 1948, Hempel and Oppenheim were looking forward to the time when theories of explanation dealing with all four boxes would be available.

9 What are laws of nature?

Hempel and Oppenheim emphasized the crucial role played by laws in scientific explanation; in fact, the D-N pattern is often called the *covering-law model*. As we will see, laws play a central part in other conceptions of scientific explanation as well. Roughly speaking, a *law* is a regularity that holds throughout the universe, at all places and all times. A *law-statement* is simply a statement to the effect that such a regularity exists. A problem arises immediately. Some regularities appear to be lawful and others do not. Consider some examples of laws:

(i)   All gases, kept in closed containers of fixed size, exert greater pressure when heated.

(ii) In all closed systems the quantity of energy remains constant.

(iii) No signals travel faster than light.

Contrast these with the following:

(iv)   All of the apples in my refrigerator are yellow.
(v)   All Apache basketry is made by women.
(vi)   No golden spheres have masses greater than 100,000 kilograms.

Let us assume, for the sake of argument, that all of the statements (i)–(vi) are true. The first thing to notice about them is their generality. Each of them has the overall form, "All *A* are *B*" or "No *A* are *B*." Statements having these forms are known as *universal generalizations*. They mean, respectively, "*Anything* that is an *A* is also a *B*" and '*Nothing* that is an *A* is also a *B*." Nevertheless, statements (i)–(iii) differ fundamentally from (iv)–(vi). Notice, for example, that none of the statements (i)–(iii) makes any reference to any particular object, event, person, time, or place. In contrast, statement (iv) refers to a particular person (me), a particular object (my refrigerator), and a particular time (now). This statement is not completely general since it singles out certain particular entities to which it refers. The same remark applies to statement (v) since it refers to a particular limited group of people (the Apache).

Laws of nature are generally taken to have two basic capabilities. First, they support counter-factual inferences. A *counterfactual statement* is a conditional statement whose antecedent is false. Suppose, for example, that I cut a branch from a tree and then, immediately, burn it in my fireplace. This piece of wood was never placed in water and never will be. Nevertheless, we are prepared to say, without hesitation, that *if it had been placed in water, it would have floated*. This italicized statement is a counterfactual conditional. Now, a law-statement, such as (i), will support a counterfactual assertion. We can say, regarding a particular sample of some gas, held in a closed container of fixed size but not actually being heated, that if it were heated it *would* exert greater

pressure. We can assert the counterfactual because we take statement (i) to be a statement of a law of nature.

When we look at statement (iv) we see that it does not support any such counterfactual statement. Holding a red delicious apple in my hand, I cannot claim, on the basis of (iv), that this apple would be yellow if it were in my refrigerator.

A second capability of laws of nature is to support *modal* statements of physical necessity and impossibility. Statement (ii), the first law of thermodynamics, implies that it is impossible to create a perpetual motion machine of the first kind – that is, a machine that does useful work without any input of energy from an external source. In contrast, statement (v) does not support the claim that it is impossible for an Apache basket to be made by a male. It is physically possible that an Apache boy might be taught the art of basket making, and might grow up to make a career of basketry.

When we compare statements (iii) and (vi) more subtle difficulties arise. Unlike statements (iv) and (v), statement (vi) does not make reference to any particular entity or place or time.[8] It seems clear, nevertheless, that statement (vi) – even assuming it to be true – cannot support either modal statements or counterfactual conditionals. Even if we agree that nowhere in the entire history of the universe – past, present, or future – does there exist a gold sphere of mass greater than 100,000 kilograms, we would not be justified in claiming that it is *impossible* to fabricate a gold sphere of such mass. I once made a rough calculation of the amount of gold in the oceans of the earth, and it came to about 1,000,000 kilograms. If an incredibly rich prince were determined to impress a woman passionately devoted to golden spheres it would be physically possible for him to extract a little more than 100,000 kilograms from the sea to create a sphere that massive.

Statement (vi) also lacks the capacity to support counterfactual conditionals. We would not be justified in concluding that, if two golden hemispheres, each of 50,001 kilogram mass, were put together, they would not form a golden sphere of mass greater than 100,000 kilograms. To appreciate the force of this point, consider the following statement:

(vii) No enriched uranium sphere has a mass greater than 100,000 kilograms.

This *is* a lawful generalization, because the critical mass for a nuclear chain reaction is just a few kilograms. If 100,000 kilograms of enriched uranium were to be assembled, we would have a gigantic nuclear explosion. No comparable catastrophe would ensue, as far as we know, if a golden sphere of the same mass were put together.

Philosophers have often claimed that we can distinguish true generalizations that are *lawful* from those that are *accidental*. Even if we grant the truth of (vi), we must conclude that it is an accidental generalization. Moreover, they have maintained that among universal generalizations, regardless of truth, it is possible to distinguish *lawlike generalizations* from those that are not lawlike. A lawlike generalization is one that has all of the qualifications for being a law except, perhaps, being true.

It is relatively easy to point to the characteristic of statements (iv) and (v) that makes them nonlawlike, namely, that they make reference to particular objects, persons, events, places, or times. The nonlawlike character of statement (vi) is harder to diagnose. One obvious suggestion is to apply the criteria of supporting counterfactual and/or modal statements. We have seen that (vi) fails on that score. The problem with that approach is that it runs a serious risk of turning out to be circular. Consider statement (ii). Why do we consider it *physically impossible* to build a

perpetual motion machine (of the first type)? Because to do so would violate a law of nature, namely (ii). Consider statement (vi). Why do we consider it *physically possible* to fabricate a golden sphere whose mass exceeds 100,000 kilograms? Because to do so would not violate a law of nature. It appears that the question of what modal statements to accept hinges on the question of what regularities qualify as laws of nature.

A similar point applies to the support of counterfactual conditionals. Consider statement (i). Given a container of gas that is not being heated, we can say that, if it were to be heated, it would exert increased pressure on the walls of its container – sufficient in many cases to burst the container. (I learned my lesson on this as a Boy Scout heating an unopened can of beans in a camp fire.) The reason that we can make such a counterfactual claim is that we can infer from statement (i) what would happen, and (i) states a law of nature. Similarly, from (iii) we can deduce that if something travels faster than light it is not a signal – that is, it cannot transmit information. You might think that this is vacuous because, as the theory of relativity tells us, nothing can travel faster than light. However, this opinion is incorrect. Shadows and various other kinds of "things" can easily be shown to travel faster than light. We can legitimately conclude that, if something does travel faster than light, it is not functioning as a signal, because (iii) is, indeed, a law of nature.

What are the fundamental differences between statement (vi) on the one hand and statements (i)–(iii) and (vii) on the other? The main difference seems to be that (i)–(iii) and (vii) are all deeply embedded in well-developed scientific theories, and that they have been, directly or indirectly, extensively tested. This means that (i)–(iii) and (vii) have a very different status within our body of scientific knowledge than do (iv)–(vi). The question remains, however, whether the regularities described by (i)–(iii) and (vii) have a different status in the physical universe than do (iv)–(vi).

At the very beginning of this chapter, we considered the explanation of the fact that the moon assumes a coppery hue during total eclipse. This is a regularity found in nature, but is it a lawful regularity? Is the statement, "The moon turns a coppery color during total eclipses," a law-statement? The immediate temptation is to respond in the negative, for the statement makes an explicit reference to a particular object, namely, our moon. But if we reject that statement as a lawful generalization, it would seem necessary to reject Kepler's laws of planetary motion as well, for they make explicit reference to our solar system. Galileo's law of falling bodies would also have to go, for it refers to things falling near the surface of the earth. It would be unreasonable to disqualify all of them as laws.

We can, instead, make a distinction between basic and derived laws. Kepler's laws and Galileo's law can be derived from Newton's laws of motion and gravitation, in conjunction with descriptions of the solar system and the bodies that make it up. Newton's laws are completely general and make no reference to any particular person, object, event, place, or time. The statement about the color of the moon during total eclipse can be derived from the laws of optics in conjunction with a description of the earth's atmosphere and the configuration of the sun, moon, and earth when an eclipse occurs. The statement about the color of the moon can also be taken as a derivative law. But what about statements (iv) and (v)? The color of the apples in my refrigerator can in no way be derived from basic laws of nature in conjunction with a description of the refrigerator. No matter how

fond I may be of golden delicious apples, there is no physical impossibility of a red delicious getting into my refrigerator. Similarly, there are no laws of nature from which, in conjunction with descriptions of the Apache and their baskets, it would be possible to derive that they can only be made by women.

10 Problems for the D-N pattern of explanation

Quite remarkably the classic article by Hempel and Oppenheim received virtually no attention for a full decade. Around 1958, however, a barrage of criticism began and a lively controversy ensued. Much of the criticism was brought into sharp focus by means of counterexamples that have, themselves, become classic. These examples fall into two broad categories. The first consists of arguments that fulfill all of the requirements for D-N explanation, yet patently fail to qualify as bona fide explanations. They show that the requirements set forth by Hempel and Oppenheim are not *sufficient* to determine what constitutes an acceptable scientific explanation. The second consists of examples of allegedly bona fide explanations that fail to fulfill the Hempel-Oppenheim requirements. They are meant to show that it is not *necessary* to fulfill those requirements in order to have correct explanations. We must treat this second category with care, for Hempel and Oppenheim never asserted that all correct explanations fit the D-N pattern. They explicitly acknowledged that legitimate statistical explanations can be found in science. So, statistical explanations are not appropriate as counterexamples. However, the attempt has been to find examples that are clearly not statistical, but which fail to fulfill the Hempel-Oppenheim criteria. Let us look at some counterexamples of each type.

CE-1. The flagpole and its shadow.[9] On a flat and level piece of ground stands a flagpole that is 12′ tall. The sun, which is at an elevation of 53.13° in the sky, shines brightly. The flagpole casts a shadow that is 9′ long. If we ask why the shadow has that length, it is easy to answer. From the elevation of the sun, the height of the flagpole, and the rectilinear propagation of light, we can deduce, with the aid of a bit of trigonometry, the length of the shadow. The result is a D-N explanation that most of us would accept as correct. So far, there is no problem.

If, however, someone asks why the flagpole is 12′ tall, we could construct essentially the same argument as before. But instead of deducing the length of the shadow from the height of the flagpole and the elevation of the sun, we would deduce the height of the flagpole from the length of the shadow and the elevation of the sun. Hardly anyone would regard that argument, which satisfies all of the requirements for a D-N explanation, as an adequate explanation of the height of the flagpole.

We can go one step farther. From the length of the shadow and the height of the flagpole, using a similar argument, we can deduce that the sun is at an elevation of 53.13°. It seems most unreasonable to say that the sun is that high in the sky because a 12′ flagpole casts a 9′ shadow. From the fact that a 12′ flagpole casts a 9′ shadow we can infer *that* the sun is that high in the sky, but we cannot use those data to explain *why* it is at that elevation. Here we must be sure to remember the distinction between confirmation and explanation (discussed in Section 1). The explanation of the elevation rests upon the season of the year and the time of day.

The moral: The reason it is legitimate to explain the length of the shadow in terms of the height of the flagpole and the elevation of the sun is that the shadow is the effect of those two causal factors. We can explain effects by citing their causes. The reason it is illegitimate to explain the

height of the flagpole by the length of the shadow is that the length of the shadow is an effect of the height of the flagpole (given the elevation of the sun), but it is no part of the cause of the height of the flagpole. We cannot explain causes in terms of their effects. Furthermore, although the elevation of the sun is a crucial causal factor in the relation between the height of the flagpole and the length of the shadow, the flagpole and its shadow play no causal role in the position of the sun in the sky.

CE-2. The barometer and the storm. Given a sharp drop in the reading on a properly functioning barometer, we can predict that a storm will shortly occur. Nevertheless, the reading on the barometer does not explain the storm. A sharp drop in atmospheric pressure, which is registered on the barometer, explains both the storm and the barometric reading.

The moral: Many times we find two effects of a common cause that are correlated with one another. In such cases we do not explain one effect by means of the other. The point is illustrated also by diseases. A given illness may have many different symptoms. The disease explains the symptoms; one symptom does not explain another.

CE-3. A solar eclipse. From the present positions of the earth, moon, and sun, using laws of celestial mechanics, astronomers can predict a future total eclipse of the sun. After the eclipse has occurred, the very same data, laws, and calculations provide a legitimate D-N explanation of the eclipse. So far, so good. However, using the same laws and the same positions of the earth, moon, and sun, astronomers can retrodict the previous occurrence of a solar eclipse. The argument by which this retrodiction is made fulfills the requirements for a D-N explanation just as fully as does the prediction of the eclipse. Nevertheless, most of us would say that, while it is possible to explain an eclipse in terms of antecedent conditions, it is not possible to explain an eclipse in terms of subsequent conditions.

The moral: We invoke earlier conditions to explain subsequent facts; we do not invoke later conditions to explain earlier facts. The reason for this asymmetry seems to lie in the fact that causes, which have explanatory import, precede their effects – they do not follow their effects.

CE-4. The man and the pill. A man explains his failure to become pregnant during the past year on the ground that he has regularly consumed his wife's birth control pills, and that any man who regularly takes oral contraceptives will avoid getting pregnant.

The moral: This example shows that it is possible to construct valid deductive arguments with true premises in which some fact asserted by the premises is actually irrelevant. Since men do not get pregnant regardless, the fact that this man took birth control pills is irrelevant. Nevertheless, it conforms to the D-N pattern.

Counterexamples CE-l–CE-4 are all cases in which an argument that fulfills the Hempel-Oppenheim requirements manifestly fails to constitute a bona fide explanation. They were designed to show that these requirements are too weak to sort out the illegitimate explanations. A natural suggestion would be to strengthen them in ways that would rule out counterexamples of these kinds. For example, CE-1 and CE-2 could be disqualified if we stipulated that the antecedent conditions cited in the explanans must be causes of the explanandum. CE-3 could be eliminated by insisting that the so-called antecedent conditions must actually obtain prior to the explanandum. And CE-4 could be ruled out by stipulating that the antecedent conditions must be relevant to the explanandum. For various reasons Hempel declined to strengthen the requirements for D-N explanation in such ways.

The next counterexample has been offered as a case of a legitimate explanation that does not meet the Hempel-Oppenheim requirements.

CE-5. The ink stain. On the carpet, near the desk in Professor Jones's office, is an unsightly black stain. How does he explain it? Yesterday, an open bottle of black ink stood on his desk, near the corner. As he went by he accidentally bumped it with his elbow, and it fell to the floor, spilling ink on the carpet. This seems to be a perfectly adequate explanation; nevertheless, it does not incorporate any laws. Defenders of the D-N pattern would say that this is simply an incomplete explanation, and that the laws are tacitly assumed. Michael Scriven, who offered this example, argued that the explanation is clear and complete as it stands, and that any effort to spell out the laws and initial conditions precisely will meet with failure.

The moral: It is possible to have perfectly good explanations without any laws. The covering law conception is not universally correct.

The fifth counterexample raises profound problems concerning the nature of causality. Some philosophers, like Scriven, maintain that one event, such as the bumping of the ink bottle with the elbow, is obviously the cause of another event, such as the bottle falling off of the desk. Moreover, they claim, to identify the cause of an event is all that is needed to explain it. Other philosophers, including Hempel, maintain that a causal relation always involves (sometimes explicitly, sometimes implicitly) a general causal law. In the case of the ink stain, the relevant laws would include the laws of Newtonian mechanics (in explaining the bottle being knocked off the desk and falling to the floor) and some laws of chemistry (in explaining the stain on the carpet as a result of spilled ink).

11 Two patterns of statistical explanation

Anyone who is familiar with any area of science – physical, biological, or social – realizes, as Hempel and Oppenheim had already noted, that not all explanations are of the deductive-nomological variety. Statistical laws play an important role in virtually every branch of contemporary science and statistical explanations – those falling into the two lower boxes in Table 18.1 – are frequently given. In 1965b Hempel published a comprehensive essay, "Aspects of Scientific Explanation," in which he offered a theory of statistical explanation encompassing both types.

In the first type of statistical explanation, the *deductive-statistical (D-S) pattern*, statistical regularities are explained by deduction from more comprehensive statistical laws. Many examples can be found in contemporary science. For instance, archaeologists use the radiocarbon dating technique to ascertain the ages of pieces of wood or charcoal discovered in archaeological sites. If a piece of wood is found to have a concentration of C^{14} (a radioactive isotope of carbon) equal to one-fourth that of newly cut wood, it is inferred to be 11,460 years old. The reason is that the half-life of C^{14} is 5730 years, and in two half-lives it is extremely probable that about three-fourths of the C^{14} atoms will have decayed. Living trees replenish their supplies of C^{14} from the atmosphere; wood that has been cut cannot do so. Here is the D-S explanation:

(5) Every C^{14} atom (that is not exposed to external radiation[10]) has a probability of 1/2 of disintegrating within any period of 5730 years.

In any large collection of C^{14} atoms (that are not exposed to external radiation)

approximately three-fourths will *very probably* decay within 11,460 years.

This derivation constitutes a deductive explanation of the probabilistic generalization that stands as its conclusion.

Deductive-statistical explanations are very similar, logically, to D-N explanations of generalizations. The only difference is that the explanation is a statistical law and the explanans must contain at least one statistical law. Universal laws have the form "All A are B" or "No A are B"; statistical laws say that a certain proportion of A are B.[11] Accordingly, the problem that plagued D-N explanations of universal generalizations also infects D-S explanations of statistical generalizations. Consider, for instance, one of the statistical generalizations in the preceding example – namely, that the half-life of C^{14} is 5730 years. There is a bona fide explanation of this generalization from the basic laws of quantum mechanics in conjunction with a description of the C^{14} nucleus. However, this statistical generalization can also be deduced from the conjunction of itself with Kepler's laws of planetary motion. This deduction would not qualify as any kind of legitimate explanation; like the case cited in Section 8, it would simply be a pointless derivation of the generalization about the half-life of C^{14} from itself.

Following the 1948 article, Hempel never returned to this problem concerning explanations of laws; he did not address it in Hempel (1965a), which contains characterizations of all four types of explanation represented in Table 18.1. This leaves both boxes on the right-hand side of Table 18.1 in a highly problematic status. Nevertheless, it seems clear that many sound explanations of both of these types can be found in the various sciences.

The second type of statistical explanation – the *inductive-statistical (I-S) pattern* – explains particular occurrences by subsuming them under statistical laws, much as D-N explanations subsume particular events under universal laws. Let us look at one of Hempel's famous examples. If we ask why Jane Jones recovered rapidly from her streptococcus infection, the answer is that she was given a dose of penicillin, and almost all strep infections clear up quickly upon administration of penicillin. More formally:

(6)   Almost all cases of streptococcus infection clear up quickly after the administration of penicillin.

Jane Jones had a streptococcus infection.

Jane Jones received treatment with penicillin.

== [r]

Jane Jones recovered quickly.

This explanation is an argument that has three premises (the explanans); the first premise states a statistical regularity – a statistical law – while the other two state antecedent conditions. The conclusion (the explanandum) states the fact-to-be-explained. However, a crucial difference exists between explanations (3) and (6): D-N explanations subsume the events to be explained deductively, while I-S explanations subsume them inductively. The single line separating the premises from the conclusion in (3) signifies a relation of deductive entailment between the premises and conclusion. The double line in (6) represents a relationship of inductive support, and the attached variable r stands for the strength of that support. This strength of support may be expressed exactly, as a numerical value of a probability, or vaguely, by means of such phrases as "very probably" or "almost certainly."

An explanation of either of these two kinds can be described as an argument to the effect that *the event to be explained was to be expected by virtue of certain explanatory facts*. In a D-N explanation, the event to be explained is deductively certain, given the explanatory facts; in an I-S explanation the event to be explained has high inductive probability relative to the explanatory facts. This feature of expectability is closely related to the *explanation-prediction symmetry thesis* for explanations of particular facts. According to this thesis any acceptable explanation of a particular fact is an argument, deductive or inductive, that could have been used to predict the fact in question if the facts stated in the explanans had been available prior to its occurrence.[12] As we shall see, this symmetry thesis met with serious opposition.

Hempel was not by any means the only philosopher in the early 1960s to notice that statistical explanations play a significant role in modern science. He was, however, the first to present a detailed account of the nature of statistical explanation, and the first to bring out a fundamental problem concerning statistical explanations of particular facts. The case of Jane Jones and her quick recovery can be used as an illustration. It is well known that certain strains of the streptococcus bacterium are penicillin-resistant, and if Jones's infection were of that type, the probability of her quick recovery after treatment with penicillin would be small. We could, in fact, set up the following inductive argument:

(7)   Almost no cases of penicillin-resistant streptococcus infection clear up quickly after the administration of penicillin.

Janes Jones had a penicillin-resistant streptococcus infection.

Jane Jones received treatment with penicillin.

== [q]

Jane Jones did not recover quickly.

The remarkable fact about arguments (6) and (7) is that their premises are mutually compatible – they could all be true. Nevertheless, their conclusions contradict one another. This is a situation that can never occur with deductive arguments. Given two valid deductions with incompatible conclusions, their premises must also be incompatible. Thus, the problem that has arisen in connection with I-S explanations has no analog in D-N explanations. Hempel called this *the problem of ambiguity of I-S explanation*.

The source of the problem of ambiguity is a simple and fundamental difference between universal laws and statistical laws. Given the proposition that all *A* are *B*, it follows immediately that all things that are both *A* and *C* are *B*. If all humans are mortal, then all people who are over six feet tall are mortal. However, even if almost all humans who are alive now will be alive five years from now, *it does not follow* that almost all living humans with advanced cases of pancreatic cancer will be alive five years hence. As we noted in Section 5, there is a parallel fact about arguments. Given a valid deductive argument, the argument will remain valid if additional premises are supplied, as long as none of the original premises is taken away. Deduction is erosion-proof. Given a strong inductive argument − one that supports its conclusion with a high degree of probability − the addition of one more premise may undermine it completely. For centuries Europeans had a great body of inductive evidence to support the proposition that all swans are white, but one true report of a black swan in Australia completely refuted that conclusion. Induction is not erosion-proof.

Hempel sought to resolve the problem of ambiguity by means of his *requirement of maximal specificity* (RMS). It is extremely tricky to state RMS with precision, but the basic idea is fairly simple. In constructing I-S explanations we must include all relevant knowledge we have that would have been available, in principle, prior to the explanandum-fact. If the information that Jones's infection is of the penicillin-resistant

type is available to us, argument (6) would not qualify as an acceptable I-S explanation.[13]

In Section 8 we stated Hempel and Oppenheim's four conditions of adequacy for D-N explanations. We can now generalize these conditions so that they apply both to D-N and I-S explanations as follows:

1   The explanation must be an argument having correct (deductive or inductive) logical form.
2   The explanans must contain at least one general law (universal or statistical), and this law must actually be required for the derivation of the explanandum.
3   The explanans must have empirical content; it must be capable, at least in principle, of test by experiment or observation.
4   The sentences constituting the explanans must be true.
5   The explanation must satisfy the requirement of maximal specificity.[14]

The theory of scientific explanation developed by Hempel in his "Aspects" essay won rather wide approval among philosophers of science. During the mid-to-late 1960s and early 1970s it could appropriately be considered *the received view of scientific explanation*. According to this view, every legitimate scientific explanation must fit the pattern corresponding to one or another of the four boxes in Table 18.1.

12 Criticisms of the I-S pattern of scientific explanation

We noticed in Section 10 that major criticisms of the D-N pattern of scientific explanation can be posed by means of well-known counterexamples. The same situation arises in connection with the I-S pattern. Consider the following:

CE-6. Psychotherapy. Suppose that Bruce Brown has a troublesome neurotic symptom. He undergoes psychotherapy and his symptom disappears. Can we explain his recovery in terms of the treatment he has undergone? We could set out the following inductive argument, in analogy with argument (6):

(8)   Most people who have a neurotic symptom of type N and who undergo psychotherapy experience relief from that symptom.

Bruce Brown had a symptom of type N and he underwent psychotherapy.

================================= [r]

Bruce Brown experienced relief from his symptom.

Before attempting to evaluate this proffered explanation we should take account of the fact that there is a fairly high spontaneous remission rate – that is, many people who suffer from that sort of symptom get better regardless of treatment. No matter how large the number r, if the rate of recovery for people who undergo psychotherapy is no larger than the spontaneous remission rate, it would be a mistake to consider argument (8) a legitimate explanation. A high probability is not *sufficient* for a correct explanation. If, however, the number r is not very large, but is greater than the spontaneous remission rate, the fact that the patient underwent psychotherapy has at least some degree of explanatory force. A high probability is not *necessary* for a sound explanation.

Another example reinforces the same point.

CE-7. Vitamin C and the common cold.[15] Suppose someone were to claim that large doses of vitamin C would produce rapid cures for the common cold. To ascertain the efficacy of vitamin C in producing rapid recovery from colds, we should note, it is *not* sufficient to establish that most people recover quickly; most colds disappear within a few days regardless of treatment. What is required is a double-blind controlled experiment[16] in which the rate of quick recovery for those who take vitamin C is compared with the rate of quick recovery for those who receive only a placebo. If there is a significant difference in the probability of quick recovery for those who take vitamin C and for those who do not, we may conclude that vitamin C has some degree of causal efficacy in lessening the duration of colds. If, however, there is no difference between the two groups, then it would be a mistake to try to explain a person's quick recovery from a cold by constructing an argument analogous to (6) in which that result is attributed to treatment with vitamin C.

The moral: CE-6 and CE-7 call attention to the same point as CE-4 (the man and the pill). All of them show that something must be done to exclude irrelevancies from scientific explanations. If the rate of pregnancy among men who consume oral contraceptives is the same as for men who do not, then the use of birth control pills is causally and explanatorily irrelevant to pregnancy among males. Likewise, if the rate of relief from neurotic symptoms is the same for those who undergo psychotherapy as it is for those who do not, then psychotherapy is causally and explanatorily irrelevant to the relief from neurotic symptoms. Again, if the rate of rapid recovery from common colds is the same for those who do and those who do not take massive doses of vitamin C, then consumption of massive doses of vitamin C is causally and explanatorily irrelevant to rapid recovery from colds.[17] Hempel's requirement of maximal specificity was designed to insure that *all* relevant information (of a suitable sort) is included in I-S explanations. What is needed in addition is a requirement insuring that *only* relevant information is included in D-N or I-S explanations.

CE-8. Syphilis and paresis. Paresis is a form of tertiary syphilis which can be contracted only by people who go through the primary, secondary, and latent forms of syphilis without treatment with penicillin. If one should ask why a particular person suffers from paresis, a correct answer is that he or she was a victim of untreated latent syphilis. Nevertheless, only a small proportion of those with untreated latent syphilis – about 25% – actually contract paresis. Given a randomly selected member of the class of victims of untreated latent syphilis, one should predict that that person *will not* develop paresis.

The moral: there are legitimate I-S explanations in which the explanans *does not* render the explanandum highly probable. CE-8 responds to the explanation-prediction symmetry thesis – the claim that an explanation is an argument of such a sort that it could have been used to predict the explanandum if it had been available prior to the fact-to-be-explained. It is worth noting, in relation to CE-6 and CE-7, that untreated latent syphilis is highly relevant to the occurrence of paresis, although it does not make paresis highly probable, or even more probable than not.

CE-9. The biased coin. Suppose that a coin is being tossed, and that it is highly biased for heads – in fact, on any given toss, the probability of getting heads is 0.95, while the probability of tails is 0.05. The coin is tossed and comes up heads. We can readily construct an I-S explanation fitting all of the requirements. But suppose it comes up tails. In this case an I-S explanation is out of the question. Nevertheless, to the degree that we understand the mechanism involved, and consequently the probable outcome of heads, to that same degree we understand the improbable outcome, even though it occurs less frequently.

The moral: If we are in a position to construct statistical explanations of events that are highly probable, then we also possess the capability of framing statistical explanations of events that are extremely improbable.

13 Determinism, indeterminism, and statistical explanation

When we look at an I-S explanation such as (6), there is a strong temptation to regard it as incomplete. It may, to be sure, incorporate all of the relevant knowledge we happen to possess. Nevertheless, we may feel, it is altogether possible that medical science will discover enough about streptococcus infections and about penicillin treatment to be able to determine precisely which individuals with strep infections will recover quickly upon treatment with penicillin and which individuals will not. When that degree of knowledge is available we will not have to settle for I-S explanations of rapid recoveries from strep infections; we will be able to provide D-N explanations instead. Similar remarks can also be made about several of the counter-examples – in particular, examples CE-6–CE-9.

Consider CE-8, the syphilis-paresis example. As remarked above, with our present state of knowledge we can predict that about 25% of all victims of untreated latent syphilis contract paresis, but we do not know how to distinguish those who will develop paresis from those who will not. Suppose Sam Smith develops paresis. At this stage of our knowledge the best we can do by way of an I-S explanation of Smith's paresis is the following:

(9)   25% of all victims of untreated latent
      syphilis develop paresis.
      Smith had untreated latent syphilis.
      ══════════════════════════════════ [.25]
      Smith contracted paresis.

This could not be accepted as an I-S explanation because of the weakness of the relation of inductive support.

Suppose that further research on the causes of paresis reveals a factor in the blood – call it the P-factor – which enables us to pick out, with fair reliability – say 95% – those who will develop paresis. Given that Smith has the P-factor, we can construct the following argument:

(10) 95% of all victims of untreated latent syphilis who have the P-factor develop paresis.
Smith had untreated latent syphilis.
Smith had the P-factor.
$$=============[.95]$$
Smith developed paresis.

In the knowledge situation just described, this would count as a pretty good I-S explanation, for 0.95 is fairly close to 1.

Let us now suppose further that additional medical research reveals that, among those victims of untreated latent syphilis who have the P-factor, those whose spinal fluid contains another factor Q invariably develop paresis. Given that information, and the fact that Smith has the Q-factor, we can set up the following explanation:

(11) All victims of untreated latent syphilis who have the P-factor and the Q-factor develop paresis.
Smith had untreated latent syphilis.
Smith had the P-factor.
Smith had the Q-factor.

Smith developed paresis.

If the suppositions about the P-factor and the Q-factor were true, this argument would qualify as a correct D-N explanation. We accepted (10) as a correct explanation of Smith's paresis only because we were lacking the information that enabled us to set up (11).

Determinism is the doctrine that says that everything that happens in our universe is completely determined by prior conditions.[18] If this thesis is correct, then each and every event in the history of the universe – past, present, or future – is, in principle, deductively explainable. If determinism is true, then every sound I-S explanation is merely an incomplete D-N explanation. Under these circumstances, the I-S pattern is not really a stand-alone type of explanation; all fully correct explanations fit the D-N pattern. The lower left-hand box of Table 18.1 would be empty. This does *not* mean that I-S explanations – that is, incomplete D-N explanations – are useless, only that they are incomplete.

Is determinism true? We will not take a stand on that issue in this chapter. Modern physics – quantum mechanics in particular – seems to offer strong reasons to believe that determinism is false, but not everyone agrees with this interpretation. However, we will take the position that indeterminism *may* be true, and see what the consequences are with respect to statistical explanation.

According to most physicists and philosophers of physics, the spontaneous disintegration of the nucleus of an atom of a radioactive substance is a genuinely indeterministic happening. Radioactive decay is governed by laws, but they are fundamentally and irreducibly statistical. Any C^{14} atom has a fifty-fifty chance of spontaneously disintegrating within the next 5730 years and a fifty-fifty chance of not doing so. Given a collection of C^{14} atoms, the probability is overwhelming that some will decay and some will not in the next 5730 years. However, no way exists, even in principle, to select in advance those that will. No D-N explanation of the decay of any such atom can possibly be constructed; however, I-S explanations can be formulated. For example, in a sample containing 1 milligram of C^{14} there are approximately 4×10^{19} atoms. If, in a period of 5730 years, precisely half of them decayed, approximately 2×10^{19} would remain intact. It is *extremely unlikely* that *exactly* half of them would disintegrate in that period, but it is *extremely likely* that *approximately* half would decay. The following argument – which differs from (5) by referring to one particular sample S – would be a strong I-S explanation:

(12) S is a sample of C^{14} that contained one milligram 5730 years ago.

S has not been exposed to external radiation.[19]

The half-life of C^{14} is 5730 years.

$$=======================[r]$$

S now contains one-half milligram ($\pm 1\%$) of C^{14}.

In this example, r differs from 1 by an incredibly tiny margin, but is not literally equal to 1. In a world that is not deterministic, I-S explanations that are not merely incomplete D-N explanations can be formulated.

14 The statistical relevance (S-R) model of explanation

According to the received view, scientific explanations are arguments; each type of explanation in Table 18.1 is some type of argument satisfying certain conditions. For this reason, we can classify the received view as an *inferential conception* of scientific explanation. Because of certain difficulties, associated primarily with I-S explanation, another pattern for statistical explanations of particular occurrences was developed. A fundamental feature of this model of explanation is that it *does not* construe explanations as arguments.

One of the earliest objections to the I-S pattern of explanation – as shown by CE-6 (psychotherapy) and CE-7 (vitamin C and the common cold) – is that statistical relevance

rather than high probability is the crucial relationship in statistical explanations. Statistical relevance involves a relationship between two different probabilities. Consider the psychotherapy example. Bruce Brown is a member of the class of people who have a neurotic symptom of type N. Within that class, regardless of what the person does in the way of treatment or nontreatment, there is a certain probability of relief from the symptom (R). That is the *prior probability* of recovery; let us symbolize it as "$Pr(R/N)$." Then there is a probability of recovery in the class of people with that symptom who undergo psychotherapy (P); it can be symbolized as "$Pr(R/N.P)$." If

$$Pr(R/N.P) > Pr(R/N)$$

then psychotherapy is positively relevant to recovery, and if

$$Pr(R/N.P) < Pr(R/N)$$

then psychotherapy is negatively relevant to recovery. If

$$Pr(R/N.P) = Pr(R/N)$$

then psychotherapy is irrelevant to recovery. Suppose psychotherapy is positively relevant to recovery. If someone then asks why Bruce Brown, who suffered with neurotic symptom N, recovered from his symptom, we can say that it was because he underwent psychotherapy. That is at least an important part of the explanation.

Consider another example. Suppose that Grace Green, an American woman, suffered a serious heart attack. In order to explain why this happened we search for factors that are relevant to serious heart attacks – for example, smoking, high cholesterol level, and body weight. If we find that she was a heavy cigarette smoker, had a serum cholesterol level above 300, and was seriously overweight, we have at least a good part of an explanation, for all of those factors are positively relevant to serious heart attacks. There are, of course, other relevant factors, but these three will do for purposes of illustration.

More formally, if we ask why this member of the class *A* (American women) has characteristic H (serious heart attack), we can take the original reference class *A* and subdivide or *partition* it in terms of such relevant factors as we have mentioned: S (heavy cigarette smokers), C (high cholesterol level), and W (overweight). This will give us a partition with eight cells (where the dot signifies conjunction and the tilde "~" signifies negation):

S.C.W	~S.C.W
S.C.~W	~S.C.~W
S.~C.W	~S.~C.W
S.-C.~W	~S.~C.~W

An S-R explanation of Green's heart attack has three parts:

1. The prior probability of H, namely, $Pr(H/A)$.
2. The posterior probabilities of H with respect to each of the eight cells, $Pr(H/S.C.W)$, $Pr(H/S.C.~W), \ldots, Pr(H/~S.~C.~W)$.
3. The statement that Green is a member of S.C.W.

It is stipulated that the partition of the reference class must be made in terms of all and only the factors relevant to serious heart attacks.

Clearly, an explanation of that sort is not an argument; it has neither premises nor conclusion. It does, of course, consist of an explanans and an explanandum. Items 1–3 constitute the explanans; the explanandum is Green's heart attack. Moreover, no restrictions are placed on the size of the probabilities – they can be high, middling, or low. All that is required is that these

probabilities differ from one another in various ways, because we are centrally concerned with relations of statistical relevance.

Although the S-R pattern of scientific explanation provides some improvements over the I-S model, it suffers from a fundamental inadequacy. It focuses on statistical relevance rather than causal relevance. It may, as a result, tend to foster a confusion of causes and correlations. In the vitamin C example, for instance, we want a controlled experiment to find out whether taking massive doses of vitamin C is *causally relevant* to quick recovery from colds. We attempt to find out whether taking vitamin C is *statistically relevant* to rapid relief because the statistical relevance relation is evidence regarding the presence or absence of *causal relevance*. It is causal relevance that has genuine explanatory import. The same remark applies to other examples as well. In the psychotherapy example we try to find out whether such treatment is statistically relevant to relief from neurotic symptoms in order to tell whether it is causally relevant. In the case of the heart attack, many clinical studies have tried to find statistical relevance relations as a basis for determining what is causally relevant to the occurrence of serious heart attacks.

15 Two grand traditions

We have been looking at the development of the received view, and at some of the criticisms that have been leveled against it. The strongest intuitive appeal of that view comes much more from explanations of laws than from explanations of particular facts. One great example is the *Newtonian synthesis*. Prior to Newton we had a miscellaneous collection of laws including Kepler's three laws of planetary motion and Galileo's laws of falling objects, inertia, projectile motion, and pendulums. By invoking three simple laws of motion and one law of gravitation, Newton was able to explain these laws – and in some cases correct them. In addition, he was able to explain many other regularities, such as the behavior of comets and tides, as well. Later on, the molecular-kinetic theory provided a Newtonian explanation of many laws pertaining to gases. Quite possibly the most important feature of the Newtonian synthesis was the extent to which it systematized our knowledge of the physical world by subsuming all sorts of regularities under a small number of very simple laws. Another excellent historical example is the explanation of light by subsumption under Maxwell's theory of electromagnetic radiation.

The watchword in these beautiful historical examples is *unification*. A large number of specific regularities are unified in one theory with a small number of assumptions or postulates. This theme was elaborated by Michael Friedman (1974) who asserted that our comprehension of the universe is increased as the number of independently acceptable assumptions we require is reduced. I would be inclined to add that this sort of systematic unification of our scientific knowledge provides a comprehensive world picture or worldview. This, I think, represents one major aspect of scientific explanation – it is the notion that we understand what goes on in the world if we can fit it into a comprehensive worldview. As Friedman points out, this is a *global* conception of explanation. The value of explanation lies in fitting things into a universal pattern, or a pattern that covers major segments of the universe.[20]

As we look at many of the criticisms that have been directed against the received view, it becomes clear that causality is a major focus. Scriven offered his ink stain example, CE-5, to support the claim that finding the explanation amounts, in many cases, simply to finding the causes. This is clearly explanation on a very *local* level. All we need to do, according to Scriven, is to get a handle on events in an extremely limited spacetime region that led up, causally, to the stain on the carpet, and we have adequate understanding of that particular fact. In this connection, we should also recall CE-1 and CE-2. In the first of these we sought a local causal explanation for the length of a shadow, and in the second we wanted a causal explanation for a particular storm. Closely related noncausal "explanations" were patently unacceptable. In such cases as the Chernobyl accident and the Challenger space-shuttle explosion we also seek causal explanations, partly in order to try to avoid such tragedies in the future. Scientific explanation has its practical as well as its purely intellectual value.

It often happens, when we try to find causal explanations for various occurrences, that we have to appeal to entities that are not directly observable with the unaided human senses. For example, to understand AIDS (Acquired Immunodeficiency [sic] Syndrome), we must deal with viruses and cells. To understand the transmission of traits from parents to offspring, we become involved with the structure of the DNA molecule. To explain a large range of phenomena associated with the nuclear accident at Three Mile Island, we must deal with atoms and subatomic particles. When we try to construct causal explanations we are attempting to discover the mechanisms – often hidden mechanisms – that bring about the facts we seek to understand. The search for causal explanations, and the associated attempt to expose the hidden workings of nature, represent a second grand tradition regarding scientific explanation. We can refer to it as *the causal-mechanical tradition*.

Having contrasted the two major traditions, we should call attention to an important respect in which they overlap. When the search for hidden mechanisms is successful, the result is often to reveal a small number of basic mechanisms that underlie wide ranges of phenomena. The explanation of diverse phenomena in terms of the same mechanisms constitutes theoretical unification. For instance, the kinetic-molecular theory of gases unified thermodynamic phenomena with Newtonian particle mechanics. The discovery of the double-helical structure of DNA, for another example, produced a major unification of biology and chemistry.

Each of the two grand traditions faces certain fundamental problems. The tradition of explanation as unification – associated with the received view – still faces the problem concerning explanations of laws that was pointed out in 1948 by Hempel and Oppenheim. It was never solved by Hempel in any of his subsequent work on scientific explanation. If the technical details of Friedman's theory of unification were satisfactory, it would provide a solution to that problem. Unfortunately, it appears to encounter serious technical difficulties (see Kitcher 1976 and Salmon 1989).

The causal-mechanical tradition faces a longstanding philosophical difficulty concerning the nature of causality that had been posed by David Hume in the eighteenth century. The problem – stated extremely concisely – is that we seem unable to identify the *connection* between cause and effect, or to find the *secret power* by which the cause brings about the effect. Hume is able to find certain *constant conjunctions* – for instance, between fire and heat – but he is unable to find the connection. He is able to see the spatial contiguity of events we identify as cause and effect, and the temporal priority of the cause to the effect – as in collisions of billiard balls, for instance – but still no *necessary connection*. In the end he locates the connection in the human imagination – in the psychological expectation we feel with regard to the effect when we observe the cause.[21]

Hume's problem regarding causality is one of the most recalcitrant in the whole history of philosophy. Some philosophers of science have tried to provide a more objective and robust concept of causality, but none has enjoyed widespread success. One of the main reasons the received view was reticent about incorporating causal considerations in the analysis of scientific explanation was an acute sense of uneasiness about Hume's problem. One of the weaknesses of the causal view, as it is handled by many philosophers who espouse it, is the absence of any satisfactory theory of causality.[22]

16   The pragmatics of explanation

As we noted in Section 4, the term "explanation" refers sometimes to linguistic entities – that is, collections of statements of facts – and sometimes to nonlinguistic entities – namely, those very facts. When we think in terms of the human activity of explaining something to some person or group of people, we are considering linguistic behavior. Explaining something to someone involves uttering or writing statements. In this section we look at some aspects of this *process of explaining*. In this chapter, up to this point, we have dealt mainly with the *product* resulting from this activity, that is, the explanation that was offered in the process of explaining.

When philosophers discuss language they customarily divide the study into three parts: syntax, semantics, and pragmatics. Syntax is concerned only with relationships among the symbols, without reference to the meanings of the symbols or the people who use them. Roughly speaking, syntax is pure grammar; it deals with the conventions governing combinations and manipulations of symbols. Semantics deals with the relationships between symbols and the things to which the symbols refer. Meaning and truth are the major semantical concepts. Pragmatics deals with the relationships among symbols, what they refer to, and the users of language. Of particular interest for our discussion is the treatment of the context in which language is used.

The 1948 Hempel-Oppenheim essay offered a highly formalized account of D-N explanations of particular facts, and it characterized such explanations in syntactical and semantical terms alone. Pragmatic considerations were not dealt with. Hempel's later characterization of the other types of explanations were given mainly in syntactical and semantical terms, although I-S explanations are, as we noted, relativized to knowledge situations. Knowledge situations are aspects of the human contexts in which explanations are sought and given. Such contexts have other aspects as well.

One way to look at the pragmatic dimensions of explanation is to start with the question by which an explanation is sought. In Section 3 we touched briefly on this matter. We noted that many, if not all, explanations can properly be requested by *explanation-seeking why-questions*. In many cases, the first pragmatic step is to clarify the question being asked; often the sentence uttered by the questioner depends upon contextual clues for its interpretation. As Bas van Fraassen, one of the most important contributors to the study of the pragmatics of explanation, has shown, the emphasis with which a speaker poses a question may play a crucial role in determining just what question is being asked. He goes to the Biblical story of the Garden of Eden to illustrate. Consider the following three questions:

(i)   Why did Adam eat *the apple?*
(ii)   Why did *Adam* eat the apple?
(iii) Why did Adam *eat* the apple?

Although the words are the same – and in the same order – in each, they pose three very different questions. This can be shown by considering what van Fraassen calls the *contrast class.* Sentence (i) asks why Adam ate the apple instead of a pear, a banana, or a pomegranate. Sentence (ii) asks why Adam, instead of Eve, the serpent, or a goat, ate the apple. Sentence (iii) asks why Adam ate the apple instead of throwing it away, feeding it to a goat, or hiding it somewhere. Unless we become clear on the question being asked, we can hardly expect to furnish appropriate answers.

Another pragmatic feature of explanation concerns the knowledge and intellectual ability of the person or group requesting the explanation. On the one hand, there is usually no point in including in an explanation matters that are obvious to all concerned. Returning to (3) – our prime example of a D-N explanation of a particular fact – one person requesting an explanation of the sudden dramatic increase in the skater's rate of rotation might have been well aware of the fact that she drew her arms in close to her body, but unfamiliar with the law of conservation of angular momentum. For this questioner, knowledge of the law of conservation of angular momentum is required in order to understand the explanandum-fact. Another person might have been fully aware of the law of conservation of angular momentum, but failed to notice what the skater did with her arms. This person needs to be informed of the skater's arm maneuver. Still another person might have noticed the arm maneuver, and might also be aware of the law of conservation of angular momentum, but failed to notice that this law applies to the skater's movement. This person needs to be shown how to apply the law in the case in question.

On the other hand, there is no point in including material in an explanation that is beyond the listeners' ability to comprehend. To most schoolchildren, for example, an explanation of the darkness of the night sky that made reference to the non-Euclidean structure of space or the mean free path of a photon would be inappropriate. Many of the explanations we encounter in real-life situations are incomplete on account of the explainer's view of the background knowledge of the audience.

A further pragmatic consideration concerns the interests of the audience. A scientist giving an explanation of a serious accident to a congressional investigating committee may tell the members of Congress far more than they want to know about the scientific details. In learning why an airplane crashed, the committee might be very interested to find that it was because of an accumulation of ice on the wing, but totally bored by the scientific reason why ice-accumulations cause airplanes to crash.

Peter Railton (1981) has offered a distinction that helps considerably in understanding the role of pragmatics in scientific explanation. First, he introduces the notion of an *ideal explanatory text*. An ideal explanatory text contains *all* of the facts and *all* of the laws that are relevant to the explanandum-fact. It details *all* of the causal connections among those facts and *all* of the hidden mechanisms. In most cases the ideal explanatory text is huge and complex. Consider, for example, an explanation of an automobile accident. The *full* details of such items as the behavior of both drivers, the operations of both autos, the condition of the highway surface, the dirt on the windshields, and the weather, would be unbelievably complicated. That does not really matter, for the ideal explanatory text is seldom, if ever, spelled out fully. What is important is to have the ability to illuminate portions of the ideal text as they are wanted or needed. When we do provide knowledge to fill in some aspect of the ideal text we are furnishing *explanatory information*.

A request for a scientific explanation of a given fact is almost always — if not literally always — a request, not for the ideal explanatory text, but for explanatory information. The ideal text contains all of the facts and laws pertaining to the explanandum-fact. These are the completely objective and nonpragmatic aspects of the explanation. If explanatory information is to count as legitimate it must correspond to the objective features of the ideal text. The ideal text determines what is *relevant* to the explanandum-fact. Since, however, we cannot provide the whole ideal text, nor do we want to, a selection of information to be supplied must be made. This depends on the knowledge and interests of those requesting and those furnishing explanations. The information that satisfies the request in terms of the interests and knowledge of the audience is *salient* information. The pragmatics of explanation determines salience — that is, what aspects of the ideal explanatory text are appropriate for an explanation in a particular context.

17 Conclusion

Several years ago, a friend and colleague — whom I will call *the friendly physicist* — was sitting on a jet airplane awaiting takeoff. Directly across the aisle was a young boy holding a helium-filled balloon by a string. In an effort to pique the child's curiosity, the friendly physicist asked him what he thought the balloon would do when the plane accelerated for takeoff. After a moment's thought the boy said that it would move toward the back of the plane. The friendly physicist replied that *he* thought it would move toward the front of the cabin. Several adults in the vicinity became interested in the conversation, and they insisted that the friendly physicist was wrong. A flight attendant offered to wager a miniature bottle of Scotch that he was mistaken — a bet that he was quite willing to accept. Soon thereafter the plane accelerated, the balloon moved forward, and the friendly physicist enjoyed a free drink.[23]

Why did the balloon move toward the front of the cabin? Two explanations can be offered, both of which are correct. First, one can tell a story about the behavior of the molecules that made up the air in the cabin, explaining how the rear wall collided with nearby molecules when it began its forward motion, thus creating a pressure gradient from the back to the front of the cabin. This pressure gradient imposed an unbalanced force on the back side of the balloon, causing it to move forward with respect to the walls of the cabin.[24] Second, one can cite an extremely general physical principle – Einstein's *principle of equivalence* – according to which an acceleration is physically equivalent, from the standpoint of the occupants of the cabin, to a gravitational field. Since helium-filled balloons tend to rise in the atmosphere in the earth's gravitational field, they will move forward when the airplane accelerates, reacting just as they would if a massive object were suddenly placed behind the rear wall.

The first of these explanations is causal-mechanical. It appeals to unobservable entities, describing the causal processes and causal inter-actions involved in the explanandum phenom-enon. When we are made aware of these explanatory facts we understand how the phenomenon came about. This is the kind of explanation that advocates of the causal-mechanical tradition find congenial. The second explanation illustrates the unification approach. By appealing to an extremely general physical principle, it shows how this odd little occur-rence fits into the universal scheme of things. It does not refer to the detailed mechanisms. This explanation provides a different kind of under-standing of the same fact.

Which of these explanations is correct? Both are. Both of them are embedded in the ideal explanatory text. Each of them furnishes valu-able explanatory information. It would be a serious error to suppose that any phenomenon has only one explanation. It is a mistake, I believe, to ask for *the* explanation of any occur-rence. Each of these explanations confers a kind of scientific understanding. Pragmatic considera-tions might dictate the choice of one rather than the other in a given context. For example, the explanation in terms of the equivalence

principle would be unsuitable for a ten-year-old child. The same explanation might be just right in an undergraduate physics course. But both are bona fide explanations.

As we noted in Section 10, the 1948 Hempel-Oppenheim essay attracted almost no attention for about a decade after its publication. Around 1959 it became the focus of intense controversy, much of it stemming from those who saw causality as central to scientific explanation. The subsequent thirty years have seen a strong oppo-sition between the advocates of the received view and the proponents of causal explanation. Each of the two major approaches has evolved considerably during this period – indeed, they have developed to the point that they can peace-fully coexist as two distinct aspects of scientific explanation. Scientific understanding is, after all, a complicated affair; we should not be surprised to learn that it has many different aspects. Exposing underlying mechanisms and fitting phenomena into comprehensive pictures of the world seem to constitute two important aspects. Moreover, as remarked above, we should remember that these two types of understanding frequently overlap. When we find that the same mechanisms underlie diverse types of natural phenomena this *ipso facto* constitutes a theoretical unification.

On one basic thesis there is nearly complete consensus. Recall that in the early decades of the twentieth century many scientists and philoso-phers denied that there can be any such thing as scientific explanation. Explanation is to be found, according to this view, only in the realms of theology and metaphysics. At present it seems virtually unanimously agreed that, however it may be explicated, there is such a thing as scien-tific explanation. Science *can* provide deep understanding of our world. We do *not* need to appeal to supernatural agencies to achieve understanding. Equally importantly, we can contrast the objectively based explanations of contemporary science with the pseudounder-standing offered by such flagrantly unscientific approaches as astrology, creation science, and scientology. These are points worth remembering in an age of rampant pseudoscience.

Notes

1 Confirmation will be treated in Chapter 2 of this book. [sic]

2 The Doppler effect is the lengthening of waves emitted by a source traveling away from a receiver and the shortening of waves emitted by a source approaching a receiver. This effect occurs in both light and sound, and it can be noticed in the change of pitch of a whistle of a passing train.

3 Because of certain logical technicalities, there are valid deductive arguments that have no premises at all, but arguments of this sort will not be involved in our discussion.

4 The familiar slogan, "Garbage in, garbage out," does not accurately characterize deductive arguments.

5 As James Lennox points out in Chapter 7 [sic], teleological explanations are anthropomorphic only if they appeal to human-like purposes. In evolutionary biology – and other scientific domains as well – there are teleological explanations that are not anthropomorphic.

6 In this example we may ignore the friction of the skate on the ice, and the friction of the skater's body in the surrounding air.

7 Hempel and Oppenheim provide, in addition to these conditions of adequacy, a precise technical definition of "explanation." In this book we will not deal with these technicalities.

8 If the occurrence of the kilogram in (vi) seems to make reference to a particular object – the international prototype kilogram kept at the international bureau of standards – the problem can easily be circumvented by defining mass in terms of atomic mass units.

9 The counterexample was devised by Sylvain Bromberger, but to the best of my knowledge he never finished it.

10 This qualification is required to assure that the disintegration is spontaneous and not induced by external radiation.

11 As James Lennox remarks in Chapter 7 [sic] on philosophy of biology, Darwin's principle of natural selection is an example of a statistical law.

12 This thesis was advanced for D-N explanation in Hempel-Oppenheim (1948, 249), and reiterated, with some qualifications, for D-N and I-S explanations in Hempel (1965a, Sections 2.4, 3.5).

13 Nor would (6) qualify as an acceptable I-S explanation if we had found that Jones's infection was of the non-penicillin-resistant variety, for the probability of quick recovery among people with that type of infection is different from the probability of quick recovery among those who have an unspecified type of streptococcus infection.

14 D-N explanations of particular facts automatically satisfy this requirement. If all A are B, the probability that an A is a B is one. Under those circumstances, the probability that an A which is also a C is a B is also one. Therefore, no partition of A is relevant to B.

15 Around the time Hempel was working out his theory of I-S explanation, Linus Pauling's claims about the value of massive doses of vitamin C in the prevention of common colds was receiving a great deal of attention. Although Pauling made no claims about the ability of vitamin C to cure colds, it occurred to me that a fictitious example of this sort could be concocted.

16 In a controlled experiment there are two groups of subjects, the experimental group and the control group. These groups should be as similar to one another as possible. The members of the experimental group receive the substance being tested, vitamin C. The members of the control group receive a placebo, that is, an inert substance such as a sugar pill that is known to have no effect on the common cold. In a blind experiment the subjects do not know whether they are receiving vitamin C or the placebo. This is important, for if the subjects knew which treatment they were receiving, the power of suggestion might skew the results. An experiment is double-blind if neither the person who hands out the pills nor the subjects

know which subject is getting which type of pill. If the experiment is not double-blind, the person administering the pills might, in spite of every effort not to, convey some hint to the subject.

17 It should be carefully noted that I am claiming *neither* that psychotherapy is irrelevant to remission of neurotic symptoms *nor* that vitamin C is irrelevant to rate of recovery from colds. I *am* saying that that is the point at issue so far as I-S explanation is concerned.

18 Determinism is discussed in detail in Chapter 6 [*sic*].

19 This qualification is required to assure that the disintegrations have been spontaneous and not induced by external radiation.

20 The unification approach has been dramatically extended and improved by Philip Kitcher (1976, 1981, and 1989).

21 Hume's analysis of causation is discussed in greater detail in Chapter 2, Part II [*sic*].

22 I have tried to make some progress in this direction in Salmon (1984, Chapters 5–7).

23 This little story was previously published in Salmon (1980). I did not offer an explanation of the phenomenon in that article.

24 Objects that are denser than air do not move toward the front of the cabin because the pressure difference is insufficient to overcome their inertia.

References

Braithwaite, R.B. (1953), *Scientific Explanation: A Study of the Function of Theory, Probability and Law in Science*. Cambridge, England: Cambridge University Press.

Friedman, Michael (1974), "Explanation and Scientific Understanding", *The Journal of Philosophy* 71: 5–19.

Frohlich, Cliff (1980), "The Physics of Somersaulting and Twisting," *Scientific American* 242 (3): 154–164.

Hempel, Carl G. (1962), "Deductive-Nomological vs. Statistical Explanation," in Feigl, H. and Maxwell, G. (eds.), *Minnesota Studies in the Philosophy of Science* Volume 3. *Scientific Explanation, Space and Time*. Minneapolis: University of Minnesota Press, pp. 98–169.

Hempel, Carl G. (1965a), "Aspects of Scientific Explanation," in Hempel (1965b), pp. 331–496.

Hempel, Carl G. (1965b), *Aspects of Scientific Explanation and Other Essays in the Philosophy of Science*. New York: Free Press.

Hempel, Carl G. and Oppenheim, Paul (1948), "Studies in the Logic of Explanation," *Philosophy of Science* 15: 135–175. Reprinted in Hempel (1965b), pp. 245–290, with a 1964 Postscript added.

Kitcher, Philip (1976), "Explanation, Conjunction, and Unification," *The Journal of Philosophy* 73: 207–212.

Kitcher, Philip (1981), "Explanatory Unification," *Philosophy of Science* 48: 507–531.

Kitcher, Philip (1989), "Explanatory Unification and the Causal Structure of the World," in Kitcher, P. and Salmon, W.C. (eds.) (1989), *Minnesota Studies in the Philosophy of Science*. Volume 13, *Scientific Explanation*. Minneapolis: University of Minnesota Press, pp. 410–505.

Nagel, Ernest (1949), "The Meaning of Reduction in the Natural Sciences," in Stauffer, R.C. (ed.) *Science and Civilisation*. Madison, WI: University of Wisconsin Press, pp. 97–135.

Popper, Karl R. ([1935] 1959), *The Logic of Scientific Discovery*. New York: Basic Books.

Railton, Peter (1981), "Probability, Explanation, and Information," *Synthese* 48: 233–256.

Salmon, Wesley C. (1980), *Space, Time, and Motion: A Philosophical Introduction*. 2d ed. Minneapolis: University of Minnesota Press.

Salmon, Wesley C. (1984), *Scientific Explanation and the Causal Structure of the World*. Princeton: Princeton University Press.

Salmon, Wesley C. (1989), "Four Decades of Scientific Explanation," in Kitcher, P. and Salmon, W.C. (eds.) (1989), *Minnesota Studies in the Philosophy of Science*. Volume 13, *Scientific Explanation*. Minneapolis: University of Minnesota Press, pp. 3–219.

Chapter 19

Bas van Fraassen

THE PRAGMATICS OF EXPLANATION[1]

If cause were non-existent everything would have been produced by everything and at random. Horses, for instance, might be born, perchance, of flies, and elephants of ants; and there would have been severe rains and snow in Egyptian Thebes, while the southern districts would have had no rain, unless there had been a cause which makes the southern parts stormy, the eastern dry.

Sextus Empiricus, *Outlines of Pyrrhonism* III, V, 1

A theory is said to have explanatory power if it allows us to explain; and this is a virtue. It is a pragmatic virtue, albeit a complex one that includes other virtues as its own preconditions. After some preliminaries in Section 1, I shall give a frankly selective history of philosophical attempts to explain explanation. Then I shall offer a model of this aspect of scientific activity in terms of why-questions, their presuppositions, and their context-dependence. This will account for the puzzling features (especially asymmetries and rejections) that have been found in the phenomenon of explanation, while remaining compatible with empiricism.

1 The language of explanation

One view of scientific explanation is encapsulated in this argument: science aims to find

explanations, but nothing is an explanation unless it is true (explanation requires true premisses); so science aims to find true theories about what the world is like. Hence scientific realism is correct. Attention to other uses of the term 'explanation' will show that this argument trades on an ambiguity.

1.1 *Truth and grammar*

It is necessary first of all to distinguish between the locutions 'we have an explanation' and 'this theory explains'. The former can be paraphrased 'we have a theory that explains' – but then 'have' needs to be understood in a special way. It does not mean, in this case, 'have on the books', or 'have formulated', but carries the conversational implicature that the theory tacitly referred to is acceptable. That is, you are not warranted in saying 'I have an explanation' unless you are warranted in the assertion 'I have a theory *which is acceptable* and which explains'. The important point is that the mere statement 'theory T explains fact E' does not carry any such implication: not that the theory is true, not that it is empirically adequate, and not that it is acceptable.

There are many examples, taken from actual usage, which show that truth is not presupposed by the assertion that a theory explains something. Lavoisier said of the phlogiston hypothesis that it is too vague and consequently 's'adapte à toutes les explications dans lesquelles on veut le faire entrer'.[2] Darwin explicitly allows explanations by false theories when he says 'It can hardly be supposed that a false theory would explain, in so satisfactory a manner as does the theory of natural selection, the several large classes of facts above specified.'[3] Gilbert Harman, we recall, has argued similarly: that a theory explains certain phenomena is part of the evidence that leads us to accept it. But that means that the explanation-relation is visible before we believe that the theory is true. Finally, we criticize theories selectively: a discussion of celestial mechanics around the turn of the century could surely contain the assertion that Newton's theory does explain many planetary phenomena. Yet it was also agreed that the advance in the perihelion of Mercury seems to be inconsistent with the theory, suggesting therefore that the theory is not empirically adequate – and hence, is false – without this agreement undermining the previous assertion. Examples can be multiplied: Newton's theory explained the tides, Huygens's theory explained the diffraction of light, Rutherford's theory of the atom explained the scattering of alpha particles, Bohr's theory explained the hydrogen spectrum, Lorentz's theory explained clock retardation. We are quite willing to say all this, although we will add that, for each of these theories, phenomena were discovered which they could not only not explain, but could not even accommodate in the minimal fashion required for empirical adequacy.

Hence, to say that a theory explains some fact or other, is to assert a relationship between this theory and that fact, which is independent of the question whether the real world, as a whole, fits that theory.

Let us relieve the tedium of terminological discussion for a moment and return to the argument displayed at the beginning. In view of the distinctions shown, we can try to revise it as follows: science tries to place us in a position in which we have explanations, and are warranted in saying that we do have. But to have such warrant, we must first be able to assert with equal warrant that the theories we use to provide premises in our explanations are true. Hence science tries to place us in a position where we have theories which we are entitled to believe to be true.

The conclusion may be harmless of course if 'entitled' means here only that one can't be convicted of irrationality on the basis of such a belief. That is compatible with the idea that we have warrant to believe a theory only because, and in so far as, we have warrant to believe that it is empirically adequate. In that case it is left open that one is at least as rational in believing merely that the theory is empirically adequate.

But even if the conclusion were construed in this harmless way, the second premise will have to be disputed, for it entails that someone who merely accepts the theory as empirically adequate, is not in a position to explain. In this second premiss, the conviction is perhaps expressed that having an explanation is not to be equated with having an acceptable theory that explains, but with having a true theory that explains.

That conviction runs afoul of the examples I gave. I say that Newton could explain the tides, that he had an explanation of the tides, that he did explain the tides. In the same breath I can add that this theory is, after all, not correct. Hence I would be inconsistent if by the former I meant that Newton had a true theory which explained the tides – for if it was true then, it is true now. If what I meant was that it was true *then* to say that Newton had an acceptable theory which explains the tides, that would be correct.

A realist can of course give his own version: to have an explanation means to have 'on the books' a theory which explains and which one is entitled to believe to be true. If he does so, he will agree that to have an explanation does not require a true theory, while maintaining his contention that science aims to place us in a position to give *true* explanations. That would bring us back, I suppose, to our initial disagreement, the detour through explanation having brought no benefits. If you can only be entitled to assert that the theory is true because, and in so far as, you are entitled to assert that it is empirically adequate, then the distinction drawn makes no practical difference. There would of course be a difference between *believe* (to-be-true) and *accept* (believe-to-be-empirically-adequate) but no real difference between be-entitled-to-believe and be-entitled-to-accept. A realist might well dispute this by saying that if the theory explains facts then that gives you an *extra* good reason (over and above any evidence that it is empirically adequate) to believe that the theory is true. But I shall argue that this is quite impossible, since explanation is not a special additional feature that can give you good reasons for belief in addition to evidence that the theory fits the observable phenomena. For 'what more there is to' explanation is something quite pragmatic, related to the concerns of the user of the theory and not something new about the correspondence between theory and fact.

So I conclude that (a) the assertion that theory T explains, or provides an explanation for, fact E does not presuppose or imply that T is true or even empirically adequate, and (b) the assertion that we have an explanation is most simply construed as meaning that we have 'on the books' an acceptable theory which explains. I shall henceforth adopt this construal.

To round off the discussion of the terminology, let us clarify what sorts of terms can be the grammatical subjects, or grammatical objects, of the verb 'to explain'. Usage is not regimented: when we say 'There is the

explanation!', we may be pointing to a fact, or to a theory, or to a thing. In addition, it is often possible to point to more than one thing which can be called 'the explanation'. And, finally, whereas one person may say that Newton's theory of gravitation explained the tides, another may say that Newton used that theory to explain the tides. (I suppose no one would say that the hammer drove the nail through the wood; only that the carpenter did so, using the hammer. But today people do sometimes say that the computer calculated the value of a function, or solved the equations, which is perhaps similar to saying that the theory explained the tides.)

This bewildering variety of modes of speech is common to scientists as well as philosophers and laymen. In Huygens and Young the typical phrasing seemed to be that phenomena may be explained *by means of* principles, laws, and hypotheses, or *according to* a view.[4] On the other hand, Fresnel writes to Arago in 1815 'tous ces phénomènes . . . sont réunis et expliqués par la même théorie des vibrations', and Lavoisier says that the oxygen hypothesis he proposes *explains* the phenomena of combustion.[5] Darwin also speaks in the latter idiom: 'In scientific investigations it is permitted to invent any hypothesis, and if it explains various large and independent classes of facts it rises to the rank of a well-grounded theory'; though elsewhere he says that the facts of geographical distribution are *explicable* on the theory of migration.[6]

In other cases yet, the theory assumed is left tacit, and we just say that one fact explains another. For example: the fact that water is a chemical compound of oxygen and hydrogen explains why oxygen and hydrogen appear when an electric current is passed through (impure) water.

To put some order into this terminology, and in keeping with previous conclusions, we can regiment the language as follows. The word 'explain' can have its basic role in expressions of form 'fact E explains fact F relative to theory T'. The other expressions can then be parsed as: 'T explains F' is equivalent to: 'there are facts which explain F relative to T'; 'T was used to explain F' equivalent to 'it was shown that there are facts which explain F relative to T'; and so forth. Instead of 'relative to T' we can sometimes also say 'in T'; for example, 'the gravitational pull of the moon explains the ebb and flow of the tides in Newton's theory'.

After this, my concern will no longer be with the derivative type of assertion that we *have* an explanation. After this point, the topic of concern will be that basic relation of explanation, which may be said to hold between facts relative to a theory, quite independently of whether the theory is true or false, believed, accepted, or totally rejected.

1.2 Some examples

Philosophical discussion is typically anchored to its subject through just a few traditional examples. The moment you see 'Pegasus', 'the king of France', or 'the good Samaritan', in a philosophical paper, you know exactly to what problem area it belongs. In the philosophical discussion of explanation, we also return constantly to a few main examples: paresis, the red shift, the flagpole. To combat the increasing sense of unreality this brings, it may be as well to rehearse, briefly, some workaday examples of scientific explanation.

(1) Two kilograms of copper at 60 °C are placed in three kilograms of water at 20 °C. After a while, water and copper reach the same temperature, namely 22.5 °C, and then cool down together to the temperature of the surrounding atmosphere.

There are a number of facts here for which we may request an explanation. Let us just ask why the equilibrium temperature reached is 22.5 °C.

Well, the specific heats of water and copper are 1 and 0.1, respectively. Hence if the final temperature is T, the copper loses $0.1 \times 2 \times (60 - T)$ units of heat and the water gains $1 \times 3 \times (T - 20)$. At this point we appeal to the principle of the Conservation of Energy, and conclude that the total amount of heat neither increased nor diminished. Hence,

$$0.1 \times 2 \times (60 - T) = 1 \times 3 \times (T - 20)$$

from which T = 22.5 can easily be deduced.

(2) A short circuit in a power station results in a momentary current of 10^6 amps. A conductor, horizontally placed, 2 metres in length and 0.5 kg in mass, is warped at that time.

Let us ask why the conductor was warped. Well, the earth's magnetic field at this point is not negligible; its vertical component is approximately $5/10^5$ tesla. The theory of electromagnetism allows us to calculate the force exerted on the conductor at the time in question:

$$(5/10^5) \times 2 \times 10^6 = 100 \text{ newtons}$$

which is directed at right angles to the conductor in the horizontal plane. The second law of Newton's mechanics entails in turn that, at that moment, the conductor has an acceleration of

$$100 \div 0.5 = 200 \text{ m/sec}^2$$

which is approximately twenty times the downward acceleration attributable to gravity (9.8 m/sec^2) – which allows us to compare in concrete terms the effect of the short circuit on the fixed conductor, and the normal effect of its weight.

(3) In a purely numerological way, Balmer, Lyman, and Paschen constructed formulae fitting frequency series to be found in the hydrogen spectrum, of the general form:

$$f^n_m = R\left(\frac{1}{m^2} - \frac{1}{n^2}\right)$$

where Balmer's law had $m = 2$, Lyman's had $m = 1$, and Paschen's $m = 3$; both m and n range over natural numbers.

Bohr's theory of the atom explains this general form. In this theory, the electron in a hydrogen atom moves in a stable orbit, characterized by an angular momentum which is an integral multiple of $h/2\pi$. The associated energy levels take the form

$$E_n = -E_0/n^2$$

where E_0 is called the ground state energy.

When the atom is excited (as when the sample is heated), the electron jumps into a higher energy state. It then spontaneously drops down again, emitting a photon with energy equal to the energy lost by that electron in its drop. So if the drop is from level E_n to level E_m, the photon's energy is

$$E = E_n - E_m = (-E_0/n^2) - (-E_0/m^2)$$

$$= E_0/m^2 - E_0/n^2$$

The frequency is related to the energy by the equation

$$E = hf$$

so the frequencies exhibited by the emitted photons are

$$f^n_m = \frac{E}{h} = \frac{E_0}{h}\left(\frac{1}{m^2} - \frac{1}{n^2}\right)$$

which is exactly of the general form found above, with E_0/h being the constant R.

The reader may increase this stock of examples by consulting elementary texts and the *Science*

Digest. It should be clear at any rate that scientific theories are used in explanation, and that how well a theory is to be regarded, depends at least in part on how much it can be used to explain.

2   A biased history

Current discussion of explanation draws on three decades of debate, which began with Hempel and Oppenheim's 'Studies in the Logic of Explanation' (1948).[7] The literature is now voluminous, so that a retrospective must of necessity be biased. I shall bias my account in such a way that it illustrates my diagnoses of the difficulties and points suggestively to the solution I shall offer below.

2.1   *Hempel: grounds for belief*

Hempel has probably written more papers about scientific explanation than anyone; but because they are well known I shall focus on the short summary which he gave of his views in 1966.[8] There he lists two criteria for what is an explanation:

> *explanatory relevance:* 'the explanatory information adduced affords good grounds for believing that the phenomenon did, or does, indeed occur.'

> *testability:* the statements constituting a scientific explanation must be capable of empirical test.'

In each explanation, the information adduced has two components, one ('the laws') information supplied by a theory, and the other ('the initial or boundary conditions') being auxiliary factual information. The relationship of providing good grounds is explicated separately for statistical and non-statistical theories. In the latter, the information implies the fact that is explained; in the former, it *bestows high probability* on that fact.

As Hempel himself points out, the first criterion does not provide either sufficient or necessary conditions for explanation. This was established through a series of examples given by various writers (but especially Michael Scriven and Sylvain Bromberger) and which have passed into the philosophical folklore.

First, giving good grounds for belief does not always amount to explanation. This is most strikingly apparent in examples of the asymmetry of explanation. In such cases, two propositions are strictly equivalent (relative to the accepted background theory), and the one can be adduced to explain why the other is the case, but not conversely. Aristotle already gave examples of this sort (*Posterior Analytics*, Book 1, Chapter 13). Hempel mentions the phenomenon of the *red shift*: relative to accepted physics, the galaxies are receding from us if and only if the light received from them exhibits a shift toward the red end of the spectrum. While the receding of the galaxies can be cited as the reason for the red shift, it hardly makes sense to say that the red shift is the reason for their motion. A more simple-minded example is provided by the *barometer*, if we accept the simplified hypothesis that it falls exactly when a storm is coming, yet does not explain (but rather, is explained by) the fact that a storm is coming. In both examples, good grounds of belief are provided by either proposition for the other. The flagpole is perhaps the most famous asymmetry. Suppose that a flagpole 100 feet high, casts a shadow 75 feet long. We can explain the length of the shadow by noting the angle of elevation of the sun, and appealing to the accepted theory that light travels in straight lines. For given that angle, and the height of the pole, trigonometry enables us to deduce the length of the base of the right-angled triangle formed by pole, light ray, and shadow. However, we can similarly deduce the length of the pole from the length of the shadow plus the angle of elevation. Yet if someone asks us why the pole is 100 feet high, we cannot explain that fact by saying 'because it has a shadow 75 feet long'. The most we could explain that way is how we *came to know*, or how he might himself verify the claim, that the pole is indeed so high.

Second, not every explanation is a case in which good grounds for belief are given. The famous example for this is *paresis*: no one contracts this dreadful illness unless he had latent, untreated syphilis. If someone asked the doctor to explain to him why he came down with this disease, the doctor would surely say: 'because you had latent syphilis which was left untreated'. But only a low percentage of such cases are followed by paresis. Hence if one knew of someone that he might have syphilis, it would be reasonable to warn him that, if left untreated, he might contract paresis – but not reasonable to expect him to get it. Certainly we do not have here the high probability demanded by Hempel.

It might be replied that the doctor has only a partial explanation, that there are further factors which medical science will eventually discover. This reply is based on faith that the world is, for macroscopic phenomena at least, deterministic or nearly so. But the same point can be made with examples in which we do not believe that there is further information to be had, even in principle. The half-life of uranium U^{238} is $(4.5) . 10^9$ years. Hence the probability that a given small enough sample of uranium will emit radiation in a specified small interval of time is low. Suppose, however, that it does. We still say that atomic physics explains this, the explanation being that this material was uranium, which has a certain atomic structure, and hence is subject to spontaneous decay. Indeed, atomic physics has many more examples of events of very low probability, which are explained in terms of the structure of the atoms involved. Although there are physicists and philosophers who argue that the theory must therefore be incomplete (one of them being Einstein, who said 'God does not play with dice') the prevalent view is that it is a contingent matter whether the world is ultimately deterministic or not.

In addition to the above, Wesley Salmon raised the vexing problem of *relevance* which is mentioned in the title of the first criterion, but does not enter into its explication. Two examples which meet the requirements of providing good grounds are:

> John Jones was almost certain to recover from his cold because he took vitamin C, and almost all colds clear up within a week of taking vitamin C.

> John Jones avoided becoming pregnant during the past year, for he has taken his wife's birth control pills regularly, and every man who takes birth control pills avoids pregnancy.[9]

Salmon assumed here that almost all colds spontaneously clear up within a week. There is then something seriously wrong with these 'explanations', since the information adduced is wholly or partly irrelevant. So the criterion would have to be amended at least to read: 'provides good and *relevant* grounds'. This raises the problem of explicating relevance, also not an easy matter.

The second criterion, of testability, is met by all scientific theories, and by all the auxiliary information adduced in the above examples, so it cannot help to ameliorate these difficulties.

2.2 Salmon: *statistically relevant factors*

A number of writers adduced independent evidence for the conclusion that Hempel's criterion is too strong. Of these I shall cite three. The first is Morton Beckner, in his discussion of evolution. This is not a deterministic theory, and often explains a phenomenon only by showing how it could have happened – and indeed, might well have happened in the presence of certain describable, believable conditions consistent with the theory.

> Selectionists have devoted a great deal of effort to the construction of models that are aimed at demonstrating that some observed or suspected phenomena are possible, that is, that they are compatible with the established or confirmed biological hypotheses . . . These models all state strongly that if conditions were (or are) so and so, then, the laws of genetics being what they are, the phenomena in question must occur.[10]

Thus evolution theory explains, for example, the giraffe's long neck, although there was no independent knowledge of food shortages of the requisite sort. Evolutionists give such explanations by constructing models of processes which utilize only genetic and natural selection mechanisms, in which the outcome agrees with the actual phenomena.

In a similar vein, Putnam argued that Newton's explanations were *not* deductions of the facts that had to be explained, but rather demonstrations of compatibility. What was demonstrated was that celestial motions could be as they were, given the theory and certain possible mass distributions in the universe.[11]

The distinction does not look too telling as long as we have to do with a deterministic theory. For in that case, the phenomena E are *compatible* with theory T if and only if there are possible preceding conditions C such that C plus T imply E. In any case, deduction and merely logical consistency cannot be what is at issue, since to show that T is logically compatible with E it would suffice to show that T is irrelevant to (has nothing to say about) E – surely not sufficient for explanation.

What Beckner and Putnam are pointing to are demonstrations that tend to establish (or at least remove objections to) claims of empirical adequacy. It is shown that the development of the giraffe's neck, or the fly-whisk tail fits a model of evolutionary theory; that the observed celestial motions fit a model of Newton's celestial mechanics. But a claim of empirical adequacy does not amount to a claim of explanation — there must be more to it.

Wesley Salmon introduced the theory that an explanation is not an argument, but an assembly of statistically relevant factors. A fact *A* is statistically relevant to a phenomenon E exactly if the probability of E *given A* is different from the probability of E *simpliciter*:

$$P(E/A) \neq P(E)$$

Hempel's criterion required $P(E/A)$ to be high (at least greater than ½). Salmon does not require this, and he does not even require that the information *A* increases the probability of E. That Hempel's requirement was too strong, is shown by the paresis example (which fits Salmon's account very well), and that $P(E/A)$ should not be required to be higher than $P(E)$ Salmon argues independently.

He gives the example of an equal mixture of uranium-238 atoms and polonium-214 atoms, which makes the Geiger counter click in interval $(t, t + m)$. This means that one of the atoms disintegrated. Why did it? The correct answer will be: because it was a uranium-238 atom; if that is so — although the probability of its disintegration is much higher relative to the previous knowledge that the atom belonged to the described mixture.[12] The problem with this argument is that, on Salmon's criterion, we can explain not only why there was a disintegration, but also why the disintegration occurred, let us say, exactly half-way between t and t + m. For the information is statistically relevant to that occurrence. Yet would we

not say that this is the sort of fact that atomic physics leaves unexplained?

The idea behind this objection is that the information is statistically relevant to the occurrence at t + (m/2), but does not favour that as against various other times in the interval. Hence, if E = (a disintegration occurred) and E_x = (a disintegration occurred at time x), then Salmon bids us compare $P(E_x)$ with $P(E_x/A)$, whereas we naturally compare *also* $P(E_x/A)$ with $P(E_y/A)$ for other times y. This suggests that mere statistical relevance is not enough.

Nancy Cartwright has provided several examples to show that Salmon's criterion of statistical relevance also does not provide necessary or sufficient conditions for explanation.[13] As to sufficiency, suppose I spray poison ivy with defoliant which is 90 per cent effective. Then the question 'Why is *this* poison ivy now dead?' may correctly be answered 'Because it was sprayed with the defoliant.' About 10 per cent of the plants are still alive, however, and for those it is true that the probability that they are still alive was not the same as the probability that they are still alive *given* that they were sprayed. Yet the question 'Why is *that* plant now alive?' cannot be correctly answered 'Because it was sprayed with defoliant.'

Nor is the condition necessary. Suppose, as a medical fiction, that paresis can result from either syphilis or epilepsy, and from nothing else, and that the probability of paresis given either syphilis or epilepsy equals 0.1. Suppose in addition that Jones is known to belong to a family of which every member has either syphilis or epilepsy (but, fortunately, not both), and that he has paresis. Why did *he* develop this illness? Surely the best answer *either* is 'Because he had syphilis' *or* is 'Because he had epilepsy', depending on which of these is true. Yet, with all the other information we have, the probability that Jones would get paresis is already established as 0.1, and this probability is not changed if we are told in addition, say, that he has a history of syphilis. The example is rather similar to that of the uranium and polonium atoms, except that the probabilities are equal – and we still want to say that in this case, the explanation of the paresis is the fact of syphilis.

Let me add a more general criticism. It would seem that if either Hempel's or Salmon's approach was correct, then there would not really be more to explanatory power than empirical adequacy and empirical strength. That is, on these views, explaining an observed event is indistinguishable from showing that this event's occurrence does not constitute an objection to the claim of empirical adequacy for one's theory, and in addition, providing significant information entailed by the theory and relevant to that event's occurrence. And it seems that Salmon, at that point, was of the opinion that there really cannot be more to explanation:

> When an explanation . . . has been provided, we know exactly how to regard any A with respect to the property B . . . We know all the regularities (universal or statistical) that are relevant to our original question. What more could one ask of an explanation?[14]

But in response to the objections and difficulties raised, Salmon, and others, developed new theories of explanation according to which there is more to explanatory power. I shall examine Salmon's later theory below.

2.3 Global properties of theories

To have an explanation of a fact is to have (accepted) a theory which is acceptable, and which explains that fact. The latter relation must indubitably depend on what that fact is, since a theory may explain one fact and not another. Yet the following may also be held: it is a necessary condition that the theory, considered as a whole,

has certain features beyond acceptability. The relation between the theory and this fact may be called a *local* feature of the theory, and characters that pertain to the theory taken as a whole, *global* features.

This suggestive geometric metaphor was introduced by Michael Friedman, and he attempted an account of explanation along these lines. Friedman wrote:

> On the view of explanation that I am proposing, the kind of understanding provided by science is global rather than local. Scientific explanations do not confer intelligibility on individual phenomena by showing them to be somehow natural, necessary, familiar, or inevitable. However, our overall understanding of the world is increased . . .[15]

This could be read as totally discounting a specific relation of explanation altogether, as saying that theories can have certain overall virtues, at which we aim, and because of which we may ascribe to them explanatory power (with respect to their primary domain of application, perhaps). But Friedman does not go quite as far. He gives an explication of the relation *theory T explains phenomenon P*. He supposes (p. 15) that phenomena, i.e. general uniformities, are represented by lawlike sentences (whatever those may be); that we have as background a set K of accepted lawlike sentences, and that the candidate S (law, theory, or hypothesis) for explaining P is itself representable by a lawlike sentence. His definition has the form:

> S explains P exactly if P is a consequence of S, relative to K, and S 'reduces' or 'unifies' the set of its own consequences relative to K.

Here A is called a consequence of B relative to K exactly if A is a consequence of B and K together. He then modifies the above formula, and explicates it in a technically precise way. But as he explicates it, the notion of reduction cannot do the work he needs it to do, and it does not seem that anything like his precise definition could do.[16] More interesting than the details, however, is the form of the intuition behind Friedman's proposal. According to him, we evaluate something as an explanation relative to an assumed background theory K. I imagine that this theory might actually include some auxiliary information of a non-lawlike character, such as the age of the universe, or the boundary conditions in the situation under study. But of course K could not very well include all our information, since we generally know that P when we are asking for an explanation of P. Secondly, relative to K, the explanation implies that P is true. In view of Salmon's criticisms, I assume that Friedman would wish to weaken this Hempel-like condition. Finally, and here is the crux, it is the character of K plus the adduced information together, regarded as a complex theory, that determines whether we have an explanation. And the relevant features in this determination are global features, having to do with all the phenomena covered, not with P as such. So, whether or not K plus the adduced information provides new information about facts other than those described in P, appears to be crucial to whether we have an explanation of P.

James Greeno has made a similar proposal, with special reference to statistical theories. His abstract and closing statement says:

> The main argument of this paper is that an evaluation of the overall explanatory power of a theory is less problematic and more relevant as an assessment of the state of knowledge than an evaluation of statistical explanations of single occurrences. . .[17]

Greeno takes as his model of a theory one which specifies a single probability space Q as the correct one, plus two partitions (or random variables) of which one is designated *explanandum* and the other *explanans*. An example: sociology cannot explain why Albert, who lives in San Francisco and whose father has a high income, steals a car. Nor is it meant to. But it does explain delinquency in terms of such other factors as residence and parental income. The degree of explanatory power is measured by an ingeniously devised quantity which measures the information I the theory provides of the *explanandum* variable M on the basis of *explanans* S. This measure takes its maximum value if all conditional probabilities $P(M_i/S_j)$ are zero or one (D–N case), and its minimum value zero if S and M are statistically independent.

But it is not difficult to see that Greeno's way of making these ideas precise still runs into some of the same old difficulties. For suppose S and M describe the behaviour of barometers and storms. Suppose that the probability that the barometer will fall (M_1) equals the probability that there will be a storm (S_1), namely 0.2, and that the probability that there is a storm *given* that the barometer falls equals the probability that the barometer falls *given* that there will be a storm, namely 1. In that case the quantity I takes its maximum value – and indeed, does so even if we interchange M and S. But surely we do not have an explanation in either case.

2.4 The difficulties: asymmetries and rejections

There are two main difficulties, illustrated by the old paresis and barometer examples, which none of the examined positions can handle. The first is that there are cases, clearly in a theory's domain, where the request for explanation is nevertheless rejected. We can explain why John, rather than his brothers, contracted paresis, for he had syphilis; but not why he, among all those syphilitics, got paresis. Medical science is incomplete, and hopes to find the answer some day.

But the example of the uranium atom disintegrating just then rather than later, is formally similar and we believe the theory to be complete. We also reject such questions as the Aristotelians asked the Galileans: why does a body free of impressed forces retain its velocity? The importance of this sort of case, and its pervasive character, has been repeatedly discussed by Adolf Grünbaum. It was also noted, in a different context, by Thomas Kuhn.[18] Examples he gives of explanation requests which were considered legitimate in some periods and rejected in others cover a wide range of topics. They include the qualities of compounds in chemical theory (explained before Lavoisier's reform, and not considered something to be explained in the nineteenth century, but now again the subject of chemical explanation). Clerk Maxwell accepted as legitimate the request to explain electromagnetic phenomena within mechanics. As his theory became more successful and more widely accepted, scientists ceased to see the lack of this as a shortcoming. The same had happened with Newton's theory of gravitation which did not (in the opinion of Newton or his contemporaries) contain an explanation of gravitational phenomena, but only a description. In both cases there came a stage at which such problems were classed as intrinsically illegitimate, and regarded exactly as the request for an explanation of why a body retains its velocity in the absence of impressed forces. While all of this may be interpreted in various ways (such as through Kuhn's theory of paradigms) the important fact for the theory of explanation is that not everything in a theory's domain is a legitimate topic for why-questions; and that what is, is not determinable a priori.

The second difficulty is the asymmetry revealed by the barometer, the red shift, and the flagpole examples: even if the theory implies that one condition obtains when and only when another does, it may be that it explains the one

in terms of the other and not vice versa. An example which combines both the first and second difficulties is this: according to atomic physics, each chemical element has a characteristic atomic structure and a characteristic spectrum (of light emitted upon excitation). Yet the spectrum is explained by the atomic structure, and the question why a substance has that structure does not arise at all (except in the trivial sense that the questioner may need to have the terms explained to him).

To be successful, a theory of explanation must accommodate, and account for, both rejections and asymmetries. I shall now examine some attempts to come to terms with these, and gather from them the clues to the correct account.

2.5 Causality: the conditio sine qua non

Why are there no longer any Tasmanian natives? Why are the Plains Indians now living on reservations? Of course it is possible to cite relevant statistics: in many areas of the world, during many periods of history, upon the invasion by a technologically advanced people, the natives were displaced and weakened culturally, physically, and economically. But such a response will not satisfy: what we want is the story behind the event.

In Tasmania, attempts to round up and contain the natives were unsuccessful, so the white settlers simply started shooting them, man, woman, and child, until eventually there were none left. On the American Plains, the whites systematically destroyed the great buffalo herds on which the Indians relied for food and clothing, thus dooming them to starvation or surrender. There you see the story, it moves by its own internal necessity, and it explains why.

I use the word 'necessity' advisedly, for that is the term that links stories and causation. According to Aristotle's *Poetics*, the right way to write a story is to construct a situation which, after the initial parameters are fixed, moves toward its conclusion with a sort of necessity, inexorably – in retrospect, 'it had to end that way'. This was to begin also the hallmark of a causal explanation. Both in literature and in science we now accept such accounts as showing only how the events could have come about in the way they did. But it may be held that, to be an explanation, a scientific account must still tell a story of how things did happen and how the events hang together, so to say.

The idea of causality in modern philosophy is that of a relation among events. Hence it cannot be identified even with efficient causation, its nearest Aristotelian relative. In the modern sense we cannot say, correctly and non-elliptically, that the salt, or the moisture in the air, caused the rusting of the knife. Instead, we must say that certain events caused the rusting: such events as dropping of the salt on the knife, the air moistening that salt, and so on. The exact phrasing is not important; that the *relata* are events (including processes and momentary or prolonged states of affairs) is very important.

But what exactly is that causal relation? Everyone will recognize Hume's question here, and recall his rejection of certain metaphysical accounts. But we do after all talk this way, we say that the knife rusted because I dropped salt on it – and, as philosophers, we must make sense of explanation. In this and the next subsection I shall discuss some attempts to explicate the modern causal relation.

When something is cited as a cause, it is not implied that it was sufficient to produce (guarantee the occurrence) of the event. I say that this plant died because it was sprayed with defoliant, while knowing that the defoliant is only 90 per cent effective. Hence, the tradition that identifies the cause as the *conditio sine qua non*: had the plant not been sprayed, it would not have died.[19]

There are two problems with restating this as: a cause is a necessary condition. In the first place, not every necessary condition is a cause; and secondly, in some straightforward sense, a cause may not be necessary, namely, alternative causes could have led to the same result. An example for the first problem is this: the existence of the knife is a necessary condition for its rusting, and the growth of the plant for its dying. But neither of these could be cited as a cause. As to the second, it is clear that the plant could have died some other way, say if I had carefully covered it totally with anti-rust paint.

J. L. Mackie proposed the definition: a cause is an insufficient but necessary part of an unnecessary but sufficient condition.[20] That sufficient condition must precede the event to be explained, of course; it must not be something like the (growth-plus death-plus rotting) of the plant if we wish to cite a cause for its death. But the first problem still stands anyway, since the existence of the knife is a necessary part of the total set of conditions that led to its rusting. More worrisome is the fact that there may be no sufficient preceding conditions at all: the presence of the radium is what caused the Geiger counter to click, but atomic physics allows a non-zero probability for the counter not clicking at all under the circumstances.

For this reason (the non-availability of sufficient conditions in certain cases), Mackie's definition does not defuse the second problem either.

David Lewis has given an account in terms of counterfactual conditionals.[21] He simply equates '*A* caused *B*' with 'if *A* had not happened, *B* would not have happened'. But it is important to understand this conditional sentence correctly, and not to think of it (as earlier logicians did) as stating that *A* was a necessary condition for the occurrence of *B*. Indeed, the 'if . . . then' is not correctly identified with any of the sorts of implication traditionally discussed in logical theory, for those obey the law of *Weakening*:

> 1. If *A* then *B*
> *hence*
> if *A* and *C* then *B*.

But our conditionals, in natural language, typically do not obey that law:

> 2. If the match is struck it will light
> *hence* (?)
> if the match is dunked in coffee and struck, it will light;

the reader will think of many other examples. The explanation of why that 'law' does not hold is that our conditionals carry a tacit *ceteris paribus* clause:

> 3. If the plant had not been sprayed
> (*and all else had been the same*)
> then it would not have died.

The logical effect of this tacit clause is to make the 'law' of Weakening inapplicable.

Of course, it is impossible to spell out the exact content of *ceteris paribus*, as Goodman found in his classic discussion, for that content changes from context to context.[22] To this point I shall have to return. Under the circumstances, it is at least logically tenable to say, as David Lewis does, that whenever '*A* is the (a) cause of (or: caused) *B*' is true, it is also true that if *A* had not happened, neither would *B* have.

But do we have a sufficient criterion here? Suppose David's alarm clock goes off at seven a.m. and he wakes up. Now, we cite the alarm as the cause of the awakening, and may grant, if only for the sake of argument, that if the alarm had not sounded, he would not (then) have woken up. But it is also true that if he had not gone to sleep the night before, he would not have woken in the morning. This does not seem sufficient reason to say that he woke up because he had gone to sleep.

The response to this and similar examples is that the counterfactuals single out all the nodes in the causal net on lines leading to the event (the awakening), whereas 'because' points to specific factors that, for one reason or other, seem especially relevant (*salient*) in the context of our discussion. No one will deny that his going to sleep was one of the events that 'led up' to his awakening, that is, in the relevant part of the causal net. That part of the causal story is objective, and which specific item is singled out for special attention depends on the context – *every* theory of causation must say this.

Fair enough. That much context-dependence everyone will have to allow. But I think that much more context-dependence enters this theory through the truth-conditions of the counterfactuals themselves. So much, in fact, that we must conclude that there is nothing in science itself – nothing in the objective description of nature that science purports to give us – that corresponds to these counterfactual conditionals.

Consider again statement (3) about the plant sprayed with defoliant. It is true in a given situation exactly if the 'all else' that is kept 'fixed' is such as to rule out death of the plant for other reasons. But who keeps what fixed? The speaker, in his mind. There is therefore a contextual variable – determining the content of that tacit *ceteris paribus* clause – which is crucial to the truth-value of the conditional statement. Let us suppose that I say to myself, *sotto voce*, that a certain fuse leads into a barrel of gunpowder, and then say out loud, 'If Tom lit that fuse there would be an explosion.' Suppose that before I came in, you had observed to yourself that Tom is very cautious, and would not light any fuse before disconnecting it, and said out loud, 'If Tom lit that fuse, there would be no explosion.'

Have we contradicted each other? Is there an objective right or wrong about keeping one thing rather than another firmly in mind when uttering the antecedent 'If Tom lit that fuse . . .'? It seems rather that the proposition expressed by the sentence depends on a context, in which 'everything else being equal' takes on a definite content.

Robert Stalnaker and David Lewis give truth-conditions for conditionals using the notion of similarity among possible worlds. Thus, on one such account, 'if A then B' is true in world w exactly if B is true in the most similar world to w in which A is true. But there are many similarity relations among any set of things. Examples of the sort I have just given have long since elicited the agreement that the relevant similarity relation changes from context to context. Indeed, without that agreement, the logics of conditionals in the literature are violated by these examples.

One such example is very old: Lewis Carroll's puzzle of the three barbers. It occurs in *The Philosophy of Mr. B*rtr*nd R*ss*ll* as follows:

Allen, Brown, and Carr keep a barber's shop together; so that one of them must be in during working hours. Allen has lately had an illness of such a nature that, if Allen is out, Brown must be accompanying him. Further, if Carr is out, then, if Allen is out, Brown must be in for obvious business reasons.[23]

The above story gives rise to two conditionals, if we first suppose that Carr is out:

1   If Allen is out then Brown is out
2   If Allen is out then Brown is in

the first warranted by the remarks about Allen's illness, the second by the obvious business reasons. Lewis Carroll, thinking that 1 and 2 contradict each other, took this as a *reductio ad*

absurdum of the supposition that Carr is out. R*ss*ll, construing 'if A then B' as the material conditional ('either B or not A') asserts that 1 and 2 are both true if Allen is not out, and so says that we have here only a proof that if Carr is out, then Allen is in. ('The odd part of this conclusion is that it is the one which common-sense would have drawn', he adds.)

We have many other reasons, however, for not believing the conditional of natural language to be the material conditional. In modal logic, the strict conditional is such that 1 and 2 imply that it is not possible that Allen is out. So the argument would demonstrate 'If Carr is out then it is not possible that Allen is out.' This is false; if it looks true, it does so because it is easily confused with 'It is not possible that Carr is out and Allen is out.' If we know that Carr is out we can conclude that it is false that Allen is out, not that it is impossible.

The standard logics of counterfactual conditionals give exactly the same conclusion as the modal logic of strict conditionals. However, by noting the context-dependence of these statements, we can solve the problem correctly. Statement 1 is true in a context in which we disregard business requirements and keep fixed the fact of Allen's illness; statement 2 is true if we reverse what is fixed and what is variable. Now, there can exist contexts c and c' in which 1 and 2 are true respectively, only if their common antecedent is false; thus, like R*ss*ll, we are led to the conclusion drawn by common sense.

Any of the examples, and any general form of semantics for conditionals, will lend themselves to make the same point. What sort of situation, among all the possible unrealized ones, is more like ours in the fuse example: one in which nothing new is done except that the fuse is lit, or one in which the fuse is lit after being disconnected? It all depends – similar in what respect? Similar in that no fuse is disconnected or similar in that no one is being irresponsible? Quine brought out this feature of counterfactuals – to serve another purpose – when he asked whether, if Verdi and Bizet had been compatriots, would they have been French or Italian? Finally, even if someone feels very clear on what facts should be kept fixed in the evaluation of a counterfactual conditional, he will soon realize that it is not merely the facts but the description of the facts – or, if you like, facts identified by non-extensional criteria – that matter: Danny is a man, Danny is very much interested in women, i.e. (?) in the opposite sex – if he had been a woman would he have been very much interested in men, or a Lesbian?

These puzzles cause us no difficulty, if we say that the content of 'all else being equal' is fixed not only by the sentence and the factual situation, but also by contextual factors. In that case, however, the hope that the study of counterfactuals might elucidate science is quite mistaken: scientific propositions are not context-dependent in any essential way, so if counterfactual conditionals are, then science neither contains nor implies counterfactuals.

The truth-value of a conditional depends in part on the context. Science does not imply that the context is one way or another. Therefore science does not imply the truth of any counterfactual – except in the limiting case of a conditional with the same truth-value in all contexts. (Such limiting cases are ones in which the scientific theory plus the antecedent strictly imply the consequent, and for them logical laws like Weakening and Contraposition are valid, so that they are useless for the application to explanation which we are at present exploring.)

There was at one point a hope, expressed by Goodman, Reichenbach, Hempel, and others, that counterfactual conditionals provide an objective criterion for what is a law of nature, or at least, a lawlike statement (where a law is a true lawlike statement). A merely general truth was to be distinguished from a law because the latter, and not the former, implies counterfactuals. This idea must be inverted: if laws imply counterfactuals then, because counterfactuals are context-dependent, the concept of law does not point to any objective distinction in nature.

If, as I am inclined to agree, counterfactual language is proper to explanation, we should conclude that explanation harbours a significant degree of context-dependence.

2.6 Causality: Salmon's theory

The preceding subsection began by relating causation to stories, but the accounts of causality it examined concentrated on the links between particular events. The problems that appeared may therefore have resulted from the concentration on 'local properties' of the story. An account of causal explanation which focuses on extended processes has recently been given by Wesley Salmon.[24]

In his earlier theory, to the effect that an explanation consists in listing statistically relevant factors, Salmon had asked 'What more could one ask of an explanation?' He now answers this question:

> What does explanation offer, over and above the inferential capacity of prediction and retrodiction . . .? It provides knowledge of the mechanisms of *production* and *propagation* of structure in the world. That goes some distance beyond mere recognition of regularities, and of the possibility of subsuming particular phenomena thereunder.[25]

The question, what is the causal relation? is now replaced by: what is a causal process? and, what is a causal interaction? In his answer to these questions, Salmon relies to a large extent on Reichenbach's theory of the common cause, which we encountered before. But Salmon modifies this theory considerably.

A process is a spatio-temporally continuous series of events. The continuity is important, and Salmon blames some of Hume's difficulties on his picture of processes as chains of events with discrete links.[26] Some processes are causal, or genuine processes, and some are pseudo-processes. For example, if a car moves along a road, its shadow moves along that road too. The series of events in which the car occupies successive points on that road is a genuine causal process. But the movement of the shadow is merely a pseudo-process, because, intuitively speaking, the position of the shadow at later times is not caused by its position at earlier times. Rather, there is shadow *here* now because there is a car here now, and not because there was shadow *there* then.

Reichenbach tried to give a criterion for this distinction by means of probabilistic relations.[27] The series of events A_r is a causal process provided

the probability of A_{r+s} given A_r is greater than or equal to the probability of A_{r+s} given A_{r-t}, which is in turn greater than the probability of A_{r+s} simpliciter.

This condition does not yet rule out pseudo-processes, so we add that each event in the series *screens off* the earlier ones from the later ones:

the probability of A_{r+s} given both A_r and A_{r-t} is just that of A_{r+s} given A_r

and, *in addition*, there is no other series of events B_r which screens off A_{r+s} from A_r for all r. The idea in the example is that if A_{r+s} is the position of the shadow at time $r + s$, then B_r is the position of the car at time $r + s$.

This is not satisfactory for two reasons. The first is that (2) reminds one of a well-known property of stochastic processes, called the Markov property, and seems to be too strong to

go into the definition of causal processes. Why should not the whole history of the process up to time r give more information about what happens later than the state at time r does by itself? The second problem is that in the addition to (1) we should surely add that B_r must itself be a genuine causal process? For otherwise the movement of the car is not a causal process either, since the movement of the shadow will screen off successive positions of the car from each other. But if we say that B_r must be a genuine process in this stipulation, we have landed in a regress.

Reichenbach suggested a second criterion, called the *mark method* and (presumably because it stops the threatened regress) Salmon prefers that.

If a fender is scraped as a result of a collision with a stone wall, the mark of that collision will be carried on by the car long after the interaction with the wall occurred. The shadow of a car moving along the shoulder is a pseudo-process. If it is deformed as it encounters a stone wall, it will immediately resume its former shape as soon as it passes by the wall. It will not transmit a mark or modification.[28]

So if the process is genuine then interference with an earlier event will have effects on later events in that process. However, thus phrased, this statement is blatantly a causal claim. How shall we explicate 'interference' and 'effects'? Salmon will shortly give an account of causal interactions (see below) but begins by appealing to his 'at-at' theory of motion. The movement of the car consists simply in being *at* all these positions *at* those various times. Similarly, the propagation of the mark consists simply in the mark being there, in those later events. There is not, over and above this, a special propagation relation.

However, there is more serious cause for worry. We cannot define a genuine process as one that *does* propagate a mark in this sense. There are features which the shadow carries along in that 'at-at' sense, such as that its shape is related, at all times, in a certain topologically definable way to the shape of the car, and that it is black. Other special marks are not always carried – imagine part of a rocket's journey during which it encounters nothing else. So what we need to say is that the process is genuine if, *were* there to be a given sort of interaction at an early stage, there *would be* certain marks in the later stages. At this point, I must refer back to the preceding section for a discussion of such counterfactual assertions.

We can, at this point, relativize the notions used to the theory accepted. About some processes, our theory *implies* that certain interactions at an early stage will be followed by certain marks at later stages. Hence we can say that, *relative to the theory* certain processes are classifiable as genuine and others as pseudo-processes. What this does not warrant is regarding the distinction as an objective one. However, if the distinction is introduced to play a role in the theory of explanation, and if explanation is a relation of theory to fact, this conclusion does not seem to me a variation on Salmon's theory that would defeat its purpose.[29]

Turning now to causal interactions, Salmon describes two sorts. These interactions are the 'nodes' in the causal net, the 'knots' that combine all those causal processes into a causal structure. Instead of 'node' or 'knot' Reichenbach and Salmon also use 'fork' (as in 'the road forks'). Reichenbach described one sort, the *conjunctive fork* which occurs when an event C, belonging to two processes, is the *common cause* of events A and B, in those separate processes, occurring after C. Here common cause is meant in Reichenbach's original sense:

(3) $P(A \& B/C) = P(A/C) \cdot P(B/C)$
(4) $P(A \& B/\overline{C}) = P(A/\overline{C}) \cdot P(B/\overline{C})$
(5) $P(A/C) > P(A/\overline{C})$
(6) $P(B/C) > P(B/\overline{C})$

which, as noted in Chapter 2 [*sic*], entails that there is a positive correlation between A and B.

In order to accommodate the recalcitrant examples (see Chapter 2) [*sic*] Salmon introduced in addition the *interactive fork*, which is like the preceding one except that (3) is changed to

(3*) $P(A \& B/C) > P(A/C) \cdot P(B/C)$

These forks then combine the genuine causal processes, once identified, into the causal net that constitutes the natural order.

Explanation, on Salmon's new account, consists therefore in exhibiting the relevant part of the causal net that leads up to the events that are to be explained. In some cases we need only point to a single causal process that leads up to the event in question. In other cases we are asked to explain the confluence of events, or a positive correlation, and we do so by tracing them back to forks, that is, common origins of the processes that led up to them.

Various standard problems are handled. The sequence, barometer falling–storm coming, is not a causal process since the relevance of the first to the second is screened off by the common cause of atmospheric conditions. When paresis is explained by mentioning latent untreated syphilis, one is clearly pointing to the causal process, whatever it is, that leads from one to the other – or to their common cause, whatever that is. It must of course be a crucial feature of this theory that ordinary explanations are 'pointers to' causal processes and interactions which would, if known or described in detail, give the full explanation.

If that is correct, then each explanation must have, as cash-value, some tracing back (which is possible in principle) of separate causal processes to the forks that connect them. There are various difficulties with this view. The first is that to be a causal process, the sequence of events must correspond to a continuous spatio-temporal trajectory. In quantum mechanics, this requirement is not met. It was exactly the crucial innovation in the transition from the Bohr atom of 1913 to the new quantum theory of 1924, that the exactly defined orbits of the electrons were discarded. Salmon mentions explicitly the limitation of this account to macroscopic phenomena (though he does discuss Compton scattering). This limitation is serious, for we have no independent reason to think that explanation in quantum mechanics is essentially different from elsewhere.

Secondly, many scientific explanations certainly do not look as if they are causal explanations in Salmon's sense. A causal law is presumably one that governs the temporal development of some process or interaction. There are also 'laws of coexistence', which give limits to possible states or simultaneous configurations. A simple example is Boyle's law for gases (temperature is proportional to volume times pressure, at any given time); another, Newton's law of gravitation; another, Pauli's exclusion principle. In some of these cases we can say that they (or their improved counterparts) were later deduced from theories that replaced 'action at a distance' (which is not action at all, but a constraint on simultaneous states) with 'action by contact'. But suppose they were not so replaceable – would that mean that they could not be used in genuine explanations?

Salmon himself gives an example of explanation 'by common cause' which actually does not seem to fit his account. By observations on Brownian motion, scientists determined Avogadro's number, that is, the number of molecules in one mole of gas. By quite different observations, on the process of electrolysis, they determined the number of electron charges equal to one Faraday, that is, to the amount of electric charge needed to deposit one mole of a monovalent metal. These two numbers are equal. On the face of it, this equality is astonishing; but physics can explain this equality by deducing it from the basic theories governing both sorts of phenomena. The common cause Salmon identifies here is the basic mechanism – atomic and molecular structure – postulated to account for these phenomena. But surely it is clear that, however much the adduced explanation may deserve the name 'common cause', it does not point to a relationship between events (in Brownian motion on specific occasions and in electrolysis on specific occasions) which is traced back via causal processes to forks connecting these processes. The explanation is rather that the number found in experiment *A* at time *t* is the same as that found in totally independent experiment B at *any* other time *t'*, because of the *similarity* in the physically independent causal processes observed on those two different occasions.

Many highly theoretical explanations at least look as if they escape Salmon's account. Examples here are explanations based on principles of least action, based on symmetry considerations, or, in relativistic theories, on information that relates to space–time as a whole, such as specification of the metric or gravitational field.

The conclusion suggested by all this is that the type of explanation characterized by Salmon, though apparently of central importance, is still at most a subspecies of explanations in general.

2.7 The clues of causality

Let us agree that science gives us a picture of the world as a net of interconnected events, related to each other in a complex but orderly way.

The difficulties we found in the preceding two sections throw some doubt on the adequacy of the terminology of cause and causality to describe that picture; but let us not press this doubt further. The account of explanation suggested by the theories examined can now be restated in general terms as follows:

(1) Events are enmeshed in a net of causal relations,

(2) What science describes is that causal net,

(3) Explanation of why an event happens consists (typically) in an exhibition of salient factors in the part of the causal net formed by lines 'leading up to' that event,

(4) Those salient factors mentioned in an explanation constitute (what are ordinarily called) the *cause(s)* of that event.

There are two clear reasons why, when the topic of explanation comes up, attention is switched from the causal net as a whole (or even the part that converges on the event in question) to 'salient factors'. The first reason is that any account of explanation must make sense of common examples of explanation – especially cases typically cited as scientific explanations. In such actual cases, the reasons cited are particular prior events or initial conditions or combinations thereof. The second reason is that no account of explanation should imply that we can never give an explanation – and to describe the whole causal net in any connected region, however small, is in almost every case impossible. So the least concession one would have to make is to say that the explanation need say no more than that *there is* a structure of causal relations of *a certain sort*, which could *in principle* be described in detail: the salient features are what picks out the 'certain sort'.

Interest in causation as such focuses attention on (1) and (2), but interest in explanation requires us to concentrate on (3) and (4). Indeed,

from the latter point of view, it is sufficient to guarantee the truth of (1) and (2) by *defining*

> the causal net = whatever structure of relations science describes

and leaving to those interested in causation as such the problem of describing that structure in abstract but illuminating ways, if they wish.

Could it be that the explanation of a fact or event nevertheless resides solely in that causal net, and that *any* way of drawing attention to it explains? The answer is *no*; in the case of causal explanation, the *explanation* consists in drawing attention to certain ('special', 'important') features of the causal net. Suppose for example that I wish to explain the extinction of the Irish elk. There is a very large class of factors that preceded this extinction and was statistically relevant to it – even very small increases in speed, contact area of the hoof, height, distribution of weight in the body, distribution of food supply, migration habits, surrounding fauna and flora – we know from selection theory that under proper conditions any variation in these can be decisive in the survival of the species. But although, if some of these had been different, the Irish elk would have survived, they are not said to provide the explanation of why it is now extinct. The explanation given is that the process of sexual selection favoured males with large antlers, and that these antlers were, in the environment where they lived, encumbering and the very opposite of survival-adaptive. The other factors I mentioned are not spurious causes, or screened off by the development of these huge and cumbersome antlers, because the extinction was the total effect of many contributing factors; but those other factors are not the salient factors.

We turn then to those salient features that are cited in explanation – those referred to as 'the cause(s)' or 'the real cause(s)'. Various philosophical writers, seeking for an objective account of explanation, have attempted to state criteria that single out those special factors. I shall not discuss their attempts. Let me just cite a small survey of their answers: Lewis White Beck says that the cause is that factor over which we have most control; Nagel argues that it is often exactly that factor which is not under our control; Braithwaite takes the salient factors to be the unknown ones; and David Bohm takes them to be the factors which are the most variable.[30]

Why should different writers have given such different answers? The reason was exhibited, I think, by Norwood Russell Hanson, in his discussion of causation.

> There are as many causes of *x* as there are explanations of *x*. Consider how the cause of death might have been set out by a physician as 'multiple haemorrhage', by the barrister as 'negligence on the part of the driver', by a carriage-builder as 'a defect in the brakeblock construction', by a civic planner as 'the presence of tall shrubbery at that turning'.[31]

In other words, the salient feature picked out as 'the cause' in that complex process, is salient to a given person because of his orientation, his interests, and various other peculiarities in the way he approaches or comes to know the problem – contextual factors.

It is important to notice that in a certain sense these different answers cannot be combined. The civic planner 'keeps fixed' the mechanical constitution of the car, and gives his answer in the conviction that regardless of the mechanical defects, which made a fast stop impossible, the accident need not have happened. The mechanic 'keeps fixed' the physical environment; despite the shrubbery obscuring vision, the accident need not have happened if the brakes had been better. What the one varies, the other keeps fixed, and you cannot do both at once. In other words, the selection of the salient causal factor is not simply a matter of pointing to the most interesting one, not like the selection of a tourist attraction; it is a matter of *competing* counterfactuals.

We must accordingly agree with the Dutch philosopher P. J. Zwart who concludes, after examining the above philosophical theories,

> It is therefore not the case that the meaning of the sentence 'A is the cause of B' depends on the nature of the phenomena A and B, but that this meaning depends on the context in which this sentence is uttered. The nature of A and B will in most cases also play a role, indirectly, but it is in the first place the orientation or the chosen point of view of the speaker that determines what the word cause is used to signify.[32]

In conclusion, then, this look at accounts of causation seems to establish that explanatory factors are to be chosen from a range of factors which are (or which the scientific theory lists as) objectively relevant in certain special ways – but that the choice is then determined by other factors that vary with the context of the explanation request. To sum up: no factor is explanatorily relevant unless it is scientifically relevant; and among the scientifically relevant factors, context determines explanatorily relevant ones.

2.8 *Why-questions*

Another approach to explanation was initiated by Sylvain Bromberger in his study of why-questions.[33] After all, a why-question is a request for explanation. Consider the question:

1. Why did the conductor become warped during the short circuit?

This has the general form

2. Why (is it the case that) P?

where P is a statement. So we can think of 'Why' as a function that turns statements into questions.

Question 1 *arises*, or *is in order*, only if the conductor did indeed become warped then. If that is not so, we do not try to answer the question, but say something like: 'You are under a false impression, the conductor became warped much earlier,' or whatever. Hence Bromberger calls the statement that P the *presupposition* of the question *Why P?* One form of the rejection of explanation requests is clearly the denial of the presupposition of the corresponding why-question.

I will not discuss Bromberger's theory further here, but turn instead to a criticism of it. The following point about why-questions has been made in recent literature by Alan Garfinkel and Jon Dorling, but I think it was first made, and discussed in detail, in unpublished work by Bengt Hannson circulated in 1974.[34] Consider the question

3.   Why did Adam eat the apple?

This same question can be construed in various ways, as is shown by the variants:

3a.   Why was it Adam who ate the apple?
3b.   Why was it the apple Adam ate?
3c.   Why did Adam *eat* the apple?

In each case, the canonical form prescribed by Bromberger (as in 2 above) would be the same, namely

4.   Why (is it the case that) (Adam ate the apple)?

yet there are three different explanation requests here.

The difference between these various requests is that they point to different contrasting alternatives. For example, 3b may ask why Adam ate the *apple* rather than some other fruit in the garden, while 3c asks perhaps why Adam *ate* the apple rather than give it back to Eve untouched. So to 3b, 'because he was hungry' is not a good answer, whereas to 3c it is. The correct general, underlying structure of a why-question is therefore

5.   Why (is it the case that) P in *contrast to* (other members of) X?

where X, the *contrast-class*, is a set of alternatives. P may belong to X or not; further examples are:

Why did the sample burn green (rather than some other colour)?

Why did the water and copper reach equilibrium temperature 22.5 °C (rather than some other temperature)?

In these cases the contrast-classes (colours, temperatures) are 'obvious'. In general, the contrast-class is not explicitly described because, *in context*, it is clear to all discussants what the intended alternatives are.

This observation explains the tension we feel in the paresis example. If a mother asks why her eldest son, a pillar of the community, mayor of his town, and best beloved of all her sons, has this dread disease, we answer: because he had latent untreated syphilis. But if that question is asked about this same person, immediately after a discussion of the fact that everyone in his country club has a history of untreated syphilis, *there is no answer*. The reason for the difference is that in the first case the contrast-class is the mother's sons, and in the second, the members of the country club, contracting paresis. Clearly, an answer to a question of form 5 must adduce information that *favours* P in *contrast to* other members of X. Sometimes the availability of such information depends strongly on the choice of X.

These reflections have great intuitive force. The distinction made is clearly crucial to the paresis example and explains the sense of ambiguity and tension felt in earlier discussion of such examples. It also gives us the right way to explicate such assertions as: individual events are never explained, we only explain a particular event *qua* event of a certain kind. (We can explain *this* decay of a uranium atom *qua* decay of a uranium atom, but not *qua* decay of a uranium atom at *this* time.)

But the explication of what it is for an answer to favour one alternative over another proves difficult. Hannson proposed: answer *A* is a good answer to (Why *P* in contrast to *X*?) exactly if the probability of *P* given *A* is higher than the average probability of members of *X* given *A*. But this proposal runs into most of the old difficulties. Recall Salmon's examples of irrelevancy: the probability of recovery from a cold *given* administration of vitamin C is nearly one, while the probability of not recovering *given* the vitamins is nearly zero. So by Hannson's criterion it would be a good answer – even if taking vitamin C has no effect on recovery from colds one way or the other.

Also, the asymmetries are as worrisome as ever. By Hannson's criterion, the length of the shadow automatically provides a good explanation of the height of the flagpole. And 'because the barometer fell' is a good answer to 'why is there a storm?' (upon selection of the 'obvious' contrast-classes, of course). Thus it seems that reflection on the contrast-class serves to solve some of our problems, but not all.

2.9 The clues elaborated

The discussions of causality and of why-questions seem to me to provide essential clues to the correct account of explanation. In the former we found that an explanation often consists in listing salient factors, which point to a complete story of how the event happened. The effect of this is to eliminate various alternative hypotheses about how this event did come about, and/or eliminate puzzlement concerning how the event could have come about. But salience is context-dependent, and the selection of the correct 'important' factor depends on the range of alternatives contemplated in that context. In N. R. Hanson's example, the barrister wants this sort of weeding out of hypotheses about the death relevant to the question of legal accountability; the carriage-builder, a weeding out of hypotheses about structural defects or structural limitations under various sorts of strain. *The context*, in other words, *determines relevance* in a way that goes well beyond the statistical relevance about which our scientific theories give information.

This might not be important if we were not concerned to find out exactly how having an explanation goes beyond merely having an acceptable theory about the domain of phenomena in question. But that is exactly the topic of our concern.

In the discussion of why-questions, we have discovered a further contextually determined factor. The range of hypotheses about the event which the explanation must 'weed out' or 'cut down' is not determined solely by the interests of the discussants (legal, mechanical, medical) but also by a range of contrasting alternatives to the event. This *contrast-class* is also determined by context.

It might be thought that when we request a *scientific* explanation, the relevance of possible hypotheses, and also the contrast-class are automatically determined. But this is not so, for both the physician and the motor mechanic are asked for a scientific explanation. The physician explains the fatality *qua* death of a human organism, and the mechanic explains it *qua* automobile crash fatality. To ask that their explanations be scientific is only to demand that they rely on scientific theories and experimentation, not on old wives' tales. Since any explanation of an individual event must be an explanation of that event *qua* instance of a certain kind of event, nothing more can be asked.

The two clues must be put together. The description of some account as an explanation of a given fact or event, is incomplete. It can only be an explanation with respect to a certain *relevance relation* and a certain *contrast-class*. These are contextual factors, in that they are determined neither by the totality of accepted scientific theories, nor by the event or fact for which an explanation is requested. It is sometimes said that an Omniscient Being would have a complete explanation, whereas these contextual factors only bespeak our limitations due to which we can only grasp one part or aspect of the complete explanation at any given time. But this is a mistake. If the Omniscient Being has no specific interests (legal, medical, economic; or just an interest in optics or thermodynamics rather than chemistry) and does not abstract (so that he never thinks of Caesar's death *qua* multiple stabbing, or *qua* assassination), then no why-questions ever arise for him in any way at all – and he does not have any explanation in the sense that we have explanations. If he does have interests, and does abstract from individual peculiarities in his thinking about the world, then his why-questions are as essentially context-dependent as ours. In either case, his advantage is that he always has all the information needed to answer any specific explanation request. But that information is, in and by itself, not an explanation; just as a person cannot be said to be older, or a neighbour, except in relation to others.

3 Asymmetries of explanation: a short story

3.1 *Asymmetry and context: the Aristotelian sieve*

That vexing problem about paresis, where we seem both to have and not to have an explanation, was solved by reflection on the contextually supplied contrast-class. The equally vexing, and much older, problem of the asymmetries of explanation, is illuminated by reflection on the other main contextual factor: contextual relevance.

If that is correct, if the asymmetries of explanation result from a contextually determined relation of relevance, then it must be the case that these asymmetries can at least sometimes be reversed by a change in context. In addition, it should then also be possible to account for specific asymmetries in terms of the interests of questioner and audience that determine this relevance. These considerations provide a crucial test for the account of explanation which I propose.

Fortunately, there is a precedent for this sort of account of the asymmetries, namely in Aristotle's theory of science. It is traditional to understand this part of his theory in relation to his metaphysics; but I maintain that the central aspects of his solution to the problem of asymmetry of explanations are independently usable.[35]

Aristotle gave examples of this problem in the *Posterior Analytics* I, 13; and he developed a typology of explanatory factors ('the four causes'). The solution is then simply this. Suppose there are a definite (e.g. four) number of types of explanatory factors (i.e. of relevance relations for why-questions). Suppose also that relative to our background information and accepted theories, the propositions A and B are equivalent. It may then still be that these two propositions describe factors of different types. Suppose that in a certain context, our interest is in the mode of production of an event, and 'Because B' is an acceptable answer to 'Why A?'. Then it may well be that A does not describe any mode of production of anything, so that, *in this same context*, 'Because A' would not be an acceptable answer to 'Why B?'.

Aristotle's lantern example (*Posterior Analytics* II, 11) shows that he recognized that in different contexts, verbally the same why-question may be a request for different types of explanatory factors. In modern dress the example would run as follows. Suppose a father asks his teenage son, 'Why is the porch light on?' and the son replies 'The porch switch is closed and the electricity is reaching the bulb through that switch.' At this point you are most likely to feel that the son is being impudent. This is because you are most likely to think that the sort of answer the father needed was something like: 'Because we are expecting company.' But it is easy to imagine a less likely question context: the father and son are re-wiring the house and the father, unexpectedly seeing the porch light on, fears that he has caused a short circuit that bypasses the porch light switch. In the second case, he is *not* interested in the human expectations or desires that led to the depressing of the switch.

Aristotle's fourfold typology of causes is probably an over-simplification of the variety of interests that can determine the selection of a range of relevant factors for a why-question. But in my opinion, appeal to some such typology will successfully illuminate the asymmetries (and also the rejections, since no factor of a *particular* type may lead to a telling answer to the why-question). If that is so then, as I said before, asymmetries must be at least sometimes reversible through a change in context. The story which follows is meant to illustrate this. As in the lantern (or porch light) example, the relevance changes from one sort of efficient cause to another, the second being a person's desires. As in all explanations, the correct answer consists in the exhibition of a single factor in the causal net, which is made salient in that context by factors not overtly appearing in the words of the question.

3.2 'The tower and the shadow'

During my travels along the Saône and Rhône last year, I spent a day and night at the ancestral home of the Chevalier de St. X . . ., an old friend of my father's. The Chevalier had in fact been the French liaison officer attached to my father's brigade in the first war, which had − if their reminiscences are to be trusted − played a not insignificant part in the battles of the Somme and Marne.

The old gentleman always had *thé à l'Anglaise* on the terrace at five o'clock in the evening, he told me. It was at this meal that a strange incident occurred; though its ramifications were of course not yet perceptible when I heard the Chevalier give his simple explanation of the length of the shadow which encroached upon us there on the terrace. I had just eaten my fifth piece of bread and butter and had begun my third cup of tea when I chanced to look up. In the dying light of that late afternoon, his profile was sharply etched against the granite background of the wall behind him, the great aquiline nose thrust forward and his eyes fixed on some point behind my left shoulder. Not understanding the situation at first, I must admit that to begin with, I was merely fascinated by the sight of that great hooked nose, recalling my father's claim that this had once served as an effective weapon in close combat with a German grenadier. But I was roused from this brown study by the Chevalier's voice.

'The shadow of the tower will soon reach us, and the terrace will turn chilly. I suggest we finish our tea and go inside.'

I looked around, and the shadow of the rather curious tower I had earlier noticed in the grounds, had indeed approached to within a yard from my chair. The news rather displeased me, for it was a fine evening: I wished to remonstrate but did not well know how, without overstepping the bounds of hospitality. I exclaimed,

'Why must that tower have such a long shadow? This terrace is so pleasant!'

His eyes turned to rest on me. My question had been rhetorical, but he did not take it so.

'As you may already know, one of my ancestors mounted the scaffold with Louis XVI and Marie Antoinette. I had that tower erected in 1930 to mark the exact spot where it is said that he greeted the Queen when she first visited this house, and presented her with a peacock made of soap, then a rare substance. Since the Queen would have been one hundred and seventy-five years old in 1930, had she lived, I had the tower made exactly that many feet high.'

It took me a moment to see the relevance of all this. Never quick at sums, I was at first merely puzzled as to why the measurement should have been in feet; but of course I already knew him for an Anglophile. He added drily, 'The sun not being alterable in its course, light travelling in straight lines, and the laws of trigonometry being immutable, you will perceive that the length of the shadow is determined by the height of the tower.' We rose and went inside.

I was still reading at eleven that evening when there was a knock at my door. Opening it I found the housemaid, dressed in a somewhat old-fashioned black dress and white cap, whom I had perceived hovering in the background on several occasions that day. Courtseying prettily, she asked, 'Would the gentleman like to have his bed turned down for the night?'

I stepped aside, not wishing to refuse, but remarked that it was very late – was she kept on duty to such hours? No, indeed, she answered, as she deftly turned my bed covers, but it had occurred to her that some duties might be pleasures as well. In such and similar philosophical reflections we spent a few pleasant hours together, until eventually I mentioned casually how silly it seemed to me that the tower's shadow ruined the terrace for a prolonged, leisurely tea.

At this, her brow clouded. She sat up sharply. 'What exactly did he tell you about this?' I replied lightly, repeating the story about Marie Antoinette, which now sounded a bit far-fetched even to my credulous ears.

'The *servants* have a different account', she said with a sneer that was not at all becoming, it seemed to me, on such a young and pretty face. 'The truth is quite different, and has nothing to do with ancestors. That tower marks the spot where he killed the maid with whom he had been in love to the point of madness. And the height of the tower? He vowed that shadow would cover the terrace where he first proclaimed his love, with every setting sun – that is why the tower had to be so high.'

I took this in but slowly. It is never easy to assimilate unexpected truths about people we think we know – and I have had occasion to notice this again and again.

'Why did he kill her?' I asked finally.

'Because, sir, she dallied with an English brigadier, an overnight guest in this house.' With these words she arose, collected her bodice and cap, and faded through the wall beside the doorway.

I left early the next morning, making my excuses as well as I could.

4 A model for explanation

I shall now propose a new theory of explanation. An explanation is not the same as a proposition, or an argument, or list of propositions; it is an *answer*. (Analogously, a son is not the same as a man, even if all sons are men, and every man is a son.) An explanation is an answer to a why-question. So, a theory of explanation must be a theory of why-questions.

To develop this theory, whose elements can all be gleaned, more or less directly, from the preceding discussion, I must first say more about some topics in formal pragmatics (which deals with context-dependence) and in the logic of questions. Both have only recently become active areas in logical research, but there is general agreement on the basic aspects to which I limit the discussion.

4.1 Contexts and propositions[36]

Logicians have been constructing a series of models of our language, of increasing complexity and sophistication. The phenomena they aim to save are the surface grammar of our assertions and the inference patterns detectable in our arguments. (The distinction between logic and theoretical linguistics is becoming vague, though logicians' interests focus on special parts of our language, and require a less faithful fit to surface grammar, their interests remaining in any case highly theoretical.) Theoretical entities introduced by logicians in their models of language (also called 'formal languages') include domains of discourse ('universes'), possible worlds, accessibility ('relative possibility') relations, facts and propositions, truth-values, and, lately, contexts. As might be guessed, I take it to be part of empiricism to insist that the adequacy of these models does not require all their elements to have counterparts in reality. They will be good if they fit those phenomena to be saved.

Elementary logic courses introduce one to the simplest models, the languages of sentential and quantificational logic which, being the simplest, are of course the most clearly inadequate. Most logic teachers being somewhat defensive about this, many logic students, and other philosophers, have come away with the impression that the over-simplifications make the subject useless. Others, impressed with such uses as elementary logic does have (in elucidating classical mathematics, for example), conclude that we shall not understand natural language until we have seen how it can be regimented so as to fit that simple model of horseshoes and truth tables.

In elementary logic, each sentence corresponds to exactly one proposition, and the truth-value of that sentence depends on whether the proposition in question is true in the actual world. This is also true of such extensions of elementary logic as free logic (in which not all terms need have an actual referent), and normal modal logic (in which non-truth functional connectives appear), and indeed of almost all the logics studied until quite recently.

But, of course, sentences in natural language are typically context-dependent; that is, which proposition a given sentence expresses will vary with the context and occasion of use. This point was made early on by Strawson, and examples are many:

'I am happy now' is true in context x exactly if the speaker in context x is happy at the time of context x,

where a context of use is an actual occasion, which happened at a definite time and place, and in which are identified the speaker (referent of 'I'), addressee (referent of 'you'), person discussed (referent of 'he'), and so on. That contexts so conceived are idealizations from real contexts is obvious, but the degree of idealization may be decreased in various ways, depending on one's purposes of study, at the cost of greater complexity in the model constructed.

What must the context specify? The answer depends on the sentence being analysed. If that sentence is

Twenty years ago it was still possible to prevent the threatened population explosion in that country, but now it is too late

the model will contain a number of factors. First, there is a set of possible worlds, and a set of contexts, with a specification for each context of the world of which it is a part. Then there will be for each world a set of entities that exist in that world, and also various relations of relative possibility among these worlds. In addition there is time, and each context must have a time of occurrence. When we evaluate the above sentence we do so relative to a context and a world. Varying with the context will be the referents of 'that country' and 'now', and perhaps also the relative possibility relation used to interpret 'possible', since the speaker may have intended one of several senses of possibility.

This sort of interpretation of a sentence can be put in a simple general form. We first identify certain entities (mathematical constructs) called propositions, each of which has a truth-value in each possible world. Then we give the context as its main task the job of selecting, for each sentence, the proposition it expresses 'in that context'. Assume as a simplification that when a sentence contains no indexical terms (like 'I', 'that', 'here', etc.), then all contexts select the same proposition for it. This gives us an easy intuitive handle on what is going on. If *A* is a sentence in which no indexical terms occur, let us designate as $|A|$ the proposition which it expresses in every context. Then we can generally (though not necessarily always) identify the proposition expressed by any sentence in a given context as the proposition expressed by some indexical-free sentence. For example:

> In context *x*, 'Twenty years ago it was still possible to prevent the population explosion in that country' expresses the proposition 'In 1958, it is (tenseless) possible to prevent the population explosion in India'

To give another example, in the context of my present writing. 'I am here now' expresses the proposition that Bas van Fraassen is in Vancouver, in July 1978.

This approach has thrown light on some delicate conceptual issues in philosophy of language.

Note for example that 'I am here' is a sentence which is true no matter what the facts are and no matter what the world is like, and no matter what context of usage we consider. Its truth is ascertainable a priori. But the proposition expressed, that van Fraassen is in Vancouver (or whatever else it is) is not at all a necessary one: I might not have been here. Hence, a clear distinction between a priori ascertainability and necessity appears.

The context will generally select the proposition expressed by a given sentence *A* via a selection of referents for the terms, extensions for the predicates, and functions for the functors (i.e. syncategorematic words like 'and' or 'most'). But intervening contextual variables may occur at any point in these selections. Among such variables there will be the assumptions taken for granted, theories accepted, world-pictures or paradigms adhered to, in that context. A simple example would be the range of conceivable worlds admitted as possible by the speaker; this variable plays a role in determining the truth-value of his modal statements in that context, relative to the 'pragmatic presuppositions'. For example, if the actual world is really the only possible world there is (which exists) then the truth-values of modal statements in that context but *tout court* will be very different from their truth-values relative to those pragmatic presuppositions – and only the latter will play a significant role in our understanding of what is being said or argued in that context.

Since such a central role is played by propositions, the family of propositions has to have a fairly complex structure. Here a simplifying hypothesis enters the fray: propositions can be uniquely identified through the worlds in which they are true. This simplifies the model considerably, for it allows us to identify a proposition with a set of possible worlds, namely, the set of worlds in which it is true. It allows the family of propositions to be a complex structure, admitting of interesting operations, while keeping the structure of each individual proposition very simple.

Such simplicity has a cost. Only if the phenomena are simple enough, will simple models fit them. And sometimes, to keep one part of a model simple, we have to complicate another part. In a number of areas in philosophical logic it has already been proposed to discard that simplifying hypothesis, and to give propositions more 'internal structure'. As will be seen below, problems in the logic of explanation provide further reasons for doing so.

4.2 Questions

We must now look further into the general logic of questions. There are of course a number of approaches; I shall mainly follow that of Nuel Belnap, though without committing myself to the details of his theory.[37]

A theory of questions must needs be based on a theory of propositions, which I shall assume given. A *question* is an abstract entity; it is expressed by an *interrogative* (a piece of language) in the same sense that a proposition is expressed by a declarative sentence. Almost anything can be an appropriate response to a question, in one situation or another; as 'Peccavi' was the reply telegraphed by a British commander in India to the question how the battle was going (he had been sent to attack the province of Sind).[38] But not every response is, properly speaking, an answer. Of course, there are degrees; and one response may be more or less of an answer than another. The first task of a theory of questions is to provide some typology of answers. As an example, consider the following question, and a series of responses:

Can you get to Victoria both by ferry and by plane?

(a)   Yes.

(b)   You can get to Victoria both by ferry and by plane.

(c)   You can get to Victoria by ferry.

(d)   You can get to Victoria both by ferry and by plane, but the ferry ride is not to be missed.

(e)   You can certainly get to Victoria by ferry, and that is something not to be missed.

Here (b) is the 'purest' example of an answer: it gives enough information to answer the question completely, but no more. Hence it is called a *direct answer*. The word 'Yes' (a) is a *code* for this answer.

Responses (c) and (d) depart from that direct answer in opposite directions: (c) says properly less than (b) − it is implied by (b) − while (d), which implies (b), says more. Any proposition implied by a direct answer is called a *partial answer* and one which implies a direct answer is a *complete answer*. We must resist the temptation to say that therefore an answer, *tout court*, is any combination of a partial answer with further information, for in that case, every proposition would be an answer to any question. So let us leave (e) unclassified for now, while noting it is still 'more of an answer' than such responses as 'Gorilla!' (which is a response given to various questions in the film *Ich bin ein Elephant, Madam*, and hence, I suppose, still more of an answer than some). There may be some quantitative notion in the background (a measure of the extent to which a response really 'bears on' the question) or at least a much more complete typology (some more of it is given below), so it is probably better not to try and define the general term 'answer' too soon.

The basic notion so far is that of direct answer. In 1958, C. L. Hamblin introduced the thesis that a question is uniquely identifiable through its answers.[39] This can be regarded as a simplifying hypothesis of the sort we come across for propositions, for it would allow us to identify a question with the set of its direct answers. Note that this does not preclude a good deal of complexity in the determination of exactly what question is expressed by a given interrogative. Also, the hypothesis does not identify the question with the disjunction of its direct answers. If that were done, the clearly distinct questions

> Is the cat on the mat?
>> *direct answers:* The cat is on the mat.
>> The cat is not on the mat.
> Is the theory of relativity true?
>> *direct answers:* The theory of relativity is true.
>> The theory of relativity is not true.

would be the same (identified with the tautology) if the logic of propositions adopted were classical logic. Although this simplifying hypothesis is therefore not to be rejected immediately, and has in fact guided much of the research on questions, it is still advisable to remain somewhat tentative towards it.

Meanwhile we can still use the notion of direct answer to define some basic concepts. One question Q may be said to *contain* another, Q′, if Q′ is answered as soon as Q is – that is, every complete answer to Q is also a complete answer to Q′. A question is *empty* if all its direct answers are necessarily true, and *foolish* if none of them is even possibly true. A special case is the *dumb* question, which has no direct answers. Here are examples:

1   Did you wear the black hat yesterday or did you wear the white one?
2   Did you wear a hat which is both black and not black, or did you wear one which is both white and not white?
3   What are three distinct examples of primes among the following numbers: 3, 5?

Clearly 3 is dumb and 2 is foolish. If we correspondingly call a necessarily false statement foolish too, we obtain the theorem *Ask a foolish question and get a foolish answer*. This was first proved by Belnap, but attributed by him to an early Indian philosopher mentioned in Plutarch's *Lives* who had the additional distinction of being an early nudist. Note that a foolish question contains all questions, and an empty one is contained in all.

Example 1 is there partly to introduce the question form used in 2, but also partly to introduce the most important semantic concept after that of direct answer, namely presupposition. It is easy to see that the two direct answers to 1 ('I wore the black hat', 'I wore the white one') could both be false. If that were so, the respondent would presumably say 'Neither', which is an answer not yet captured by our typology. Following Belnap who clarified this subject completely, let us introduce the relevant concepts as follows:

> a *presupposition*[40] of question Q is any proposition which is implied by all direct answers to Q.
>
> a *correction* (or *corrective answer*) to Q is any denial of any presupposition of Q.
>
> the *(basic) presupposition* of Q is the proposition which is true if and only if some direct answer to Q is true.

In this last notion, I presuppose the simplifying hypothesis which identifies a proposition through the set of worlds in which it is true; if that hypothesis is rejected, a more complex definition needs to be given. For example 1, 'the' presupposition is clearly the proposition that the addressee wore either the black hat or the white one. Indeed, in any case in which the number of direct answers is finite, 'the' presupposition is the disjunction of those answers.

Let us return momentarily to the typology of answers. One important family is that of the partial answers (which includes direct and complete answers). A second important family is that of the corrective answer. But there are still more. Suppose the addressee of question 1 answers 'I did not wear the white one.' This is not even a partial answer, by the definition given: neither direct answer implies it, since she might have worn both hats yesterday, one in the afternoon and one in the evening, say. However, since the questioner is presupposing that she wore at least one of the two, the response is *to him* a complete answer. For the response plus the presupposition together entail the direct answer that she wore the black hat. Let us therefore add:

> a *relatively complete answer* to Q is any proposition which, together with the presupposition of Q, implies some direct answer to Q.

We can generalize this still further: a complete answer to Q, relative to theory T, is something which together with T, implies some direct answer to Q – and so forth. The important point is, I think, that we should regard the introduced typology of answers as open-ended, to be extended as needs be when specific sorts of questions are studied.

Finally, which question is expressed by a given interrogative? This is highly context-dependent, in part because all the usual index-ical terms appear in interrogatives. If I say 'Which one do you want?' the context deter-mines a range of objects over which my 'which one' ranges – for example, the set of apples in the basket on my arm. If we adopt the simpli-fying hypothesis discussed above, then the main task of the context is to delineate the set of direct answers. In the 'elementary questions' of Belnap's theory ('whether-questions' and 'which-questions') this set of direct answers is specified through two factors: a *set of alternatives* (called the *subject* of the question) and *request* for a selection among these alternatives and, possibly, for certain information about the

selection made ('distinctness and completeness claims'). What those two factors are may not be made explicit in the words used to frame the interrogative, but the context has to determine them exactly if it is to yield an interpretation of those words as expressing a unique question.

4.3 A theory of why-questions

There are several respects in which why-questions introduce genuinely new elements into the theory of questions.[41] Let us focus first on the determination of exactly what question is asked, that is, the contextual specifi-cation of factors needed to understand a why-interrogative. After that is done (a task which ends with the delineation of the set of direct answers) and as an independent enterprise, we must turn to the evaluation of those answers as good or better. This evaluation proceeds with reference to the part of science accepted as 'background theory' in that context.

As example, consider the question 'Why is this conductor warped?' The questioner implies that the conductor is warped, and is asking for a reason. Let us call the proposition that the conductor is warped the *topic* of the question (following Henry Leonard's terminology, 'topic of concern'). Next, this question has a *contrast-class*, as we saw, that is, a set of alternatives. I shall take this contrast-class, call it X, to be a class of propositions which includes the topic. For this particular interrogative, the contrast could be that it is *this* conductor rather than *that* one, or that this conductor has warped rather than retained its shape. If the question is 'Why does this material burn yellow?' the contrast-class could be the set of propositions: this material burned (with a flame of) colour x.

Finally, there is the respect-in-which a reason is requested, which determines what shall count as a possible explanatory factor, the relation of *explanatory relevance*. In the first example, the request might be for *events* 'leading up to' the warping. That allows as relevant an account of human error, of switches being closed or moisture condensing in those switches, even spells cast by witches (since the evaluation of what is a good answer comes later). On the other hand, the events leading up to the warping might be well known, in which case the request is likely to be for the standing conditions that made it possible for those events to lead to this warping: the presence of a magnetic field of a certain strength, say. Finally, it might already be known, or considered immaterial, exactly how the warping is produced, and the question (possibly based on a misunderstanding) may be about exactly what function this warping fulfils in the operation of the power station. Compare 'Why does the blood circulate through the body?' answered (1) 'because the heart pumps the blood through the arteries' and (2) 'to bring oxygen to every part of the body tissue'.

In a given context, several questions agreeing in topic but differing in contrast-class, or conversely, may conceivably differ further in what counts as explanatorily relevant. Hence we cannot properly ask what is relevant to this topic, or what is relevant to this contrast-class. Instead we must say of a given proposition that it is or is not relevant (in this context) to the topic with respect to that contrast-class. For example, in the same context one might be curious about the circumstances that led Adam to eat the apple rather than the pear (Eve offered him an apple) and also about the motives that led him to eat it rather than refuse it. What is 'kept constant' or 'taken as given' (that he ate the fruit; that what he did, he did to the apple) which is to say, the contrast-class, is not to be dissociated entirely from the respect-in-which we want a reason.

Summing up then, the why-question Q expressed by an interrogative in a given context will be determined by three factors:

The topic P_k
The *contrast-class* $X = \{P_1, \ldots, P_k, \ldots\}$
The *relevance relation* R

and, in a preliminary way, we may identify the abstract why-question with the triple consisting of these three:

$$Q = \langle P_k, X, R \rangle$$

A proposition A is called *relevant to* Q exactly if A bears relation R to the couple $\langle P_k, X \rangle$.

We must now define what are the direct answers to this question. As a beginning let us inspect the form of words that will express such an answer:

(*) P_k in contrast to (the rest of) X because A

This sentence must express a proposition. What proposition it expresses, however, depends on the same context that selected Q as the proposition expressed by the corresponding interrogative ('Why P_k?'). So some of the same contextual factors, and specifically R, may appear in the determination of the proposition expressed by (*).

What is claimed in answer (*)? First of all, that P_k is true. Secondly, (*) claims that the other members of the contrast-class are not true. So much is surely conveyed already by the question — it does not make sense to ask why Peter rather than Paul has paresis if they both have it. Thirdly, (*) says that A is true. And finally, there is that word 'because': (*) claims that A is a *reason*.

This fourth point we have awaited with bated breath. Is this not where the inextricably modal or counterfactual element comes in? But not at all; in my opinion, the word 'because' here signifies only that A is relevant, in this context, to this question. Hence the claim is merely that A bears relation R to $\langle P_k, X \rangle$. For example, suppose you ask why I got up at seven o'clock this morning, and

I say 'because I was woken up by the clatter the milkman made'. In that case I have interpreted your question as asking for a sort of reason that at least includes events-leading-up-to my getting out of bed, and my word 'because' indicates that the milkman's clatter was that sort of reason, that is, one of the events in what Salmon would call the causal process. Contrast this with the case in which I construe your request as being specifically for a motive. In that case I would have answered 'No reason, really. I could easily have stayed in bed, for I don't particularly want to do anything today. But the milkman's clatter had woken me up, and I just got up from force of habit I suppose.' In this case, I do not say 'because' for the milkman's clatter does not belong to the relevant range of events, as I understand your question.

It may be objected that 'because *A*' does not only indicate that *A* is *a* reason, but that it is *the* reason, or at least that it is a good reason. I think that this point can be accommodated in two ways. The first is that the relevance relation, which specifies what sort of thing is being requested as answer, may be construed quite strongly: 'give me a motive strong enough to account for murder', 'give me a statistically relevant preceding event not screened off by other events', 'give me a common cause', etc. In that case the claim that the proposition expressed by *A* falls in the relevant range, is already a claim that it provides a telling reason. But more likely, I think, the request need not be construed that strongly; the point is rather that anyone who answers a question is in some sense tacitly claiming to be giving a good answer. In either case, the determination of whether the answer is indeed good, or telling, or better than other answers that might have been given, must still be carried out, and I shall discuss that under the heading of 'evaluation'.

As a matter of regimentation I propose that we count (*) as a direct answer *only if*

A is relevant.[42] In that case, we don't have to understand the claim that *A* is relevant as explicit part of the answer either, but may regard the word 'because' solely as a linguistic signal that the words uttered are intended to provide an answer to the why-question just asked. (There is, as always, the tacit claim of the respondent that what he is giving is a good, and hence a relevant answer – we just do not need to make this claim part of the answer.) The definition is then:

B is a *direct answer* to question Q = P_k, X, R exactly if there is some proposition *A* such that *A* bears relation R to P_k, X and B is the proposition which is true exactly if (P_k; *and for all* i ≠ k, *not* P_i; *and A*) is true

where, as before, X = {P_1,.... P_k, ... }. Given this proposed definition of the direct answer, what does a why-question presuppose? Using Belnap's general definition we deduce:

a why-question *presupposes* exactly that
- (a) its topic is true
- (b) in its contrast-class, only its topic is true
- (c) at least one of the propositions that bears its relevance relation to its topic and contrast-class, is also true.

However, as we shall see, if all three of these presuppositions are true, the question may still not have a *telling* answer.

Before turning to the evaluation of answers, however, we must consider one related topic: when does a why-question arise? In the general theory of questions, the following were equated: question Q arises, all the presuppositions of Q are true. The former means that Q is not to be rejected as mistaken, the latter that Q has some true answer.

In the case of why-questions, we evaluate answers in the light of accepted background theory (as well as background information) and it seems to me that this drives a wedge between the two concepts. Of course, sometimes we reject a why-question because we think that it has no true answer. But as long as we do not think that, the question does arise, and is not mistaken, regardless of what is true.

To make this precise, and to simplify further discussion, let us introduce two more special terms. In the above definition of 'direct answer', let us call proposition A the *core* of answer B (since the answer can be abbreviated to '*Because A*'), and let us call the proposition that (P_k *and for all* $i \neq k$, *not* P_i) the *central presupposition* of question Q. Finally, if proposition A is relevant to $\langle P_k, X \rangle$ let us also call it relevant to Q.

In the context in which the question is posed, there is a certain body K of accepted background theory and factual information. This is a factor in the context, since it depends on who the questioner and audience are. It is this background which determines whether or not the question arises; hence a question may arise (or conversely, be rightly rejected) in one context and not in another.

To begin, whether or not the question genuinely *arises*, depends on whether or not K implies the central presupposition. As long as the central presupposition is not part of what is assumed or agreed to in this context, the why-question does not arise at all.

Secondly, Q presupposes *in addition* that one of the propositions A, relevant to its topic and contrast-class, is true. Perhaps K does not imply that. In this case, the question will still arise, provided K does not imply that all those propositions are false.

So I propose that we use the phrase 'the question arises in this context' to mean exactly this: K implies the central presupposition, and K does not imply the denial of any presupposition. Notice that this is very different from 'all

the presuppositions are true', and we may emphasize this difference by saying 'arises in context'. The reason we must draw this distinction is that K may not tell us which of the possible answers is true, but this *lacuna* in K clearly does not eliminate the question.

4.4 Evaluation of answers

The main problems of the philosophical theory of explanation are to account for legitimate rejections of explanation requests, and for the asymmetries of explanation. These problems are successfully solved, in my opinion, by the theory of why-questions as developed so far.

But that theory is not yet complete, since it does not tell us how answers are evaluated as telling, good, or better. I shall try to give an account of this too, and show along the way how much of the work by previous writers on explanation is best regarded as addressed to this very point. But I must emphasize, first, that this section is not meant to help in the solution of the traditional problems of explanation; and second, that I believe the theory of why-questions to be basically correct as developed so far, and have rather less confidence in what follows.

Let us suppose that we are in a context with background K of accepted theory plus information, and the question Q arises here. Let Q have topic B, and contrast-class X = {B, C, . . ., N}. How good is the answer *Because A*?

There are at least three ways in which this answer is evaluated. The first concerns the evaluation of A itself, as acceptable or as likely to be true. The second concerns the extent to which A *favours* the topic B as against the other members of the contrast-class. (This is where Hempel's criterion of giving reasons to expect, and Salmon's criterion of statistical relevance may find application.) The third concerns the comparison of *Because A* with other possible answers to the same question; and this has three aspects. The first is whether *A* is more probable (in view of *K*); the second whether it favours the topic to a greater extent; and the third, whether it is made wholly or partially irrelevant by other answers that could be given. (To this third aspect, Salmon's considerations about *screening off* apply.) Each of these three main ways of evaluation needs to be made more precise.

The first is of course the simplest: we rule out *Because A* altogether if *K* implies the denial of *A*; and otherwise ask what probability *K* bestows on *A*. Later we compare this with the probability which *K* bestows on the cores of other possible answers. We turn then to favouring.

If the question why B rather than C,. . ., N arises here, *K* must imply B and imply the falsity of C,. . ., N. However, it is exactly the information that the topic is true, and the alternatives to it not true, which is irrelevant to how favourable the answer is to the topic. The evaluation uses only that part of the background information which constitutes the general theory about these phenomena, plus other 'auxiliary' facts which are known but which do not imply the fact to be explained. This point is germane to all the accounts of explanation we have seen, even if it is not always emphasized. For example, in Salmon's first account, *A* explains B only if the probability of B given *A* does not equal the probability of B *simpliciter*. However, if I know that *A* and that B (as is often the case when I say that B because *A*), then my *personal probability* (that is, the probability given all the information I have) of *A* equals that of B and that of B given *A*, namely 1. Hence the probability to be used in evaluating answers is not at all the probability given all my background information, but rather, the probability given some of the general theories I

accept plus some selection from my data.[43] So the evaluation of the answer *Because A* to question Q proceeds with reference only to a certain part *K(Q)* of *K*. How that part is selected is equally important to all the theories of explanation I have discussed. Neither the other authors nor I can say much about it. Therefore the selection of the part *K(Q)* of *K* that is to be used in the further evaluation of *A*, must be a further contextual factor.[44]

If *K(Q)* plus *A* implies B, and implies the falsity of C,. . ., N, then *A* receives in this context the highest marks for favouring the topic B.

Supposing that *A* is not thus, we must award marks on the basis of how well *A* redistributes the probabilities on the contrast-class so as to favour B against its alternatives. Let us call the probability in the light of *K(Q)* alone the *prior* probability (in this context) and the probability given *K(Q)* plus *A* the *posterior* probability. Then *A* will do best here if the posterior probability of B equals 1. If *A* is not thus, it may still do well provided it shifts the mass of the probability function toward B; for example, if it raises the probability of B while lowering that of C,. . ., N; or if it does not lower the probability of B while lowering that of some of its closest competitors.

I will not propose a precise function to measure the extent to which the posterior probability distribution favours B against its alternatives, as compared to the prior. Two factors matter: the minimum odds of B against C,. . ., N, *and* the number of alternatives in C,. . ., N to which B bears these minimum odds. The first should increase, the second decrease. Such an increased favouring of the topic against its alternatives is quite compatible with a decrease in the probability of the topic. Imagining a curve which depicts the probability distribution, you can easily see how it could be changed quite dramatically so as to single out the topic − as the tree that stands out from the wood, so to say – even though the new advantage is only a relative one. Here is a schematic example:

> Why E_1 rather than $E_2, \ldots, E_{1000}$?
> Because *A*.
> Prob $(E_1) = \ldots = $ Prob $(E_{10}) = 99/1000 = 0.099$
> Prob $(E_{11}) = \ldots = $ Prob $(E_{1000}) = 1/99{,}000 = 0.00001$
> Prob $(E_1/A) = 90/1000 = 0.090$
> Prob $(E_2/A) = \ldots = $ Prob $(E_{1000}/A) = 910/999{,}000 \simeq 0.001$

Before the answer, E_1 was a good candidate, but in no way distinguished from nine others; afterwards, it is head and shoulders above all its alternatives, but has itself lower probability than it had before.

I think this will remove some of the puzzlement felt in connection with Salmon's examples of explanations that lower the probability of what is explained. In Nancy Cartwright's example of the poison ivy ('Why is this plant alive?') the answer ('It was sprayed with defoliant') was statistically relevant, but did not redistribute the probabilities so as to favour the topic. The mere fact that the probability was lowered is, however, not enough to disqualify the answer as a telling one.

There is a further way in which *A* can provide information which favours the topic. This has to do with what is called Simpson's Paradox; it is again Nancy Cartwright who has emphasized the importance of this for the theory of explanation (see n. 13 above). Here is an example she made up to illustrate it. Let H be 'Tom has heart disease'; S be 'Tom smokes'; and E, 'Tom does exercise'. Let us suppose the probabilities to be as follows:

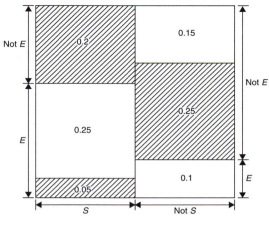

Figure 19.1

Shaded areas represent the cases in which H is true, and numbers the probabilities. By the standard calculation, the conditional probabilities are

$$\text{Prob}\,(H\,/\,S) = \text{Prob}\,(H) = \frac{1}{2}$$

$$\text{Prob}\,(H\,/\,S\,\&\,E) = \frac{1}{6}$$

$$\text{Prob}\,(H\,/\,E) = \frac{1}{8}$$

$$\text{Prob}\,(H\,/\,S\,\&\,\text{not}\,E) = 1$$

$$\text{Prob}\,(H\,/\,\text{not}\,E) = \frac{3}{4}$$

In this example, the answer 'Because Tom smokes' does favour the topic that Tom has heart disease, in a straightforward (though derivative) sense. For as we would say it, the odds of heart disease increase with smoking regardless of whether he is an exerciser or a non-exerciser, and he must be one or the other.

Thus we should add to the account of what it is for *A* to favour B as against C, . . ., N that: if $Z = \{Z_1, \ldots, Z_n\}$ is a logical partition of explanatorily relevant alternatives, and *A* favours B as against

C,..., N if any member of Z is added to our background information, then A does favour B as against C,..., N.

We have now considered two sorts of evaluation: how probable is A itself? *and*, how much does A favour B as against C,..., N? These are independent questions. In the second case, we know what aspects to consider, but do not have a precise formula that 'adds them all up'. Neither do we have a precise formula to weigh the importance of how likely the answer is to be true, against how favourable the information is which it provides. But I doubt the value of attempting to combine all these aspects into a single-valued measurement.

In any case, we are not finished. For there are relations among answers that go beyond the comparison of how well they do with respect to the criteria considered so far. A famous case, again related to Simpson's Paradox, goes as follows (also discussed in Cartwright's paper): at a certain university it was found that the admission rate for women was lower than that for men. Thus 'Janet is a woman' appears to tell for 'Janet was not admitted' as against 'Janet was admitted'. However, this was not a case of sexual bias. The admission rates for men and women for each department in the university were approximately the same. The appearance of bias was created because women tended to apply to departments with lower admission rates. Suppose Janet applied for admission in history; the statement 'Janet applied in history' *screens off* the statement 'Janet is a woman' from the topic 'Janet was not admitted' (in the Reichenbach–Salmon sense of 'screens off': P screens off A from B exactly if the probability of B given P and A is just the probability of B given P alone). It is clear then that the information that Janet applied in history (or whatever other department) is a much more telling answer than the earlier reply, in that it makes that reply irrelevant.

We must be careful in the application of this criterion. First, it is not important if some proposition P screens off A from B if P is not the core of an answer to the question. Thus if the why-question is a request for information about the mechanical processes leading up to the event, the answer is no worse if it is statistically screened off by other sorts of information. Consider 'Why is Peter dead?' answered by 'He received a heavy blow on the head' while we know already that Paul has just murdered Peter in some way. Secondly, a screened-off answer may be good but partial rather than irrelevant. (In the same example, we know that there must be some true proposition of the form 'Peter received a blow on the head with impact x', but that does not disqualify the answer, it only means that some more informative answer is possible.) Finally, in the case of a deterministic process in which state A_i, and no other state, is followed by state A_{i+1}, the best answers to the question 'Why is the system in state A_n at time t_n?' may all have the form 'Because the system was in state A_i at time t_i', but each such answer is screened off from the event described in the topic by some other, equally good answer. The most accurate conclusion is probably no more than that if one answer is screened off by another, and not conversely, then the latter is better in some respect.

When it comes to the evaluation of answers to why-questions, therefore, the account I am able to offer is neither as complete nor as precise as one might wish. Its shortcomings, however, are shared with the other philosophical theories of explanation I know (for I have drawn shamelessly on those other theories to marshal these criteria for answers). And the traditional main problems of the theory of explanation are solved not by seeing what these criteria are, but by the general theory that explanations are answers to why-questions, which are themselves contextually determined in certain ways.

4.5 Presupposition and relevance elaborated

Consider the question 'Why does the hydrogen atom emit photons with frequencies in the general Balmer series (only)?' This question presupposes that the hydrogen atom emits photons with these frequencies. So how can I even ask that question unless I believe that theoretical presupposition to be true? Will my account of why-questions not automatically make scientific realists of us all?

But recall that we must distinguish carefully what a theory *says* from what we believe when we accept that theory (or rely on it to predict the weather or build a bridge, for that matter). The epistemic commitment involved in accepting a scientific theory, I have argued, is not belief that it is true but only the weaker belief that it is empirically adequate. In just the same way we must distinguish what the question says (i.e. *presupposes*) from what we believe when we ask that question. The example I gave above is a question which arises (as I have defined that term) in any context in which those hypotheses about hydrogen, and the atomic theory in question, are *accepted*. Now, when I ask the question, if I ask it seriously and in my own person, I imply that I believe that this question arises. But that means then only that my epistemic commitment indicated by, or involved in, the asking of this question, is exactly – no more and no less than – the epistemic commitment involved in my acceptance of these theories.

Of course, the discussants in this context, in which those theories are accepted, are conceptually immersed in the theoretical world-picture. They talk the language of the theory. The phenomenological distinction between objective or real, and not objective or unreal, is a distinction between what there is and what there is not which is drawn in that theoretical picture. Hence the questions asked are asked in the theoretical language – how could it be

otherwise? But the epistemic commitment of the discussants is not to be read off from their language.

Relevance, perhaps the other main peculiarity of the why-question, raises another ticklish point, but for logical theory. Suppose, for instance, that I ask a question about a sodium sample, and my background theory includes present atomic physics. In that case the answer to the question may well be something like: because this material has such-and-such an atomic structure. Recalling this answer from one of the main examples I have used to illustrate the asymmetries of explanation, it will be noted that, *relative to* this background theory, my answer is a proposition necessarily equivalent to: because this material has such-and-such a characteristic spectrum. The reason is that the spectrum is unique – it identifies the material as having that atomic structure. But, here is the asymmetry, I could not well have answered the question by saying that this material has that characteristic spectrum.

These two propositions, one of them relevant and the other not, are equivalent relative to the theory. Hence they are true in exactly the same possible worlds allowed by the theory (less metaphysically put: true in exactly the same models of that theory). So now we have come to a place where there is a conflict with the simplifying hypothesis generally used in formal semantics, to the effect that propositions which are true in exactly the same possible worlds are identical. If one proposition is relevant and the other not, they cannot be identical.

We could avoid the conflict by saying that of course there are possible worlds which are not allowed by the background theory. This means that when we single out one proposition as relevant, in this context, and the other as not relevant and hence distinct from the first, we do so in part by thinking in terms of worlds (or models) regarded in this context as impossible.

I have no completely telling objection to this, but I am inclined to turn, in our semantics, to a different modelling of the language, and reject the simplifying hypothesis. Happily there are several sorts of models of language, not surprisingly ones that were constructed in response to other reflections on relevance, in which propositions can be individuated more finely. One particular sort of model, which provides a semantics for Anderson and Belnap's logic of tautological entailment, uses the notion of *fact*.[45] There one can say that

It is either raining or not raining

It is either snowing or not snowing

although true in exactly the same possible situations (namely, in all) are yet distinguishable through the consideration that today, for example, the first is *made true* by the fact that it is raining, and the second is made true by quite a different fact, namely, that it is not snowing. In another sort of modelling, developed by Alasdair Urquhart, this individuating function is played not by facts but by bodies of information.[46] And still further approaches, not necessarily tied to logics of the Anderson-Belnap stripe, are available.

In each case, the relevance relation among propositions will derive from a deeper relevance relation. If we use facts, for example, the relation R will derive from a request to the effect that the answer must provide a proposition which describes (is made true by) facts of a certain sort: for example, facts about atomic structure, or facts about this person's medical and physical history, or whatever.

5 Conclusion

Let us take stock. Traditionally, theories are said to bear two sorts of relation to the observable phenomena: *description* and *explanation*. Description

can be more or less accurate, more or less informative; as a minimum, the facts must 'be allowed by' the theory (fit some of its models), as a maximum the theory actually implies the facts in question. But in addition to a (more or less informative) description, the theory may provide an explanation. This is something 'over and above' description; for example, Boyle's law describes the relationship between the pressure, temperature, and volume of a contained gas, but does not explain it – kinetic theory explains it. The conclusion was drawn, correctly I think, that even if two theories are strictly empirically equivalent they may differ in that one can be used to answer a given request for explanation while the other cannot.

Many attempts were made to account for such 'explanatory power' purely in terms of those features and resources of a theory that make it informative (that is, allow it to give better descriptions). On Hempel's view, Boyle's law does explain these empirical facts about gases, but minimally. The kinetic theory is perhaps better *qua* explanation simply because it gives so much more information about the behaviour of gases, relates the three quantities in question to other observable quantities, has a beautiful simplicity, unifies our over-all picture of the world, and so on. The use of more sophisticated statistical relationships by Wesley Salmon and James Greeno (as well as by I. J. Good, whose theory of such concepts as weight of evidence, corroboration, explanatory power, and so on deserves more attention from philosophers), are all efforts along this line.[47] If they had succeeded, an empiricist could rest easy with the subject of explanation.

But these attempts ran into seemingly insuperable difficulties. The conviction grew that explanatory power is something quite irreducible, a special feature differing in kind from empirical adequacy and strength. An inspection of examples defeats any attempt to identify the ability to explain with any complex of those more familiar and down-to-earth virtues that are used to evaluate the theory *qua* description. Simultaneously it was argued that what science is really after is understanding, that this consists in being in a position to explain, hence what science is really after goes well beyond empirical adequacy and strength. Finally, since the theory's ability to explain provides a clear reason for accepting it, it was argued that explanatory power is evidence for the *truth* of the theory, special evidence that goes beyond any evidence we may have for the theory's empirical adequacy.

Around the turn of the century, Pierre Duhem had already tried to debunk this view of science by arguing that explanation is not an aim of science. In retrospect, he fostered that explanation–mysticism which he attacked. For he was at pains to grant that explanatory power does not consist in resources for description. He argued that only metaphysical theories explain, and that metaphysics is an enterprise foreign to science. But fifty years later, Quine having argued that there is no demarcation between science and philosophy, and the difficulties of the ametaphysical stance of the positivist-oriented philosophies having made a return to metaphysics tempting, one noticed that scientific activity does involve explanation, and Duhem's argument was deftly reversed.

Once you decide that explanation is something irreducible and special, the door is opened to elaboration by means of further concepts pertaining thereto, all equally irreducible and special. The premises of an explanation have to include lawlike statements; a statement is lawlike exactly if it implies some non-trivial counterfactual conditional statement; but it can do so only by asserting relationships of necessity in nature. Not all classes correspond to genuine properties; properties and propensities figure in explanation. Not everyone has joined this return to essentialism or neo-Aristotelian realism, but

some eminent realists have publicly explored or advocated it.

Even more moderate elaborations of the concept of explanation make mysterious distinctions. Not every explanation is a scientific explanation. Well then, that irreducible explanation-relationship comes in several distinct types, one of them being scientific. A scientific explanation has a special form, and adduces only special sorts of information to explain – information about causal connections and causal processes. Of course, a causal relationship is just what 'because' must denote; and since the *summum bonum* of science is explanation, science must be attempting even to describe something beyond the observable phenomena, namely causal relationships and processes.

These last two paragraphs describe the flights of fancy that become appropriate if explanation is a relationship *sui generis* between theory and fact. But there is no direct evidence for them at all, because if you ask a scientist to explain something to you, the information he gives you is not different in kind (and does not sound or look different) from the information he gives you when you ask for a description. Similarly in 'ordinary' explanations: the information I adduce to explain the rise in oil prices, is information I would have given you to a battery of requests for description of oil supplies, oil producers, and oil consumption. To call an explanation scientific, is to say nothing about its form or the sort of information adduced, but only that the explanation draws on science to get this information (at least to some extent) and, more importantly, that the criteria of evaluation of how good an explanation it is, are being applied using a scientific theory (in the manner I have tried to describe in Section 4 above).

The discussion of explanation went wrong at the very beginning when explanation was conceived of as a relationship like description: a relation between theory and fact. Really it is a three-term relation, between theory, fact, and context. No wonder that no single relation between theory and fact ever managed to fit more than a few examples! Being an explanation is essentially relative, for an explanation is an *answer.* (In just that sense, being a daughter is something relative: every woman is a daughter, and every daughter is a woman, yet being a daughter is not the same as being a woman.) Since an explanation is an answer, it is evaluated *vis-à-vis* a question, which is a request for information. But exactly what is requested, by means of the interrogative 'Why is it the case that P?', differs from context to context. In addition, the background theory plus data relative to which the question is evaluated, as arising or not arising, depends on the context. And even what part of that background information is to be used to evaluate how good the answer is, *qua* answer to that question, is a contextually determined factor. So to say that a given theory can be used to explain a certain fact, is always elliptic for: there is a proposition which is a telling answer, relative to this theory, to the request for information about certain facts (those counted as relevant for *this* question) that bears on a comparison between this fact which is the case, and certain (contextually specified) alternatives which are not the case.

So scientific explanation is not (pure) science but an application of science. It is a use of science to satisfy certain of our desires; and these desires are quite specific in a specific context, but they are always desires for descriptive information. (Recall: every daughter is a woman.) The exact content of the desire, and the evaluation of how well it is satisfied, varies from context to context. It is not a single desire, the same in all cases, for a very special sort of thing, but rather, in each case, a different desire for something of a quite familiar sort.

Hence there can be no question at all of explanatory power as such (just as it would be silly to speak of the 'control power' of a theory, although of course we rely on theories to gain control over nature and circumstances). Nor can there be any question of explanatory success as providing evidence for the truth of a theory that goes beyond any evidence we have for its providing an adequate description of the phenomena. For in each case, a success of explanation is a success of adequate and informative description. And while it is true that we seek for explanation, the value of this search for science is that the search for explanation is *ipso facto* a search for empirically adequate, empirically strong theories.

Notes

1 This chapter is based in part on my paper by the same title, *American Philosophical Quarterly*, **14** (1977), 143–50, presented to the American Philosophical Association, Portland, March 1977, with commentary by Kit Fine and Clark Glymour.

2 A. Lavoisier, *Oeuvres* (Paris: Imp. Imperiale, 1862), vol. II, p. 640. I owe this and the other historical references to my former student, Paul Thagard.

3 C. Darwin, *On the Origin of Species* (text of 6th edn, New York: Collier, 1962), p. 476

4 Cf. Christiaan Huygens, *Treatise on Light*, trans. S.P. Thompson (New York: Dover, 1962), pp. 19f, 22, 63; Thomas Young, *Miscellaneous Works*, ed. G. Peacock (London: John Murray, 1855), vol. I, pp. 168, 170.

5 A. Fresnel, *Oeuvres complètes* (Paris: Imp. Imperiale, 1866), vol. I, p. 36 (see also pp. 254, 355); Lavoisier, op. cit., p. 233.

6 C. Darwin, *The Variation of Animals and Plants* (London: John Murray, 1868), Vol. I, p. 9; *On the Origin of the Species* (Facs. of first edition, Cambridge, Mass., 1964), p. 408.

7 C.G. Hempel and P. Oppenheim, 'Studies in the Logic of Explanation', *Philosophy of Science*, **15** (1948), 135–75

8 C.G. Hempel, *Philosophy of Natural Science* (Englewood Cliffs, N.J.: 1966), pp. 48f.; see S. Bromberger, 'Why-questions', (n. 32 below) for some of the counterexamples.

9   W.C. Salmon, *Statistical Explanation and Statistical relevance* (Pittsburgh: University of Pittsburgh Press, 1971), pp. 33f.

10 M. Beckner, *The Biological Way of Thought* (Berkeley: University of California Press, 1968), p. 165 (first pub. 1959, by Columbia University Press).

11 In a presentation at a conference at the University of Illinois; a summary of the paper may be found in F. Suppe (ed.), *The Structure of Scientific Theories* (Urbana, Ill.: University of Illinois Press, 1974).

12 Salmon, op. cit. p. 64.

13 Nancy Cartwright, 'Causal Laws and Effective Strategies', *Noûs*, **13**(4) (1979), 419–37. The examples cited had been communicated in 1976.

14 Salmon, op. cit., p. 78. For reasons of exposition, I shall postpone discussion of the *screening off* relation to Sect. 2.6 although Salmon already used it here.

15 M. Freidman, 'Explanation and Scientific Understanding', *Journal of Philosophy*, **71** (1971), 5–19.

16 See P. Kitcher, 'Explanation, Conjunction, and Unification', *Journal of Philosophy*, **73** (1976), 207–12.

17 J. Greeno, 'Explanation and Information', pp. 89–104, in W.C. Salmon, op. cit. (n. 9).

18 T. Kuhn, *The Structure of Scientific Revolutions* (Chicago: University of Chicago Press, 1970), pp 107f.

19 P.J. Zwart, *Causaliteit* (Assen: van Gorcum, 1967), p. 133.

20 J.L. Mackie, 'Causes and Conditions', *American Philosophical Quarterly*, **2** (1965), 245–64. Since then, Mackie has published a much more extensive theory of causation in *The Cement of the Universe* (Oxford: Clarendon Press, 1974). Since I am of necessity restricted to a small and selective, i.e. biased, historical introduction to my own theory of explanation, I must of necessity do less than justice to most authors discussed.

21 D. Lewis, 'Causation', *Journal of Philosophy*, **70** (1973), 556f.

22 See N. Goodman, *Fact, Fiction, and Forecast* (Cambridge, Mass.: Harvard University Press, 1955), Ch. 1. The logical theory of counterfactual conditionals has been developed, with success in basic respects but in others still subject to debate, in a number of

articles of which the first was Robert Stalnaker, 'A Theory of Conditionals', pp. 98–112, in N. Rescher (ed.) *Studies in Logical Theory* (Oxford: Blackwell, 1968), and one book (David Lewis, *Counterfactuals* (Oxford: Blackwell, 1973)). For a summary of the results, and problems, see my 'Report on Conditionals', *Teorema*, **5** (1976), 5–25.

23 *Mind*, N.S. **3** (1894), 436–8; P.E.B. Jourdain (ed.), *The Philosophy of Mr. B*rtr*nd R*ss*ll* (London: Allen and Unwin, 1918), p39. The analysis of the example which I give here is due to Richmond Thomason.

24 W.C. Salmon, 'Why ask, "Why?"?', presidential address to the Pacific Division of the American Philosophical Association, San Francisco, March 1978. Page references are to the manuscript version completed and circulated in May 1978; the paper is published in *Proceedings and Addresses of the American Philosophical Association*, **51** (1978), 683–705.

25 Op. cit, pp. 29f.

26 Ibid., pp. 14f.

27 Hans Reichenbach, *The Direction of Time* (Berkeley: University of California Press, 1956), Sects 19 and 22.

28 Salmon, op. cit., p. 13.

29 But this might defeat the use of Salmon's theory in metaphysical arguments, for example, his argument for realism at the end of this paper.

30 This survey is found in Zwart, op. cit., p. 135, n 19; references are to Beck's and Nagel's papers in H. Feigl and M. Brodbeck (eds), *Readings in the Philosophy of Science* (New York: Appleton-Century-Crofts, 1953), pp. 374 and 698, R.B. Braithwaite, *Scientific Explanation* (Cambridge: Cambridge University Press, 1953), p. 320; D. Bohm, *Causality and Chance in Modern Physics* (London: Routledge & Kegan Paul, 1957), *passim*.

31 N.R. Hanson, *Patterns of Discovery* (Cambridge: Cambridge University Press, 1958), p. 54.

32 Zwart, op. cit., p. 136; my translation.

33 S. Bromberger, 'Why Questions', pp. 86–108, in R.G. Colodny (ed.) *Mind and Cosmos* (Pittsburgh: University of Pittsburgh Press, 1966).

34 'Explanations-of-what?' mimeographed and circulated, Stanford University, 1974. The idea was independently developed, by Jon Dorling in a

paper circulated in 1974, and reportedly by Alan Garfinkel in *Explanation and individuals* (Yale University Press, forthcoming). I wish to express my debt to Bengt Hannson for discussion and correspondence in the autumn of 1975 which clarified these issues considerably for me.

35 For a fuller account of Aristotle's solution of the asymmetries, see my 'A Reexamination of Aristotle's Philosophy of Science', *Dialogue*, 1980. The story was written in reply to searching questions and comments by Professor J. J. C. Smart, and circulated in November 1976.

36 At the end of my 'The Only Necessity is Verbal Necessity', *Journal of Philosophy*, **74** (1977), 71–85 (itself an application of formal pragmatics to a philosophical problem), there is a short account of the development of these ideas, and references to the literature. The paper 'Demonstratives' by David Kaplan which was mentioned there as forthcoming, was completed and circulated in mimeo'd form in the spring of 1977; it is at present the most important source for the concepts and applications of formal pragmatics, although some aspects of the form in which he develops this theory are still controversial (see also, n. 30 to Ch. 6 below [sic]).

37 Belnap's theory was first presented in *An analysis of questions: preliminary report* (Santa Monica, Cal.: System Development Corporation, technical memorandum 7-1287-1000/00, 1963), and is now more accessible in N.D. Belnap Jr. and J.B. Steel, Jr., *The Logic of Questions and Answers* (New Haven: Yale University Press, 1976).

38 I heard the example from my former student Gerald Charlwood. Ian Hacking and J. J. C. Smart told me that the officer was Sir Charles Napier.

39 C.L. Hamblin, 'Questions', *Australasian Journal of Philosophy*, **36** (1958), 159–68.

40 The defining clause is equivalent to 'any proposition which is true if any direct answer to Q is true'. This includes, of course, propositions which would normally be expressed by means of 'metalinguistic' sentences – a distinction which, being language-relative, is unimportant.

41 In the book by Belnap and Stell (see n. 37 above), Bromberger's theory of why-questions is cast in the general form common to elementary questions. I think that Bromberger arrived at his concept of 'abnormic law' (and the form of answer exhibited by ' "Grünbaum" is spelled with an umlaut because it is an English word borrowed from German, and no English words are spelled with an umlaut except those borrowed from another language in which they are so spelled'), because he ignored the tacit *rather than* (contrast-class) in why-interrogatives, and then had to make up for this deficiency in the account of what the answers are like.

42 I call this a matter of regimentation, because the theory could clearly be developed differently at this point, by building the claim of relevance into the answer as an explicit conjunct. The result would be an alternative theory of why-questions which, I think, would equally well save the phenomena of explanation or why-question asking and answering.

43 I mention Salmon because he does explicitly discuss this problem, which he calls *the problem of reference class*. For him this is linked with the (frequency) interpretation of probability. But it is a much more general problem. In deterministic, non-statistical (what Hempel would call a deductive-nomological) explanation, the adduced information implies the fact explained. This implication is relative to our background assumptions, or else those assumptions are part of the adduced information. But clearly, our information that the fact to be explained is actually the case, and all its consequences, must carefully be kept out of those background assumptions if the account of explanation is not to be trivialized. *Mutatis mutandis* for statistical explanations given by a Bayesian, as is pointed out by Glymour in his *Theory and Evidence*.

44 I chose the notation K(Q) deliberately to indicate the connection with models of rational belief, conditionals, and hypothetical reasoning, as discussed for example by William Harper. There is, for example, something called the Ramsey test: to see whether a person with total beliefs K accepts that if A then B, he must check whether K(A) implies B, where K(A) is the 'least revision' of K that implies A. In order to 'open the question' for A, such a person must similarly shift his beliefs

from K to $K?A$, the 'least revision of K that is consistent with A; and we may conjecture that $K(A)$ is the same as $(K?A)\&A$. What I have called $K(Q)$ would, in a similar vein, be a revision of K that is compatible with every member of the contrast class of Q, and also with the denial of the topic of Q. I don't know whether the 'least revision' picture is the right one, but these suggestive similarities may point to important connections; it does seem, surely, that explanation involves hypothetical reasoning. Cf. W. Harper, 'Ramsey Test Conditionals and Iterated Belief Change', pp. 117–35, in W. Harper and C.A. Hooker, *Foundations of Probability Theory, Statistical Inferences and Statistical Theories of Science* (Dordrecht: Reidel, 1976), and his 'Rational Conceptual Change', in F. Suppe and P. Asquith (eds) *PSA 1976* (East Lansing: Philosophy of Science Association, 1977).

45 See my 'Facts and Tautological Entailment', *Journal of Philosophy*, **66** (1969), 477–87 and reprinted in A.R. Anderson and N.D. Belnap, Jr., *Entailment* (Princeton: Princeton University Press, 1975), and 'Extension, Intension, and Comprehension', M. Munitz (ed.), *Logic and Ontology* (New York: New York University Press, 1973).

46 For this and other approaches to the semantics of relevance see Anderson and Belnap, op. cit. (n. 45 above).

47 I.J. Good, 'Weight of Evidence, Corroboration, Explanatory Power, and the Utility of Experiments', *Journal of the Royal Statistical Society*, series B, **22** (1960), 319–31; and 'A Causal Calculus', *British Journal for the Philosophy of Science*, **11** (1960/61), 605–18 and **12** (1961/62), 43–51. For discussion, see W. Salmon, 'Probabilistic Causality', *Pacific Philosophical Quarterly*, **61** (1980), 61–74.

Chapter 20

Peter Lipton

EXPLANATION

Understanding explanation

Once we have made an inference, what do we do with it? Our inferred beliefs are guides to action that help us to get what we want and avoid trouble. Less practically, we also sometimes infer simply because we want to learn more about the way the world is. Often, however, we are not satisfied to discover that something is the case: we want to know *why*. Thus our inferences may be used to provide explanations, and they may themselves be explained. The central question about our explanatory practices can be construed in several ways. We may ask what principles we use to distinguish between a good explanation, a bad explanation, and no explanation at all. Or we may ask what relation is required between two things to count one to be an explanation of the other. We can also formulate the question in terms of the relationship between knowledge and understanding. Typically, someone who asks why something is the case already knows that it is the case. The person who asks why the sky is blue knows that it is blue, but does not yet understand why. The question about explanation can then be put this way: What has to be added to knowledge to get understanding?

As in the case of inference, explanation raises problems both of justification and of description. The problem of justification can be understood in various ways. It may be seen as the problem of showing whether things we take to be good explanations really are, whether they really provide understanding. The issue here, to distinguish it from the case of inference, is not whether there is any reason to believe that our putative explanations are themselves true, but whether, granting that they are true, they really explain. There is no argument against the possibility of explanation on a par with Hume's argument against induction. The closest thing is regress of why. The why-regress is a feature of the logic of explanation that many of us discovered as children, to our parents' cost. I vividly recall the moment it dawned on me that, whatever my mother's answer to my latest why-question, I could simply retort by asking 'Why?' of the answer itself, until my mother ran out of answers or patience. But if only something that is itself understood can be used to explain, and understanding only comes through being explained by something else, then the infinite chain of why's makes explanation impossible. Sooner or later, we get back to something unexplained, which ruins all the attempts to explain that are built upon it (cf. Friedman, 1974, pp. 18–19).

This skeptical argument is not very troubling. One way to stop the regress is to argue for phenomena that are self-explanatory or that can be understood without explanation. But while there may be such phenomena, this reply concedes too much to the skeptical argument. A better reply is that explanations need not themselves be understood. A drought may explain a poor crop, even if we don't understand why there was a drought; I understand why you didn't come to the party if you explain that you had a bad headache, even if I have no idea why you had a headache; the big bang explains the background radiation, even if the big bang is itself inexplicable, and so on. Understanding is not like a substance that the explanation has to possess in order to pass it on to the phenomenon to be explained. Rather than show that explanation is impossible, the regress argument brings out the important facts that explanations can be chained and that what explains need not itself be understood, and so provides useful constraints on a proper account.

The reason there is no skeptical argument against explanation on a par with Hume's argument against induction is fairly clear. Hume's argument, like all the great skeptical arguments, depends on our ability to see how our methods of acquiring beliefs could lead us into massive error. There are possible worlds where our methods persistently mislead us, and Hume exploits these possibilities by arguing that we have no non-circular way of showing that the actual world is not one of these misleading worlds. In the case of explanation, by contrast, the skeptic does not have this handle, at least when the issue is not whether the explanation is true, but whether the truth really explains. We do not know how to make the contrast between understanding and merely seeming to understand in a way that makes sense of the possibility that most of the things that meet all our standards for explanation might nonetheless not really explain. To put the matter another way, we do not see a gap between meeting our standards for the explanation and actually understanding

in the way we easily see a gap between meeting our inductive standards and making an inference that is actually correct.

It is not clear whether this is good or bad news for explanation. On the one hand, in the absence of a powerful skeptical argument, we feel less pressure to justify our practice. On the other, the absence seems to show that our grasp on explanation is even worse than our grasp on inference. We know that inferences are supposed to take us to truths and, as Hume's argument illustrates, we at least have some appreciation of the nature of these ends independently of the means we use to try to reach them. The situation is quite different for explanation. We may say that understanding is the goal of explanation, but we do not have a clear conception of understanding apart from whatever our explanations provide. If this is right, the absence of powerful skeptical arguments against explanation does not show that we are in better shape here than we are in the case of inference. Perhaps things are even worse for explanation: here we may not even know what we are trying to do. Once we know that something is the case, what is the point of asking why?

Reason, familiarity, deduction

Explanation also raises the problem of description. Whatever the point of explanation or the true nature of understanding, we have a practice of giving and judging explanations, and the problem of description is to give an account of how we do this. As with the problem of justification, the central issue here is not how we judge whether what we claim to be an explanation is true, but whether, granting that it is true, it really does explain what it purports to explain. Like our inductive practices, our explanatory practices display the gap between doing and describing. We discriminate between things we understand and things we do not, and between good explanations and bad explanations, but we are strikingly bad at giving any sort of principled account of how we do this. As before, the best way to make this claim convincing is to canvass some of the objections to various popular accounts of explanation. In this chapter, I will consider three accounts: the reason model, the familiarity model, and the deductive-nomological model.

According to the reason model, to explain a phenomenon is to give a reason to believe that the phenomenon occurs (cf. Hempel 1965, pp. 337, 364–76). On this view, an engineer's explanation of the collapse of a bridge succeeds by appealing to theories of loading, stress, and fatigue which, along with various particular facts, show that the collapse was likely. There is a germ of truth in this view, since explanations do quite often make the phenomenon likely and give us a reason to believe it occurs. A particularly satisfying type of explanation takes a phenomenon that looks accidental and shows how, given the conditions, it was really inevitable, and these deterministic explanations do seem to provide strong reasons for belief. Moreover, the reason model suggests a natural connection between the explanatory and predictive uses of scientific theories since, in both cases, the theory works by providing grounds for belief.

On balance, however, the reason model is extremely implausible. It does not account for the central difference between knowing that a phenomenon occurs and understanding why it occurs. The model claims that understanding why the phenomenon occurs is having a reason to believe that it occurs, but we already have this when we know that it occurs. We already have a reason to believe the bridge collapsed when we ask why it did, so we can not simply be asking for a reason when we ask for an explanation. Explanations may provide reasons for belief, but that is not enough. It is also often too much:

many explanations do not provide any actual reason to believe that the phenomenon occurs. Suppose you ask me why there are certain peculiar tracks in the snow in front of my house. Looking at the tracks, I explain to you that a person on snowshoes recently passed this way. This is a perfectly good explanation, even if my only reason for believing it is that I see the very tracks I am explaining. This 'self-evidencing explanation' (Hempel 1965, pp. 370–4) has a distinctive circularity: the passing person on snowshoes explains the tracks and the tracks provide the evidence for the passing. What is significant is that this circularity is *virtuous*: it ruins neither the explanation nor the justification. It does, however, show that the justification view of explanation is false, since to take the explanation to provide a reason to believe the phenomenon after the phenomenon has been used as a reason to believe the explanation would be vicious. In other words, if the reason model were correct, self-evidencing explanations would be illicit, but self-evidencing explanations may be perfectly acceptable and are ubiquitous, as we will see in chapter four [*sic*], so the reason model is wrong. Providing reasons for belief is neither necessary nor sufficient for explanation.

Another answer to the descriptive question is the familiarity model. On this view, unfamiliar phenomena call out for explanation, and good explanations somehow make them familiar (cf. Hempel, 1965, pp. 430–3; Friedman, 1974, pp. 9–11). On a common version of this view, there are certain familiar phenomena and processes that we do not regard as in need of explanation, and a good explanation of a phenomenon not in this class consists in showing it to be the outcome of a process that is analogous to the processes that yield the familiar phenomena. The kinetic theory of gases explains various phenomena of heat by showing that gases behave like collections of tiny billiard balls;

Darwin's theory of natural selection explains the traits of plants and animals by describing a mechanism similar to the mechanism of artificial selection employed by animal breeders; and a theory of electronics explains by showing that the flow of current through a wire is like the flow of water through pipes. But this is not a particularly attractive version of the familiarity view, in part because it does not help us to understand why certain phenomena are familiar in the first place, and because not all good explanations rely on analogies.

A more promising version of the familiarity view begins with the idea that a phenomenon is unfamiliar when, although we may know that it occurs, it remains surprising because it is in tension with other beliefs we hold. A good explanation shows how the phenomenon arises in a way that eliminates the tension and so the surprise (cf. Hempel, 1965, pp. 428–30). We may know that bats navigate with great accuracy in complete darkness, yet find this very surprising, since it seems in tension with our belief that vision is impossible in the dark. Finding out about echolocation shows that there is no real tension, and we are no longer surprised. The magician tells me the number I was thinking of, to my great surprise; a good explanation of the trick ruins it by making it unsurprising. This version of the familiarity view has the virtue of capturing the fact that it is very often surprise that prompts the search for explanations. It also calls our attention to the process of 'defamiliarization', which is often the precursor to asking why various common phenomena occur. In one sense, the fact that the sky is blue is a paradigm of familiarity, but we become interested in explaining this when we stop to consider how odd it is that the sky should have any color at all. Again, the fact that the same side of the moon always faces us at first seems not to call out for any interesting explanation, since it just seems to show that the moon is not spinning on its own axis. It is only when we realize that the moon must actually spin in order to keep the same side towards us, and moreover with a period that is exactly the same as the period of its orbit around the earth, that this phenomenon cries out for explanation. The transformation of an apparently familiar phenomenon into an extraordinary 'coincidence' prompts the search for an adequate explanation. The surprise version of the familiarity model also suggests that a good explanation of a phenomenon will sometimes show that beliefs that made the phenomenon surprising are themselves in error, and this is an important part of our explanatory practice. If I am surprised to see a friend at the supermarket because I expected him to be away on vacation, he will not satisfy my curiosity simply by telling me that he needed some milk, but must also say something about why my belief about his travel plans was mistaken.

There are three objections that are commonly made to a familiarity model of explanation. One is that familiarity is too subjective a notion to yield a suitably objective account of explanation. The surprise version of the familiarity model does make explanation audience relative, since what counts as a good explanation will depend on prior expectations, which vary from person to person. But it is not clear that this is a weakness of the model. Nobody would argue that an account of inference is unacceptable because it makes warranted inductions vary with prior belief. Similarly, an account that makes explanation interest relative does not thereby make explanation perniciously subjective. (We will consider the interest relativity of explanation in the next chapter [sic].) In particular, the familiarity model does not collapse the distinction between understanding a phenomenon and mistakenly thinking one does. The explanation must itself be true, and it must say something about how the phenomenon came about, but just what it says may legitimately depend on the audience's interests and expectations.

Another objection to the familiarity model is that good explanations often themselves appeal to unfamiliar events and processes. This is particularly common in the advanced sciences. Here again, however, it is unclear whether this is really a difficulty for a reasonable version of the model. Scientific explanations often appeal to exotic processes, but it is not clear that such an explanation is very good if it conflicts with other beliefs. The third and most telling objection is that we often explain familiar phenomena. The process of defamiliarization goes some way towards meeting this, but it is not enough to turn the objection completely. The rattle in my car is painfully familiar, and consistent with everything else I believe, but while I am sure there is a good explanation for it, I don't have any idea what it is. Nor do you have to convince me that, in fact, it is somehow surprising that there should be a rattle to get me interested in explaining it. Surprise is often a precursor to the search for explanation, but it is not the only motivation. A reasonable version of the familiarity theory has more going for it than many of its critics suppose, but does not by itself provide an adequate description of our explanatory practices.

The third and best known account of explanation is the deductive-nomological model, according to which we explain a phenomenon by deducing it from a set of premises that includes at least one law that is necessary for the deduction (Hempel, 1965, pp. 335–76). On this view, we can explain why a particular star has its characteristic spectrum shifted towards the red by deducing this shift from the speed at which the star is receding from us, and the Doppler law that links recession and red shift. (This Doppler effect is similar to the change in pitch of a train whistle as the train passes by.) Of course, many scientific explanations, and most lay explanations, do not meet the strict requirements of the model, since they either do not contain exceptionless laws or do not strictly entail the phenomenon, but they can be seen as 'explanation sketches' (Hempel, 1965, pp. 423–4) that provide better or worse approximations to a full deductive-nomological explanation.

The deductive-nomological model is closely related to the reason model, since the premises that enable us to deduce the phenomenon often also provide us with a reason to believe that the phenomenon occurs, but it avoids some of the weaknesses of the reason model. The explanation of the red shift in terms of the recession satisfies the deductive-nomological model even though, as it happens, the red shift is itself crucial evidence for the speed of recession that explains it. That is, the Doppler explanation is self-evidencing. Unlike the reason model, the deductive-nomological model allows for self-evidencing explanations. It also does better than the reason model in accounting for the difference between knowing and understanding. When we know that a phenomenon occurs but do not understand why, we usually do not know laws and supporting premises that entail the phenomenon. So at least a deductive-nomological argument usually gives us something new. To its credit, it also does seem that, when theories are used to give scientific explanations, these explanations often do aspire to deductive-nomological form. Moreover, the model avoids the main objection to the familiarity model, since a phenomenon may be common and unsurprising, but still await a deductive-nomological explanation.

The model nevertheless faces debilitating objections. It is almost certainly too strong: very few explanations fully meet the requirements of the model and, while some scientific explanations at least aspire to deductive-nomological status, many ordinary explanations include no laws and allow no deduction, yet are not incomplete or mere sketches. The model is also too weak. Perhaps the best known objection is that it does not account for the asymmetries of explanation (Bromberger, 1966; Friedman, 1974, p. 8; van Fraassen, 1980, p. 112). Consider again the Doppler explanation of the red shift. In that explanation, the law is used to deduce the shift from the recession, but the law is such that we could equally well use it to deduce the recession from the shift. Indeed this is how we figure out what the recession is. What follows from this is that, according to the deductive-nomological model, we can explain why the star is receding as it is by appeal to its red shift. But this is wrong: the shift no more explains the recession than a wet sidewalk explains why it is raining. The model does not account for the many cases where there is explanatory asymmetry but deductive symmetry.

The deductive-nomological model should produce a sense of *déjà vu*, since it is isomorphic to the hypothetico-deductive model of confirmation, which we considered in the last chapter [sic]. In one case we explain a phenomenon by deducing it from a law; in the other we show that the evidence confirms a hypothesis by deducing the evidence from the hypothesis. That deduction should play a role both in explanation and in inductive support is not itself suspicious, but the isomorphism of the models suggests that weaknesses of one may also count against the other, and this turns out to be the case. As we saw, the main weakness of the hypothetico-deductive model is that it is over-permissive, counting almost any datum as evidence for almost any hypothesis. The deductive-nomological model similarly makes it far too easy to explain (Hempel, 1965, p. 273, n. 33; pp. 293–4). This comes out most clearly if we consider the explanation of a general phenomenon, which is itself described by a law. Suppose, for example, that we wish to explain why the planets move in ellipses. According to the deductive-nomological model, we can 'explain' the ellipse law by deducing it from the conjunction of itself and any law you please, say a law in economics. The model also suffers from a problem analogous to the raven paradox. We may explain why an object warmed by pointing out that it was in the sun and everything warms when in the sun, but we cannot explain why an object was not in the sun by pointing out that it was not warmed. Just as the hypothetico-deductive model leaves the class of hypotheses confirmed by the available data dramatically underdetermined, so the deductive-nomological model underdetermines the class of acceptable explanations for a given phenomenon.

I conclude that none of the three models we have briefly considered gives an adequate description of our explanatory practices. Each of them captures distinctive features of certain explanations: some explanations provide reasons for belief, some make the initially unfamiliar or surprising familiar, and some are deductions from laws. But there are explanations that have none of these features, and something can have all of these features without being an acceptable explanation. I want now to consider a fourth model of explanation, the causal model. This model has its share of difficulties, but I believe it is more promising than the other three. I also can offer a modest development of the model that improves its descriptive adequacy and that will make it an essential tool for the discussion of Inference to the Best Explanation to follow [sic]. For these reasons, the causal model deserves a chapter of its own.

References

Bromberger, S. (1966) 'Why-Questions', in R.G. Colodny (ed.) *Mind and Cosmos*, Pittsburgh, University of Pittsburgh Press, 86–111.

Friedman, M. (1974) 'Explanation and Scientific Understanding', *Journal of Philosophy*, LXXI, 1–19.

Hempel, C. (1965) *Aspects of Scientific Explanation*, New York: Free Press.

van Fraassen, B.C. (1980) *The Scientific Image*, Oxford: Oxford University Press.

PART FIVE

Laws and causation

INTRODUCTION TO PART FIVE

I N PART FOUR, on Explanation, we looked at a prominent account of explanation, the covering-law model, which says that to explain a phenomenon is to show how it is an instance of a law of nature. An alternative emphasizes the bona fides of causal explanation. In this part we look at both laws and causes.

We mentioned that the empiricist approach to explanation involves either rejecting or minimizing the metaphysical component of explanation. Either explanation is a pragmatic or subjective notion, or if it is a worldly relation it is a straightforward one, such as a deductive relation with a covering law. The latter is acceptable to the empiricist so long as there is an account of law that is also acceptable to the empiricist.

What should an empiricist say about laws? The available approaches mirror those for explanation. The empiricist can reject the idea that 'law' refers to any metaphysically significant component of the world. We saw that van Fraassen denies that there really is any worldly explanation: the notion of explanation is a purely pragmatic one. Similarly, he argues that there really are no such things as laws of nature. When we consider the various properties that various philosophers have understood laws to possess, we see, argues van Fraassen, that there can be no laws. In particular, the principal requirements on something's being a law of nature are *universality* and *necessity*. The requirement of universality is that laws are general, covering everything, or everything of a certain kind. *All* matter is subject to the law of gravitation; *all* metal objects expand on heating and conduct electricity; *all* sodium chloride will dissolve in (pure) water, and so forth. However, universality is not enough, since some truths might be universally true, but not laws of nature. All persisting gold spheres have a radius of less than ten metres. That is no law of nature: at current gold prices the US would have to invest its annual GDP to create such a 10m-radius gold sphere. On the other hand, all persisting spheres of uranium-235 have a radius of less than 10m and that is a law of nature, since any sphere of that size would be above critical mass and would bring about a nuclear explosion. A natural response to these cases is to say that while there is no such gold sphere, there *could* be such a sphere, whereas there *could not* be any such uranium sphere. This is the idea of necessity: laws describe not just how things *are*, but (in some sense) how things *must* be.

Bas van Fraassen argues that any account of laws faces a dilemma, which he frames this way. The first problem relates to universality: our account of laws must make it a valid *inference* from the laws to the universal truths. The second problem is one of *identification*: which features of the world mark something out as a law? The dilemma, according to van Fraassen is this: the features that seem required to make a successful identification of laws also make it difficult to see why there should be a valid inference from laws to universal truths.

To see how van Fraassen's dilemma is supposed to work, let us start with what would seem to be an entirely acceptable account of laws according to an empiricist point of view. The simplest account would just regard the laws of nature as the universal *regularities* in the world. According to this view, there is nothing meta-physically special about laws beyond their being universal or fully general facts. The principal problem with this simple regularity view is the one already encountered: it fails to distinguish between the laws and the merely 'accidental' regularities (such as the fact that every gold sphere has a radius of less than 10m). So Lewis (1973, 72–7), in an empiricist tradition going back to Mill, articulates the idea that what distinguishes the laws from the regularities is the *systematic* nature of laws. Taken together the laws form a system that is both simple and strong – where strength is a matter of the number of actual and possible cases the laws cover. For example, 'everything is red', if true, would be a claim that is both simple and strong, since it covers everything. Newton's law of universal gravitation: $F = Gm_1m_2/r^2$ is less simple, but is also strong since it covers all matter. Lewis imagines a system of axioms that encapsulates all the particular facts there are. Find the simplest such system: the laws of nature are the axioms of this system plus any general truth deducible from the axioms.

Lewis's systematic account of laws has the advantage that it matches the way that scientists think about and look for laws of nature. Scientists look for simple, powerful theories, and they expect laws to fit and work together in explaining phenomena. For example, Newton's law of gravitation explains nothing *on its own*, but it is a powerful explanatory tool when conjoined with his laws of motion. However, those who are not empiricists, such as Dretske, hold that there are many reasons to doubt Lewis's account. Is it inevitable that laws of nature are simple, strong and systematic? Could there not be a possible world where the laws are many, complex, limited in scope and not well integrated? One feature of laws is that they support *subjunctive* and *counterfactual conditionals*. For example, the laws of nature make it true that *were* one to construct a 10m radius sphere of pure uranium-235, it *would* fail to persist (it would explode). That is one way of distinguishing laws from accidental truths, since it is *not* true that if I were to construct a 10m radius sphere of pure gold it would fail to persist. If the systematic regularity view of laws is correct, then at the root of this difference between laws and accidents is the systematicity of laws. But is that plausible? How does their being organised in a strong and simple system explain the ability of law to support these conditionals? In Part Four we pointed to a problem for the regularity account of laws when combined with the D-N model of explanation. A regularity is not much more than a summary of its many specific instances. So how can it explain those instances?

These kinds of problem suggest that the regularity theorist has failed as regards the identification – systematicity cannot be the distinguishing feature of laws: that distinguishing feature must be something that allows law to support subjunctive and counterfactual conditionals and to explain their instances. A promising idea here is that of *necessity*. Laws are not just about the way things actually are. They are also about how they *must* be. That's why they support counterfactuals and explain their instances. Fred Dretske, David Armstrong, and Michael Tooley articulate this as follows. Firstly we need to accept the existence of *universals*. The idea here is that properties, some natural properties at least, are genuine entities of some sort. The difference between an electron and a positron is a difference in one of their properties, charge. This is a genuine ontological difference according to the proponent of universals – there is some *thing* different between the two, and that thing is universal, the property of being negatively charged, which is a component of the being of the electron but not of the being of the positron (and vice versa for the universal, the property of being positively charged). Once one accepts universals one can consider that there might be relations between universals. Dretske proposes that there is a relation of *necessitation* that can hold between universals, and that the laws of nature are just cases of universals being related by this necessitation. Armstrong symbolizes laws this way: N(F,G), where 'N' denotes the relation of necessitation between the universals F and G (or, in Dretske's terms, F-ness and G-ness). Clearly this is just schematic and needs more detailed articulation for the different kinds of law, for example functional laws such as the law of gravitation.

Dretske argues that this 'nomic necessitation' view avoids the problems we raised for the regularity view. On this view, a law is a very different kind of thing from a regularity. Nonetheless, there must be *some* connection between a law and a regularity. Laws are supposed to explain regularities; non-accidental regularities are there because of laws. So, according to van Fraassen a minimal requirement on laws is that it should be a valid deduction from 'it is a law that Fs are Gs' to 'all Fs are Gs'. This is the inference problem. Clearly the regularity view has no difficulty with the inference problem. Laws just are regularities, so the inference is trivial. On the other hand, the nomic necessitation view, we have just said, avoids the problems of the regularity view by saying that laws are very different from regularities. So it is not trivial for that view to say the regularities are deducible from laws. N(F,G) is a relation of necessitation between the universals F-ness and G-ness. What in that description tells us that whenever N(F,G) holds it is also true that all Fs are Gs? It is important to recognize that just being a relation between universals does not give N this power. After all, if there is such a relation of necessitation, perhaps there is also a relation of prevention, P(H,J) whereby H-ness *prevents* J-ness, or a relation of making likely, L(S,T), whereby S-ness makes T-ness likely. The relations P and S do not entail the holding of the corresponding regularity, so the mere fact of N(F,G) being a relation between F-ness and G-ness tells us nothing about whether or not all Fs are Gs. So there must be something special about that specific relation N that makes it true that from 'N(F,G)' we can deduce 'all Fs are Gs'. But what? That is van Fraassen's inference problem. Note that the fact that we have given the relation the name 'necessitation' does not answer the problem. For, as Lewis quips, calling a man 'Armstrong' does not mean that he has mighty biceps.

Van Fraassen's view is that nothing meets the principal requirements on what a law is, thereby answering the dilemma of the identification and inference problems. There are, therefore, no laws. Other philosophers also take the view that there are no laws – *if* statements of laws are understood to be descriptions of the way the world itself is. Several philosophers take this to be shown by 'the problem of provisos'. Many, perhaps all, laws we find in science are ones that, if interpreted as statements of regularity, are not strictly true. For example, Kepler's first law says that planets travel in ellipses. But that is not exactly correct, since all planets deviate from ellipses due to the gravitational attraction of other planets, and for some planets, such as Uranus, the deviation is significant. One response is to add a proviso; really Kepler's law should state: 'planets travel in ellipses unless under the gravitational attraction of another planet'. But as Lange and others argue, that approach cannot work in general, since the list of such provisos is indefinite. On the other hand, a catch-all proviso, 'planets travel in ellipses unless they are prevented from travelling in ellipses' threatens to be vacuous. This is clearly problematic for the regularity view, as Lange argues. And it looks impossible to draw upon the covering-law model of explanation. For that, as we saw in Part Four, requires us to be able to deduce the explanandum from the law plus conditions. But if we lack a clear statement of the provisos attached to a law, we are not able to make that deduction. Other philosophers therefore regard law statements not as descriptions of the world as a whole, but as descriptions of something else: an experimental setup or nomological machine (Nancy Cartwright) or an abstract model (Ron Giere). Lange proposes that we abandon this descriptive conception of laws and replace it by a normative view. An intuitive understanding of the conditions of application of a law statement governs a scientist's use of it in relations of justification.

In Part Four we saw that one empiricist approach to causation was to try to show that causal explanation is an instance of covering-law explanation. One problem with this is that while the relevant laws are symmetrical, explanation and causation are not. Nancy Cartwright expands on this idea to show that causal laws cannot be reduced to laws of association, generalizing to include the case of probabilistic laws. The empiricist approach here has two steps: (i) regard laws as grounded only in stable frequencies, and (ii) analyze causation in terms of the probabilistic relations encapsulated in the laws. To fulfil step (i) the empiricist takes laws to be claims such as 'the probability of suffering from heart disease conditional on receiving hormone replacement therapy (HRT) is p', which are to be understood as founded just on the facts about the proportion of women receiving HRT who also suffer from heart disease. Cartwright accepts this but she rejects step (ii), which would analyze 'HRT protects against (i.e. causes a reduced chance of) heart disease' as 'the probability that a woman does not suffer from heart disease given that she has HRT is greater than the probability that a woman does not suffer from heart disease given that she does not have HRT' (i.e. having HRT is correlated with a reduced probability of heart disease). But as this example shows, such an analysis fails. For HRT *does not* confer any protection against heart disease. Then why the association with reduced heart disease? This is because HRT is more frequently used by women in the more affluent socio-economic groups, and being from a more affluent socio-economic group is itself a causal factor in reduced heart disease. Once one controls for socio-economic status, the association disappears. That is to say, if we look at women just within each socio-economic group, then we find that in no group is there a decreased probability of heart disease for women receiving HRT.

Can focusing on the subgroups help retrieve a probabilistic analysis of causation? Cartwright agrees that we must focus on subgroups to understand causation, but not that this will help give a reductive account of causation. If C causes E (e.g. smoking causes cancer) then E will be more probable when C (cancer will be more probable given smoking) in *all subgroups where causal factors for E are held constant*. For example, smoking is associated with an increased probability of cancer in groups that are homogenous with respect to socio-economic group, with respect to diet, with respect to sex, with respect to exercise and so forth, and with respect to any combination of these. But Cartwright emphasizes that the subgroups in question are those determined by all causally relevant factors *and only* such factors, not just any factor. For if all factors are relevant in determining subgroups we will be able to find some such subgroups where smoking is correlated with a *decrease* in cancer. For example, newspapers are keen on reporting cases of 90-year-olds who have spent a lifetime smoking and are in excellent health. They are not so interested in reporting stories about 90-year-old smokers who do have cancer. So it is likely to be true that if we consider the group '90-year-olds who are the focus of newspaper stories' that we find that cancer is less prevalent among the smokers in this group than it is among the non-smokers. We should ignore this subgroup because being the focus of a newspaper story is not a causal factor for cancer. However, if what determines which subgroup is relevant is whether the factors in question are *causal* factors, then clearly this approach cannot be used to provide an analysis of causation in terms solely of probabilistic laws of association.

Further reading

Armstrong, D.M. 1985 *What is a Law of Nature?* Cambridge University Press.
Bird, A. 2009 *Nature's Metaphysics: Laws and Properties*. Oxford University Press.
Giere, R.N. 1999 *Science Without Laws*. University of Chicago Press.
Lewis, D.K. 1973 'Causation', *Journal of Philosophy* 70: 556–67.
Lewis, D.K. 1973 *Counterfactuals*, Blackwell. See especially pp. 72–7.
Psillos, S. 2002 *Causation and Explanation*. Acumen.
Tooley, M. 1977 'The Nature of Laws', *Canadian Journal of Philosophy*, 7: 667–98.
Woodward, J. 2005 *Making Things Happen: A Theory of Causal Explanation*. Oxford University Press.

Chapter 21

Fred I. Dretske

LAWS OF NATURE

It is a traditional empiricist doctrine that natural laws are universal truths. In order to overcome the obvious difficulties with this equation most empiricists qualify it by proposing to equate laws with universal truths that play a certain role, or have a certain function, within the larger scientific enterprise. This view is examined in detail and rejected; it fails to account for a variety of features that laws are acknowledged to have. An alternative view is advanced in which laws are expressed by singular statements of fact describing the relationship between universal properties and magnitudes.

It is tempting to identify the laws of nature with a certain class of universal truths. Very few empiricists have succeeded in resisting this temptation. The popular way of succumbing is to equate the fundamental laws of nature with what is asserted by those universally true statements of nonlimited scope that embody only qualitative predicates.[1] On this view of things a law-like statement is a statement of the form "$(x)(Fx \supset Gx)$" or "$(x)(Fx \equiv Gx)$" where "F" and "G" are purely qualitative (nonpositional). Those law-like statements that are true express laws. "All robins' eggs are greenish blue," "All metals conduct electricity," and "At constant pressure any gas expands with increasing temperature" (Hempel's examples) are law-like statements.

If they are true, they express laws. The more familiar sorts of things that we are accustomed to calling laws, the formulae and equations appearing in our physics and chemistry books, can supposedly be understood in the same way by using functors in place of the propositional functions "Fx" and "Gx" in the symbolic expressions given above.

I say that it is tempting to proceed in this way since, to put it bluntly, conceiving of a law as having a content greater than that expressed by a statement of the form $(x)(Fx \supset Gx)$ seems to put it beyond our epistemological grasp.[2] We must work with what we are given, and what we are given (the observational and experimental data) are facts of the form: this F is G, that F is G, all examined F's have been G, and so on. If, as some philosophers have argued,[3] law-like statements express a kind of nomic necessity between events, something *more* than that F's are, as a matter of fact, always and everywhere, G, then it is hard to see what kind of evidence might be brought in support of them. The whole point in acquiring instantial evidence (evidence of the form "This F is G") in support of a law-like hypothesis would be lost if we supposed that what the hypothesis was actually asserting was some kind of nomic connection, some kind of modal relationship, between things that were F and things that were G. We would, it seems, be in the position of someone trying to confirm the *analyticity* of "All bachelors are unmarried" by collecting evidence about the marital status of various bachelors. This kind of evidence, though relevant to the *truth* of the claim that all bach-elors are unmarried, is powerless to confirm the *modality* in question. Similarly, if a hypothesis, in order to quality [*sic*] as a law, must express or assert some form of necessity between F's and G's, then it becomes a mystery how we ever manage to confirm such attributions with the sort of instantial evidence available from observation.

Despite this argument, the fact remains that laws are *not* simply what universally true statements express, not even universally true statements that embody purely qualitative predicates (and are, as a result, unlimited in scope). This is not particularly newsworthy. It is commonly acknowledged that law-like statements have some peculiarities that prevent their straightforward assimilation to universal truths. That the concept of a law and the concept of a universal truth are different concepts can best be seen, I think, by the following consideration: assume that $(x)(Fx \supset Gx)$ is true and that the predicate expressions satisfy all the restrictions that one might wish to impose in order to convert this universal statement into a statement of law.[4] Consider a predicate expression "K" (eternally) coextensive with "F"; i.e., $(x)(Fx \equiv Kx)$ for all time. We may then infer that if $(x)(Fx \supset Gx)$ is a universal truth, so is $(x)(Kx \supset Gx)$. The class of universal truths is closed under the operation of coextensive predicate substitution. Such is *not* the case with laws. If it is a law that all F's are G, and we substitute the term "K" for the term "F" in this law, the result is not necessarily a law. If diamonds have a refractive index of 2.419 (law) and "is a diamond" is coextensive with "is mined in kimberlite (a dark basic rock)" we cannot infer that it *is a law* that things mined in kimberlite have a refractive index of 2.419. Whether this is a law or not depends on whether

the coextensiveness of "is a diamond" and "is mined in kimberlite" is *itself* law-like. The class of laws is not closed under the same operation as is the class of universal truths.

Using familiar terminology we may say that the predicate positions in a statement of law are *opaque* while the predicate positions in a universal truth of the form $(x)(Fx \supset Gx)$ are *transparent*. I am using these terms in a slightly unorthodox way. It is not that when we have a law, "All F's are G," we can alter its truth value by substituting a coextensive predicate for "F" or "G." For if the statement is true, it will remain true after substitution. What happens, rather, is that the expression's status *as a law* is (or may be) affected by such an exchange. The matter can be put this way: the statement

(A) All F's are G (understood as $(x)(Fx \supset Gx)$)

has "F" and "G" occurring in transparent positions. Its truth value is unaffected by the replacement of "F" or "G" by a coextensive predicate. The same is true of

(B) It is universally true that F's are G.

If, however, we look at

(C) It is a law that F's are G.

we find that "F" and "G" occur in opaque positions. If we think of the two prefixes in (B) and (C), "it is universally true that . . ." and "it is a law that . . .," as operators, we can say that the operator in (B) does not, while the operator in (C) does, confer opacity on the embedded predicate positions. To refer to something as a statement of law is to refer to it as an expression in which the descriptive terms occupy opaque positions. To refer to something as a universal truth is to refer to it as an expression in which the descriptive terms occupy transparent positions. Hence, our concept of a law differs from our concept of a universal truth.[5]

Confronted by a difference of this sort, many philosophers have argued that the distinction between a natural law and a universal truth was not, fundamentally, an *intrinsic* difference. Rather, the difference was a difference in the *role* some universal statements played within the larger theoretical enterprise. Some universal statements are more highly integrated into the constellation of accepted scientific principles, they play a more significant role in the explanation and prediction of experimental results, they are better confirmed, have survived more tests, and make a more substantial contribution to the regulation of experimental inquiry. But, divorced from this context, stripped of these *extrinsic* features, a law is nothing but a universal truth. It has the same empirical content. Laws are to universal truths what shims are to slivers of wood and metal; the latter *become* the former by being *used* in a certain way. There is a *functional* difference, nothing else.[6]

According to this reductionistic view, the peculiar opacity (described above) associated with laws is not a manifestation of some intrinsic difference between a law and a universal truth. It is merely a symptom of the special status or function that some universal statements have. The basic formula is: law = universal truth + X. The "X" is intended to indicate the special function, status or role that a universal truth must have to qualify as a law. Some popular candidates for this auxiliary idea, X, are:

(1) High degree of confirmation,
(2) Wide acceptance (well established in the relevant community),
(3) Explanatory potential (can be used to explain its instances),
(4) Deductive integration (within a larger system of statements),
(5) Predictive use.

To illustrate the way these values of X are used to buttress the equation of laws with universal truths, it should be noted that each of the concepts appearing on this list generates an opacity similar to that witnessed in the case of genuine laws. For example, to say that it is a law that all F's are G may possibly be no more than to say that it is well established that $(x)(Fx \supset Gx)$. The peculiar opacity of laws is then explained by pointing out that the class of expressions that are well established (or highly confirmed) is not closed under substitution of coextensive predicates: one cannot infer that $(x)(Kx \supset Gx)$ is well established just because "Fx" and "Kx" are coextensive and $(x)(Fx \supset Gz)$ is well established (for no one may know that "Fx" and "Kx" *are* coextensive). It may be supposed, therefore, that the opacity of laws is merely a manifestation of the underlying fact that a universal statement, to qualify as a law, must be well established, and the opacity is a result of this epistemic condition. Or, if this will not do, we can suppose that one of the other notions mentioned above, or a combination of them, is the source of a law's opacity.

This response to the alleged uniqueness of natural laws is more or less standard fare among empiricists in the Humean tradition. Longstanding (= venerable) epistemological and ontological commitments motivate the equation: law = universal truth + X. There is disagreement among authors about the differentia X, but there is near unanimity about the fact that laws are a *species* of universal truth.

If we set aside our scruples for the moment, however, there is a plausible explanation for the opacity of laws that has not yet been mentioned. Taking our cue from Frege, it may be argued that since the operator "it is a law that . . ." converts the otherwise transparent positions of "All F's are G" into opaque positions, we may conclude that this occurs because within the context of this operator (either explicitly present or implicitly understood) the terms "F" and "G" do not have their usual referents. There is a shift in what we are talking about. To say that *it is a law* that F's are G is to say that "All F's are G" is to be understood (in so far as it expresses a law), not as a statement about the extensions of the predicates "F" and "G," but as a singular statement describing a relationship between the universal properties F-ness and G-ness. In other words, (C) is to be understood as having the form:

(6)   F-ness → G-ness.[7]

To conceive of (A) as a universal truth is to conceive of it as expressing a relationship between the extensions of its terms; to conceive of it as a law is to conceive of it as expressing a relationship between the properties (magnitudes, quantities, features) which these predicates express (and to which we may refer with the corresponding abstract singular term). The opacity of laws is merely a manifestation of this change in reference. If "F" and "K" are coextensive, we cannot substitute the one for the other in the *law* "All F's are G" and expect to preserve truth; for the law asserts a connection between F-ness and G-ness and there is no guarantee that a similar connection exists between the properties K-ness and G-ness just because all F's are K and *vice versa*.[8]

It is this view that I mean to defend in the remainder of this essay. Law-like statements are singular statements of fact describing a relationship between properties or magnitudes. Laws are the relationships that are asserted to exist by true law-like statements. According to this view, then, there is an *intrinsic* difference between laws and universal truths. Laws imply universal truths, but universal truths do not imply laws. Laws are (expressed by) *singular* statements describing the relationships that exist between universal qualities and quantities; they are not universal statements about the particular objects

and situations that exemplify these qualities and quantities. Universal truths are not transformed into laws by acquiring some of the extrinsic properties of laws, by being used in explanation or prediction, by being made to support counterfactuals, or by becoming well established. For, as we shall see, universal truths *cannot* function in these ways. They *cannot* be made to perform a service they are wholly unequipped to provide.

In order to develop this thesis it will be necessary to overcome some metaphysical prejudices, and to overcome these prejudices it will prove useful to review the major deficiencies of the proposed alternative. The attractiveness of the formula: law = universal truth + X, lies, partly at least, in its ontological austerity, in its tidy portrayal of what there is, or what there must be, in order for there to be laws of nature. The antidote to this seductive doctrine is a clear realization of how utterly hopeless, epistemologically and functionally hopeless, this equation is.

If the auxiliary ideas mentioned above (explanation, prediction, confirmation, etc.) are deployed as values of X in the reductionistic equation of laws with universal truths, one can, as we have already seen, render a satisfactory account of the opacity of laws. In this particular respect the attempted equation proves adequate. In what way, then, does it fail?

(1) and (2) are what I will call "epistemic" notions; they assign to a statement a certain epistemological status or cognitive value. They are, for this reason alone, useless in understanding the nature of a law.[9] Laws do not begin to be laws only when we first become aware of them, when the relevant hypotheses become well established, when there is public endorsement by the relevant scientific community. The laws of nature are the same today as they were one thousand years ago (or so we believe); yet, some hypotheses are highly confirmed today

that were not highly confirmed one thousand years ago. It is certainly true that we only begin to *call* something a law when it becomes well established, that we only recognize something as a statement of law when it is confirmed to a certain degree, but that something *is* a law, that some statement does in fact express a law, does not similarly await our appreciation of this fact. We discover laws, we do not invent them – although, of course, some invention may be involved in our manner of expressing or codifying these laws. Hence, the status of something as a statement of law does not depend on its epistemological status. What does depend on such epistemological factors is our ability to identify an otherwise qualified statement *as true* and, therefore, *as a statement of law*. It is for this reason that one cannot appeal to the epistemic operators to clarify the nature of laws; they merely confuse an epistemological with an ontological issue.

What sometimes helps to obscure this point is the tendency to conflate laws with the verbal or symbolic expression of these laws (what I have been calling "statements of law"). Clearly, though, these are different things and should not be confused. There are doubtless laws that have not yet (or will never) receive symbolic expression, and the same law may be given different verbal codifications (think of the variety of ways of expressing the laws of thermodynamics). To use the language of "propositions" for a moment, a law is the proposition expressed, not the vehicle we use to express it. The *use* of a sentence *as an expression of law* depends on epistemological considerations, but the law itself does not.

There is, furthermore, the fact that whatever auxiliary idea we select for understanding laws (as candidates for X in the equation: law = universal truth + X), if it is going to achieve what we expect of it, should help to account for the variety of other features that laws are

acknowledged to have. For example, it is said that laws "support" counterfactuals of a certain sort. If laws are universal truths, this fact is a complete mystery, a mystery that is usually suppressed by using the word "support." For, of course, universal statements do not *imply* counterfactuals in any sense of the word "imply" with which I am familiar. To be told that all F's are G is not to be told anything that implies that if this x were an F, it would be G. To be told that all dogs born at sea have been and will be cocker spaniels is *not* to be told that we would get cocker spaniel pups (or no pups at all) if we arranged to breed dachshunds at sea. The only reason we might *think* we were being told this is because we do not expect anyone to assert that all dogs born at sea *will be* cocker spaniels unless they know (or have good reasons for believing) that this is true; and we do not understand *how* anyone could *know* that this is true without being privy to information that insures this result – without, that is, knowing of some bizarre [sic] law or circumstance that *prevents* anything but cocker spaniels from being born at sea. Hence, if we accept the claim at all, we do so with a certain presumption about what our informant must know in order to be a serious claimant. We assume that our informant knows of certain laws or conditions that *insure* the continuance of a past regularity, and it is this presumed knowledge that we exploit in endorsing or accepting the counterfactual. But the simple fact remains that the statement "All dogs born at sea have been and will be cocker spaniels" does not *itself* support or imply this counterfactual; at best, *we* support the counterfactual (if we support it at all) on the basis of what the claimant is supposed to know in order to advance such a universal projection.

Given this incapacity on the part of universal truths to support counterfactuals, one would expect some assistance from the epistemic condition if laws are to be analyzed as well established universal truths. But the expectation is disappointed; we are *left* with a complete mystery. For if a statement of the form "All F's are G" does not support the counterfactual, "If this (non-G) were an F, it would be G," it is clear that it will not support it just because it is well established or highly confirmed. The fact that all the marbles in the bag are red does not support the contention that if this (blue) marble were in the bag, it would be red; but neither does the fact that we *know* (or it is highly confirmed) that all the marbles in the bag are red support the claim that if this marble were in the bag it would be red. And making the universal truth *more universal* is not going to repair the difficulty. The fact that all the marbles in the universe are (have been and will be) red does not imply that I *cannot* manufacture a blue marble; it implies that I *will not*, not that I cannot or that if I were to try, I would fail. To represent laws on the model of one of our epistemic operators, therefore, leaves wholly unexplained one of the most important features of laws that we are trying to understand. They are, in this respect, unsatisfactory candidates for the job.

Though laws are not merely well established general truths, there is a related point that deserves mention: laws are the *sort* of thing that can become well established prior to an exhaustive enumeration of the instances to which they apply. This, of course, is what gives laws their predictive utility. Our confidence in them increases at a much more rapid rate than does the ratio of favorable examined cases to total number of cases. Hence, we reach the point of confidently using them to project the outcome of unexamined situations while there is still a substantial number of unexamined situations to project.

This feature of laws raises new problems for the reductionistic equation. For, contrary to the argument in the third paragraph of this essay,

it is hard to see how confirmation is possible for universal truths. To illustrate this difficulty, consider the (presumably easier) case of a general truth of *finite* scope. I have a coin that you have (by examination and test) convinced yourself is quite normal. I propose to flip it ten times. I conjecture (for whatever reason) that it will land heads all ten times. You express doubts. I proceed to "confirm" my hypothesis. I flip the coin once. It lands heads. Is this evidence that my hypothesis is correct? I continue flipping the coin and it turns up with nine straight heads. Given the opening assumption that we are dealing with a fair coin, the probability of getting all ten heads (the probability that my hypothesis is true) is now, after examination of 90% of the total population to which the hypothesis applies, exactly .5. If we are guided by probability considerations alone, the likelihood of all ten tosses being heads is now, after nine favorable trials, a toss-up. After nine favorable trials it is no more reasonable to believe the hypothesis than its denial. In what sense, then, can we be said to have been accumulating evidence (during the first nine trials) that all would be heads? In what sense have we been confirming the hypothesis? It would appear that the probability of my conjecture's being true never exceeds .5 until we have exhaustively examined the entire population of coin tosses and found them *all* favorable. The probability of my conjecture's being true is either: (i) too low (≤ .5) to invest any confidence in the hypothesis, or (ii) so high (= 1) that the hypothesis is useless for prediction. There does not seem to be any middle ground.

Our attempts to confirm universal generalizations of nonlimited scope is, I submit, in exactly the same impossible situation. It is true, of course, that after nine successful trials the probability that all ten tosses will be heads is greatly increased over the initial probability that all would be heads. The initial probability (assuming a fair coin) that all ten tosses would be heads was on the order of .002. After nine favorable trials it is .5. In this sense I have increased the probability that my hypothesis is true; I have raised its probability from .002 to .5. The important point to notice, however, is that this sequence of trials did not alter the probability that the *tenth* trial would be heads. The probability that the unexamined instance would be favorable remains exactly what it was before I began flipping the coin. It was originally .5 and it is now, after nine favorable trials, still .5. I am in no better position now, after extensive sampling, to predict the outcome of the tenth toss than I was before I started. To suppose otherwise is to commit the converse of the Gambler's Fallacy.

Notice, we could take the first nine trials as evidence that the tenth trial would be heads *if* we took the results of the first nine tosses as evidence that the coin was biased in some way. Then, on *this* hypothesis, the probability of getting heads on the last trial (and, hence, on all ten trials) would be greater than .5 (how much greater would depend on the conjectured degree of bias and this, in turn, would presumably depend on the extent of sampling). This new hypothesis, however, is something quite different than the original one. The original hypothesis was of the form: $(x)(Fx \supset Gx)$, all ten tosses will be heads. Our new conjecture is that there is a physical asymmetry in the coin, an asymmetry that tends to yield more heads than tails. We have succeeded in confirming the general hypothesis (all ten tosses will be heads), but we have done so via an intermediate hypothesis involving *genuine laws* relating the physical make-up of the coin to the frequency of heads in a population of tosses.

It is by such devices as this that we create for ourselves, or some philosophers create for themselves, the *illusion* that (apart from supplementary *law-like* assumptions) general truths can be confirmed by their instances and therefore qualify, in this respect, as laws of nature. The illusion is fostered in the following way. It is assumed that confirmation is a matter of *raising the probability of a hypothesis*.[10] On this assumption any general statement of finite scope can be confirmed by examining its instances and finding them favorable. The hypothesis about the results of flipping a coin ten times can be confirmed by tossing nine straight heads, and this confirmation takes place without *any* assumptions about the coin's bias. Similarly, I confirm (to some degree) the hypothesis that all the people in the hotel ballroom are over thirty years old when I enter the ballroom with my wife and realize that *we* are both over thirty. In both cases I raise the probability that the hypothesis is true over what it was originally (before flipping the coin and before entering the ballroom). But this, of course, isn't confirmation. Confirmation is not simply raising the probability that a hypothesis is true, it is raising the probability that the unexamined cases resemble (in the relevant respect) the examined cases. It is *this* probability that must be raised if genuine confirmation is to occur (and if a confirmed hypothesis to be useful in *prediction*), and it is precisely this probability that is left unaffected by the instantial "evidence" in the above examples.

In order to meet this difficulty, and to cope with hypotheses that are *not* of limited scope,[11] the reductionist usually smuggles into his confirmatory proceedings the very idea he professes to do without: viz., a type of law that is not merely a universal truth. The general truth then gets confirmed but *only* through the mediation of these supplementary laws. These supplementary assumptions are usually introduced to *explain* the regularities manifested in the examined instances so as to provide a basis for projecting these regularities to the unexamined cases. The only way we can get a purchase on the unexamined cases is to introduce a hypothesis which, while *explaining* the data we already have, *implies* something about the data we do not have. To suppose that our coin is biased (first example) is to suppose something that contributes to the explanation of our extraordinary run of heads (nine straight) and simultaneously implies something about the (probable) outcome of the tenth toss. Similarly (second example) my wife and I may be attending a reunion of some kind, and I may suppose that the other people in the ballroom are old classmates. This hypothesis not only explains our presence, it implies that most, if not all, of the remaining people in the room are of comparable age (well over thirty). In both these cases the generalization can be confirmed, but only via the introduction of a law or circumstance (combined with a law or laws) that helps to explain the data already available.

One additional example should help to clarify these last remarks. In sampling from an urn with a population of colored marbles, I can confirm the hypothesis that all the marbles in the urn are red by extracting at random several dozen red marbles (and no marbles of any other color). This is a genuine example of confirmation, not because I have raised the probability of the hypothesis that all are red by reducing the number of ways it can be false (the same reduction would be achieved if you *showed* me 24 marbles from the urn, all of which were red), but because the hypothesis that all the marbles in the urn are red, together with the fact (law) that you cannot draw nonred marbles from an urn containing only red marbles, *explains* the result of my random sampling. Or, if this is too strong, the law that assures me that random sampling from an urn containing a substantial number of nonred marbles would reveal (in all likelihood) at least one nonred marble lends its support to my confirmation that the urn contains only (or mostly) red marbles. Without the assistance of such auxiliary laws a sample of 24 red

marbles is powerless to confirm a hypothesis about the total population of marbles in the urn. To suppose otherwise is to suppose that the *same* degree of confirmation would be afforded the hypothesis if you, whatever your deceitful intentions, showed me a carefully selected set of 24 red marbles from the urn. This *also* raises the probability that they are all red, but the trouble is that it does not (due to your unknown motives and intentions) raise the probability that the unexamined marbles resemble the examined ones. And it does not raise this probability because we no longer have, as the best available explanation of the examined cases (all red), a hypothesis that implies that the remaining (or most of the remaining) marbles are also red. Your careful selection of 24 red marbles from an urn containing many different colored marbles is an equally good explanation of the data and it does *not* imply that the remainder are red. Hence, it is not just the fact that we have 24 red marbles in our sample class (24 positive instances and no negative instances) that confirms the general hypothesis that all the marbles in the urn are red. It is this data *together with a law* that confirms it, a law that (together with the hypothesis) explains the data in a way that the general hypothesis alone cannot do.

We have now reached a critical stage in our examination of the view that a properly qualified set of universal generalizations can serve as the fundamental laws of nature. For we have, in the past few paragraphs, introduced the notion of *explanation*, and it is this notion, perhaps more than any other, that has received the greatest attention from philosophers in their quest for the appropriate X in the formula: law = universal truth + X. R. B. Braithwaite's treatment ([3]) is typical. He begins by suggesting that it is merely deductive integration that transforms a universal truth into a law of nature. Laws are simply universally true statements of the form $(x)(Fx \supset Gx)$ that are derivable from certain higher level hypotheses. To say that $(x)(Fx \supset Gx)$ is a statement of law is to say, not only that it is true, but that it is *deducible from* a higher level hypothesis, H, in a well established scientific system. The fact that it must be deducible from some higher level hypothesis, H, confers on the statement the opacity we are seeking to understand. For we may have a hypothesis from which we can derive $(x)(Fx \supset Gx)$ but from which we cannot derive $(x)(Kx \supset Gx)$ despite the coextensionality of "F" and "K." Braithwaite also argues that such a view gives a satisfactory account of the counterfactual force of laws.

The difficulty with this approach (a difficulty that Braithwaite recognizes) is that it only postpones the problem. Something is not a statement of law simple [sic] because it is true and deducible from some well established higher level hypothesis. For every generalization implies another of smaller scope (e.g. $(x)(Fx \supset Gx)$ implies $(x)(Fx \cdot Hx \supset Gx)$), but this fact has not the slightest tendency to transform the latter generalization into a law. What is required is that the higher level hypothesis *itself* be law-like. You cannot give to others what you do not have yourself. But now, it seems, we are back where we started from. It is at this point that Braithwaite begins talking about the higher level hypotheses having *explanatory force* with respect to the hypotheses subsumed under them. He is forced into this maneuver to account for the fact that these higher level hypotheses – not themselves law-like on his characterization (since not themselves derivable from still higher level hypotheses) – are capable of conferring lawlikeness on their consequences. The higher level hypotheses are laws because they explain; the lower level hypotheses are laws because they are deducible from laws. This fancy twist smacks of circularity. Nevertheless, it represents a conversion to *explanation* (instead of *deducibility*) as the fundamental feature of laws, and Braithwaite concedes this: "A hypothesis

to be regarded as a natural law must be a general proposition which can be thought to *explain* its instances" ([3], p. 303) and, a few lines later, "Generally speaking, however, a true scientific hypothesis will be regarded as a law of nature if it has an explanatory function with regard to lower-level hypotheses or its instances." Deducibility is set aside as an incidental (but, on a Hempelian model of explanation, an important) facet of the more ultimate idea of explanation.

There is an added attraction to this suggestion. As argued above, it is difficult to see how instantial evidence can serve to confirm a universal generalization of the form: $(x)(Fx \supset Gx)$. If the generalization has an infinite scope, the ratio "examined favorable cases/total number of cases" never increases. If the generalization has a finite scope, or we treat its probability as something other than the above ratio, we may succeed in raising its probability by finite samples, but it is never clear how we succeed in raising the probability that the unexamined cases resemble the examined cases without invoking laws as auxiliary assumptions. And this is the very notion we are trying to analyze. To this problem the notion of explanation seems to provide an elegant rescue. If laws are those universal generalizations that explain their instances, then following the lead of a number of current authors (notably Harman ([8], [9]); also see Brody ([4])) we may suppose that universal generalizations can be confirmed because confirmation is (roughly) the converse of explanation; E confirms H if H explains E. *Some* universal generalizations can be confirmed; they are those that explain their instances. Equating laws with universal generalizations having explanatory power therefore achieves a neat economy: we account for the confirmability of laws in terms of the explanatory power of those generalizations to which laws are reduced.

To say that a law is a universal truth having explanatory power is like saying that a chair is a breath of air used to seat people. You cannot make a silk purse out of a sow's ear, not even a very good sow's ear; and you cannot *make* a generalization, not even a purely universal generalization, explain its instances. The fact that *every* F is G fails to explain why *any* F is G, and it fails to explain it, not because its explanatory efforts are too feeble to have attracted our attention, but because the explanatory attempt is never even made. The fact that all men are mortal does not explain why you and I are mortal; it *says* (in the sense of *implies*) that we are mortal, but it does not even suggest *why* this might be so. The fact that all ten tosses will turn up heads is a fact that logically guarantees a head on the tenth toss, but it is not a fact that explains the outcome of this final toss. On one view of explanation, *nothing* explains it. Subsuming an instance under a universal generalization has exactly as much explanatory power as deriving Q from P · Q. None.

If universal truths of the form $(x)(Fx \supset Gx)$ could be *made* to explain their instances, we might succeed in making them into natural laws. But, as far as I can tell, no one has yet revealed the secret for endowing them with this remarkable power.

This has been a hasty and, in some respects, superficial review of the doctrine that laws are universal truths. Despite its brevity, I think we have touched upon the major difficulties with sustaining the equation: law = universal truth + X (for a variety of different values of "X"). The problems center on the following features of laws:

(a)   A statement of law has its descriptive terms occurring in opaque positions.

(b)   The existence of laws does not await our identification of them *as* laws. In this sense they are objective and independent of epistemic considerations.

(c)   Laws can be confirmed by their instances and the confirmation of a law raises the probability that the unexamined instances will resemble (in the respect described by the law) the examined instances. In this respect they are useful tools for prediction.

(d)   Laws are not merely summaries of their instances; typically, they figure in the explanation of the phenomena falling within their scope.

(e)   Laws (in some sense) "support" counterfactuals; to know a law is to know what would happen if certain conditions were realized.

(f)   Laws tell us what (in some sense) must happen, not merely what has and will happen (given certain initial conditions).

The conception of laws suggested earlier in this essay, the view that laws are expressed by singular statements of fact describing the relationships between properties and magnitudes, proposes to account for these features of laws in a single, unified, way: (a)–(f) are all manifestations of what might be called "ontological ascent," the shift from talking about individual objects and events, or collections of them, to the quantities and qualities that these objects exemplify. Instead of talking about green and red things, we talk about the *colors* green and blue. Instead of talking about gases that have a volume, we talk about the volume (temperature, pressure, entropy) that gases have. Laws eschew reference to the things that have length, charge, capacity, internal energy, momentum, spin, and velocity in order to talk about these quantities themselves and to describe *their* relationship to each other.

We have already seen how this conception of laws explains the peculiar opacity of law-like statements. Once we understand that a law-like statement is not a statement about the extensions of its constituent terms, but about the intensions (= the quantities and qualities to which we may refer with the abstract singular form of these terms), then the opacity of laws to *extensional* substitution is natural and expected. Once a law is understood to have the form:

(6)   F-ness → G-ness

the relation in question (the relation expressed by "→") is seen to be an *extensional* relation between *properties* with the terms "F-ness" and "G-ness" occupying *transparent* positions in (6). Any term referring to the same quality or quantity as "F-ness" can be substituted for "F-ness" in (6) without affecting its truth or its law-likeness. Coextensive terms (terms referring to the same *quantities* and *qualities*) can be freely exchanged for "F-ness" and "G-ness" in (6) without jeopardizing its truth value. The tendency to treat laws as some kind of intensional relation between extensions, as something of the form $(x)(Fx \boxed{N}\!\!\rightarrow Gx)$ (where the connective is some kind of modal connective), is simply a mistaken rendition of the fact that laws are extensional relations between intensions.

Once we make the ontological ascent we can also understand the modal character of laws, the feature described in (e) and (f) above. Although true statements having the form of (6) are not themselves *necessary* truths, nor do they describe a modal relationship between the respective qualities, the contingent relationship between properties that is described imposes a modal quality on the particular events falling within its scope. This F *must* be G. Why? Because F-ness is linked to G-ness; the one property yields or generates the other in much the way a change in the thermal conductivity of a metal yields a change in its electrical conductivity. The pattern of inference is:

(I) F-ness → G-ness

   This is F
   ――――――――
   This must be G.

This, I suggest, is a valid pattern of inference. It is quite unlike the fallacy committed in (II):

(II) $(x)(Fx \supset Gx)$

   This is F
   ――――――――
   This must be G.

The fallacy here consists in the absorption into the conclusion of a modality (entailment) that belongs to the relationship *between* the premises and the conclusion. There is no fallacy in (I), and this, I submit, is the source of the "physical" or "nomic" necessity generated by laws. It is this which explains the power of laws to tell us what *would* happen if we did such-and-such and what *could* not happen whatever we did.

I have no proof for the validity of (I). The best I can do is an analogy. Consider the complex set of legal relationships defining the authority, responsibilities, and powers of the three branches of government in the United States. The executive, the legislative, and the judicial branches of government have, according to these laws, different functions and powers. There is nothing *necessary* about the laws themselves; they could be changed. There is no law that prohibits scrapping all the present laws (including the constitution) and starting over again. Yet, given these laws, it follows that the President *must* consult Congress on certain matters, members of the Supreme Court *cannot* enact laws nor declare war, and members of Congress *must* periodically stand for election. The legal code lays down a set of relationships between the various *offices* of government, and this set of relationships (between the abstract offices) impose legal constraints on the individuals who occupy these offices – constraints that we express with such modal terms as "cannot" and "must." There are certain things the individuals (and collections of individuals – e.g., the Senate) can and cannot do. Their activities are subjected to this modal qualification whereas the framework of laws from which this modality arises is itself modality-free. The President (e.g., Ford) *must* consult the Senate on matter M, but the relationship between the *office* of the President and that *legislative body* we call the Senate that makes Gerald Ford's action obligatory is not *itself* obligatory. There is no law that says that this relationship between the office of President and the upper house of Congress must (legally) endure forever and remain indisoluble.

In matters pertaining to the offices, branches and agencies of government the "can" and "cannot" generated by laws are, of course, legal in character. Nevertheless, I think the analogy revealing. Natural laws may be thought of as a set of relationships that exist between the various "offices" that objects sometimes occupy. Once an object occupies such an office, its activities are constrained by the set of relations connecting that office to other offices and agencies; it *must* do some things, and it *cannot* do other things. In both the legal and the natural context the modality at level n is generated by the set of relationships existing between the entities at level $n + 1$. Without this web of higher order relationships there is nothing to support the attribution of constraints to the entities at a lower level.

To think of statements of law as expressing relationships (such as class inclusion) between the extensions of their terms is like thinking of the legal code as a set of universal imperatives directed to a set of particular individuals. A law that tells us that the United States President must consult Congress on matters pertaining to M is not an imperative issued to Gerald Ford, Richard Nixon, Lyndon Johnson, *et al*. The law tells us

something about the duties and obligations attending the *Presidency*; only indirectly does it tell us about the obligations of the Presidents (Gerald Ford, Richard Nixon, *et al*.). It tells us about their obligations in so far as they are occupants of this office. If a law was to be interpreted as of the form: "For all x, if x is (was or will be) President of the United States, then x must (legally) consult Congress on matter M," it would be incomprehensible why Sally Bickle, were she to be president, would have to consult Congress on matter M. For since Sally Bickle never was, and never will be, President, the law, understood as an imperative applying to *actual* Presidents (past, present and future) does not apply to her. Even if there is a possible world in which she becomes President, this does not make her a member of that class of people to which the law applies; for the law, under this interpretation, is directed to that class of people who become President in this world, and Sally is not a member of this class. But we all know, of course, that the law does not apply to individuals, or sets of individuals, in this way; it concerns itself, in part, with the offices that people occupy and only indirectly with individuals in so far as they occupy these offices. And this is why, if Sally Bickle were to become President, if she occupied this office, she would have to consult Congress on matters pertaining to M.[12]

The last point is meant to illustrate the respect and manner in which natural laws "support" counterfactuals. Laws, being relationships between properties and magnitudes, *go beyond* the sets of things in this world that exemplify these properties and have these magnitudes. Laws tell us that quality F is linked to quality G in a certain way; hence, if object O (which has neither property) were to acquire property F, it would also acquire G in virtue of this connection between F-ness and G-ness. A statement of law asserts something that allows us to entertain the prospect of alterations in the extension of the predicate expressions contained in the statement. Since they make no reference to the extensions of their constituent terms (where the extensions are understood to be the things that are F and G in this world), we can hypothetically alter these extensions in the antecedent of our counterfactual ("if this were an F . . .") and use the connection asserted in the law to reach the consequent (". . . it would be G"). Statements of law, by talking about the relevant properties rather than the sets of things that have these properties, have a far wider scope than any true generalization about the actual world. Their scope extends to those possible worlds in which the extensions of our terms differ but the connections between properties remains invariant. This is a power that no universal generalization of the form (x) (Fx ⊃ Gx) has; this statement says something about the actual F's and G's in *this* world. It says absolutely nothing about those possible worlds in which there are *additional* F's or *different* F's. For this reason it cannot imply a counterfactual. To do this we must ascend to a level of discourse in which what we talk about, and what we say about what we talk about, remains the *same* through alternations in extension. This can only be achieved through an ontological ascent of the type reflected in (6).

We come, finally, to the notion of explanation and confirmation. I shall have relatively little to say about these ideas, not because I think that the present conception of laws is particularly weak in this regard, but because its very real strengths have already been made evident. Laws figure in the explanation of their instances because they are not merely summaries of these instances. I can explain why this F is G by describing the relationship that exists between the properties in question. I can explain why the current increased upon an increase in the voltage by appealing to the relationship that exists between the flow of charge (current intensity)

and the voltage (notice the definite articles). The period of a pendulum decreases when you shorten the length of the bob, not because all pendulums do that, but because the period and the length are related in the fashion $T = 2\pi\sqrt{L/g}$. The principles of thermodynamics tell us about the relationships that exist between such quantities as energy, entropy, temperature and pressure, and it is for this reason that we can use these principles to explain the increase in temperature of a rapidly compressed gas, explain why perpetual motion machines cannot be built, and why balloons do not spontaneously collapse without a puncture.

Furthermore, if we take seriously the connection between explanation and confirmation, take seriously the idea that to confirm a hypothesis is to bring forward data for which the hypothesis is the best (or one of the better) competing explanations, then we arrive at the mildly paradoxical result that laws can be confirmed *because* they are more than generalizations of that data. Recall, we began this essay by saying that if a statement of law asserted anything more than is asserted by a universally true statement of the form $(x)(Fx \supset Gx)$, then it asserted something that was beyond our epistemological grasp. The conclusion we have reached is that *unless* a statement of law goes beyond what is asserted by such universal truths, unless it asserts something that cannot be completely verified (even with a complete enumeration of its instances), it cannot be confirmed and used for predictive purposes. It cannot be confirmed because it cannot explain; and its inability to explain is a symptom of the fact that there is not enough "distance" between it and the facts it is called upon to explain. To get this distance we require an ontological ascent.

I expect to hear charges of Platonism. They would be premature. I have not argued that there are universal properties. I have been concerned to establish something weaker, something conditional in nature: viz., universal properties exist, and there exists a definite relationship between these universal properties, if there are any laws of nature. If one prefers desert landscapes, prefers to keep one's ontology respectably nominalistic, I can and do sympathize. I would merely point out that in such barren terrain there are no laws, nor is there anything that can be dressed up to look like a law. These are inflationary times, and the cost of nominalism has just gone up.

Notes

For their helpful comments my thanks to colleagues at Wisconsin and a number of other universities where I read earlier versions of this paper. I wish, especially, to thank Zane Parks, Robert Causey, Martin Perlmutter, Norman Gillespie, and Richard Aquilla for their critical suggestions, but they should not be blamed for the way I garbled them.

1 This is the position taken by Hempel and Oppenheim ([10]).

2 When the statement is of nonlimited scope it is already beyond our epistemological grasp in the sense that we cannot *conclusively* verify it with the (necessarily) finite set of observations to which traditional theories of confirmation restrict themselves. When I say (in the text) that the statement is "beyond our epistemological grasp" I have something more serious in mind than this rather trivial limitation.

3 Most prominently, William Kneale in [12] and [13].

4 I eliminate quotes when their absence will cause no confusion. I will also, sometimes, speak of laws and statements of law indifferently. I think, however, that it is a serious mistake to conflate these two notions. Laws are what is expressed by true lawlike statements (see [1], p. 2, for a discussion of the possible senses of "law" in this regard). I will return to this point later.

5 Popper ([17]) vaguely perceives, but fails to appreciate the significance of, the same (or a similar) point. He distinguishes between the structure of terms in laws and universal generalizations, referring to their occurrence in laws as "intensional" and their occurrence in universal generalizations as "extensional." Popper fails to develop this insight, however, and continues to equate laws with a certain class of universal truths.

6 Nelson Goodman gives a succinct statement of the functionalist position: "As a first approximation then, we might say that a law is a true sentence used for making predictions. That laws are used predictively is of course a simple truism, and I am not proposing it as a novelty. I want only to emphasize the Humean idea that rather than a sentence being used for prediction because it is a law, it is called a law because it is used for prediction; and that rather than the law being used for prediction because it describes a causal connection, the meaning of the causal connection is to be interpreted in terms of predictively used laws" ([7], p. 26). Among functionalists of this sort I would include Ayer ([2]), Nagel ([16]), Popper ([17]), Mackie ([14]), Bromberger ([6]), Braithwaite ([3]), Hempel ([10], [11]) and many others. Achinstein is harder to classify. He says that laws express regularities that can be cited in providing analyses and explanations ([1], p. 9), but he has a rather broad idea of regularities: "regularities might also be attributed to properties" ([1], pages 19, 22).

7 I attach no special significance to the connective "→." I use it here merely as a dummy connective or relation. The kind of connection asserted to exist between the universals in question will depend on the particular law in question, and it will vary depending on whether the law involves quantitative or merely qualitative expressions. For example, Ohm's Law asserts for a certain class of situations a constant ratio (R) between the magnitudes E (potential difference) and I (current intensity), a fact that we use the "=" sign to represent: $E/I = R$. In the case of simple qualitative laws (though I doubt whether there are many genuine laws of this sort) the connective "→" merely expresses a link or connection between the respective qualities and may be read as "yields." If it is a

law that all men are mortal, then humanity yields mortality (humanity → mortality). Incidentally, I am not denying that we can, and do, express laws as simply "All F's are G" (sometimes this is the only convenient way to express them). All I am suggesting is that when lawlike statements are presented in this form it may not be clear what is being asserted: a law or a universal generalization. When the context makes it clear that a relation of law is being described, we can (without ambiguity) express it as "All F's are G" for it is then understood in the manner of (6).

8 On the basis of an argument concerned with the restrictions on predicate expressions that may appear in laws, Hempel reaches a similar conclusion but he interprets it differently. "Epitomizing these observations we might say that a lawlike sentence of universal nonprobabilistic character is not about classes or about the extensions of the predicate expressions it contains, but about these classes or extensions *under certain descriptions*" ([11], p. 128). I guess I do not know what being *about* something *under a description* means unless it amounts to being about the property or feature expressed by that description. I return to this point later.

9 Molnar ([15]) has an excellent brief critique of attempts to analyze a law by using epistemic conditions of the kind being discussed.

10 Brody argues that a qualitative confirmation function need not require that any E that raises the degree of confirmation of H thereby (qualitatively) confirms H. We need only require (perhaps this is also too much) that if E does qualitatively confirm H, then E raises the degree of confirmation of H. His arguments take their point of departure from Carnap's examples against the special consequence and converse consequence condition ([4], pages 414–18). However this may be, I think it fair to say that most writers on confirmation theory take a *confirmatory* piece of evidence to be a piece of evidence that *raises* the probability of the hypothesis for which it is confirmatory. How well it must be confirmed to be acceptable is another matter of course.

11 If the hypothesis is of nonlimited scope, then its scope is not known to be finite. Hence, we cannot know whether we are getting a numerical increase in the ratio: examined favorable cases/total number of cases. If an increase in the probability of a hypothesis is equated with a (known) increase in this ratio, then we cannot raise the probability of a hypothesis of nonlimited scope in the simple-minded way described for hypotheses of (known) finite scope.

12 If the law was interpreted as a universal imperative of the form described, the most that it would permit us to infer about Sally would be a counteridentical: If Sally were one of the Presidents (i.e. identical with either Ford, Nixon, Johnson, . . .), then she would (at the appropriate time) have to consult Congress on matters pertaining to M.

References

[1] Achinstein, P. *Law and Explanation*. Oxford: Clarendon Press, 1971.

[2] Ayer, A.J. "What is a Law of Nature." In [5], pages 39–54.

[3] Braithwaite, R.B. *Scientific Explanation*. Cambridge, England: Cambridge University Press, 1957.

[4] Brody, B.A. "Confirmation and Explanation." *Journal of Philosophy* 65 (1968): 282–99, Reprinted in [5], pages 410–26.

[5] Brody, B.A. *Readings in the Philosophy of Science*. Englewood Cliffs, N.J.: Prentice Hall, 1970.

[6] Bromberger, S. "Why-Questions." In [5], pages 66–87.

[7] Goodman, N. *Fact, Fiction and Forecast*. London: The Athlone Press, 1954.

[8] Harman, G. "The Inference to the Best Explanation." *Philosophical Review* 74 (1965): 88–95.

[9] Harman, G. "Knowledge, Inference and Explanation." *Philosophical Quarterly* 18 (1968): 164–73.

[10] Hempel, C.G., and Oppenheim, P. "Studies in the Logic of Explanation." In [5], pages 8–27.

[11] Hempel, C.G. "Maximal Specificity and Lawlikeness in Probabilistic Explanations." *Philosophy of Science* 35 (1968): 116–33.

[12]   Kneale, W. "Natural Laws and Contrary-to-Fact Conditionals." *Analysis* **10** (1950): 121–5.

[13]   Kneale, W. *Probability and Induction*. Oxford: Oxford University Press, 1949.

[14]   Mackie, J.L. "Counterfactuals and Causal Laws." In *Analytical Philosophy*. (First Series). Edited by R.J. Butler. Oxford: Basil Blackwell, 1966.

[15]   Molnar, G. "Kneale's Argument Revisited." *Philosophical Review* **78** (1969): 79–89.

[16]   Nagel, E. *The Structure of Science*. New York: Harcourt Brace, 1961.

[17]   Popper, K. "A Note on Natural Laws and So-Called 'Contrary-to-Fact Conditionals." *Mind* **58** (1949): 62–6.

Chapter 22

Bas van Fraassen

WHAT ARE LAWS OF NATURE?

This question has a presupposition, namely that there are laws of nature. But such a presupposition can be cancelled or suspended or, to use Husserl's apt phrase, 'bracketed'. Let us set aside this question of reality, to begin, and ask what it means for there to be a law of nature. There are a good half-dozen theories that answer this question today, but, to proceed cautiously, I propose to examine briefly the apparent motives for writing such theories, and two recent examples (Peirce, Davidson) of how philosophers write about laws of nature. Then I shall collect from the literature a number of criteria of adequacy that an account of such laws is meant to satisfy. These criteria point to two major problems to be faced by any account of laws.

1 The importance of laws

What motives could lead a philosopher today to construct a theory about laws of nature? We can find three. The first comes from certain traditional arguments, which go back at least to the realist-nominalist controversy of the fourteenth century. The second concerns science. And the last comes from a reflection on philosophical practice itself; for while in the seventeenth century it was scientific treatises that relied on the notion of law, today it is philosophical writings that do so.

The motive provided by the traditional arguments I shall spell out in the next section, drawing on the lectures of Charles Sanders Peirce.

The second and much more fashionable motive lies in the assertion that laws of nature are what science aims to discover. If that is so, philosophers must clearly occupy themselves with this subject. Thus Armstrong's *What Is a Law of Nature?* indicates in its first section, 'the nature of a law of nature must be a central ontological concern for the philosophy of science'.

This does indeed follow from the conception of science found among seventeenth-century thinkers, notably Descartes. Armstrong elaborates it as follows. Natural science traditionally has three tasks: first, to discover the topography and history of the actual universe; *second*, to discover what sorts of thing and sorts of property there are in the universe; and *third*, to state the laws which the things in the universe obey. The three tasks are interconnected in various ways. David Lewis expresses his own view of science in such similar comments as these:

Physics is relevant because it aspires to give an inventory of natural properties. . . . Thus the business of physics is not just to discover laws and causal explanations. In putting forward as comprehensive theories that recognize only a limited range of natural properties, physics proposes inventories of the natural properties instantiated in our world. . . . Of course, the discovery of natural properties is inseparable from the discovery of laws.[1]

But what status shall we grant this view of science? Must an account of what the laws of nature are vindicate this view – or conversely, is our view of what science is to be bound to this conception? We know whence it derives: the ideal of a metaphysics in which the sciences are unified, as parts of an explanatory, all-embracing, and coherent world-picture (recall Descartes's 'philosophy as a whole is like a tree whose roots are metaphysics, whose trunk is physics, and whose branches, . . . are all the other sciences'). But this ideal is not shared throughout Western philosophy, nor ever was.

By its fruits, of course, shall we know this tree. If, starting with this conception, philosophers succeed in illuminating the structure of science and its activities, we shall have much reason to respect it. I do not share this conception of science, and do not see prima facie reason to hold it.

On the other hand, if metaphysics ought to be developed in such a way that the sciences can be among its parts, that does indeed place a constraint on metaphysics. It will require at least a constant series of plausibility arguments – to assure us that the introduction of universals, natural properties, laws, and physical necessities do not preclude such development. But this observation yields, in itself, only a motive for metaphysicians to study science, and not a motive for philosophers of science to study metaphysics.

The third and final motive, I said, lies in our reflection on philosophical practice itself. Even in areas far removed from philosophy of science, we find arguments and positions which rely for

their very intelligibility on there being a significant distinction between laws and mere facts of nature. I can do no better than to give an example, in section 3 below, of one such philosophical discussion, by Donald Davidson, about whose influence and importance everyone is agreed.

2 Peirce on scholastic realism

The traditional arguments are two-fold: to the conclusion that there must be laws of nature, and quite independently, to the conclusion that we must believe that there are such laws. The first argues from the premiss that there are pervasive, stable regularities in nature (sometimes itself backed up by noting the success of science). But no regularity will persist by chance – there must be a reason. That reason is the existence of a law of nature.

The second argues that if the preceding be denied, we are reduced to scepticism. If you say that there is no reason for a regularity – such as that sodium salts always burn yellow – then you imply that there is no reason for the regularity to persist. But if you say there is no reason, then you can't have any reason to expect it to persist. So then you have no basis for rational expectation of the future.

Charles Sanders Peirce asserted, correctly, that the general form of such arguments appeared well before the idea of laws of nature appeared in its modern sense.[2] Arguments of this form were given by the scholastic realists of the late Middle Ages against the nominalists. Peirce himself devoted the first section of his fourth lecture, 'The Reality of Thirdness', in his 1903 lecture series at Harvard, to his own variant of these arguments.[3] The lecture starts with the assertion that something quite beyond what nominalists acknowledge, is operative in nature. Dramatically opening his hand to the audience, Peirce displayed a stone (piece of writing chalk?):

Suppose we attack the question experimentally. Here is a stone. Now I place that stone where there will be no obstacle between it and the floor, and I will predict with confidence that as soon as I let go my hold upon the stone it will fall to the floor. I will prove that I can make a correct prediction by actual trial if you like. But I see by your faces that you all think it will be a very silly experiment.

Why silly? Because we all know what will happen.

But how can we know that? In words to be echoed later by Einstein, Podolsky, and Rosen, he answers 'If I *truly know* anything, that which I know must be real.' The fact that we know that this stone will fall if released, 'is the proof that the formula, or uniformity [which] furnish[es] a safe basis for prediction, is, or if you like it better, *corresponds to*, a reality'. A few sentences later he names that reality as a law of nature (though for him that is not the end of the story).

Do we have here the first or the second argument, or both? We very definitely have the second, for Peirce clearly implies you have no right to believe that the phenomena will continue the same in the future, unless you believe in the reality in question. But the reality cannot be a mere regularity, a fact about the future 'ungrounded' in the present and past, for that could not be known. Peirce did recognize chance, and agreed that anything at all could come about spontaneously, by chance, without such underlying reasons. Therefore he does not subscribe to the validity of 'There is a regularity, therefore there must be a reason for it, since no regularity could come about without a reason.' However he does not allow that we can know the premiss of that argument to be true, unless we also know the conclusion — nor to believe the premiss unless we believe the conclusion. This is a subtle point but important.

He gives the example of a man observed to wind his watch daily over a period of months, and says we have a choice: (*a*) 'suppose that some *principle* or *cause* is *really* operative to *make* him wind his watch daily' and predict that he will continue to do so; or else (*b*) 'suppose it is mere chance that his actions have hitherto been regular; and in that case regularity in the past affords you not the slightest reason for expecting its continuance in the future'. It is the same with the operations of nature, Peirce goes on to say, and the observed regularity of falling stones leaves us only two choices. We can suppose the regularity to be a matter of chance only, and declare ourselves in no position to predict future cases — or else insist that we can predict because we regard the uniformity with which stones have been falling as 'due to some *active general principle*'.

There is a glaring equivocation in this reasoning, obscured by a judicious choice of examples. Sometimes 'by chance' is made to mean 'due to no reason', and sometimes 'no more likely to happen than its contraries'. Of course, I cannot logically say that certain events were a matter of chance in the second sense, and predict their continuation with any degree of certainty. That would be a logical mistake. Nor do I think that a person winds his watch for no reason at all, unless he does it absent-mindedly; and absent-mindedness is full of chance fluctuations. But I can quite consistently say that all bodies maintain their velocities unless acted upon, and add that this is just the way things are. That is consistent; it asserts a regularity and denies that there is some deeper reason to be found. It would be strange and misleading to express this opinion by saying that this is the way things are by chance. But that just shows that the phrase 'by chance' is tortured if we equate it to 'for no reason'.

Perhaps we should not accuse Peirce of this equivocation, but attribute to him instead the tacit premiss that whatever happens either does so for a reason or else happens no more often than its contraries. But that would mean that a universe without laws – if those are the reasons for regularities – would be totally irregular, chaotic. That assertion was exactly the conclusion of the first argument. Hence if this is how we reconstruct Peirce's reasoning, we have him subscribing to the first argument as well. His indeterminism would then consist in the view that individual events may indeed come about for no reason, but not regularities.[4]

Peirce knew well the contrary tradition variously labelled 'nominalist' and 'empiricist', which allows as rational also simple extrapolation from regularities in past experience to the future. He saw this represented most eminently by John Stuart Mill, and attacked it vigorously. The following argument appears in Peirce's entry 'Uniformity' in Baldwin's Dictionary (1902).[5] Of Mill, Peirce says that he 'was apt to be greatly influenced by Ockham's razor in forming theories which he defended with great logical acumen; but he differed from other men of that way of thinking in that his natural candour led to his making many admissions without perceiving how fatal they were to his negative theories' (ibid. 76).

Mill had indeed mentioned the characterization of the general uniformity of nature as the 'fact' that 'the universe is governed by general laws'.[6] (He did not necessarily endorse that form of language as the most apt, though he does again use it in the next paragraph.) Any particular uniformity may be arrived at by induction from observations. The peculiar difficulty of this view lies in the impression that the rule of induction gives, of presupposing some prior belief in the uniformity of nature itself. Mill offered a heroic solution:

> the proposition that the course of nature is uniform is the fundamental principle, or

general axiom, of Induction. It would yet be a great error to offer this large generalization as any explanation of the inductive process. On the contrary I hold it to be itself an instance of induction, and induction by no means of the most obvious kind. (*Collected Works*, 392)

According to Peirce, Mill used the term 'uniformity' in his discussions of induction, to avoid the use of 'law', because that signifies an element of reality no nominalist can admit. But if his 'uniformity' meant merely regularity, and implied no real connection between the events covered, it would destroy his argument. Thus Peirce writes:

> It is, surely, not difficult to see that this theory of uniformities, far from helping to establish the validity of induction, would be, if consistently admitted, an insuperable objection to such validity. For if two facts, A and B, are entirely independent in their real nature, then the truth of B cannot follow, either necessarily or probably, from the truth of A. (*Collected Papers*, 77)

But this statement asserts exactly the point at issue: why should A, though bearing in itself no special relation to B, not be invariably or for the most part be followed by B? It is true that there would be no logical necessity about it, nor any probability logically derivable from descriptions of A and B in and by themselves. But why should all that is true, or even all that is true and important to us, be logically derivable from some internal connection or prior circumstance?

The convictions expressed by Peirce are strong, and have pervaded a good half of all Western philosophy. Obviously we shall be returning to these convictions, in their many guises, in subsequent chapters [*sic*]. A law must be conceived as *the reason which accounts for* uniformity in nature, not the mere uniformity or regularity itself. And the law must be conceived as something real, some element or aspect of reality quite independent of our thinking or theorizing – not merely a principle in our preferred science or humanly imposed taxonomy.

3 A twentieth-century example: Davidson

Concepts developed or analysed in one part of philosophy tend to migrate to others, where they are then mobilized in arguments supporting one position or another. From the roles they are expected to play in such auxiliary deployment, we should be able to cull some criteria for their explication. A good example is found in recent philosophy of mind.

Is there mind distinct from matter? Peter felt a sudden fear for his safety, and said 'I know him not'. The first was a mental event, the second at least in part a physical one. But materialists say that the mental event too consisted solely in Peter's having a certain neurological and physiological state – so that it too was (really) physical. Donald Davidson brought a new classification to this subject, by focusing on the question whether there are psychophysical laws. Such a law, if there is one, might go like this: every human being in a certain initial physiological state, if placed in certain circumstances, will feel a sudden fear for his or her safety. Davidson denies that there are such laws, yet asserts that all mental events are physical.

It may make the situation clearer to give a fourfold classification of theories of the relation between mental and physical events that emphasizes the independence of claims about laws and claims of identity. On the one hand there are those who assert, and those who deny, the existence of psychophysical laws;

on the other hand there are those who say mental events are identical with physical and those who deny this. Theories are thus divided into four sorts: *nomological monism*, which affirms that there are correlating laws and that the events correlated are one (materialists belong in this category); *nomological dualism*, which comprises various forms of parallelism, interactionism, and epiphenomenalism; *anomalous dualism*, which combines ontological dualism with the general failure of laws correlating the mental and the physical (Cartesianism). And finally there is *anomalous monism*, which classifies the position I wish to occupy.[7]

This last position is that every strict law is a physical law, and most if not all events fall under some such law – which they can obviously do only if they admit of some true physical description. Therefore most if not all events are physical. This is consistent provided that, although every individual mental event has some physical description, we do not assert that a class of events picked out by some mental description – such as 'a sudden feeling of fear' – must admit some physical description which appears in some strict law.

This point of consistency is easy enough to see once made. It does not at all depend on what laws are. But whether the position described even could be, at once, non-trivial and true – that does depend on the notion of law. If, for example, there were no distinction between laws and true statements in general, then there obviously are psychophysical laws, even if no interesting ones. Imagine an omniscient being, such as Laplace envisaged in his discussion of determinism, but capable also of using mental descriptions. Whatever class of events we describe to It, this being can list all the actual members of this class, and hence all the states of the universe in which these members appear. It can pick out precisely, for example, the set of conditions of the universe under which at least one of these states is realized within the next four years. Davidson must object that what It arrives at in such a case is in general not a law, although it is a true general statement.[8]

The form of objection could be anthropomorphic: although It could know that, we humans could not. Then the cogency of the objection would hinge on the notion of law involving somehow this distinction between what is and is not accessible (knowable, confirmable, . . .) to humans. The position of anomalous monism would no longer have the corollary 'Therefore most if not all events are physical', but rather something like: every event which we humans could cover in some description that occurs in a humanly accessible (knowable, or confirmable, or . . .) general regularity, is physical. In that case the position would seem to have no bearing at all on the usual mind-body problems, such as whether the mental 'supervenes' on the physical (which means, whether our mental life being otherwise would have required the physical facts to be otherwise).

Davidson's objection to the story about this omniscient genie would therefore need to be non-anthropomorphic. It would have to insist on a distinction between what the laws are and truths in general, independent of human limitations. The reason this being would not automatically arrive at a law, by reflection on just any class of events we mentioned to It, would have to be due to a law being a special sort of fact about the universe.

Davidson himself notes this presupposition of his argument, and places the burden of significance squarely on the notion of law. What he then goes on to say about laws is unfortunately in part predicated on the logical positivists' very unsuccessful approach to the subject, and in part deliberately non-committal: 'There is (in my view) no non-question begging criterion of the lawlike, which is not to say that there are no reasons in particular cases for a judgment' (*Essays*, 217). This statement, which begins his discussion of laws, itself presupposes the positivists' idea that laws are simply the truths among a class of statements (the 'lawlike' ones) singled out by some common element of form or meaning, rather than by what the world is like. (Davidson comments 'nomologicality is much like analyticity, as one might expect since both are linked to meaning' (p. 218). This presumption was later strongly criticized, for example by Dretske; at this point we should note only that it is dubitable, and not innocuous. I do not mean to go further into how Davidson discusses laws here; the point I wanted to make should now be clear.

The assumptions involved are that there is a significant concept of natural law, that the distinction between laws and truths in general is non-anthropomorphic and concerns what the world is like, and that the correct account of laws must do justice to all this. These are indispensable to Davidson's classification of philosophical positions on mind and matter, to the arguments for his position, and for the significance of that position.[9] This is a striking illustration of how general philosophy had, by our century, learned to rely on this notion of law.

4 Criteria of adequacy for accounts of laws

If we do have the concept of a law of nature, this must mean at least that we have some clear intuitions about putative examples and counter-examples. These would be intuitions, for example, about what is and what is not, or what could be and what could not be, a law of nature, if some sufficiently detailed description of the world is supposed true. It does not follow that we have intuitions of a more general sort about what laws are like. But when we are offered ideas of this more general sort, we can test them against our intuitions about specific examples.

The use of such examples and our intuitive reactions to them serves at least to rule out overly simplistic or naïve accounts of laws of nature. Their use has also led to a number of points on which, according to the literature, all accounts of laws must agree. None of these points is entirely undisputed, but all are generally respected.

Disagreements about the criteria should not dismay us at the outset. As Wittgenstein taught, many of our concepts are 'cluster concepts' – they have an associated cluster of criteria, of which only *most* need be satisfied by any instance. The more of the criteria are met, the more nearly we have a 'clear case'. This vagueness does not render our concepts useless or empty – our happiness here as elsewhere depends on a properly healthy tolerance of ambiguity.

In what follows I shall discuss about a dozen criteria found in the literature. Some are less important, or more controversial than others. We can use them to dismiss some naïve ideas, especially cherished by empiricists – and in subsequent chapters [*sic*] bring them to bear on the main remaining accounts of law. Nowhere should we require that all the criteria be met; but any account should respect this cluster as a whole.

Universality

The laws of nature are universal laws, and universality is a mark of lawhood. This criterion has been a great favourite, especially with empiricists, who tend to be wary of nearly all the criteria we shall discuss subsequently. There is indeed nothing in the idea of universality that should make philosophical hackles rise, nor would there be in the idea of law if a law stated merely what happens always and everywhere. The hope that this may be so must surely account

for the curiously uncritical attitude toward this notion to be found in even the most acute sceptics:

> Whitehead has described the eighteenth century as an age of reason based upon faith – the faith in question being a confidence in the stability and regularity of the universal frame of Nature. Nothing can better illustrate Hume's adherence to this faith, and its separation in his mind from his philosophical scepticism, than his celebrated Essay *Of Miracles.* The very man who proves that, for all we can tell, anything may be the 'cause' of anything, was also the man who disproved the possibility of miracles because they violated the invariable laws of Nature.[10]

That does not make Hume inconsistent. If what a law is concerns only what is universal and invariable, the faith in question could hardly impugn Hume's scepticism about mysterious connections in nature beyond or behind the phenomena. For in that case it would merely be a faith in matters of fact, which anyone might have, and which would not – unlike the 'monkish virtues' – bar one from polite society (the standard Hume himself so steadfastly holds out to us).

Unfortunately this mark of universality has lately fallen on hard times, and that for many reasons. Let us begin with the point that universality is not enough to make a truth or law of nature. No rivers past, present, or future, are rivers of Coca-Cola, or of milk. I think that this is true; and it is about the whole world and its history. But we have no inclination to call it a law.[11] Of course we can cavil at the terms 'river', 'Coca-Cola', or 'milk'. Perhaps they are of earthly particularity. But we have no inclination to call this general fact a law because we regard it as a merely incidental or accidental truth. Therefore we will have the same intuition, regardless of the terms employed. This is brought out most strikingly by parallel examples, which employ exactly the same categories of terms, and share exactly the same logical form, yet evoke different responses when we think about what could be a law. The following have been discussed in various forms by Reichenbach and Hempel:[12]

1. All solid spheres of enriched uranium (U235) have a diameter of less than one mile.
2. All solid spheres of gold (Au) have a diameter of less than one mile.

I guess that both are true. The first I'm willing to accept as putatively a matter of law, for the critical mass of uranium will prevent the existence of such a sphere. The second is an accidental or incidental fact − the earth does not have that much gold, and perhaps no planet does, but the science I accept does not rule out such golden spheres. Let us leave the reasons for our agreement to one side − the point is that, if 1 could be law, if only a little law, and 2 definitely could not, it cannot be due to a difference in universality.

Another moral that is very clear now is that laws cannot be simply the true statements in a certain class characterized in terms of syntax and semantics. There is no general syntactic or semantic feature in which the two parallel examples differ. So we would go wrong from the start to follow such writers as Goodman, Hempel, and Davidson in thinking of the laws as the true 'lawlike' statements.

We can agree in the intuitions invoked above, before any detailed analysis of universality. But we have also already discerned some reason to think that the analysis would not be easy. In fact, it is extremely difficult to make the notion precise without trivializing it. The mere linguistic form 'All ... are ...' is not a good guide, because it does not remain invariant under logical transformations. For example,

'Peter is honest' is in standard logic equivalent to the universal statement 'Everyone who is identical with Peter, is honest.' To define generality of content turns out to be surprisingly difficult. In semantics, and philosophy of science, these difficulties have appeared quite poignantly.[13] Opinions in the literature are now divided on whether laws must indeed be universal to be laws. Michael Tooley has constructed putative counterexamples.[14] David Armstrong's account requires universality, but he confesses himself willing to contemplate amendment.[15] David Lewis's account does not require it.[16] In Part III we shall find an explication of generality allied to concepts of symmetry and invariance [sic]. While I regard this as important to the understanding of science, the generality we shall find there is theory-relative.

The criterion of universality, while still present in discussion of laws, is thus no longer paramount.

Relations to necessity

In our society, one must do what the laws demand, and may do only what they do not forbid. This is an important part of the positive analogy in the term 'laws of nature'.

Wood burns when heated, because wood must burn when heated. And it must burn because of the laws which govern the behaviour of the chemical elements of which wood and the surrounding air are composed. Bodies do not fall by chance; they must fall because of the law of gravity. In such examples as these we see a close connection between 'law' and 'must,' which we should stop to analyse.

Inference. The most innocuous link between law and necessity lies in two points that are merely logical or linguistic. The first is that if we say that something is a law, we endorse it as being true. The inference

(1)   It is a law of nature that *A* Therefore, *A*

is warranted by the meaning of the words. This point may seem too banal to mention – but it turns out, surprisingly, to be a criterion which some accounts of law have difficulty meeting. One observes of course that the inference is not valid if 'of nature' is left out, since society's laws are not always obeyed. Nor does it remain valid if we replace 'law of nature' by 'conjecture' or even 'well-confirmed and universally accepted theory'. Hence the validity must come from the special character of laws of nature. In Chapter 5, the problem of meeting this criterion will be called the problem of inference [*sic*].

Intensionality.   The second merely logical point is that the locution 'It is a law that' is *intensional*. Notice that the above inference pattern (1) does remain valid if we replace 'a law of nature' by 'true'. But something important has changed when we do, for consider the following argument:

(2)   It is true that all mammals have hair.
        All rational animals are mammals.
        Therefore, it is true that all rational
        animals have hair.

This is certainly correct, but loses its validity if we now replace 'true' again by 'a law of nature'. Another example would be this: suppose that it is a law that diamonds have a refraction index >2, and that as a matter of fact all mankind's most precious stones are diamonds. It still does not follow that it is a law that all mankind's most precious stones have a refraction index >2. Here we see the distinction between law and mere truth or matter of fact at work.

Our first two criteria are therefore merely points of logic, and I take them to be entirely uncontroversial.

Necessity bestowed.   The moon orbits the earth and must continue to do so, because of the law of gravity. This illustrates the inference from *It is a law that A* to *It is necessary that A*; but this must be properly understood.

The medievals distinguished the *necessity of the consequence* from the *necessity of the consequent*. In the former sense it is quite proper to say 'If all mammals have hair then whales must have hair, because whales are mammals.' The 'must' indicates only that a certain consequence follows from the supposition. For law this point was therefore already covered above. The criterion of necessity bestowed is that there is more to it: if *It is a law that A* is true then also, rightly understood, *It is necessary that A* is true. This necessity is then called physical necessity or nomological necessity (and is now often generalized to physical probability).

Empiricists and nominalists have always either rejected this criterion or tried to finesse it. For they believe that necessity lies in connections of words or ideas only, so ultimately the only necessity there can be lies in the necessity of the consequence. This is not altogether easy to maintain, while acknowledging the preceding points of logic. Yet their persistent attempts to reconstrue the criterion of necessity bestowed, so that it is fulfilled if 'properly' understood, show the strength of the intuition behind it.[17]

Necessity inherited.   There is a minority opinion that what the laws are is itself necessary.[18] This point definitely goes beyond the preceding, for logic does not require what is necessary to be necessarily necessary. More familiar is the idea that there are many different ways the world could have been, including differences in its laws governing nature. If gravity had obeyed an inverse cube law, we say, there would have been no stable solar system – and we don't think we are contemplating an absolute impossibility. But we could be wrong in this.

Of course, if laws are themselves necessary truths, their consequences would inherit this necessity. Therefore the strong criterion of *necessity inherited* entails that of *necessity bestowed*. And since what is necessary must be actual, the criterion of *necessity bestowed* entails that of *inference*. The entailments do not go in the opposite direction. So three of the criteria we have formulated here form a logical chain of increasing strength.

Explanation

Such writers as Armstrong insist that laws are needed to explain the phenomena, and indeed, that there are no explanations without laws. This is not in accord with all philosophical theories of explanation.[19] A more moderate requirement would be that laws must be conceived as playing an indispensable role in some important or even pre-eminent pattern of explanation.

There does indeed appear to be such a pattern, if there is an intimate connection between laws and necessity (and objective probability). It may even be the pre-eminent pattern involved in all our spontaneous confrontations with the world. Witness that Aristotle made it the key to narrative and dramatic structure in tragedy:

> And these developments must grow out of the very structure of the plot itself, in such a way that on the basis of what has happened previously this particular outcome follows either by necessity or in accordance with probability; for there is a great difference in whether these events happen because of those or merely after them. (*Poetics*, 52ª 17–22)

This account of tragedy bears a striking resemblance to Aristotle's account of how science must depict the world, in his *Physics*.[20] The parallel is no accident, though one must admit that Aristotle's demands upon our understanding of nature persisted longer than those he made upon our appreciation of literature.

What exactly is this criterion, that laws must explain the phenomena? When a philosopher – as so many do – raises explanation to pre-eminence among the virtues, the good pursued in science and all natural inquiry, he or she really owes us an account of why this should be so. What is this pearl of great price, and why is it so worth having? What makes laws so well suited to secure us this good? When laws give us 'satisfying' explanations, in what does this warm feeling of satisfaction consist? There are indeed philosophical accounts of explanation, and some mention laws very prominently; but they disagree with each other, and in any case I have not found that they go very far toward answering *these* questions.[21]

Hence we should not get very far with this criterion for accounts of laws, if its uses depended greatly on the philosophical opinions of what explanation is. Fortunately there are two factors which keep us from being incapacitated here. The first factor is the very large measure of agreement on what counts as explanation when we are confronted with specific, concrete examples. The other factor is the great degree of abstraction which characterizes many discussions of law. In Chapter 6, for example [sic], we shall be able to take up Dretske's and Armstrong's arguments concerning what is for them the crucial argument form of Inference to the Best Explanation – and its relation to laws – without ever having to reproach them for the fact that they nowhere tell us what an explanation is.

We shall encounter a certain tension between the criteria regarding the connections of law with necessity on the one hand, and with science on the other. Here the concept of explanation could perhaps play an important mediating role: If explanation is what we look for in science, while necessity is crucial to explanation and law crucial to necessity, then that tension may perhaps be '*aufgehoben*' in a higher unity. We shall have to see.

Prediction and confirmation

That there is a law of gravity is the reason why the moon continues to circle the earth. The premiss that there is such a law is therefore a good basis for prediction. The second traditional argument which I briefly sketched above – and illustrated from Peirce's lecture – goes on: and if we deny there is such a reason, then we can also have no reason for making that prediction. We shall have no reason to expect the phenomenon to continue, and so be in no position to predict.

If there is a problem with this argument today, it is surely that we cannot be so ready to equate *having reason to believe that A* with *believing that there is a reason for A*. Linguistic analysis in philosophy makes us very wary of such pretty rhetoric. But the equation might perhaps hold for the special case of empirical regularities and laws. Certainly, a form of this second traditional argument is found very prominently in Armstrong's book. After canvassing some views on what laws are, he notes a possibility which he says was brought to his attention by Peter Forrest:

> There is one truly eccentric view. . . . This is the view that, although there are regularities in the world, there are no laws of nature. . . . This Disappearance view of law can nevertheless maintain that inferences to the unobserved are reliable, because although the world is not law-governed, it is, by luck or for some other reason, regular.[22]

Armstrong replies immediately that such a view cannot account for the fact that we can have *good reasons to think* that the world is regular. He gives an argument for this, which I shall discuss in Chapter 6 [sic].

A little of the recent history of confirmation appeared along the way in the article by Davidson which we discussed above. While Davidson attempts no definition or theory of what laws are, he says among other things that laws are general statements which are confirmed by their instances – while this is not always so for general statements. This makes sense if laws are the truths among lawlike statements, and if in addition we (who assess the evidence) can distinguish lawlike statements from other generalities. For else, how can instances count for greater confirmation?

But this idea receives rather a blow from the parallel gold and uranium examples we discussed above. These parallel examples are so parallel in syntactic form and semantic character that the independent prior ability to distinguish lawlike from other general statements is cast into serious doubt.

We should also observe that for writers on laws there is – and perhaps must be – a crucial connection between confirmation and explanation. For consider the following argument: that it is a law that P could be supported by claims either of successful explanation or of successful prediction (or at least, successful fitting of the data). But prediction cannot be enough, for the second sort of claim works equally well for the bare statement that *A: It is a law that A* entails or fits factual data only in so far as, and because, *A* does. Hence confirmation for the discriminating claim *It is not only true but a law that A* can only be on the basis of something in addition to conforming evidence. One traditional candidate for this something extra is successful explanation.

This observation gives us, I think, the best explanation of why advocates of laws of nature typically make Inference to the Best Explanation the cornerstone of their epistemology.

Counterfactuals and objectivity

Philosophy, being a little other-worldly, has always been fascinated with the conditional form If (*antecedent*) then (*consequent*). When the antecedent is false ('the conditional is contrary to fact' or 'counterfactual') what speculative leaps and fancies are not open to us? If wishes were horses then beggars would ride; if gravity had been governed by an inverse cube law there would have been no stable solar system; if Caesar had not crossed the Rubicon, ... Being also a little prosaic, the philosopher sets out to find the bounds of fancy: when must such a conditional be true, when false?

There is one potentially large class of cases where the answer is clear. If B follows from A with necessity, then *If A then B* is true and *If A then not B* is false. Thus if iron must melt at 2000 °C, it follows that this iron horse-shoe would melt if today it were heated to that temperature . . . this is clear even if the horseshoe remains at room temperature all day. At midnight we will be able to say, with exactly the same warrant, that it would have melted if it had been heated to 2000 °C. Many other such conditionals command our intuitive assent: Icarus' father too would have fallen if his wings had come loose, and so would I if I had stepped off the little platform when I went up the cathedral tower in Vienna. We observe that in all such cases we intuitively agree also to describe our warrant in terms of laws. These facts about iron and gravity are matters of law; if it is a law then it must be so; and if it must be so then it will or would be so if put to the test.

This large class of cases falls therefore very nicely under the previous criterion of necessity bestowed. But the requirement that laws be the sort of thing that warrant counterfactuals, has a much greater prominence in the literature. Is there more to it?

In the mid-1940s, Nelson Goodman and Roderick Chisholm made it clear that there are mysteries to the counterfactual conditional, which had escaped their logical treatment so far. This treatment did indeed fit necessary implications. Typical sanctioned argument patterns include

Whatever is A must be B.
Therefore, if this thing is (were) both A and C, then it is (would be) B.

But can all conditionals derive from necessities in this way? Consider: if I had struck this match, then it would have lit. It does not follow that if I had struck this match, and it had been wet at the time, then it would have lit. Nor, if I agree that the latter is false, do I need to retract the former. I can just say: well, it wasn't wet. We see therefore that counterfactual conditionals violate the principles of reasoning which govern 'strict' or necessary conditionals.

How are we to explain this mystery? Goodman did not explain it, but related it to laws.[23] We can, he said, support a counterfactual claim by citing a law. We cannot similarly support it by merely factual considerations, however general. For example it is a fact (but not a matter of law) that all coins in Goodman's pocket were silver. We cannot infer from this that if this nickel had been in his pocket then, it would have been silver. On the other hand it is also a fact and a matter of law that silver melts at 960.5 °C. Therefore if this silver had been heated to that degree, it would have melted. This observation, Goodman thought, went some way toward clearing up the mystery of counterfactual conditionals. The mystery was not thereby solved, so the connection was inverted: giving warrant for counterfactual conditionals became the single most cited criterion for lawhood in the post-war literature.

But the mystery was solved in the mid-1960s by the semantic analysis due to Robert Stalnaker and extended by David Lewis. Unfortunately for laws, this analysis entails that the violations of those principles of inference that work perfectly well for strict conditionals are due to context-dependence. The interesting counterfactuals which do not behave logically like the strict ones do not derive from necessities alone, but also from some contextually fixed factual considerations. Hence (I have argued elsewhere) science by itself does not imply these more interesting counterfactuals; and if laws did then they would have to be context-dependent.[24] Robert Stalnaker has recently replied to this that science does imply counterfactuals, in the same sense that it implies indexical statements.[25] An example would be:

> Science implies that your materialist philosophy is due to a dietary deficiency.

This is a context-dependent sense of 'implies' (not of course the sense which I had in mind), because the referent of 'you' depends on context. Stalnaker's point is quite correct. But it leads us to conclude at best that the speaker may believe that some law is the case, and holds its truth-value fixed in a tacit *ceteris paribus* clause (which gives the counterfactual sentence its semantic content in this context). This is certainly correct, but is equally correct for any other sort of statement, and cannot serve to distinguish laws from mere truths or regularities. I suspect that the real use of Goodman's requirement concerned counterfactuals considered true in cases where the corresponding physical necessity statement is also implied. If so, the requirement coincided in philosophical practice with the requirement of bestowed necessity.

Context-independence. In view of the above, however, it is important to isolate the sense in which law statements cannot be context-dependent. Stalnaker's sort of example leaves us with the requirement:

> If the truth value of statement *A* is context-independent, then so is that of *It is a law that A*.

Related to the context-independence of the locution 'It is a law that', but not at all the same,

is the point that laws are to be conceived of as objective.

Objectivity. Whether or not something is a law is entirely independent of our knowledge, belief, state of opinion, interests, or any other sort of epistemological or pragmatic factor. There have definitely been accounts of law that deny this. But they have great difficulty with such intuitively acceptable statements as that there may well be laws of nature which not only have not been discovered and perhaps never will be, but of which we have not even yet conceived.[26]

Relation to science

We come now to a final criterion which is of special importance. Laws of nature must, on any account, be the sort of thing that science discovers. This criterion is crucial, given the history of the concept and the professed motives of its exponents.

This criterion too is subject to a number of difficulties. First of all, there is no philosophically neutral account of what science discovers, or even what it aims to discover. Secondly, although the term 'law' has its use in the scientific literature, that use is not without its idiosyncrasies. We say: Newton's laws of motion, Kepler's laws of planetary motion, Boyle's law, Ohm's law, the law of gravity. But Schroedinger's equation, or Pauli's exclusion principle, which are immensely more important than, for example, Ohm's law, are never called laws. The epithet appears to be an honour, and often persists for obscure historical reasons.

Attempts to regiment scientific usage here have not been very successful. Margenau and Lindsay note disapprovingly that other writers speak of such propositions as *copper conducts electricity* as laws.[27] They propose that the term should be used to denote any precise numerical equation describing phenomena of a certain kind.[28]

That would make Schroedinger's equation, but not Pauli's exclusion principle, a law. Even worse: it would be quite easy to make up a quantitative variant of the rivers of Coca-Cola example which would meet their criterion trivially.

Faced with this situation, some writers have reserved 'law' for low-level, empirical regularities, thus classifying the law of conservation of energy rather than Boyle's law as terminological idiosyncrasy. To distinguish, these low-level laws are also called phenomenological laws, and contrasted with basic principles which are usually more theoretical. Science typically presents the phenomenological laws as only approximate, strictly speaking false, but useful. According to Nancy Cartwright's stimulating account of science, the phenomenological laws are applicable but always false; the basic principles accurate but never applicable.[29] It is therefore not so easy to reconcile science as it is with the high ideals of those who see it as a search for the true and universal laws of nature.

The criterion of adequacy, that an account of laws must entail that laws are (among) what science aims to discover, is therefore not easy to apply. Certainly it cannot be met by reliance on a distinction embodied in what scientists do and do not call a law. Nor, because of serious philosophical differences, can it rely on an uncontroversial notion of what the sciences (aim to) discover. The criterion of objectivity we listed earlier, moreover, forbids identification of the notion of law with that of a basic principle or any other part of science, so identified. For if there are laws of nature, they would have been real, and just the same even if there had been no scientists and no sciences.

It appears therefore that in accounts of law, we must try to discern simultaneously a view of what science is and of what a law is, as well as of how the two are related. These views must then be evaluated both independently and in terms of this final criterion, that they should stand in a significant relationship.

Earlier in this century, the logical positivists and their heirs discussed laws of nature, and utilized that concept in their own explications and polemics. I shall not examine those discussions in any detail. If we look at their own efforts to analyse the notion of law, we find ourselves thoroughly frustrated. On the one hand we find their own variant of the sin of psychologism. For example, there is a good deal of mention of natural laws in Carnap's *The Logical Syntax of Language*. But no sooner has he started on the question of what it means to say that it is a law that all *A* are *B*, than he gets involved in the discussion about how we could possibly verify any universal statement. On the other hand there is the cavalier euphoria of being involved in a philosophical programme all of whose problems are conceived of as certain to be solved some time later on. Thus in Carnap's much later book *Philosophical Foundation of Physics*[30], we find him hardly nearer to an adequate analysis of laws or even of the involved notions of universality or necessity – but confident that the necessary and sufficient conditions for lawhood are sure to be formulated soon. The culmination of Carnap's, Reichenbach's, and Hempel's attempts, which is found in Ernest Nagel's *The Structure of Science*, was still strangely inconclusive. In retrospect it is clear that they were struggling with modalities which they could not reduce, saw no way to finesse, could not accept unreduced, and could not banish.

Having perceived these failures of logical empiricism, some philosophers have in recent years taken a more metaphysical turn in their accounts of laws. I shall focus my critique on those more recent theories.

5 Philosophical accounts: the two main problems

Of the above criteria, never uniformly accepted in the literature, five seem to me pre-eminent.

They are those relating to necessity, universality, and objectivity and those requiring significant links to explanation and science.

But apart from the more or less piecemeal evaluation these allow, of all proffered philosophical accounts of laws, there will emerge two major problems. I shall call these the *problem of inference* and the *problem of identification*. As we shall see, an easy solution to either spells serious trouble from the other.

The problem of inference is simply this: that it is a law that *A*, should *imply* that *A*, on any acceptable account of laws. We noted this under the heading of necessity. One simple solution to this is to equate *It is a law that A* with *It is necessary that A*, and then appeal to the logical dictum that necessity implies actuality. But is 'necessary' univocal? And what is the ground of the intended necessity, what is it that makes the proposition a necessary one? To answer these queries one must identify the relevant sort of fact about the world that gives 'law' its sense; that is the problem of identification. If one refuses to answer these queries − by consistent insistence that necessity is itself a primitive fact − the problem of identification is evaded. But then one cannot rest irenically on the dictum that necessity implies actuality. For 'necessity', now primitive and unexplained, is then a mere label given to certain facts, hence without logical force − Bernice's Hair does not grow on anyone's head, whatever be the logic of 'hair'.

The little dialectic just sketched is of course too elementary and naïve to trip up any philosopher. But it illustrates in rudimentary fashion how the two problems can operate as dilemma. We shall encounter this dangerous duet in its most serious form with respect to objective chance (irreducibly probabilistic laws), but it will be found somehow in many places. In the end, almost every account of laws founders on it.

Besides this dialectic, the most serious recurring problem concerns the relation between laws and science. The writers on laws of nature by and large do not so much develop as presuppose a philosophy of science. Its mainstay is the tenet that laws of nature are the sciences' main topic of concern. Even if we do not require justification for that presupposition, it leaves them no rest. For they are still required to show that science aims to find out laws *as construed on their account*. This does not follow from the presupposition, even if it be sacrosanct.

While I cannot possibly examine all extant accounts of laws, and while new ones could spring up like toadstools and mushrooms every damp and gloomy night, these problems form the generic challenge to *all* philosophical accounts of laws of nature. In the succeeding three chapters, we shall see the three main extant sorts of account founder on them [sic].

Notes

1 D. Lewis, 'New Work for a Theory of Universals', *Australasian Journal of Philosophy*, **61** (1983), 343–77; citations from 356–357, 364, 365.

2 For Peirce's discussion of the history of this idea, see his essay on Hume's 'Of Miracles'; *Collected Papers of Charles Sanders Peirce*, ed. C. Hartshorne and P. Weiss (Cambridge, Mass., 1931–5, vol. VI, bk. Ii, ch. 5, especially pp. 364–6. See also the section 'Letters to Samuel P. Langley, and "Hume on Miracles and Laws of Nature"' in P. Weiner (ed.), *Values in a World of Chance: Selected Writings of C. S. Peirce* (New York, 1966).

3 *Collected Papers*, vol. v, sect 5.93–5.119 (pp. 64–76). That first section bears the title 'Scholastic Realism'.

4 This may be compared with Reichenbach's conception of indeterminism which I discussed in my 'The Charybdis of Realism; Epistemological Implications of Bell's Inequalities', *Synthese*, **5** (1982), 25–38; see also ch. 5, sect. 6 below [sic] and *The Scientific Image*, ch. 2, sect. 5.

5 Reprinted Hartshorne and Weiss vi. 75–85; see esp. 76–9.

6 *A System of Logic* (London 1846), vol. I, I; bk. iii, ch. 3; J.S. Mill, *Collected Works* (Toronto; 1963), 306.

7 D. Davidson, *Essays on Actions and Events* (Oxford, 1980), 213–14. I want to thank Mark Johnston for helpful conversations; I have also drawn on his 'Why Having a Mind Matters', in E. LePore and B. McLaughlin (eds), *Actions and Events* (Oxford, 1985), 408–26.

8 The statement is general provided only that the universe satisfies the principle of identity of indiscernibles: for in that case, whatever class of events we describe will have some uniquely identifying description.

9 The same can be said of Davidson's analysis of causation, relied on in this article. See 'Causal Relations' in his *Essays on Actions and Events* (Oxford, 1980), 149–62.

10 R. Willey, *The Eighteenth Century Background* (London, 1940), 126–7. This is not an accurate assessment if taken as a description of Hume's argument; as Peirce pointed out, that concerned probability alone. But that Hume expressed this faith in uniformity is certainly evident.

11 Some writers do not balk; see e.g. N. Swartz, *The Concept of Physical Law* (Cambridge, 1985).

12 David Lewis tells me that such parallel pairs were discussed at UCLA in the 1960s, and I speculate this was the heritage of Reichenbach's discussion of such examples.

13 See my 'Essence and Existence' in *American Philosophical Quarterly* Monograph Series No. 12 (1978); J. Earman 'The Universality of Laws', *Philosophy of Science*, **45** (1978), 173–81.

14 M. Tooley, 'The Nature of Law', *Canadian Journal of Philosophy*, **7** (1977), 667–98 (see esp. 686).

15 D. Armstrong, *What is a Law of Nature?* (Cambridge: Cambridge University Press, 1983), 26.

16 See Earman, 'The Universality of Laws'; he does add that laws in the sense of Lewis are 'likely' to be universal.

17 My own attempts, utilizing formal pragmatics, are found in 'The only Necessity is Verbal Necessity', *Journal of Philosophy*, **74** (1977), 71–85, and 'Essences and Laws of Nature', in R. Healey (ed.), *Reduction, Time and Reality* (Cambridge: Cambridge University Press, 1981).

18 Cf. Armstrong, *What is a Law of Nature?*, ch. 11; C. Swoyer, 'The Natural Laws', *Australasian Journal of Philosophy*, **60** (1982), 203–23.

19 For example, it is not in accord with my own, cf. *The Scientific Image*, ch. 5. An entirely different, but also contrary, point is made by McMullen, 1984: 'In the new sciences, lawlikeness is not an explainer, it is what has to be explained' (p. 51).

20 Perhaps the resemblance goes further: according to the *Posterior Analytics*, science deals only with the truly universal; and in the *Poetics*, Aristotle writes 'poetry is something more philosophic and of graver import than history, since its statements are of the nature of universals' (1451^7–9). The parallel can not be pushed too far, but remains striking nevertheless, if we are at all puzzled by our insistent craving for reasons why.

21 My own account of explanation, in *The Scientific Image*, implies that explanation is not a good in itself, and is worth pursuing only as a tactical aim, to bring us to empirical strength and adequacy if it can.

22 Armstrong, *What is a Law of Nature?*, 5; see P. Forrest, 'What Reasons do we have for Believing There are Laws of Nature?', *Philosophical Inquiry Quarterly*, **7** (1985), 1–12.

23 N. Goodman, *Fact, Fiction, and Forecast* (Atlantic Highlands, NJ, 1954).

24 See *The Scientific Image*, p. 118, and 'Essences and Laws of Nature'.

25 R. Stalnaker, *Inquiry* (Cambridge, Mass., 1984), 149–50.

26 Cf. F. Dretske, 'Laws of Nature', *Philosophy of Science*, **44** (1977) 24, 278–68; see esp. 251–2 and 254–5.

27 H. Margenau and R.B. Lindsay, *Foundations of Physics* (New York, 1957), 20.

28 That is certainly in accordance with one use of the term, which may indeed be quite common in some idiolects. Thus P. Abbot and H. Marshall, *National Certificate Mathematics*, 2nd edn (London, 1960) designed 'for students taking mechanical or electrical engineering courses' has a chapter entitled 'Determination of Laws' which teaches how to find equations that fit given sets of data.

29 N. Cartwright, *How the Laws of Physics Lie* (Oxford, 1983).

30 Ed. M. Gardner (New York, 1966).

Chapter 23

Marc Lange

NATURAL LAWS AND THE PROBLEM OF PROVISOS

I

According to the regularity account of physical law – versions of which have been advocated by Ayer (1963), Braithwaite (1953, ch. 9), Goodman (1983, pp. 17–27), Hempel (1965a, pp. 264ff.), Lewis (1973, pp. 72–77; 1986), Mackie (1962, pp. 71–73), Nagel (1961, pp. 58ff.), and Reichenbach (1947, ch. 8), among others – laws of nature are regularities among events or states of affairs and a law-statement, the linguistic expression of a law, is a description of a regularity that is a law. The familiar challenge faced by this account is to distinguish those descriptions of regularities that are law-statements from those that are accidental generalizations. I wish to consider a more fundamental problem: that many a claim we believe to describe no regularity at all, nomological or accidental, we nevertheless accept as a law-statement. This problem arises from what Hempel (1988) calls "the problem of provisos."

Consider the familiar statement of the law of thermal expansion: "Whenever the temperature of a metal bar of length L_0 changes by ΔT, the bar's length changes by $\Delta L = k \cdot L_0 \cdot \Delta T$, where k is a constant characteristic of that metal." This statement states a relation between L_0, ΔT, and ΔL that does not obtain; it may be violated, for instance, if someone is hammering the rod inward at one end. Since this statement does not describe a regularity, it is not a law-statement, on the regularity account.

Hempel (1988) would hold that a complete statement of the law includes the condition ". . . if the end of the rod is not being hammered in." On this view, whereas the familiar "law-statement" takes the form $(x) (Fx \supset Gx)$, and the only premise one must add to yield the conclusion Ga is Fa, the genuine law-statement takes the form $(x) (Fx \& Px \supset Gx)$, and one needs to add Pa as well as Fa to infer Ga. It is the regularity account of laws that leads Hempel to characterize such conditions $(x) \ldots$ if Px as "essential" to law-statements (and the corresponding premises Pa as "essential" to inferences); without these conditions, the claims would be false and so, on the regularity account, would not be law-statements. Following Hempel (1988, p. 23), I'll refer to those conditions (and premises) that, by this reasoning, are necessary to law-statements (and to inferences from law-statements), but are "generally unstated," as "provisos."

Provisos pervade scientific practice. By Hempel's reasoning, Snell's law of refraction – when a beam of light passes from one medium to another, $\sin i / \sin r = $ constant, where i is the angle of incidence of the beam upon the second medium, r is the angle of refraction in that medium, and the constant is characteristic of the two types of media – must require particular temperatures and pressures of the media as well as the absence of any magnetic or electrical potential difference across the boundary, uniform optical density and transparency and non-double-refractivity in the two media, and a monochromatic beam; these conditions are provisos. Likewise, the law of freely falling bodies – the distance a body falls to earth in time t is $(1/2)gt^2$ – must specify when fall qualifies as "free"; while the law can remain approximately true away from the height at which g is measured, its predictions may be drastically wrong when electromagnetic forces, air resistance, or other collisions affect the falling body.

On Hempel's proposal, it becomes impossible to state very many genuine law-statements since, as Giere puts it (1988, p. 40), "the number of provisos implicit in any law is indefinitely large." To state the law of thermal expansion, for instance, one would need to specify not only that no one is hammering the bar inward at one end, but also that the bar is not encased on four of its six sides in a rigid material that will not yield as the bar is heated, and so on. For that matter, not all cases in which the bar is hammered upon constitute exceptions to $\Delta L = k \cdot L_0 \cdot \Delta T$; the bar may be hammered upon so softly and be on such a frictionless surface that the hammering produces translation rather than compression of the bar. One is driven to say that the only way to utter a complete law-statement is to employ some such condition as ". . . in the absence of other relevant factors." But Hempel deems such an expression inadmissible in a law-statement: On the regularity account, a law-statement states a particular relation (which must obtain and be a law), but a claim (x) $(Fx$ & there are no other relevant factors $\supset Gx)$ does not assert any determinate relation at all because it fails to specify which other factors count as relevant, i.e., which specific premises are to be

added to Fa and the law-statement to infer Ga. Such a claim is no better than "The relation $\Delta L = k \cdot L_0 \cdot \Delta T$ holds when it holds," and so is further from being a law-statement than is the familiar statement of the law of thermal expansion, which at least ascribes a particular (albeit false) relation to various quantities.

In short, Hempel sees the existence of provisos as posing a dilemma: For many a claim that we commonly accept as a law-statement, either that claim states a relation that does not obtain, and so is false, or is shorthand for some claim that states no relation at all, and so is empty. In either case, the regularity account must admit that many claims commonly accepted as law-statements are neither complete law-statements themselves nor even colloquial stand-ins for complete law-statements. If we continue to regard those familiar claims as law-statements, then we violate the regularity account. This is the problem of provisos. (Prior to Hempel (1988), versions of this problem were discussed by Canfield and Lehrer (1961) and Coffa (1968), as well as by Hempel himself (1965b, p. 167), and the issue, in general terms, was anticipated by Scriven (1961). Difficulties similar to the problem of provisos have also been noted in ethics, e.g., with regard to Ross's (1930) definition of a "prima facie duty".)

One may be tempted to reject this problem by insisting that genuine law-statements lack provisos; on this view, that many familiar "law-statements" are actually neither law-statements nor abbreviations for law-statements only goes to show that we have discovered very few genuine laws (or nomological explanations). To yield to this temptation would, I think, be unjustified. An account of laws must accommodate the fact that scientists show no reluctance to use these familiar claims in the manner distinctive of law-statements, e.g., in explanations and in support of counterfactuals. This fact would be difficult to explain if we held that scientists do not consider them to express laws. To insist nevertheless that only claims without provisos are genuine law-statements is to hold that whether a claim must or must not be captured by an analysis of what it is to believe a claim to state a physical law is not determined by whether scientists treat that claim as able or unable to perform those functions that distinguish law-statements from accidental generalizations. But, then, on what basis is the adequacy of an account of law to be evaluated? Scientific practice is the only phenomenon that exists for an account of law to save; if an account is tested against not actual science but science as idealized to conform to that account, the test is circular. If accounts of law are not free to disregard the fact that scientists treat "All gold cubes are smaller than one cubic mile" as an accidental generalization and "All cubes of Uranium-235 are smaller than one cubic mile" as expressing a law, even though this fact is troublesome for many proposed accounts, I see no reason why an account of law should be permitted to ignore the fact that scientists treat as law-statements many claims involving provisos.

Another temptation is to argue that although some law-statements involve provisos, these derive from other, more fundamental law-statements, which themselves need no provisos to describe regularities. For instance, the law of falling bodies follows in classical physics from the fundamental laws of motion and gravitation along with information about the earth; the proviso ". . . so long as the body is falling freely" restricts the law to those cases in which the gravitational-force law applied to this information about the earth accounts for all of the forces acting on the body. However, this temptation should also be resisted. It merely pushes the burden onto the fundamental laws: What regularity is described by the gravitational-force law? If it described a regularity between the masses and separation of two bodies and the *total* force

that each exerts on the other, it would be false unless it included the proviso ". . . so long as the bodies exert no other forces upon each other." But to regard this proviso as part of this "fundamental" law not only conflicts with the law's applicability to a case in which two charged and massive bodies interact, but also raises the familiar problem: The proviso fails to specify the circumstances in which other forces are present, and so prevents the gravitational-force law from setting forth a particular relation. If each of the other force-laws is supposed to specify when a given non-gravitational force is present and thereby help to determine the relation asserted by the gravitational-force law, then each of these other force-laws would have to apply to a body affected by many types of forces and so could not describe a regularity involving the *total* force exerted on the body. Alternatively, for the gravitational-force law to describe a regularity between the masses and separation of two bodies and the *gravitational* (rather than total) force that each exerts on the other, a component gravitational force would have to be a real entity that conforms to certain regularities. That component forces are real is a controversial contention (see, e.g., Cartwright 1983), and in any event, it seems to me that the nomic status of the gravitational-force law does not depend on it; after all, we use the Coriolis-force law in the manner distinctive of law-statements even though we believe there to be no Coriolis force to figure in a regularity that the law-statement might describe.

Perhaps one would be justified in setting the problem of provisos aside if one had some account of physical law that, except for this problem, were entirely successful. But since I know of no such account, I think it worth investigating whether greater progress toward one can be made by reflecting on provisos than by disregarding them. In this paper, I'll offer a response to the problem of provisos that ultimately undermines the regularity account of physical law and suggests an alternative, normative conception of law-statements according to which they specify the claims we ought to use, in various contexts, to justify certain other claims. I'll argue that some claims are properly adopted as law-statements although they are believed not to describe regularities, because one who believes that "It is F" ought to be used to justify "It is G" need not believe that some regularity, such as that all Fs are G, obtains.

In Section II, I'll explain the problem of provisos more fully, and in Section III, I'll attack Hempel's view of what a "law-statement" must be to qualify as complete, on which the problem depends. The more liberal criterion of completeness that I'll defend permits us to avoid Hempel's dilemma by enabling us to regard familiar "law-statements" as law-statements. I'll argue in section IV, however, that this response to the problem of provisos requires the rejection of any regularity account of physical law because many a claim commonly accepted as a law-statement describes no regularity; a normative conception of law-statements then suggests itself. In Section V, I'll maintain that this strategy can be used to argue against many other conceptions of physical law besides the regularity account, such as those of Armstrong (1983) and Kneale (1949). Finally, I'll argue that my response to the problem of provisos is superior to that offered by Giere (1988).

II

Hempel does not explicitly present the problem of provisos as posing the dilemma that "law-statements" are either false or empty. Yet this dilemma certainly stands behind his discussion. It must be because he believes a "law-statement" without provisos would be false, and so would not be a law-statement, that Hempel defines provisos as "essential." He goes on to point out

that if the complete law-statement includes the proviso $(x) \ldots$ if Px, then among the premises of an inference from the law-statement to testable predictions must be Pa, as well as other auxiliary hypotheses. This might at first appear to reduce at least part of the problem of provisos to a special case of the Duhem–Quine problem: a law-statement is not falsifiable (at least, not in a straightforward sense) because to make a prediction from it that can be tested, one must use auxiliary hypotheses, so one can preserve the law-statement, if the prediction fails, by rejecting an auxiliary hypothesis. If provisos are merely additional auxiliary hypotheses, distinguished from others only by the fact that they generally go unstated in scientific practice, then (it might appear) they do not represent a novel kind of threat to the falsifiability of individual hypotheses.

Hempel (1988, pp. 25f.) emphasizes, however, that the existence of provisos presents some obstacle to falsifiability beyond that posed by the Duhem–Quine problem. The Duhem–Quine problem assumes that the law-statements and auxiliary hypotheses are jointly sufficient to entail the testable prediction. But, Hempel maintains, if there are provisos among the auxiliary hypotheses, then this assumption often fails because, for many a familiar "law-statement," one can state neither the complete set of proviso conditions $(x) \ldots$ if Px needed to make that "law-statement" true nor the complete set of auxiliary hypotheses Pa needed to infer the testable prediction from the law-statement. Therefore, Hempel says that in comparison to the Duhem–Quine problem, "[t]he argument from provisos leads rather to the stronger conclusion that even a comprehensive system of hypotheses or theoretical principles will not entail any [testable predictions] because the requisite deduction is subject to provisos" – that is to say, always remains subject to provisos, no matter how many auxiliary premises one adds to try to exclude all factors disturbing to the law.

Hence, it is only because he considers complete law-statements impossible, since "the number of provisos implicit in any law is indefinitely large" (Giere 1988, p. 40), that Hempel sees provisos as presenting a difficulty that is distinct from the Duhem–Quine problem. But suppose Hempel believed that a condition such as ". . . in the absence of other relevant factors" could appear in a law-statement, and likewise that "There are no other relevant factors" could function as an auxiliary hypothesis. Then he would have to admit that any law *can* be completely stated by a claim that includes only a finite number of conditions $(x) \ldots$ if Px, one of which might be ". . . if there are no disturbing factors," and that any inference to a testable prediction includes only a finite number of auxiliary hypotheses Pa, one of which might be "There are no disturbing factors." The existence of provisos would then add nothing new to the Duhem–Quine problem. Hempel therefore must regard the expression ". . . in the absence of other relevant factors" as inappropriate for a law-statement. Though he does not explain why this is, the regularity account of laws, implicit throughout his discussion, suggests an answer: The sentence "$\Delta L = k \cdot L_0 \cdot \Delta T$ obtains in the absence of factors that disturb it" states no definite relation and so cannot be a law-statement. This worry is evident in Giere's remark (1988, p. 40): "The problem is to formulate the needed restrictions without rendering the law completely trivial."

III

By Hempel's definition, a proviso $(x) \ldots$ if Px usually is omitted from a statement of the law, and the corresponding premise Pa usually is not mentioned in inferences involving that law. Hempel regards these inferences as enthymemes and these familiar "law-statements" as incomplete. But why are we able to make do with incomplete law-statements? And is Hempel correct in considering them incomplete? I'll now argue that Hempel's criterion of completeness is motivated by an incorrect view of what is necessary in order for a sentence to state a determinate relation. I'll argue that familiar law-statements, which include clauses such as "in the absence of other relevant factors," are complete as they stand.

That proviso premises are distinguished by their absence from ordinary conversation is not an incidental feature of Hempel's definition. As I've explained, it is bound up with the fact that the number of provisos is "indefinitely large," which makes it impossible to offer them *all* as premises. But why do scientists find it unnecessary to mention *any* of the proviso premises in order to put an end to demands for the justification of their conclusions? Why is it that although it is known that when someone is hammering on the bar, $\Delta L = k \cdot L_0 \cdot \Delta T$ need not obtain, in actual practice a claim concerning k, L_0 and ΔT is recognized as sufficient, without "No one is hammering on the bar," to put an end to demands for the justification of a claim concerning ΔL?

The answer is that in practice, when no one is hammering on the bar, nearly all of those who demand justifications of claims concerning ΔL already believe that this is so. (Likewise, to consider a different proviso example, it is widely understood by workers in many fields of physics that no cases will involve velocities approaching that of light.) Of course, to someone who presents a claim concerning k, L_0 and ΔT to justify a claim concerning ΔL, one *could* object, "You have not told me that no one is hammering on the bar, and this you must do because if someone is, then (you will agree) your conclusion may well be false even though your premise is true." But apparently, that no one is hammering on the bar would, in nearly any actual case in which it is true, be believed in advance by those who might demand the justification of some claim concerning ΔL. In nearly all cases, then, someone who demands a justification for a claim concerning ΔL should regard a claim concerning k, L_0 and ΔT as a sufficient response.

Attention to this kind of shared background not only explains why scientists needn't in practice give any of the "indefinitely large" number of proviso premises in order to justify their conclusions, but also reveals why Hempel's standard of completeness is too high. Hempel apparently considers a "law-statement" complete only if it suffices, in the absence of any background understanding, to inform one of what it takes for nature to obey the corresponding law. This ideal of completeness requires that the complete law-statement include all of the proviso conditions (x) . . . if Px and, more importantly, that none of these conditions be "in the absence of disturbing factors," because this condition plainly appeals to background understanding. A generalization "All Fs are G, except when disturbing factors are present" does not indicate, in a manner intelligible to one who doesn't know already what constitutes a disturbing factor, some determinate regularity to which nature conforms. In exactly the same way, someone who is told to follow a rule "Conform to the regularity . . . when it is appropriate to apply this rule" can understand what it would take to follow this rule only if she already knows when this rule is appropriately applied.

But to require that a rule be intelligible in the absence of implicit background understanding of how to apply it is not a reasonable criterion of completeness because no rule can satisfy it. As Wittgenstein (1958) suggests, one can always conceive of alternative interpretations of a rule that recognize the same actual past actions as conforming to the rule but regard different hypothetical actions as what it would take to

follow the rule. It is futile to try to avoid this by including in the rule an expression that specifies explicitly how to apply the rest of the rule, for alternative interpretations of that expression are likewise conceivable. In the same way, a law-statement specifies a determinate relation only by exploiting implicit background understanding of what it would take for nature to obey this law. This point applies to *any* law-statement, whether or not it *blatantly* appeals to implicit background understanding by referring to "disturbing factors."

That the proper way to apply a rule is not itself specified by any rule, intelligible without implicit background understanding of how to follow that rule, does not imply, as Hempel seems to think, that nothing counts as a violation of the given rule. The background understanding, albeit implicit, enables the rule to impose determinate requirements. This implicit understanding must be capable of being taught and of being made the explicit subject of discussion if disputes over it ever arise. Some fortune tellers explain away your failure to make accurate predictions by using their rules as the result of your having misapplied the claims they believe to be law-statements, of your having ignored some clause they say they neglected to mention. It is doubtful that they have undertaken any determinate commitments at all by adopting those "law-statements." It is as if someone says, "I can run a four-minute mile," but with each failure reveals a proviso that she had not stated earlier: ". . . except on this track," ". . . except on sunny Tuesdays in March," and so on. It quickly becomes apparent that this person will not acknowledge having committed herself to any claim by asserting "I can run a four-minute mile." Science is distinguished from such bunk neither by the explicit inclusion in scientific law-statements of all conditions Hempel would deem "essential" nor by the absence of implicit background understanding of how to apply those law-statements. What is noteworthy about science is that this background understanding is genuine background *understanding*. In general, all researchers identify the same testable claims as those to which one would become committed by adding a given lawlike hypothesis to a certain store of background beliefs. Because they agree on how to apply the hypothesis, it is subject to honest test.

IV

I'll now argue that this attractive response to the problem of provisos is incompatible with the view that law-statements describe regularities of a certain kind. This account of provisos leads instead to a conception of law-statements as specifying the claims we ought to respect, in a certain context, as able to justify certain other claims.

What, according to the regularity account, is the regularity stated by the familiar expression of the law of thermal expansion? Presumably, it is that a bar's length changes by $k \cdot L_0 \cdot \Delta T$ whenever the bar has a certain composition, its initial length is L_0, its temperature changes by ΔT, and there are no disturbing factors. That nearly all of us agree on whether "There are no disturbing factors" is appropriately said of a given case saves this expression, and so the law-statement, from emptiness; we share an implicit understanding of which predictions the law-statement underwrites, of when it is properly applied. But whether certain scientists are correct in saying of a given case that there are no disturbing factors depends not just on the physical features of this case but also on their purposes, e.g., on the degree of approximation they can tolerate considering the use they intend to make of this prediction. Even if the regularity account can countenance as laws some uniformities involving the concerns of scientists, the law of thermal expansion was surely not supposed to be such a uniformity; somehow, the subject has changed

from a law of physics to a law of the science of scientific activity.

It gets worse for the regularity account. Suppose that according to the regularity account, the familiar expression of the law of thermal expansion states that a bar's length changes by $k \cdot L_0 \cdot \Delta T$ whenever the bar has a certain composition, its initial length is L_0, its temperature changes by ΔT, no one is hammering on the bar hard enough to cause deviations from $\Delta L = k \cdot L_0 \cdot \Delta T$ great enough (given our interests) to matter to us, and so on. Nevertheless, this claim cannot qualify as a law-statement according to the regularity account, for the relation it states does not obtain. It is violated, for example, when someone is hammering on the bar hard enough to cause the actual change in the bar's length to depart from $k \cdot L_0 \cdot \Delta T$ but lightly enough for this departure to be irrelevant to the investigator's concerns. While in this case it is not true that the actual length of the bar changes by $k \cdot L_0 \cdot \Delta T$, it is proper for such an investigator to predict that the length of the bar will change by $k \cdot L_0 \cdot \Delta T$. The relation stated by the law involves not the bar's actual change in length but rather the change in length one is justified in predicting.

This suggests that the law of thermal expansion doesn't describe a regularity among events or states of affairs but concerns the way a claim concerning ΔL ought to be justified. The response I've advocated to the problem of provisos leads to a conception of the law of thermal expansion as the objective fact that under certain (partly pragmatic) circumstances, a premise about the bar's initial length, its change in temperature, and its composition ought to be used to justify a certain claim about its change in length. On this view, a law-statement has a normative element because it says that under certain circumstances, certain claims ought to be used to justify certain other claims.

To succeed, this account would have to show that law-statements are able to explain their instances, to support counterfactuals, and to be confirmed inductively by their instances *because* they specify the roles that certain descriptions should play in justifications. This account would likewise have to show that *because* accidental generalizations lack this prescriptive import, they cannot be used in these ways. To show this would be to break the familiar unilluminating circle of analysis from a law's explanatory power, to its physical necessity, to its capacity for counterfactual support, to its lawlikeness, to its capacity to be inductively confirmed by its instances, to its explanatory power. This task is well beyond the scope of this paper; I begin it in my (1993). My concern here is to show how this normative account of law-statements arises from a plausible response to the problem of provisos.

Let me summarize the argument. Contrary to Hempel, a law-statement that includes "so long as there are no disturbing factors" is not thereby rendered trivial. Like any other expression, this condition derives its content from an implicit shared understanding of how one should use it. Hence, the relation that a "law-statement" that includes this condition claims to obtain, which is determined by the appropriate way to apply this statement, ultimately depends on proprieties not codified explicitly. All there is to give meaning to "so long as there are no disturbing factors," and thereby save the "law-statement" from triviality, is how scientists consider the "law-statement" properly applied; because there is near unanimity on this point, the "law-statement" is not empty. But since the statement is properly applied to a given physical circumstance only when investigators have certain interests, the regularity that supposedly constitutes the law described by this statement must involve not only physical events but also investigators' concerns. Even if the regularity account

regarded such uniformities as laws of physics, these uniformities do not obtain anyway and so, according to the regularity account, cannot be described by law-statements. For with regard to a given physical situation, it sometimes is and sometimes is not appropriate for us to say that ΔL will be $k \cdot L_0 \cdot \Delta T$, depending on our concerns. But surely in a given physical situation, there is only one real amount by which the bar expands; the actual behavior of the bar does not depend on our interests. Once the law-statement, in stating a relation involving "no one is hammering on the bar hard enough or otherwise disturbing it enough to matter to us," turns out to involve our interests, the other relatum, "ΔL," is found to be infected by our interests as well. It refers to the change in the bar's length according to the claim that ought to be considered justified by these premises, which depends on the purpose for which we intend to use this claim, rather than to the bar's actual change in length, which does not. So the law-statement expresses a norm rather than a regularity involving the bar's real length. It states a relation between the presence of certain conditions (some having to do with our interests) and the way one ought to predict the bar's change in length.

A law-statement, then, informs an audience already able to tell whether there are "disturbing factors" that if there are none, they ought to use a given claim to justify another. On this view, a proviso is not "essential" *tout court*, contrary to Hempel. There is no fact of the matter to whether a law-statement has or lacks provisos. There is only whether those who would typically discuss that law-statement can learn, merely from reading it, what it prescribes they do (as is the case for the familiar expression of Newton's second law), or whether making the proviso explicit is essential *for that audience*.

Moreover, a normative conception of law-statements does not deny that there are regularities in nature. It denies only that law-statements, when performing their distinctive functions, are describing some of them. Laws, in the sense that the regularity account envisions them, need not be invoked to understand what law-statements say.

V

The foregoing argument, if successful, can be used to undermine not only any regularity account of law but also any account that takes a law-statement to describe a state of affairs that necessarily presupposes such a regularity. For instance, it has been suggested (e.g., by Kneale 1949) that law-statements describe not regularities of a certain kind but non-Humean connections of physical necessity. Such an account implies that the non-Humean connection supplements a regularity among circumstances. But with this regularity, the problem of provisos takes hold. The same reasoning can be deployed against recent accounts (see Armstrong, 1983, and Dretske, 1977) according to which law-statements describe relations among universals, such as the property of lengthening by a given amount. On these accounts, the law-statement of thermal expansion entails that events conform to a certain relation "so long as there are no disturbing factors." Precisely what this comes to, i.e., whether some factor qualifies as "disturbing" or not, must be fixed by the law-statement. But, I have argued, the law-statement appeals to a determinate set of disturbing conditions only because it states a relation involving not the property of expanding in length by a given amount but the property of being able to serve as the subject of a justified claim attributing expansion by a given amount. In short, if the above reasoning goes through, it constitutes a recipe for an argument against any account according to which a law is or requires a regularity among events or states of affairs.

A law-statement concerning a particular influence, such as the Coriolis-force law or Newton's two-body gravitational-force law, suffices to tell persons *having comparatively little background understanding* how they ought to justify a certain claim. This is because the magnitude of the influence covered by the law-statement does not depend on which other influences are at work; hence, while there are provisos, there is none that demands significant background understanding, such as "in the absence of disturbing factors" would. The gravitational-force law, for example, specifies how one should justify a claim concerning the gravitational force between two bodies, whatever the other influences with which this subtotal should be combined to reach, say, the total force on a given body.

However, Giere's (1988) response to the problem of provisos does not capture the fact that to be committed to the gravitational-force law is to be committed to treating a certain inference to the component gravitational force as correct, no matter what the other relevant influences. In light of the problem of provisos, Giere holds certain familiar law-statements to be false as claims about the physical world, but he contends that, as law-statements, they must nevertheless function as descriptions of something. He therefore tries to find something that they describe. While this search has led Armstrong and Dretske to the exotic realm of universals, Giere (along with Cartwright, 1983, whose account is similar in all relevant respects) maintains that a law-statement describes a scientific model. One may hold that the relevant behavior of a given real system can be predicted by using some model; Giere terms such a claim a "theoretical hypothesis." Since the law-statement describes the model, not reality, it needs no qualification by provisos to be accurate.

Consider, then, what Giere says (1988, p. 44) about two laws, each concerning a single influence, that are combined to account for a magnetically influenced pendulum:

> We have discovered a new kind of pendulum . . . in which the force of gravity is supplemented by a magnetic force directed toward a point below the point of rest. Constructing a theoretical model that does apply to such systems is a fairly easy problem in physics.

If the gravitational-force law specifies how one ought to calculate a subtotal (the gravitational influence of one body on another) no matter what the other influences present, and the magnetic-force law does likewise, then this is indeed an easy problem. Having already accepted these law-statements, and having recognized the proper way to add forces and to use the net force on a body to infer its acceleration, we are committed to a particular procedure for predicting the bob's motion.

But Giere takes these law-statements not to prescribe which models to use (which is the job of theoretical hypotheses) but merely to describe certain models. Thus, Giere must admit that by accepting that the gravitational-force law describes certain models, we are not committed to saying that one ought to use a model it describes to predict the bob's motion. Moreover, one who accepts the theoretical hypothesis that this law should be used to calculate the gravitational force exerted by the earth on a non-magnetic bob is not thereby committed to the theoretical hypothesis that this law should be used to calculate the gravitational force exerted by the earth on a magnetically influenced bob.

On my view, in contrast, to accept this law-statement is to recognize the way one should justify claims concerning the component gravitational force, whatever the other relevant influences. This view accounts for what Giere's view obscures: That we are committed to some common element in our treatments of ordinary and magnetically augmented pendula (namely, an identical way of justifying a certain subtotal) in virtue of which the magnetically augmented pendulum is an easy problem. We became committed to elements of its solution when we adopted solutions to other problems.

I have argued against Hempel's contention that a complete law-statement must specify its own range without depending on implicit background understanding in order to do so. I have also argued against Giere's (and Cartwright's) alternative claim that a law-statement says nothing about its range, leaving it for theoretical hypotheses to specify. I have defended the view that a law-statement specifies its range in a fashion that may involve blatant appeal to implicit proprieties of use. I have thereby tried to offer a way around the problem of provisos, at the price of abandoning the regularity account of law in favor of a normative analysis. Further discussion of that proposal must await another occasion.

References

Armstrong, D.: 1983, *What Is a Law of Nature?*, Cambridge University Press, Cambridge.

Ayer, A.J.: 1963, "What is a Law of Nature?", in *The Concept of a Person and Other Essays*, Macmillan, London, pp. 209–234.

Braithwaite, R.B.: 1953, *Scientific Explanation*, Cambridge University Press, Cambridge.

Canfield, J. and Lehrer, K.: 1961, 'A Note on Prediction and Deduction', *Philosophy of Science* **28**, 204–208.

Cartwright, N.: 1983, *How the Laws of Physics Lie*, Clarendon Press, Oxford.

Coffa, J.: 1968, 'Discussion: Deductive Predictions', *Philosophy of Science* **35**, 279–283.

Dretske, F.: 1977, 'Laws of Nature', *Philosophy of Science* **44**, 248–268.

Giere, R.: 1988, 'Laws, Theories, and Generalizations', in A. Grünbaum and W. Salmon (eds), *The Limits of Deductivism*, University of California Press, Berkeley, pp. 37–46.

Goodman, N.: 1983, *Fact, Fiction, and Forecast*, 4th ed., Harvard University Press, Cambridge.

Hempel, C.G.: 1965a, 'Studies in the Logic of Explanation', in *Aspects of Scientific Explanation*, The Free Press, New York, pp. 245–295.

Hempel, C.G.: 1965b, 'Typological Methods in the Natural and the Social Sciences', ibid., pp. 155–172.

Hempel, C.G.: 1988, 'Provisos', in A. Grünbaum and W. Salmon (eds). *The Limits of Deductivism*, University of California Press, Berkeley, pp. 19–36.

Kneale, W.: 1949, *Probability and Induction*, Oxford University Press. Oxford.

Lange, M.B.: 1993, 'Lawlikeness', *Noûs*, forthcoming.

Lewis, D.K.: 1973, *Counterfactuals*, Harvard University Press. Cambridge.

Lewis, D.K.: 1986, 'Postscript to "A Subjectivist's Guide to Objective Chance", in *Philosophical Papers*: Volume 2, Oxford University Press. Oxford, pp. 121–126.

Mackie, J.L.: 1962, 'Counterfactuals and Causal Laws', in R.S. Butler (ed.). *Analytic Philosophy*, Barnes and Noble, New York, pp. 66–80.

Nagel, E.: 1961, *The Structure of Science*, Harcourt, Brace, and World, New York.

Reichenbach, H.: 1947, *Elements of Symbolic Logic*, Macmillan, New York.

Ross, W.D.: 1930, *The Right and The Good*, Oxford University Press, Oxford.

Scriven, M.: 1961, 'The Key Property of Physical Laws – Inaccuracy', in H. Feigl and G. Maxwell (eds), *Current Issues in the Philosophy of Science*, Holt, Rinehart, and Winston, New York, pp. 91–104.

Wittgenstein, L.: 1958, *Philosophical Investigations*, 3rd ed., trans. G.E.M. Anscombe, Macmillan, New York.

Chapter 24

Nancy Cartwright

CAUSAL LAWS AND EFFECTIVE STRATEGIES

0 Introduction

There are at least two kinds of laws of nature: laws of association and causal laws. Laws of association are the familiar laws with which philosophers usually deal. These laws tell how often two qualities or quantities are co-associated. They may be either deterministic – the association is universal – or probabilistic. The equations of physics are a good example: whenever the force on a classical particle of mass m is f the acceleration is f/m. Laws of association may be time indexed, as in the probabilistic laws of Mendelian genetics, but, apart from the asymmetries imposed by time indexing, these laws are causally neutral. They tell how often two qualities co-occur; but they provide no account of what makes things happen.

Causal laws, by contrast, have the word 'cause' – or some causal surrogate – right in them. Smoking causes lung cancer; perspiration attracts wood ticks; or, for an example from physics, force causes change in motion: to quote Einstein and Infeld, 'The action of an external force changes the velocity ... such a force either increases or decreases the velocity according to whether it acts in the direction of motion or in the opposite direction.'[1]

Bertrand Russell argued that laws of association are all the laws there are, and that causal principles cannot be derived from the causally symmetric laws of association.[2] I shall here argue in support of Russell's second claim, but against the first. Causal principles cannot be reduced to laws of association; but they cannot be done away with.

The argument in support of causal laws relies on some facts about strategies. They are illustrated in a letter which I recently received from TIAA-CREF, a company that provides insurance for college teachers. The letter begins:

> It simply wouldn't be true to say,
> 'Nancy L. D. Cartwright ... if you own a TIAA life insurance policy you'll live longer.'
> But it is a fact, nonetheless, that persons insured by TIAA do enjoy longer lifetimes, on the average, than persons insured by commercial insurance companies that serve the general public.

I will take as a starting point for my argument facts like those reported by the TIAA letter: it wouldn't be true that buying a TIAA policy would be an effective strategy for lengthening one's life. TIAA may, of course, be mistaken; it could after all be true. What is important is that their claim is, as they suppose, the kind of claim which is either true or false. There is a pre-utility sense of goodness of strategy; and what is and what is not a good strategy in this pre-utility sense is an objective fact. Consider a second example. Building the canal in Nicaragua, the French discovered that spraying oil on the swamps is a good strategy for stopping the spread of malaria, whereas burying contaminated blankets is useless. What they discovered was true, independent of their theories, of their desire to control malaria, or of the cost of doing so.

The reason for beginning with some uncontroversial examples of effective and ineffective strategies is this: I claim causal laws cannot be done away with, for they are needed to ground the distinction between effective strategies and ineffective ones. If indeed, it isn't true that buying a TIAA policy is an effective way to lengthen one's life, but stopping smoking is, the difference between the two depends on the causal laws of our universe, and on nothing weaker. This will be argued in Part 2. Part 1 endorses the first of Russell's claims, that causal laws cannot be reduced to laws of association.

1 Statistical analyses of causation

I will abbreviate the causal law, 'C causes E' by C ↳ E. Notice that C and E are to be filled in by general terms, and not names of particulars; for example, 'Force causes motion' or 'Aspirin relieves headache'. The generic law 'C causes E' is not to be understood as a universally quantified law about particulars, even about particular causal facts. It is generically true that aspirin relieves headache even though some particular aspirins fail to do so. I will try to explain what causal laws assert by giving an account of how causal laws relate on the one hand to statistical laws, and on the other to generic truths about strategies. The first task is not straightforward; although causal laws are intimately connected with statistical laws, they cannot be reduced to them.

A primary reason for believing that causal laws cannot be reduced to probabilistic laws is

broadly inductive: no attempts so far have been successful. The most notable attempts recently are by the philosophers Patrick Suppes[3] and Wesley Salmon[4] and, in the social sciences, by a group of sociologists and econometricians working on causal models, of whom Herbert Simon and Hubert Blalock[5] are good examples.

It is not just that these attempts fail, but rather why they fail that is significant. The reason is this. As Suppes urges, a cause ought to increase the frequency of its effect. But this fact may not show up in the probabilities if other causes are at work. Background correlations between the purported cause and other causal factors may conceal the increase in probability which would otherwise appear. A simple example will illustrate.

It is generally supposed that smoking causes heart disease (S ↳ H). Thus we may expect that the probability of heart disease on smoking is greater than otherwise. (We can write this as either $Prob(H/S) > Prob(H)$, or $Prob(H/S) > Prob(H/\neg S)$, for the two are equivalent.) This expectation is mistaken. Even if it is true that smoking causes heart disease, the expected increase in probability will not appear if smoking is correlated with a sufficiently strong preventative, say exercising. (Leaving aside some niceties, we can render 'Exercise prevents heart disease' as X ↳ ¬H.) To see why this is so, imagine that exercising is more effective at preventing heart disease than smoking at causing it. Then in any population where smoking and exercising are highly enough correlated,[6] it can be true that $Prob(H/S) = Prob(H)$, or even $Prob(H/S) < Prob(H)$. For the population of smokers also contains a good many exercisers, and when the two are in combination, the exercising tends to dominate.

It is possible to get the increase in conditional probability to reappear. The decrease arises from looking at probabilities that average over both exercisers and non-exercisers. Even though in the general population it seems better to smoke than not, in the population consisting entirely of exercisers, it is worse to smoke. This is also true in the population of non-exercisers. The expected increase in probability occurs not in the general population but in both sub-populations.

This example depends on a fact about probabilities known as Simpson's paradox,[7] or sometimes as the Cohen–Nagel–Simpson paradox, because it is presented as an exercise in Morris Cohen's and Ernest Nagel's text, *An Introduction to Logic and Scientific Method*.[8] Nagel suspects that he learned about it from G. Yule's *An Introduction to the Theory of Statistics* (1904), which is one of the earliest textbooks written on statistics; and indeed it is discussed at length there. The fact is this: any association – $\mathrm{Prob}(A/B) = \mathrm{Prob}(A)$; $\mathrm{Prob}(A/B) > \mathrm{Prob}(A)$; $\mathrm{Prob}(A/B) < \mathrm{Prob}(A)$ – between two variables which holds in a given population can be reversed in the sub-populations by finding a third variable which is correlated with both.

In the smoking–heart disease example, the third factor is a preventative factor for the effect in question. This is just one possibility. Wesley Salmon[9] has proposed different examples to show that a cause need not increase the probability of its effect. His examples also turn on Simpson's paradox, except that in his cases the cause is correlated, not with the presence of a negative factor, but with the absence of an even more positive one.

Salmon considers two pieces of radioactive material, uranium 238 and polonium 214. We are to draw at random one material or the other, and place it in front of a Geiger counter for some time. The polonium has a short half-life, so that the probability for some designated large number of clicks is .9; for the long-lived uranium, the probability is .1. In the situation described, where one of the two pieces is drawn at random,

the total probability for a large number of clicks is ½(.9)+ ½(.1)=.5. So the conditional probability for the Geiger counter to click when the uranium is present is less than the unconditional probability. But when the uranium has been drawn and the Geiger counter does register a large number of clicks, it is the uranium that causes them. The uranium decreases the probability of its effect in this case. But this is only because the even more effective polonium is absent whenever the uranium is present.

All the counter examples I know to the claim that causes increase the probability of their effects work in this same way. In all cases the cause fails to increase the probability of its effects for the same reason: in the situation described the cause is correlated with some other causal factor which dominates in its effects. This suggests that the condition as stated is too simple. A cause must increase the probability of its effects; but only in situations where such correlations are absent.

The most general situations in which a particular factor is not correlated with any other causal factors are situations in which all other causal factors are held fixed, that is situations that are homogeneous with respect to all other causal factors. In the population where everyone exercises, smoking cannot be correlated with exercising. So, too, in populations where no-one is an exerciser. I hypothesize then that the correct connection between causal laws and laws of association is this:

'C causes E' if and only if C increases the probability of E in every situation which is otherwise causally homogeneous with respect to E.

Carnap's notion of a state description[10] can be used to pick out the causally homogeneous situations. A complete set of causal factors for E is the set of all C_i such that either $C_i \hookrightarrow + E$ or $C_i \hookrightarrow \neg E$. (For short $C_i \hookrightarrow \pm E$). Every possible arrangement of the factors from a set which is complete except for C picks out a population homogeneous in all causal factors but C. Each such arrangement is given by one of the 2^n state descriptions $K_j = \wedge \pm C_i$ over the set $\{C_i\}$ (i ranging from 1 to n) consisting of all alternative causal factors. These are the only situations in which probabilities tell anything about causal laws. I will refer to them as *test* situations for the law $C \hookrightarrow E$.

Using this notation the connection between laws of association and causal laws is this:

$CC: C \hookrightarrow E$ iff $\text{Prob}(E/C.K_j) > \text{Prob}(E/K_j)$ for all state descriptions K_j over the set $\{C_i\}$, where $\{C_i\}$ satisfies

  (i) $C_i \in \{C_i\} \Rightarrow C_i \hookrightarrow \pm E$
  (ii) $C \notin \{C_i\}$
  (iii) $\forall D\, (D \hookrightarrow \pm E \Rightarrow D = C \text{ or } D \in \{C_i\})$
  (iv) $C_i \in \{C_i\} \Rightarrow \neg(C \hookrightarrow C_i)$.

Condition (iv) is added to ensure that the state descriptions do not hold fixed any factors in the causal chain from C to E. It will be discussed further in the section after next.

Obviously CC does not provide an analysis of the schema $C \hookrightarrow E$, because exactly the same schema appears on both sides of the equivalence. But it does impose mutual constraints, so that given sets of causal and associational laws cannot be arbitrarily conjoined. CC is, I believe, the strongest connection that can be drawn between causal laws and laws of association.

1.1 Two advantages for scientific explanation

C. G. Hempel's original account of inductive-statistical explanation[11] had two crucial features which have been given up in later accounts, particularly in Salmon's: (1) an explanatory factor must increase the probability of the fact to

be explained; (2) what counts as a good explanation is an objective, person-independent matter. Both of these features seem to me to be right. If we use causal laws in explanations, we can keep both these requirements and still admit as good explanations just those cases that are supposed to argue against them.

(i) Hempel insisted that an explanatory factor increase the probability of the phenomenon it explains. This is an entirely plausible requirement, although there is a kind of explanation for which it is not appropriate. In one sense, to explain a phenomenon is to locate it in a nomic pattern. The aim is to lay out all the laws relevant to the phenomenon; and it is irrelevant to this aim whether the phenomenon has high or low probability under these laws. Although this seems to be the kind of explanation that Richard Jeffrey describes in 'Statistical Explanation vs. Statistical Inference',[12] it is not the kind of explanation that other of Hempel's critics have in mind. Salmon, for instance, is clearly concerned with causal explanation.[13] Even for causal explanation Salmon thinks the explanatory factor may decrease the probability of the factor to be explained. He supports this with the uranium–polonium example described above.

What makes the uranium count as a good explanation for the clicks in the Geiger counter, however, is not the probabilistic law Salmon cites (Prob(clicks/uranium) < Prob(clicks)), but rather the causal law – 'Uranium causes radioactivity'. As required, the probability for radioactive decay increases when the cause is present, for *every test situation*. There is a higher level of radioactivity when uranium is added both for situations in which polonium is present, and for situations in which polonium is absent. Salmon sees the probability decreasing because he attends to a population which is not causally homogeneous.

Insisting on increase in probability across all test situations not only lets in the good cases of explanation which Salmon cites; it also rules out some bad explanations that must be admitted by Salmon. For example, consider a case which, so far as the law of association is concerned, is structurally similar to Salmon's uranium example. I consider eradicating the poison oak at the bottom of my garden by spraying it with defoliant. The can of defoliant claims that the spray is 90 per cent effective; that is, the probability of a plant's dying given that it is sprayed is .9, and the probability of its surviving is .1. Here in contrast to the uranium case only the probable outcome, and not the improbable, is explained by the spraying. One can explain why some plants died by remarking that they were sprayed with a powerful defoliant; but this will not explain why some survive.[14]

The difference is in the causal laws. In the favourable example, it is true both that uranium causes high levels of radioactivity and that uranium causes low levels of radioactivity. This is borne out in the laws of association. Holding fixed other causal factors for a given level of decay, either high or low, it is more probable that that level will be reached if uranium is added than not. This is not so in the unfavourable case. It is true that spraying with defoliant causes death in plants, but it is not true that spraying also causes survival. Holding fixed other causes of death, spraying with my defoliant will increase the probability of a plant's dying; but holding fixed other causes of survival, spraying with that defoliant will decrease, not increase, the chances of a plant's surviving.

(ii) All these explanations are explanations by appeal to causal laws. Accounts, like Hempel's or Salmon's or Suppes's, which instead explain by appeal to laws of association, are plagued by the reference class problem. All these accounts allow that one factor explains another just in case some privileged statistical relation obtains between them. (For Hempel the probability of the first factor on the second must be high; for Suppes it must be higher than when the second factor is

absent; Salmon merely requires that the probabilities be different.) But whether the designated statistical relation obtains or not depends on what reference class one chooses to look in, or on what description one gives to the background situation. Relative to the description that either the uranium or the polonium is drawn at random, the probability of a large number of clicks is lower when the uranium is present than it is otherwise. Relative to the description that polonium and all other radioactive substances are absent, the probability is higher.

Salmon solves this problem by choosing as the privileged description the description assumed in the request for explanation. This makes explanation a subjective matter. Whether the uranium explains the clicks depends on what information the questioner has to hand, or on what descriptions are of interest to him. But the explanation that Hempel aimed to characterize was in no way subjective. What explains what depends on the laws and facts true in our world, and cannot be adjusted by shifting our interest or our focus.

Explanation by causal law satisfies this requirement. Which causal laws are true and which are not is an objective matter. Admittedly certain statistical relations must obtain; the cause must increase the probability of its effect. But no reference class problem arises. In how much detail should we describe the situations in which this relation must obtain? We must include all and only the other causally relevant features. What interests we have, or what information we focus on, is irrelevant.

I will not here offer a model of causal explanation, but certain negative theses follow from my theory. Note particularly that falling under a causal law (plus the existence of suitable initial conditions) is neither necessary nor sufficient for explaining a phenomenon.

It is not sufficient because a single phenomenon may be in the domain of various causal laws, and in many cases it will be a legitimate question to ask, 'Which of these causal factors actually brought about the effect on this occasion?' This problem is not peculiar to explanation by causal law, however. Both Hempel in his inductive-statistical model and Salmon in the statistical relevance account sidestep the issue by requiring that a 'full' explanation cite all the possibly relevant factors, and not select among them.

Conversely, under the plausible assumption that singular causal statements are transitive, falling under a causal law is not necessary for explanation either. This results from the fact that (as CC makes plain) causal laws are not transitive. Hence a phenomenon may be explained by a factor to which it is linked by a sequence of intervening steps, each step falling under a causal law, without there being any causal law that links the explanans itself with the phenomenon to be explained.

1.2 Some details and some difficulties

Before carrying on to Part 2, some details should be noted and some defects admitted.

(a) *Condition (iv)*. Condition (iv) is added to the above characterization to avoid referring to singular causal facts. A test situation for $C \leftrightarrow E$ is meant to pick out a (hypothetical, infinite) population of individuals which are alike in all causal factors for E, except those which on that occasion are caused by C itself. The test situations should not hold fixed factors in the causal chain from C to E. If it did so, the probabilities in the populations where all the necessary intermediate steps occur would be misleadingly high; and where they do not occur, misleadingly low. Condition (iv) is added to except factors caused by C itself from the description of the test situation. Unfortunately it is too strong. For condition (iv) excepts any factor which *may* be caused by C even on those particular occasions

when the factor occurs for other reasons. Still, (iv) is the best method I can think of for dealing with this problem, short of introducing singular causal facts, and I let it stand for the nonce.

(b) *Interactions.* One may ask, 'But might it not happen that $\text{Prob}(E/C) > \text{Prob}(E)$ in *all* causally fixed circumstances, and still C not be a cause of E?' I do not know. I am unable to imagine convincing examples in which it occurs; but that is hardly an answer. But one kind of example is clearly taken account of. That is the problem of spurious correlation (sometimes called 'the problem of joint effects'). If two factors E_1 and E_2 are both effects of a third factor C, then it will frequently happen that the probability of the first factor is greater when the second is present than otherwise, over a wide variety of circumstances. Yet we do not want to assert $E_1 \leftrightarrow E_2$. According to principle CC, however, $E_1 \leftrightarrow E_2$ only if $\text{Prob}(E_1/E_2) > \text{Prob}(E_1)$ both when C obtains, and also when C does not obtain. But the story that E_1 and E_2 are joint effects of C provides no warrant for expecting either of these increases.

One may have a worry in the other direction as well. Must a cause increase the probability of its effect in *every* causally fixed situation? Might it not do so in some, but not in all? I think not. Whenever a cause fails to increase the probability of its effect, there must be a reason. Two kinds of reasons seem possible. The first is that the cause may be correlated with other causal factors. This kind of reason is taken account of. The second is that interaction may occur. Two causal factors are interactive if in combination they act like a single causal factor whose effects are different from at least one of the two acting separately. For example, ingesting an acid poison may cause death; so too may the ingestion of an alkali poison. But ingesting both may have no effect at all on survival.

In this case, it seems, there are three causal truths: (1) ingesting acid without ingesting alkali causes death; (2) ingesting alkali without ingesting acid causes death; and (3) ingesting both alkali and acid does not cause death. All three of these general truths should accord with CC.

Treating interactions in this way may seem to trivialize the analysis; anything may count as a cause. Take any factor that behaves sporadically across variation of causal circumstances. May we not count it as a cause by looking at it separately in those situations where the probability increases, and claim it to be in interaction in any case where the probability does not increase? No. There is no guarantee that this can always be done. For interaction is always interaction with some other *causal factor*; and it is not always possible to find some other factor, or conjunction of factors, which obtain just when the probability of E on the factor at issue decreases, and which itself satisfies principle CC relative to all other causal factors.[15] Obviously, considerably more has to be said about interactions; but this fact at least makes it reasonable to hope they can be dealt with adequately, and that the requirement of increase in probability across all causal situations is not too strong.

(c) *0, 1 probabilities and threshold effects.* Principle CC as it stands does not allow $C \hookrightarrow E$ if there is even a single arrangement of other factors for which the probability of E is one, independent of whether C occurs or not. So CC should be amended to read:

$$C \hookrightarrow E \text{ iff } (\forall j) \{ \text{Prob}(E/C.K_j) > \text{Prob}(E/K_j) \text{ or } \text{Prob}(E/K_j) = 1 = \text{Prob}(E/C.K_j) \} \text{ and } (\exists j) \{ \text{Prob}(E/K_j) \neq 1 \}.$$

It is a consequence of the second conjunct that something that occurs universally can be the consequent of no causal laws. The alternative is to let anything count as the cause of a universal fact.

There is also no natural way to deal with threshold effects, if there are any. If the probability of some phenomenon can be raised just so high, and no higher, the treatment as it stands allows no genuine causes for it.

(d) *Time and causation.* CC makes no mention of time. The properties may be time indexed; taking aspirins at t causes relief at $t + \Delta t$, but the ordering of the indices plays no part in the condition. Time ordering is often introduced in statistical analyses of causation to guarantee the requisite asymmetries. Some, for example, take increase in conditional probability as their basis. But the causal arrow is asymmetric, whereas increase in conditional probability is symmetric: $\text{Prob}(E/C) > \text{Prob}(E)$ iff $\text{Prob}(C/E) > \text{Prob}(C)$. This problem does not arise for CC, because the set of alternative causal factors for E will be different from the set of alternative causal factors for C. I take it to be an advantage that my account leaves open the question of backwards causation. I doubt that we shall ever find compelling examples of it; but if there were a case in which a later factor increased the probability of an earlier one in all test situations, it might well be best to count it a cause.

2 Probabilities in decision theory

Standard versions of decision theory require two kinds of information. (1) How desirable are various combinations of goals and strategies and (2) how effective are various strategies for obtaining particular goals. The first is a question of utilities, which I will not discuss. The second is a matter of effectiveness; it is generally rendered as a question about probabilities. We need to know what may roughly be characterized as 'the probability that the goal will result if the strategy is followed.' It is customary to measure effectiveness by the conditional probability. Following this custom, we could define

!S is an *effective strategy* for G iff $\text{Prob}(G/S) > \text{Prob}(G)$.

I have here used the volative mood marker ! introduced by H. P. Grice,[16] to be read 'let it be the case that'. I shall refer to S as *the strategy state*. For example, if we want to know whether the defoliant is effective for killing poison oak, the relevant strategy state is 'a poison oak plant is sprayed with defoliant'. On the above character-ization, the defoliant is effective just in case the probability of a plant's dying, given that it has been sprayed, is greater than the probability of its dying given that it has not been sprayed. Under this characterization the distinction between effective and ineffective strategies depends entirely on what laws of association obtain.

But the conditional probability will not serve in this way, a fact that has been urged by Allan Gibbard and William Harper.[17] Harper and Gibbard point out that the increase in conditional probability may be spurious, and that spurious correlations are no grounds for action. Their own examples are somewhat complex because they specifically address a doctrine of Richard Jeffrey's not immediately to the point here. We can illus-trate with the TIAA case already introduced. The probability of long life given that one has a TIAA policy is higher than otherwise. But, as the letter says, it would be a poor strategy to buy TIAA in order to increase one's life expectancy.

The problem of spurious correlation in deci-sion theory leads naturally to the introduction of counterfactuals. We are not, the argument goes, interested in how many people have long lives among people insured by TIAA, but rather in the probability that one *would have* a long life if one *were* insured with TIAA. Apt as this suggestion is, it requires us to evaluate the probability of coun-terfactuals, for which we have only the begin-nings of a semantics (via the device of measures over possible worlds)[18] and no methodology, much less an account of why the methodology is suited to the semantics. How do we test claims about probabilities of counterfactuals? We have no answer, much less an answer that fits with our nascent semantics. It would be preferable to have a measure of effectiveness that requires only probabilities over events that can be tested in the actual world in the standard ways. This is what I shall propose.

The Gibbard and Harper example, an example of spurious correlation due to a joint cause, is a special case of a general problem. We saw that the conditional probability will not serve as a mark of causation in situations where the putative cause is correlated with other causal factors. Exactly the same problem arises for effectiveness. For whatever reason the correlation obtains, the conditional probability is not a good measure of effectiveness in any populations where the strategy state is correlated with other factors causally relevant to the goal state. Increase in conditional probability is no mark of effective-ness in situations which are causally heteroge-neous. It is necessary, therefore, to make the same restrictions about test situations in dealing with strategies that we made in dealing with causes:

> !S is an *effective strategy* for obtaining G in situ-ation L iff $\mathrm{Prob}(G/S.K_L) > \mathrm{Prob}(G/K_L)$.

Here K_L is the state description true in L, taken over the complete set $\{C_i\}$ of causal factors for G, barring S. But L may not fix a unique state descrip-tion. For example L may be the situation I am in when I decide whether to smoke or not, and at the time of the decision it is not determined whether I will be an exerciser. In that case we should compare not the actual values $\mathrm{Prob}(G/S.K_L)$ and $\mathrm{Prob}(G/K_L)$, but rather their expected values:

> SC: !S is an *effective strategy* for obtaining G in L iff

$$\sum_j \mathrm{Prob}(G/S.K_j)\mathrm{Prob}(K_j) > \sum_j \mathrm{Prob}(G/K_j)\mathrm{Prob}(K_j),$$

where j ranges over all K_j consistent with L.[19]

This formula for computing the effectiveness of strategies has several desired features: (1) it is a function of the probability measure, Prob, given by the laws of association in the actual world; and hence calculable by standard methods of statistical inference. (2) It reduces to the conditional probability in cases where it ought. (3) It restores a natural connection between causes and strategies.

(1) SC avoids probabilities over counterfactuals. Implications of the arguments presented here for constructing a semantics for probabilities for counterfactuals will be pointed out in section 2.2.

(2) Troubles for the conditional probability arise in cases like the TIAA example in which there is a correlation between the proposed strategy and (other) causal factors for the goal in question. When such correlations are absent, the conditional probability should serve. This follows immediately: when there are no correlations between S and other causal factors, $\text{Prob}(K_j/S) = \text{Prob}(K_j)$; so the left-hand side of SC reduces to $\text{Prob}(G/S)$ in the situation L and the right-hand side to $\text{Prob}(G)$ in L.

(3) There is a natural connection between causes and strategies that should be maintained; if one wants to obtain a goal, it is a good (in the pre-utility sense of good) strategy to introduce a cause for that goal. So long as one holds both the simple view that increase in conditional probability is a sure mark of causation and the view that conditional probabilities are the right measure of effectiveness, the connection is straightforward. The arguments in Part 1 against the simple view of causation break this connection. But SC re-establishes it, for it is easy to see from the combination of CC and SC that if $X \hookrightarrow G$ is true, then !X will be an effective strategy for G in any situation.

2.1 Causal laws and effective strategies

Although SC joins causes and strategies, it is not this connection that argues for the objectivity of sui generis causal laws. As we have just seen, one could maintain the connection between causes and strategies, and still hope to eliminate causal laws by using simple conditional probability to treat both ideas. The reason causal laws are needed in characterizing effectiveness is that they pick out the right properties on which to condition. The K_j which are required to characterize effective strategies must range over all and only the causal factors for G.

It is easy to see, from the examples of Part 1, why the K_j must include all the causal factors. If any are left out, cases like the smoking-heart disease example may arise. If exercising is not among the factors which K_j fixes, the conditional probability of heart disease on smoking may be less than otherwise in K_j, and smoking will wrongly appear as an effective strategy for preventing heart disease.

It is equally important that the K_j not include too much. $\{K_j\}$ partitions the space of possible situations. To partition too finely is as bad as not to partition finely enough. Partitioning on an irrelevancy can make a genuine cause look irrelevant, or make an irrelevant factor look like a cause. Earlier discussion of Simpson's paradox shows that this is structurally possible. Any association between two factors C and E can be reversed by finding a third factor which is correlated in the right way with both. When the third factor is a causal factor, the smaller classes are the right ones to use for judging causal relations between C and E. In these, whatever effects the third factor has on E are held fixed in comparing the effects of C versus those of ¬C. But when the third factor is causally irrelevant to E – that is, when it has no effects on E – there is no reason for it to be held fixed, and holding it fixed gives wrong judgements both about causes and about strategies.

I will illustrate from a real life case.[20] The graduate school at Berkeley was accused of discriminating against women in their admission policies, thus raising the question 'Does being a woman cause one to be rejected at Berkeley?' The accusation appeared to be borne out in the probabilities: the probability of acceptance was much higher for men than for women. Bickel, Hammel, and O'Connell[21] looked at the data more carefully, however, and discovered that this was no longer true if they partitioned by department. In a majority of the eighty-five departments, the probability of admission for women was just about the same as for men, and in some even higher for women than for men. This is a paradigm of Simpson's paradox. Bickel, Hammel and O'Connell accounted for the paradoxical reversal of associations by pointing out that women tended to apply to departments with high rejection rates, so that department by department women were admitted in about the same ratios as men; but across the whole university considerably fewer women, by proportion, are admitted.

This analysis seems to exonerate Berkeley from the charge of discrimination. But only because of the choice of partitioning variable. If, by contrast, the authors had pointed out that the associations reversed themselves when the applicants were partitioned according to their roller skating ability that would count as no defence.[22] Why is this so?

The difference between the two situations lies in our antecedent causal knowledge. We know that applying to a popular department (one with considerably more applicants than positions) is just the kind of thing that causes rejection. But without a good deal more detail, we are not prepared to accept the principle that being a good roller skater causes a person to be rejected by the Berkeley graduate school, and we make further causal judgements accordingly. If the increased probability for rejection among women disappears when a causal variable is held fixed, the hypothesis of discrimination in

admissions is given up; but not if it disappears only when some causally irrelevant variable is held fixed.

The Berkeley example illustrates the general point: only partitions by causally relevant variables count in evaluating causal laws. If changes in probability under causally irrelevant partitions mattered, almost any true causal law could be defeated by finding, somewhere, some third variable that correlates in the right ways to reverse the required association between cause and effect.

2.2 Alternative accounts which employ 'true probabilities' or counterfactuals

One may object: once all causally relevant factors have been fixed, there is no harm in finer partitioning by causally irrelevant factors. Contrary to what is claimed in the remarks about roller skating and admission rates, further partitioning will not change the probabilities. There is a difference between true probabilities and observed relative frequencies. Admittedly it is likely that one can always find some third, irrelevant, variable which, on the basis of estimates from finite data, appears to be correlated with both the cause and effect in just the ways required for Simpson's paradox. But we are concerned here not with finite frequencies, or estimates from them, but rather with true probabilities. You misread the true probabilities from the finite data, and think that correlations exist where they do not.

For this objection to succeed, an explication is required of the idea of a true probability, and this explication must make plausible the claim that partitions by what are pre-analytically regarded as non-causal factors do not result in different probabilities. It is not enough to urge the general point that the best estimate often differs from the true probability; there must in addition be reason to think that that is happening in every case where too-fine partitioning seems to generate implausible causal hypotheses. This is not an easy task, for often the correlations one would want to classify as 'false' are empirically indistinguishable from others that ought to be classified 'true'. The misleading, or 'false', correlations sometimes pass statistical tests of any degree of stringency we are willing to accept as a general requirement for inferring probabilities from finite data. They will often, for example, be stable both across time and across randomly selected samples.

To insist that these stable frequencies are not true probabilities is to give away too much of the empiricist programme. In the original this programme made two assumptions. First, claims about probabilities are grounded only in stable frequencies. There are notorious problems about finite versus infinite ensembles, but at least this much is certain: what probabilities obtain depends in no way, either epistemologically or metaphysically, on what causal assumptions are made. Secondly, causal claims can be reduced completely to probabilistic claims, although further empirical facts may be required to ensure the requisite asymmetries.

I attack only the second of these two assumptions. Prior causal knowledge is needed along with probabilities to infer new causal laws. But I see no reason here to give up the first, and I think it would be a mistake to do so. Probabilities serve many other concerns than causal reasoning and it is best to keep the two as separate as possible. In his *Grammar of Science* Karl Pearson taught that probabilities should be theory free, and I agree. If one wishes nevertheless to mix causation and probability from the start then at least the arguments I have been giving here show some of the constraints that these 'true probabilities' must meet.

Similar remarks apply to counterfactual analyses. One popular kind of counterfactual analysis would have it that

!S is effective strategy for G in L iff
$$\text{Prob}(S \,\square\!\!\rightarrow G/L) > \text{Prob}(\neg S \,\square\!\!\rightarrow G/L)^{23}$$

The counterfactual and the causal law approach will agree, only if

A: $\text{Prob}(\alpha \,\square\!\!\rightarrow G/X) = \text{Prob}(G/\alpha.K_x)$

where K_x is the maximal *causal* description (barring α) consistent with X. Assuming the arguments here are right, condition A provides an adequacy criterion for any satisfactory semantics of counterfactuals and probabilities.

3 How some worlds could not be Hume worlds[24]

The critic of causal laws will ask, what difference do they make? A succinct way of putting this question is to consider for every world its corresponding Hume world – a world just like the first in its laws of association, its temporal relations, and even in the sequences of events that occur in it. How does the world that has causal laws as well differ from the corresponding Hume world? I have already argued that the two worlds would differ with respect to strategies.

Here I want to urge a more minor point, but one that might go unnoticed: not all worlds could be turned into Hume worlds by stripping away their causal laws. Given the earlier condition relating causal laws and laws of association, many worlds do not have related Hume worlds. In fact no world whose laws of association provide any correlations could be turned into a Hume world. The demonstration is trivial. Assume that a given world has no causal laws for a particular kind of phenomenon E. The earlier condition tells us to test for causes of E by looking for factors that increase the probability of E in maximal causally homogeneous sub-populations. But in the Hume world there are no causes, so every sub-population is homogeneous in all causal factors, and the maximal homogeneous population is the whole population. So if there is any C such that Prob(E/C) > Prob(E), it will be true that C causes E, and this world will not be a Hume world after all.

Apparently the laws of association under-determine the causal laws. It is easy to construct examples in which there are two properties, P and Q, which could be used to partition a popu-lation. Under the partition into P and ¬P, C increases the conditional probability of E in both sub-populations; but under the partition into Q and ¬Q, Prob(E/C) = Prob(E). So relative to the assumption that P causes E, but Q does not, 'C causes E' is true. It is false relative to the assump-tion that Q ⑄ E, and P ⑄ E. This suggests that, for a given set of laws of association, any set of causal laws will do. Once some causal laws have been settled, others will automatically follow, but any starting point is as good as any other. This suggestion is mistaken. Sometimes the causal laws are underdetermined by the laws of association, but not always. Some laws of asso-ciation are compatible with only one set of causal laws. In general laws of association do not entail causal laws: but in particular cases they can. Here is an example.

Consider a world whose laws of association cover three properties, A, B, and C; and assume that the following are implied by the laws of association:

(1) Prob(C/A) > Prob(C)
(2) Prob(C/B & A) > Prob(C/A);
      Prob(C/B & ¬A) > Prob(C/¬A)
(3) Prob(C/B) = Prob(C)

In this world, B ⑄ C. The probabilities might for instance be those given in Chart 24.1. From just the probabilistic facts (1), (2), and (3), it is possible to infer that both A and B are causally relevant to C. Assume B ⑄ ±C. Then by (1), A ⑄ C,

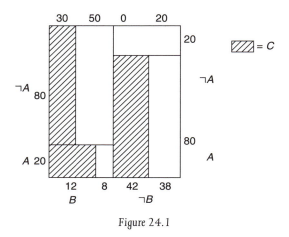

Figure 24.1

since the entire population is causally homoge-neous (barring A) with respect to C and hence counts as a test population for A's effects on C. But if A ⑄ C, then by (2), B ⑄ ±C. Therefore B ⑄ ±C. But from (3) this is not possible unless A is also relevant, either positively or negatively, to C. In the particular example pictured in the chart, A and B are both positively relevant to C.

This kind of example may provide solace to the Humean. Often Humeans reject causal laws because they have no independent access to them. They suppose themselves able to determine the laws of association, but they imagine that they never have the initial causal information to begin to apply condition C. If they are lucky, this initial knowledge may not be necessary. Perhaps they live in a world that is not a Hume world; it may nevertheless be a world where causal laws can be inferred just from laws of association.

4 Conclusion

The quantity Prob(E/C.K_j), which appears in both the causal condition of Part 1 and in the measure of effectiveness from Part 2, is called by statisticians the partial conditional probability of E on C, holding K_j fixed; and it is used in ways similar to the ways I have used it here. It forms the foundation for regression analyses of causation and it is applied by both Suppes and Salmon to treat the problem of joint effects. In decision theory the formula SC is structurally identical to one proposed by Brian Skyrms in his deft solution to Newcomb's paradox; and elaborated further in his book *Causal Necessity*.[25] What is especially significant about the partial conditional probabilities which appear here is the fact that these hold fixed all and only causal factors.

The choice of partition, $\{K_j\}$, is the critical feature of the measure of effectiveness proposed in SC. This is both (a) what makes the formula work in cases where the simple conditional probability fails; and (b) what makes it necessary to admit causal laws if you wish to sort good strategies from bad. The way you partition is crucial. In general you get different results from SC if you partition in different ways. Consider two different partitions for the same space, $K_1, \ldots, K_n$. and $I_1, \ldots, I_s$, which cross-grain each other – the K_i are mutually disjoint and exhaustive, and so are the I_j. Then it is easy to produce a measure over the field ($\pm G$, $\pm C$, $\pm K_i$, $\pm I_j$) such that

$$\sum_{i=1}^{n} \mathrm{Prob}(G\,/\,C.K_i)\mathrm{Prob}(K_j) \neq \sum_{i=1}^{s} \mathrm{Prob}(G\,/\,C.I_j)\mathrm{Prob}(I_j).$$

What partition is employed is thus essential to whether a strategy appears effective or not. The right partition – the one that judges strategies to be effective or ineffective in accord with what is objectively true – is determined by what the causal laws are. Partitions by other factors will give other results; and, if you do not admit causal laws, there is no general procedure for picking out the right factors. The objectivity of strategies requires the objectivity of causal laws.

Notes

1 Albert Einstein and Leopold Infeld, *The Evolution of Physics* (Cambridge: Cambridge University Press, 1971), p. 9.

2 Bertrand Russell, 'On the Notion of Cause', *Proceedings of the Aristotelian Society* **13** (1912–13), pp. 1–26.

3 See Patrick Suppes, *A Probabilistic Theory of Causality* (Amsterdam: North-Holland Publishing Co., 1970).

4 See Wesley Salmon, 'Statistical Explanation' in Salmon, Wesley (ed.), *Statistical Explanation and Statistical Relevance* (Pittsburgh: University of Pittsburgh Press, 1971).

5 See H.M. Blalock, Jr., *Causal Models in the Social Sciences* (Chicago: Aldine-Atherton, 1971).

6 Throughout, '*A* and *B* are correlated' will mean $\mathrm{Prob}(A/B) \neq \mathrm{Prob}(A)$.

7 E.H. Simpson, 'The Interpretation of Interaction in Contingency Tables', *Journal of the Royal Statistical Society*, Ser. B. **13** (1951), pp. 238–41.

8 See Morris R. Cohen and Ernest Nagel, *An Introduction to Logic and Scientific Method* (New York: Harcourt, Brace and Co., 1934).

9 See Wesley Salmon, op. cit.

10 See Rudolf Carnap, *The Continuum of Inductive Methods* (Chicago: University of Chicago Press, 1952).

11 See C.G. Hempel, *Aspects of Scientific Explanation* (New York: Free Press, 1965).

12 See Richard C. Jeffrey, 'Statistical Explanation vs. Statistical Inference', in Wesley Salmon, op. cit.

13 This is explicitly stated in Salmon's later papers (see 'Theoretical Explanation' in S. Körner, *Explanation* (Oxford: Basil Blackwell 1975), but it is already clear from the treatment in 'Statistical Explanation' that Salmon is concerned with causal explanations, otherwise there is no accounting for his efforts to rule out 'spurious' correlations as explanatory.

14 This example can be made exactly parallel to the uranium–polonium case by imagining a situation in which we choose at random between this defoliant and a considerably weaker one that is only 10 per cent effective.

15 See section 3.

16 H.P. Grice, 'Some Aspects of Reason', the Immanuel Kant Lectures, Stanford University, 1977.

17 See Allan Gibbard and William Harper, 'Counterfactuals and Two Kinds of Expected Utility'. Discussion Paper No. 194, Center for Mathematical

Studies in Economics and Management Science Northwestern University, January 1976.

18 See William Harper, Robert Stalnaker, and Glenn Pearce (eds), *University of Western Ontario Series in Philosophy of Science* (Dordrecht: D. Reidel Publishing Co., 1981).

19 I first derived this formula by reasoning about experiments. I am especially grateful to David Lewis for pointing out that the original formula was mathematically equivalent to the shorter and more intelligible one presented here.

20 Roger Rosenkrantz and Persi Diaconis first pointed out to me that the feature of probabilities described here is called 'Simpson's Paradox', and the reference for this example was supplied by Diaconis.

21 See Peter J. Bickel, Eugene A. Hammel and J. William O'Connell, 'Sex Bias in Graduate Admissions: Data from Berkeley', in William B. Fairley and Frederick Mosteller, *Statistics and Public Policy* (Reading, Mass: Addison-Wesley, 1977).

22 William Kruskal discusses the problem of choosing a partition for these data briefly in correspondence following the Bickel, Hammel, and O'Connell article referred to in footnote 21.

23 See articles in Harper, op. cit.

24 I learned the term 'Hume World' from David Lewis. Apparently it originated with Frank Jackson and other Australian philosophers.

25 Brian Skyrms, *Causal Necessity* (New Haven: Yale University Press, 1980).

PART SIX

Science and medicine

INTRODUCTION TO PART SIX

I S MEDICINE A science or an art? On the one hand there is clearly a vast amount of science in medicine. Huge amounts of effort and money are put into scientific research that seeks the causes of diseases, develops drugs or tests, investigates the effects on health of environment and lifestyle, and assesses the effectiveness of therapies. There is no doubt that this is science if anything is. When Barry Marshall showed that the bacterium *H. pylori* is the principal cause of stomach ulcers, Françoise Barré-Sinoussi and Luc Montagnier discovered HIV, and Richard Doll and Austin Bradford Hill linked smoking to lung cancer, their research was science. Likewise the development of antibiotics from the work of Fleming, Chain, Florey and Heatley on penicillin was a matter of laboratory science followed by scientifically organized therapeutic trials (in most cases – penicillin itself was an exception; it was very effective and urgently needed in the Second World War, and was never trialled).

On the other hand, medicine aims, for the most part, at treating individual patients for their particular ailments. This might look very different from science for at least two reasons. First, the individual nature of diagnosis, treatment, and care contrasts with the generalized, statistical nature of medical research. Science can tell us about what will tend to happen in a certain proportion of cases. But that knowledge cannot always determine what is best for the individual. Secondly, effective diagnosis, prognosis, deciding upon and administering a treatment regime and assisting a patient through it depend on a number of factors that fall outside the province of science. These require judgements made on the basis of a clinician's experience, of both the disease at issue and also of the individual patient, and will require sensitivity to the patient's particular circumstances (for example, their cultural background). A good 'bedside manner' is not important just to make the patient feel cared for – although that is important in ensuring compliance with the treatment regime and for its significant positive placebo effect – a good bedside or surgery relationship between physician and patient can significantly assist the physician in making a correct diagnosis. The individual nature of care and its dependence on clinical judgement suggest that medicine retains the character of an art, an acquired skill. Helping the ailing patient is at the heart of medicine.

Ronald Munson distinguishes medicine from science on different but related grounds. According to him, science and medicine have different aims. Science aims at giving us knowledge and understanding, whereas medicine aims at the treatment and prevention of disease. Medicine may be scientific, in that it employs scientific methods and knowledge in achieving its aims, but that does not suffice to show that medicine *is* a science (policing can be scientific, but that does not make policing a science).

What makes medical research *medical* is that it holds out the promise of making a difference to the health of individuals. Munson acknowledges that he does not discuss the possibility that medicine is an applied science. However, it might be that his argument can still be maintained. An engineer might work on the problem of calculating the forces within a structure of a certain design. That might well be regarded as an interesting problem in engineering science, even if there is no prospect of using that knowledge to build a stronger or lighter structure. On this view, an applied science is a science whose problems are generated by certain areas of human activity where science can be put to practical use. What makes applied science *science* is that its aim is the generation of knowledge irrespective of use. So while much applied biological science is materially the same as medical research, it is *medical* research only insofar as it aims at some practical health benefit. Munson also invokes a moral dimension: the physician has a moral responsibility, as a physician, for the health of her patient. From this he concludes that medicine has an inherent moral principle, whereas science lacks such a principle. Whether this argument is successful may depend on whether medicine is just what it is that individual doctors do with individual patients. It might appear that there is a lot of medicine that is not concerned with the doctor–patient relationship and does not carry those moral responsibilities. Public health is one such area of medicine. For example, a doctor might engage in a campaign of vaccination less out of concern for the individual recipients of the vaccine but rather in order to halt the spread of a disease or in order to protect those who cannot be vaccinated (e.g. the very young or those with allergies). The fact that the doctor is not concerned solely or primarily with the individual being treated, whatever other moral issues this may raise, does not itself mean that this doctor is not practicing medicine. That said, it could be argued that they still have a moral concern with the population, so something analogous to Munson's point stands.

One motivation for the view that medicine is a science is the thought that all the important knowledge in medicine can in principle be reduced to more basic, non-medical sciences, such as the various branches of biology. While that would not answer Munson's points relating to ethics, the reduction of medical science would encourage the view that the distinctive features of medicine are superficial or are just a matter of relating the hard science to individual treatment. On the other hand the irreducibility of medicine would leave room for the view that medicine is sui generis and not really a science at all. Quite what constitutes a reduction of one scientific theory to another is a difficult and contended question in the philosophy of science. Kenneth Schaffner describes not only the principal views on reduction in general but also develops a theory about how reduction works – or does not work – when applied to medicine.

The difference between medicine-as-science and medicine-as-art has been commonly reflected in the distinction between 'medical science' and 'clinical practice'. We might say that the latter is an art informed and aided by the former (perhaps as music is an art informed by the science and technology of acoustics). Yet, such a distinction is coming increasingly under pressure. 'Translational medicine' aims to bring the results of research in the biomedical sciences much closer to clinical medicine, translating science into clinical practice much more effectively than hitherto. Translational medicine may be seen as an implementation of the 'Evidence-Based Medicine' movement (EBM).

According to David Sackett, one of the leading lights of the EBM movement, 'Evidence-based medicine, whose philosophical origins extend back to mid-19th century Paris and earlier, is the conscientious, explicit and judicious use of current best evidence in making decisions about the care of individual patients.' (Sackett, 1997) When described like this, it may be difficult to see how there could be anything contentious about EBM. One can get some insight into what Sackett has in mind by considering his reference to 'mid-19th century Paris': this is Pierre-Charles-Alexandre Louis and his use of the 'numerical method' in medicine. Louis decided to test efficacy of bloodletting, a hugely popular treatment for fevers and other forms of illness. Using numerical data concerning the treatment of 77 patients with pneumonia, Louis concluded that bloodletting was rather less effective than had previously been thought (although he accepted that it could help in certain kinds of case). What is significant about this piece of history is the fact that the evidence produced by thinking in terms of populations (i.e. thinking about a large number of patients) contradicts what the large majority of physicians were prescribing for the individuals they were treating. If Louis's reasoning is correct, most of what was being done to patients (with pneumonia at least) was inefficacious and quite possibly harmful. Yet those physicians were not being entirely unreasonable. No doubt many of their patients did recover from their illness after bloodletting; and if others did not, that could be ascribed to the severity of the illness, the particular susceptibility of the patient or other conditions, such as the climate. Furthermore, bloodletting was what they were taught and it was promoted by the great physicians of the day, for example the great revolutionary doctor François-Joseph-Victor Broussais, whose new theory of disease explained why bloodletting would be effective. If any doctor had doubts he would find that his colleagues were confident in the value of bloodletting and would be able to cite numerous cases where it would appear to have cured the patient.

The supporter of EBM would point out that we now accept that evidence of the sort produced by Louis shows that the accepted practice, expert advice and reasoning from individual cases employed by advocates of bloodletting were all mistaken – and consequently harmful for patients. That EBM enthusiast would then go on to argue that although there is today much more (and better) evidence available of the kind that Louis produced, it does not have the influence on clinical practice that it ought. Physicians are still relying too heavily on what they were taught and on expert opinion (which might be out of date). In particular, there is a tendency to over-emphasize the value of individual clinical judgement as the fruit of experience. According to EBM, different kinds of evidence have different evidential weight: the evidence from data systematically collected over numerous patients has more value than anecdotes about a small number of cases. EBM advocates the careful evaluation of distinct forms of evidence and this leads to the EBM hierarchy of evidence: systematic reviews of clinical trials are better sources of evidence than cohort studies, which are better sources of evidence than expert opinion (see the reading 'Hierarchy of Evidence' from the School of Health and Related Research, University of Sheffield). It is true that many medical conditions and treatments cannot be understood without considering large populations rather than individual cases, if only because some conditions are so rare that even an experienced clinical practitioner cannot be expected to have encountered them. A particularly important reason for emphasizing large-scale studies is that for many conditions we have effective drugs, yet marginal improvements would still save many lives. Such improvements would only be detectable with large-scale studies.

Even so, EBM is not universally accepted. Many physicians resist its downplaying of clinical expertize and judgement. No systematic review can tell one how to treat an individual patient, since no treatment works for every patient and many have side-effects. The doctor must decide whether the treatment is likely to benefit *this* patient (recall Munson's claim above that the physician has a moral obligation to act in the interests of each patient individually). Furthermore, clinical trials usually take as their subjects patients who do not have health problems other than the one under study. But a large proportion of patients suffer from more than one complaint. The study will not tell us how such patients will fare with the treatment, especially if they are taking drugs for their other illnesses. This is an instance of the problem of *external* validity. *Internal* validity concerns the reliability of the study in showing a causal link between the drug given to the subjects in the study and a certain health improvement in *those* subjects. External validity concerns an inference from the results of the study to the larger population of patients outside the study who suffer from the disease in question and who might be potential beneficiaries of the treatment. Clinical judgement may be required to estimate the likely degree of external validity of a clinical trial. (More moderate articulations of the aims of EBM take this point on board, and talk of the integration of best evidence with clinical judgement.) A distinct complaint, made by John Worrall, is that over-emphasizing the importance of controlled clinical trials (trials involving a 'control' group for comparison that does not receive the new treatment) and downplaying clinical judgement, informed by knowledge of pathology and physiology, can lead to unnecessary and even unethical studies being carried out and delays to the implementation of new therapies.

According to the EBM hierarchy of evidence the best kind of investigation is the randomized controlled trial (RCT). Systematic reviews and meta-analyses come higher, but these are reviews or amalgamations of other studies, preferably RCTs. The key feature of an RCT is the comparison of two groups (or in some trials more than two groups). One group, the control group, will be the group against which the others are being compared. The other (test or intervention) group(s) will receive the treatment(s) under study. The test group must be as like the control group as possible, with the sole exception of the treatment being studied. If that is the case, then Mill's Method of Difference tells us that we can regard the cause of any difference in outcome between the two groups (such as faster recovery in the test group) as the effect of the treatment. This may require giving the control group a placebo, for example a sugar pill that looks just like the drug being studied except that it does not contain the active ingredient. Without a placebo we may not be able to attribute the improved health outcome to the active ingredient of the test drug rather than to the positive psychological effects of being given a treatment. Other precautions, such as blinding the physicians as to which patients are receiving the treatment and which are receiving the placebo, are put in place to remove other possible sources of difference between the two groups.

A standard feature of such trials is that the allocation of patients to test or control groups is determined by a random process, such as tossing a coin or use of a computerized random number generator. The use of randomization is philosophically interesting. Why should leaving allocation to chance be a good way of determining who receives treatment and who is in the control group? There are three answers that one can find in the literature.

First, following the great statistician Sir Ronald Fisher, it is argued that in order to use certain standard ('classical') statistical inferences techniques, random allocation is required. When we carry out an RCT we might find that there is a difference in health outcomes between the test and control group – the average time to recovery in the first group was 5.3 days whereas the average recovery time in the second was 4.6 days. That looks favourable to the drug. On the other hand we know that recovery from illness is itself something of a 'stochastic' process (it involves an element of chance). Perhaps the difference between the two groups arose by chance, and is nothing to do with the drug? This is where classical statistical technique of 'significance testing' comes in. Given certain information (such as the size of the groups, and the degree to which recovery rates can vary) one can calculate the probability that this difference between the two groups could arise by chance alone. If the probability is low, say less than 5%, then one can conclude that the difference between two groups is 'statistically significant' ('at the 5% level') and so one is in a good position to attribute the difference to the efficacy of the drug. The classical justification of this inference of statistical significance requires that the sampling (in this case assignment to the two groups) is done at random. According to the classical approach, if this 'random sampling assumption' is not true, then one cannot infer that the difference is statistically significant, in which case we cannot infer a correlation between taking the drug and an improved recovery, and so, furthermore, we cannot infer the causal efficacy of the drug.

Secondly, one can argue that randomization is required in order to make the inference from a statistically significant correlation to the causal efficacy of the drug. 'Correlation is not causation' is a favourite mantra of epidemiologists. This is because a correlation between two factors A and B can be brought about even if A and B have no causal influence on one another. That can happen when some third factor is the common cause of both A and B. For example several studies seemed to show that a beneficial side-effect of hormone replacement therapy (HRT) was that it reduced the chances of heart disease. Further studies showed, however, that although there was a correlation between HRT and reduced heart disease, this was entirely due to the fact that the women who chose HRT tended to have higher socio-economic status than those who did not. That higher socio-economic status was itself causally linked to lower risk of heart disease (through lower smoking rates, healthier diet, etc.). So in order to infer causation from correlation we must be able to exclude such common causes (or 'confounders' as they are known). We might be able to exclude some potential confounders by careful examination of the data, but that is possible only for factors that we *know* might be possible common causes. What about factors we don't know about? According to its second defence, only randomization can exclude confounding by both known and unknown possible common causes. Papineau's own justification of randomization is a version of this argument.

The third argument for randomization points to a very specific kind of problem that can arise with controlled trials. Let us say that you are testing a new drug you are hoping will prove to be an effective cure. You are allocating patients to the test and control groups. You might give in to the temptation to put the healthier-looking patients in the test group and the sicker ones in the control group. The better outcome in the test group might suggest that the drug is efficacious, but given the biased method of selecting patients, that clearly cannot be inferred. Randomized allocation removes the possibility of selection bias. One of the first and most important clinical trials to use randomization was the 1947 Medical Research Council trial of the antibiotic streptomycin in the treatment of tuberculosis. It has been widely assumed that the reason for employing randomization was the fact that the statistician on the trial, Austin Bradford Hill, was influenced by Ronald Fisher's work on the design of experiments and so must have used randomization for the first justification mentioned above. But the reports of the trial's organizers, who included Philip D'Arcy Hart, suggest that the principal reason was to avoid the possibility of selection bias. Worrall accepts this as the only possible justification for randomization.

Further reading

Cartwright, N. 2007 'Are RCTs the gold standard?', *BioSocieties*, 2: 11–20.

Collins, H.M. and Pinch, T.J. 2005 *Dr. Golem: How to Think About Medicine*. University of Chicago Press.

Guyatt G, Cook D and Haynes B. 2004 'Evidence based medicine has come a long way', *British Medical Journal*, 329(7473): 990–1.

Hacking, I. 1988 'Telepathy: Origins of randomization in experimental design', *Isis*, 79: 427–51.

Papineau, D. 1994 'The Virtues of Randomization', *British Journal for the Philosophy of Science*, 45(2): 437–450.

Reference

Sackett, D. 'Evidence-based medicine', *Seminars in Perinatology*, (1997) vol. 21, pp. 3–5.

Chapter 25

Ronald Munson

WHY MEDICINE CANNOT BE A SCIENCE

A widely-shared belief about contemporary medicine is that it is well on the way to becoming a science. The sources of this belief are not difficult to locate. The new understanding of disease processes and pharmacological mechanisms contributed by biochemistry and molecular biology, the establishment of therapies on the basis of experimental research, and the development of new modes of acquiring and utilizing precise laboratory data are just a few of the factors that contribute to the sense of medicine's being rapidly transformed into science.

Furthermore, quite apart from the impression that such factors produce, some observers claim that there are good reasons for saying that medicine has already become a science. They argue that contemporary medicine possesses features that satisfy explicit criteria that justify our including it among the sciences.

I believe these views are mistaken. In contrast, I want to argue here that medicine is not a science, nor can it ever become a science. Medicine can, of course, be *scientific* in ways that are easily specified, and medicine can participate in scientific research and contribute to scientific understanding. In the final analysis, however, medicine is fundamentally different from science. This is not, I want to suggest, a weakness of medicine produced by some fatal flaw. Rather, it is a reflection of the features medicine possesses by being the sort of enterprise that it is.

The particular thesis that medicine cannot be a science is a consequence of a more general position. I hold that medicine is an autonomous discipline with its own aims, constraints, and framework of basic commitments. However, to make this view plausible and to defend it from potential objections would require a more comprehensive treatment than is possible in a single paper. Accordingly, I want to restrict myself to considering just one implication – that science and medicine are inherently different. This means that several related and quite significant issues will not be dealt with. In particular, I shall not attempt to evaluate the following specific claims:

(1) Medicine is an *applied science* (analogous to automotive engineering);
(2) Medicine is a *craft* (analogous to automobile mechanics);
(3) Medicine is a *practical art* (analogous to interior decorating).

To establish my general position, I would have to show that medicine possesses features that justify our rejecting each of these claims. I must, however, leave these tasks to a later occasion.

It is worth stressing at the outset, though, that the question of whether medicine can be a science merits serious attention in its own right. An adequate reason for dealing with the matter lies in the fact that medicine is presently regarded by many to be a science and arguments supporting this view have been presented. If there are grounds for believing this view to be mistaken, this is a sound academic reason for bringing them forward.

More important, however, attempting to answer the question satisfactorily challenges us to explicate aspects of the fundamental nature of medicine, and this promises benefits that are more than purely intellectual. To mention just one, we require an understanding of what medicine really is in order to determine what we can legitimately demand of it. Whether we consider medicine a science (actually or potentially) is sure to affect the decisions we make about establishing research programs, meeting social needs, and conducting clinical practice.

For example, is it legitimate to fund a medical research program that holds out the hope of eventually understanding the mechanisms of a particular disease, but offers no direct promise of finding a means of controlling the disease? The answer given to such a question will depend, to some extent, upon whether the aim of medicine and the aim of science are considered identical. If they are the same, the program is justified by appealing to the scientific goal of understanding. If they are not, then it is necessary to ask if the aim of medicine can be secured by such research.

We clearly expect certain kinds of products and performances from science. If medicine is a science, it would be reasonable to expect these of medicine and to justify medicine by the same standards we apply to the sciences. By contrast, if medicine is not a science, our demands, expectations, and standards of success might be different. However, these are complex matters

and we cannot even begin to discuss them in a useful practical way until we understand the character of medicine much better than we do at present. Examining the status of medicine with respect to science will not take us to the end of this road, but it is the first step in a journey worth undertaking.

In the first section of this paper, I shall consider the most serious effort I know of to establish the thesis that medicine is a science. I shall show that there are adequate reasons for regarding the attempt as unsuccessful. In the second section, I will compare science and medicine with respect to their internal aims and criteria of success and argue that in each instance the two are fundamentally different. Furthermore, I shall argue that medicine possesses an inherently moral aspect that distinguishes it from science. In the third section, I shall consider the issue of reduction and argue that while the cognitive content of medicine may be reducible to scientific theories, medicine itself cannot be. Finally, all these considerations permit me to conclude that medicine is an autonomous enterprise quite unlike science, and I shall end by pointing out some of the consequences of this result.

1 The claim that medicine is a science

L. A. Forstram objects to the common assumption that medicine should be counted among the traditional crafts. Actually, Forstram argues, "clinical medicine has a rightful place among the sciences," because it "conforms to what is ordinarily understood by 'science'" (1977, p. 8). Forstram argues that not only is clinical medicine a science, it is one that differs fundamentally from all the established sciences. This second claim I shall have to ignore as not relevant to my purpose here. The first claim is completely relevant, however, for if *any* branch of medicine is a science, then my thesis that medicine cannot be a science is completely wrong.[1]

To establish his primary thesis, Forstram draws upon R. B. Braithwaite's characterization of the special sciences (physics, biology, etc.) and finds there are two features which he offers as criteria for counting something a science. An activity is a science if it: (1) has a "natural domain" and (2) has an "investigative function" which commits it to seeking generalizations and to increasing knowledge of phenomena in its domain (1977, p. 9). Clinical medicine, Forstram claims, satisfies both criteria.

These criteria are not particularly clear or well-specified. (Exactly what is a "natural domain," for example?) Yet rather than raising general questions about them, it is more productive to consider how Forstram employs them in connection with medicine. We will examine what he says about each criterion and then look at the result of using them in the way he suggests.

Forstram begins by merely telling us that clinical medicine has a domain: "The domain of clinical medicine is the human organism in its manifold environmental contexts, in health and disease" (1977, p. 9). It is immediately clear to him that this domain is much too large, since it would make all aspects of human behavior the proper concern of clinical medicine. Accordingly, he limits clinical medicine to concerns "having to do with the natural history of human diseases and the factors affecting these, including the various forms of medical intervention" (1977, p. 9).

Forstram apparently fails to realize that his altered characterization is still much too broad and so fails to pick out a unique area as the "natural domain" of clinical medicine. Medical sociology, epidemiology, bacteriology, biochemistry, and social work are just a few of the disciplines that are also concerned with "the natural history of human diseases and the factors affecting them." Some of these disciplines are also concerned with "the various forms of medical intervention." Accordingly, various endeavors might legitimately claim the area described by Forstram as belonging to their "natural domain."

Of course, Forstram might ascribe this domain to clinical medicine alone, if the undertaking of medical interventions is unique to it. Yet if this step is taken, the "natural domain" that Forstram wishes to assign to clinical medicine is acquired only at the cost of circularity. We are left with the truism that the domain of clinical medicine is the domain of medical intervention undertaken by clinical medicine. Forstram clearly fails to show that clinical medicine satisfies his first criterion for being a science.

Clinical medicine supposedly meets the second criterion because it "works importantly in the enlargement of the body of theory and knowledge to which the term 'medical science' corresponds" (1977, p. 10). Forstram goes on to point out that clinical medicine involves active inquiry in which observations, testing, and so on play the same role as in recognized sciences like physics and biology. Furthermore, Forstram considers the activity of seeking knowledge an essential aspect of clinical medicine and he warns us that we must not make the "mistake" of identifying clinical medicine with its "social functions," namely "the diagnosis and treatment of human disease" (1977, p. 10).

In explaining why he believes this second criterion is satisfied, much of what Forstram says is indeed true of clinical medicine. It does involve inquiry and the seeking of knowledge, and it does employ methods typical of the sciences. However, two critical points are worth making.

First, the use of 'scientific methods' does not itself make an activity a science. If barbers conducted experiments, made observations, and confirmed generalizations about cutting and styling hair, we would still not consider barbering a science. We might speak of 'scientific barbering' in recognition of the methods employed to establish claims to knowledge and to determine effective and satisfactory techniques, but we would still insist that being a science requires more than using certain methods of inquiry. In particular, we would call attention to the fact that the aims and products of inquiry in barbering differ significantly from those of the sciences. Similarly, the fact that medicine employs the methods of science and works to expand knowledge is sufficient to justify our speaking of 'scientific medicine,' but it is not sufficient to demonstrate that medicine is a science. (Later I will show that even when medicine is indistinguishable from science in its methods, research activities, and contributions to knowledge there are still reasons for saying that medicine has not 'become' science.)

Second, and more seriously, I believe Forstram has fallen victim to a means–end confusion. He is right to stress that medicine is concerned with acquiring knowledge, and it is also true that medicine is often concerned with establishing and testing general laws and theories. However, Forstram is mistaken in identifying the acquisition of knowledge as the primary "function" of clinical medicine.

Medicine, as medicine, seeks knowledge as a *means* of performing more effectively its task of dealing with disease. (In confusing the end of medicine with the means of achieving it, Forstram is like someone who starts out to build a wall and, while moving stones to the site, forgets about the wall and comes to believe that his purpose is to move stones.) If this is correct (and in the next section I shall argue that it is), then the diagnosis and treatment of diseases is not merely a secondary, "social function" which we "mistakenly" identify with clinical medicine. Rather, it is the distinctive *aim* of clinical medi-

cine and Forstram's second criterion is not met.[2]

Having pointed out some flaws in Forstram's attempt to show that clinical medicine satisfies his two criteria for being a science, let us look at the results of accepting the criteria. It is easy to see that even if clinical medicine satisfied them, we still would not be warranted in considering it a science. The problem is with the criteria themselves – they are too inclusive.

In relying upon them, a good case could be made for counting as sciences such activities as cooking, dog training, leather tanning, horse breeding, and so on. Thus, as Creson (1978, p. 257) has pointed out, the very crafts Forstram hoped to distinguish from medicine would become sciences along with it.

The more important point, of course, is that criteria that lead us to include such activities in the class of sciences are clearly defective. They have failed to capture and reconstruct the significant differences that we antecedently recognize between, e.g., farming and physics. Thus, we cannot trust the criteria to help us in cases where there is doubt or disagreement.

Although I reject Forstram's thesis, his discussion is valuable in calling attention to some of the ways in which medicine and science are related. I believe that two of those ways, in particular, justify our speaking of "scientific medicine." First, as Forstram noted, medicine involves research directed towards establishing general theories by employing the same experimental methods and techniques as those typical of the sciences. Second, medicine makes use of concepts and generalizations (e.g. those concerning antigen-antibody reactions) borrowed from the sciences and established, in part, by clinical research. Methodologically and conceptually, then, contemporary medicine is *scientific*. Although this is a point that must not be ignored, I now want to show that, for all that, medicine is not, and cannot be, a science.

2 Differences between medicine and science

The first difficulty that faces anyone who addresses the question of whether medicine can be a science is the fact that we have no generally accepted analysis of what it is for a discipline or an enterprise to be a science. Thus, it is not possible to solve the problem of medicine by consulting a set of explicit criteria and then making a decision. Furthermore, I think it is pointless to step into this darkness and declare more or less by fiat that certain conditions are necessary or sufficient for being a science. Science is simply too diverse, various, and changeable to be summed up by a few rules of language use. (The failure of Forstram's criteria illustrates this, but other instances might easily be offered.)

I see no reason for despair over this difficulty, however. We must simply pay heed to Aristotle's dictum about ethics and not demand a greater degree of precision than our subject matter will allow. We can start by recognizing that we do have commonly accepted or core cases of sciences – physics, chemistry, biology, psychology, and so on. Consequently, to claim that a discipline is a science is to assert that it possesses features that make it reasonable to include it with these core cases. By contrast, to claim that the discipline is not a science is to assert that it has features that make it significantly different from the core cases and thus make it unreasonable to include it with them.

The features selected as relevant for comparison should be fundamental ones. They should be so connected with the character of the discipline that, if they were altered, we would be justified in saying that the entire nature of the discipline has changed and would, accordingly, react to the discipline differently, alter our expectations about it, and so on. The following features are ones that I believe are fundamental

in comparing medicine with science: (1) the internal aims; (2) the internal criteria of success, and (3) the internal principles regulating the conduct of the discipline's activities. I now want to show how medicine and science differ with respect to each feature.

2.1 The aims

To avoid confusion in comparing the aims of science and medicine, it is useful to recognize at the outset that both are social activities and both are intellectual activities. Accordingly, we can distinguish between their respective *external* and *internal* aim.[3] (A similar distinction is required for the other two features as well.)

Science as an organized social enterprise can be portrayed as seeking to secure a number of aims. Thus, we speak of science as working towards increasing national prestige, discovering ways of increasing our food supply, providing cheap energy, and so on. Like other social institutions, science justifies itself in terms of the desirable functions it serves.

From the same social point of view, scientists as people are motivated in their careers by a variety of personal aims. Like other people, scientists seek recognition, money, power, prestige, and the satisfactions of work well done.

The external aims of science and scientists are no different in kind from those of other social institutions and their constituent members. Similar social and personal goals can also be ascribed to religious, financial, educational, and legal enterprises. In social criticism, policy making, and sociological explanations, reference to external aims plays a definite role, and no picture of science is complete unless it acknowledges science's social character.

Yet science is also an intellectual endeavor, and it is pursued in accordance with its own internal aims. Scientists *as scientists* are no more concerned with the external aim of, e.g., increasing national prestige than writers who buy pencils are concerned with contributing to the GNP. Rather, scientists seek such internal goals as designing successful experiments, formulating hypotheses and theories, developing accurate models, and so on.

These are relatively specific aims, but I think it is also correct to say that science has one general and fundamental aim with respect to which the others are subservient. The basic internal aim of science is *the acquisition of knowledge and understanding of the world and the things that are in it.*

The knowledge sought by science can sometimes be turned to useful purpose, i.e., be 'applied' to solve practical or technological difficulties. Furthermore, social needs may pose a problem and provide a motive for pursuing a certain kind of research. Nevertheless, this does not alter the fact that what we sometimes call 'pure' science does not seek knowledge for the sake of its potential usefulness.[4] The internal aim of science is the epistemic one of theoretical knowledge, and it is with respect to this aim that particular scientific activities are (internally) justified and evaluated. Thus, certain research projects are branded as trivial, pointless, or 'merely applied' if they do not promise to contribute to our knowledge of what the world is like.

Medicine, too, is a social institution, and many of its external aims do not differ from those of science, the law, and so on. Yet medicine also has a unique external aim: *to promote the health of people through the prevention or treatment of disease.* It is primarily with respect to this aim that medicine justifies itself to the society.

A crucially important feature of medicine is that this external aim is also its fundamental *internal* aim. That is, this aim defines and determines the activity of medicine in the same way as the aim of acquiring knowledge defines and determines the activity of science. In short, the basic internal aim of medicine is not identical with the basic internal aim of science.

The most obvious objection to this view is that medicine also has the aim of acquiring knowledge. After all, contemporary medical research is often identical with scientific research. A physician-researcher and a biochemist, for example, might both be engaged in investigating the biological properties of cell membranes. Thus, it makes no sense to say one is doing medicine and the other science, for they are doing the same thing.

It is undeniable that medical researchers engage in scientific inquiry, and when they do, their specific aims may be no different from their scientist colleagues (say, understanding the membranes). But what the above argument neglects (as did Forstram) is that research can be justified as *medical* research only when it holds some promise of helping to secure the aim of promoting health. The research need not be specifically 'targeted,' and the promise need be no more than a possibility. But if there is no promise at all, then the research cannot legitimately be described as 'medical.'

Thus, a physician who engaged in paleontological inquiry might make a genuine contribution to scientific knowledge, but if he wished his work to be considered medical research, then he would have to undertake to show how it might contribute to the aim of promoting health. By contrast, a biochemist, to justify his work as a scientist, need only demonstrate that it is likely to increase our knowledge. The medical researcher must, in effect, present a dual justification: (1) the work will increase our knowledge; (2) the knowledge will be relevant to the aim of medicine.

The reason that medical and scientific research are often identical lies in the fact (as I suggested earlier) that medicine has discovered that its goal can be most effectively achieved by means of knowledge and understanding. It is this recognition that has led medicine to choose deliberately what Claude Bernard called the "permanent scientific path" (1957, p. 1). Had the path not been chosen, medicine would still be recognizable as the enterprise we know today, for it would continue to be identified by its commitment to the prevention and treatment of disease. Because it would not be *scientific* medicine, it would almost certainly not be as effective as it is now, but that is a different matter.

It is worth pointing out that even now a physician who refused to employ relevant scientific methods and knowledge, when they are available and would ordinarily be employed, would still be considered to be practicing medicine. Of course, we might call him incompetent or say he is practicing 'bad medicine.' What this shows is that the scientific aspects of contemporary medicine can serve as standards for evaluating the *quality* of medical practice, but they do not determine what *counts as* medical practice. Once again, it is the aim of promoting health that does that.

2.2 Criteria of success

When is science successful? In general, it is successful when it achieves its aim of providing us with knowledge or understanding of the world. Obviously, however, science as a whole is not successful or unsuccessful. Rather, it is the theories, laws, models, and hypotheses of science that are, for they are the cognitive products of scientific inquiry that purport to tell us what the world is like. It is through them that what I call the reality-revealing function of science is exercised.

If we take theories as the most important of the cognitive products of science (or, at least, as representative of them), we can ask, "When is a theory considered successful within science?" The traditional answer is that a theory is successful when we have adequate reasons for considering it to be true or as likely to be true. If the aim of science is to provide us with knowledge of the world, then truth is necessarily the internal criterion of success for cognitive claims. In short, theories are successful if, and only if, they give us a correct account of what there is.

However, this is not to say that theories known to be false are not employed within science. Sometimes seriously flawed theories are the best available, and sometimes false theories are used for such limited purposes as making convenient calculations. Furthermore, to claim that truth is the criterion of success is not to say that we can categorize each current theory as true or not. We may even believe some theories are false and yet keep them in the hope that something can be salvaged and the theory developed into one we can accept as true.

The traditional view that truth is the criterion of success for theories has been criticized in recent years, and various alternatives to it (or construals of it) have been suggested. Thomas Kuhn (1970), for example, has claimed that truth is always relative to a "paradigm" or constellation of beliefs about the nature of reality. Consequently, scientific progress cannot mean coming closer to the truth. Rather, it must consist in finding more and better solutions to "puzzles" generated by successive paradigms. Similarly, Larry Laudan (1977) has argued that theories should be judged by their problem-solving effectiveness in connection with an on-going research program.

For my purpose, it is not necessary to attempt to resolve the conflict among these competing views. (To do so would require that we become enmeshed in epistemological issues about the nature of truth, scientific progress, and the character of scientific theories.) It is enough to note that all the views agree in recognizing that what we demand of a scientific theory is that it provide us with a reliable account of whatever aspect of the world that the theory deals with. Whether truth or some sort of progress in puzzle-solving is the best way to analyze this demand, we need not consider. Thus, while admitting I am dodging this question, I will continue to speak of truth as the criterion of scientific success.

We can now ask of medicine the question we raised about science: When is medicine successful? In general, medicine is successful when it achieves its aim of promoting health through the prevention and treatment of disease. Like science, however, medicine as a whole is neither successful nor unsuccessful. Medicine is successful only when its particular therapeutic and diagnostic rules, causal theories, preventive measures, and so on are successful, for it is through them that the aim of medicine is realized.

The picture of medicine is complicated somewhat by the fact that its cognitive contents include theories and laws of the natural sciences. Some of these are borrowed ones (e.g. the laws of chemical bonding), while others are the product of joining clinical-scientific research (e.g., theories of immunology). In either case, the theories are judged by the criterion of success employed in science. Medicine looks to such theories to provide explanations of disease processes and to serve as a basis for developing means of prevention and treatment. (To mention only a famous example, the use of insulin in the management of diabetes was the direct result of its isolation by Banting and Best in 1921 and of their recognition of its role in glucose metabolism.[5])

This aspect of medicine illustrates the point I stressed earlier, namely, that contemporary medicine is committed to the notion that effectiveness in seeking its goal is best achieved through science. It is unquestionable that medicine prefers to have its rules of diagnosis, treatment, and prevention grounded in scientific knowledge of causal mechanisms. Not only does such knowledge offer a means of developing successful modes of intervention, but it also provides a way of determining the limits of appropriateness, reliability, and effectiveness of the standard rules.

Yet medicine's preference for science-based rules should not obscure the fact that medicine's criterion of success is not the epistemic one of truth. Rather, the basic standard of evaluation in medicine is instrumental or practical success. Since the aim of medicine is to promote health through the prevention or treatment of disease, cognitive formulations are acceptable within medicine when they can serve as the basis for successful practical action in achieving this aim.

Let me put the point another way. In seeking knowledge, science can be described as a quest for truth about the world. In seeking to promote health, medicine can be described as a quest for *control* over factors affecting health. Knowledge or understanding of biological processes is important to medicine, because it leads to control. But where understanding is lacking, medicine will seek control in other ways. In particular, it will rely upon empirical rules that are validated (at least partially) by practical success.

That practical success is medicine's standard of evaluation is illustrated most clearly by examples from therapy. It is often the case that drugs have a demonstrable therapeutic effect, even though an understanding of the biochemical and biological mechanisms of their mode of operation (and of the disease itself) is lacking. The chlorpromazines, for instance, are generally effective in treating the symptoms of schizophrenia. Although knowledge of the neurophysiological processes through which they work is far from complete, their use is warranted by their comparative success. This, in turn, warrants a therapeutic rule (a generalization) specifying the conditions under which the use of a chlorpromazine drug is indicated. (See Baldessarini, 1977, pp. 12–56.)

The same criterion justifies measures that are taken to control the symptoms of other diseases with unknown causes. Narcolepsy, for example, is a syndrome characterized by sudden loss of muscle tone and recurrent attacks of sleep. Episodes may be triggered by even mild emotional excitement, but virtually nothing is known about the process that produces them. Consequently, therapy is 'empirical' and is aimed at preventing the attacks by the use of drugs such as amphetamines that have been found to be effective in clinical experience.

Innumerable historical examples also testify to the existence and wisdom of using control or practical success as a criterion in medicine. For example, although the cholera bacillus was not identified by Koch until 1883, empirically-based hygienic measures had already brought the disease under control by then (Singer and Underwood, 1962, p. 391). Or again, in 1846 Semmelweis showed how cleansing the hands of examining physicians could dramatically reduce the incidence of puerperal fever, and ten years later Joseph Lister introduced the antiseptic method into surgery (Singer and Underwood, 1962, pp. 360–361). Both events took place before Pasteur demonstrated the theoretical basis for them by working out his germ theory of disease in the 1860s.

It should be no surprise that contemporary medicine, despite its commitment to science, continues to employ practical success or control as its basic criterion for evaluating rules, procedures, and causal claims. Medicine is an eminently practical enterprise that must attempt to meet immediate and urgent demands. It cannot afford to wait for the acquisition of the appropriate scientific knowledge but must do the best it can in the face of ignorance and uncertainty. We do not look to medicine to tell us what the world is like. Rather, we count on medicine to act against disease and suffering.

Science, by contrast, is a leisured pursuit. It may be prodded by external demands to solve practical problems, but its internal standard of success continues to be truth. It can afford to wait and work until this is met. For medicine, the constant external demand to promote health is identical with medicine's internal aim, and as a result, practical success is both the external and internal standard. What we expect of medicine is what it expects of itself.

2.3 Internal principles and the moral aspect of medicine

Internal to the activities of both medicine and science are certain principles or commitments that are recognized (at least implicitly) as required for making the enterprise possible. That is, the principles must be acknowledged in practice or the entire enterprise either fails or becomes a different kind of enterprise.

In science, perhaps the most obvious of these principles is that of giving honest reports of observations or experimental results. Since science is a cooperative endeavor in which reported data are often used as the basis for theory development or additional investigation, being able to trust (in general) the reports of fellow investigators is necessary for the conduct of the enterprise. This is the primary reason that faked data and false results are such crimes in science. Not only do they in themselves violate the aim of science, but all claims and projects based on them are compromised. Hence, science itself is threatened.[6]

Besides honesty in reporting, there are no doubt a number of other internal principles basic to the practice of science. A preference for quantitative results, a commitment to accuracy and precision, an acknowledgement in practice of certain criteria for acceptable explanations, preferred types of theories, and significant problems are perhaps some that might be specified. None of them need concern us here, however.

Medicine also requires honesty in the conduct of its enterprise, and very likely it also acknowledges other principles embodied in the conduct of science. However, implicit in the practice of medicine is a principle that is not found in science and that radically distinguishes medicine from science. It is one connected with medicine's aim of promoting health, but it is not identical with this aim. The principle is this: a physician, acting in his or her capacity as a physician, is committed to promoting the health of any individual accepted as patient. Since the principle enjoins an obligation (although a conditional one), I shall say that medical practice has an inherently moral aspect.[7]

The principle is not an obvious part of medicine and may seem prima facie implausible. But I believe an explanation of the character of the principle and of how and when it operates will show that it is in fact tacitly recognized. Perhaps the following three remarks will serve this purpose.

First, let me make it clear why the moral principle is not identical with medicine's aim. In so far as one is doing medical work at all, one must be working to promote health. However, it is possible to do this in a variety of ways. A medical researcher, for example, may never deal with a patient and yet be actively pursuing medicine's aim through the expected consequences of her research. Similarly, a physician involved in public health or health education may contribute significantly to preventing diseases without ever participating in a clinical encounter. In such instances, the concern with promoting health is expressed in a general or even statistical way. That is, it is the health of a *population* of people and not that of certain individuals that is sought.

The situation is quite different in medical practice. Medicine has no internal principles that oblige a physician to practice medicine nor, if he chooses to do so, to accept just anyone as a patient. However, if he does accept a person as a patient, then he enters into a certain kind of

medical relationship. He implicitly commits himself to seeking the aim of medicine in a special way, namely, by promoting the health of the *patient*. Medicine's aim is then, so to speak, manifested in the patient, and to achieve it the physician as a practitioner must work for the sake of the patient. Thus, the commitment of a physician to the welfare of a patient is not identical with the commitment to promoting health but arises only within the special circumstances of medical practice.

Second, I think the presence of an inherently moral aspect of medical practice can be demonstrated clearly by considering circumstances in which we believe it is legitimate to doubt the propriety of the actions of a physician in providing care. The nature of medical practice is such that we consider it justifiable for a patient to assume that the actions taken by his physician are for *the patient's own good* and are not taken solely for the purpose of acquiring knowledge, benefiting others, increasing the physician's income or reputation, or anything of the sort. Such results might be produced or even sought, but they must not be the physician's primary aim. Thus, when there is reason to believe that anything other than the good of the patient directs the physician's actions, then we consider this to constitute grounds for moral doubt or even condemnation by both the patient and the medical community. In not acting to promote his patient's health, the physician violates the principle of medical practice. He might be said to have acted wrongly or to have practiced (a different kind of) 'bad medicine' or even not have been practicing medicine at all. (I will discuss this more fully below.) In any event, the possibility of such evaluations shows the presence in medical practice of an internal principle that can be violated.

Along similar lines, that medical practice has such a moral aspect can be clearly seen by recognizing the role it plays in producing a familiar dilemma that faces clinical researchers. As practicing physicians, they are required to seek the aim of medicine by promoting the health of their patients as individuals, while as medical researchers they are committed to promoting the health of patients as a population. Thus, if they act primarily for the benefit of individual patients, they may lose the opportunity to acquire knowledge that might benefit the population, and if they act primarily for the sake of acquiring such knowledge, then they may not be acting in the best interest of their individual patients.

The dilemma results from a conflict between two equally legitimate ways of seeking the aim of medicine, and the conflict itself calls attention to the presence of the commitment to the individual that is inherent in medical practice. Notice, too, that the dilemma is not the product of the principles of some external moral theory applied to medicine. Rather, it is a result of a conflict between internal factors, and I suspect that it cannot be resolved by any internal norms. This leads me to the additional suspicion (although I shall not argue for it here) that additional principles of an external ethical theory are required to justify choosing one option over the other in cases in which the dilemma is generated.

Finally, I want to stress that it is medical *practice* that possesses this particular moral aspect. Medical practice, roughly speaking, is the application of medical information (or knowledge) to particular cases. By "medical information" I mean the body of rules, beliefs, concepts, and theories that forms the cognitive content of medicine. Medical information has no more a moral aspect to it than does, say, the body of knowledge associated with physics.

Since medical practice is committed to promoting the health of the patient, actions by a physician that violate this principle are *wrong* or *bad*. Of course, a physician who has observed this principle and yet failed in the attempt to secure the patient's health is not necessarily to be blamed. Acting in accordance with the principle and trying to achieve the aim of practice is all that is required to make the actions medically legitimate. In this sense, the physician has fulfilled an obligation and acted morally.

Similar principles that impose an obligation are found in end-directed human relationships other than medical practice. A business manager for an actor, for example, is committed to the principle of safeguarding and advancing the actor's financial interests. This is an obligation inherent in the role of business manager. If the manager is careless with the money turned over to him or if he uses it for his own purposes, then he has violated the principle. Thus, his actions are wrong or bad in a way exactly analogous to the actions of a physician who has sought some end other than the health of his patient.

Neither the business manger nor (as I mentioned above) the physician need be successful in achieving their aims in order for their actions to be legitimate. The manager may lose his client's money and the physician may fail to save his patient's life. Yet so long as each has acted to promote the end inherent in their respective relationships, neither can be criticized for violating the obligation imposed by the relationship. (Of course, they may be open to other sorts of criticism – negligence, incompetence, etc.)

A principle of the sort I have discussed is present only when an activity is both directed towards a general practical end and also involves a form of practice in which the end sought becomes manifested in individuals. I have called such a principle "moral" because it enjoins an obligation on one who engages in the practice. In fact, however, this is not an adequate justification for considering the obligation moral. The obligation results from a role relationship – whoever accepts the role is obliged to attempt to secure the end. This would mean that an engineer who undertakes to design an electric chair commits himself to seeing to it that the device kills its victims. Although he acquires an obligation to produce an effective design, it would here be peculiar to call it moral. I suggest, then, that such an obligation is moral only when the general aim of the enterprise is to promote the interest (the good) of the class of clients in some respect. This is not the general aim of engineering, although it may sometimes be a specific aim as part of a design problem. (The designer of an aircraft has the problem of seeing to the safety and comfort of the passengers.) It is, however, an aim of activities such as financial management, clinical psychology, athletic training, teaching, and medicine.

Science, unlike medicine, lacks an inherent moral principle. As I pointed out, such a principle is inherent only in role relationships in which a general practical aim is instantiated by individuals and the enterprise itself aims at promoting the interest of a class of clients. No such principle is found in science, because no such role relationships are found in science. Science is goal-directed, but its end is theoretical, not practical, and science is not committed to promoting any interest at all.

All of this means that there cannot be instances of scientific practice of the appropriate sort. The relationship between a scientist and a subject involves attempting to *learn* something for its *own* sake, not attempting to *do* something for the *subjects'* sake. The case is quite otherwise in medicine, and it is through medicine's aim and the principle implicit in the physician-patient relationship that medicine acquires a moral character not found in science.

3 Medicine and reduction

As we noticed earlier, the close relationship between medicine and the biological sciences encourages the view that medicine itself, either actually or potentially, is a science. What is more, this relationship makes it reasonable to ask whether it is not possible for medicine to be *reduced to* biology.

What would such a reduction involve? The continuing discussion of the possibility of reducing biology itself to the physical sciences is helpful in answering this question. Generally speaking, one science or discipline is reduced to another when the fundamental theories of the first can be derived from theories of the second. The exact requirements for accomplishing such a reduction are still a matter of controversy, but at least two of the conditions stipulated (in different versions) by Schaffner (1967) and Nagel (1961, pp. 433–435) are generally accepted as necessary:

(1) The basic terms (concepts) of the reduced theory must be defined in terms of the reducing theory. The terms need be equivalent only in reference, not in meaning.

(2) The basic principles of the reduced theory must be deducible from the principles of the reducing theory, supplemented by the definitions.

(Schaffner's statement of these two conditions, in addition to his other conditions, takes into account the fact that the principles actually deduced may constitute a 'corrected' version of the original theory. Furthermore, the reducing theory must make it possible to explain why the original version of the reduced theory was imprecise, restricted, or otherwise faulty.)

I want to suggest that there are no a priori reasons for supposing that the cognitive content of medicine – its concepts, laws, and theories – cannot in principle satisfy such conditions. At the same time, however, I want to argue that the features of medicine I called attention to in the preceding section eliminate the possibility of reducing medicine to biology or any other science.

My basic claim is that medicine as an enterprise cannot be reduced, but that it may well be possible for medicine's fundamental concepts to be replaced by biological ones. Thus I reject those attempts to establish the autonomy of medicine by claiming that medicine employs special kinds of concepts.

It is frequently asserted, for example, that the concept of disease is both fundamental to medicine and impossible to explicate in biological terms. To identify some condition as a disease is to invoke implicitly social norms and values, and such value judgments cannot be replaced by biological descriptions.[8]

As an argument against reduction, this view depends on demonstrating that disease (or health) is a normative or social concept. I believe this notion is mistaken. To show that it is would require developing a non-normative account of disease in terms of biological functioning, and I cannot undertake that here.[9] In terms of a non-normative account (disease realism), it may be true as a matter of fact that different societies may identify different conditions as diseases or that what is considered to be a disease at one time (masturbation, for example) may not be considered to be such at a later time. Yet such facts merely call attention to the possibility of error. Physicians in the nineteenth century were just *wrong* to think masturbation was a disease. Thus if disease normativism is wrong and disease realism is correct, then the concept of disease is no barrier to reducing the content of medicine to biology.

Even without deciding this issue, however, the normative view does not rule out the possibility of reduction. Even if we grant that the concept of disease cannot be replaced by biological concepts, this does not show that it is impossible to explain *diseases* in biological terms

only. We consider the collapse of an apartment building in which people are killed to be a disaster, yet we look to the laws of physics to explain the collapse. It is true that those laws do not explain the collapse *as* a disaster but only as a certain kind of physical event involving masses and forces. We have to look to norms and values to understand 'disaster' as a description.

But the reduction of one discipline to another does not require that the terms of the reducing theory be equivalent in meaning to those of the reduced theory. It is not required that the reducing theory explain the events covered by the reduced theory under the same description. In the physics example, all we need be able to do to explain physical disasters is to assimilate every case of one to appropriate physical descriptions so that the theories of physics will apply.

What is true of physics is also true of medicine. Biology cannot explain why the Japanese once considered *osmidrosis axillae* (armpit odor) a disease (Wing, 1978, pp. 16–17). But it can explain the condition itself, and it can explain why some ten percent of the population display the trait. Similarly, biological explanations of 'real' diseases like pericarditis or various forms of pneumonia are both in principle possible and are avidly sought by medical researchers themselves.

It seems reasonable to believe that, in principle, every somatic condition identified as a disease can be provided with an appropriate biological description and explained by means of a biological theory. If this is so, then the only autonomy guaranteed to medicine by the normative disease concept is the power of deciding what condition is to be considered a disease. Thus what initially appears to be a strong a priori barrier to the *explanation* of medical (disease) phenomena by biological laws is no more than an insistence that we recognize the fact that the social use of the term 'disease' is governed by norms and values. In fact, the normative concept of disease does not at all rule out the possibility of a reduction that satisfies the conditions mentioned above.

There seem to be no compelling a priori reasons standing in the way of reducing the cognitive content of medicine to biology. Medicine seeks its own aim of securing health through the use and development of scientific theories. Its cognitive content includes such theories, but it also includes empirical generalizations that serve as the basis for diagnostic or therapeutic rules. Often such generalizations are imprecise and unreliable and are employed only for the lack of anything better.

From the standpoint of medicine's practical aim, a precise and reliable scientific theory is preferable. Consequently, to the extent that the reduction of medical concepts and generalizations promises to provide a more effective means of preventing and treating diseases, then reduction may be one of the legitimate goals of medical research. Indeed, there seems to be much of medical value to be gained from a successful reduction. For example, we now know why the empirical rules followed by Lister were successful in reducing the incidence of postoperative infection. Similarly, an account in terms of a biological theory of the processes and mechanisms involved in narcolepsy might well be able to explain why the usual empirical therapies are sometimes successful and sometimes not.

There is some reason for saying that, for scientific medicine, the explanation of all normal functioning, disease processes, and therapeutic procedures in terms of biological theories is a goal dictated by attempting to secure the aim of medicine in the most effective way. Thus the reduction of the concepts, generalizations, and empirical rules of medicine might be sought in service of that aim.

Yet a reduction of the cognitive content of medicine would in no way entail the loss of medicine's autonomy. The barrier that stands in the way of reducing medicine to biology has nothing to do with medicine's use of special concepts, methods, or kinds of generalizations. Rather, it has everything to do with the fact that medicine is an *enterprise* as well as a discipline. Even if it is possible to reduce the information content of medicine to biological theories, medicine will still retain an independent status.

The enterprise of medicine, as I indicated earlier, is defined by its commitment to the aim of preventing and treating disease by whatever means is found effective and by its implicit commitment to the welfare of people, either collectively or as individuals. Thus, although medicine has a cognitive content, it is not identical with its content.

Medicine as an enterprise is a social activity, and in this it does not differ from science. But since the external and internal aim of medicine is the same, unlike science, medicine is an inherently social activity. That is, being a social activity is basic to the character of medicine. The effective pursuit of its aim requires that medicine be deeply involved with science, even to the extent of engaging in research that is, in itself, indistinguishable from research undertaken for the sake of a purely scientific aim. Yet the way in which medicine pursues its aim does not alter the fact that its aim, its criteria of success, and its moral commitment are factors that distinguish it from science and establish it as an enterprise.

Because medicine is not identical with its cognitive content, it is possible to acknowledge the possibility of reducing that content to biology, and yet insist that it is inappropriate to talk about reducing *medicine* to biology. It is inappropriate in the same way it would be to talk about reducing hopscotch to physics. As a game, hopscotch must be understood in terms of rules and aims. The pattern of activity displayed in playing the game gains its significance from a network of social conventions.

The 'game' of medicine is likewise understandable only in terms of the basic commitments that regulate its activities. The cognitive contents are acquired and used for the sake of achieving the aim of preventing and treating disease, and it is this aim, realized in the individual, that governs medical practice. Thus, the reason that it is impossible to reduce medicine as a whole to biology or any other science is that medicine as an activity, as an enterprise, is an inappropriate subject for reduction, although its content is not.

4 Summary and conclusions

I began by showing that an argument to the effect that medicine (or part of it) is already a science is unsuccessful in its own terms. Next, I argued that medicine and science differ both in their aims and their criteria for success: the aim of medicine is to promote health through the prevention and treatment of disease, while the aim of science is to acquire knowledge; medicine judges its cognitive formulations by their practical results in promoting health, while science evaluates its theories by the criterion of truth. I then demonstrated that medicine (as medical practice) has a moral aspect that is not present in science. Finally, I tried to show that while the content of medicine may be reducible to biology, medicine itself cannot be.

Have I shown that medicine and science are fundamentally different? I believe I have. I have made it clear how medicine can be *scientific* (in its methods, use of scientific concepts and theories, etc.) and why in its pursuit of its goal medicine engages in scientific research. I have, that is, shown the legitimate place of those aspects of medicine that incline some to identify medicine with science either actually or potentially. Without denying these aspects, I have demonstrated that medicine possesses features that distinguish it from science in such basic ways as

to make it unreasonable to include medicine among the sciences. I have shown why it is that medicine cannot remain medicine and yet become a science.

Perhaps, in a sense, such detailed considerations were not required to establish my claim. I might merely have appealed to history or to imagination. After all, medicine existed before science, and it is easy to imagine it existing without science. Yet such appeals are not wholly persuasive, for contemporary medicine has such a close relationship with science that it is necessary to investigate whether medicine has not merged with science or might not be expected to do so eventually. To show why such an outcome could not happen required a close examination of the basic character of medicine.

By way of a conclusion, I wish to indicate briefly four results that I believe emerge from our examination of some of the fundamental differences between medicine and science.

(1) I think it is important to note that the attempt to identify medicine with science poses a potential danger by obscuring the basic character of medicine. The claim of identity can be made plausible only by focusing on medicine's scientific components and commitments and downgrading (á la Forstram) medicine's basic aim into a secondary "social function." Thus, by implication, physicians and researchers are encouraged to regard the treatment and prevention of disease as less significant than the acquisition of knowledge.

Whether or not this is true from a broad philosophical point of view, it is surely wrong from the *medical* point of view. To formulate policies and to conduct practice on the assumption that medicine is just a special kind of science would be certain to have undesirable social consequences. Indeed, a medicine valuing theoretical knowledge of disease over its effective treatment and prevention would cease to be medicine at all. Both its intellectual and social character would be destroyed and society would be left without a scientific medicine.

(2) We should recognize that medicine must continue to use empirical rules in conducting its business. It is not reasonable to expect or demand that all therapeutic or diagnostic rules satisfy the same standards of corroboration required in the sciences before serving as a basis for intervention. Certainly some standards must be met, and it is clearly an ideal of scientific medicine to ground its rules in an understanding of fundamental biological processes. In principle, it may well be possible to achieve this ideal, but because medicine is a practical enterprise operating under the pressure of time and immediate need, we must recognize that it frequently must base decisions about intervention on purely empirical probabilistic generalizations.

(3) Medical researchers must recognize that the different aims of medicine and science make the demand that they justify their work as *medical* a legitimate one. Because promoting health is both the external and internal goal of medicine, it is not enough for an investigator to say merely that her research increases our knowledge. Perhaps an argument can be made to support the claim that all "basic research" furthers the aim of medicine, but in any case, the obligation to make it falls on the medical researcher. Otherwise, one would be open to the charge that one is *only* "doing science" and thus, one would lose one's intellectual (and perhaps financial) base.

(4) We should recognize that the advent of the physician-researcher generates an internal conflict within scientific medicine that requires the use of external ethical principles to resolve. Thus, physicians ought not consider the introduction of ethics into medicine as something forced upon them by society or by law. The activities and aims of the discipline itself generate the need.

More practically, there is a need to work for the design and general acceptance of procedures that guarantee the patient that the researcher *qua* physician is working primarily for his or her welfare, while assuring the physician *qua* researcher the opportunity to further medical knowledge through clinical investigation. The line between clinical inquiry and patient care – between practice and research – may often be a blurred one, but this does not mean that the conditions for requiring consent cannot be made sharp.

As my last word here, I want to stress that the claim that medicine is fundamentally distinct from science neither implies nor supports the conclusion that medicine is in some qualitative way inferior to science. In particular, I am not arguing that medicine is just a crude or imperfect form of science. There is good reason to believe that medicine will continue to become more scientific, operate more on the basis of scientific knowledge and methods, and thereby become more effective. Yet medicine will continue to be a distinct enterprise, pursuing its own aim, measuring success in its own terms, and regulating its practice by its own principle of obligation. To insist that medicine can never become a science is only to insist on medicine's continuing uniqueness.

Notes

1   For an argument similar to Forstram's, see McWhinney (1978).

2   See Creson (1978, p. 257) for a similar, though not identical, objection.

3   For a fuller discussion of internal and external 'norms' in medicine, see Gorovitz and MacIntyre (1976, pp. 52–55).

4   For a discussion of the relation between pure and applied science, see Bunge (1974, pp. 28–30).

5   For details, see Singer and Underwood (1962, pp. 554–555).

6 We might imagine a situation in which honesty was not recognized and a scientist could trust only his own results. Perhaps some kind of science would still be possible, but it would no longer resemble science as we know it now.

7 Because science requires honesty, it might be said that it too has a moral aspect. I have no objection to this, but what I shall show is that medical practice involves an obligation to individuals.

8 Specific versions of this thesis are defended by, among others, Engelhardt (1975, 1976), King (1954), and Wing (1978).

9 For a criticism of the normative view and a defense of disease judgments as value-neutral, see Boorse (1977). In my opinion, Boorse's analysis of biological function, which is essential to his analysis, is mistaken. For another line of criticism, see Bunzl (1980).

References

Baldessarini, R.J.: 1977, *Chemotherapy in Psychiatry*, Harvard University Press, Cambridge, pp. 12–56.

Bernard, C.: 1957, *An Introduction to the Study of Experimental Medicine*, Dover Publications, New York.

Boorse, C.: 1977, 'Health as a theoretical concept', *Philosophy of Science* **44**, 542–573.

Bunge, M.: 1974, 'Towards a philosophy of technology', in A.C. Michalos (ed.), *Philosophical Problems of Science and Technology*, Allyn and Bacon, Boston, pp. 28–46.

Bunzl, M.: 1980, 'Comment on "Health as a theoretical concept"', *Philosophy of Science* **47**, 116–118.

Creson, D.L.: 1978, 'Why science: A rejoinder', *Journal of Medicine and Philosophy* **3**, 256–261.

Engelhardt, H.T., Jr.: 1975, 'The concepts of health and disease', in H.T. Engelhardt and S.F. Spicker (eds), *Evaluation and Explanation in the Biomedical Sciences*, D. Reidel, Dordrecht, pp. 125–141.

Engelhardt, H.T., Jr.: 1976, 'Ideology and etiology', *Journal of Medicine and Philosophy* **1**, 256–268.

Forstram, L.A.: 1977, 'The scientific autonomy of clinical medicine', *Journal of Medicine and Philosophy* **2**, 8–19.

Gorovitz, S. and A. MacIntyre: 1976, 'Towards a theory of medical fallibility', *Journal of Medicine and Philosophy* **1**, 51–71.

King, L.S.: 1954, 'What is disease?', *Philosophy of Science* **21**, 193–202.

Kuhn, T.S.: 1970, *The Structure of Scientific Revolutions* (2nd ed.), University of Chicago Press, Chicago.

Laudan, L.: 1977, *Progress and Its Problems*, University of California Press, Los Angeles.

McWhinney, E.R.: 1978, 'Family medicine as a science', *Journal of Family Practice* **7**, 53–58.

Nagel, E.: 1961, *The Structure of Science*, Harcourt, Brace, and World, New York, pp. 429–446.

Schaffner, K.F.: 1967, 'Approaches to reduction', *Philosophy of Science* **34**, 137–147.

Singer, C. and E.A. Underwood: 1962, *A Short History of Medicine* (2nd ed.), Oxford University Press, New York.

Wing, J.K.: 1978, *Reasoning About Madness*, Oxford University Press, New York.

Chapter 26

Kenneth F. Schaffner

PHILOSOPHY OF MEDICINE

Philosophical analyses of medicine have a long pedigree (see Engelhardt 1986; Pellegrino 1976, 1986; Pellegrino and Thomasma 1981, Chapter 1). Nonetheless, only in the last dozen years or so have professional philosophers and philosophically sensitive physicians paid any sustained attention to a philosophy of medicine. In part this renewed interest can be traced to the burgeoning impact of medical ethics – a branch of philosophy of medicine broadly conceived. Part of this interest is also dependent on the dynamic growth of studies in the philosophy of biology which furnish insights into the sciences that are traditionally viewed as lying at the core of medicine.

In this chapter we will examine two related topics in the philosophy of medicine: the scientific status of medicine and the nature of reduction in the biomedical sciences. A satisfactory answer to the question whether medicine is a science will ultimately depend heavily on the account given of the nature of reduction, whence the rationale for the inclusion of this topic.

1 Is medicine a science?

We begin by considering a general question: "What is the nature of 'medicine'?" and in particular focus on the more specific question: "Is medicine a *science?*" By attempting to answer these questions we have an opportunity to examine both the reach of science as well as its limitations.

It should be clear to all without reciting any extensive number of examples that medicine is *at least in part grounded on* the biomedical sciences. The development of the germ theory of disease, the extent to which micro- and molecular biological techniques are used in the development of new antibiotics, and the manner in which biochemical tests are employed in diagnosis are testimony to the scientific nature of modern medicine. Moreover, significant progress has been made in rationalizing medical decision making through the application of Bayesian probabilistic reasoning techniques to clinical problems (Zarin and Pauker 1984), as well as other important advances in computerizing medical diagnosis (Miller, Pople, and Myers 1982; Schaffner 1985). However, medicine is not simply biology, as every patient will recognize. The doctor-patient interaction is a human interchange, and as Engelhardt has noted:

One cannot effectively treat patients without attending to their ideas and values. One cannot humanely treat patients without recognizing that their lives are realized within a geography of cultural expectations, of ideas and images of how to live, be ill, suffer, and die. (1986, 3)

The great nineteenth-century pathologist Virchow (1849) observed that medicine may best be conceived of as a *social science*, in part because of these types of complexities. But other philosophers have argued in recent years that medicine is not – in its essence – a science at all. A review of some of these arguments and an elaboration of what we have come to learn about the intricate nature of medicine by discussing a brief example of a particular patient problem will help us understand the character of the subject better.

2 Problems with characterizing medicine as a science

Continuous debates have been in the pages of *The Journal of Medicine and Philosophy* and in its sister publication *Theoretical Medicine* about the nature of medicine as a science. In addition a vigorous debate has taken place within one medical discipline, psychiatry, about the scientific status of that subject (see Grünbaum 1984). In his article, Munson (1981) challenged Forstrom's (1977) earlier claim that medicine (in general) was a science by proposing we examine universally recognized sciences such as physics, chemistry and biology along three fundamental dimensions and compare medicine with these paradigm sciences. Munson suggested that we consider the fundamental *aims*, *criteria of success*, and *principles regulating the disciplines* of recognized sciences and medicine. In each of these three categories Munson claimed that significant differences exist between the recognized sciences and medicine, and then, on the basis of an additional further argument based on construing medicine "as an *enterprise*" (1981,

183), concluded that "medicine is not, and cannot be, a science" (1981, 189). Let us examine Munson's three rubrics for comparing medicine and other sciences.

The first important dimension along which science and medicine are contrasted is the *aim* of the discipline. For Munson, the basic (internal) aim of science is "*the acquisition of knowledge and understanding of the world and the things that are in it.*" (1981, 190) For medicine, however, the basic aim is "*to promote the health of people through the prevention or treatment of disease*" (1981, 191). Acquiring knowledge and understanding of diseases is subservient to disease prevention and treatment. The second rubric in Munson's comparison involves the *criteria of success* of the disciplines. Science is successful when it achieves true (or approximately true) knowledge, whereas medicine is successful when it achieves its aim of preventing or ameliorating disease regardless of the truth of its cognitive content. Finally, Munson compares the basic moral commitments of science and medicine. Science must give *honest* reports of observations or experimental results; without such honesty science as a cooperative endeavor cannot succeed. Medicine in contrast is primarily committed to "promoting the health of any individual accepted as patient" (1981, 196). This is not to assert that medicine as an enterprise can encourage dishonesty, nor that physicians ought not be honest with their patients, but only that health promotion for the physician's individual patient is primary and essential.

These are useful and largely correct points made by Munson and they suggest at the minimum that important *differences* exist between the traditional sciences and medicine. By appealing to one further argument, however, Munson believes that he can demonstrate that medicine cannot *ever* be *reduced to* – in the sense of replaced by – the biomedical sciences, which are so heavily employed by medicine. This additional argument is important since, as we see in the later parts of this chapter, *apparent differences* between sciences cannot be taken as compelling evidence that a reduction cannot be accomplished between those sciences. Thus the differences Munson specifies are but the first of two steps he must take to demonstrate that medicine cannot be a science.

Munson appears to accept something like the general reduction model to be described in Section 6 as prescribing the requirements for achieving a reduction. These requirements, which are considered in more detail later, essentially demand that we define all the terms in medicine in vocabulary that belongs to the biomedical *sciences*, and that we then be able to derive medicine as a special case from the biomedical sciences. On the basis of that type of account, however, Munson contends that medicine cannot be reduced to its conceptual content – the biomedical sciences – because it is an *enterprise* – an inherently social activity – as well as a discipline Though the *content* of medicine might someday be reducible to biology, it would be "inappropriate to talk about reducing *medicine* to biology" (1981, 203). Munson provides an analogy:

> It is inappropriate in the same way it would be to talk about reducing hopscotch to physics. As a game, hopscotch must be understood in terms of rules and aims. The pattern of activity displayed in playing the game gains its significance from a network of social conventions. (1981, 203)

Munson's new argument rests on two new claims: (1) medicine has an essentially social character to it; and (2) as such, it is irreducible to biology because networks of social conventions (and interactions) are not reducible to biology. Munson's argument here bears some similarities to a view urged by Pellegrino and

Thomasma, "medicine is not reducible to biology, physics, chemistry, or psychology; . . . we argue that medicine is a form of unique relationship" (1981, xiv). This relationship – if we understand it correctly – is that between the healer and the patient in which the patient participates as subject and object simultaneously. In the next section we examine the nature of the social aspect of medicine, and in the conclusion re-examine Munson's second claim, relating it to the material in the earlier sections of this chapter.

3 Medicine and a biopsychosocial model

As noted in Section 1, it was Virchow who first suggested that medicine should be conceived of as a "social science." Other scholars have proposed similar ideas, but the writer who has probably been the most articulate proponent of significance of the social – and psychological – components of medicine has been George Engel. Engel has propounded the notion of a "biopsychosocial" model for medicine to be contrasted with a more reductionistic "biomedical" model (1977, 198).

In his writings, Engel suggests that the striking successes of twentieth-century medicine such as the discovery of insulin and the antibiotics, as well as new breakthroughs in molecular biology, can result in tunnel vision on the part of the physician, nurse, or other health-care professional. Such successes incline healers toward a biomedical vision of medicine, in which the patient is "nothing but" a bag of chemicals. For such a doctor or nurse, sensitivity to factors which can have marked influences on illness, such as psychosocial stresses, becomes weakened. Such healers may overlook important predisposing causes of disease, or may fail to incorporate significant psychosocial dimensions into their healing art.

Engel (1981) recounts the story of a patient suffering from a heart attack who is being examined in a hospital's emergency room. The patient has an inexperienced intern who attempts to take a sample of his arterial blood as part of a standard diagnostic workup for presumptive heart attack patients. This arterial puncture is a simple procedure which, however, can become painful if it is not performed deftly. Repeated unsuccessful attempts generate pain and anxiety which result in a second heart attack for the patient. Reassuring discussion with the patient, and a request for assistance with the arterial puncture, could have achieved a significantly better outcome for this patient. Figure 26.1 is taken from Engel's essay and graphically depicts

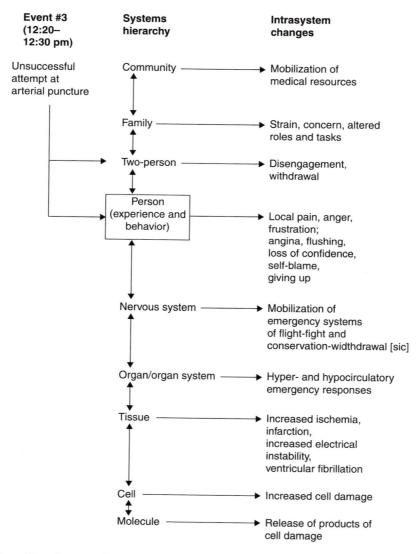

Figure 26.1 *Event #3 in the course of patient's diagnosis of a myocardial infarction or heart attack, depicting the interlevel causal interactions. See text for details. From Engel 1981, with permission.*

the hierarchy of perspectives that are found in medicine and the complex ways that these "levels" interact in one time-slice of a patient's encounter with the health-care system.

Engel's suggestion is that better health care would be delivered by healers mindful of the psychosocial as well as the biological dimensions of illness. The appreciation of the complexity of a human being's reality and a joint analysis of the interactions among various levels of causation, from the molecular through the organ-level to the intellectual, emotional, familial, and economic, will both permit a better understanding of how illness arises, as well as provide a richer armamentarium for the physician. This richer, more complex perspective is what Engel terms the *biopsychosocial model of medicine*. Two fundamental questions to be addressed are whether such suggestions (1) are sound; and (2) count against the reducibility of medicine to biology.

Engel's proposals that medicine needs to be conceptualized as a complex multilevel science receive support from recent studies examining the interactions of the neurological and immunological systems. Thus Engel's suggestions can be viewed as sound, and also can envisage an important role for the social dimension of medicine at the core of its fundamental aim as a healing practice. We are still left, however, with the question whether the acceptance of these "levels" as working heuristics in a sense – and this seems to be all that we are committed to thus far – is any argument against in-principle reducibility. In an important sense, we may stop temporarily at many levels in the course of bringing about a reduction; we see how this works in the account of Kandel and his associates' research discussed in later sections. Another example to be examined explores the interaction between neurological and immunological systems – and the endocrine system as well – which may assist us in understanding how a

"biopsychosocial" model may be implemented in the framework of a particular disease, such as breast cancer, and whether including such concepts as "emotion" into our medical ontology counts for or against the prospect of reduction.

4 Reduction in the biomedical sciences

Reduction for biomedical scientists often suggests that biological entities are "nothing but" aggregates of physicochemical entities. This can be termed "ontological reductionism." A closely related but more *methodological* position is that sound scientific generalizations are available only at the level of physics and chemistry. The contrary of the first *ontological* position is antireductionism, which often is further distinguished into a strong *vitalistic* position and a second weaker *emergentist* view. From the perspective of ontological antireductionism, the biomedical sciences are autonomous. The contrary of the second *methodological* reductionist position is a kind of methodological emergentism. Here too we can distinguish a strong position taken by such distinguished biologists as Simpson (1964) and Weiss (1969), and a weaker form that attributes the necessity to work with higher level entities to current gaps in our knowledge (Grobstein 1965).

The vitalistic thesis, championed earlier in this century by Hans Driesch, proposed the existence of special forces and irreducible causes which were peculiar to living entities. The vitalist thesis led to no testable hypothesis and is essentially a dead issue in contemporary biology.

Analysis of the emergentist position requires further distinctions. One type of emergentism, *in principle emergentism*, holds that the biomedical sciences cannot, regardless of progress in biology and the physicochemical sciences, be reduced to physics and chemistry. Generally this view is based on what its proponents perceive to be certain general timeless *logical* features of biomedical entities and modes of explanation. A similar thesis has been applied to other areas of experience, and we have seen such a view proposed for "medicine" as an enterprise by Munson (1981) and by Pellegrino and Thomasma (1981).

The weaker sense of "emergentism" refers to the *current* status of the relation between the biomedical and the physicochemical sciences (or between a science of social interactions and the biomedical sciences). It should be clear from the views developed thus far that not all biomedical entities, such as cells, are *at present* reducible to molecular terms and chemical theories. However, some *portions* of the biomedical sciences *are* so reduced. Such reductions, partial though they may be, are important accomplishments in the biomedical sciences and constitute the subject matter of biophysics, biochemistry, and parts of "molecular" biology. *Why* the accomplishments are important, and specifically, what conditions are and should be met in reductions will occupy us in the next few sections.

A major component of reductionism to be elucidated is often referred to as *intertheoretic* reductionism. Much of the discussion by scientists and philosophers has focused on this notion, though the term "intertheoretic" is somewhat misleading. Essentially intertheoretic reduction is the *explanation* of a theory initially formulated in one branch of science by a theory from another discipline. Sometimes this type of reduction is referred to as *branch* reduction when it is felt that emphasis on *theories* is not appropriate (Kemeny and Oppenheim 1956). This is a useful corrective, since what can be captured by general high-level abstract theory in the biomedical sciences is often considerably different from what can be so represented in physics. In the next section, a notion of branch reduction is developed and then discussed in the context of the various senses of reduction and emergence outlined above.

5 Reduction and the human mind

An analysis of a complex concept such as "reduction in the biomedical sciences" will not be fruitful unless it can be shown how – at least in a schematic way – it works in practice in those sciences. This suggests that we will need to look at some substantive area in which reduction has occurred or may be occurring. There are a number of plausible candidates, given molecular biological developments in the past decade, including genetics and the neurosciences. Hull (1974, 1976) and Schaffner (1967, 1969, 1976, 1977) have engaged in a debate over the reducibility of classical genetics to molecular genetics, with important contributions to the discussion by Wimsatt (1976) and Kitcher (1984). Rosenberg (1985) has discussed both sickle-cell anemia and recombinant DNA in the context of reductionism, and has also developed a position on the relation of classical and molecular genetics. Also see Balzer and Dawe (1986) who have recently examined the reduction of classical to molecular genetics.

An examination, however, of the way in which reductionistic explanations of learning behavior are being developed by the neuroscientists is more relevant to the themes in this chapter. The explanation of mental processes through appeals to physical or biological events has traditionally been controversial, and remains so today. What has altered in the past decade are significant advances in the application of molecular biology to the neurosciences. In previous sections, "medicine" has been characterized as possessing important psychological (and social) dimensions, and it has been indicated that to adequately address these issues we needed an account of reduction which would be sufficiently strong to give promise of reducing such psychosocial dimensions to the biomedical sciences. Furthermore, significant advances are occurring in neurobiology which have immediate clinical applications, for example, neural tissue transplants for Parkinson's disease. Accordingly, it may be most fruitful to explore reduction at the leading edges of molecular neurobiology. Thus an account of reduction in the biomedical sciences is developed in the context of neuroscientists' molecular explanations of learning behavior and, ultimately, of the mind.

Questions concerning the substantial nature of the mind are as old as philosophy (and religion). Today the ontological status of "the mental" is coming under renewed scrutiny in the neurobiological sciences. Some of the major contributors to the science have argued that *molecular biological approaches* are superior to alternative means of studying neurobiology. In the Preface to a recent volume on *Molecular Neurobiology in Neurology and Psychiatry* (1987), Eric R. Kandel wrote:

> This volume reflects the impact of molecular biology on neural science and particularly on neurology and psychiatry. These new approaches have accelerated the growth of neurobiology. The resulting increase in knowledge has brought with it two unanticipated consequences that have changed the ways in which clinical researchers and practitioners can now view the findings that came from basic science.
>
> The first consequence is a new unity, a greater coherence, in biology as a whole, as studies move from the level of the cell to that of the molecule.... [T]he second consequence of our increased knowledge ... [is that a]s science becomes more powerful it becomes more ambitious – it becomes bolder.... Many molecular biologists now frankly admit that the ultimate object of their

interest is not simply the system with which they work. It is not simply *lambda*, the T4 phages, or *E. coli*. It is not even *C. elegans*, *Drosophila*, or *Aplysia*. It is human biology. And some biological researchers are so bold as to see their ultimate interest as the function of the human mind. (1987, vii–viii)

In Section 7 we look closely at the nature of some of these molecular biological explanations of neurobiological phenomena, and most specifically at learning and at memory. In what follows we will present a brief historical sketch of some of the main features of philosophical analyses of reduction, and then go on both to criticize those accounts and to sketch the outlines of what we take to be a more adequate perspective on reduction. We believe that the account to be a proposed has direct application in molecular neurobiology. In point of fact we will elaborate a detailed example from molecular neurobiology, but we also think that its field of application is broader. We think the analysis to be presented in the present chapter covers not only neurobiology but also the biological, psychological, and social sciences, and that it can be applied and tested in a wide domain.

6 A brief historical account of reduction in philosophy of science

As mentioned in Section 4, much of the debate about reductionism has focused on the relationship between *theories*. The traditional analysis of a scientific theory in philosophy of science understood a theory to be a collection of a few sentences, typically each of them scientific laws. These sentences could be taken to be the basic *axioms* or leading premises from which theorems could be proven, much as in any high school geometry text. These axioms had both logical terms, such as the equal sign, and non-logical terms, such as "photon" and "gene." Philosophers of science also believed that casting those sentences into a logical formalism would aid in the clarification of the nature of a scientific theory, especially since the explicit relations between the axioms and the theorems – other "derived" scientific laws or descriptions of experimental results – could be examined with formal precision. More recently philosophers of science have thought of scientific theories in different ways, such as viewing them as "predicates" (see Chapter 3 [sic]). This is often referred to as a "semantic" conception of scientific theories, as distinguished from the older "syntactic" conception. Such newer views are compatible with the older notion of a theory as a small collection of sentences. In fact, simply conjoining the collection of sentences (the axioms) together is sufficient to define the predicate of interest. (See Chapter 3 [sic].)

The *locus classicus* of philosophical analyses of theory reduction can be found in the work of Ernest Nagel (1949, 1961). Nagel envisaged reduction as a relation between *theories* in science, and also assimilated it to a generalization of the classical Hempel deductive-nomological model of explanation (see Chapter 1 [sic]). A theory in biology, say, was reducible to a theory in chemistry, if and only if (1) all the nonlogical terms appearing in the biological theory were *connectable* with those in the chemical theory, for example, gene had to be connected with DNA; and (2) with the aid of these *connectability assumptions*, the biological theory could be *derived* from the chemical theory (with the additional aid of general logical principles). Later connectability came to be best seen as representing a kind of "synthetic identity," for example, gene = DNA sequence, (Schaffner 1967, Sklar 1967, Causey 1977).

An interesting alternative and more semantic approach to reduction was developed by Suppes and Adams:

Many of the problems formulated in connection with the question of reducing one science to another may be formulated as a series of problems using the notion of a representation theorem for the models of a theory. For instance, the thesis that psychology may be reduced to physiology would be for many people appropriately established if one could show that for any model of a psychological theory it was possible to construct an isomorphic model within physiological theory. (Suppes, 1967, 59)

Another example of this type of reduction is given by Suppes when he states:

To show in a sharp sense that thermodynamics may be reduced to statistical mechanics, we would need to axiomatize both disciplines by defining appropriate set-theoretical predicates, and then show that given any model T of thermodynamics we may find a model of statistical mechanics on the basis of which we may construct a model isomorphic to T. (1957, 271)

This model-theoretic approach allows a somewhat complementary but essentially equivalent way to approach the issues, to be discussed.

In contrast to Suppes and Adams (and thus also later work by Sneed, Stegmuller, and their followers), most writers working on reduction dealt with the more syntactic requirements of Nagel: connectability and derivability. The *derivability* condition (as well as the connectability requirement) was strongly attacked in influential criticisms by Popper (1957a), Feyerabend (1962), and Kuhn (1962). Feyerabend, citing Watkins, suggested but did not agree with the proposal that perhaps some form of reduction could be preserved by allowing approximation (1962, 93). Schaffner (1967) elaborated a modified reduction model designed to preserve the strengths of the Nagel account but flexible enough to accommodate the criticisms of Popper, Feyerabend, and Kuhn. The model, termed the *general reduction model*, has been criticized (Hull 1974, Wimsatt 1976, Hooker 1981), defended (Schaffner 1976, Ruse 1976), further developed (Wimsatt 1976, Schaffner 1977, Hooker 1981), and recriticized recently by Kitcher (1984) and Rosenberg (1985). A somewhat similar approach to reduction was developed by Paul Churchland in several books and papers (1979, 1981, 1984).

An approach close to that of the original general reduction model has quite recently been applied in the area of neurobiology by Patricia Churchland (1986). She offers the following concise statements of that general model:

Within the new, reducing theory T_B, construct an *analogue* T_R^* of the laws, etc., of the theory that is to be reduced, T_R. The analogue T_R^* can then be logically deduced from the reducing theory T_B plus sentences specifying special conditions (e.g., frictionless surfaces, perfect elasticity). Generally, the analogue will be constructed with a view to mapping expressions of the old theory onto expressions of the new theory, laws of the old theory onto sentences (but not necessarily *laws*) of the new. Under these conditions the old theory reduces to the new. When reduction is successfully achieved, the new theory will explain the old theory, it will explain why the old theory worked as well as it did, and it will explain much where the old theory was buffaloed. (Ibid., 282–283)

Churchland goes on to apply this notion of reduction to the sciences of psychology (and also what is termed "folk psychology") which are the sciences *to be reduced*, and to the rapidly evolving neurosciences, which are the *reducing sciences*. In so doing she finds she needs to relax

the model even further, to accommodate, for example, cases not of reduction but of replacement, and of partial reduction. She never explicitly reformulates the model to take such modifications into account, however, and it would seem useful, given the importance of such modifications, to say more as to how this might be accomplished. The most general model must also allow for those cases in which the T_R is NOT modifiable into a T_R^* but rather is REPLACED by T_B or a T_B^*. Though not historically accurate, a reduction of phlogiston theory by a combination of Lavoisier's oxidation theory and Dalton's atomic theory would be such a replacement. Replacement of a demonic theory of disease with a germ theory of disease, but with retention, say of the detailed observations of the natural history of the diseases and perhaps preexisting syndrome clusters as well, is another example of reduction with replacement (see Schaffner 1977).

In the simplest replacement situation we have the essential *experimental arena* of the previous T_R (but not the theoretical premises) directly connected via new correspondence rules associated with T_B (or T_B^*) to the reducing theory. A correspondence rule is here understood as a telescoped causal sequence linking (relatively) theoretical processes to (relatively) observable ones. (A detailed account of this interpretation of correspondence rules can be found in Schaffner 1969.) Several of these rules would probably suffice, then, to allow for further explanation of the experimental results of T_R's subject area by a T_B (or T_B^*). In the more complex but realistic case, we also want to allow for partial reduction, that is, the possibility of a partially adequate component of T_R being maintained together with the entire *domain* (or even only part of the domain) of T_R'. (The sense given to "domain" here is that of Shapere (1974): a domain is a complex of experimental results which either are accounted for by T_R and/or *should be* accounted for by T_R when (and if) T_R is or becomes completely and adequately developed.) Thus, the possibility arises of a continuum of reduction relations in which T_B (or T_B^*) can participate. (In those cases where only one of T_B or T_B^* is the reducing theory, we use the expression $T_B^{(*)}$.) To allow for such a continuum, T_R must be construed not only as a completely integral theory but also as a theory dissociable into weaker versions of the theory, and also associated with an experimental subject area(s) or domain(s). Interestingly the Suppes-Adams model might lend some additional structure through the use of model-theoretic terminology to this notion of partial reduction, though it will need some (strengthened) modifications to that original schema, to be discussed.

We may conceive of "weaker versions of the theory" either (1) as those classes of models of the theory in which not all the assumptions of the theory are satisfied or (2) as a restricted subclass of all the models of the reduced and/or reducing theory. The first weakening represents a restriction of assumption, the second a restriction in the applied scope of the theory. As an example of the first type of restriction consider a set of models in which the first and second laws of Newton are satisfied but which is silent or which deny the third law. The application to reduction is straightforward, since in point of fact there are models of optical and electromagnetic theories in the nineteenth century which satisfy Newton's first two laws but which violate the third (see Schaffner 1972, 65). As an example of the second type of restriction, consider a restriction of scope of statistical mechanical models that eliminates (for the purpose of achieving the reduction) those peculiar systems (of measure zero) in which entropy does not increase, for example, a collection of particles advancing in a straight line. This second type of restriction is unfortunately *ad hoc*.

Under some reasonable assumptions, the general reduction model introduced in the quotation from Patricia Churchland above could be modified into the general reduction-replacement model, characterized by the conditions given in the following text box. (Text in boxes as well as all the figures with the exception of Figure 26.1 are optional material.) These conditions are of necessity formulated in somewhat technical language, but the concepts involved should be reasonably clear from the discussion above. (The italicized *ors* should be taken in the weak, inclusive sense of "or," i.e., and/or.)

General reduction-replacement (GRR) model

T_B – the reducing theory/model
T_B^* – the "corrected" reducing theory/ model
T_R – the original reduced theory/model
T_R^* – the "corrected" reduced theory/ model

Reduction in the most general sense occurs if and only if:

(1)(a) All primitive terms of T_R^* are associated with one or more of the terms of $T_B^{(*)}$ such that:

(i) T_R^* (entities) = function $(T_B^{(*)}$ (entities))
(ii) T_R^* (predicates) = function $(T_B^{(*)}$(predicates))
or
(1)(b) The domain of T_R^* be connectable with $T_B^{(*)}$ via new correspondence rules. (Condition of generalized connectability.)

(2)(a) Given fulfillment of condition (1)(a), that T_R^* be derivable from $T_B^{(*)}$ supplemented with (1)(a)(i) and (1)(a)(ii) functions.

or

(2)(b) Given fulfillment of condition (1)(b) the domain of T_R be derivable from $T_B^{(*)}$ supplemented with the new correspondence rules. (Condition of generalized derivability.)

(3)   In case (1)(a) and (2)(a) are met, T_R^* corrects T_R, that is, T_R^* makes more accurate predictions. In case (1)(b) and (2)(b)

are met, it may be the case that $T_B^{(*)}$ makes more accurate predictions in T_R's domain than did T_R.

(4)(a) T_R is explained by $T_B^{(*)}$ in that T_R and T_R^* are strongly analogous, and $T_B^{(*)}$ indicates why T_R worked as well as it did historically.

or

(4)(b) T_R's domain is explained by $T_B^{(*)}$ even when T_R is replaced.

Such a model has as a limiting case what we have previously characterized as the general reduction model, which in turn yields Nagel's model as a limiting case. The use of the weak sense of *or* in conditions (1), (2) and (4) allows the "continuum" ranging from reduction as subsumption to reduction as explanation of the experimental domain of the replaced theory. Though in this latter case we do not have intertheoretic reduction, we do maintain the "branch" reduction previously mentioned. This flexibility of the general reduction-replacement model is particularly useful in connection with discussions concerning current theories that may explain "mental" phenomena.

Though these are useful expansions of the traditional model, we have yet to see a reexamination of one of the basic premises that reduction is best conceived of as a relation between *theories* (though we have allowed a reduction to occur between a theory and the *domain* of a previous theory). Furthermore, nowhere has it yet been seriously questioned whether the notion of *laws* or a collection of laws is the appropriate reductandum and reductans.

Wimsatt (1976) criticized Schaffner's earlier approach to theory reduction (Schaffner 1967,

1969, 1976) for among other things as not taking "mechanisms" as the focus of the reducing science. He seems to have been motivated in part by some of Hull's (1974) observations, and especially by Salmon *et al.*'s (1971) model of explanation. Wimsatt construes one of Salmon *et al.*'s important advances as shifting our attention away from statistical *laws* to statistically relevant *factors* and to underlying *mechanisms* (Wimsatt 1976, 488). Wimsatt's suggestions are very much to the point. The implications of them are considered further below in terms of the causal gloss placed on reductions as explanations.

The question of the relevance of "laws" in reductions in the biomedical sciences in general, and in connection with genetics in particular, was also raised more recently by Philip Kitcher. In his paper on reduction in molecular biology, Kitcher (1984) argued that the Nagel model of reduction and all analogous accounts suffered from several problems. One of those problems was that when one applies such accounts to biology in general, and to genetics in particular, it is hard to find collections of sentences which are the "laws" of the theory to be reduced.

Kitcher notes that finding many laws about genes is difficult, as opposed to the way that one finds various gas laws peppering the gas literature, or the extent to which there are explicit laws in optics, such as Snell's law, Brewster's law, and the like. Why this is the case is complex (see Schaffner 1980 and 1986), but it has interesting implications for reduction and thus requires some discussion. As Kitcher adds, however, two principles or laws have received that term and are extractable from Mendel's work and its rediscoverers. These are the famous laws of segregation and independent assortment. They continue to be cited even in very recent genetics texts such as Watson *et al.* (1987), and a statement of them can be found in Chapter 7 [sic]. These laws were the subject of a searching historical inquiry by Olby (1966).

Olby believes that Mendel's laws are important and, in their appropriate context, still accurate. In answer to the question whether Mendel's law of independent assortment still holds, Olby writes:

> . . . yes; but, like any other scientific law, it holds only under prescribed conditions. Mendel stated most of these conditions, but the need for no linkage, crossing-over, and polyploidy were stated after 1900. (1966, 140)

In his essay, Kitcher gleans a different lesson from the post-1900 discoveries of the limitations of Mendel's law. It is not that one could not save Mendel's law by specifying appropriate restrictions or articulating a suitable approximation and reducing it, but rather that "Mendel's second law, amended or unamended, simply becomes irrelevant to subsequent research in classical genetics" (Kitcher 1984, 342) What seems to concern Kitcher is that Mendel's second law is only interesting and relevant if it is embedded in a cytological perspective which assigns Mendelian genes to chromosomal segments within cells:

> What figures largely in genetics after Morgan is the technique [of using cytology], and this is hardly surprising when we realize that one of the major research problems of classical genetics has been the problem of discovering the distribution of genes *on the same chromosome*, a problem which is beyond the scope of the amended law. (1984, 343)

We think Kitcher is on the right track here but that he has not fully identified the units that are at work in actual scientific reductions. His later elaboration of what he terms "practices" – which is a kind of analogue of a Kuhnian paradigm, a Lakatosian research program, or a Laudanian research tradition (see Chapter 4 [sic]) – though interesting in its own right does not really address the issue of *reduction*, but is more suited to clarify problems in the domain of scientific progress. What we take from Wimsatt's arguments and Kitcher's discussion is that a focus on the notion of a theory as a collection of general laws may be problematic in the biological sciences. Any biomedical generalization needs to be embedded in a broader context of similar and overlapping theories in order to adequately capture the nature of reduction in these sciences. We will see how this works in neurobiology in Section 8.

What is the import of this view for reduction? To assess that, it will be appropriate to turn to an extended example of reduction in neurobiology, after which we can in the light of the philosophical discussion just presented, *generalize* the example and examine what reduction in neurobiology might look like from a philosophical point of view.

7 Short-term and long-term learning in *Aplysia*

In a series of critically important papers in neurobiology, Kandel, Schwartz, and their colleagues have been deciphering the complex neurobiological events underlying the primitive forms of learning which the marine mollusc *Aplysia* exhibits. This invertebrate organism, sometimes called the "sea hare" (see Figure 26.2 for a diagram of its gross structure and note its similarity to a rabbit) exhibits simple reflexes which can be altered by environmental stimuli. The nervous system of *Aplysia* is known to consist of discrete aggregates of neurons called ganglia containing several thousand nerve cells. These nerve cells can be visualized under a microscope, and can have microelectrodes inserted into them for monitoring purposes. Individual sensory as well as motor nerve cells have been identified, and are essentially identical within the species. This has permitted the tracing of the synaptic connections among the nerve cells as well as the identification of the organs which they innervate. "Wiring diagrams" of the nervous system have been constructed from this information.

It has been a surprise to many neuroscientists that such simple organisms as shellfish, crayfish, and fruit flies can be shown to exhibit habituation, sensitization, classical conditioning, and operant conditioning. Both short-term and long-term memory by these organisms can be demonstrated and serve as the basis for model approaches to understanding the molecular basis of such learning and memory. Though it is not expected that a *single universal* mechanism for learning will be found that holds for all organisms including humans, it is felt that basic mechanisms will have certain fundamental relationships to one another, such as reasonably strong analogies.

1. Sensitization and short-term memory. Sensitization is a simple form of nonassociative learning in which an organism such as *Aplysia* learns to respond to a usually noxious stimulus (e.g., a squirt of water from a water pik) with a strengthening of its defensive reflexes against a previously neutral stimulus. A well-known defensive reflex in *Aplysia* known as a *gill-siphon withdrawal reflex* has been studied intensively by Kandel and Schwartz and their colleagues. When the siphon or mantle shelf of *Aplysia* is stimulated by light touch, the siphon, mantle shelf, and gill all contract vigorously and withdraw into the mantle cavity (see Figure 26.2

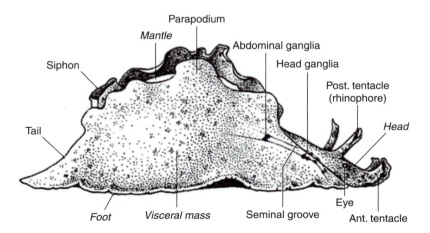

Figure 26.2 *Gross structure of Aplysia californica, also known as the "sea hare." The gill structure is not visible in this view, but is located adjacent to the mantle behind the parapodium and in front of the siphon. From Cellular Basis of Behavior by E. R. Kandel copyright © 1976 by W.H. Freeman and Company. Reprinted with permission.*

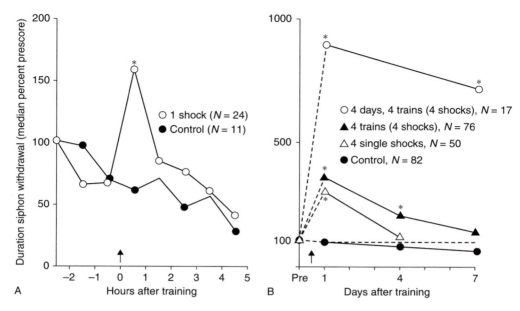

Figure 26.3 A and B: Short-term and long-term sensitization of the withdrawal reflex in Aplysia, after Kandel and others (1987), p. 117, with permission. The arrow indicates a single strong electric shock to the tail or the neck of Aplysia, resulting in the heightened response as shown in the open circles in the graphs. Filled-in circles are controls.

for these structures). Short-term and long-term sensitization of the withdrawal reflex is shown in Figures 26.3A and 26.3B (from Kandel *et al.* 1987).

The nerve pathways that mediate this reflex are known, and a simplified wiring diagram of the gill component of the withdrawal reflex is shown in Figure 26.4. The sensitizing stimulus applied to the sensory receptor organ (in this case the head or the tail) activates facilitatory interneurons that in turn act on the follower cells of the sensory neurons to increase neurotransmitter release. Three types of cells apparently use different neurotransmitters; serotonin (= 5-HT) and two small interrelated peptides known as SCP_A and SCP_B are identified transmitters in two of the cell types. This explains sensitization at the cellular level to some extent, but Kandel and his colleagues have pushed this investigation to a deeper level, that of the molecular mechanisms involved.

For short-term sensitization, all the neurotransmitters have a common mode of action: each activates an enzyme known as adenylate cyclase which increases the amount of cyclic AMP – a substance known as the second messenger – in the sensory neuronal cells. This cyclic AMP then turns on (or turns up) the activity of another enzyme, a protein kinase, which acts to modify a "family of substrate proteins to initiate a broad cellular program for short-term synaptic plasticity" (Kandel *et al.* 1987, 118). The program involves the kinase that phosphorylates (or adds a phosphate group onto) a K^+-channel protein, closing one class of K^+ channels that normally would restore or repolarize the neuron's action potential to the original level. This channel closing increases the excitability of the neuron and also prolongs its action potential, resulting in more Ca^{++} flowing into the terminals, and permitting more neurotransmitter to be released. (There may also be another component to this

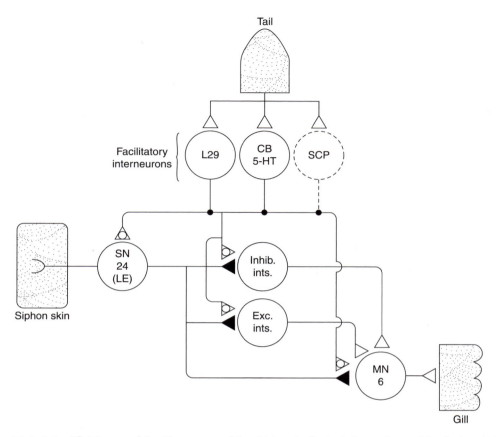

Figure 26.4 *A simplified diagram of the gill component of the withdrawal reflex in Aplysia after Kandel and others (1987),
p. 118, with permission. The 24 mechanoreceptor neurons transmit the information from the siphon skin to the inhibitory and
excitatory interneurons, as well as to the gill motorneurons (MN). Stimulation of the tail or the neck excites some of the
facilitory interneurons increasing the strength of the signal between the sensory neurons and the cells they stimulate. The facilitory
interneurons use three different types of neurotransmitters, two of which have been identified as shown.*

mechanism involving a movement of a C-kinase
to the neuron's membrane where it may enhance
mobilization and sustain release of neurotransmitter.) The diagram in Figure 26.5 should make
this sequence of causally related events clearer.

2. Long-term memory in Aplysia. A similar type
of explanation can be provided for long-term
memory for sensitization, but we will not go
into it in any detail here, save to point out that
the mechanism can best be described as exhibiting important *analogies* with short-term

memory. Suffice it to say that the long-term
mechanism is somewhat more complex, though
it shares similarities with the short-term mechanisms such as the adenylate cyclase – cyclic AMP
– protein kinase cascade. Since protein synthesis
occurs in connection with long-term memories,
additional different mechanisms need to be
invoked, including, almost certainly, new genes
being activated. Figure 26.6 provides us with
one of at least two possible overviews of this
more complex – and at present more speculative
– mechanism.

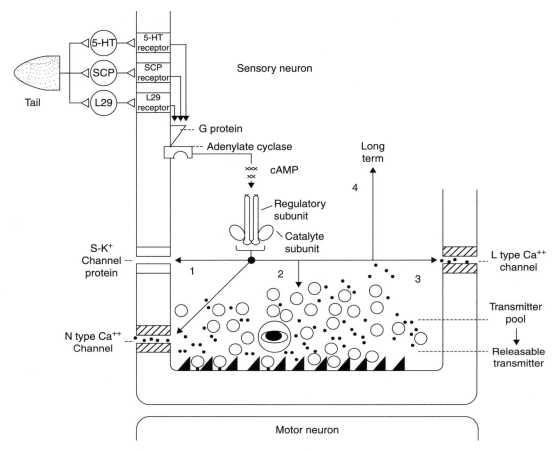

Figure 26.5 *The biochemical mechanism of short-term sensitization in Aplysia updated from Kandel and others (1987), p. 119, with permission.*

Additional Complexity and Parallel Processing in Aplysia. The above description of Kandel and his associates' work focuses on an account which has "had the advantage of allowing a relatively detailed analysis of the cellular and molecular mechanisms underlying one component important for both short- and long-term memory for sensitization" (Frost *et al.* 1988, 298). As these authors note in their more recent analysis, however, "the siphon withdrawal response exhibits a complex choreography with many different components ..." (ibid.).

Further work on both the gill-component and the siphon component of this reflex has suggested that memory for sensitization in *Aplysia* – as well as other organisms – appears to involve "parallel processing," an expression which is "similar" to a "view, called 'parallel distributed processing' [which] has emerged from theoretical studies in artificial intelligence and cognitive psychology and is based on the idea that common computational processes recur at many sites within a network ..." (Frost *et al.* 1988, 297–298).

In this chapter we will not have an opportunity to explore the complex interactions which Kandel and his colleagues have found even in this comparatively simple form of short-term memory. Suffice it to say that they have found it involves at least four circuit sites, each involving a different type of neuronal plasticity, which are shown in Figure 26.7: (1) presynaptic facilitation of the central sensory neuron connections as described earlier in this section; (2) presynaptic inhibition made by L30 onto circuit interneurons including L29 shown; (3) posttetanic potentiation (PTP) of the synapses made by L29, onto the siphon motor neurons, and (4) increases in the tonic firing rate of the LFS motor neurons, leading to neuromuscular facilitation (Frost *et al.* 1988, 299).

Frost *et al.* (1988) anticipated such complexity but were also surprised to find that though the PTP discovered at L29 was a homosynaptic process, the other three components appeared to be coordinately regulated by a common modulatory transmitter, serotonin, and that the common second-messenger system, cyclic AMP, was involved in each (pp. 323–324). Additional modulatory transmitters and other second-messenger systems are explicitly not ruled out, however (see 1988, 324).

In some other recent work, Kandel has in fact reported that tail stimuli in *Aplysia* can lead to transient inhibition (in addition to prominent facilitation). Studies of this inhibitory component indicate that the mechanism is presynaptic inhibition in which the neurotransmitter is the peptide FMRFamide, and that the "second messenger" in this component is not cyclic AMP; rather, inhibition is mediated by the lipoxygenase pathway of arachidonic acid. This "unexpected richness" as

Kandel has characterized the existence of two balancing pathways, constitutes still further evidence for the philosophical views to be discussed below.

8 Implications of this example for reduction in neurobiology

Several points are important to note in the account given in Section 7 which gives a molecular explanation of a behavioral, in a sense "psychological," explanandum. First, the GRR model will be discussed in the light of the specific *Aplysia* example, then in the following section more general issues regarding the GRR model will be considered, ultimately leading to two forms of this account, one (comparatively) *simple*, and one both *partial* and more *complex*.

When we examine the *Aplysia* sensitization exemplar, we do not see anything fully resembling the "laws" we find in physics explanations. The generalizations that exist are of varying generality – for example, protein kinase enzymes act by phosphorylating molecules, and cyclic AMP is a (second) messenger – which we could extract from this account, though they are typically left implicit. Such generalizations are, however, introduced explicitly in introductory chapters in neuroscience texts. In addition, however, there are usually also generalizations of narrow scope, such as the "two balancing pathways" found in Kandel's very recent work. Further, these generalizations are typically borrowed from a very wide-ranging set of models (such as protein synthesis models, biochemical models, and neurotransmitter models). The set in biology is so broad it led Morowitz's (1985) committee to invoke the notion of a many-many modeling in the form of a complex biomedical "matrix." But even when such generalizations, narrow or broad, are made explicit, they need to be supplemented with the details of specific

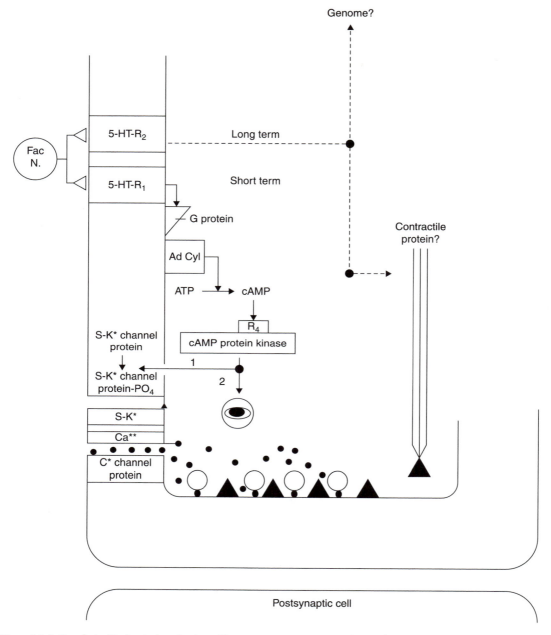

Figure 26.6 *Speculative biochemical mechanism of long-term sensitization in Aplysia after Kandel and others* (1987), p. 126, with permission.

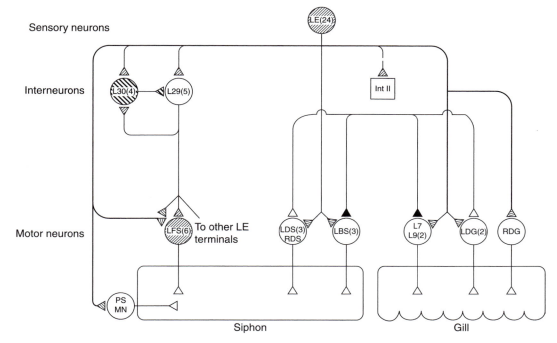

Figure 26.7 *Parallel processing in the gill and siphon withdrawal reflex circuit showing the sites found to be modified by sensitizing stimuli (shaded structures). From Frost, W.N., Clark, G.A. and Kandel. E.R. (1988), "Parallel Processing of Short-term Memory for Sensitization in Aplysia" in* Journal of Neurobiology **19**:297–334. Copyright © 1988 John Wiley & Sons, Inc. Reprinted by permission of John Wiley & Sons, Inc. See textbox 8.2 for details [sic].

connections in the system in order to constitute an explanation, such as the linkage of three-fold L-29 − SCP − 5-HT receptor to the G protein shown in Figure 26.5. Are these connections playing the role of "initial conditions" in traditional philosophy of science? This might not be the case since we do not have a set of powerful generalizations such as Newton's laws of motion to which we can add minor details concerning the force function and the initial position and momentum, and generate explanations across a wide range of domains. The generalizations we can glean from such molecular mechanisms have a variable scope, and typically need to be changed to *analogous* mechanisms as one changes organism or behavior. The brief account given of the relation between short-term and long-term mechanisms for sensitization in *Aplysia* is a case in

point, as is the further complexity found in short-term sensitization in the more recent "parallel processing" account previously given.

What we appear to have are intricate *systems* using both broad and narrow **causal generalizations** which are typically *not* framed in purely biochemical terminology, but which are characteristically *interlevel*. Such systems often *explain* by providing a temporal (and often causal) sequence as part of their models (see Schaffner 1980, 1986, and 1993).

This view of the explaining system as a combination of broad and narrow generalizations that may be framed at a variety of different levels of aggregation suggests that *theory reduction* needs to consider not only general summary statements of entities' behavior which might be said to function at what we call the "γ (for very general) level of theory axiomatization" (e.g., the general definition of learning by sensitization), but also needs to note the significance of the finer-structured elements of a theory – for example, the cytological or what we term the "σ level of specification." (This σ, level of specification represents a means of *realizing* a higher level γ principle.) One can also in addition introduce a more fine-structured level than σ that we call the "δ-level," which can represent molecular level details realizing in turn σ-level mechanisms. We reexamine the implications of these levels of specification for reduction further below in Section 9. The introduction of these levels of specification or realizations (γ, σ, and δ) very roughly track the general process, cell implementation, and molecular implementation levels, but are more intended to capture the level of *generality* than the level of aggregation (though empirically they *may* track both simultaneously.) For the utility of such levels as parts of "extended theories" in accounting for scientific change see Schaffner (1993, Chapter 5).

What we appear to have in the neuroscience example is an instance of a multilevel system employing causal generalizations of both broad and narrow scope. Moreover, as an explanation of a behavioral phenomenon like sensitization is given in molecular terms, one maps the phenomenon into a neural vocabulary. Sensitization becomes not just the phenomenon shown earlier in Figure 26.3, but it is *reinterpreted as* an instance of neuronal excitability and increased transmitter release, that is as *enhanced synaptic transmission*. Thus something like Nagel's condition of connectability is found. Is this explanation also in accord with explanation by derivation from a theory? Again, the answer may be yes, but this time with some important differences. As argued above, we do not have a very general set of sentences (the laws) which can serve as the premises from which we can deduce the conclusion. Rather, what we have is a set of causal sentences of varying degrees of

generality, many of them specific to the system in question. In some distant future all of these causal sentences of narrow scope, such as "This phosphorylation closes one class of K+ channels that normally repolarize the action potential" (Kandel *et al.* 1987, 120–121), may be explainable by general laws of protein chemistry, but it is not the case at present. In part this is because we cannot even fully infer the three-dimensional structure of a protein like the kinase enzyme or the K+ channels mentioned from a complete knowledge of the amino acids which make up the proteins. Fundamental and very general principles will have to await a more developed science than we will have for some time. Thus the explanans, the explaining generalizations in such an account, will be a complex web of interlevel causal generalizations of varying scope, and will typically be expressed in terms of an idealized system of the type shown in Figure 26.5, with some textual elaboration on the nature of the casual sequence leading through the system.

This then is a kind of **partial model reduction with largely implicit generalizations**, often of narrow scope, licensing the temporal sequence of causal propagation of events through the model. It is not unilevel reduction, that is, to biochemistry, but it *is* characteristically what is termed a *molecular biological explanation*. The model effecting the "reduction" is typically interlevel, mixing different levels of aggregation from cell to organ back to molecule, and the reducing model may be further integrated into another model, as the biochemical model is integrated into or seen as a more detailed expansion of the neural circuit model for the gill-siphon reflex. The model or models also may not be robust across this organism or other organisms; it may well have a narrow domain of application in contrast to what we typically encounter in physical theories.

In some recent discussion on the theses developed in the present essay, Wimsatt (personal communication) has raised the question of why not refer to the reduction as being accomplished by "mechanisms" in the sense of Wimsatt (1976), and simply forget about the more syntactic attempt to examine the issues in terms of "generalizations"? Though Wimsatt's suggestion (and his position) are fruitful, "mechanism" should not be taken as an unanalyzed term. Furthermore, it seems that we do (and, as will be argued, *must*) have "generalizations" of varying scope at work in these "molecular biological explanations" which are interlevel and preliminary surrogates for a unilevel reduction, and that it is important to understand the varying scope of the generalizations and how they can be applied "analogically" within and across various biological organisms. This point of view relates closely to the question of "theory structure" in the biomedical sciences, an issue mentioned earlier (also see Morowitz's National Academy of Science Report, Committee on Models for Biomedical Research 1985 and Schaffner 1987). This said, however, there is no reason that the *logic* of the relation between the explanans and the explanandum cannot be *cast* in deductive form. Typically this is not done because it requires more formalization than it is worth, but some fairly complex engineering circuits, such as a full adder shown in Figure 26.8 and described in the following text box, can effectively be represented in the first-order predicate calculus and useful deductions made from the premises, which in the adder example number some 30.

thus consists of 26 objects: 6 components or subcomponents and 20 ports.

The structure and operation of this circuit can be captured in first order predicate logic (FOL), and a theorem prover used with the logical formulas to generate answers to questions about the device's operations. A full treatment is beyond the scope of this chapter (but can be found in Genesereth and Nilsson's (1987, 29–32; 78–84). Here we will give only a few of the axioms used to characterize this circuit in FOL. After the vocabulary is introduced, some 30 logic sentences are needed for the full description.

Vocabulary examples: Xorg(x) means that **x** is an *xor* gate; **I(i,x)** designates the **i**th (**1, 2,** or **3**) input port of device **x**; **Conn (x,y)** means that port **x** is connected to port **y**; **V (x,z)** means that the value of port **x** is **z**; **1** and **0** designate the high and low signals respectively.

Connectivity and behavior of components examples:

. . .

(7) $Conn(I(1,F1), I(1,X1))$

(8) $Conn(I(2,F1), I(2,X1))$

(26) $\forall x \, \forall n \, (Org(x) \land V(I(n,x),1) \Rightarrow V(O(1,x),1))$

. . .

(30) $\forall x \forall y \forall z \, (Conn(x,y) \land V(x,z) \Rightarrow V(y,z))$

(As noted, these assumptions represent only 4 of the 30 needed.)

The full adder shown in Figure 26.8 is an integer processing circuit consisting of five subcomponents called "gates." There are two exclusive "or" (*xor*) gates x_1 and x_2, two "and" gates a_1 and a_2, and an inclusive "or" gate o_1. There are input ports on the left side of the box and output ports on the right as well as ports into and out of the subcomponents. The universe of discourse

The picture of reduction that emerges from any detailed study of molecular biology as it is practiced is not an elegant one. A good overall impression of reduction in practice can be

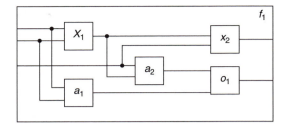

Figure 26.8 A circuit diagram of the full adder described in text box, after © 1987 Morgan Kaufmann publishers. Reprinted with permission from Geneserth and Nilsson, "Logical Foundations of Artificial Intelligence."

obtained if one reads through such a text as Watson *et al.*'s version of his classic *Molecular Biology of the Gene* (1987) or Lewin's recent *Genes-IV* (1990). An alternative similar overall impression can be found in those intermediary metabolism charts that hang on laboratory walls, displaying wheels and cycles of complex reactions feeding into more cycles and pathways of complex reactions (Bechtel 1986). This material found in chapters of Watson *et al.* (1987), in Lewin (1990), and in the intermediary metabolism charts plays the role of the explaining theories and initial conditions of molecular biology and biochemistry. Unfortunately, as already remarked, it is not possible to usefully separate out a small core of general laws and add a list of initial conditions to the core to generate an explanation.

The extensive complexity found in molecular biology has tended to generate pessimism regarding the reduction of such theories as Mendelian genetics by molecular genetics (Hull 1974, Kitcher 1984, Rosenberg, 1985), though both Kitcher (1989) and Rosenberg (1985) acknowledge that different forms of knowledge representation may alleviate this pessimism, Rosenberg writing that "[T]he 'cannot' [involved in obtaining a derivation about a regularity in transmission genetics] is not a logical one; it has to do with the limitations on our powers to express and manipulate symbols ..." (1985, 106). Kitcher's solution is to develop his notion of "practice" in some detail so that we might see how scientific research programs are accomplishing molecular explanations ("reductions") (Kitcher 1989). Rosenberg's approach is perhaps more sensitive to the complexity of molecular explanation than is Kitcher's, but his own solution, appealing to "supervenience," (1985, 111–117) appears to sidestep the real issue of knowledge representation and derivation (see Hull 1981, 135).

9 The relation between the causal-mechanical account of reduction and the GRR model; two versions of the GRR model – simple and complex

In the previous section, the discussion is strongly focused on various causal features involved in (what is at least a partial) reduction. At several points, however, we addressed the relations of that account to Nagel's conditions of connectability and derivability which continue to figure prominently in the GRR model introduced earlier in this chapter. In the present section, we consider those relations more systematically.

It appears that the GRR, account, which might be termed following Salmon's (1989) account as a more "epistemic" approach to reduction, may *best* capture the relations that exist in a "clarified" science where a unilevel reductans has been fully developed. We might term this a *simple interpretation* of the GRR model; it allows as to construe it as constituting a *set of conditions* which must be met if a *complete* reduction is to have been judged as effected. Under this interpretation the GRR model falls into what might be termed the justificatory or evaluative sphere. The requirements of connectability and derivability are seen in this light to be both reasonable and benign: connectability *must* be established if two languages are to be related, and derivability is just the implementation of a well understood truth-preserving mode of inference – one that carries impeccable credentials.

This *simple interpretation* of the GRR model does not, however, without elaboration and interpretation of its tenets along the lines proposed in Section 6, adequately capture the heavy reliance of scientific explanation on causal mechanisms of the type that Wimsatt (1976) and Hull (through personal communication) have stressed. In terms of active scientific investigation and of the language in scientific research reports (and textbooks), a causal-mechanical analysis (to use Salmon's [1989] term) *seems* much more suitable.

Along similar lines to those made by Salmon (1989) on explanation about the possibility of a "rapprochement" between causal and unificatory analyses, it is thus *tempting* to think of the causal reduction analysis presented above (in terms of interlevel causal generalizations that become progressively better characterized at the detailed δ levels of aggregation and specificity) as perhaps *another sense* of reduction, by analogy with Carnap's probability$_1$ and probability$_2$ (and especially Salmon's explanation$_1$ and explanation$_2$). On such a view, there would be a reduction$_{grr}$ and a reduction$_{cm}$ standing for the epistemic "general reduction-replacement" model at one extreme and for the ontic "causal mechanism" at the other, perhaps with some overlap (mixed) analyses between these two extreme types.

Though tempting, this view might be mistaken without keeping in mind four additional clarificatory comments. First, any appeal to "mechanisms" also requires appeals to the "laws of working" (the term is Mackie's 1974) that are implemented in the mechanisms and which provide part of the analysis and grounding of the "causal" aspect of the causal-mechanical approach (see Schaffner 1993, Chapter 6 for

details). Second, if we appeal to *any* general notion of a theory (i.e., to a semantic or a syntactic analysis) it would seem that we would have to deal with "generalizations" and with connecting the generalizations' entities and predicates, as in the GRR model. For example, if we consider the semantic conception of a scientific theory as it might be involved in reductions, such "theories" are set-theoretic (or other language) predicates which are essentially constituted by component generalizations. Thus if we were to think about attempting to characterize a "causal-mechanical" approach to reduction with the aid of semantic construal of theories, we would still require recourse to traditional generalizations. Third, appeals to either a semantic or a syntactic notion of theories still requires that entity and predicate terms be used (in addition to the generalizations) to express the theories which we consider. Thus some form of connectability assumptions relating terms such as "sensitization" and "enhanced synaptic transmission" will be required to effect a reduction. Fourth and finally, in those cases in which we wish to verify the *reasoning* supporting some *causal consequence* of a reducing theory (whether it be a general result or a particular one), we can do no better than utilize the principles of *deductive* reasoning, other forms of logic and reasoning being significantly more suspect. These four points support the characterization of reduction presented in the GRR account previously provided.

If we keep these caveats in mind, however, it does seem that we *can* effect a "rapprochement" of sorts between the two CM and GRR senses of reduction by construing reductions as having two aspects: (1) ongoing advances that occur in *piecemeal* ways, for example, as some features of a model are progressively elaborated at the molecular level or perhaps a new "mechanism" is added to the model, and (2) *assessments* of the explanatory fit between two theories viewed as a *collection* of models or even as two reasonably complete *branches* of science. Failure to keep these two aspects distinct has led to confusion about the adequacy of the GRR model (by, for example, Wimsatt 1976). Aspect (1) is akin to what Mayr (1982) has termed explanatory reduction, a notion that has been further elaborated by Sarkar (1989, 1992). Aspect (2) is the more traditional notion of intertheoretic reduction.

As scientists further develop reducing theories in a domain, they frequently elaborate various aspects of the theories both by (a) modifying a δ assumption or proposing a δ-level assumption that accounts for (part of) a σ-level process, and (b) describing these new assumptions in *causal* terms. Thus the result of such advances are typically complex interlevel connections with causal processes appealed to as part of the connections. (Part-whole or constituent relationships may also be appealed to in such situations.) This task is typically done by scientists, and reflects what Salmon (1989) calls the *ontic* dimension. This ongoing process will establish connections at several different levels, and, prior to arriving at something like a unilevel theory in a clarified science, many weblike and bushy connections can be expected. These reticulate relations however do not count *against* a reduction; they can be thought of as part of the process leading *towards* a "simplified" unilevel reduction. Something very much like this progressive process is described in Culp and Kitcher's (1989) recent essay on the discovery of enzymatic RNA, though they prefer to characterize the process more as an "embedding of a problem-solving schemata of one field of science in those of another" (1989, 479) than as a complex partial reduction.

These various theoretical developments and attendant connections – bushy and weblike though they may be – frequently have the ability to "explain" at least partially some features of a higher level domain, for example, a phenomenon such as sensitization in *Aplysia*. The more

traditional question of reduction, however, is a question of the adequacy of the explanatory relationship in a more *global* or gross sense which appeals (usually) to preexisting disciplines or fields (or theories). Working out this type of relation is typically done by the philosopher or by a scientist interested in reviewing very broad questions of domain relations. Such a relation(s) is much more *systematic* than the ontic dimension(s) referred to in the previous paragraph.

In connection with this global question, the various features of the simple GRR model can be appealed to as providing the *logical* conditions that must be satisfied, but usually satisfied in a post hoc evaluative manner. Thus one looks for (1) synthetic identities, which introduce basic ontological economies, and (2) derivability of the corrected reduced science (with the help of causal generalizations in the reducing science), which insures that security of relation that only truth-preserving nonampliative inference can provide. A representation of intertheoretic relations using the GRR model thus is a kind of "executive summary" of much scientific work that has gone on to establish interdisciplinary connections; such analysis also typically falls into the justificatory arena of philosophy of science. The GRR model in what may be considered its *simple* interpretation is thus useful in providing a kind of systematic summary and regulatory ideal, but it should not in general be confused with the *process* of establishing reductions and in the ongoing elaboration of the complex web of connections that typically unite reduced and reducing theories.

Though the GRR model, particularly in its limiting case as the general reduction model, can be conceived of as a summary and a regulatory ideal for unilevel reductions, it can also accommodate two interfield or interlevel theories. This would *not* be the *simple interpretation* of the GRR model discussed above, but there is no reason in principle why a theory which was *interlevel but primarily cellular* could not be explained by a theory which was *interlevel but primarily molecular*. To an extent, this feature of the GRR model will also rely on its ability to accommodate *partial* reductions. Insofar as the explanation was adequate, this type of theory would be akin to a type of "homogeneous" but now interlevel reduction discussed by Nagel (1961). Perhaps it might better be termed a "complex homogeneous" reduction. The GRR model could accommodate this form of reduction, but its reduction functions will reflect the *mixed interlevel* character of the theories. Such a *complex interpretation* of the GRR model may be more realistic than the *simple interpretation* of the GRR model discussed earlier where unilevel theories functioned as reductans and reductandum.

Certain general methodological principles are likely to govern the implementation of this *complex* GRR model. For example, it seems likely that

- T_R^{*}'s γ-central hypotheses will be preserved unless the relation falls on the replacement side of the reduction-replacement continuum.
- T_R^{*}'s σ hypotheses will either be replaced or amplified.
- T_R^{*}'s δ hypotheses will either be replaced or amplified.

Kandel's group's recent amplifications of *Aplysia's* cellular and molecular mechanisms accounting for sensitization appear to fit these suggestions. (In addition, in Schaffner 1993, Chapter 5, an extensive example that shows how these suggestions work in practice is provided in terms of the development of the clonal selection theory from 1957–1967. Current theories of antibody diversity generation and of T-lymphocyte stimulation (see Watson *et al.* 1987, Chapter 23) suggest that these types of changes are taking

place in those fields studying the molecular biology of the immune response.)

The GRR model is also an appropriate framework in terms of which deeper logical questions, such as predicate analysis, the nature and means of establishing synthetic identities, and the metaphysical implications of such identities can be pursued (see Schaffner 1976, and 1993). As such, the GRR framework forces an actual ongoing reduction into a "rational reconstructive" mode. It should be added as a final point, however, that occasionally a focus on the satisfaction of even the simple unilevel interpretation of the GRR conditions can illuminate a scientific debate.

Accordingly the GRR model, in spite of its syntactic (epistemic?) flavor, and though it may prima facie appear to be intuitively less in accord with the pervasive causal flavor of middle range examples of theories (whether used in explanatory or reduction contexts) presented in this chapter is a strong and defensible account of reduction. Though the causal-mechanical alternative is a valuable approach representing an important aspect of reductions, it is not the whole story by any means. It seems that deeper analysis of what a causal-mechanical alternative involves *leads us back* to just those assumptions which constitute the GRR model.

As noted, reduction in the biomedical sciences is complex, and causal and (re)interpretative connections are found at a number of loci. Before closing this section, it would be well to introduce one additional point of clarification to guard against the misinterpretation of interpretative connections as causal connections.

An examination of the system studied by Kandel and others (1987) indicates that the **behavioral** level of analysis (represented by Figures 26.3A and 26.3B) deals with such entities as shocks administered to organisms, gross reflex descriptions, and a time course of hours or days. Comparison with the **neural network** level indicates that these entities are now viewed at a higher level of detail; this level of detail is further increased at the **biochemical** level. As one descends levels from the behavioral to the biochemical, one does *not* traverse a causal sequence in reverse, with behavior viewed as effects and biochemical mechanisms as causes, rather one traces a **perspectival** sequence. **Causal** sequences can be found at any level of aggregation *and between levels of aggregation*, but a description of a set of parts of a whole does not automatically entail a causal account relating the parts as causes to the whole as an effect.

What frequently is the case is that the causal sequence between cause and effect is best understood at the biochemical level. Thus stimuli can be **interpreted** as propagating action potentials, and neurotransmitters can be invoked as chemical messengers. Biochemical cascades can be examined as fine-structure connecting networks, leading to an event to be explained which can be **reinterpreted** so as to be identified with an original explanandum. What should not be done, however, is to conflate perspectival reinterpretation with causation for it can only lead to confusion in an already extraordinarily complex set of relations.

We thus finally have a response to Kandel's optimistic assessment of the impact of reductionism on neurobiology: The pursuit of reductionism in molecular biology, and that includes the molecular neurosciences, will result in systematicity and unity *only in comparison* with the even more bizarre complexity that biologists would have to deal with without some occasional broad generalizations obtainable from the molecular approach. Reduction is also bound to be patchy. Portions of learning theory as understood to be applied to specific types of model organisms will be reduced, but it is unlikely that any *general theory* will emerge which is universal and simple that does anything useful for neuroscientists – or for behavioral scientists. Thus *partial reductions* as well as some reductions which

may well have the effect of *replacing* rather than explaining previous material is likely, given the complexity of the domains of interest and the lack of sound foundations. Thus we will most likely see progress in the neurosciences where features of the reduction-replacement model described are implemented implicitly, but with a rather baroque set of models playing the role of the reducing theories.

10 Relations between reduction and explanation

In the (comparatively) simple example from Kandel and his associates' analysis of learning in *Aplysia*, we appear to encounter a paradigm case of *both* a reduction *and* a scientific explanation. This suggests the question of how the account of explanation given in association with the Kandel example relates to extant models of scientific explanation discussed in Chapter 1 of this book [sic].

The most useful approach may be to refer back to the contrast between two very general analyses of scientific explanation noted by Wesley Salmon in Chapter 1 [sic] as well as in his *Scientific Explanation and the Causal Structure of the World* (1984). Salmon contrasts the "epistemic" approach to explanation with the "ontic" approach. The former, in its "inferential" interpretation, is best represented by the Hempel-Oppenheim model of explanation. The "ontic" approach, on the other hand, is one which Salmon has articulated and defended under the rubric of the "causal-mechanical" tradition. Salmon believes that the epistemic approach also characterizes the "theoretical unification" analyses of explanation developed by Friedman (1974) and Kitcher (1981, 1989).

Salmon was unwilling, however, to completely yield "unification" to the epistemic approach and suggested that it could be accommodated in a way within the causal-mechanical tradition:

The ontic conception looks upon the world, to a large extent at least, as a black box whose workings we want to understand. Explanation involves laying bare the underlying mechanisms that connect the observable inputs to the observable outputs. We explain events by showing how they fit into the causal nexus. Since there seem to be a small number of fundamental causal mechanisms, and some extremely comprehensive laws that govern them, the ontic conception has as much right as the epistemic conception to take the unification of natural phenomena as a basic aspect of our comprehension of the world. *The unity lies in the pervasiveness of the underlying mechanisms* upon which we depend for explanation. (Salmon 1984, 276)

In (Salmon 1992), Salmon introduces two briefly characterized biomedical examples to illustrate his approach. He mentions that "to understand AIDS, we must deal with viruses and cells. To understand the transmission of traits from parents to offspring, we become involved with the structure of the DNA molecule. . . . When we try to construct causal explanations we are attempting to discover the mechanisms – often hidden mechanisms – that bring about the facts we seek to understand" (p. 34).

Salmon adds to the contrast between the two major traditions that in "an important respect" these two traditions "overlap":

When the search for hidden mechanisms is successful, the result is often to reveal a small number of basic mechanisms that underlie wide ranges of phenomena. The explanation of diverse phenomena in terms of the same mechanism constitutes theoretical unification. For instance . . . [t]he discovery of the double-helical structure of DNA . . . produced a major unification of biology and chemistry, (p. 34)

The biomedical sciences in general, and the neurosciences, AIDS virology, and molecular genetics in particular, tend, with a few important exceptions, to propose what Salmon terms causal-mechanical explanations. The Kandel example detailed above offers clear evidence to support his point. As such, the appeal to "a small number of basic mechanisms that underlie wide ranges of phenomena" is the way both explanation and reduction is achieved, though an important caveat is in order.

The caveat has to do with the expression "small number of mechanisms." In point of fact, close analysis of the biomedical sciences discloses an extensive variety of mechanisms, some with comparatively narrow scope and some with almost universal scope. In a number of areas, a range of mechanisms which bear close analogies to each other are found. This, given the evolutionary backdrop, is not unexpected, but the subtle variation biologists continue to encounter requires them to be attentive to changes in biological processes as they analyze different types of organisms or even as they analyze different forms of entities in the same organism. Thus though the genetic code is (almost) universal, messenger RNA is processed quite differently in prokaryotes (e.g., bacteria) in contrast with eukaryotes (e.g., multicellular organisms such as *Aplysia*). Furthermore, variation can be found in the same organisms, for example, in the human where the actions of muscle fibers in skeletal muscle are regulated importantly differently than in cardiac muscle, and serious errors can result if the differences are not kept in mind.

Thus the biomedical sciences display a rather complex and partially attenuated theoretical unification: some mechanisms are nearly universal, many have evolutionarily fixed idiosyncrasies built into them, and some are highly individualized. Almost a spectrum of scope is encountered in the diversity of the life sciences.

Nevertheless, generalizations and mechanisms are available to be used in explanations and to be tested in different laboratories: the variable scope breadth does not negate the nomic force of the generalizations and the mechanisms.

This picture of explaining-reducing generalizations of variable scope instantiated in a series of overlapping mechanisms is congruent with the account of theory structure proposed by Schaffner (1980, 1986), and further developed in Schaffner (1987). The analysis also comports with the GRR model of reduction developed in Section 6. It would take us beyond the scope of this introductory book to pursue these issues in any further depth. Suffice it to say, then, that reduction (and explanation) can be located in that "overlap" range discussed by Salmon where (comparatively) few "hidden" mechanisms account for a (comparatively) wide range of phenomena by exhibiting (complex) causal pathways.

11 The neuro-immune-endocrine connection as exemplifying the biopsychosocial model

We have now traced the details of how a reduction occurs in a part of neurobiology, and it remains to apply those results back to the issues raised in our treatments of Munson and Engel in Sections 2 and 3. Over the past half-dozen years, a number of biomedical scientists have discovered that important connections exist between the nervous system and the immune system in both animals and in humans. That the nervous system interacted with the endocrine system – the group of glands that produce hormones – has been appreciated for some time, but the mutual influence of the immune system and the neuroendocrine apparatus is a relatively new idea.

Scientists have found bidirectional communication between the systems. The cells of the immune system have been found to synthesize biologically active neural hormones and the immune system's cells have receptors on their surface for such hormones. In addition, the central nervous system can affect immune responses by inducing the release of adrenal and other hormones, which can substantially modulate immune reactivity since immune cells also have receptors for adrenal hormones on their surface. Thus each system can influence the other via a variety of chemical substances. A recently published book on *The Neuro-Immune-Endocrine Connection* (Cotman et al. 1987) demonstrates that well characterized *causal pathways* exist whereby certain forms of behavioral stress can be shown to result in a decreased immune response. Animals experiencing such stress – often referred to as an analogue of a human's helpless-hopeless situation – become susceptible to bacterial and viral illnesses, and also appear to have a higher susceptibility of developing cancers.

This research on the neuro-immune-endocrine connection is based in part on an idea that was at one time controversial, but which today seems well-founded. This is the notion that one of the functions of the body's immune system is to identify at an early stage those cells which have become cancerous and to eliminate them before they can divide and overwhelm the body's defenses. This hypothesis was championed by the late Dr. MacFarlane Burnet who termed it "immunological surveillance." The increase of cancers in patients who have been artificially immunosuppressed because of receiving organ transplants, as well as the appearance of a rare form of cancer known as Kaposi's sarcoma in patients with the Acquired Immunodeficiency Syndrome or AIDS, supports this idea.

Recent studies done at the Pittsburgh Cancer Institute at the University of Pittsburgh by Dr. Sandra Levy and Dr. Ronald Herberman and their colleagues also suggests that psychological status including emotions may play an important role in accounting for different prognoses of cancer patients. These scientists followed the immunological and psychological status of 75 breast cancer patients who had undergone surgery and chemotherapy-radiation therapy. An important predictor of these patients' prognoses was the activity level of their natural killer (NK) or cancer-fighting immunological cells. Drs Levy and Herberman found that they could explain about 51% of the decreased NK activity on the basis of three distress indicators: the patient's poor adjustment, the lack of social support, and their fatigue-depression symptoms. A similar but less pronounced effect was also noted after a period of three months. More recent work by these investigators, carried out on a second sample of 125 early stage breast cancer patients, showed that the perception of high quality emotional support from the patient's spouse was a more potent predictor of NK activity than the endocrinological character of the tumor itself, as determined by its endocrine receptor status. These investigators concluded that these central nervous system mediated factors thus accounted for a significant effect on the cancer's prognosis.

Other attempts to correlate emotions and cancer prognosis have been less successful, but these linkages have been given a rational basis by the neuro-immunological-endocrine connection discussed. Human beings are very complex and it is very difficult to isolate the extensive interactions just indicated so as to produce unequivocal experimental conclusions. Research in cancer immunology is proceeding at an increasing pace as investigators realize that the more traditional forms of chemotherapy have likely gone as far as possible, and the new direction is likely to involve means of boosting or stimulating the immune system via interferons, interleukins, and other active immunological modifiers to help the body fight cancer. Augmentation and fine tuning of the immunological response via the neuroendocrine connection is in the process of exploration.

These considerations force us to take seriously that multiple perspectives, in the sense of multiple interacting systems, play significant roles in disease and in the healing process. To this point, however, the advances described both in the early sections and in the previous section suggest that biomolecular advances are discovering just *how* such interlevel causal connections as indicated in Figure 26.1 are implemented *at the biomolecular level*. Thus such progress as is occurring seems to support the reducibility of medicine to biology, and would confirm the characterization of medicine as a science. There is, however, one additional at least prima facie difficulty for such a reductionistic approach. As suggested in Engelhardt's remarks at the beginning of Section 1, and in Munson's comments about the essential moral features of medicine, medicine may contain intrinsic normative or ethical features that make a reductionistic approach difficult to defend. It is to this issue that we now turn.

12 The ethical dimension of medicine

Ethical problems in medicine have become an important part of the public consciousness in recent years. As medicine has developed its technical prowess, it has generated hard choices concerning the use of its capacity to help individuals with their health. The costs associated with leading-edge innovations such as organ transplants pose difficult allocation decisions for individuals and for society. The potential to keep terminally ill and/or comatose patients alive for indefinite periods has led to a vigorous ethical and legal debate about active and passive euthanasia. These choices and debates are not external to medicine but in a very real sense are part of health-care delivery.

The situation does not change whether we focus our attention on so-called "macro-" issues involving societal determinations of what portion of the gross national product to commit to health care or on "micro-" problems such as whether to withdraw an artificial respirator from a particular patient. In both types of deliberation, normative principles need to be utilized in order to reach an ethical conclusion of what it is "best" to do. The literature of medical ethics has grown enormously over the past fifteen or so years, with no lack of approaches which offer normative frameworks to assist with these matters.

Many ethicists have found that what can be described as a "principle" approach to medical ethics is helpful in clarifying and resolving clinical ethical problems. Such a perspective can be both flexible and sufficiently philosophically eclectic that it does not generate additional problems of communication by dogmatic adherence to any one philosophical or religious system. The approach has its roots in several Federal Commission reports (including the *Belmont Report* and the many volumes published by the President's Commission for the Study of Ethical Problems in Medicine) and has been articulated systematically by Beauchamp and Childress (1989).

It will not be feasible in this chapter to develop any significant medical ethical analysis utilizing this approach. Suffice it to say that the "principle" approach does introduce several prima facie *values* such as individual "self-determination" and "well-being" which represent what is taken to be ethically desirable features to be maximized in any given situation. Unfortunately such values frequently conflict and may require a more general and comprehensive ethical theory such as "utilitarianism" or Rawlsian social contract theory to resolve

them. Critical for our inquiry in this chapter is the question whether such normative principles could be in any interesting sense "reduced" to the biological sciences. A major barrier to such a reduction is an observation made by the philosopher David Hume several hundred years ago and sometimes referred to as the "is-ought" distinction.

Hume pointed out that no logical rule of inference exists by which a sentence expressing (moral) obligation, e.g., "You ought not steal," can be *derived* from a set of premises that are descriptive of the world or of human nature ([1739–1740] 1978). Such premises, including all standard scientific statements among them, state what *is* the case, not what *ought* to be the case. Recalling the conditions of the general reduction model of Section 6, what would be necessary to achieve a reduction of ethics would be a "connectability assumption" defining an ethical property in terms of a descriptive property. This, however, seems to be exactly against what Hume has cautioned us.

The situation is somewhat more complex though not more encouraging in terms of pointing toward a clear solution. A number of philosophers, John Stuart Mill (the utilitarian) among them, contend that the "good" – what is ethically desirable – *can* be specified in what are prima facie nonethical terms. For Mill, the "good" or "*summum bonum*" was "the greatest happiness" ([1861] 1979, 1,3). Mill offered a complex theory of happiness as well as a number of subtle arguments for his identifying the "good" with "the greatest happiness for the greatest number," none of which we can discuss here. Other philosophers, however, including Kant, have contended that such a reduction or *naturalizing* of ethical notions is faulty and incorrect. Thus we encounter in our discussion of the reducibility of medicine – itself a difficult question – one of the perennial philosophical problems about the nature of ethics.

While no major new insights will be offered here concerning this issue, the concluding section suggests a tentative position which has the virtue of permitting us to continue with further inquiry, and which may point toward a more satisfactory resolution of the problem.

13 Summary and conclusion

In this chapter we have examined two inter-related problems in the philosophy of medicine. The first issue addressed was the status of medicine as a science and of the reducibility of medicine to its constituent biological sciences. There the significant and essential social (and psychological) dimensions of medicine were emphasized.

In the second part we examined the nature of reduction in the biomedical sciences in general, and reduction involving what has often been called "mental" to the "physical" in particular. Those sections showed both how complex such a reduction is, as well as how fragmentary the results have been in spite of major advances in molecular neurobiology. On the way to these conclusions, a complex model of reduction was also developed – the GRR account – which was suggested to have special utility in those areas involving the relation of the mental and the biological.

Finally, the inquiry closed by suggesting that though on the other hand molecular biology may be furthering ways to reduce the biopsychosocial intricacies of medicine to biochemistry, on the other hand the normative aspects of the biopsychosocial dimensions of medicine seemed to strongly resist any such reduction.

One way to come to a reasonable conclusion regarding these different tendencies is to acknowledge that *for the present and the foreseeable future* medicine will not be a reducible science. Whether medicine can be conceived of as science at all will also wait on a resolution of the reducibility of the inherently normative components which constitute part of it. This is a position described in Section 4. as "weak emergentism." Close analysis of the example of *Aplysia* sensitization will affirm that even this paradigm of the "reduction" of a behavioral event to a set of molecular mechanisms involves conjoint appeals to multi-level concepts including the cellular as well as the molecular. On the other hand, the GRR model as well as the *Aplysia* example suggests that translation and reinterpretation – and *possible* replacement – of some fundamental entities, properties, or processes may occur as science advances. This is a thesis that Paul and Patricia Churchland have argued may even involve basic concepts in "folk psychology" such as "beliefs" and "desires" (see Churchland 1986, 1988). A very powerful molecular learning theory may well indicate why we possess and utilize such normative concepts as self-determination, and the role that such ethical analyses may play in individual and collective homeostasis. Would such a theory commit the "naturalistic fallacy" of attempting to derive an "ought" from an "is" that Hume ([1739–1740] 1978) and Moore ([1903] 1962) said was not possible, or might it be conceived more in the spirit of a Millean attempt at a rational reconstruction of a naturalistic foundation for ethics? Such possibilities are at present mere (and weak) speculations but as the molecular neurosciences advance rapidly, their salience – and importance – are highly likely to confront us with increased urgency.

References

Balzer, W. and Dawe, C. M. (1986), "Structure and Comparison of Genetic Theories: (1) Classical

Genetics; (2) the Reduction of Character-Factor Genetics to Molecular Genetics," *The British Journal for the Philosophy of Science* **37**: 55-69 and 177-191.

Beauchamp, T.L. and Childress, J.F (1989), *Principles of Biomedical Ethics*. 3rd ed. New York: Oxford University Press.

Bechtel, William (1986), "The Nature of Scientific Integration," in Wiliam Bechtel (ed.), *Integrating Scientific Disciplines*. Dordrecht: Martinus Nijhoff, pp. 3-52.

Causey, Robert L. (1977), *Unity of Science*. Dordrecht: Reidel.

Churchland, Patricia Smith (1986), *Neurophilosophy: Toward a Unified Science of the Mind/Brain*. Cambridge, MA: Bradford/MIT Press.

Churchland, Patricia Smith (1988), "The Significance of Neuroscience for Philosophy," *Trends in Neurosciences* **11**: 304-307.

Churchland, Paul M. (1979), *Scientific Realism and the Plasticity of Mind*. Cambridge, England: Cambridge University Press.

Churchland, Paul M. (1981), "Eliminative Materialism and the Propositional Attitudes," *The Journal of Philosophy* **78**: 67-90.

Churchland, Paul M. (1984), *Matter and Consciousness: A Contemporary Introduction to the Philosophy of Mind*. Cambridge, MA: Bradford/MIT Press.

Cohen, R.S.; Hooker, C.A.; Michalos, A.C.; van Evra, J.W. (eds.) (1976), *PSA 1974*. Dordrecht: Reidel.

Committee on Models for Biomedical Research (1985), *Models for Biomedical Research:A New Perspective*.Washington, DC: National Academy Press.

Cotman, Carl W.; Brinton, Roberta E.; Galaburda, A.; McEwen, B.; and Schneider, Diana M. (eds.) (1987), *The Neuro-Immune-Endocrine Connection*. New York: Raven Press.

Culp, Sylvia and Kitcher, Philip (1989), "Theory Structure and Theory Change in Contemporarty Molecular Biology," *The British Journal for the Philosophy of Science* **40**: 459-483.

Engel, George L. (1977), "The Need for a New Medical Model: A Challenge for Biomedicine," *Science* **196**: 129-136.

Engel, George L. (1981), "The Clinical Application of the Biopsychosocial Model," *The Journal of Medicine and Philosophy* **6**: 101-123.

Engelhardt, Jr., H. Tristram (1986), "From Philosophy and Medicine to Philosophy of Medicine," *The Journal of Medicine and Philosophy* **11**: 3-8.

Feyerabend, Paul K. (1962), Explanation, Reduction, and Empiricism," in Feigl, H. and Maxwell, G. (eds), *Minnesota Studies in the Philosophy of Science,Volume 3, Scientific Explanation, Space and Time*. Minneapolis: University of Minnesota Press, pp. 28-97.

Forstrom, Lee A. (1977), "The Scientific Autonomy of Clinical Medicine," *The Journal of Medicine and Philosophy* **2**: 8-19.

Friedman, Michael (1974) "Explanation and Scientific Understanding", *The Journal of Philosophy* **71**: 5-19.

Frost, W.N., Clark, G.A. and Kandel, E.R. (1988), "Parallel Processing of Short-term Memory for Sensitization in *Aplysia*," in Journal of Neurobiology **19**: 297–334.

Genesereth, Michael R. and Nilsson, Nils J. (1987), *Logical Foundations of Artificial Intelligence*. Los Altos, CA: Morgan Kaufmann.

Grobstein, Clifford (1965), *The Strategy of Life*. San Francisco: Freeman.

Grünbaum, Adolf (1984), *The Foundations of Psychoanalysis: A Philosophical Critique*. Berkeley and Los Angeles: University of California Press.

Hooker, C.A. (1981), "Towards a General Theory of Reduction. Part I: Historical and Scientific Setting. Part II: Identity in Reduction. Part III: Cross-Categorial Reduction," *Dialogue* **20**: 38-59, 201-236, 496-529.

Hull, David L. (1974), *Philosophy of Biological Science*. Englewood Cliffs, NJ: Prentice-Hall.

Hull, David L. (1976), "Informal Aspects of Theory Reduction," in Cohen et al., pp. 653-670.

Hull, David L. (1981), "Reduction and Genetics," *The Journal of Medicine and Philosophy* **6**:125-143.

Hume, David ([1739–1740] 1978), *A Treatise of Human Nature*, Edited by L.A. Selby-Bigge. 2d ed. Oxford: Clarendon Press.

Kandel, Eric R. (1976), *Cellular Basis of Behavior:An Introduction to Behavioral Neurobiology*. W.H. Freeman and Company.

Kandel, Eric R. (ed.) (1987), *Molecular Neurobiology in Neurology and Psychiatry*. New York: Raven Press.

Kandel, Eric R.; Castellucci, Vincent F.; Goelet, Philip; and Schacher Samuel (1987), "Cell-Biological Interrelationships between Short-Term and Long-Term Memory," in Kandel (1987), pp.111–132.

Kemeny, John G. and Oppenheim, Paul (1956),"On Reduction," *Philosophical Studies* **7**: 6–19.

Kitcher, Philip (1981), "Explanatory Unification," *Philosophy of Science* **48**: 507–531.

Kitcher, Philip (1984), "1953 and All That. A Tale of Two Sciences," *The Philosophical Review* **93**: 335–373.

Kitcher, Philip (1989), "Explanatory Unification and the Causal Structure of the World," in Kitcher, P. and Salmon, W.C. (eds.) (1989), *Minnesota Studies in the Philosophy of Science*. Volume 13, *Scientific Explanation*. Minneapolis: University of Minnesota Press, pp. 410–505

Kuhn, T. (1962) *The Structure of Scientific Revolutions*. Chicago: University of Chicago Press.

Lewin, Benjamin (1990), *Genes IV*. Oxford: Oxford University Press.

Mackie, J. (1974), *The Cement of the Universe: A Study of Causation*. Oxford: Clarendon Press.

Mayr, Ernst (1982), *The Growth of Biological Thought: Diversity, Evolution, and Inheritance*. Cambridge, MA: Belknap Press of Harvard University Press.

Mill, John Stuart ([1861] 1979), *Utilitarianism*. Edited by George Sher. Indianapolis: Hackett.

Miller, Randolph A.; Pople, Jr., Harry E.; and Myers, Jack D. (1982), "Internist-I, an Experimental Computer-Based Diagnostic Consultant for General Internal Medicine," *The New England Journal of Medicine* **307**: 468–476.

Moore, George Edward ([1903] 1962), *Principia Ethica*. Cambridge, England: Cambridge University Press.

Morowitz (1985), see Committee on Models for Biomedical Research.

Munson, Ronald (1981), "Why Medicine Cannot Be a Science," *The Journal of Medicine and Philosophy* **6**: 183–208.

Nagel, Ernest (1949), "The Meaning of Reduction in the Natural Sciences," in Stauffer, R.C. (ed.) *Science and Civilisation*. Madison, WI: University of Wisconsin Press, pp. 97–135.

Nagel, Ernest (1961), *The Structure of Science: Problems in the Logic of Scientific Explanation*. New York: Harcourt, Brace, & World.

Olby, Robert C. (1966), *Origins of Mendelism*. London: Constable.

Pellegrino, Edmund D. (1976), "Philosophy of Medicine: Problematic and Potential," *The Journal of Medicine and Philosophy* **1**: 5–31.

Pellegrino, Edmund D. (1986), "Philosophy of Medicine: Towards a Definition," *The Journal of Medicine and Philosophy* **11**: 9–16.

Pellegrino, Edmund D. and Thomasma, David C. (1981), *A Philosophical Basis of Medical Practice: Toward a Philosophy and Ethic of the Healing Professions*. New York: Oxford University Press.

Popper, Karl R. (1957a), "The Aim of Science," *Ratio* **1**: 24–35.

Rosenberg, Alexander (1985), *The Structure of Biological Science*. Cambridge, England: Cambridge University Press.

Ruse, Michael (1976), "Reduction in Genetics," in Cohen et al., pp. 633–651.

Salmon, Wesley C. (1984), *Scientific Explanation and the Causal Structure of the World*. Princeton: Princeton University Press.

Salmon, Wesley C. (1989), "Four Decades of Scientific Explanation," in Kitcher, P. and Salmon, W.C. (eds.) (1989), *Minnesota Studies in the Philosophy of Science*. Volume 13, *Scientific Explanation*. Minneapolis: University of Minnesota Press, pp. 3–219.

Salmon, Wesley C.; Jeffrey, Richard C.; and Greeno, James G. (1971), *Statistical Explanation and Statistical Relevance*. Pittsburgh: University of Pittsburgh Press.

Salmon, Wesley C. (1992), "Scientific Explanation", in Merrilee H Salmon et al. (eds.), *Introduction to the Philosophy of Science*. Englewood Cliffs, NJ: Prentice-Hall Inc.

Sarkar, Sahotra (1989), "Reductionism and Molecular Biology: A Reappraisal." Ph.D. Dissertation, University of Chicago.

Sarkar, Sahotra (1992), "Models of Reduction and Categories of Reductionism," *Synthese* **91**: 167–194.

Schaffner, Kenneth F. (1967), "Approaches to Reduction," *Philosophy of Science* **34**: 137–147.

Schaffner, Kenneth F. (1969), "Correspondence Rules," *Philosophy of Science* **36**: 280–290.

Schaffner, Kenneth F. (1972), *Nineteenth-Century Aether Theories*. Oxford: Pergamon Press.

Schaffner, Kenneth F. (1976), "Reductionism in Biology: Prospects and Problems," in Cohen et al., pp. 613–632.

Schaffner, Kenneth F. (1977), "Reduction, Reductionism, Values, and Progress in the Biomedical Sciences," in Colodny, Robert G. (ed.), *Logic, Laws, and Life: Some Philosophical Complications*. Pittsburgh: University of Pittsburgh Press, pp. 143–171.

Schaffner, Kenneth F. (1980), "Theory Structure in the Biomedical Sciences," *The Journal of Medicine and Philosophy* **5**: 57–97.

Schaffner, Kenneth F. (ed.) (1985), *Logic of Discovery and Diagnosis in Medicine*. Berkeley: University of California Press.

Schaffner, Kenneth F. (1986), "Exemplar Reasoning About Biological Models and Diseases: A Relation Between the Philosophy of Medicine and Philosophy of Science," *The Journal of Medicine and Philosophy* **11**: 63–80.

Schaffner, Kenneth F. (1987), "Computerized Implementation of Biomedical Theory Structures: An Artificial Intelligence Approach," in Fine, Arthur and Machamer, Peter (eds.) *PSA 1986*. East Lansing, MI: Philosophy of Science Association, pp. 17–32.

Schaffner, Kenneth F. (1993), *Discovery and Explanation in Biology and Medicine*. Chicago: University of Chicago Press.

Shapere, Dudley (1974), "Scientific Theories and Their Domains," in Suppe, Frederick (ed.), *The Structure of Scientific Theories*. Urbana: University of Illinois Press, pp. 518–565.

Simpson, George Gaylord (1964), *This View of Life: The World of an Evolutionist*. New York: Harcourt, Brace, & World.

Sklar, Lawrence (1967), "Types of Inter-Theoretic Reduction," *The British Journal for the Philosophy of Science* **18**: 109–124.

Suppes, Patrick (1957), *Introduction to Logic*. Princeton: Van Nostrand.

Suppes, Patrick (1967), "What is a Scientific Theory?" in Morgenbesser, S. (ed.), *Philosophy of Science Today*. New York: Basic Books, pp. 55–67.

Virchow, R. (1849), "Wissenschaftliche Methode und therapeutische Standpunkte," *Archiv für pathologische Anatomie und Physiologie und für klinische Medicin* **2**: 37.

Watson, James D. et al. (1987), *Molecular Biology of the Gene*. 4th ed. Two volumes. Menlo Park, CA: Benjamin/Cummings.

Wimsatt, William C. (1976), "Reductive Explanation: A Functional Account," in Cohen et al., pp. 671–710.

Zarin, Deborah A. and Pauker, Stephen G. (1984), "Decision Analysis as a Basic for Medical Decision Making: The Tree of Hippocrates," *The Journal of Medicine and Philosophy* **9**: 181–213.

Chapter 27

School of Health and Related Research, University of Sheffield

HIERARCHY OF EVIDENCE

For many years scientists have recognized two types of research:

- Primary: original studies, based on observation or experimentation on subjects.
- Secondary: reviews of published research, drawing together the findings of two or more primary studies.

In biomedical science there is general agreement over a hierarchy: the higher up a methodology is ranked, the more robust and closer to objective truth it is assumed to be. The orthodox hierarchy looks something like Table 27.1.

You may note that this ranking has an evolutionary order, moving from simple observational methods at the bottom through to increasingly sophisticated and statistically refined methodologies. To some extent, the physical sciences are consistent: you can repeatedly measure the acceleration due to gravity, in different places and with different techniques, but it will always come out close to a constant value. But, biological systems are multivariate and less predictable: if you looked at the effect of dietary supplements on growth in different groups even from the same species, you would get very mixed results. Social phenomena are even more complex and difficult to isolate. Many biomedical discoveries have had to await the

development of techniques capable of dealing with populations in a statistical manner. For instance:

- Simple observation may prove that sewing up a torn vein stops patients bleeding to death
- In the 1950s cohort studies established the link between smoking and lung cancer
- Nowadays, to detect small but important differences in survival rates with different combinations of drugs for cancer, large multicentre randomised controlled trials (RCTs) may be needed.

James Lind hit lucky with his work on scurvy: he picked a disease that was nearly universal in sailors on long voyages, but where practically all cases could be cured by citrus fruit juice, or more accurately, the vitamin C it contained. Such discoveries were often empirical, and pre-dated the theories that would back them up. Another 150 years passed before vitamins were isolated and described, but that did not stop mariners protecting themselves from scurvy with lemon and lime juice!

Finally, you should remember two points in relation to this hierarchy:

1. Techniques which are lower down the ranking are not always superfluous. For instance,

Table 27.1

Rank	Methodology	Description
1	Systematic reviews and meta-analyses	Systematic review: review of a body of data that uses explicit methods to locate primary studies, and explicit criteria to assess their quality.
		Meta-analysis: A statistical analysis that combines or integrates the results of several independent clinical trials considered by the analyst to be "combinable" usually to the level of re-analysing the original data, also sometimes called: pooling, quantitative synthesis. Both are sometimes called "overviews."
2	Randomised controlled trials (finer distinctions may be drawn within this group based on statistical parameters like the confidence intervals)	Individuals are randomly allocated to a control group and a group who receive a specific intervention. Otherwise the two groups are identical for any significant variables. They are followed up for specific end points.
3	Cohort studies	Groups of people are selected on the basis of their exposure to a particular agent and followed up for specific outcomes.
4	Case-control studies	"Cases" with the condition are matched with "controls" without, and a retrospective analysis used to look for differences between the two groups.
5	Cross sectional surveys	Survey or interview of a sample of the population of interest at one point in time
6	Case reports	A report based on a single patient or subject; sometimes collected together into a short series
7	Expert opinion	A consensus of experience from the good and the great.
8	Anecdotal	Something a bloke told you after a meeting or in the bar.

if you wanted to study risk factors for an illness, any ethics committee would quite rightly reject an RCT whereby you proposed exposing half your subjects to some noxious agent! Here, you need a cohort, where you find a group exposed to the agent by chance or their own choice and compare how they fare with another group who were not exposed. As we shall see later in the discussion of qualitative and quantitative methods, it is not that one method is better: for a given type of question there will be an appropriate method.

2. This hierarchy is not fixed in tablets of stone: the ranking may change, and there is debate over the relative positions of systematic reviews and large RCTs. The future may also see

changes in the hierarchy of scientific evidence. Traditionally, the randomised controlled trial has been regarded as the most objective method of removing bias and producing comparable groups. But, the technique is often slow, expensive and produces results that are difficult to apply to messier, multivariate "real situations." In the future, expanding computer capacity may make observational studies with multiple variables possible.

In some ways, the wheel has come full circle. Printing presses with moveable type arrived in Europe in the fifteenth century, and provided the technology necessary to produce reading material quickly and cheaply. It was another two centuries before the rise of scientific societies and an interest in both performing and disseminating experimental work generated a need for technical periodicals. Now, three hundred years later, the rate of technological change, the arrival of the electronic database and the Internet, the volume of research and demands for services like healthcare to be justified by research evidence are driving the evolution of the scientific press into the twenty-first century.

Chapter 28

John Worrall

WHAT EVIDENCE IN EVIDENCE-BASED MEDICINE?

1 Introduction

The usual reaction from outside observers on being told that there is a (relatively) new movement called "Evidence-Based Medicine" is "What on earth was medicine based on before?" Telling clinicians that they ought to operate in accordance with EBM sounds about as controversial as telling people that they ought to operate in accordance with virtue.

However, just as everyone agrees that people should act in accordance with virtue, but disagrees about what virtue precisely is, so in the case of EBM, disagreements soon emerge once we get to the details. The idea that clinicians ought to base practice on best evidence surely ought to win widespread acceptance, but what exactly counts as best evidence? How persuasive are different kinds of evidence (or rather, how persuasive ought they to be)? What evidential role, if any, is played by 'clinical experience' or 'clinical expertise'? EBM needs, but I shall argue does not yet possess, a fully coherent, articulated and detailed account of the correct relationship between the evidence and various therapeutic and causal claims that would answer questions such as these from general first principles. This seems to me an area where philosophers of science can, for once, be of real practical value. After all, the topic of the relationship between

theory and evidence has, of course, long been a central one in the philosophy of science.

There are two main areas in which EBM has yet to produce a fully defensible account of the view of evidence that it recommends. The first concerns the role and evidential power of randomization; the second concerns the role and evidential power of clinical judgment and expertise. In the present paper I concentrate exclusively on the first of these.

2 EBM and RCTS

It is widely believed in the medical profession that the only truly scientifically "valid" evidence to be obtained from clinical trials is that obtained from trials employing randomized controls. This view derives from the frequentist statisticians. Tukey (1977, 684) for example asserts that: "the *only* source of reliable evidence about the usefulness of almost any sort of therapy . . . is that obtained from well-planned and carefully conducted randomized . . . clinical trials." While Sheila Gore (1981, 1958) writes: "Randomized trials remain the reliable method for making specific comparisons between treatments."

While it is often supposed that EBM endorses this view, closer attention to the EBM literature reveals a much more qualified account.[1] For example, the 1996 attempt to clarify the position ("Evidence-Based Medicine: What it is and What it isn't".) is quite explicit that:

EBM is not restricted to randomised trials and meta-analyses. It involves tracking down the best external evidence with which to answer our clinical questions. To find out about the accuracy of a diagnostic test, we need to find proper cross sectional studies of patients clinically suspected of harbouring the relevant disorder, not a randomised trial. For a question about prognosis, we need proper follow up studies of patients assembled at a uniform, early point in the clincial [sic] course of their disease. And sometimes the evidence we need will come from the basic sciences such as genetics or immunology. It is when asking questions about therapy that we should try to avoid the non-experimental approaches, since these routinely lead to false positive conclusions about efficacy. Because the randomised trial, and especially the systematic review of several randomised trials, is so much more likely to inform us and so much less likely to mislead us, it has become the "gold standard" for judging whether a treatment does more good than harm. However some questions about therapy do not require randomised trials (successful interventions for otherwise fatal conditions) or cannot wait for the trials to be conducted. And if no randomised trial has been carried out for our patient's predicament, we must follow the trail to the next best external evidence and work from there.

(Sackett *et al.* 1996, 72)

Moreover, in the selection criteria for articles to be abstracted in the journal *Evidence-Based Medicine*, randomization is again required only for therapeutic trials, while an explicitly more open policy is declared toward studies of causation:

Criteria for studies of causation: a clearly identified comparison group for those at risk for, or having, the outcome of interest (whether from randomised, quasi-randomised, or nonrandomised controlled trials; cohort-analytic studies with case-by-case matching or statistical adjustment to create comparable groups; or case-control studies); masking of observers of outcomes to exposures (assumed to be met if the outcome is objective [e.g. all-cause mortality or an objective test]); observers of exposures masked to outcomes for case-control studies OR masking of subjects to exposure for all other study designs.

(*Evidence-Based Medicine*, **1**, 1, 2)

Finally, in a 1995 article titled "Clinical Practice is Evidence-Based," randomized trials are explicitly deemed inessential for "Group 2 interventions." These are defined as follows:

Intervention with convincing non-experimental evidence – Interventions whose face validity is so great that randomised trials were unanimously judged by the team to be both unnecessary, and, if a placebo would have been involved, unethical. Examples are starting the stopped hearts of victims of heart attacks and transfusing otherwise healthy individuals in haemorrhagic shock. A self-evident intervention was judged effective for the individual patient when we concluded that its omission would have done more harm than good.

(Sackett *et al.* 1995, 408–409)

In sum,

1   RCTs are not required except for trials of therapy.

2   Even in the case of therapy, RCTs are sometimes unnecessary – for example, in

"successful interventions for otherwise fatal conditions" (notice, by the way, that this seems clearly to imply that this is not just a pragmatic matter, we can properly judge an intervention "successful" independently of an RCT).

3   Moreover, even presumably outside of such cases, RCTs may be deemed – presumably again *properly* deemed – unnecessary in the case of interventions with "convincing non-experimental evidence" defined to be those whose "face-validity" is agreed on unanimously by "the [presumably unusually competent] team."

4   In the case of therapy, the RCT undoubtedly represents the "gold standard," while other non-randomized trials "routinely lead to false positive conclusions about efficacy." But despite this, and in general (and so in particular in the case of trials of therapy), "no RCT" should not be taken to entail "no scientific evidence" – instead "we must follow the trail to the next best external evidence and work from there." (And, of course, EBM supplies a hierarchy of strength of evidence – starting always with RCTs as the highest and working down toward clinical experience.)

No one, of course, disputes that EBM needed a more qualified account of evidence in medicine – the claim that the only real scientific evidence is that obtained from an RCT may be clear, clean, and crisp, but then it is clearly untenable. The problem is not that the position just quoted is qualified, but that the qualifications are not explained. Several justificatory questions emerge once an attempt is made to think through the various claims and concessions. These include:

(1) What exactly does the view on the special power of randomization amount to once it is agreed that, even for therapeutic claims, nonrandomized controlled evidence can sometimes be (effectively) conclusive? (Note that everyone, even the staunchest advocate of the virtues of randomization, in the end admits this – even if only in the small print.[2] After all, everyone agrees that there is no doubt that aspirin is effective for minor headaches, that penicillin is effective for pneumonia, that appendectomy may be beneficial in the case of acute appendicitis and so on and so on – yet none of these therapies has ever been subjected to an RCT. There is, note, no hint of "second-best" here – the effectiveness of these therapies is regarded, surely correctly, as at least as well established as that of therapies that *have been* successfully subjected to RCTs.)

(2) The effectiveness of no therapy is "self-evident." In calling a therapy's effectiveness "self evident" what is presumably meant is that that effectiveness is properly established by evidence we already have. But then, since that is, by definition, *pre*-RCT evidence, this in fact again concedes that other evidence may be, at least to all intents and purposes, compelling. So, again, why such an emphasis on RCTs now?

(3) Why, if randomization is not specially privileged in the case of studies of causation, should it have this "highly preferred, if not strictly necessary" status concerning trials of therapy?

(4) What justifies the hierarchy of evidence involved in EBM and just how far down that hierarchy are scientific clinicians supposed to go in search of the "next best" evidence – presumably there should be some point at which we ought to admit that there is no real evidence at all, but only unjustified opinion?

Contrary, perhaps, to certain fashionable views in philosophy of science about the inevitable (and welcome) "disunity" of methods, it must surely be a good idea to at least attempt to find some unified, general "first principles" perspective from which to answer these questions, and hence to supply some sort of general rationale for the complex position summarised in points 1 to 4 (or, more likely, for some modified version of that position). This is, of course, a very tall order and I make no pretence to meet it fully here. But some important first steps can be made. These stem from reexamining the main arguments for the special power of RCTs. We shall see that at least some of the tensions in the complex EBM position on evidence may result from a continuing overestimation of the epistemic power of the RCT.

3 Why randomize?

There have traditionally been three answers to this question – to which, as we will see, a fourth answer of a reliabilist kind was added later. (My account of the first two of the traditional answers follows the earlier treatments of Peter Urbach – from which I have also taken a number of quotations from other authors.[3])

3a The Fisherian argument from the logic of significance testing

Fisher argued that the logic of the classical statistical significance test requires randomization. Fisher wrote that it is only "[t]he full method of randomisation by which the validity of the test of significance can be guaranteed" (1947, 19), and Fisher's claim is repeatedly echoed by classical frequentist statisticians.[4]

An argument that some observed outcome of a trial was "statistically significant" at, say, the 95% level, can be valid, so this line of reasoning goes, only if the division between control and experimental groups was made by some random process so that any given individual in the trial had the same probability of landing in either group. Only then might the observed data imply that an outcome has happened that has only a 5% chance or less of happening if there is no real difference in therapeutic effect between the

two treatments (standard and experimental or placebo and experimental) involved in the trial.

I shall not consider this often-examined argument in any detail here (it is in any event not the one that has carried most persuasive force sociologically speaking). I just report first that it is not in fact clear that the argument is convincing even on its own terms;[5] and *secondly* that there are, of course, many – not all of them convinced Bayesians – who regard the whole of classical signficance-testing as having no epistemic validity, and hence who would not be persuaded of the need for randomisation even if it *had* been convincingly shown that the justification for a significance test presupposes randomization.

3b Randomization "controls for all variables, known and unknown"

The *second* traditional argument for the power of randomization is the one that seems chiefly to have persuaded the medical community that RCTs supply the "gold standard."

The basic logic behind *controlling* trials is, at least superficially, clear. First the *post hoc ergo propter hoc* fallacy must be avoided – the fact that, to take a hackneyed example, a large group of people suffering from a cold all recovered within a week when given regular vitamin C would constitute no sort of evidence for the efficacy of vitamin C for colds without evidence from a "control group" who were given some other treatment (perhaps none) and whose colds proved more tenacious. But not just any control group will do. The effects of the factor whose effect is being investigated must be "shielded" from other possible confounding factors. Suppose those in the experimental group taking vitamin C recovered from their colds much better on average than those in the control group. This would still constitute no sort of evidence for the efficacy of vitamin C if, say, the general state of health of the members of the experimental group was considerably more robust than that of members of the control group. The control and experimental groups could be deliberately *matched* relative to some features, and, despite the qualms of some avid randomizers, surely ought to be matched with respect to factors that there is some good reason to think may play a role in recovery from, or amelioration of the symptoms of, the condition at issue. But even laying aside issues about the practicality of matching with respect to any reasonable number of factors, it is of course *in principle* impossible to match for all *possible* "confounding" factors. At most we can match for all the "known" (possibly) confounding factors. (This really means all those it is reasonable to believe, on the basis of background knowledge, might play a role.) There is, however, clearly an indefinite number of unknown factors that *might* play a causal role. Even a pair of experimental and control groups matched perfectly with respect to all "known" confounding factors might of course be significantly skewed with respect to one or more unknown factors. Thus the possibility is always open that any observed positive effect might be due, not to the treatment at issue, but to the greater representation of patients with unknown factor X within one or the other group.

This is where, according to this very influential line of reasoning, randomization, and randomization alone, can come to the rescue. It is often supposed that by dividing the patients in a study into experimental and control groups by some random process *all* possible confounding factors, both known *and unknown*, are controlled for at once.

Ron Giere says exactly as much: randomized groups "are automatically controlled for ALL other factors, even those no one suspects." (1979, 296). And so did Fisher (1947, 19):

> The full procedure of randomisation [is the method] by which the test of significance

may be guaranteed against corruption by the causes of disturbance which have not been eliminated [that is, not deliberately controlled for ahead of the randomisation].

This claim is of course, if taken literally, trivially unsustainable. It is perfectly possible that a properly applied random process might "by chance" produce a division between control and experimental groups that is significantly skewed with respect to some uncontrolled prognostic factor that in fact plays a role in therapeutic outcome. Giere, and above all Fisher, of course knew this and so presumably what they meant, despite what they say, is something weaker – only that the randomization controls for all factors, known or unknown, "in some probabilistic sense."

And in fact most of those who advocate RCTs choose their words more carefully in line with this weaker formulation. But what exactly might that weaker claim amount to? Schwartz *et al.* suggest that

> Allocating patients to treatments A and B by randomisation produces two groups which are alike *as possible* with respect to all their characteristics, both known and unknown.
>
> (1980, 7; emphasis supplied).

While Byar *et al.* in their highly influential paper claim that

> randomisation *tends to* balance treatment groups in covariates (prognostic factors), whether or not these variables are known. This balance means that the treatment groups being compared will in fact *tend* to be truly comparable.
>
> (1976, 75; the emphases are mine)

And Sheila Gore (1981, 1958) talks of randomisation as supplying a "long run insurance against possible bias".

Presumably what is being claimed here is that if the division between experimental and control group is made at random then, *with respect to any one given possible unknown prognostic factor*, it is improbable that its distribution between the two groups is very skewed compared to the distribution in the population as a whole – the improbability growing with the degree of skewedness and with the size of the trial. Hence, if the randomization were performed indefinitely often, the number of cases of groups skewed with respect to that factor would be very small. The fact, however, is that a given RCT has not been performed indefinitely often but only once. Hence it is of course possible that, "unluckily," the distribution even of the one unknown prognostic factor we are considering is significantly skewed between the two groups. And indeed, the advice given by even the staunchest supporters of RCTs is that, should it be noticed after randomization that the two groups are unbalanced with respect to a variable that may, on reflection, play a role in therapeutic outcome, then one should either re-randomize or employ some suitable adjustment technique to control for that variable post hoc. (This of course again gives the lie to the idea, not seriously held but nonetheless highly influential, that randomization *guarantees* comparability of experimental and control groups. It also seems to me to render even more doubtful the advice that one quite often hears from advocates of RCTs that one should, in the interests of simplicity and pragmatic efficiency, explicitly control for few, if indeed any, variables – relying on the randomization to control for all variables. Surely any telling trial *must* be deliberately controlled for all factors that it seems plausible to believe, in the light of background knowledge, may well play a role in therapeutic outcome.[6])

But, moreover, whatever may be the case with respect to *one* possible unknown "confounder," there is, as the Bayesian Dennis Lindley among others has pointed out (1982), a major difficulty once we take into account the fact that there are indefinitely many possible confounding factors. Even if there is only a small probability that an individual factor is unbalanced, given that there are indefinitely many possible confounding factors, then it would seem to follow that the probability that there is some factor on which the two groups are unbalanced (when remember randomly constructed) might for all anyone knows be high. Prima facie those frequentist statisticians who argue that randomization "tends" to balance the groups in all factors commit a simple quantificational fallacy.

3c Selection bias

The third argument for the value of randomized controls is altogether more down to earth (and tends to be the one cited by the advocates of RCTs when their backs are against the wall).[7] If the clinicians running a trial are allowed to determine which arm a particular patient is assigned to then, whenever they have views about the comparative merits and comparative risks of the two treatments – as they standardly will, there is a good deal of leeway for those clinicians to affect the outcome of the trial. (The motive here might be simply, and worthily, to protect the interests of their patients; it may also, more worryingly, be that, because of funding considerations, they have a vested interest in producing a 'positive' result.) Clinicians may for example – no doubt subconsciously – predominantly direct those patients they think are most likely to benefit to the new treatment or, in other circumstances, they may predominantly direct those whom they fear may be especially badly affected by any side-effects of the new treatment to the control group (which will generally mean that those patients who are frailer will be over-represented in the control group, which is of course likely to overestimate the effectiveness of the therapy under test). Moreover, since the eligibility criteria for a trial are always open to interpretation, if a clinician is aware of the arm of trial that a given patient will go into (as she will be in unblinded studies), then there is room for that clinician's views about whether or not one of the therapies is likely to be more beneficial to affect whether or not that patient is declared eligible. (Remember that anyone declared ineligible for the trial will automatically receive "standard treatment.")

The fact that the investigator chooses the arm of the trial that a particular patient joins also means, or at any rate usually means, that the trial is at best single blind. This in turn opens up the possibility that the doctor's expectations about the likely success or failure may subconsciously play a role in affecting the patient's attitude toward the treatment s/he receives, which may in turn affect the outcome – especially where the effect expected is comparatively small. Finally, performing the trial single-blind also means that the doctor knows which arm the patient was on when coming to assess whether or not there was any benefit from whichever treatment was given – the doctor's own prior beliefs may well affect this judgment whenever the outcome measure is at any rate partially subjective.

It is undeniable that selection bias may play a role. Because it provides an alternative explanation for positive outcomes (at any rate for small positive effects),[8] we need to control for such bias before declaring that the evidence favors the efficacy of the treatment. One way to control is by standard methods of randomization – applied after the patient has been declared eligible for the trial. The arm that a given patient is assigned to will then be determined by the toss of a coin (or equivalent) and not by any clinician.

This seems to me a cast-iron argument for randomization: far from facing methodological difficulties, it is underwritten by the simple but immensely powerful general principle that one should test a hypothesis against plausible alternatives before pronouncing it well supported by the evidence. The theory that any therapeutic effect, whether negative or positive, observed in the trial is caused (or "largely caused") by selection bias is always a plausible alternative theory to the theory that the effect is produced by the characteristic features of the therapeutic agent itself. Notice however that randomization as a way of controlling for selection bias is very much a means to an end, rather than an end in itself. It is blinding (of the clinician) that does the real methodological work – randomization is simply one method of achieving this.

3d Observational studies are "known" to exaggerate treatment effects

A quite different argument for the superior evidential weight of results from randomized trials is based on the claim that, whatever the methodological rights and wrongs of the randomization debate, we just know on an empirical basis that random allocation leads to more trustworthy results. This is because there are studies that "show" that non-randomized trials routinely exaggerate the real effect. For example, the 1996 clarification of the EBM position (Sackett *et al.* 1996, 72) cites the following reason for downgrading "observational studies" (at least when it comes to assessing therapy): "It is when asking questions about therapy that we should try to avoid the non-experimental approaches, *since these routinely lead to false positive conclusions about efficacy*" (emphasis supplied).[9]

This claim stems from work in the 1970s and 1980s[10] which looked at cases where some single treatment had been assessed using *both* randomized *and* non-randomized trials – in fact the latter usually involved "historical controls." These studies found that, in the cases investigated, the historically controlled trials tended to produce more "statistically significant" results and more highly positive point-estimates of the effect than RCTs on the same intervention.

This issue merits careful examination; but I here simply make a series of brief points:

(1) The claim that these studies, even if correct, show that historically controlled trials exaggerate the true effect follows *only* if the premise is added that RCTs measure that true effect (or at least can be reliably assumed to come closer to it than trials based on other methods). Without that premise, the data from these studies is equally consistent with the claim that RCTs consistently *underestimate* the true effect.[11]

(2) It is of course possible that the particular historically controlled trials that Chalmers *et al.* compared to the RCTs were comparatively poorly done; and indeed they themselves argued that the control and experimental groups in the trials they investigated were "maldistributed" with respect to a number of plausible prognostic factors. (Notice that if such maldistributions can be recognized, then it is difficult to see any reason why they should not be controlled for post hoc, by standard adjustment techniques.)

(3) More recent studies of newer research in which some therapeutic intervention has been assessed using both RCTs and "observational" (non-randomized) trials have come to quite different conclusions than those arrived at by Chalmers *et al.* Kunz and Oxman (1998, 1188) found that

> Failure to use random allocation and conceal-ment of allocation were associated with rela-tive increases in estimates of effects of 150% or more, relative decreases of up to 90%, inversion of the estimated effect and, in some cases, no difference.[12]

Benson and Hartz (2000), comparing RCTs to "observational" trials with concurrent but non-randomly selected control groups, found, still more significantly,

> little evidence that estimates of treatment effects in observational studies reported after 1984 are either consistently larger than or qualitatively different from those obtained in randomized, controlled trials (1878).

And they suggested that the difference between their results and those found earlier by Chalmers *et al.* may be due to the more sophisticated methodology underlying the "observational studies" investigated: "Possible methodologic improvements include a more sophisticated choice of data sets and better statistical methods. Newer methods may have eliminated some systematic bias" (Benson and Hartz 2000, 1878).

In the same issue of the *New England Journal of Medicine*, Concato, Shah, and Horwitz argue that "[t]he results of well-designed observational studies . . . do not systematically overestimate the magnitude of the effects of treatment as compared with those in RCTs on the same topic" (2000, 1887).

These authors emphasize that their findings "challenge the current [EBM-based] consensus about a hierarchy of study designs in clinical research." The "summary results of RCTs and observational studies were remarkably similar for each clinical topic [they] examined"; while investigation of the spread of results produced by single RCTs and by observational studies on the same topic revealed that the RCTs produced much the greater variability. Moreover, the different observational studies despite some variability of outcome none the less all pointed in the same direction (treatment effective or ineffective); while, on the contrary, the exami-nation of cases where several RCTs had been performed on the same intervention produced several "paradoxical results" – that is, cases of individual trials pointing in the opposite direction to the "overall" result (produced by techniques of meta-analysis).

(4) This last point is in line with the result of the 1997 study by Lelorier *et al.* who found, contrary at least to what clinicians tend to believe when talking of RCTs as the "gold standard," that

> the outcomes of ... large randomised, controlled trials that we studied were not predicted accurately 35% of the time by the meta-analyses published previously on the same topics.
>
> (Lelorier *et al.* 1997, 536)

(That is, 35% of the individual RCTs came to a different judgment as to whether the treatment at issue was effective or not than that delivered by the meta-analysis of all such trials on that treatment.)

Unless I have missed something, these more recent (meta-)results have completely blown the reliabilist argument for RCTs out of the water.

4 Toward a unified account of the evidential weight of therapeutic trials

So, recall the overall project: I embarked on a critical examination of the arguments for randomization, not for its own sake, but with a view to finding a general explanation from first principles of the complex account of clinical evidence given by EBM, or of some modified version of it. Here is the "first principles" view that *seems to* be emerging from that critical examination.

Of course clinical practice should be based on best evidence. Best evidence for the positive effect of a therapeutic intervention arises when plausible alternative explanations for the difference in outcomes between experimental and control groups have been eliminated. This means

controlling for plausible alternatives. There are, of course, indefinitely many possible alternative causal factors to the characteristic features of the intervention under test. But "background knowledge" indicates which of these are *plausible* alternatives. It is difficult to see how we can do better than control, whether in advance or post hoc, for all plausible alternatives. The idea that randomization controls all at once for known and unknown factors (or even that it "tends" to do so) is a will-o'-the-wisp. The only solid argument for randomization appears to be that standard means of implementing it have the side-effect of blinding the clinical experimenter and hence controlling for a known factor – selection bias. But if selection bias can be eliminated or controlled for in some other way, then why should randomization continue to be thought essential.

Notice that my analysis supplies no reason to take a more negative attitude toward the results of RCTs, but it does seem clearly to indicate a more positive attitude toward the results of carefully conducted (i.e. carefully controlled) non-randomized studies.[13] If something like the line that emerges from that critical survey is correct, then we do indeed move toward a more unified overall account of clinical evidence than that summarized in points 1 to 4 considered in section 2 above.

Notes

I am especially indebted to Dr. Jennifer Worrall, for extensive discussions of the issues raised in this paper. Unlike myself, Dr. Worrall has first hand experience both of clinical trials and of the problems of trying to practice medicine in an evidence-based way – and hence her contributions to this paper have been invaluable. I am also greatly indebted to my former colleague, Dr. Peter Urbach, who first brought me to question the orthodoxy on the value of randomization (see Urbach 1985, 1993 and Howson and Urbach 1993 – as noted Urbach's work forms the basis for some parts of my

1 In fact the advocates of RCTs in general, whether explicit EBM-ers or not, tend to hide a much more guarded view behind slogans like those just quoted. The "fine print view" tends to be that, at least under some conditions, some useful and "valid" information can sometimes be gleaned from studies that are not randomized; but that randomized trials are undoubtedly epistemically *superior*. So, Stuart Pocock for example writes: "it is now generally accepted that the *randomised controlled trial* is the most reliable method of conducting clinical research" (1983, 5). Or Grage and Zelen (1982) assert that the randomized trial "is not only an elegant and pure method to generate reliable statistical information, but most clinicians regard it as the most trustworthy and unsurpassed method to generate the unbiased data so essential in making therapeutic decisions" (24).

2 See, for example, Doll and Peto 1980.

3 See Urbach 1985, 1993 and Howson and Urbach 1993, chap. 11. A fuller treatment than I can give here would respond to the criticisms of Urbach's views by David Papineau (1994).

4 Byar et al. 1976 for example assert, explicitly in connection with RCTs, that "randomisation guarantees the statistical tests of significance that are used to compare the treatments." Or, in the same paper: "It is the process of randomisation that generates the significance test."

5 See, for example, Lindley 1982 and Howson and Urbach 1993, chap. 11.

6 Peto et al. (1976), for instance, hold that stratification (a form of matching) "is an unnecessary elaboration of randomisation." Stuart Pocock (1983) holds that while, to the contrary, in some circumstances "stratification would seem worthwhile," this is not true in the case of larger trials.

7 For example, Doll and Peto (1980) write that the main objection to historically controlled trials and the main reason why RCTs are superior is "that the criteria for selecting patients for treatment with an exciting new agent or method may differ from the criteria used for selecting the control patients."

8 Doll and Peto (1980) claim that selection bias "cannot plausibly give rise to a *tenfold* artefactual difference in disease outcome . . . [but it may and often does] easily give rise to *twofold* artefactual differences. Such twofold biases are, however, of critical importance, since most of the really important therapeutic advances over the past decade or so have involved recognition that some particular treatment for some common condition yields a *moderate but important* improvement in the proportion of favourable outcomes."

9 Or, as Brian Haynes suggested to me, the superiority of random allocation is "not merely a matter of logic, common sense or faith: non-random allocation usually results in more optimistic differences between intervention and control groups than does random allocation."

10 See in particular Chalmers, Matta, Smith, and Kunzler 1977 and Chalmers, Celano, Sacks, and Smith 1983.

11 And indeed there are now some claims that they do exactly that; see e.g. Black 1996.

12 Kunz and Oxman take themselves to be looking at the variety of "distortions" that can arise from not randomizing (or concealing). They explicitly concede, however, that "we have assumed that evidence from randomised trials is the reference standard to which estimates of non-randomised trials are compared." Their subsequent admission that "as with other gold standards, randomised trials are not without flaws and this assumption is not intended to imply that the true effect is known, or that estimates derived from randomised trials are always closer to the truth than estimates from non-randomised trials" leaves their results hanging in thin air. Indeed their own results showing the variability of the results of randomized and non-randomized on the same intervention seems intuitively to tell strongly against their basic assumption. (They go on to make the interesting suggestion that "it is possible that randomised controlled trials can sometimes underestimate the effectiveness of an intervention in routine practice by forcing healthcare professionals and patients to acknowledge their uncertainty and thereby reduce the strength of the placebo effect.")

13 Right from the beginning EBM-ers seemed to have been ready to move from consideration of some spectacularly flawed non-randomized study or studies (where alternative explanations for the observed effect leap out at you) to the conclusion that such studies are inherently flawed (and therefore that we need an alternative in the form of the RCT). A notable example is provided by Cochrane (1972, chap. 4), who is very much a father figure to the movement.

References

Benson, K. and A.J. Hartz (2000), "A Comparison of Observational Studies and Randomised, Controlled Trials", *New England Journal of Medicine* **342**: 1878–1886.

Black, N. (1996), "Why We Need Observational Studies to Evaluate the Effectiveness of Health Care", *British Medical Journal* **312**: 1215–1218.

Byar, D.P. *et al.* (1976), "Randomized Clinical Trials: Perspectives on Some Recent Ideas", *New England Journal of Medicine* **295** (2): 74–80.

Chalmers T.C., R.J. Matta, H. Smith, Jr., and A.M. Kunzler (1977), "Evidence Favoring the Use of Anticoagulants in the Hospital Phase of Acute Myocardial Infarction" *New England Journal of Medicine* **297**: 1091–1096.

Chalmers, T.C., P. Celano, H.S. Sacks, and H. Smith, Jr. (1983), "Bias in Treatment Assignment in Controlled Clinical Trials", *New England Journal of Medicine* **309**: 1358–1361.

Cochrane, A.I. (1972), *Effectiveness and Efficiency. Random Reflections on the Health Service.* Oxford: The Nuffield Provincial Hospitals Trust.

Concato, J., N. Shah, and R.I. Horwitz (2000), "Randomised Controlled Trials, Observational Studies, and the Hierarchy of Research Designs", *New England Journal of Medicine* **342**: 1887–1892.

Doll, R. and R. Peto (1980), "Randomised Controlled Trials and Retrospective Controls", *British Medical Journal* **280**: 44.

Fisher, R.A. (1947), *The Design of Experiments*, 4th ed. Edinburgh: Oliver and Boyd.

Giere, R.N. (1979), *Understanding Scientific Reasoning.* New York: Holt, Rinehart and Winston.

Gore, S.M. (1981), "Assessing Clinical Trials – Why Randomise?", *British Medical Journal* **282**: 1958–1960.

Grage, T.B. and M. Zelen (1982), "The Controlled Randomised Trial in the Evaluation of Cancer Treatment – the Dilemma and Alternative Designs", *UICC Tech. Rep. Ser.* **70**: 23–47.

Howson, C. and P.M. Urbach (1993), *Scientific Reasoning – the Bayesian Approach*, 2nd ed. Chicago and La Salle: Open Court.

Kunz, R. and A.D. Oxman (1998), "The Unpredictability Paradox: Review of Empirical Comparisons of Randomised and Non-Randomised Clinical Trials", *British Medical Journal* **317**: 1185–1190.

Lelorier, J. et al. (1997), "Discrepancies between Meta-Analyses and Subsequent Large Randomised Controlled Trials", *Journal of the American Medical Association* **337**: 536–542.

Lindley, D.V. (1982), "The Role of Randomisation in Inference", *PSA 1982*, vol. 2. East Lansing, MI: Philosophy of Science Association, 431–446.

Papineau, D. (1994). "The Virtues of Randomisation", *British Journal for the Philosophy of Science* **45**: 451–466.

Peto, R. et al. (1976), "Design and Analysis of Randomised Clincial Trials Requiring Prolonged Observation of Each Patient: I. Introduction and Design", *British Journal of Cancer* **34**: 585–612.

Pocock, S.J. (1983), *Clinical Trials – A Practical Approach.* New York: Wiley.

Sackett, D.L. et al. (1995), "Clinical Practice is Evidence-Based", *The Lancet* **350**: 405–410.

—— (1996), "Evidence-Based Medicine: What It Is and What It Isn't", *British Medical Journal* **312**: 71–72.

Schwarz, D. et al. (1980), *Clinical Trials.* London: Academic Press.

Tukey, J.W. (1977), "Some Thoughts on Clinical Trials, Especially Problems of Multiplicity", *Science* **198**: 679–684.

Urbach, P.M. (1985), "Randomisation and the Design of Experiments", *Philosophy of Science* **52**: 256–273.

—— (1993), "The Value of Randomisation and Control in Clinical Trials", *Statistics in Medicine* **12**: 1421–1431.

PART SEVEN

Probability in action: forensic science

INTRODUCTION TO PART SEVEN

PROBABILITY THEORY is a relatively recent addition to mathematics. This is surprising because it has aspects that seem to be elementary. Games of chance were played in antiquity but it was only in the sixteenth and seventeenth centuries that the mathematical theory of probability began. Since then there has been a massive growth in the subject, and probability theory is now a fundamental and indispensable part of science. Probability has two main aspects: the first is uncertainty and the second is frequency. These are closely related since, if a certain kind of event may happen but we are uncertain about whether it will, then matching our degree of uncertainty about it to the frequency of that kind of event would seem appropriate. So, for example, if one is uncertain whether one will roll a six on a die, then given that sixes occur once every six throws on average, one should proportion one's belief that one will roll a six accordingly. The exact relationship between what are called 'degrees of belief' and probability is controversial. 'Bayesians' hold that probabilities just are such degrees of belief in the face of uncertainty, while others argue that probabilities must always be tied to frequencies.

Statistics delivers information about frequencies and other facts about populations relevant to the estimation of probabilities, so probability theory and statistics are intimately related in theory and in practice. Both are essential to the functioning of the courts in many countries because forensic science can only give uncertain although perhaps highly probable evidence. This was so even when fingerprint evidence was the limit of identification technology. A fingerprint expert might look at a partial print from a crime scene and only be able to say that it is very unlikely that it did not come from a suspect. Such evidence is always prone to abuse, and biases about statistical evidence and fallacies of probabilistic reasoning may systematically favour prosecution or defendant or neither. In any case, it is clearly of grave importance that the foundations of forensic science are clear. It is remarkable therefore that the rules used for the interpretation of even fingerprint evidence are not unanimously agreed upon. For example, in the UK a fingerprint expert will compare a print from a suspect with a print from a crime scene and log the similarities before arriving at a view as to whether the suspect left the print at the crime scene. In the Netherlands, the expert is required to first look at the crime scene print and to make a list of distinguishing features before looking at the suspect's print to see if the same features are in evidence. The difference in methodology may be important because it is well known that people primed to look for similarities between things are more likely to ignore features that differ between them (this is called 'confirmation bias'). The expert who first logs the distinguishing features of the crime scene print is therefore more likely to include features that do not match the features of the suspect's print when they are different.

The problem with uncertainty and probability is that it often cannot be carried over into decision making. In the legal context, people are either convicted or acquitted (in most legal systems no more nuanced judgments are possible), and someone cannot be 96% sent to jail. The fallacies of probabilistic reasoning we will discuss are easy to make and they could result in someone being wrongly convicted or wrongly acquitted. Unfortunately, it seems that people are extremely bad at probabilistic reasoning without training, and even those who are trained still make elementary errors. This is well illustrated by two fallacies that people are very prone to make.

The base rate fallacy

The base rate fallacy is illustrated by the case of testing for a disease. Suppose the condition is rare so that only one person in 1000 has it. Suppose that the test is accurate so that only 2% of those tested will test positive when they don't have the disease (that is, only a 2% false positive rate), and suppose that the false negative rate is zero so all of those who have the disease will test positive. Suppose that a person tests positive; what is the probability that they have the disease? Is it very likely, likely, evens, unlikely or very unlikely? Most people seem to think that the answer is at least 'likely' but in fact it remains very unlikely that he or she has the disease. This is because the overall probability that they have the disease must take into account the 'base rate' of one in 1000 as well as the reliability of the test. To see this, consider the fact that two in 100 is 20 in 1000; so, in a population of 1000, although only one person will have the disease another 20 will test positive for it. Hence, overall, only one person in 21 who tests positive has the disease; and so the probability that any given individual who tests positive has the disease is only about 5%, which is still very unlikely although much higher than the background chance of 0.1%.

The 'Monty Hall' problem

Imagine you are on a game show and that you are allowed to choose one of three doors. Behind one will be a prize and behind the other two is nothing or a booby prize. The game show host tells you that he knows where the prize is and that after you have chosen one of the doors he will open another door behind which the prize will definitely not be. You chose, say, door 1, and the host opens another door, say, door 2. You are now allowed to change your choice of door if you wish. The question is, 'should you do so?' The answer is, 'yes you should'. When this question and the solution were published in a column in a magazine in 1990, around 10,000 readers, including about 1000 with PhDs and including a number of professors of statistics, wrote in to say that the solution was wrong and that it is not rationally required that one switch one's choice. To see why you should, consider the fact that when you originally chose you had a 1/3 chance of being right; hence the probability that the prize is behind one of the other two doors is 2/3. But if it is behind one of the other two doors, it will not be behind the one that the host opens. So if you switch, your chance of getting the prize goes from 1/3 to 2/3.

The reading from Thompson and Schumann introduces the idea of 'incidence rate'. Suppose we are interested in a population of people, some of whom are lawyers and some of whom carry briefcases, and suppose that most lawyers carry briefcases. If one observes someone with a briefcase and wants to estimate the probability that he or she is a lawyer, one must know both the number of lawyers in the population – the base rate – and also the prevalence of briefcase carrying in the population – the incidence rate. By analogy, the incidence rate in a criminal trial will be the proportion of people in the population who share some characteristic – say, a kind of fingerprint pattern, or genetic marker – with those found at the crime scene. The 'Prosecutor's Fallacy' explained in the reading is very similar to the base rate fallacy above. Suppose that there are very few lawyers and that all carry briefcases, and that only one person in ten carries a briefcase. If one observes someone with a briefcase it does not follow that there is a one in ten chance that he or she is a lawyer. Similarly, even if only a small proportion of people in a population have a certain genetic marker, and even if that marker is found at a crime scene and in a suspect, it does not follow that it is very likely that the suspect is guilty because the background probability that the suspect is guilty may otherwise be very low. The 'Defense Attorney's Fallacy' is the converse error of assuming that if a rare blood type or genetic marker is found at the crime scene and in a suspect then, since a rare characteristic may still be found among many people, this information is irrelevant.

The particularly worrying thing about the way probabilities are used in criminal trials is that how prone people are to committing fallacies when reasoning about them has been shown to depend on how the information is presented. For example, in the case of the test for the disease discussed above one can present the information about the false positive rate either as a frequency – two in 100 people will test positive even though they lack the disease – or as a 'conditional probability' – the probability that you test positive, given that you lack the disease, is 2%. In consideration of criminal cases it seems that conditional probabilities make people more likely to commit the Prosecutor's Fallacy. One can see why. Returning to the disease example, if we say that there is only a 2% chance that you will test positive if you do not have the disease, this makes it seem that the chance of having the disease if you do test positive is 98%. When it is put in terms of frequencies as we saw above, the false positive rate means that 20 in 1000 people will test positive even though only one in 1000 will have the disease. The point is that the probability of testing positive in the absence of the disease being 2% does not mean that the probability of having the disease if tested positive is 98%. Similarly, though it may be true that there is only a 2% probability of a suspect having blood or genetic markers that match the crime scene if he or she is innocent, it does not follow that if they do match they are 98% likely to be guilty.

People are correspondingly more likely to commit the Defence Attorney's Fallacy if they are presented with frequency information. They are neglecting the fact that, in the case of the base rate fallacy, the positive test result does raise the probability that the person has the disease from 0.1% to something close to 5%, which is a very significant increase.

The second reading, by Never Sesardić, takes us deeper into the fraught business of using statistics to estimate probabilities for consumption by judges and juries. The author argues that eminent statisticians commenting on a criminal trial involving alleged homicide that could be attributed instead to sudden infant death syndrome made a number of errors and wrongly accused a key prosecution witness of elementary errors. The paper uses a Bayesian analysis.

Chapter 29

William C. Thompson and Edward L. Schumann

INTERPRETATION OF STATISTICAL EVIDENCE IN CRIMINAL TRIALS: THE PROSECUTOR'S FALLACY AND THE DEFENSE ATTORNEY'S FALLACY

In criminal cases where the evidence shows a match between the defendant and the perpetrator on some characteristic, the jury often receives statistical evidence on the incidence rate of the "matching" characteristic. Two experiments tested undergraduates' ability to use such evidence appropriately when judging the probable guilt of a criminal suspect based on written descriptions of evidence. Experiment 1 varied whether incidence rate statistics were presented as conditional probabilities or as percentages, and found the former promoted inferential errors favoring the prosecution while the latter produced more errors favoring the defense. Experiment 2 exposed subjects to two fallacious arguments on how to interpret the statistical evidence. The majority of subjects failed to detect the error in one or both of the arguments and made judgments consistent with fallacious reasoning. In both experiments a comparison of subjects' judgments to Bayesian norms revealed a general tendency to underutilize the statistical evidence. Theoretical and legal implications of these results are discussed.

Introduction

Crime laboratories often play an important role in the identification of criminal suspects (Saferstein, 1977; Schroeder, 1977; Giannelli, 1983). Laboratory tests may show, for example, that blood shed by the perpetrator at the scene of the crime matches the suspect's blood type (Jonakait, 1983), that a hair pulled from the head of the perpetrator matches samples of the suspect's hair (Note, 1983), or that carpet fibers found on the victim's body match the carpet in the suspect's apartment (Imwinkelried, 1982b). Testimony about "matches" found through these comparisons is called associative evidence (Stoney, 1984). It is increasingly common in criminal trials where the defendant's identity is at issue (Imwinkelried, 1982b; Schroeder, 1977; Peterson et al., 1984).

Associative evidence is sometimes accompanied by statistical testimony about the *incidence rate* of the "matching" characteristic. Where tests show the defendant and perpetrator share the same blood type, for example, an expert may provide information on the percentage of people in the general population who possess that blood type (e.g., Grunbaum, Selvin, Pace, and Black, 1978). Where microscopic comparisons reveal a match between the defendant's hair and samples of the perpetrator's hair, the expert may provide information on the incidence rate of such "matches" among hairs drawn from different individuals (Gaudette and Keeping, 1974). During the past 15 years, forensic scientists have devoted much effort to studying the incidence rate of various characteristics of hair (Gaudette and Keeping, 1974), soil (Saferstein, 1977, pp. 63–64), glass (Fong, 1973; Davis and DeHaan, 1977), paint (Pearson *et al.*, 1971), and bodily fluids (Owens and Smalldon, 1975; Briggs, 1978; Gettinby, 1984). Statistical data from this literature are increasingly presented in criminal trials (Imwinkelried, 1982b; Note, 1983). One legal commentator, discussing research on blood typing, concluded that "our criminal justice system is now at the threshold of an explosion in the presentation of mathematical testimony" (Jonakait, 1983, p. 369).

The reaction of appellate courts to this type of evidence has been divided. The conflict stems largely from differing assumptions about the way jurors respond to incidence rate statistics. A few appellate courts have rejected such evidence on the grounds that jurors are likely to greatly overestimate its value (*People v. Robinson*, 1970; *People v. Macedonio*, 1977; *State v. Carlson*, 1978; *People v. McMillen*, 1984). The majority of jurisdictions, however, admit such evidence on the grounds that jurors are unlikely to find it misleading and will give it appropriate weight (Annotation, 1980; Jonakait, 1983). Legal commentary on the issue appears divided between those who argue that statistical evidence may have an exaggerated impact on the jury (Tribe, 1971), and those who argue that statistical evidence is likely to be underutilized (Finkelstein and Fairley, 1970; Saks and Kidd, 1980).

Which of these positions is correct? Although no studies have examined people's evaluation of incidence rate statistics directly, research does exist on people's reactions to similar types of statistical information. A number of studies have shown, for example, that when people are asked to judge the likelihood of an event they often ignore or underutilize statistics on the base rate frequency of that event (for reviews see Bar-Hillel, 1980; Borgida and Brekke, 1981). When judging whether a man is a lawyer or an engineer, for example, people tend to give less weight than they should to statistics on the relative number of lawyers and engineers in the relevant population (Kahneman and Tversky, 1973). This error has been labeled the base rate fallacy.

Because base rate statistics are similar to incidence rate statistics, one is tempted to assume incidence rates will be underutilized as well. There are important differences between the two types of statistics, however, which render this generalization problematic. Base rate statistics indicate the frequency of a target outcome in a relevant population, while incidence rate statistics indicate the frequency of a trait or characteristic that is merely diagnostic of the target outcome. Suppose one is judging the likelihood Joan, who works in a tall building and owns a briefcase, is a lawyer. The percentage of women in Joan's building who are lawyers is a base rate statistic. The percentage of women in the building who own briefcases is an incidence rate statistic. In a criminal trial, the percentage of defendants in some relevant comparison population who are guilty is a base rate statistic,[1] while the percentage of some relevant population who possess a characteristic linking the defendant to the crime is an incidence rate statistic. Because incidence rate statistics are likely to play a different role in people's inferences than base rate statistics, people's tendency to underutilize the latter may not generalize to the former.

Research on the "pseudodiagnosticity phenomenon" (Doherty, Mynatt, Tweney, and Schiavo, 1979; Beyth-Marom and Fischhoff, 1983) has examined people's reactions to a form of statistical data more closely analogous to incidence rate statistics. In one series of studies, Beyth-Marom and Fischhoff asked people to judge the likelihood that a man, drawn from a group consisting of university professors and business executives, is a professor (rather than an executive) based on the fact he is a member of the Bears Club. These researchers were interested in how people respond to information about the percentage of professors and business executives who are "Bears." Data on the percentage of business executives who are Bears are most analogous to incidence rate statistics because they speak to the probability the man would be a Bear if he is not a professor, just as incidence rate statistics speak to the probability a defendant would possess a "matching" characteristic if he is not guilty.

When subjects in these studies were asked what information they would require to evaluate the probability that the man was a professor based on the fact he is a "Bear," most were interested primarily in knowing the percentage of professors who are Bears; only half expressed an interest in knowing the percentage of executives who are "Bears," although the two types of information are equally important (Beyth-Marom and Fischhoff, 1983, Experiment 1). Moreover, those subjects who expressed an interest in the latter percentage often did so based on mistaken or illogical reasoning (Beyth-Marom and Fischhoff, 1983, Experiment 4). Nevertheless, when subjects were informed of the respective percentages, most subjects considered both and adjusted their beliefs in the proper direction (Beyth-Marom and Fischhoff, 1983, Experiment 5). Beyth-Marom and Fischhoff conclude that "people are much better at using [statistical] information . . . than they are at seeking it out . . . or articulating reasons for its usage" (p. 193).

Although the findings of Beyth-Marom and Fischhoff are hopeful, anecdotal evidence suggests people sometimes make serious errors when evaluating incidence rate statistics. One of the authors recently discussed the use of incidence rate statistics with a deputy district attorney. This experienced prosecutor insisted that one can determine the probability of a defendant's guilt by subtracting the incidence rate of a "matching" characteristic from one. In a case where the defendant and perpetrator match on a blood type found in 10% of the population, for example, he reasoned that there is a 10% chance the defendant would have this blood type if he were innocent and therefore concluded there is a 90% chance he is guilty. This assessment is misguided because it purports to determine the defendant's probability of guilt based solely on the associative evidence, ignoring the strength of other evidence in the case. If a prosecutor falls victim to this error, however, it is possible that jurors do as well.

The fallacy in the prosecutor's logic can best be seen if we apply his analysis to a different problem. Suppose you are asked to judge the probability a man is a lawyer based on the fact he owns a briefcase. Let us assume all lawyers own a briefcase but only one person in ten in the general population owns a briefcase. Following the prosecutor's logic, you would jump to the conclusions that there is a 90% chance the man is a lawyer. But this conclusion is obviously wrong. We know that the number of nonlawyers is many times greater than the number of lawyers. Hence, lawyers are probably outnumbered by briefcase owners who are not lawyers (and a given briefcase owner is more likely to be a nonlawyer than a lawyer). To draw conclusions about the probability the man is a lawyer based on the fact he owns a briefcase, we must consider not just the incidence rate of briefcase ownership, but also the a priori likelihood of being a lawyer. Similarly, to draw conclusions about the probability a criminal suspect is guilty based on evidence of a "match," we must consider not just the percentage of people who would match but also the a priori likelihood that the defendant in question is guilty.[2]

The prosecutor's misguided judgmental strategy (which we shall call the Prosecutor's Fallacy) could lead to serious error, particularly where the other evidence in the case is weak and therefore the prior probability of guilt is low. Suppose, for example, that one initially estimates the suspect's probability of guilt to be only .20, but then receives additional evidence showing that the defendant and perpetrator match on a blood type found in 10% of the population. According to Bayes' theorem, this new evidence should increase one's subjective probability of guilt to .71, not .90.[3]

There is also anecdotal evidence for a second error, which we first heard voiced by a criminal defense attorney and therefore call the Defense Attorney's Fallacy. Victims of this fallacy assume associative evidence is irrelevant, regardless of the rarity of the "matching" characteristic. They reason that associative evidence is irrelevant because it shows, at best, that the defendant and perpetrator are both members of the same large group. Suppose, for example, that the defendant and perpetrator share a blood type possessed by only 1% of the population. Victims of the fallacy reason that in a city of 1 million there would be approximately 10,000 people with this blood type. They conclude there is little if any relevance in the fact that the defendant and perpetrator both belong to such a large group. What this reasoning fails to take into account, of course, is that the great majority of people with the relevant blood type are not suspects in the case at hand. The associative evidence drastically narrows the group of people who are or could have been suspects, while failing to exclude the defendant, and is therefore highly probative, as a Bayesian analysis shows. The Defense Attorney's Fallacy is not limited to defense attorneys. Several appellate justices also appear to be victims of this fallacy (See, e.g., *People v. Robinson*, 1970). If defense attorneys and appellate justices fall victim to this fallacy, it is quite possible that some jurors do as well, thereby giving less weight to associative evidence than it warrants.

Whether people fall victim to the Prosecutor's Fallacy, the Defense Attorney's Fallacy, or neither may depend on the manner in which incidence rate statistics are presented and explained. In criminal trials, forensic experts often present information about incidence rates in terms of the conditional probability the defendant would have a particular characteristic *if he were innocent* (Jonakait, 1983). Where 1% of the population possess a blood type shared by the defendant and perpetrator, for example, experts often present only the conclusory statement that there is one chance in 100 that the defendant would have this blood type if he were innocent. This type of testimony seems especially likely to lead jurors to commit the Prosecutor's Fallacy. On the other hand, a defense tactic used in some actual cases is to point out that, notwithstanding its low incidence rate, the characteristic shared by the defendant and perpetrator is also possessed by thousands of other individuals. Where 1% of the population possess a blood type shared by the defendant and perpetrator, for example, the expert might be forced to admit during cross examination that in a city of one million people, approximately 10,000 individuals would have the "rare" blood type. Statements of this type may reduce the tendency toward the Prosecutor's Fallacy but induce more errors consistent with the Defense Attorney's Fallacy.

To test these hypotheses, Experiment 1 had subjects estimate the likelihood a criminal suspect was guilty based, in part, on statistical evidence concerning the incidence rate of a characteristic shared by the defendant and perpetrator. The part of the evidence concerning the incidence rates was presented in two different ways. In one condition the forensic expert presented only the conditional probability that an innocent person would have the "matching" characteristic. In a second condition the expert stated the percentage of the population who possess the relevant characteristic and the approximate number of people who possess this characteristic in the city where the crime occurred.

Experiment 1

Method

Subjects

All subjects in this and the following experiment were volunteers from a university human subjects pool who were given extra credit in coursework as an incentive to participate. Subjects (N = 144) were run in groups of about ten in sessions lasting one half hour. Each subject was randomly assigned to one of the two experimental conditions.

Procedure

On arriving for the experiment, subjects were given a five-page packet of stimulus materials. The first page, containing instructions, stated (a) that the experiment was designed to test people's ability to draw reasonable conclusions from evidence involving probabilities, (b) that subjects would be asked to read a description of a criminal case and to indicate their estimate of the likelihood of the suspect's guilt by writing a percentage between 0 and 100, and (c) that when making these estimates subjects should disregard the concept of reasonable doubt and indicate the likelihood the suspect "really did it," rather than the likelihood a jury would convict the suspect. The experimenter reviewed these instructions with subjects and answered any questions, then subjects read the description of the criminal case.

The case involved the robbery of a liquor store by a man wearing a ski mask. The store clerk was able to describe the robber's height, weight and clothing, but could not see his face or hair. The police apprehended a suspect near the liquor store who matched the clerk's description but the suspect did not have the ski mask or the stolen money. In a trash can near where the suspect was apprehended, however, the police found the mask and the money.

At this point, subjects made an initial estimate of the probability of the suspect's guilt based only on the information they had received to that point.

Next subjects read a summary of testimony by a forensic expert who reported that samples of the suspect's hair were microscopically indistinguishable from a hair found inside the ski mask. The expert also described an empirical study that yielded data on the probability that two hairs drawn at random from different individuals would be indistinguishable. The expert's testimony was modeled on that of the actual prosecution experts in *U.S. v. Massey* (1979) and *State v. Carlson* (1978). The empirical study described in the stimulus materials was similar but not identical to that reported by Gaudette and Keeping (1974).

The experimental manipulation was the way in which the expert described the incidence rate of matching hair. In the Conditional Probability condition, the expert reported the incidence rate as a conditional probability, stating that the study indicated there "is only a two percent chance the defendant's hair would be indistinguishable from that of the perpetrator if he were innocent. . . ." In the Percentage and Number condition, the expert reported that the study indicated only 2% of people have hair that would be indistinguishable from that of the defendant and stated that in a city of 1,000,000 people there would be approximately 20,000 such individuals. Half the subjects were assigned to each condition.

After reading all of the evidence, subjects made a final judgment of the probability of the suspect's guilt.

Results

Fallacious judgments

About one quarter of the subjects made final judgments of guilt consistent with their having fallen victim to one of the fallacies described above. Overall, 19 subjects (13.2%) were coded as victims of the Prosecutor's Fallacy because they estimated the probability of guilt to be exactly .98, which is the probability one would obtain by simply subtracting the incidence rate of the "matching" characteristic from one. Eighteen subjects (12.5%) were coded as victims of the Defense Attorney's Fallacy because their final judgment of guilt was the same as their initial judgment of guilt, which indicates they gave no weight to the associative evidence. The remaining subjects (74.3%) were coded as victims of neither fallacy because their final judgments of guilt were higher than their initial judgments (indicating they gave some weight to the associative evidence) but were less than .98.

The manner in which the statistical information was presented significantly influenced the likelihood subjects would make judgments consistent with fallacious reasoning. Table 29.1 shows the number of subjects in each condition who were coded as victims of the Prosecutor's Fallacy, Defense Attorney's Fallacy, or neither fallacy. As predicted, more subjects made judgments consistent with the Prosecutor's Fallacy when the incidence rate was presented in the form of a conditional probability (22.2%) than when it was presented as a percentage and number (4.2%; $\chi^2(2,N = 144) = 10.21$, $p < .01$). Fewer subjects committed the Defense

Table 29.1 Number of subjects rendering judgments consistent with fallacious reasoning (Experiment 1)

Fallacy committed	Mode of presentation		
	Conditional probability	Percentage and number	Total
Prosecutor's Fallacy	16	3	19
Defense Attorney's Fallacy	6	12	18
Neither fallacy	50	57	107
Total	72	72	144

Attorney's Fallacy when the incidence rate was presented as a conditional probability (8.3%) than when it was presented as a percentage and number (16.7%), but this difference was only marginally significant ($\chi^2(2,N = 144) = 5.81$, $p < .10$).[4]

Initial and final judgments

After receiving only preliminary information about the arrest of a suspect, the mean judgment of the suspect's probability of guilt was .25. Subjects in the two experimental conditions had received identical preliminary information and their initial judgments did not significantly differ [Conditional Probability condition M = .27, Percentage and Number condition M = .22; $t(142) = 1.53$, $p = .13$]. Subjects' *final* judgments of probable guilt, which took into account the "match" between the suspect's and perpetrator's hair and the reported low incidence rate of such hair, were significantly higher than initial judgments [M = .63; $t(143) = 16.10$, $p < .001$, paired comparison]. More importantly, final judgments of subjects in the Conditional Probability condition (M = .72) were significantly higher than those of subjects in the Percentage and Number condition (M = .53; $t(142) = 4.34$, $p < .001$). An analysis of covariance, using initial judgments as a covariate, confirmed that this effect was significant after controlling for any initial differences [$F(1,143) = 16.14$, $p < .001$]. This finding indicates that the manner in which incidence rate statistics were presented had an important effect on subjects' judgments of probable guilt.

Comparing subjects' final judgments to model Bayesian judgments

Between their initial and final judgments, subjects learned that the suspect's hair "matched" that of the perpetrator and that the incidence rate of such hair was only 2%. To determine whether subjects gave this information the weight it would be accorded by Bayes' theorem, subjects' final judgments were compared to "model" judgments computed by revising each subject's initial judgment in accordance with the Bayesian formula in note 2. For each subject, $p(D/H)$ was assumed to be 1.00 and $p(D/H)$ was assumed to be .02. Each subject's own initial judgment of probability of guilt was used as $p(H)$. These probabilities were combined using the Bayesian formula to yield a posteriori probability guilt for each subject. The model Bayesian judgments ($M = .93$) were significantly higher than subjects' final judgments [$M = .63$; $t(143) = 9.64$, $p < .001$, paired comparison]. This finding is consistent with the general tendency of people to be more conservative than Bayes' theorem when revising judgments in light of new information (Edwards, 1968).

Discussion

The results confirm suspicions, arising from anecdotal evidence, that people can make serious errors when judging guilt based on associative evidence and incidence rate statistics. About one quarter of subjects made judgments consistent with their having fallen victim to the Prosecutor's Fallacy or Defense Attorney's Fallacy. Furthermore, the number of subjects who were apparent victims of the fallacies, and mean judgments of guilt, were significantly affected by a subtle manipulation in the way incidence rate statistics were presented.

It is important to note that subjects in the two conditions did not receive different information. In both conditions subjects learned that the suspect and perpetrator matched on a characteristic found in 2% of the population. But this information was presented in ways that focused attention on different, though rather straightforward, implications of the data. Subjects in the conditional probability condition were told there was only a 2% chance the suspect would "match" if he was innocent. Presenting the data in this manner probably led more subjects to commit the Prosecutor's Fallacy because they falsely assumed the conditional probability of a "match" given innocence is the complement of the conditional probability of guilt given a "match."[5] In any case, reflection on the low probability that the match could have occurred by chance probably promoted the impression that the suspect is likely to be guilty. Subjects in the percentage and number condition were told that 2% of the population would "match" and that in a city of one million people there would be 20,000 such individuals. Presenting the data in this manner probably made subjects less likely to begin thinking about the low likelihood that the suspect would "match" if he was innocent. Instead, this presentation encourages thoughts about the large number of other individuals who also would match. This line of thinking seems more conducive to the Defense Attorney's Fallacy – and, in fact, a larger number of subjects made judgments consistent with this fallacy in the percentage and number condition, though this difference was not significant. It also creates the general impression that "a lot of people could have done it," which probably accounts for the lower estimates of probable guilt in the percentage and number condition.

Although most subjects realized that the "match" between the suspect and perpetrator was diagnostic of guilt, they may not have fully appreciated the strength of the evidence. Most subjects revised their judgments in the right direction but not by as much as Bayes' theorem would dictate given subjects' initial estimates of probability of guilt and the low incidence rate of the "matching" hair. The most obvious explanation for this apparent conservatism is that subjects gave less weight to the match between the suspect's and perpetrator's hair than this evidence deserves. One must be cautious about drawing this conclusion, however, because there are other possible explanations that are not ruled out by this design. One possibility is that subjects' initial judgments of guilt overstated their perception of the strength of the nonstatistical evidence in the case (perhaps due to a tendency to avoid the extreme lower end of the response scale). In other words, subjects' final judgments may appear conservative because a response bias in initial judgments caused the "model" Bayesian judgments to be too high, rather than because under-utilization of the associative evidence caused subjects' final judgments to be too low (cf. DuCharme, 1970; Slovic and Lichtenstein, 1971).

The practical significance of Experiment 1 is difficult to judge without more information. One key limitation of the experiment is that subjects were not exposed to any arguments about how much weight the statistical evidence deserved. Because jurors in actual cases may hear such arguments from attorneys or other jurors, it is important to consider how people respond to these arguments. Do they recognize the flaws in an argument for a fallacious position, or are they persuaded by fallacious reasoning? Is the impact of a fallacious argument neutralized by hearing an argument for a contrary position? Experiment 2 addresses these questions.

Experiment 2

Method

The procedure of Experiment 2 was similar to that of Experiment 1. Undergraduate subjects (N = 73) read a description of a murder case in which the killer's identity was unknown but the victim was known to have wounded the killer with a knife. The police find some of the killer's blood at the scene of the crime and laboratory tests indicate it is a rare type found in only one person in 100. While questioning the victim's

neighbors, a detective notices that one man is wearing a bandage. Based on his overall impression of this man, the detective estimates the probability of his guilt to be .10. Later the detective receives some new evidence indicating that this suspect has the same rare blood type as the killer.

The subjects' task was to decide whether the detective should revise his estimate of the probability of the suspect's guilt in light of this new evidence and, if so, by how much. To help them make this judgment they read two arguments regarding the relevance of the blood type evidence. The "Prosecution argument" advocated the Prosecutor's Fallacy as follows:

> The blood test evidence is highly relevant. The suspect has the same blood type as the attacker. This blood type is found in only 1% of the population, so there is only a 1% chance that the blood found at the scene came from someone other than the suspect. Since there is only a 1% chance that someone *else* committed the crime, there is a 99% chance the suspect is guilty.

The "Defense argument" advocated the Defense Attorney's Fallacy:

> The evidence about blood types has very little relevance for this case. Only 1% of the population has the "rare" blood type, but in a city . . . [like the one where the crime occurred] with a population of 200,000 this blood type would be found in approximately 2000 people. Therefore the evidence merely shows that the suspect is one of 2000 people in the city who might have committed the crime. A one-in-2000 chance of guilt (based on the blood test evidence) has little relevance for proving *this* suspect is guilty.

Half of the subjects first received the Prosecution argument and then received the Defense argument (Pros–Def Condition). The remaining subjects received the arguments in reverse order (Def–Pros Condition). After reading each argument, subjects answered three questions. First, they indicated whether they believed the reasoning and logic of the argument was correct or incorrect. Then they indicated whether they thought the detective should revise his estimate of the suspect's probable guilt in light of the blood type evidence. Finally, they indicated what they thought the detective's estimate of probability of guilt should be after receiving the blood type evidence.

Notice that Experiment 2 differs from Experiment 1 in the method used to establish the suspect's initial or prior probability of guilt. In Experiment 1 subjects established the prior themselves by estimating the suspect's initial probability of guilt. They then revised their own initial estimates in light of the associative evidence. This approach left open the possibility that the conservatism of subjects' final judgments was caused by subjects' overstating their initial judgments rather than underestimating the strength of the associative evidence. In Experiment 2 the prior is established by telling subjects about the detective's initial estimate of the suspect's likelihood of guilt. Subjects are then asked how much the detective should revise this estimate in light of the "match." Asking subjects to make this judgment "in the second person" rules out a response bias in subjects' initial judgments as an explanation for conservatism.

Results and discussion

Recognition of fallacious arguments

A substantial number of subjects failed to recognize that the fallacious arguments were incorrect. As Table 29.2 indicates, the Defense argument was more convincing than the Prosecution argument: Overall 50 subjects

Table 29.2 Number of subjects rating the Prosecution argument and Defense argument correct and incorrect (Experiment 2)

Prosecution argument	Defense argument		
	Correct	Incorrect	Total
Correct	14	7	21
Incorrect	36	16	52
Total	50	23	73

(68.5%) labeled the Defense argument "correct" while 21 (28.8%) labeled the Prosecution argument "correct" (McNemar $\chi^2_c(1, N = 73) = 18.23$, $p < .0001$). Only 16 subjects (22.2%) recognized that both arguments are incorrect. It is unclear, of course, how much subjects' perceptions of the correctness of each argument depend on the specific wordings used here.

The order in which the arguments were presented did not significantly affect ratings of correctness. The distribution of subjects across the four categories shown in Table 29.2 was not significantly different in the Pros–Def condition than in the Def–Pros condition ($\chi^2(3, N = 73) = 5.93$, n.s.).

Fallacious judgments

To determine whether subjects were responding to the associative evidence in a manner consistent with fallacious reasoning, we examined responses to the questions asking whether and how much the detective should revise his estimate of the suspect's probability of guilt. These responses were divided into three categories. Judgments that the detective should increase his estimate of probable guilt to exactly .99 were coded as consistent with the Prosecutor's Fallacy because .99 is the probability one would obtain by subtracting the incidence rate of the "matching" characteristic from one. Judgments that the detective should not revise his estimate of probable guilt in light of the associative evidence were coded as consistent with the Defense Attorney's Fallacy. Judgments that the detective should increase his estimate of probable guilt to some level other than .99 were coded as consistent with neither fallacy. Each subject judged whether and how much the detective should revise his estimate at two points – once after reading each argument. Hence, the 73 subjects made a total of 146 codable responses.

Overall, responses consistent with the Prosecutor's Fallacy were rare, but responses consistent with the Defense Attorney's Fallacy were surprisingly common: only four responses (3%) were consistent with the Prosecutor's Fallacy while 82 (56%) were consistent with the Defense Attorney's Fallacy and 60 (41%) were consistent with neither fallacy. The order in which the arguments were presented did not affect the distribution of responses across the three response categories.

Seventy percent of subjects made at least one response consistent with fallacious reasoning. Three subjects (4%) made one or more judgments consistent with the Prosecutor's Fallacy, 48 subjects (66%) made one or more judgments consistent with the Defense Attorney's Fallacy, and only 22 subjects (30%) made no judgments consistent with fallacious reasoning. The order in which the arguments were presented did not significantly affect the distribution of subjects across these three categories $[\chi^2(2,N = 73) = 3.80,$ n.s.$]$.

Forty-eight percent of subjects made two responses consistent with fallacious reasoning. Thirty-four subjects (47%) made two judgments consistent with the Defense Attorney's Fallacy and one made two judgments consistent with the Prosecutor's Fallacy; none mixed fallacies.

We had expected that the likelihood subjects would respond in a manner consistent with the Prosecutor's Fallacy or the Defense Attorney's Fallacy would depend on which argument they had read. Surprisingly, the findings did not support this prediction. A between group comparison, looking at responses of subjects who had read only the first of the two arguments, revealed no significant difference between those who had read only the Prosecution argument and those who had read only the Defense argument in the distribution of the two groups across the three response categories. The number of subjects coded as victims of the Prosecutor's Fallacy, Defense Attorney's Fallacy and neither fallacy was 3, 18, and 16, respectively among those who read only the Prosecution argument, and 0, 21, and 18 among those who read only the Defense argument $[\chi^2(2,N = 73) = 3.08,$ n.s.$]$. Nor was there any difference between these two groups in their responses after reading the second of the two arguments $[\chi^2(2,N = 73) = 1.06,$ n.s.$]$. Regardless of which argument they read, about half or more of the subjects made judgments consistent with the Defense Attorney's Fallacy.

It is unclear why the Prosecutor's Fallacy was so much less prevalent and the Defense Attorney's Fallacy so much more prevalent in Experiment 2 than in Experiment 1. Certainly the results indicate the Defense argument was more persuasive than the Prosecution argument. Whether this finding depends on the specific arguments used here or not remains to be seen. It is interesting to note that even among subjects who had read only the Prosecution argument, nearly half responded in a manner consistent with the Defense Attorney's Fallacy. Perhaps these subjects detected something "fishy" about the Prosecution argument and therefore decided to disregard the associative evidence altogether. If this is the case, the weak Prosecution argument may actually have promoted the Defense Attorney's Fallacy.

Judgments of probable guilt

Although the arguments did not affect the number of subjects making judgments consistent with fallacious reasoning, they did affect subjects' judgments of probable guilt. As Table 29.3 shows, subjects thought the detective's subjective probability of guilt should be higher after reading the Prosecution argument (M = .31) than after reading the Defense argument (M = .24; $F(1,71) = 7.89$, $p < .01$). Order of presentation also influenced these judgments. Subjects in the Pros–Def condition thought the detective's estimate of guilt should be higher than did subjects in the Def–Pros condition, $F(1,71) = 8.44$, $p < .005$. The order effect is probably due to a simple anchoring phenomenon. Subjects who received the Prosecution argument first made higher initial judgments than subjects who received the Defense argument first. The initial judgments then served as an anchor point for subjects' second judgments. It is interesting to note, however, that a 2 × 2 analysis of variance examining the effects of type of argument (Prosecution vs. Defense) and order of presentation (Pros–Def vs. Def–Pros), revealed a significant argument by order interaction, indicating there was less variation by type of argument in the Def–Pros condition than in the Pros–Def condition [$F(1,71) = 4.43$, $p < .05$]. One way of looking at this finding is that the Defense argument, when received first, "anchored" subjects' judgments more firmly than did the Prosecution argument. This interpretation is, of course, consistent with the previously noted finding that the Defense argument was more persuasive than the Prosecution argument.

A comparison of judgments of probable guilt to a Bayesian model showed the same conservative bias that was evident in Experiment 1. According to a Bayesian analysis, the detective's subjective probability of guilt should increase from .10 to .92 after receiving the associative evidence. As Table 29.3 indicates, however, subjects thought the detective's revised estimate of probable guilt should be considerably lower (overall M = .28). Mean judgments were low, in part, because a substantial percentage of subjects, apparent victims of the Defense Attorney's Fallacy, indicated that the detective's estimate of probable guilt should remain at .10. Even among subjects who thought the detective's estimate should increase, however, mean judgments were well below what Bayes' theorem dictates (M = .43). These findings provide additional confirmation of subjects' conservatism when revising an initial judgment of probable guilt in light of associative evidence. Because subjects were indicating how much the detective should revise his initial estimate of probable guilt, rather than revising their own initial estimates, subjects' conservatism cannot be attributed to a tendency to avoid the lower end of the response scale when making initial judgments. Because subjects' final judgments were, on average, below the midpoint of the response scale, subjects' conservatism also cannot be attributed to an artifactual tendency to avoid the upper end of the response scale (DuCharme, 1970). The most likely explanation is that subjects simply gave less weight to the associative evidence than it deserves.

Table 29.3 Mean final estimate of probability of guilt (Experiment 2)

		Order of presentation	
Argument		Prosecution defense[a]	Defense prosecution[b]
Prosecution argument	M	.42	.20
	(SD)	(.34)	(.18)
Defense argument	M	.29	.18
	(SD)	(.29)	(.14)

a Prosecution defense condition, n = 37.
b Defense-prosecution condition, n = 36.

General discussion

Theoretical implications

These experiments indicate that people are not very good at drawing correct inferences from associative evidence and incidence rate statistics. They are strongly influenced by subtle and logically inconsequential differences in how the statistics are presented (Experiment 1). They are unable to see the error in crude arguments for fallacious interpretations of the evidence, and their judgments of probable guilt are strongly influenced by such arguments (Experiment 2). It appears, then, that people generally lack a clear sense of how to draw appropriate conclusions from such evidence and that, as a result, judgments based on such evidence are highly malleable.

People's responses to the evidence were far from uniform. A relatively small percentage (13% in Experiment 1; 4% in Experiment 2) gave responses consistent with the simple but erroneous judgmental strategy we have labeled the Prosecutor's Fallacy. A larger percentage (12% in Experiment 1; 66% in Experiment 2) gave responses consistent with another judgmental error, which we call the Defense Attorney's Fallacy. The remaining subjects responded in a manner that was consistent with neither fallacy but that suggested a tendency to underestimate the value of associative evidence. We will discuss each type of response in turn.

Prosecutor's Fallacy

The Prosecutor's Fallacy probably results from confusion about the implications of conditional probabilities. In the cases used in these experiments, the incidence rate statistics established the conditional probability that the suspect would "match" *if he was innocent*. Victims of the fallacy may simply have assumed that this probability is the complement of the probability the suspect would be guilty *if he matched*.

The Prosecutor's Fallacy is similar to a fallacy documented by Eddy (1982) in physicians' judgments of the probability that a hypothetical patient had a tumor. When physicians were told there is a 90% chance a particular test will be positive *if the patient has a tumor*, most of them jumped to the conclusion that there is a 90% chance the patient has a tumor *if the test result is positive*. This judgment is, of course, erroneous, except where the prior probability of a tumor and the prior probability of a positive test result are equal. But physicians made this judgment even when this condition clearly was not met. The difference between the physicians' error and the Prosecutor's Fallacy is most easily explained in formal terms: where H and $\bar{H}$ indicate that the matter being judged is true and false, respectively, and D is evidence relevant to that judgment, the physicians responded as if $p(H/D) = p(D/H)$; victims of the Prosecutor's Fallacy respond as if $p(H/D) = 1 - p(D/\bar{H})$.

The Prosecutor's Fallacy is clearly inappropriate as a general strategy for assigning weight to associative evidence because it fails to take into account the strength of other evidence in the case. Particularly where the other evidence against the defendant is weak, it can lead to errors. On the other hand, there are some circumstances in which judgments based on the Prosecutor's Fallacy will closely approximate Bayesian norms. Where the incidence rate of a "matching" characteristic is extremely low (e.g., below 1%), for example, the posterior probabilities of guilt dictated by Bayes' theorem and by the Prosecutor's Fallacy will converge at the upper end of the scale and may, for practical purposes, be indistinguishable. There is also a convergence where the prior probability of guilt is near .50.[6] Whether people actually fall victim to the Prosecutor's Fallacy under all of these circumstances is, of course, speculative at this point. For present purposes it is sufficient to note that the practical consequences of the Prosecutor's

Fallacy, if it occurs, are likely to be most significant where the prior probability is not close to .50 and the incidence rate is greater than .01.

Defense Attorney's Fallacy

Perhaps the most surprising finding of this research was how easily people can be persuaded to give *no* weight to associative evidence. The associative evidence presented in the two experiments was quite powerful: a match between the suspect and perpetrator on a characteristic found in only 2% (Experiment 1) or 1% (Experiment 2) of the population. According to Bayes' theorem, a person who initially thought the suspect's probability of guilt was .10 should revise that estimate upward to .85 and .93, respectively, in light of this associative evidence. Of course, one need not rely on Bayes' theorem to conclude that the Defense Attorney's Fallacy is inappropriate. It is difficult to imagine any normative model of judgment that would give no weight to associative evidence. Yet many people in these experiments were persuaded that because a large number of individuals other than the suspect would also "match" on the relevant characteristic, the "match" is uninformative with regard to the suspect's likelihood of guilt. As noted earlier, what this reasoning ignores is that the overwhelming majority of the other people who possess the relevant characteristic are not suspects in the case at hand. The associative evidence drastically narrows the class of people who could have committed the crime, but fails to eliminate the very individual on whom suspicion has already focused.

The argument favoring the Defense Attorney's Fallacy was particularly persuasive. In Experiment 2, over 60% of people who heard the argument were persuaded that the associative evidence deserved no weight. Finding some way to combat this powerful fallacy is clearly an imperative for future research.

Underutilization of associative evidence

Among subjects who thought the associative evidence deserved some weight, final judgments of guilt still tended to be significantly lower than a Bayesian analysis suggests they should have been. Of course efforts to compare human judgments to Bayesian models are problematic (DuCharme, 1970; Slovic and Lichtenstein, 1971). But, as discussed earlier, the design and results of the two experiments appear to rule out the most obvious possible artifacts. Hence, these findings lend support to the claim that people underutilize associative evidence (Saks and Kidd, 1980).

Legal implications

Because the present research has some important limitations, conclusions about its legal implications are best viewed as preliminary. It is unclear, for example, how much the *individual* judgmental errors documented by these studies affect the *group* decisions of juries. We do not know what happens when victims of opposing fallacies encounter each other in deliberation. Perhaps in the crucible of deliberation fallacious reasoning is detected and the jurors, as a group, adopt a more appropriate evaluation of the evidence. On the other hand, deliberation may simply cause the most persuasive line of fallacious reasoning to dominate, reinforcing the biases of the majority of jurors. These intriguing issues await further research. Another limitation of the present research is that its findings are largely based on people's estimates of probabilities rather than their decision to convict or acquit. It is important that future research confirm that the tendency to misuse associative evidence, suggested by these findings, goes beyond the articulation of numbers and actually influences the sorts of decisions juries are called upon to make.

Nevertheless, people's tendency to draw erroneous conclusions from descriptions of evidence closely modeled on evidence from actual cases is troubling. College undergraduates are unlikely to be worse at evaluating such evidence than actual jurors, so the findings suggest such evidence may well be misinterpreted and misused by juries.

As noted earlier, the legal debate over the admissibility of incidence rate statistics in connection with associative evidence stems largely from differing assumptions about the way jurors are likely to respond to such evidence. The findings of the present research cast some initial light on this issue. The finding that some people make judgments consistent with the Prosecutor's Fallacy lends support to the argument of some appellate courts and commentators (e.g., Tribe, 1971) that statistical evidence may have an exaggerated impact on the jury. On the other hand, the powerful tendency to commit the Defense Attorney's Fallacy, particularly after reading arguments, and the general tendency to make conservative judgments in light of associative evidence, suggest that underutilization of such evidence is the more serious problem.

From a legal point of view, the primary danger of admitting statistical evidence is that it will be overutilized. There is little harm in admitting it if jurors give it no more weight than it deserves. If jurors underutilize such evidence or even ignore it they may reach the wrong verdict, but they are not more likely to err if the evidence is admitted than if it is not. The major danger of admitting this evidence is therefore the possibility that juries will commit the Prosecutor's Fallacy.

Judgments consistent with the Prosecutor's Fallacy were more common when the statistical evidence was presented as a conditional probability than when it was presented as a "percentage and number" (Experiment 1). This finding suggests that courts that admit incidence rate

statistics can reduce the likelihood that jurors will commit the Prosecutor's Fallacy by forbidding experts to present incidence rates as conditional probabilities, requiring them to state the incidence rate as a percentage and requiring them to provide an estimate of the number of people in the area who would also have the relevant characteristic. Lawyers' arguments may also counteract the tendency to commit the Prosecutor's Fallacy. In Experiment 2 the argument for the Defense Attorney's Fallacy proved considerably more persuasive than the argument for the Prosecutor's Fallacy. An attorney worried about jurors committing the Prosecutor's Fallacy might do well to fight fallacy with fallacy. Of course, these tactics for preventing the Prosecutor's Fallacy have a price – they make it much more likely that jurors will underutilize the associative evidence or ignore it altogether. When deciding how incidence rates should be presented to the jury, then, the key issue is not whether jurors will draw appropriate conclusions from it or not, but whether one type of error will be more likely than another.

In recent years a number of scholars have suggested that jurors be instructed in the use of "decision aids" based on Bayes' theorem in cases in which they must deal with statistical evidence (Finkelstein and Fairley, 1970; Feinberg and Kadane, 1983; Lindley, 1977). These proposals have been severely criticised by legal scholars, who argue that the proposed cure is worse than the alleged disease (e.g., Tribe, 1971; Brilmayer and Kornhauser, 1978; Callen, 1982; Cullison, 1979). Indeed, the legal community's reaction to these proposals has been so hostile that it appears unlikely any of these proposals will be adopted in the foreseeable future. Nevertheless, the present findings lend support to the claim that jurors are likely to misuse statistical evidence if they are not provided with decision aids (Saks and Kidd, 1980). Whether jurors would evaluate statistical evidence better with decision aids than without is not an issue addressed by this research, but is a worthy topic for future study.

Another possible remedy for misuse of statistical evidence is to rely on cross-examination and arguments by the attorneys. The arguments used in Experiment 2, for example, seemed to counteract the Prosecutor's Fallacy (though they may have promoted the Defense Attorney's Fallacy). Perhaps clever lines of cross-examination or argument exist or could be developed that are effective in counteracting misuse of statistical evidence (see, e.g., Imwinkelried, 1982b). Future research might examine the way experienced attorneys actually deal with statistical evidence and might, through simulation experiments, test the effectiveness of those and other techniques.

The use of mathematical evidence is likely to increase dramatically in the near future (Jonakait, 1983) and legal professionals will increasingly face difficult choices about how to deal with it. Because their choices will turn, in part, on assumptions about the way people respond to mathematical evidence, now is an opportune time for social scientists to begin exploring this issue. Our hope is that social scientists, building on studies like those reported here, will be able to answer the key underlying behavioral questions so that lawyers and judges may base decisions about mathematical evidence on empirical data rather than unguided intuitions.

Notes

This research was supported by grants to the first author from the National Science Foundation (No. SES 86-05323) and the UCI Academic Senate Committee on Research. The authors wish to thank Karen Rook and Robyn M. Dawes for comments on earlier versions of this article and John Van-Vlear for help in collecting data.

1   Base rate statistics, when used in a trial, are sometimes called "naked statistical evidence" (Kaye, 1982). Where a person is struck by a bus of unknown ownership, evidence that a particular company operates 90% of the buses on that route is "naked statistical evidence" on who owns the bus. Where a man possessing heroin is charged with concealing an illegally imported narcotic, evidence that 98% of heroin in the US is illegally imported is "naked statistical evidence" on whether the heroin possessed by the defendant was illegally imported. Courts have generally treated "naked statistical evidence" differently from incidence rate statistics. Although the majority of jurisdictions admit incidence rate statistics, courts almost universally reject "naked statistical evidence" (see, e.g., *Smith v. Rapid Transit*, 1945), though there are a few exceptions where its admissibility has been upheld (e.g., *Turner v. U.S.*, 1970; *Sindell v. Abbott Labs*, 1980). For general discussions of "naked statistical evidence" see Kaye (1982), Cohen (1977), and Tribe (1971).

2   Bayes' theorem may be used to calculate the amount one should revise one's prior estimate of the probability of a suspect's guilt after receiving associative evidence accompanied by incidence rate statistics (for general discussions of the use of Bayes' theorem to model legal judgments, see Kaplan, 1968; Lempert, 1977; Lempert and Saltsburg, 1977; Lindley, 1977; Schum, 1977b; Schum and Martin, 1982; Kaye, 1979). Where H and $\overline{H}$ designate the suspect's guilt and innocence respectively, and D designates associative evidence showing a match between the suspect and perpetrator on some characteristic, Bayes' theorem states:

$$p(H/D) =$$

$$p(H)p(D/H)/[p(H)p(D/H) + p(\overline{H})p(D/\overline{H})].$$

The term $p(H)$ is called the prior probability and reflects one's initial estimate of the probability the suspect is guilty in light of everything that is known *before* receiving D. The term $p(H/D)$ is called the posterior probability and indicates what one's revised estimate of probable guilt should be in light of everything that is known *after* receiving D. The formula indicates that the associative evidence, D, should cause one to revise one's opinion of the suspect's guilt to the extent $p(D/H)$ differs from $p(D/\overline{H})$. If one believes the suspect and perpetrator are certain to match if the suspect is guilty, $p(D/H)$ = 1.00. If one believes an innocent suspect is no more likely than anyone else to possess the "matching" characteristic, $p(D/\overline{H})$ is equal to the incidence rate of the matching characteristic. This model assumes, of course, that the sole issue determining guilt and innocence is the suspect's identity as the perpetrator (rather than, say, his mental state).

3 The prior probability of guilt, $p(H)$, is equal to .20, and because the defendant must be either guilty or innocent, $p(\overline{H})$ = .80. Because the defendant is certain to have the relevant genetic markers if he is guilty, $p(D/H)$ = 1.00; and because the defendant is no more likely than anyone else to have the genetic markers if he is not guilty, $p(D/\overline{H})$ = .10, the incidence rate of the blood markers. These probabilities may be plugged into the Bayesian formula in note 2, allowing us to solve for $p(H/D)$, which, in this case equals .71.

4 A log-likelihood ratio chi-square on the 2 × 3 table revealed a significant overall difference between the conditional probability condition and the percentage and number condition [$\chi^2(2,N = 144) = 12.26, p < .01$]. Multiple comparisons among the three response categories, using a simultaneous test procedure recommended by Gabriel (1966), indicated that the two conditions differed mainly in the number of subjects committing the Prosecutor's Fallacy versus neither fallacy [$\chi^2(2,N = 126) = 4.97, p < .01$] and in the number committing the Prosecutor's Fallacy versus the Defense Attorney's Fallacy [$\chi^2(2,N = 37) = 5.24, p < .10$], rather than the number committing the Defense Attorney's Fallacy versus neither fallacy [$\chi^2(2,N = 125) = .57$, n.s.]. This pattern of results allows a statistically reliable inference that the two conditions are heterogeneous with respect to the number of subjects falling in the Prosecutor's Fallacy category (compared to at least one of the other categories) [$\chi^2(2,N = 144) = 10.21, p < .01$] and that the two conditions differ marginally in the number of subjects falling in the Defense Attorney's Fallacy category (compared to at least one of the other categories) [$\chi^2(2,N = 144) = 5.81, p < .10$].

5 During debriefing subjects were asked to explain their final judgments. Among those who judged the suspects' probability of guilt to be .98, two rationales were common. Some, like the district attorney discussed earlier, argued that a 98% chance of guilt is an obvious or direct implication of the 2% incidence rate. Others argued that if 2% of the population have hair that would "match" the perpetrator's there is only a 2% chance that someone other than the suspect committed the crime and therefore a 98% chance the suspect is guilty.

6 In a criminal case in which the defendant and perpetrator match on a blood type found in one person in ten, for example, a victim of the Prosecutor's Fallacy would conclude that the probability of the defendant's guilt is .90. Bayes' theorem dictates a nearly identical probability (.909) when the prior probability is .50. To the extent the prior probability is lower or higher than about .50, however, the Prosecutor's Fallacy produces results that diverge from Bayesian norms. When the prior probability is .20 and the incidence rate is .10, Bayes' theorem dictates a posterior probability of .71 while the Prosecutor's Fallacy produces a posterior of .90. By contrast, when the prior is above .50, the Prosecutor's Fallacy may actually favor the defense. When the prior probability is .80 and the incidence rate is .10, the Prosecutor's Fallacy, as before, produces a posterior of .90, which is lower than the Bayesian posterior of .98.

References

Annotation (1980). Admissibility, weight and sufficiency of blood grouping tests in criminal cases. *American Law Reports, Fourth*, **2**, 500.

Bar-Hillel, M. (1980). The base-rate fallacy in probability judgments. *Acta Psychologica*, **44**, 211–233.

Beyth-Maron, R., and Fischhoff, B. (1983). Diagnosticity and pseudodiagnosticity. *Journal of Personality and Social Psychology*, **45**, 1185–1195.

Borgida, E., and Brekke, N. (1981). The base-rate fallacy in attribution and prediction. *New Directions in Attribution Research*, Vol. 3. Hillsdale, New Jersey: Erlbaum.

Braun, L.J. (1982). Quantitative analysis and the law: Probability theory as a tool of evidence in criminal trials. *Utah Law Review*, **4**, 41–87.

Briggs, T.J. (1978). The probative value of bloodstains on clothing. *Medicine, Science and the Law*, **18**, 79–83.

Brilmayer, L., and Kornhauser, L. (1978). Review: Quantitative methods and legal decision. *University of Chicago Law Review*, **46**, 116–153.

Callen, C.R. (1982). Notes on a grand illusion: Some limits on the use of Bayesian theory in evidence law. *Indiana Law Journal*, **57**, 1–44.

Chaperlin, K., and Howarth, P.S. (1983). Soil comparison by the density gradient method – A review and evaluation. *Forensic Science International*, **23**, 161–177.

Cohen, L.J. (1977). *The Probable and the Provable*. London: Oxford University Press.

Cullison, A.D. (1979). Identification by probabilities and trial by mathematics (a lesson for beginners in how to be wrong with greater precision). *Houston Law Review*, **6**, 471–518.

Davis, R.J., and DeHaan, J.D. (1977). A survey of men's footwear. *Journal of Forensic Science Society*, **17**, 271–283.

Doherty, M.E., Mynatt, C.R., Tweney, R.D., and Schiavo, M.D. (1979). Pseudodiagnosticity. *Acta Psychologica*, **43**, 111–121.

Ducharme, W.M. (1970). A response bias explanation of conservatism in human inference. *Journal of Experimental Psychology*, **85**, 66–74.

Eddy, D.M. (1982). Probabilistic reasoning in clinical medicine. In D. Kahneman, P. Slovic, and A. Tversky (eds), *Judgments Under Uncertainty: Heuristics and Biases*. Cambridge: Cambridge University Press.

Edwards, W. (1968). Conservatism in human information processing. In B. Kleinmuntz (ed.), *Formal Representation of Human Judgment*. New York: Wiley.

Feinberg, S.E., and Kadane, J.B. (1983). The presentation of Bayesian statistical analysis in legal proceedings. *The Statistician*, **32**, 88–98.

Finklestein, M.O., and Fairley, W.B. (1970). A Bayesian approach to identification evidence. *Harvard Law Review*, **83**, 489–517.

Fong, W. (1973). Value of glass as evidence. *Journal of Forensic Science*, **18**, 398–404.

Gabriel, K.R. (1966). Simultaneous test procedures for multiple comparisons on categorical data. *Journal of the American Statistical Association*, **14**, 1081–1096.

Gaudette, B.D., and Keeping, E.S. (1974). An attempt at determining probabilities in human scalp hair comparison. *Journal of Forensic Science*, **19**, 599–605.

Gettinby, G. (1984). An empirical approach to estimating the probability of innocently acquiring bloodstains of different ABO groups on clothing. *Journal of Forensic Science Society*, **24**, 221–227.

Giannelli, P.C. (1983). Frye v. U.S. Background paper prepared for the national conference of lawyers and scientists, *Symposium of Science and the Rules of Evidence. National Conference of Lawyers and Scientists*, April 29–30.

Grunbaum, B.W., Selvin, S., Pace, N., and Black, D.M. (1978). Frequency distribution and discrimination probability of twelve protein genetic variants in human blood as functions of race, sex, and age. *Journal of Forensic Sciences*, **23**, 577–587.

Imwinkelried, E.J. (1981). A new era in the evolution of scientific evidence: A primer on evaluating the weight of scientific evidence. *William & Mary Law Review*, **23**, 261–287.

Imwinkelried, E.J. (1982a). Forensic hair analysis: The case against the underemployment of scientific evidence. *Washington and Lee Law Review*, **39**, 41–67.

Imwinkelried, E.J. (1982b). *The Methods of Attacking Scientific Evidence*. Charlottesville, Virginia: The Michie Company.

Imwinkelried, E.J. (1983). The standard for admitting scientific evidence: A critique from the perspective of juror psychology. *Villanova Law Review*, **28**, 554–571.

Jonakait, R.N. (1983). When blood is their argument: Probabilities in criminal cases, genetic markers, and once again, Bayes' theorem. *University of Illinois Law Review*, 1983, 369–421.

Kahneman, D. and Tversky, A. (1973). On the psychology of prediction. *Psychological Review*, **80**, 237–251.

Kaplan, J. (1968). Decision theory and the factfinding process. *Stanford Law Review*, **20**, 1065.

Kaye, D. (1979). The law of probability and the law of the land. *University of Chicago Law Review*, **47**, 34–56.

Kaye, D. (1982). The limits of the preponderance of the evidence standard: Justifiably naked statistical evidence and multiple causation. *American Bar Foundation Research Journal*, **2**, 487–516.

Lempert, R.L. (1977). Modeling relevance. *Michigan Law Review*, **75**, 1021–1075.

Lempert, R. and Saltzburg, S. (1977). *A Modern Approach to Evidence*. St. Paul, Minnesota: West Publishing.

Lindley, D.V. (1977). Probability and the law. *The Statistician*, **26**, 203–212.

Note. (1983). Admissibility of mathematical evidence in criminal trials. *American Criminal Law Review*, **21**, 55–79.

Owens, G.W., and Smalldon, K.W. (1975). Blood and semen stains on outer clothing and shoes not related to crime: Report of a survey using presumptive tests. *Journal of Forensic Science*, **20**, 391–403.

Pearson, E.F., May, R.W., and Dobbs, M.D.G. (1971). Glass and paint fragments found in men's outer clothing – Report of a survey. *Journal of Forensic Science*, **16**(3), 283–299.

Peterson, J.L., Mihajlovic, S., and Gilliland, M. (1984). *Forensic Evidence and the Police: The Effects of Scientific Evidence on Criminal Investigations*. Washington, DC: US Department of Justice, National Institute of Justice.

Saferstein, R. (1977). *Criminalistics: An Introduction to Forensic Science*. Englewood Cliffs, New Jersey: Prentice-Hall.

Saks, M.J., and Kidd, R.F. (1980). Human information processing and adjudication: Trial by heuristics. *Law and Society Review*, **15**, 123–160.

Saks, M.J., and Van Duizend, R. (1983). *The Use of Scientific Evidence in Litigation*. Williamsburg, Virginia: National Center for State Courts.

Schroeder, O. (1977). *Assessment of the Forensic Sciences Profession: A Legal Study Concerning the Forensic Sciences*. Washington, DC: Forensic Science Foundation.

Schum, D.A. (1977a). Contrast effects in inference: On the conditioning of current evidence by prior evidence. *Organizational Behavior and Human Performance*, **18**, 217–253.

Schum, D.A. (1977b). The behavioral richness of cascaded inference models: Examples in jurisprudence. In N.J. Castellan, D.B. Pisoni, and G.R. Potts (eds), *Cognitive Theory*. Vol. II. Hillsdale, New Jersey: Lawrence Erlbaum.

Schum, D.A., and DuCharme, W.M. (1971). Comments of the relationship between the impact and the reliability of evidence. *Organizational Behavior and Human Performance*, **6**, 111–131.

Schum, D.A., and Martin, A.W. (1982). Formal and empirical research on cascaded inference in jurisprudence. *Law and Society Review*, **17**, 105–151.

Slovic, P., Fischoff, B., and Lichtenstein, S. (1977). Behavioral decision theory. *Annual Review of Psychology*, **28**, 1–39.

Slovic, P., and Lichtenstein, S. (1971). Comparison of Bayesian and regression approaches to the study of information processing in judgment. *Organizational Behavior and Human Performance*, **6**, 649–744.

Stoney, D. (1984). Evaluation of associative evidence: Choosing the relevant question. *Journal of the Forensic Science Society*, **24**, 473–482.

Tippet, C.R, Emerson, V.J., Fereday, M.J., Lawton, R., Richardson, A., Jones, L.T., and Lampert, S.M. (1968). The evidential value of the comparison of paint flakes from sources other than vehicles. *Forensic Science Society Journal*, **8**, 61–65.

Tribe, L.H. (1971). Trial by mathematics: Precision and ritual in the legal process. *Harvard Law Review*, **84**, 1329–1393.

Tversky, A., and Kahneman, D. (1982). Evidential impact of base-rates. In D. Kahneman, A. Slovic, and A Tversky (eds), *Judgment Under Uncertainty: Heuristic Biases*. Cambridge: Cambridge University Press.

Cases

State v. Carlson, 267 N.W. 170 (Minn. 1978).

People v. Macedonio, 42 N.Y.2d 944, 366 N.E.2d 1355 (1977).

United States v. Massey, 594 F.2d 676 (8th Cir. 1979).

People v. McMillen, 126 Mich.App. 203, 337 N.W.2d 48 (1984).

People v. Robinson, 27 N.Y.2d 864, 265 N.E.2d 543, 317 N.Y.S.2d 19 (1970).

Sindell v. Abbott Labs, 26 Cal.3d 588, 607 P.2d 924, 163 Cal.Rptr. 132 (1980).

Smith v. Rapid Transit, 317 Mass. 469, 58 N.E.2d 754 (1945).

Turner v. U.S., 396 US 398, reh'g denied, 397 US 958 (1970).

Chapter 30

Neven Sesardić

SUDDEN INFANT DEATH OR MURDER? A ROYAL CONFUSION ABOUT PROBABILITIES

1 Introduction

There has been a lot of publicity recently about several women in the United Kingdom who were convicted of killing their own children after each of these mothers had two or more of their infants die in succession and under suspicious circumstances. (A few of these convictions were later overturned on appeal.) The prosecutor's argument and a much discussed opinion of a crucial expert witness in all these court cases relied mainly on medical evidence but probabilistic reasoning also played a (minor) role.

It was this latter, probabilistic aspect of the prosecutor's case that prompted a number of statisticians to issue a general warning about what they regarded as the expert witness's serious statistical error, which they thought might have had a profound influence on the jury. Among those who intervened and tried to instruct the public and the courts about the correct use of probability were the Royal Statistical Society (RSS), the President of the Royal Statistical Society, the President of the Mathematical Association and several prominent academics, including two professors of statistics (from Oxford and University College, London). In my opinion, their advice about how to proceed with statistical reasoning is

actually problematic in each of its three inferential steps.

For reasons that do not concern us, it was one of these court cases (R v. Clark) that caught most of the attention of the media. The solicitor Sally Clark's child, Christopher, died in her presence in November of 1996, 11 weeks after birth. The death was ascribed to respiratory infection. About a year later, her new baby, Harry, also died while he was alone with his mother and when he was 8 weeks old. At this point Clark was accused of murdering both infants and then in 1999 sentenced to life imprisonment. Her first appeal in 2000 was unsuccessful but on the second appeal in 2003 she was acquitted, principally because the judge decided that the surfacing of some previously undisclosed micro-biological information that was never considered at trial had made the conviction 'unsafe.'

I have no intention here to argue about Sally Clark's guilt or innocence. My only goal is to evaluate the recommendations of some leading statisticians about how to make the appropriate probability judgment in this kind of situation. I am aware, of course, that many a reader will have little patience for listening to a little-known philosopher who announces that the Royal Statistical Society is wrong about − statistics! Well, I must confess to having a strange feeling myself about being engaged in such a quixotic enterprise, but I can only hope that this will not make you stop reading and throw this article away.

2 Setting the stage: Bayes's theorem

This section should be rather uncontroversial. It introduces a basic equation from the elementary probability theory (called 'Bayes's theorem'), which is relevant for discussing different views on Sally Clark's case. The question at issue has often been phrased as deciding between the following two alternative explanations of the deaths of the two children: (i) that both cases were instances of sudden infant death syndrome (SIDS) or (ii) that the children were killed by their mother. This does not correspond to the way the debate was actually conducted in court, but since the statisticians I will critique framed the central issue in these terms (Dawid [2002]; Joyce [2002]; Hill [2004]), I will follow their presentation.

If we call these two hypotheses 2S (double SIDS) and 2M (double murder) then with E being the relevant empirical evidence presented in the court, the 'odds form' of Bayes's theorem gives the following relation between the posterior probabilities of the two rival hypotheses:

$$\frac{p(2S/E)}{p(2M/E)} = \frac{p(2S)}{p(2M)} \times \frac{p(E/2S)}{p(E/2M)} \quad (1)$$

Now according to the standard formula for calculating the probability of joint events, the probability of two cases of SIDS in one family is obtained by multiplying the probability of SIDS in general with the probability of a second SIDS, *given that one SIDS already happened*. That is:

$$p(2S) = p(S) \times p(S_2/S) \quad (2)$$

Obviously, the same logic applies to a double murder scenario, for which we can then write:

$$p(2M) = p(M) \times p(M_2/M) \quad (3)$$

Substituting Equations (2) and (3) into (1) yields:

$$\frac{p(2S/E)}{p(2M/E)} = \frac{p(S)}{p(M)} \times \frac{p(S_2/S)}{p(M_2/M)} \times \frac{p(E/2S)}{p(E/2M)} \quad (4)$$

Equation (4) is very useful because it shows that the ratio of the overall probabilities of the two competing hypotheses (2S and 2M) depends entirely on three ratios: (i) the ratio of prior probabilities of SIDS and murder, that is, $p(S)/p(M)$; (ii) the ratio of probabilities of repeated SIDS and repeated murder, that is, $p(S_2/S)/ p(M_2/M)$; and (iii) the ratio of the so-called 'likelihoods' of the two hypotheses, that is, $p(E/2S)/p(E/2M)$. The likelihood of a hypothesis is a term commonly used for the probability that a hypothesis confers upon the evidence (i.e., for how likely the evidence is on the assumption that the hypothesis is true).

How to estimate these three ratios? After carefully analyzing each of them separately I have concluded that a counsel about these matters coming from, or under the auspices of, the Royal Statistical Society is seriously misguided.

3 Prior probabilities of single SIDS and single homicide

It should be clear at the outset that 'prior' probabilities of SIDS and murder do not refer to probabilities of these events that are completely a priori. They are prior only in the sense that they represent probabilities *before* the concrete evidence pertaining to a particular case (say, Sally Clark) is taken into account. In other words, prior probabilities are also based on empirical evidence, but of a nonspecific kind (mostly data from population statistics).

Hence a very natural suggestion seems to be that the ratio of prior probabilities of SIDS and murder, $p(S)/p(M)$, should be obtained in a standard way, by first taking the recorded incidence of SIDS and of infanticide in the relevant population, and then simply dividing the former magnitude by the latter. Indeed, several statisticians follow this procedure, apparently not being aware that in this particular context the empirically observed frequencies cannot be so straightforwardly translated into probabilities. The problem with this approach is that it leads to a wrong estimate of both probabilities, $p(S)$ and $p(M)$, and, to make things worse, the two mistakes go in opposite directions: $p(S)$ is overestimated while $p(M)$ is underestimated. As a result, the ratio of the two prior probabilities is seriously overstated.

Let us first see how Philip Dawid (Professor of Statistics at University College London) derived estimates for $p(S)$ and $p(M)$. His route to $p(S)$ uses one of the most carefully conducted empirical studies of SIDS. According to the widely cited and praised Confidential Enquiry for Stillbirths and Deaths in Infancy (CESDI study), in families like Clark's (affluent, parents nonsmokers, mother over 26) the incidence of SIDS in the period 1993–1996 was 1 in 8,543 (Fleming *et al.* [2000], p. 92). Dawid takes this as the prior probability of SIDS for the case at hand.

How about $p(M)$? Here Dawid relies on the data from the Office of National Statistics for 1997, and claims that in that year there were seven cases in England and Wales in which children were killed in their first year of life. He then says that, given the total of 642,093 live births that year, 'this yields an estimate of the probability of being murdered, for one child, of 1.1×10^{-5}' (Dawid [2002], p. 76). This is one chance in about 92,000. So, the ratio of the prior probabilities that we are looking for, $p(S)/p(M)$, would be around 11. In other words, the conclusion is that before the specific evidence in the case is taken into account, the probability of one of Clark's children dying of

SIDS was eleven times higher than the probability of being murdered.

I found Dawid's figure of seven infant homicides in 1997 questionable because other scholars (e.g., Marks and Kumar [1993], p. 329; Fleming *et al.* [2000], p. 128; Levene and Bacon [2004], p. 443) say that the number of such cases per year in England and Wales has been consistently around 30. Therefore I checked Dawid's source, the report on mortality statistics for 1997 (Office of National Statistics [1998], pp. 174–7), and the mystery was easily resolved.

In the categorization of the Office of National Statistics (ONS), there are two codes for causes of death that are relevant here. First, E960–E969 refers to 'homicide and injury purposely inflicted on by other persons,' and second, E980–E989 refers to 'injury by other and unspecified means, undetermined whether accidentally or purposely inflicted.' In 1997, there were seven cases of infant deaths that were subsumed under the former code and 17 cases that were assigned the latter code. It therefore appears that Dawid arrived at his estimate of the probability of infant homicide by including only the first group in his calculation and disregarding the second group completely. This is odd. The code E960–E969 covers the *established* homicides (i.e., those officially confirmed by the coroner or criminal court), and the code E980–E989 comprises *suspected* homicides (most of these cases are so classified pending the decision of the relevant authorities). Now it should be immediately clear even to a statistically unsophisticated person that at least some of the suspected homicides will eventually end up in the category of established homicides. Dawid's apparent omission of these 17 cases is bound to underestimate the true probability of infant homicide.

The effect of this 'misunderestimation' (pardon the Bushism) is far from negligible. Dawid's own source, the Office of National

Statistics, actually recommends that in order to get the best estimate of the number of homicides, the suspected homicides should virtually all be treated as real homicides: 'A better estimate of homicides in years 1993 to 1998 may be obtained by *combining* ICD codes E960–E969, E980–E989 with inquest verdict "pending".' (Office of National Statistics [2005], italics added) The same advice comes from the CESDI study, which Dawid used for assessing the probability of SIDS: 'Deaths may be placed in the latter category [E980–E989] pending a coroner's verdict, *and most are eventually reassigned to the heading of homicide* . . . (Fleming *et al.* [2000], p. 128, italics added).

So, this would mean that the number of infant homicides in 1997 should be increased from 7 to 24 (as the authors of the CESDI study also explicitly recommend). But even after this, a further correction is needed. ONS did not include homicides that happened in the first month of baby's life, so it follows that it still gives a too low figure for the infanticide frequency per year.[1] Besides, many scholars faced with all these uncertainties of estimation think that it is best to go directly to the official Home Office crime statistics, which lists the straightforward number of homicides per year as directly recorded by the police. According to this source (Home Office [2001], p. 87), there were 33 *recorded* killings of less than one-year old children in 1997, while the average number of these homicides per year in England and Wales in the period between 1990 and 2001 was 30. Since this figure is cited (among others) by the authors of the CESDI study, I also decided to rely on it in my further calculations.[2] On this basis, Dawid's estimate for p(M) turns out to be wrong by a factor of 4.

Mathematician Ray Hill starts off on a better foot, citing the statistic of around 30 infant homicides a year among 650,000 births each year in England and Wales. But he also proceeds too quickly to translate the observed frequency into a probability statement: 'the chances of an infant being a homicide victim are about 1 in 21,700' (Hill [2004], p. 321). This estimate is also accepted by Helen Joyce in her paper ([2002]), in which she undertook to explain to the wider public how to use correctly the probabilistic reasoning in Clark's case. (Joyce was at the time the editor of *Significance*, a quarterly magazine published by the Royal Statistical Society.)

Are we ready now to obtain the ratio of prior probabilities, by dividing the probability of SIDS (Dawid's figure of 1 in 8,543) with the probability of infant homicide (Hill's figure of 1 in 21,700)? No, there are still two problems.

First, the observed frequencies from which the two probabilities are derived do not refer to the same group. Recall that Dawid's figure for SIDS refers to families like Clark's (affluent, parents nonsmokers, mother over 26). Hill's figure for infant homicide, however, refers to the whole population indiscriminately. It is well known that, like SIDS, infant homicide is also less frequent in the type of family defined by the three characteristics, but unfortunately I could not get a reliable statistic of infant homicide for that group. Therefore, in order to make the data for SIDS and infant homicide mutually comparable I decided to use the frequency of SIDS in the general population, instead of the more narrow category used by Dawid. According to the already mentioned study by Fleming *et al.*, the incidence of SIDS in the general population was 1 in 1,300.

Second, a serious methodological problem arises from the fact that SIDS is not a single well-defined syndrome, disease or type of event. It is a term of exclusion, or a 'dustbin diagnostic' (Emery [1989]). An essential component of the concept of SIDS is the ignorance of what caused the death. Simply, when the etiology of infant death is unknown then, if some additional conditions are satisfied, the death is classified as SIDS. In the most widely used formulation, SIDS is 'the sudden death of an infant under 1 year of age, which *remains unexplained* after a thorough case investigation, including performance of a complete autopsy, examination of the death scene, and review of the clinical history' (American Academy of Pediatrics [2000], p. 650, italics added).

This clearly opens up a possibility that a certain proportion of those cases that are labeled as SIDS are in fact homicides. Of course, the opposite possibility also exists, that some cases labeled as homicides are in fact SIDS, but for two reasons the latter kind of mistake is going to be much less frequent than the former. First, since a death will be classified as a homicide only if homicide is proved beyond reasonable doubt, miscategorizations will happen more rarely with cases labeled as homicide than with those labeled as SIDS, where inclusion is based largely on ignorance.[3] Second, a person who is typically the sole witness of a child death in those ambiguous 'SIDS or homicide' situations and whose report tends to be one of the most important pieces of evidence in the case will almost always try to prove the SIDS scenario. To put it bluntly, people who killed their children will argue that it was SIDS whenever there is minimal plausibility to that claim, but people whose children died of SIDS will seldom claim that they murdered them. This asymmetry will constitute another ground for taking the officially declared statistics of SIDS with some reservation and for not rushing into accepting it as an automatic reflection of the true probability of SIDS.

These are not just theoretical speculations. There are quite a number of cases that were classified as SIDS at first, but later ended in conviction for murder and even outright confession. Surely, there must also be cases of undiscovered murder that are wrongly treated as SIDS. How often does this happen?

It is hard to tell, but we can immediately reject the assumption implicitly made by Dawid, Hill and Joyce that the empirically collected data give an unbiased estimate of the actual incidence of SIDS and infanticide. They all take the recorded frequencies at face value, without taking into account the high probability that an unknown proportion of cases classified as SIDS are in fact covert homicides. To get a minimally reasonable assessment of the ratio of prior probabilities of S and M we would have to find a way to estimate the percentage of SIDS that are not true SIDS but something more sinister. This is not easy to do for the reasons explained above, but an additional difficulty is that in the case of infants even a very detailed autopsy cannot distinguish a SIDS death from a death caused by suffocation with a soft object.

Nevertheless, scholars have tried to get around the problem and obtain a reasonable ball-park figure for the proportion of homicides that are wrongly subsumed under SIDS.[4] John Emery suggested that perhaps between 10% and 20% of deaths attributed to SIDS may be unnatural deaths ([1993], p. 1099). The authors of the CESDI study considered the question and came up with the figure of 14% of SIDS cases in which maltreatment was thought to be either the main cause of death or a secondary or alternative cause of death (Fleming *et al.* [2000], pp. 126–7). In a recent article devoted to this very topic, it is suggested that, most plausibly, around 10% of SIDS are in reality homicides, while the proportion of covert homicides among deaths registered as 'unascertained' is probably even higher (Levene and Bacon [2004], p. 444). One of the latest comprehensive studies of homicide gives a figure of 20% as the best guess of the proportion of SIDS that are in reality nonnatural deaths (Brookman [2005], p. 21). Apparently, many pediatric pathologists and forensic pathologists say *in private conversations* that 'parental or adult intervention may have occurred in 20–40% of the cases of so called sudden infant death syndrome with which they are involved' (Green [1999], p. 697) but the publicly expressed opinions always give lower figures. Faced with this area of indeterminacy I decided to use estimates of p(S) and p(M) within a range of misclassified SIDS from 0% to 20%. The assumption of 0% error (all SIDS are true SIDS) is quite unrealistic, whereas many experts think that the assumption of 5–10% of misdiagnosis is unlikely to be a significant overestimate of the incidence of covert homicide.

The reader might think that a 10% error will only minimally affect our estimates of probabilities, and that it is hence somewhat pedantic to insist on introducing this kind of correction. Although this may be true about p(S) it is not at all true about p(M). Since there are so many more cases of reported SIDS than recorded infant homicides, even a small percentage of cases transferred from S to M will substantially increase the number of M cases. For instance, even if only 10% of SIDS are misdiagnosed, the correction will result in more than doubling of the probability of M. Look at Table 30.1 for details.

Let us run quickly through this table and explain its contents.

The first row (ΣS) gives the number of all SIDS cases (as recorded in the CESDI study) in its first column, and then in the subsequent columns this number is decreased by subtracting a percentage of false SIDS under each of the four different 'false SIDS scenarios' (5%, 10%, 15% and 20%).

The second row (ΣM) is based on the number of infant homicides per year from the Home Office crime statistics. Again, there is no correction in the first column because the assumption here is that there are zero cases of (covert) homicide wrongly subsumed under the SIDS rubric. In the subsequent columns, however, the number of homicides is increased by adding the corresponding number of incorrectly labeled SIDS that are in reality murders.

The third row (ΣM-corr) introduces another modification. Since the population which represents the reference group for the murder statistics (642,093 births in England and Wales in 1997) is considerably larger than the population for SIDS statistics (472,823 births covered in the CESDI study) it is necessary to add a correction factor to compensate for the difference in size between the two groups. Therefore, in each of the four cases of nonzero percentage of false SIDS, the number of estimated covert homicides that is added to the officially reported figure of recognized homicides has to be multiplied by 642,093/472,823, which is 1.36.

The fourth row gives the prior probability of S, by simply dividing ΣS (the first row) by 472,823.

Table 30.1 Frequencies and prior probabilities of S and M

	0%	5%	10%	15%	20%
ΣS	363	345	327	309	290
ΣM	30	48	66	84	103
ΣM-corr	30	55	79	104	129
p(S)	0.000768	0.000729	0.000691	0.000653	0.000614
p(M)	0.000047	0.000085	0.000124	0.000162	0.0002
p(S)/p(M)	16.4	8.6	5.6	4	3.1

The fifth row gives the prior probability of M, by dividing ΣM-corr (the third row) by 642,093.

The sixth row reveals the magnitude we are looking for in this section: the ratio of prior probabilities of S and M.

Several comments are in order here. If the reader is wondering why those two different sources were chosen for estimating probabilities (the CESDI study for sudden infant death and the government data for homicides) the answer is simple. It was not me who picked out these sources. They were used by the statisticians I critique. Recall that in this paper I am *not* trying to arrive at the most reasonable estimates of probabilities of S and M, all things considered, but rather to assess the statisticians' conclusions about these probabilities. My view is that their conclusions are badly off the mark *even when they are inferred from their own data.*

How wrong are the statisticians about $p(S)/p(M)$? Ray Hill overestimates the ratio by a factor of 2 or 3, if we accept the figure of 5 or 10% as the percentage of covert homicides passing as SIDS. Philip Dawid's error is more serious. Like Hill, he also takes at face value the statistical data about SIDS (neglecting the probability that an unknown proportion of case labeled as SIDS are homicides). So when his additional miscalculation of incidence of infant homicides (1 in 92,000) in the general population is combined with the comparable estimate of $p(S)$ as 1 in 1,300 in the general population, this yields the ratio of 70 for $p(S)/p(M)$, an assessment that is wrong by a factor of 8 or 12 (again, on the assumption that the ratio of misclassified SIDS is 5 or 10%).

A clear result of our analysis is that $p(S)$ is indeed significantly higher than $p(M)$ under any of the percentage scenarios of misdiagnosed SIDS. In other words, the first component of the right-hand side of Equation (4) makes the probability ratio favorable for the double SIDS hypothesis.

Now after having obtained our first ratio, $p(S)/p(M)$, we are ready to plug it into Equation (4), as the first step in determining the overall ratio of posterior probabilities of 2S and 2M. For concreteness, I will select the 10% assumption of misdiagnosed SIDS, and then for each stage of the discussion I will show how Equation (4) looks under that assumption. I chose this percentage because, as mentioned above, it comes from the most recent and most detailed analysis in the literature (Levene and Bacon [2004]). But remember that the figure of 10% is just an estimate, used as an illustration for those who have a preference for specific comparisons. For a more synoptic view, look at the graphs later in the text that contain more comprehensive information. (Also bear in mind the important warning spelled out in note 4).

With this caveat, here is how Equation (4) is modified after the first stage of our debate, with the ratio $p(S)/p(M)$ replaced by the value 5.6 from Table 30.1 (the 10% column):

$$\frac{p(2S/E)}{p(2M/E)} = 5.6 \times \frac{p(S_2/S)}{p(M_2/M)} \times \frac{p(E/2S)}{p(E/2M)} \quad (5)$$

4 Prior probabilities of the recurrence of SIDS and homicide

Now it is time to consider the second component of Equation (4), the ratio of prior probabilities of repeated SIDS and of repeated murder: $p(S_2/S)/p(M_2/M)$. It is in this context that the statisticians complained most about the treatment of probabilities in the Sally Clark court case. The focus was on Roy Meadow, an expert witness for the prosecution, who suggested that the probability of a double SIDS can be obtained by just squaring the probability of a single SIDS. In this way, starting with the odds of 1 in 8,543 for a *single* SIDS in the relevant type of family (affluent, parents nonsmokers, mother over 26), Meadow claimed that the probability of a *double* SIDS in that kind of family is 1 in 73 million (8,543 × 8,543).

This squaring of the single SIDS probability caused an outrage among statisticians. First, the President of the Royal Statistical Society was approached by a Fellow who was involved in the campaign to clear Sally Clark and who asked 'that the Society consider issuing a statement on the misuse of statistics in the court case' (Royal Statistical Society [2002a], p. 503). And indeed, the RSS decided to make a public statement about 'invalid probabilistic reasoning in court,' which referred to Meadow's handling of probabilities. What the Royal Statistical Society called 'a serious statistical error' was soon described as a 'howler' (Matthews [2005]), 'an elementary statistical mistake, of the kind that a first year student should not make' (Dalrymple [2003]), 'infamous statistic' (Hill [2005]), 'blunder' (Marshall [2005]), 'poor science' (the Oxford statistician Peter Donnelly, quoted in Sweeney [2000]), 'disgrace' (the geneticist Brian Lowry, quoted in Batt [2004], p. 244) and even 'the most infamous statistical statement ever made in a British courtroom' (Joyce [2002]).

Meadow was demonized in print to such an extent that, when he appeared before the disciplinary panel of the General Medical Council (GMC) in July of 2005, people might have been surprised to notice, as a journalist put it, that the 'disgraced' and 'discredited' 'child-snatcher in chief' walked on feet rather than cloven hooves (Gornall [2005]). The GMC promptly decided to strike Meadow off the medical register,[5] in the face of the isolated voice of opposition of the editor of *The Lancet*, who argued that 'facts and fairness demand that Prof Roy Meadow be found not guilty of serious professional misconduct' (Horton [2005]). A court of appeals overturned the GMC verdict in February 2006, but the appeal of GMC to that decision was partly successful in October 2006.

Returning to probabilities, what exactly was the statistical sin that Meadow was guilty of? Here is the explanation of the Royal Statistical Society:

This approach (the squaring of the single SIDS probability) is, in general, statistically invalid. It would only be valid if SIDS cases arose independently within families, an assumption that would need to be justified empirically. Not only was no such empirical justification provided in the case, but there are very strong a priori reasons for supposing that the assumption will be false. There may well be unknown genetic or environmental factors that predispose families to SIDS, so that a second case within the family becomes much more likely.

(Royal Statistical Society [2001])

The same criticism is expressed in an open letter that the President of the Royal Statistical Society sent to the Lord Chancellor on January 23, 2002 (Royal Statistical Society [2002b]).

Now it is undeniable that in inferring the probability of two SIDS in the same family Roy Meadow proceeded as if $p(S_2/S)$ were equal to $p(S_2)$, that is, as if the probability of a second child dying of SIDS was not affected by the fact that the first child died of SIDS. Let us call this assumption that $p(S_2/S) = p(S_2)$, the independence hypothesis (IH). RSS is certainly right that Meadow's squaring of probability is valid only if IH is true. It seems to me that there are two possible explanations for why Meadow squared the probability: he either did it because (i) he did not realize that it is necessary to assume the truth of IH (thus making an elementary mistake in probability reasoning), or because (ii) he actually had reasons to assume that IH is true and relied on its truth in his inference. In case (ii) there is no statistical fallacy at all, and then the only question would be whether his empirical assumption is in fact correct or not.

It is very odd that neither RSS nor the President of RSS did as much as mention possibility (ii). They immediately opted for diagnosis (i), apparently for two reasons: first, because Meadow provided no empirical justification for IH, and second, because they thought there are strong grounds to believe that IH is in reality false.

But neither of these two reasons supports explanation (i). First, as an expert witness, Meadow was surely under no immediate obligation to provide specific empirical justification for his opinion about this issue, especially since he was not asked by anyone to elaborate, and since his view was not even challenged by the defense when he expressed it.

Second, when the statisticians assert that there are 'very strong *a priori* reasons' (RSS) or 'strong reasons' (the president of RSS) for supposing that IH is false, they really venture forth onto thin ice. They are right that 'there may well be unknown genetic or environmental factors that predispose families to SIDS, so that a second case within the family becomes much more likely.' But then again, there are also 'very strong *a priori* reasons' that pull in the opposite direction, making a second SIDS case within the family *less* likely. For instance, common sense tells us that the parents whose first child died of SIDS are likely to take extreme precautions in their care of the next child, be attentive to the smallest signs of child's discomfort, never or rarely leave it alone, change their own habits (e.g., quit smoking), strictly follow the doctor's advice about minimizing the risk of recurrence, etc., which would all have the effect of making the probability of a second SIDS in the same family *lower* than the probability of SIDS in the general population.

This shows that RSS was wrong when it claimed, on the basis of 'a priori strong reasons,' that the error resulting from accepting IH is likely to be 'in one particular direction' (i.e., that

it will underestimate the probability of repeated SIDS). There is simply no way to know in advance which of the two kinds of a priori reasons pulling in opposite directions will carry more weight.

More importantly, however, since the truth-value of IH is an empirical issue, why should the discussion about it be conducted in non-empirical terms? Why start criticizing a very distinguished pediatrician's testimony about pediatrics by disputing his claims with a priori reasons or by imputing to him an elementary mistake in the logic of probability? If the expert's expressed view in his own field of competence implicitly assumes the truth of IH, does not minimal fairness dictate that, before his commitment to IH is dismissed as being the result of sloppy and fallacious thinking, we first seriously consider the possibility that he actually possessed *good empirical reasons* to believe that IH is true? The statisticians who intervened in the Sally Clark case by attacking Meadow's testimony left this avenue completely unexplored.

An indication that Meadow's reliance on IH was not an ill-considered and precipitate judgment is that he defended it in his published writings as well: 'There do not seem to be specific genetic factors entailed [for SIDS] and so *there is no reason to expect recurrence within a family*' (Meadow [1997], p. 27, italics added). Is this a solitary opinion of a maverick expert, or is it shared by some other scholars in the field? Let us see.

The authors of a carefully designed 16-year study of the SIDS recurrence statistics in the state of Washington concluded: 'Parents of SIDS victims can be advised, with considerably more confidence than in the past, that their risk of loss of a child in the future is virtually the same as that among families of like size and mother's age' (Peterson *et al.* [1986], p. 914) The authors of another similar study conducted in Norway over 14 years and on a population of more than

800,000 children claimed that 'there is no significant increased risk [of SIDS] in subsequent siblings' (Irgens and Peterson [1988]). One of the leading SIDS researchers, J. Bruce Beckwith, writes: 'Until more satisfactory information is provided, I will continue to reassure families of these SIDS victims whose family history is negative, whose clinical and post-mortem findings are classic for STDS, and who are not burdened with an excess of risk factors that their recurrence risk is not increased significantly over that of the general population' (Beckwith [1990], p. 514). In a widely cited study on the subject it is suggested that 'the chance of recurrence [of SIDS] is very small and probably no greater than the general occurrence of such deaths' (Wolkind *et al.* [1993], p. 876). The editor of the main journal in the field of child medicine says resolutely: 'Some physicians still believe SIDS runs in families. It does not – murder does' (Lucey [1997]). It is the opinion of the American Academy of Pediatrics that 'the risk for SIDS in subsequent children is not likely increased' (American Academy of Pediatrics [2001], p. 439). A chapter on 'SIDS and Infanticide' in a recent authoritative collection of articles about SIDS contains the following statement: 'No evidence has ever been presented to demonstrate that families that have suffered one SIDS death are legitimately at a greater risk of a second, to say nothing of a third, fourth, or fifth. On the contrary: anecdotal evidence supports the suggestion that, as Linda Norton MD, a medical examiner specializing in child abuse, said: "SIDS does not run in families – murder does" (personal communication)' (Firstman and Talan [2001], p. 296). In a textbook of forensic pathology that is called 'the most comprehensive, definitive, and practical medicolegal textbook of forensic pathology today' we can read: 'SIDS deaths appear to occur in families at random. There is no evidence of a genetic etiology. Siblings of SIDS victims have

the same risk as the general population' (DiMaio and DiMaio [2001], p. 327). Thomas G. Keens, professor of pediatrics at the University of California who specializes in SIDS research, gave an overview of the relevant empirical studies and concluded: 'Thus, the SIDS recurrence risk for subsequent siblings of one previous SIDS victim is probably not increased over the SIDS risk for the general population' (Keens [2002], p. 21).

I am not implying that all scholars would agree with the opinion voiced in these quotations, but obviously many leading SIDS experts believe that IH is actually supported by empirical evidence and that it is true. This completely undermines the RSS's condemnation of Meadow's squaring of probabilities. Recall that it was on the basis of some a priori reasons (which *they* regarded as cogent) that the statisticians thought that IH is most probably false, and then on that assumption they concluded that the squaring of probabilities, which depended on the truth of IH, must have been the result of a crude mathematical error. But, as the above list of quotations proves, the statisticians are simply wrong in their judgment that it is so easy to dismiss IH on intuitive grounds. Ironically, the statisticians accused a pediatrician of venturing into a field outside of his area of expertise and committing a serious statistical fallacy, when what really happened was that the statisticians themselves left their own area of competence and issued a rather dogmatic and ill-informed pronouncement about a topic – in pediatrics![6]

The authority of the Royal Statistical Society was sufficient to make this misconceived claim spread quickly in all directions and gain uncritical acceptance, sometimes in quite unexpected quarters. For instance, although many experts do argue that SIDS cases *are* independent (as documented above), the categorical but unsubstantiated statement that 'there is no scientific evidence' for the independence of SIDS deaths was even defended in the British Academy Annual Lecture that was given by the Governor of the Bank of England (King [2004], p. 10).

The intervention of the statisticians was entirely unnecessary because the legitimacy of squaring the probabilities in the Sally Clark case was never really a mathematical or statistical issue. The multiplication stands or falls just depending on whether IH is true. And as we saw, many scholars in the relevant disciplines do accept IH, which shows that there was nothing atrociously wrong or irresponsible in Meadow's testimony invoking the odds of 1 in 73 million.

True, other experts reject IH, but even if IH is false, and if consequently $p(S_2/S)$ is higher than $p(S)$, the Royal Statistical Society is still wrong in its statement that in that case the figure of 1 in 73 million would involve an error that 'is likely to be *very large*' and that 'the true frequency of families with two cases of SIDS may be *very much less* incriminating than the figure presented to the jury at trial' (Royal Statistical Society [2001], italics added). The phrases 'very large' and 'very much less' seem to suggest that we have a good reason to expect a drastic increase of probability of SIDS, given a previous SIDS case. But the best studies in the literature do not support such dramatic jumps in SIDS risk at all. For instance, in a recent policy statement the American Academy of Pediatrics concluded that 'the risk of [SIDS] recurrence in siblings, if present, is most likely *exceedingly low*' (Blackmon *et al.* [2003], p. 215, italics added). Furthermore, to mention the most widely cited research reports, it has been suggested that the risk of SIDS in subsequent siblings of a SIDS victim is increased by a factor of 3.7 (Irgens *et al.* [1984]), 1.9 (Peterson *et al.* [1986]) or 4.8 (Gunderoth *et al.* [1990]). The most recent study on the topic gives the estimate of a 6-fold increase of the risk (Carpenter *et al.* [2005]), but as a perceptive critic pointed out (Bacon [2005]), this larger figure is probably an overestimate that resulted

from two serious methodological flaws: first, the sample was not randomly selected (some families opted out from the investigation for unknown reasons), and second, some cases with a number of suspicious characteristics (possible homicides) were uncritically subsumed under the SIDS category.[7] The second objection actually also applies to the previously mentioned three studies, so that, all things considered, it could be argued that it is very unlikely that $p(S_2/S)$ is more than 4 times higher than $p(S)$.

Applying this estimate to the Sally Clark case, the prior probability of two infants dying of SIDS in that kind of family is obtained by multiplying 1/8,543 (the CESDI figure) with 1/2,136 (four times the CESDI figure), which gives the required probability as 1 in 18 million. Although this estimate (1 in 18 million) as opposed to odds of 1 in 73 million is clearly higher, it is still an astronomically low probability and it must be asked whether this would really be regarded as 'very much less incriminating,' as the Royal Statistical Society suggested.

In our search for the ratio $p(S_2/S)/p(M_2/M)$ we are now in the position to weigh the number in the numerator: $p(S_2/S)$ is either the same as $p(S)$, or it is up to 4 times higher than $p(S)$. We will make use of both of these estimates later, at the final stage of the analysis in this section.

But first we have to try to get some idea about the magnitude of the denominator. Dawid has an odd proposal about how to proceed here. He says that since Meadow's obtaining the probability of double SIDS by squaring the single SIDS probability was based on the 'unrealistic' assumption of independence, then one might as well treat the probability of double murder 'in the same (admittedly dubious) way as the corresponding probability for SIDS, and square it to account for the two deaths' (Dawid [2002], p. 76; cf. Dawid [2004], p. 2). In this way he derives the conclusion that a double SIDS is much more likely than double murder.

Dawid's basic point is that the simple multiplication of probabilities is dubious in both cases, and that, therefore, if it is allowed in calculating the probability of double SIDS then it could, equivalently, be also applied in the case of double murder.

But in estimating the repetition probabilities in cases like Sally Clark's, the two scenarios that we are considering (double SIDS and double murder) are not equivalent at all. The crucial asymmetry consists in the fact that the first child's death was initially diagnosed as being due to natural causes, which means SIDS (in the perspective of 'either-SIDS-or-murder' that we adopted from the beginning). Now if it was really SIDS, the mere classification of death as SIDS could in no way increase the probability of the next sibling dying from SIDS. But if the first child's death was not a true SIDS but homicide, then the fact that it was *wrongly* classified as SIDS *would indeed substantially increase the probability of the next sibling being killed.* Why? Well, the wrong classification would have as a direct consequence that after the *undiscovered* homicide of the first child the next sibling would continue living together with the mother that already killed one baby, and under these circumstances the new child would be under grave risk.

So, squaring the single-case probability in order to get the recurrence probability is not equally problematic in the two cases (SIDS and murder). With SIDS, many scholars in the relevant area of empirical research think that the procedure is entirely legitimate. In the case of murder, however, most people would think that in the situation under discussion such a procedure would be manifestly wrong. It stands to reason that the probability of a child being killed *if it is in the care of a parent who already proved to be capable of killing a baby* must be significantly higher than the probability of a child being killed in general.

There will be exceptions, of course, in which an infant homicide may lead to, say, a Raskolnikovian guilt and to a character transformation of a murderer into a caring and devoted parent. Or a parent might be afraid that she would be caught if she did it again,[8] and she could refrain from killing again for that reason. This is all true but we are talking about probabilities here. Although the very logic of the situation makes it impossible to get access to the relevant statistical data (how could we know the recurrence rate of child homicide in families with a previous *undiscovered* infanticide?), indirect reasoning can help. Since it is well known that child abuse has a strong tendency to repeat itself, why expect that this suddenly stops with the ultimate abuse? Also, since it is well known that previous homicide offenders are much more likely to commit a homicide than people in the general population, why expect that this recidivist tendency abruptly disappears when victims of homicide are babies? Or to make the point with a rhetorical question: if the only thing you knew about persons A and B was that A intentionally killed her own child whereas B did not, which of the two would you prefer as a babysitter for your own son or daughter?

On this basis, we have to reject as misguided Dawid's idea (repeated and endorsed in Aitken and Taroni [2004], pp. 211-3) that if one squares the probability of a single SIDS to obtain the probability of a double SIDS, this somehow justifies doing the 'same' thing with the probability of murder. It is not the same thing. Inferring the probability of double S by squaring $p(S)$ may well be regarded as controversial, but in the context that interests us, squaring $p(M)$ to obtain the probability of double M is just blatantly wrong.

How should we quantify $p(M_2/M)$? We need some numerical value, however approximate, which we can plug in to obtain the ratio $p(S_2/S)/p(M_2/M)$. Mathematician Ray Hill, who was actively involved in the efforts to free Sally Clark, estimated that $p(M_2/M)$ is 0.0078 (Hill

[2004], pp. 322–3). Since for reasons of space I cannot go into a detailed criticism of his derivation, let me briefly adumbrate four basic reasons why I do not find his figure credible. First, Hill used the data from a study that was based on voluntary participation, which raises serious concerns that the sample was not representative and that the biased selection led to the underestimate of the probability in question. Second, his calculation relied on the values of $p(S)$ and $p(M)$ that were directly estimated from observed frequencies, without correcting for a number of homicide cases that are expected to have been misdiagnosed as SIDS (the mistake that is extensively explained in Section 3). Third, Hill uncritically treated all cases classified as double SIDS as *really* being double SIDS, again disregarding an even more pronounced risk of wrong classification. Fourth, there are reasons to think that the probability we are looking for will be considerably higher than Hill's figure of 0.0078. Namely, if a child lives with a parent who already killed another baby, it is arguable that the child's life is exposed to 'a clear and present danger' to such a degree that it is not adequately represented by the probability of its being killed that is as low as less than 1%.

Helen Joyce starts with an intuitive estimate of 0.1 for $p(M_2/M)$. Although she says that this figure 'is almost certainly overestimating the incidence of double murder' (Joyce [2002]), she nevertheless regards it as one of several 'reasonable estimates of the likelihoods of relevant events'. I will adopt Joyce's suggestion for $p(M_2/M)$ because I agree that it is not an unreasonable estimate, and because in this way I will follow my basic strategy of accepting those of the statisticians' premises that are not patently wrong and then assessing whether or not the conclusions they derive from them do actually follow.

Now we are ready to obtain the second component in Equation (4), the ratio $p(S_2/S)/p(M_2/M)$.

On the assumption of statistical independence of SIDS cases in the same family, $p(S_2/S)$ is equal to $p(S)$. As before, for illustration purposes let us take $p(S)$ under the scenario of the 10% rate of misdiagnosed SIDS (covert homicide). In that case, $p(S)$ is 0.00069. If we divide this by Joyce's figure for $p(M_2/M)$, which is 0.1, we get 0.0069. This is the ratio of the repetition probabilities of SIDS and murder. If we plug this number into Equation (5), and multiply it with the earlier obtained ratio $p(S)/p(M)$, which was 5.6, the result is presented in the following Equation (6):

$$\frac{p(2S/E)}{p(2M/E)} = 5.6 \times 0.0069 \times \frac{p(E/2S)}{p(E/2M)}$$

$$= 0.04 \times \frac{p(E/2S)}{p(E/2M)} \tag{6}$$

What is the meaning of 0.04 here? It represents the ratio of *prior* probabilities of the two hypotheses we are discussing (double SIDS and double murder), that is, *before* the specific empirical evidence in any specific case is considered. Put differently, the ratio tells us that a priori (before consulting the particular evidence pertaining to the case) the double murder hypothesis is 25 times more probable than the double SIDS hypothesis. Moreover, supposing that one of the two hypotheses must be true about a given unresolved child death, we can immediately obtain the prior probabilities of both hypotheses. The prior probability of 2M is 25/26, or approximately 0.96, whereas the prior probability of 2S is 1/26, or approximately 0.04.[9]

Of course, these particular probabilities apply only to a very specific situation, that is, under the assumptions that were made about $p(S)$, $p(M)$, $p(S_2/S)$ and $p(M_2/M)$. To get a broader perspective on how these probabilities change when values of some of these parameters are altered, take a look at Figures 30.1(a) and 30.1(b). Both figures represent the situations in which SIDS and homicide are the only alternatives, so that $p(S) = 1 - p(M)$, and $p(2S) = 1 - P(2M)$.

Figure 30.1(a) shows how prior probabilities of S, M, 2S and 2M depend on the proportion of prima facie SIDS cases that are in reality covert homicides. The assumption in Figure 30.1(a) is that separate occurrences of SIDS are statistically independent, that is, $p(S_2/S) = p(S_2) = p(S)$.

In Figure 30.1(b), though, the assumption that IH is true is abandoned. In accordance with the earlier review of the literature that disputes IH, it is supposed here that $p(S_2/S) = 4 \times p(S)$.

It is interesting to note that in both Figures 30.1(a) and 30.1(b) the prior probability of *single* SIDS is consistently higher than

the prior probability of single homicide (independently of the percentage of misdiagnosed SIDS). In contrast, however, in both figures the prior probability of *double* SIDS is consistently lower than the probability of double homicide. Furthermore, apart from the implausible zero percentage scenario of misclassified SIDS, it transpires that 2M is always *considerably more probable* than 2S. This accords well with the view, expressed by many physicians, forensic scientists and criminologists, that any isolated death classified as SIDS is probably a true SIDS but that two (or more) alleged SIDS deaths in the same family should raise concerns about possible child abuse and would justify more detailed examination.

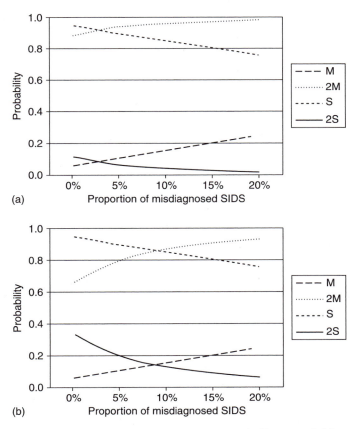

Figure 30.1 (a) Prior probabilities of SIDS and murder (SIDS independence). (b) Prior probabilities of SIDS and murder (SIDS dependence).

The inclination to take seriously the possibility of foul play in such situations is opposed by allegedly 'very strong a priori reasons' (Royal Statistical Society [2001]; Royal Statistical Society [2002b]) for supposing that certain genetic or environmental factors predispose some families to SIDS, and that, consequently, an index of suspicion should not be raised much under these conditions. This is not a purely academic issue. Let me give two real-life examples in which reliance on what the Royal Statistical Society called 'very strong reasons' led to a tragic outcome.

In the first case (described in Firstman and Talan [2001], p. 298), a child was many times rushed to the hospital from an early age, with her mother reporting that the baby had stopped breathing and turned blue. A curious thing was that an older sibling had had the same problems but only until the younger child was born, when the symptoms in some miraculous way 'transferred' to her. Some doctors started to suspect that the mother had the so-called 'Munchhausen syndrome by proxy' (MSBP), a disorder where a parent (usually the mother) tries to draw attention to herself by intentionally making her child sick and often eventually killing it.[10] But despite these publicly expressed suspicions, a closer investigation by child welfare authorities and police was blocked by a senior doctor who had an alternative theory that the child had a heart defect, and so he decided to perform surgery and install a pacemaker. Some time later the child was brought to the hospital unconscious and died soon afterwards. (The pacemaker was functioning properly.) The mother's report about what happened was refuted by a forensic pathologist and she failed the polygraph. But although the district attorney had no doubts about her guilt (like most others involved in the case) he decided not to prosecute because he thought that with the prospect of the senior, highly esteemed doctor defending

the SIDS diagnosis he could not convince the jury.

The second case is more dramatic and it involves one of the most widely cited papers about SIDS in the pediatric literature. In 1972 Alfred Steinschneider published an article (Steinschneider [1972]) in the leading journal in the field, in which he analyzed the case of Waneta Hoyt, a woman from New York State, whose three very young children died of unknown causes. Apparently guided by the 'very strong a priori reasons' for believing that certain factors predispose some families to SIDS, Steinschneider hypothesized that the children must have suffered from the same birth abnormality that caused 'apneic episodes' and ultimately death. His suggestion that this disorder might actually be behind many SIDS deaths was soon accepted by a number of medical experts, and the theory stayed at the very center of SIDS research for twenty years. Furthermore, a massive program that included specially designed equipment was established for monitoring the sleep of children who were considered to be at risk. In fact, Waneta Hoyt herself had two more children (beside the three who died), and Steinschneider described in his paper how he put them on 'apnea monitors' in the hospital in order to minimize the probability of the recurrence of SIDS. But despite these measures the two children died one after the other (a year apart), each of them having been only two months old and with the death in each case happening just a day after the child was discharged from the hospital. The diagnosis for both was an apnea-related SIDS. The paper that described the two cases in detail was widely interpreted as strong evidence that SIDS runs in families.

Enter Linda Nelson, a forensic pathologist from Texas. In 1986 she was contacted by a New York State assistant prosecutor for help with another case involving the death of several children in a family. In the course of the conversation about multiple infant deaths she mentioned that she had read Steinschneider's paper and told her colleague that 'the victims in that paper were all homicide victims, in my opinion' (Pinholster [1994], p. 199). This triggered a long and laborious criminal investigation. In Steinschneider's paper, Hoyt's children, Molly and Noah, were referred to only by initials, but this proved sufficient to identify them through death records. Eventually, the first suspicion turned out to be corroborated, and Waneta Hoyt was accused of killing all of her five children. She confessed and explained that she simply could not stand the children's crying: 'They just kept crying and crying [. . .]. I just picked up Julie and I put her into my arm, in between my arm and my neck like this [. . .] and I just kept squeezing and squeezing and squeezing . . .' (Firstman and Talan [1997], p. 435). Hoyt retracted the confession later, but this was unconvincing (except, predictably, to Steinschneider himself). She was convicted in 1995 and later died in jail.

The editor of *Pediatrics* wrote in 1997: 'We never should have published this article [. . .] Huge amounts of time and money have been wasted over the last 25 years in a useless attempts to "do something" for SIDS victims by monitoring' (Lucey [1997]). Everyone would agree, no doubt, that something much more valuable than time and money was wasted in the process. Some of Waneta Hoyt's children might be alive today if more people had kept an open mind, instead of uncritically trusting their mere hunches and faulty intuitions, which they mistook for 'a priori very strong reasons.'

5 Likelihoods of double SIDS and double homicide

What about the third probability ratio in Equation (4)? The ratio $p(E/2S)/p(E/2M)$ tells us about the import of empirical evidence (E) in

deciding between the two hypotheses, 2S and 2M. The question here is whether the concrete evidence available in the given situation is more likely on the assumption that 2M is true or that 2S is true. To address that issue, we have to leave the rarefied area of aprioristic reasoning and descend into the specifics of a particular controversial case (Sally Clark), which has become something of a test case for the role of probabilities in this kind of context.[11]

Obviously, before discussing how the two hypotheses relate to empirical evidence (E), we have first to determine what E actually was in that case.

Helen Joyce describes E in a way that is nothing short of astonishing. She says that *all* the relevant evidence in the Sally Clark case just reduces to the fact that 'both children are dead' (Joyce [2002]). From this she concludes that, trivially, both $p(E/2M)$ and $p(E/2S)$ are equal to 1. (Indeed, the probability that the children are dead, assuming that they were murdered, is 1. Likewise, the probability that the children are dead, assuming that they died of SIDS, is of course 1.)

By reducing the entire empirical evidence (E) to the mere fact that 'the children are dead,' Joyce is in effect suggesting that Sally Clark was convicted of murder *without any specific evidence against her being considered.* Joyce even explicitly states that the reason why Sally Clark went to prison was 'because the prosecution argued, and the jury accepted, that lightning does not strike twice' (ibid.). In other words, she says that the guilty verdict was here based exclusively on the prior improbability of two SIDS in the same family.

This is a blatant distortion of what really happened in the court proceedings.[12] In truth, the days of the trial were completely dominated by extensive and detailed discussions of *pathological evidence* of various kinds, while the rarity of double SIDS was just briefly mentioned by Meadow in an aside. The point was not regarded important even by the defense lawyers, who did not think, at the time, that it deserved comments or rebuttal.

What totally discredits Joyce's claim that the jury's decision was based exclusively on their belief that lightning does not strike twice is the fact that the judge actually gave the jury a clear and stern warning that they should *not* rely only on evidence from population statistics: 'I should I think, members of the jury, just sound a note of caution about the statistics. However compelling you may find those statistics to be, we do not convict people in these courts on statistics. It would be a terrible day if that were so.' (Court of Appeal [2000], para. 128)

I am not saying that Meadow's figure of 1 in 73 million did not influence the jury's decision at all. Rather, what I am saying is that Joyce's claim that the specific empirical evidence in the case is exhausted by the fact that 'the children are dead' is a ridiculous misrepresentation of the actual state of affairs.

But what exactly were these other pieces of empirical evidence that went unmentioned by Joyce? Dawid says that 'there was additional medical evidence, including hemorrhages to the children's brains and eyes' ([2002], p. 78). This is still far from satisfactory. The word 'including' indicates that there were yet some other relevant details that Dawid for whatever reason chose not to specify. Clearly, we need all these facts in the open in order to estimate p(E/2M) and p(E/2S), be it even informally and in a very approximate manner. Here are some of these data as they were summarized in the judgment of the Court of Appeal in 2000:[13] (i) inconsistent statements that Sally Clark gave to the paramedics and later at the hospital about where one of the babies was when it died,[14] (ii) a child's previous unusual nosebleed unsatisfactorily explained, (iii) a torn frenulum, (iv) extensive fresh bleeding around the spine, (v) hypoxic damage to the brain which occurred at least 3 hours

before death, (vi) fracture of the second rib which was some 4 weeks old for which there was no natural explanation, (vii) dislocation of the first rib, (viii) petechial hemorrhages to the eyelid that were acknowledged even by the defense expert to be 'a worrying feature . . .'.[15]

All these things, especially when taken together, are not what one expects in a legitimate SIDS death. But in an infant homicide, they are not surprising. For if a mother kills her baby, then there is nothing particularly odd if the medical investigation finds post mortem signs of child abuse, like bone fractures and other injuries. In normal situations, however, as the judge in the first appeal said, 'young, immobile infants do not sustain injury without the carer having a credible history as to how the injury was caused.' (Court of Appeal [2000], para. 239) No credible history was offered by Clark's defense.[16]

It might seem that rib fractures could have easily been the result of resuscitation efforts, but this is in fact extremely implausible. Here is the explanation by John L. Emery, a scholar who is held in such a high esteem that he is referred to as 'the master of pediatric pathology' and 'a pediatric pathologist par excellence' (Barness and Debich-Spicer [2005]):

We and others have gone through the movements of resuscitation on cadavers and have found that it is extremely difficult to fracture ribs in an infant by pressing on the chest or by any of the usual methods of artificial respiration. Fractures of the ribs, however, can be relatively easily produced by abnormal grasping of the child's thorax. The presence of fractures in any site in a child younger than 1 year should be considered as caused by abuse unless proven otherwise.

(Emery [1993], p. 1099)

Coming back to the Sally Clark case, it is telling that (according to the judgment of the first

Court of Appeal in 2000) even one of the defense's own experts thought that 'there were features in both deaths that gave rise to *very great concern*' (Court of Appeal [2000], para. 77, italics added). Moreover, the defense itself actually 'accepted that there were *worrying* and *unusual* features, but submitted that the evidence amounted to no more than suspicion' (ibid., para. 10, italics added). What do the italicized phrases mean? Well, the most natural interpretation of 'worrying' and 'unusual' is that the features in question were worrying and unusual *for the double SIDS hypothesis*. In other words, these features were easier to explain on the assumption that 2M is true than on the assumption that 2S is true. Or, alternatively, if these features are labeled 'E', then $p(E/2M) \gg p(E/2S)$. Therefore, it appears that even Sally Clark's defense team realized that, given the evidence presented to the court, it would not sit well with the jury to deny the obvious (the existence of suspicious circumstances) and to insist, *à la* Joyce, that $p(E/2M) = p(E/2S)$.

After it is conceded that $p(E/2M)$ is higher than $p(E/2S)$, the next question is: how much higher is it actually? We need a number for the ratio $p(E/2S)/p(E/2M)$, which could then be plugged into Equation (6) to yield the final answer to our central question (about the ratio of *posterior* probabilities of 2S and 2M). Dawid, confronted with the same need to come up with a numerical estimate, suggests that a reasonable figure for the likelihood ratio is 1/5 (Dawid [2002], p. 78). That is, he proposes that the 'worrying' evidence of the kind that was found in the Sally Clark case is five times more probable if 2M is true than if 2S is true.

In my opinion, there are good reasons to think that Dawid underestimates the difference between the two probabilities. Think about it this way: if it turns out, as it may well do, that the suspicious circumstances of the aforementioned kind are discovered in, say, about 25% of

infant homicides (this is a purely hypothetical figure), would you under these circumstances then really expect that, in accordance with Dawid's ratio of 1/5, the disturbing features like babies' broken bones, other strange injuries and the parents' incoherent reports about the death scene would be present in as many as 5% cases of true SIDS? I doubt that you would. If you agree, it is arguable that $p(E/2M)$ is more than five times higher than $p(E/2S)$.

Nevertheless, I do not intend to press this objection. Rather, I will work here with Dawid's ratio, in the spirit of my attempt to evaluate the statisticians' conclusions on the basis of their own premises. Now, with the estimation of the last unknown (the ratio of likelihoods of 2S and 2M), the ground is cleared for the final analysis.

6 Posterior probabilities of double SIDS and double homicide

If Dawid's proposal of 1/5 for the ratio $p(E/2S)/p(E/2M)$ is substituted into Equation (6), the result is:

$$\frac{p(2S/E)}{p(2M/E)} = 5.6 \times 0.0069 \times 0.2 = 0.008 \quad (7)$$

This is the probability ratio of the two rival hypotheses (double SIDS and double murder) after *all* the relevant evidence is taken into account. The probability of 2M is 125 times higher than the probability of 2S. Assuming that 2M and 2S are the only two possibilities, the ratio of these two probabilities straightforwardly yields the information about the probabilities of the two hypotheses. All things considered, after the statisticians' reasoning is corrected for numerous errors it turns out that, on their own premises, the statistical probability that Sally Clark killed her two babies would be higher than 0.99 (125/126), whereas the complementary probability that the children died of SIDS would be lower than 0.01 (1/126).[17]

We have to remember, though, that the result in Equation (7) is obtained under two specific assumptions: (a) that the proportion of homicides that are falsely diagnosed as SIDS is 10%, and (b) that IH is true, that is, $p(S_2/S) = p(S_2) = p(S)$. Let us now drop these assumptions and see how the posterior probabilities of 2S and 2M are affected by changes in these two parameters. The more general picture emerges in Figures 30.2(a) and 30.2(b).

Figure 30.2(a) shows the posterior probabilities of 2M and 2S under five different percentage scenarios of misdiagnosed SIDS, but in a world in which IH is true. Figure 30.2(b), on the other hand, represents a world in which IH is false, and in which the probability of SIDS in a family with a previous SIDS case is significantly higher than the probability of SIDS in the general population. In accordance with our earlier discussion, we assume that $p(S_2/S) = 4 \times p(S)$.

Conspicuously, the probability of 2M is *always* much higher than 2S. Moreover, under any of the ten scenarios pictured in the two figures, the

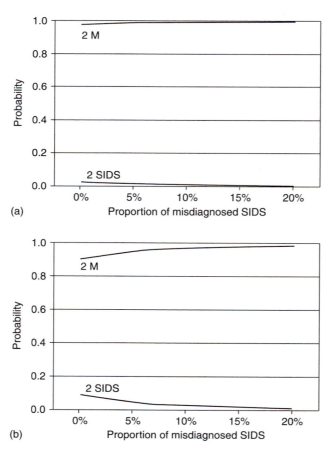

Figure 30.2 (a) Posterior probabilities of double SIDS and double murder (SIDS independence), (b) Posterior probabilities of double SIDS and double murder (SIDS dependence).

probability of double murder is greater than 0.9. This is in jarring discrepancy with the estimates of the statisticians. Joyce's final conclusion is that p(2M/E) is 0.33, while Dawid judged it to be as low as 0.04. What is the source of their underestimates?

It seems to me that Joyce's two big mistakes were, first, putting a too high value for $p(S_2/S)$, and second, totally disregarding the case-specific empirical evidence. Nevertheless, despite this fallacious analysis Joyce's explanation of probabilities appears to have been endorsed by the Royal Statistical Society, as well as praised by cosmologist Hermann Bondi in his letter in *Nature* (Bondi [2004]) and by mathematician Barry Lewis in his presidential address as the President of the Mathematical Association (Lewis [2003], pp. 424–5).

Dawid's first major error, in my view, was that he started with a seriously wrong numerical value for p(M). Second, he thought, incorrectly, that if it is permitted to square p(S) to obtain p(2S), then it must also be OK to square p(M) to obtain p(2M). But these two cases are quite different. In estimating the prior probability that Sally Clark's children died of SIDS, squaring p(S) may be controversial but it is certainly not manifestly wrong. (The procedure is actually regarded as perfectly legitimate by many scholars.) In contrast, to do the same when estimating the prior probability that the children's deaths were homicides would fly in the face of basic common sense. Speaking about $p(M_2/M)$, it is grossly implausible to claim that the probability that a child is killed, *given that it is in care of the mother who already killed one child*, is identical to the probability of an infant homicide in the general population.

7 Conclusion

The statisticians have with one voice condemned a particular treatment of probabilities in the Sally Clark case. After proposing their alternative account they have argued that statistical evidence should be presented in the court 'only by appropriately qualified statistical experts' (Royal Statistical Society [2001]) and that 'interpretation of statistical evidence cannot be safely left in the hands of the uninitiated' (Dawid [2005], p. 8). The problem, however, is that the arguments defended by the 'initiated' also suffer from serious flaws and prove on closer inspection to be logically unsound. Therefore, although it is undoubtedly true that court experts, judges and juries could benefit from a better understanding of probabilistic reasoning, this paper shows that even among the 'appropriately qualified statistical experts' there is room for improvement in this respect. *Statistice, cura te ipsum!*

Notes

I would like to thank Paisley Livingston, Samir Okasha, James Andrew Rice and Elliott Sober for useful comments on the first draft.

1 Speaking about the imperfect temporal match of our two databases (the CESDI source on SIDS and the ONS source on infant homicide), it should be mentioned that the CESDI study used the definition of SIDS that excludes infant deaths in the first *week* of life, but this 'shortening' of a 1-year period of data collection by 1/52 of its duration (less than 2%) is safely ignored. Introducing a correction for such a small discrepancy would just unnecessarily complicate our mathematics and yet the final result of our calculations would remain virtually the same.

2 Strangely enough, although Dawid has only words of praise for the authors of the CESDI study (which he calls 'a professionally executed study', Dawid [2002]), he does not even mention their advice to use the Home Office criminal statistics nor their decision to work with a more than three or four times higher figure for the incidence of infant homicide than his unreasonably low assessment.

3 As John Emery said, SIDS is 'very easy to diagnose, as in general the less found the more certain the diagnosis' (Emery [1993], p. 1097).

4 Notice that many actual but unsuspected homicides will be wrongly subsumed not only under the SIDS rubric but under *other* causes of death as well (accidents, various illnesses, etc.), and that this will lead to a further (and more massive) underestimation of the incidence of infant homicide: 'it is *generally acknowledged* that the known recorded cases of infant homicide are an underestimate of the actual number of infants killed' (Brookman and Nolan [2006], p. 870, italics added); '*Most* commentators claim that the official statistics probably *considerably underestimate* the true incidence of child homicide' (Wilczynski [1997], p. 25, cf. p. 36, italics added). I could not think of a sound way to assess the impact of this misclassification error. Therefore, the reader should keep in mind that my acceptance of a too low value for p(M) will have the direct effect of making my final estimate of the ratio p(S)/p(M) unduly inflated to an unknown extent.

5 In marked contrast, Solicitors Disciplinary Tribunal made an unprecedented decision in 2001 that Sally Clark should *not* be struck off the Roll of Solicitors, although at the time her conviction for infant homicide remained in place, as her second (successful) appeal had not yet been heard.

6 In a talk in which he discussed the Sally Clark case the Oxford statistician Peter Donnelly also made a bare assertion that IH is 'palpably false' (Donnelly [2005]), without providing any actual empirical evidence against IH. Apparently he was unaware that he was thereby overstepping his authority as a statistician and that, moreover, his unsupported contention was resolutely rejected by many of the leading experts in SIDS research. Worse still, Donnelly failed to heed his own good advice from the same talk: 'We need to understand where our competence is and is not.'

7 For example, the authors even classified as SIDS those situations where there were mental health concerns about parents, there was confirmed child abuse, and in 25% of all of these cases 'there were pathological findings compatible with an asphyxial component to the death.'

8 Notice, however, that there would not be much reason for such a fear in an environment in which multiple SIDS is regarded as a frequent and expected phenomenon.

9 An attentive reader will ask: how come that if we divide p(S) by p(M) the result turns out to be 1/24 (0.04/0.96 = 1/24), rather than our previously calculated ratio of 1/25? The reason for this discrepancy is that the numerical values of these two probabilities are rounded off to two decimal places.

10 MSBP was first described by Roy Meadow, so it was to be expected that with his fall from grace and with 'the mob currently in full pursuit' (Fitzpatrick [2004]) this syndrome would also soon come under a barrage of attacks from vehement, though not necessarily well-informed, critics. And this indeed happened. For instance, although there is a huge specialist literature that analyzes and describes hundreds and hundreds of empirical cases in which MSBP appears to be manifested (which includes even the video recordings of perpetrators caught in *flagrante delicto*), the philosopher Simon Blackburn resolutely asserts that it is 'quite false' that science has discovered that condition. He is equally adamant that there is no reason to believe that mothers ever harm or kill their children merely to gain attention (Blackburn [2005], p. 95). How does he know that? In fact, he gives no justification that would even remotely connect with the debate about MSBP in medical publications. Instead he provides some negative rhetoric, like saying that this syndrome 'merely served to direct power and prestige' to expert witnesses, and led to 'the conviction of many innocent mothers.' There is no argument here, philosophical or otherwise. His intervention only amounted to adding fuel to the fire of the anti-Meadow public sentiment that was already out of control.

11 Let me stress again that, although I will briefly go into some empirical details pertaining to Sally Clark's case here, I will not debate the question of her actual innocence or guilt. As I said earlier, my exclusive goal is to analyze the statisticians' reasoning. In that spirit, when it comes to estimating the import of empirical evidence I will eventually just follow the opinion of one of the statisticians (Dawid).

12 Strangely enough, it appears that Joyce's bad advice about how to deal with probabilities in this context is actually endorsed by the Royal Statistical Society, because when Sally Clark is mentioned on the RSS website <www.rss.org.uk/main.asp?page=1225>, there is a direct link to Joyce's paper *and to no other source*. Joyce, in turn, links her paper to the website of the Campaign to free Sally Clark, and she even invites the reader to become involved and to 'write to your MP or to Sally Clark herself' (Joyce [2002]). So, the Royal Statistical Society chose to make its webpage with its pronouncement about probabilities only two clicks away from an Internet site that was entirely devoted to partisan efforts to defend an accused person in a criminal court case. The RSS did not direct its web visitors to other sources of information for a more balanced perspective.

13 I give the snapshot of the relevant evidence at this point in time because the statisticians that I criticize intervened in the case on the basis of what was known *at that stage*. Needless to say, this is not meant to be a full account of the arguments given by both sides, but rather a summary of some of the points presented by the prosecution. My aim here, as stated before, is not to argue the guilt or innocence of Sally Clark, but merely to demonstrate that a substantial amount of empirical evidence was considered by the court, and that if taken at face value this evidence would reduce the ratio $p(E/2S)/p(E/2M)$ from Joyce's value of 1. Of course, the defense and prosecution strongly disagreed about the exact probative force that should be attached to that evidence – but a comprehensive consideration of all the arguments made by both sides would lead us too far afield. For a fuller account of the empirical evidence considered at the trial, the reader should consult (Court of Appeal [2000], [2003]).

14 The incoherence is briefly summarized in a fine article written by two law professors from the London School of Economics: 'She claimed during questioning that one of the babies had died in a baby chair in a position that was not physically possible. Later she refused to answer questions. Then at trial she gave evidence that the child had

not died in the chair.' (Nobles and Schiff [2004], p. 236).

15 Let me repeat once more that I am *not* mentioning these things here in order to argue for Sally Clark's guilt. My point is merely that, when the statisticians want to estimate *overall* probabilities of guilt and innocence by applying the Bayesian inference to this particular case, the only correct way to do this is to also include in their calculations the import of the specific empirical evidence admitted to the court. (Besides, bringing these facts to attention might help to dispel the widespread but completely wrong belief that Clark was convicted *only* on the basis of the very low probability of two SIDS in the same family.)

16 That these facts indeed have *some* evidential relevance to the issue at hand is illustrated in the following apposite comment: 'The Court of Appeal's conclusion that Sally Clark had had an unfair trial did not (legally) mean that she was factually innocent. And even if the undisclosed evidence did indicate a probable cause of death other than by the infliction of physical injuries, the evidence of such injuries, and the lack of any explanation as to how they could have been sustained, point to unanswered questions.' (Nobles and Schiff [2004], pp. 242–3)

17 Again, although our calculated ratio of the two probabilities, $p(2S/E)/p(2M/E)$, is $1/125$ (0.008), it does not come out exactly right if we divide $p(2S/E)$ with $p(2M/E)$: $0.01/0.99 = 1/99$. The reason for this is the same as explained in note 9: the numerical values for these two probabilities are not exact but are rounded off to two decimal places.

References

Aitken, C.G.G. and Taroni, F. [2004]: *Statistics and the Evaluation of Evidence for Forensic Scientists*, Chichester: John Wiley.

American Academy of Pediatrics [2000]: 'Changing Concepts of Sudden Infant Death Syndrome: Implications for Infant Sleeping Environment and Sleep Position', *Pediatrics*, **105**, pp. 650–6.

American Academy of Pediatrics [2001]: 'Distinguishing Sudden Infant Death Syndrome from Child Abuse Fatalities', *Pediatrics*, **107**, pp. 437–41.

Bacon, C. [2005]: 'Repeat Sudden Unexpected Infant Deaths', *Lancet*, **365**, p. 113.

Barness, E.G. and Debich-Spicer, D.E. (eds). [2005]: *Handbook of Pediatric Autopsy Pathology*, Totowa, NJ: Humana Press.

Batt, J. [2004]: *Stolen Innocence: A Mother's Fight for Justice, The Story of Sally Clark*, London: Ebury Press.

Beckwith, J.B. [1990]: 'Sibling Recurrence Risk of Sudden Infant Death', *Journal of Pediatrics*, **117**, pp. 513–4.

Blackburn, S. [2005]: *Truth: A Guide for the Perplexed*, London: Allen Lane.

Blackmon, L.R., Batton, D.G., Bell, E.F., Engle, W.A., Kanto, W.P., Martin, G.I., Rosenfeld, W.N., Stark, A.R. and Lemons, J.A. [2003]: 'Apnea, Sudden Infant Death Syndrome, and Home Monitoring', *Pediatrics*, **111**, pp. 914–7.

Bondi, H. [2004]: 'Statistics Don't Support Cot-Death Murder Theory', *Nature*, **428**, p. 799.

Brookman, F. [2005]: *Understanding Homicide*, London: Sage.

Brookman, F. and Nolan, J. [2006]: 'The Dark Figure of Infanticide in England and Wales: Complexities of Diagnosis', *Journal of Interpersonal Violence*, **21**, pp. 869–89.

Carpenter, R.G., Waite, A., Coombs, R.C., Daman-Willems, C, McKenzie, A., Huber, J. and Emery, J.L. [2005]: 'Repeat Sudden Unexpected and Unexplained Infant Deaths: Natural or Unnatural?', *Lancet*, **365**, pp. 29–35.

Court of Appeal [2000]: Clark, R v [2000] EWCA Crim 54 (2nd October, 2000), <www.bailii.org/ew/cases/EWCA/Crim/2000/54.html>

Court of Appeal [2003]: Clark, R v [2003] EWCA Crim 1020 (11 April 2003), <www.bailii.org/ew/cases/EWCA/Crim/2003/1020.html>

Dalrymple, T. [2003]: 'Much Maligned', *Daily Telegraph* (December 14th).

Dawid, A.P. [2002]: 'Bayes's Theorem and Weighing Evidence by Juries', in R. Swinburne (ed.), *Bayes's Theorem*, Oxford: Oxford University Press.

Dawid, A.P. [2004]: *Probability and Statistics in the Law* (Research report No. 243), Department of Statistical Science, University College London.

Dawid, A.P. [2005]: 'Statistics on Trial', *Significance*, **2**, pp. 6–8.

DiMaio, V.J. and DiMaio, D. [2001]: *Forensic Pathology*, 2nd edition, Boca Raton: CRC Press.

Donnelly, P. [2005]: 'How juries are fooled by statistics', video available online at <www.ted.com/index.php/talks/view/id/67>

Emery, J.L. [1989]: 'Is Sudden Infant Death Syndrome a Diagnosis?', *British Medical Journal*, **299**, p. 1240.

Emery, J.L. [1993]: 'Child Abuse, Sudden Infant Death Syndrome and Unexpected Infant Death', *American Journal of Diseases in Childhood*, **147**, pp. 1097–100.

Firstman, R. and Talan, J. [1997]: *The Death of Innocents: A True Story of Murder, Medicine, and High-Stake Science*, New York: Bantam.

Firstman, R. and Talan, J. [2001]: 'SIDS and Infanticide', in R.W. Byard and H.F. Krous (eds), 2001, *Sudden Infant Death Syndrome: Problems, Progress and Possibilities*, London: Arnold.

Fitzpatrick, M. [2004]: 'The Cot Death Controversy', <www.spiked-online.com/Printable/0000000CA3D8.htm>

Fleming, P.J., Blair, P.S., Bacon, C. and Berry, J. (eds). [2000]: *Sudden Unexpected Deaths in Infancy: The CESDI SUDI Studies 1993–1996*, London: The Stationery Office.

Gornall, J. [2005]: 'The Devil You Don't Know', *The Times* (June 24th).

Green, M. [1999]: 'Time to Put "Cot Death" to Bed', *British Medical Journal*, **319**, pp. 697–8.

Guntheroth, W.G., Lohmann, R. and Spiers, P.S. [1990]: 'Risk of Sudden Infant Death Syndrome in Subsequent Siblings', *Journal of Pediatrics*, **116**, pp. 520–4.

Hill, R. [2004]: 'Multiple Sudden Infant Deaths–Coincidence or beyond Coincidence?' *Paediatric and Perinatal Epidemiology*, **18**, pp. 320–6.

Hill, R. [2005]: 'Reflections on the Cot Death Cases', *Significance*, **2**, pp. 13–6.

Home Office [2001]: *Criminal Statistics: England and Wales 2000*, Home Office.

Horton, R. [2005]: 'In Defense of Roy Meadow', *Lancet*, **366** (July 2), pp. 3–5.

Irgens, L.M., Skjaerven, R. and Peterson, D.R. [1984]: 'Prospective Assessment of Recurrence Risk in Sudden Infant Death Syndrome Siblings', *Journal of Pediatrics*, **104**, pp. 349–51.

Irgens, L.M. and Peterson, D.R. [1988]: 'Sudden Infant Death Syndrome and Recurrence in Subsequent Siblings', *Journal of Pediatrics*, **111**, p. 501.

Joyce, H. [2002]: 'Beyond Reasonable Doubt', *Plus*, Issue 21, <plus.maths.org/issue21/features/clark/>.

Keens, T.G. [2002]: *Sudden Infant Death Syndrome*, <www.californiasids.com/UploadedFiles/Forms/SIDS%20Overview.pdf>

King, M. [2004]: *What Fates Impose: Facing Up To Uncertainty (The Eighth British Academy Annual Lecture)*, <www.britac.ac.uk/pubs/src/_pdf/king.pdf>

Levene, S. and Bacon, C.J. [2004]: 'Sudden Unexpected Death and Covert Homicide in Infancy', *Archives of Disease in Childhood*, **89**, pp. 443–7.

Lewis, B. [2003]: 'Taking Perspective', *Mathematical Gazette*, **87**, pp. 418–31.

Lucey, J.F. [1997]: 'Why All Pediatricians Should Read This Book', *Pediatrics*, **100**, p. A77.

Marks, M.N. and Kumar, R. [1993]: 'Infanticide in England and Wales', *Medicine, Science and the Law*, **33**, pp. 329–39.

Marshall, E. [2005]: 'Flawed Statistics in Murder Trial May Cost Expert His Medical License', *Science*, **322**, p. 543.

Matthews, R. [2005]: 'Matt's Stats: Does Faith Have a Prayer in Making a Difference?' *Daily Telegraph* (July 27th).

Meadow, R. [1997]: *ABC of Child Abuse*, 3rd edition, London: BMJ.

Nobles, R. and Schiff, D. [2004]: 'A Story of Miscarriage: Law in the Media', *Journal of Law and Society*, **31**, pp. 221–44.

Office of National Statistics [1998]: *Mortality Statistics: Cause (Review of the Registrar General on Deaths by Cause, Sex and Age, in England and Wales, 1997)*, London: Stationery Office.

Office of National Statistics [2005]: *Homicide Numbers and Rates, 1901–1998*, <www.statistics.gov.uk/StatBase/xsdataset.asp?vlnk=1880&More=Y>

Peterson, D.R., Sabotta, E.E. and Daling, J.R. [1986]: 'Infant Mortality among Subsequent Siblings of Infants Who Died of Sudden Infant Death Syndrome', *Journal of Pediatrics*, **108**, pp. 911–4.

Pinholster, G. [1994]: 'SIDS Paper Triggers a Murder Charge', *Science*, **264**, pp. 198–9.

Royal Statistical Society [2001]: *News Release: Royal Statistical Society Concerned by Issues Raised in Sally Clark Case*, October 23.

Royal Statistical Society [2002a]: 'Report of the Council for the Session 2001–2002', *The Statistician*, **51**, pp. 485–563.

Royal Statistical Society [2002b]: *Letter from the President to the Lord Chancellor Regarding the Use of Statistical Evidence in Court Cases*, January 23.

Steinschneider, A. [1972]: 'Prolonged Apnea and the Sudden Infant Death Syndrome: Clinical and Laboratory Observations', *Pediatrics*, **50**, pp. 646–54.

Sweeney, J. [2000]: *73 Million to One* (BBC Radio Five Documentary), July 15.

Wilczynski, A. [1997]: *Child Homicide*, London: Greenwich Medical Media.

Wolkind, S., Taylor, E.M., Waite, A.J., Dalton, M. and Emery, J.L. [1993]: 'Recurrence of Unexpected Infant Death', *Acta Paediatrica*, **82**, pp. 873–6.

PART EIGHT

Risk, uncertainty and science policy

INTRODUCTION TO PART EIGHT

IDENTIFYING AND QUANTIFYING risk is a vital part of personal, corporate and state planning and decision-making. While much may be risky or uncertain, 'risk' always refers to a bad outcome; we do not risk winning a prize in a lottery but we do risk wasting our money on a ticket. In medicine, 'risk' may refer to the chances of contracting a condition or to the chances of being subject to the side-effects of a drug. In economics 'risk' refers to adverse future events such as, for example, recession or high inflation. In everyday life we may ponder the risks associated with pursuing certain sports and activities and seek to minimize them. In every case then, risk is connected to the *probability* of a bad outcome. Sometimes, this probability is an objective chance, such as the risk that a volcano will erupt or the risk that a terrorist group will strike, and sometimes the probability represents our degree of belief that some event will happen, as for example with the risk that an innocent person will be wrongly convicted if the jury is not appropriately informed about probabilistic reasoning when considering forensic evidence. In this latter case the person is either guilty or not and the risk solely concerns what philosophers call 'epistemic probability', that is probability that pertains to our degrees of belief because of our information and ignorance. However, in most cases, 'risk' refers to both chances in the world and our uncertainty about them. One of the most important and controversial cases of the latter kind is the risk associated with anthropogenic climate change. Here, there are both chances of different phenomena occurring that are dependent on objective features of the world, but there is also our uncertainty about what those chances are and how to estimate them. Clearly the consequences of anthropogenic climate change could be very bad indeed for the world's population, but there are also costs associated with the various forms of action intended to prevent it. Since the stakes are so high, the chances so complex and interdependent and the uncertainty so great, saying anything precise and accurate about the risks is as difficult as it is important.

One suggestion about how to deal with risk is to obey what is called the 'precautionary principle'. This comes in various forms but the rough idea is that we should err on the side of caution and, in particular, when an action may cause a very bad outcome, we should avoid the action altogether even if the risk of it occurring is judged to be small; equivalently, when we can act in a way that would prevent something very bad happening we should so act even if the chance of it happening is small. This principle has been applied to the climate change case; the idea is that the negative effects of large-scale climate change would be so catastrophic that we ought to act to prevent it even though we are uncertain of how likely such effects are. It is well-known to economists and psychologists that many individual human beings are 'risk-averse' in the sense that they will prefer choices that do not involve the risk of negative consequences even when the choices also involve a high chance of very good outcomes. Such behaviour often violates the norms of rational decision making that are codified in decision theory. Similarly, the precautionary principle is often regarded as contradicting the other norms governing decision making. In the first paper of the section Neil Manson explains and assesses a variety of versions of the precautionary principle and argues that many formulations of it cited in the context of public policy and regulation are unsound. Whether some other version of it is defensible is left open.

Our next reading (by several authors) defends the precautionary principle against five charges that are often levelled against it. The authors argue that it can be formulated in a way that makes it compatible with science and the canons of rational decision theory. The final reading by Sven Ove Hansson addresses the broader issues that this part raises about the relationship between risk and moral value. It is sometimes suggested that science can tell us how to act, but of course it cannot. In the case of rational decision theory, it makes prescriptions about action based on our judgements of the 'value' of the outcomes we predict will follow from those actions. For example, no amount of decision theory can tell me whether to watch a movie or go to the park without the input about my preferences in respect of these activities. In the end, science policy is not science, though it should be scientifically well-informed. One of the things that the public and politicians need to be informed about is the degree to which scientific reasoning is uncertain. As we have seen throughout this book, it is almost always the case that scientists must reach conclusions based on the balance of probabilities rather than deductive logic. In the end, the data and the evidence do not determine what we should believe, and we are often only able say that a certain hypothesis is likely or unlikely to be true, not that it is definitely true or false. This is especially true when it comes to hypotheses that make predictions about the future or when we are considering a particular case falling under a statistical law. However, the public and policy-makers often want more than this; they want scientists to tell them what will happen and what is true. Principles like the precautionary principle are proposed in an attempt to mitigate this problem but the problem remains. While scientific reasoning concerns shades of grey between a theory being proven true and proven false, everyday life and collective action often do not admit of third options; either a child is vaccinated or not, either a carbon tax is imposed or not, either a drug is licensed as safe and effective or not and so on. This mismatch between the science and its application is a fact of life.

Chapter 31

Neil A. Manson

FORMULATING THE PRECAUTIONARY PRINCIPLE

The precautionary principle, enshrined in the laws of various nations and the words of various international treaties, is heralded by its proponents as embodying a radically different approach to environmental decision-making. Given the importance accorded to it, the lack of uniformity regarding its formulation comes as a surprise. Versions of the precautionary principle are many, both in terms of wording and in terms of surface syntactic structure.[1] While it may seem obvious to all of the participants in environmental disputes what is meant by "the precautionary principle," from the perspective of an outsider the content of the precautionary principle can be far from clear.

I first attempt to enhance our grasp of the possible meanings of the precautionary principle by distinguishing its core structure from the details of its various formulations. The framework provided highlights several concepts which stand in need of clarification. Second, I examine a particular (but often-invoked) version of the precautionary principle; I call it "the catastrophe principle." Arguments based on the catastrophe principle are self-defeating (for reasons familiar to critics of Pascal's Wager), which suggests that the precautionary principle must be formulated with care lest it suffer the same fate.

The structure of the precautionary principle

One of the most famous statements of the precautionary principle is Principle 15 of the 1992 Rio Declaration of the UN Conference on Environment and Development:

> In order to protect the environment, the precautionary approach shall be widely applied by States according to their capabilities. Where there are threats of serious or irreversible damage, lack of full scientific certainty shall not be used as a reason for postponing cost-effective measures to prevent environmental degradation.

As this passage indicates, the precautionary principle is supposed to provide guidance with respect to cases in which our scientific knowledge of the harmful effects of a proposed activity is significantly incomplete. The central idea is that even if the normal scientific standards for establishing causal connections are not met in the case of the relationship between an industrial/technological activity and a given harm to the environment, precaution warrants the regulation of that activity. This idea is supposed to run counter to standard decision making procedures (e.g., cost-benefit analysis), in which possible but unproven causal connections do not count. As one author notes, "the precautionary principle . . . aspires to achieve a radical breakthrough in the dominant and ineffectual pattern of balancing of risks, costs of regulatory measures and benefits of activities that cause risks."[2]

Because there is a tangle of competing official formulations layered atop this core idea, it is useful to develop a framework within which these various formulations might be understood. A helpful first step is to identify the generic elements of any statement worthy of the appellation "the precautionary principle." With those generic elements in mind, we can then identify a logical structure common to the various competing formulations. It may be that no single formulation of the precautionary principle emerges from the provision of this framework. Indeed, such a result may not even be desirable. Instead, the worth of such a framework is in helping both proponents and critics of any particular version of the precautionary principle see more clearly what the commitments are of those who endorse that particular version.

To begin the analysis, note that for a given *activity* that may have a given *effect* on the environment, the precautionary principle is supposed to indicate a *remedy*. For the sake of brevity, we can refer to these generic elements as "e-activities," "e-effects," and "e-remedies" respectively. E-activities include things such as commercial fishing, burning fossil fuels, developing land, releasing genetically modified organisms, using nuclear power, generating electrical fields, and disposing of toxic chemicals. E-effects are outcomes such as the depletion of fish stocks, global warming, species extinction, a loss of biodiversity, nuclear contamination, cancer, and birth defects. E-remedies include measures such as an outright ban on the e-activity, a moratorium on it, strict regulation of it, and further research into it.

A review of the many statements of the precautionary principle indicates there is a general, three-part structure shared by every version. The first part is the damage condition; it specifies the characteristics of an e-effect in virtue of which precautionary measures should be considered. The second part is the knowledge condition; it specifies the status of knowledge regarding the causal connections between the e-activity and the e-effect. The third part specifies the e-remedy that decision makers should take in response to the e-activity. The three-part structure shared by all versions of the precautionary principle is the following conditional statement: if the e-activity meets the damage condition and if the link between the e-activity and the e-effect meets the knowledge condition, then decision makers ought to enact the specified e-remedy.

As the chart shown in Table 31.1 indicates, the core structure is very general indeed, allowing considerable room for variation. Of course, relevant possible damage conditions, knowledge conditions, and e-remedies may not be included in the corresponding lists of possibilities, but nothing prevents additions to those lists. The important worry is whether the logical structure of the precautionary principle is being adequately represented. Proponents of the precautionary principle will have to decide this concern for themselves. They should note that the generic elements and logical structure of the precautionary principle have been identified in light both of actual usage and suggested applications (as gleaned from various laws, treaties, protocols, etc.). Because these are the primary guides an outside observer has to the meaning of any term, those who object that the suggested framework cannot capture what they mean by "the precautionary principle" are obliged to articulate [what] they do mean.

A benefit of this framework is that it serves to highlight the distinct particulars that can be

Table 31.1 Three part structure of the precautionary principle

Suggested damage conditions	Suggested knowledge conditions	Suggested e-remedies
1. Serious	1. Possible	1. Ban or otherwise prevent the e-activity
2. Harmful	2. Suspected	2. Put a moratorium on the e-activity
3. Catastrophic	3. Indicated by a precedent	3. Postpone the e-activity
4. Irreversible	4. Reasonable to think	4. Encourage research alternatives to the e-activity
5. Such as to destroy something irreplaceable	5. Not proven with certainty that it is not the case	5. Try to reduce uncertainty about the causal relationship between the e-activity and the e-effect
6. Such as to reduce or eliminate biodiversity	6. Not proven beyond a shadow of a doubt that it is not the case	6. Search for ways to diminish the consequences of the e-effect
7. Such as to violate the rights of members of future generations	7. Not proven beyond a reasonable doubt that it is not the case	

substituted for the generic elements of the skeleton precautionary principle. For example, whatever irreversibility is, it is not the same as irreplaceability; irreversibility is a property of processes (e.g., ozone depletion) while irreplaceability is a property of concrete items (e.g. the ozone layer). That this framework calls attention to such distinctions points to a further benefit of it. To the extent that the concept of, say, irreversibility is clearer than that of irreplaceability, a version specifying irreversibility in the damage condition will be a better guide to action than one specifying irreplaceability. Assuming that one desideratum of formulating the precautionary principle is the ability to guide action, we will have a criterion for evaluating competing versions of the precautionary principle, for saying one version is better than another.

What is needed here is a program of conceptual clarification regarding the various potential component concepts at work in the precautionary principle. For example, if we survey the many characteristics the possession of which have been alleged to make an e-effect meet the damage condition, we see plenty of job

opportunities for the environmental philosopher. The precautionary principle is variously said to be activated when the consequences in question are serious, harmful to humans, catastrophic, irreversible, such as to result in the loss of something irreplaceable, such as to reduce or eliminate biodiversity, or such as to violate the rights of members of future generations. Doubtless other properties could be added to the list of damage conditions. But what, exactly, are these properties? Though they bear some loose connections (as those, for example, between irreversibility, irreplaceability, and the loss of biodiversity), they are nonetheless distinct.[3] Though some of these concepts have received considerable attention (for example, the concept of the rights and interests of future generations), many have not.[4]

To take just one example of the need for conceptual clarification, the concept of irreversibility at work in debates about environmental decision making is surely not the concept at work in contemporary physics. The latter is well-defined but designed for application in statistical mechanics, in connection with the definition of entropy and the attempt to solve the problem of the "arrow of time."[5] Some have criticized the precautionary principle on the assumption that irreversibility is to be understood in the physicist's sense.

> . . . all change (and hence all damage) is irreversible in the strict sense that the precise structure of the world that pertained before cannot once again come into being. This is a consequence of the second law of thermodynamics, wherein it is observed that the state of disorder (or entropy) of the universe is constantly increasing. . . . This ultimately negates the utility of including "irreversible" as a criterion distinct from "serious."[6]

This criticism is uncharitable. Surely proponents of the precautionary principle don't mean that any e-activity which, when engaged in at a particular time, leads to the universe's being different at a later time must therefore merit an e-remedy of some sort. If we restrict ourselves to the biosphere (as opposed to the universe), the second law of thermodynamics does not apply, because the biosphere is not a closed system; it receives a constant flow of energy from the sun.

Yet, such misunderstandings are invited if no sense other than the physicist's is given to the notion of irreversibility. Consider a decision maker confronted with the proposal to dam a river. He or she knows that the dam will result in the death of all of the native trout, but also knows that in the future the dam can be removed and the river re-stocked with non-native trout. Is the e-effect confronting him or her irreversible or not? Those who promote versions of the precautionary principle which appeal to irreversibility should provide a definition of irreversibility that enables such questions to be answered. Similar cases can be constructed which would test the notion of an e-effect's being catastrophic, irreplaceable, and so on.

Environmental philosophers need to roll up their sleeves and start analyzing these concepts.

Remember also that our skeletal precautionary principle is a statement about what ought to be done; it says that if the damage and knowledge conditions are met, then the decision makers ought to impose the e-remedy. It is crucial, then, that any proposed damage condition be connected with disvalue. Few would care to deny that if an outcome is catastrophic, harmful to humans, or in violation of the rights of future generations of humans, then it is, in that respect, a bad thing. But what about irreversibility? If an e-effect is irreversible, does that mean it is, in that respect, a bad thing? Perhaps a case can be made that irreversibility is always (or at least in most cases) a bad feature of an e-effect, but the position remains in want of an argument. Again, there's work here for the philosopher.

As with the damage condition, there is considerable variety concerning the knowledge condition. A source of this diversity is disagreement about whether the precautionary principle mandates shifting the burden of proof from regulators to industrialists and potential polluters. Many advocates of the precautionary principle frame it this way, but not all do.[7] Clearly, questions of burden of proof and of presumption are highly relevant within a legal and political framework. These questions explain why so many environmental defenders favor versions of the precautionary principle in which the burden of proof rests on potential polluters. Those who prefer such versions variously suggest that the polluters must show with certainty, or beyond a shadow of a doubt, or beyond a shadow of a reasonable doubt, that the e-activity does not cause the e-effect.

Yet, not all advocates of the precautionary principle hasten to put the burden of proof on potential polluters. They seem to think that so long as they make sufficiently weak the conditions that must be satisfied in order for the precautionary principle to be activated, the desired effect (blocking the activities of potential polluters) remains the same. They are happy to admit that it is up to the opponents of any given e-activity to make a positive case that it causes the e-effect, but (they say) the standards these opponents need to meet are not those normally demanded in the policy arena. Regarding the claim that an e-activity causes an e-effect, they need only establish its bare possibility, or have a hunch that it is true, or point to a precedent for thinking that it is true, or have reasonable grounds for concern that it is true. They need not prove it with full scientific certainty, or prove it beyond a shadow of a doubt, or have any scientific evidence for it at all.

What these various specifications actually amount to – for example, what it is to have a reasonable doubt – requires some careful thinking, although here the resources for conceptual clarification are much richer thanks to legal tradition. From the perspective of an advocate of precautionary action, whether it is better to place a high burden of proof on those who would engage in the e-activity rather than a low burden of proof on those who would regulate the e-activity is entirely an issue of pragmatics rather than principles. Which formulation of the precautionary principle is most favorable to the pro-precaution side in any given situation will depend on the particular political and legal context in which that formulation is to be employed.

The last component of the precautionary principle is the e-remedy. In most versions of the precautionary principle, the e-remedy is simply the prohibition of the e-activity. Sometimes, however, other actions are said to be mandated by the precautionary principle. These include encouraging research on alternatives to the e-activity, trying to reduce uncertainty about the causal relationship between the e-activity and the e-effect, and seeing if there are ways of diminishing the negative consequences of the e-effect. Advocates of precaution are well-advised to build into their formulation of the precautionary principle reference to such follow-up actions. Doing so is an important gesture to fairness, for otherwise the precautionary principle risks imposing an obligation that proponents of the e-activity can never discharge. For example, suppose the fact that regulators are unsure of the population of a particular species of marine animal is used as a reason to play it safe by treating that species as endangered. If harvesting of the species is forbidden unless knowledge of its population is gained, then, by implication, gaining the appropriate information about the population should result in permission being granted for harvesting of the species (other things being equal). Another way of making essentially the same point is that the precautionary principle should not be formulated in such a way that it encourages and rewards ignorance.

Note that, within the above framework, any version of the precautionary principle will effectively be absolute and unconditional. The only restriction on the scope of the sorts of activity to which it will apply will be the extremely weak one that the action be such as to count as an e-activity (an activity that affects the environment). Furthermore, if the damage and knowledge conditions are met, the e-remedy is not merely supported or suggested, but made obligatory. Some may object that this approach is not the right way to understand the precautionary principle, but it is the understanding suggested by the printed word and by the contrast drawn between the precautionary approach and traditional cost-benefit analysis. "In several treaties, the precautionary principle is formulated in absolutist terms. It stipulates that once a risk of a certain magnitude is identified, preventive efforts to erase that risk are mandatory."[8] If the impression that the precautionary principle is absolute and unconditional is a mistaken one, its advocates would do well to ask themselves what they are doing to generate that impression.

The catastrophe principle

With these points in mind, let us now examine in depth a particular version of the precautionary principle,[9] one in which the damage condition is that the e-effect is catastrophic and the knowledge condition is that there is a possibility the e-activity leads to the e-effect. Let us call this version of the precautionary principle "the catastrophe principle." According to it, if we can identify an e-activity and an e-effect such that the e-effect is catastrophic and it is merely possible that the e-activity causes the e-effect, then the imposition of the e-remedy is justified regardless of the probability that the e-activity causes the e-effect.

The catastrophe principle is clearly suggested in a number of arguments in favor of specific regulations. Consider the following argument for the elimination of nuclear arsenals. It is based on the nuclear winter scenario, according to which multiple nuclear explosions would create a blanket of dust and debris that would block out the sun and drop global mean surface temperature enough to result in the destruction of most life forms.[10] In this argument the e-activity is engaging in nuclear war, the e-effect is nuclear winter, and the e-remedy is eliminating nuclear arsenals.

There are real uncertainties involved in the nuclear winter predictions. They are based on models of poorly understood processes. Many of the complex scientific problems will take many years to resolve and some of the key uncertainties will remain unless there is a nuclear war. Science cannot provide certainty on this issue. However, one doesn't require certainty to take decisions about risks. . . . With nuclear winter there would be no second chance. The potential costs are so enormous that it hardly matters for our argument whether the probability that the nuclear winter predictions are basically correct is 10 per cent, 50 per cent, or 90 per cent. . . .[11]

The authors imply that it does not matter how low the probability is that the nuclear winter model is correct. Because the result of a nuclear winter would be human extinction, uncertainties about whether nuclear war causes nuclear winter should not deter us from eliminating nuclear arsenals.

Here is another application of the catastrophe principle, this time from the possibility of global warming to the drastic reduction of the production of greenhouse gases.

The IPCC [Intergovernmental Panel on Climate Change] scientists predict, based on their computer models of climate, increases of temperature more than ten times faster than life on Earth has experienced in at least 100,000 years, and probably much longer. . . . But the IPCC scientists may be wrong. . . . such are the uncertainties of the climate system that they could, nonetheless, – just conceivably – be wrong. . . . But do we want to gamble on that tiny possibility? If the world's climate scientists are in virtual unanimity that unprecedented global warming will occur if we do nothing about greenhouse-gas emissions, would we not best serve our children and theirs if we took heed – even if there are uncertainties? . . . Nobody – repeat, nobody – can deny that there is at the very least a prospect of ecological disaster on the horizon where the greenhouse effect is concerned. Those who choose to ignore the prospect, therefore, willfully elect to ignore the environmental security of future generations.[12]

In the last two sentences, the author suggests an argument for drastic reductions in greenhouse gas emissions goes through so long as we grant the mere prospect of ecological disaster (though for him that argument is only a fallback position, as he thinks the conclusions of the IPCC are almost certainly true). He says that global warming would be just such a disaster, maintaining in an earlier part of his article that "with the emergence of the global-warming threat . . . industrial and other human activities have thrown a shadow over the very future of our species." (Wry commentators might argue that human extinction would be no catastrophe, but we can flatter ourselves that the extinction of *Homo sapiens* would be as bad as it gets!)

These are not the first arguments from the mere possibility of an extreme outcome to the practical requirement of a course of action. In his *Pensees* Blaise Pascal (1623–1662) contends that one is compelled by rational self-interest to believe in God. So long as the probability that God exists is nonzero, the infinite nature of the reward if one correctly believes that and acts as if God exists makes belief in God rational – no matter how low that probability is.

> . . . in this game you can win eternal life which is eternally happy; you have one chance of winning against a finite number of chances of losing, and what you are staking is finite. That settles it: wherever there is infinity, and where there is not an infinity of chances of losing against the chance of winning, there is no room for hesitation: you must stake everything.[13]

Pascal thinks the infinite nature of the reward obviates determining the probability that one will, in fact, gain the reward. The question of one's grounds for thinking God actually exists is divorced from the rationality of acting as if God exists. Thus, Pascal thinks there is no need to argue over the evidence for God – which is good, he thinks, because such disputes will never come to an end. Let the scholars dispute God's existence, Pascal tells us. People in the real world need to act. They want to know what to do, and they must decide now, before it is too late.

Pascal's Wager has been subjected to a number of philosophical criticisms, but for the purposes of assessing the catastrophe principle we may focus on one in particular: the so-called "many gods" objection. Consider Odin. If Odin is jealous, then on Pascal's reasoning one has equal reason not to believe in God. After all, if Odin exists and God does not, and one worships God instead of Odin, one will pay an infinite price, making belief in God a risk not worth taking. Now Odin might not strike us as a very plausible deity, but if one admits it is possible that Odin exists, then, according to the logic of Pascal's Wager, one has just as much reason to believe in Odin as in God. But one cannot believe in both. Given that both Odin and God are possible deities, Pascal's reasoning leads to contradictory practical demands, and so it cannot be valid.

The "many-gods" objection brings to light the following general point regarding the catastrophe principle: even if an e-effect is catastrophic, that fact cannot rationally compel us to impose an e-remedy unless we also know that the e-remedy itself does not lead to catastrophic results; otherwise, we shall have to apply the catastrophe principle once again, negating the result of our first application. Thus, even if we grant that the e-effect is, say, human extinction, it does not follow that we should impose the e-remedy (much less that we should disregard the probability that the e-activity causes the e-effect). Why? Because it could be that the e-remedy will bring about an outcome which also leads to human extinction (or some other equally catastrophic outcome). And the same goes for any e-activity which might lead to an e-effect that, while not possessing infinite disutility, is nonetheless very, very bad. We could be doomed if we do and doomed if we don't.

Consider a wild story. The Kyoto Treaty is ratified by the US Senate and signed into law by President Bush. All signatories to the treaty abide strictly to its demands. A global economic depression results. Massive social unrest ensues. Totalitarian dictatorships arise in Russia and the United States. War starts and nuclear weapons are launched by both sides. The predictions of the nuclear winter model prove to be perfectly accurate. Within five years, cockroaches rule the planet. The moral? We had better not do anything about greenhouse gas emissions. This conclusion is absurd. To pursue the analogy with religious belief, the scenario is not a "live option." As it stands, however, the catastrophe principle dictates this conclusion, because it fails to exclude any catastrophic possibilities from its realm of application. The catastrophe principle only requires the mere possibility of catastrophe, and since mere possibilities are so easy to construct, any application of the catastrophe principle will confront a fatal problem: the reasoning it employs can be used to generate a demand for a contradictory course of action. In other words, as it stands, the catastrophe principle is useless as a guide to action.

In light of this problem, it is natural to suggest amplifying the catastrophe principle as follows: "If an e-effect is catastrophic, if it is possible that a given e-activity causes that e-effect, and if it is not possible that imposing the given e-remedy will cause some other catastrophic e-effect, then the e-remedy should be imposed." This modification, however, would render the catastrophe principle ineffectual as a tool for action, because it would be practically impossible to show that there do not exist any catastrophic outcomes that might possibly come about as result of imposing the e-remedy.

The preceding considerations fail to impugn any of the conclusions that might be reached via application of the catastrophe principle. There are cogent arguments for the view that nuclear weapons stockpiles should be eliminated: they are not worth the expense, it is morally wrong to be prepared to kill millions of people, and so on. Even if the catastrophe principle is a defective vehicle for arriving at policy destinations, this does not mean those policy destinations might not be reached in some other way. Furthermore, given the framework provided in the first part of this paper, it is clear that many other versions of the precautionary principle can be formulated besides the catastrophe principle. That there are fundamental problems with the latter fails to show that a workable version of the former cannot be constructed. The plausibility and workability of each version will have to be determined on a case-by-case basis.

Still, a broader lesson can be drawn here: if the precautionary principle is assumed to apply to any activity whatsoever that might harm the environment, then surely it is arbitrary and unreasonable to exempt e-remedies themselves from scrutiny. As recent history has shown, well-intentioned safety measures (e.g., making passenger-side airbags mandatory in new automobiles) can lead to damaging consequences. Even if regulatory activity is not directly harmful, it can shift public behavior in harmful ways. For example, prohibiting nuclear power diverts energy production into other modes of production: coal-based, hydroelectric, and so on. To the extent that these means of energy production are damaging to the environment, the prohibition of nuclear power carries a risk, though of course that risk may be greatly outweighed by that of nuclear power itself. At best it is only a general rule of thumb that regulation and prohibition are less damaging to the environment than commercial and technological development. That such a rule of thumb holds is not a sufficient reason for a blanket exemption of these activities from the scope of the precautionary principle. Such an exemption would be *ad hoc* and would expose those who use the precautionary principle to the charge of special pleading.

Conclusion

Given its large and growing role in contemporary debates about environmental decision making, a clear formulation of the precautionary principle is needed. As we have seen, however, the array of possible formulations is vast. The formulation(s) ultimately selected from this array should meet several specifications. First, the component concepts should be clearly defined. Second, the damage conditions identified should be such that a state possessing one of those features does, indeed, have great disvalue – if not always, then most of the time. Third, since the burden of proof is determined by the knowledge condition, the knowledge condition must be chosen in light of the particular political and legal systems in which the formulation will be applied. Fourth, included within the e-remedy should be some sort of pledge to continue research, for otherwise the formulation might have the effect of rewarding ignorance. Fifth, the formulation chosen must not be self-defeating – that is, it must not be such that the imposition of the e-remedy itself gets ruled out on grounds that it is in violation of the formulation. Readers may explore for themselves whether any formulation of the precautionary principle meets all of these specifications.

Notes

1 For documentation of this variety, see David Freestone and Ellen Hay, "Origins and Development of the Precautionary Principle," in David Freestone and Ellen Hay, eds, *The Precautionary Principle and International Law: The Challenge of Implementation* (The Hague: Kluwer Law International, 1996); and Julian Morris, "Defining the Precautionary Principle," in Julian Morris, ed., *Rethinking Risk and the Precautionary Principle* (Woburn, Mass.: Butterworth Heinemann, 2000).

2 Andre Nollkaemper, "'What You Risk Reveals What You Value,' and Other Dilemmas Encountered in the Legal Assaults on Risks," in Freestone and Hay, *The Precautionary Principle and International Law*, p. 75.

3 Norman Myers, "Biodiversity and the Precautionary Principle," *Ambio* 22, pts. 2–3 (1993), 74–79.

4 See Alexandre Kiss, "The Rights and Interests of Future Generations and the Precautionary Principle," in Freestone and Hay, *The Precautionary Principle and International Law*, pp. 19–28. For a stimulating and far-reaching discussion of the concept of our responsibility to future generations, see Hans Jonas, *The Imperative of Responsibility* (Chicago: University of Chicago Press, 1984).

5 See Lawrence Sklar, *Philosophy of Physics* (Boulder, Colo.: Westview Press, 1992), chap. 3 for a good introduction to the notion of irreversibility at work in modern physics.

6 Morris, "Defining the Precautionary Principle," p. 14.

7 Carl F. Cranor, "Asymmetric Information, The Precautionary Principle, and Burdens of Proof," in Carolyn Raffensperger and Joel A. Tickner (eds), *Protecting Public Health and The Environment: Implementing the Precautionary Principle* (Washington, DC: Island Press, 1999), pp. 74–99

8 Nollkaemper, "'What You Risk Reveals What You Value,'" p. 73.

9 See Stephen P. Stich, "The Recombinant DNA Debate: A Difficulty for Pascalian-Style Wagering," *Philosophy and Public Affairs* 7 (1978): 187–205 for a presentation of essentially the same point as is made in this section. An abbreviated version of the

argument made in this section can be found in Neil A. Manson, "The Precautionary Principle, The Catastrophe Argument, and Pascal's Wager," *Ends and Means: Journal of the University of Aberdeen Centre for Philosophy, Technology, and Society* 4, no. **1**, 12–16 (Autumn 1999).

10 See R.P. Turco, O.B. Toon, T.P. Ackerman, J.B. Pollack, and Carl Sagan, "Nuclear Winter: Global Consequences of Multiple Nuclear Explosions," *Science* **222**, no. 4630 (1983): 1283–92; and Paul Ehrlich *et al.*, "Long-Term Biological Consequences of Nuclear War," *Science* **222**, no. 4630 (1983):

1293–300 for detailed accounts of the nuclear winter scenario.

11 Owen Greene, Ian Percival, and Irene Ridge, *Nuclear Winter: The Evidence and the Risks* (Cambridge, England: Polity Press, 1985), pp. 154–55.

12 Jeremy Leggett, "Global Warming: A Greenpeace View," in Jeremy Leggett (ed.), *Global Warming: The Greenpeace Report* (New York: Oxford University Press, 1990), pp. 460–61.

13 Blaise Pascal, *Pascal's Pensees: Translated with an Introduction by Martin Turnell* (London: Harvill Press, 1962), pp. 202–03.

Chapter 32

Per Sandin, Martin Peterson, Sven Ove Hansson, Christina Rudén and André Juthe

FIVE CHARGES AGAINST THE PRECAUTIONARY PRINCIPLE

1 Introduction

In 1999 Gail Charnley, then president of the Society for Risk Analysis, declared in the Society's newsletter that 'the precautionary principle is threatening to take the place of risk analysis as the basis for regulatory decision making in a number of places, particularly in Europe' (Charnley, 1999). Returning to the subject in the same forum in early 2000, she described risk analysis as 'a discipline under fire', threatened by a

> serious, growing, antirisk-analysis sentiment that is challenging the legitimacy of science in general and risk analysis in particular . . . And what is it being replaced with? The so-called precautionary principle or the "better-safe-than-sorry" approach.

She sees here

> just the newest skirmish in the age-old battle between science and ideology, between evolution and creationism. It's about religion. In one corner, we have risk analysis – the practice of using science to draw conclusions about the likelihood that something bad will happen – and in the other corner, we have the belief that instead of science, the

precautionary principle will somehow solve all our problems (Charnley, 2000).

These statements from a leading risk analyst reflect a rather widespread view that the precautionary principle is in disagreement with a scientific approach to risk assessment and risk management. A survey of the literature shows that critics have declared that the precautionary principle:

- is ill-defined (Bodansky, 1991)
- is absolutist (Manson, 1999)
- leads to increased risk-taking (Nollkaemper, 1996)
- is a value judgement or an 'ideology' (Charnley, 2000)
- is unscientific or marginalises the role of science (Gray and Bewers, 1996).

In this short essay, we intend to defend the precautionary principle against these charges, and in particular to show that it is compatible with science. In section 2, we briefly review what is meant by the precautionary principle. In section 3, we respond to each of the five cited charges against this principle, and in section 4 we conclude by noting that our defence of the precautionary principle should be encouraging for its proponents, but it should also urge them to refine their principle.

2 What is the precautionary principle?

There is considerable disagreement as to what the precautionary principle means, a problem which we discuss more fully in section 3.1. Its origins, however, can be traced back at least to German environmental legislation from the 1970s (Sandin, 1999). The basic message of the precautionary principle, in the versions discussed in the present essay, is that on some occasions, measures against a possible hazard should be taken even if the available evidence does not suffice to treat the existence of that hazard as a scientific fact.[1]

3 Meeting the charges against the precautionary principle

We can now return to the five major charges against the precautionary principle that were listed in section 1.

3.1 Is the precautionary principle ill-defined?

It is often complained that the precautionary principle is vague or ill-defined. '[The precautionary principle] is too vague to serve as a regulatory standard', writes Daniel Bodansky (1991: 5; see also Bodansky, 1992). Two other authors claim that a version of the precautionary principle used in the context of marine pollution 'poses a number of fundamental problems', as the logic of the principle is unclear and its key terms are not defined (Gray and Bewers, 1996: 768). The fact that the precautionary principle has occurred in some important official documents (e.g. the Maastricht Treaty; see, for instance, Krämer, 2000: 16f and a recent EU communication, CEC, 2000) without explicit definition also fuels the perception that it is poorly defined. Furthermore, the fact that there is a number of different versions of the precautionary principle supports the impression that it – if it is even possible to speak of 'it' – is poorly defined.

This objection certainly poses a problem for proponents of the precautionary principle. In response, they might argue that lack of specificity is not unique to the precautionary principle and that the same objection can be raised against many other decision rules. Consider for instance a rule such as 'only perform those risk reductions that are scientifically justified', a rule which seems no less in need of specification than the precautionary principle (cf. Cameron and Abouchar, 1991: 23). However, both the relevance and the truth of this defence can be put in question. First, claiming that other decision rules are as ill-defined as the precautionary principle is a weak defence of the precautionary principle. It certainly does not support the view that any of these other rules should be replaced by the precautionary principle. Second, even if other decision rules are not *in principle* better defined than the precautionary principle, they might *in fact* be, in the sense that due to their long period of use there has emerged a substantial body of interpretations and practices that partly compensate for the lack of exact definitions. There are, for instance, governmental guidance documents and court cases that can be of help in interpreting these principles.

Thus, proponents of the precautionary principle should acknowledge that the absence of a clear definition is a problem. However, it is one that can be remedied. It is possible to specify more precise versions of the precautionary principle, that are more readily applicable than a general statement that a precautionary approach should be applied (Sandin, 1999 gives an indication of how this can be done).

We may begin with emphasizing the distinction between *prescriptive* and *argumentative* versions of the precautionary principle.[2] An argumentative version is found, for instance, in Principle

15 of the Rio Declaration (UNCED, 1993).[3] This version merely requires that 'lack of full scientific certainty shall not be used as a *reason* for postponing cost-effective measures to prevent environmental degradation' (italics ours). Thus, it is not a substantial principle for decisions, but a principle for what arguments are valid, i.e. a restriction on dialogue. In essence it says little more than that arguments from ignorance should not be used. Such arguments are generally regarded as trivially fallacious (Robinson, 1971; cf., however, Wreen, 1984 and Walton, 1992), and barring them from discourse does not seem very demanding. Thus, the philosophical interest of argumentative versions of the precautionary principle are rather limited. We will not dwell upon them here.

Prescriptive versions of the precautionary principle seem more prevalent in public and scholarly debate than argumentative ones, both among proponents of the precautionary principle (reflected for instance in the Wingspread Statement, see Raffensperger and Tickner, 1999) and its critics (e.g. Manson, 1999).[4]

As was shown in Sandin (1999), most prescriptive versions of the precautionary principle share four common elements or dimensions. These formulations can be recast into the following if-clause, containing the four dimensions: 'If there is (1) a threat, which is (2) uncertain, *then* (3) some kind of action (4) is mandatory.' By paying attention to these four dimensions, the precautionary principle may be made more precise. Hence, in order to make the principle operational, the following four specifications have to be made as a first step.

1   To what types of hazards does the principle apply?
2   Which level of evidence (lower than that of full scientific certainty) should be required?
3   What types of measures against potential hazards does the principle refer to?

4   With what force are these measures recommended (mandatory, merely permitted, etc.)?

(For a further discussion of some of these issues, see Sandin, 1999, and Hansson, 1999a.)

Finally, interpretations of the precautionary principle are in fact emerging, albeit slowly. For instance, the recent EC communication (CEC, 2000), the broad lines of which were endorsed by the European Council's meeting in Nice, December 2000 (European Council, 2000: Annex III), is a modest step in this direction. The Commission notes that 'it would be wrong to conclude that the absence of a definition has to lead to legal uncertainty' (p. 10), and that '[t]he Community authorities' practical experience with the precautionary principle and its judicial review make it possible to get an ever-better handle on the precautionary principle' (ibid.)

To sum up, it must be admitted that the precautionary principle is to some extent poorly defined, and that it is even worse off than such competing decision rules for which there exists a body of interpretations and practical experience that can compensate for the lack of exact definitions. However, this can (and should) be remedied. The precautionary principle may be given more precise formulations, in the way indicated above. Interpretations of the precautionary principle are also beginning to emerge from practical experience.

3.2   Is the precautionary principle absolutist?

A further accusation against the precautionary principle is that it is absolutist or 'overly rigid' (Bodansky, 1992: 4). According to one author, '[i]n several treaties, the precautionary principle is formulated in absolutist terms. It stipulates that once a risk of a certain magnitude is *identified*, preventive measures to erase that risk are *mandatory*' (Nollkaemper, 1996: 73, italics ours).

Of course, virtually every activity is associated with some risk of non-negligible damage. Therefore, under this interpretation the precautionary principle can be used to prohibit just about any human activity.

'Absolutist' here means, roughly, that the precautionary principle forces decision-makers to pay unreasonable attention to *extremely* unlikely scenarios. For an example, let us assume that there is an extremely small cancer risk associated with the food additive R&D Orange No. 17, say 1 chance in $10^{19}/70$-year life. Now, since cancer is a non-negligible harm and we can produce food looking almost as tasty without using R&D Orange No. 17, an adherent of the precautionary principle ought to conclude, it seems, that the use of R&D Orange No. 17 must be prohibited because this substance *can* cause cancer. However, this is absurd (the argument goes), since the cancer risk associated with R&D Orange No. 17 is *extremely* small and contributes almost nothing to the total cancer risk. Thus, the precautionary principle is 'absolutist' in the sense of being insensitive to scientific facts about the *probabilities* associated with different risks.[5] Hence, it is claimed, the precautionary principle would require us to prohibit everything that *might* be dangerous. A similar charge was raised by the National Association of Swedish Fishermen, in their comment on the suggested new Swedish Environmental Code. The fishermen held that the precautionary principle as applied to fisheries would mean that no fishing at all could be undertaken (Swedish Government, 1997: §481).

Under such a strict, absolutist interpretation of the precautionary principle, it would prohibit in principle *every* action. Since any action, in a sense, might have unforeseen catastrophic consequences (perhaps, due to the chaotic nature of causation, you will cause a new world war by taking a day off tomorrow, etc.), the action of carrying it out will be prohibited, and so will the action of *not* carrying it out. This

objection was raised by McKinney (1996) and Manson (1999).

The argument from absolutism is clearly based on a misconstruction of the precautionary principle. The principle requires that actions be taken when there is lack of full scientific certainty. This, however, does not mean that precautionary measures are required when there is no particular evidence, scientific or other, of the presence of a possible hazard. Indeed, we have not been able to find any authoritative formulation or interpretation of the principle that supports such an extreme requirement. In fact, some documents explicitly demand that the possibility of harm at least should be identified (European Council, 2000: Annex III, §7).

Nevertheless, the argument from absolutism helps to put into focus one of the major specifications of the precautionary principle that was referred to in section 3.1: if full scientific proof is not required before precautionary action should be taken, then what is the required degree of scientific evidence? One obvious way of answering this question is to state a degree of evidence in qualitative terms, such as 'strong scientific evidence' or 'scientifically supported strong suspicions'. Admittedly, these phrases are not very precise, but they are not obviously less precise than many other phrases that legislators leave it to courts to interpret. It is not even clear that they are more difficult to interpret than phrases such as 'full scientific proof' or 'scientific certainty'. Today, most formulations of the precautionary principle lack such specification of the required degree of scientific evidence needed to trigger precautionary action. This is, for instance, the case with the formulations reviewed in Sandin (1999).[6] Needless to say, the development of such specifications is an essential part in the operationalisation of the precautionary principle.

Another way to deal with the problem of absolutism is to apply an adaptation of the *de* minimis principle prior to application of the precautionary principle. *De minimis* is a threshold concept. 'A de minimis risk level would ... represent a cutoff, or benchmark, below which a regulatory agency could simply ignore alleged problems or hazards' (Fiksel, 1987: 4).[7] The *de minimis* principle excludes scenarios with very small probabilities from consideration.[8]

Some proponents of this principle have suggested specific numbers for fixing a *de minimis* level (Flamm *et al.*, 1987: 89; Pease, 1992: 253; FDA, 1995). Others have claimed that probabilities are *de minimis* if and only if they cannot be scientifically detected, or distinguished from random effects in randomized studies (Suter *et al.*, 1995; Stern, 1996). (See Hansson, 1999b for a critical discussion.) None of these approaches are without problems. However, we will not go into the eminently difficult question of how to interpret and determine *de minimis* levels, but merely note that a proponent of the precautionary principle may very well − and perfectly consistently − apply the *de minimis* principle, but will expectedly use a lower probability limit than proponents of a non-precautionary approach.

It might be objected that applying some sort of *de minimis* principle requires quantitative risk assessments, something which the precautionary principle does not. Thus, the *de minimis* principle would require more information than the precautionary principle alone. This is true, but it need not be a problem. Applying the *de minimis* principle does not require that the probability of the undesired event can be determined. It only requires that it can be determined whether or not its probability is below a certain level. This is of course more information than the absolutist version of the precautionary principle requires, as it only requires that the event is possible, which in this case might be interpreted as saying that its probability is > 0. On the other hand, it is significantly *less* information than

what the calculation of expected utility requires (i.e. a reasonable probability estimate).

There is also another problem. Some proponents of the precautionary principle explicitly see it as an alternative to risk-based decisions (Santillo *et al.*, 1999). Combining the precautionary principle with a *de minimis* rule, it might be argued, would mean moving the precautionary principle towards a risk-based approach, towards which certain proponents of the precautionary principle might be hostile. To this we can only reply that while this might make the precautionary principle less effective as a rhetorical device, it will probably make it more applicable as a decision rule.[9] Furthermore, we may also reduce the risk that the precautionary principle is used in a completely unreflected manner, something that might happen if the absolutist version is used as a guide for decision making.

3.3 Does the precautionary principle lead to increased risk-taking?

Let us now turn to an argument stating that the precautionary principle leads to the imposition of new risks, since cautiousness in one respect often leads to incautiousness in another. This may come about directly − when precautionary measures themselves impose new risks − or indirectly − when precautionary measures are so costly that the resultant loss of wealth imposes risks. A clear example of when precautionary measures *directly* impose risks would be the use of pesticides in a developing country. Pesticides may be a threat to the environment, and, it can be argued that for reasons of precaution, they should not be used. But that would in some cases lead to an increased risk of crop failure and consequently of famine which, for reasons of precaution, should be avoided. Substitution of hazardous chemicals poses similar problems. If neurotoxic pesticides are not to be used, there is a possibility that we are driven to use substitute chemicals that might be less neurotoxic, but may instead be carcinogenic.[10] Less obvious, but no less important, is the possibility that precautionary measures indirectly, through economic mechanisms, impose risks.

> Risk driven regulation of one industrial sector under one treaty can be a perfect implementation of the precautionary principle, but can also consume resources that cannot be spent on equal or more serious risks in other sectors. (Nollkaemper, 1996: 91)

Another possible scenario might be that precautionary measures stifle technological progress, causing the loss of benefits that would otherwise have been available (for example, more and healthier food from genetically modified plants).

A similar argument has been raised against environmental regulations and, in fact, risk reduction efforts in general: there is a strong correlation between health and wealth; the richer you are, the healthier you are. 'Richer is Safer', as the title of Wildavsky's (1980) article aptly states it. (For formal, quantitative discussions of the problem, see Keeney, 1990; 1997.) Precautionary measures impose costs, and they might therefore in the end lead to worse effects than if they had not been carried out. Cross (1995) gives an enlightening review of this argument; see also Cross (1996). Another example can be found on the web site of the American Council on Science and Health:

> [I]f we act on all the remote possibilities in identifying causes of human disease, we will have less time, less money and fewer general resources left to deal with the real public health problems which confront us. (Whelan, 2000)

This argument is attractive and might seem troublesome for defenders of the precautionary principle. However, the problem does not depend on the precautionary principle itself but on the limited framing of the decision problem to which it is applied. When delineating a decision problem, one has to draw the line somewhere, and determine a 'horizon' for the decision (Toda, 1976). If the horizon is too narrow, then decisions will be recommended that are suboptimal in a wider perspective, and this applies irrespective of what decision rule is being used. If we apply expected utility maximization to, for instance, crop protection, seen as an isolated issue, then the decision with respect to pesticides may very well be different from what it would have been if we had applied the same decision rule to a more widely defined decision problem in which effects on nutrition and health are included. The same is true if we replace expected utility maximization by the precautionary principle, or, it might be added, any other decision rule.

Here, it must be admitted that the precautionary principle, used as a rhetorical device, might in fact tempt decision-makers to focus upon a single, conspicuous threat, while disregarding countervailing risks. This, however, should not be a reason for abandoning the precautionary principle. Instead, it should urge us to apply the precautionary principle in a reasonable and reflected manner. Particularly, the precautionary principle should be applied also to the precautionary measures prescribed by the precautionary principle itself.

3.4 Is the precautionary principle a value judgement?

Some critics of the precautionary principle have argued that it is a value judgement or an 'ideology' (Charnley, 2000), not a factual judgement. It is claimed that the precautionary principle merely expresses a subjective attitude of fear against risk taking, and therefore can neither be confirmed nor falsified by scientific studies. Since science only deals with factual truths, not subjective attitudes towards risk taking, the precautionary principle simply leaves no room for a scientific approach to risk analysis. Or so the story goes.

In order to appraise this argument, let us return to the central feature of the precautionary principle that was identified in section 2, namely that precautions are required in the absence of full scientific evidence. Hence, according to the precautionary principle, the level of evidence at which precautions should be taken is situated below the level of full scientific evidence. This is clearly a value judgement. However, the alternative standpoint, that the level at which precautions should be taken coincides with the level of full scientific evidence, is to no less degree a value judgement.

It should also be observed that the notion of 'scientific proof' may not, in itself, be as value-free as it is often thought to be. In a famous paper, Richard Rudner (1953) claimed that a decision whether or not to accept a scientific hypothesis must take into account not only the available empirical evidence but also the seriousness of the two possible types of mistakes: accepting an incorrect hypothesis and rejecting a correct one. In Rudner's own words:

> But if this is so then clearly the scientist as scientist does make value judgements. For, since no scientific hypothesis is ever completely verified, in accepting a hypothesis the scientist must make the decision that the evidence is sufficiently strong or that the probability is sufficiently high to warrant the acceptance of the hypothesis. Obviously our decision regarding the evidence and respecting how strong is 'strong enough', is going to be a function of the importance, in the typically ethical sense, of making a mistake

> in accepting or rejecting the hypothesis . . . How sure we need to be before we accept a hypothesis will depend on how serious a mistake would be. (Rudner, 1953: 2)

It is certainly a matter of debate what types of values should be allowed to influence a decision to accept or not to accept a scientific hypothesis (Rudner, 1953; Churchman, 1956; Leach, 1968; Martin, 1973; Rooney, 1992; Valerioano, 1995). According to one view, only intrascientific, 'epistemic' values such as simplicity, usefulness in further inquiries, etc. should be allowed to have an influence. According to another view, moral values may also have a role in this type of decision. (For a discussion of these different views, see Martin, 1973.) At any rate, the intrascientific standards of scientific proof are based to a large degree on considerations of investigative economy. It would be a strange coincidence if the level of evidence that is appropriate according to these criteria always coincided with the level of evidence required for practical action.

The claim that the precautionary principle should not be applied, or in other words the claim that precautions should only be taken when full scientific proof of a hazard has been obtained, is value-based in essentially the same way as the precautionary principle itself. They are indeed both value-based on two levels. First, the very identification

> level of evidence necessary for action = level of evidence necessary for scientific proof

is a value judgment, based on moral values. Of course, the same applies to the inequality:

> level of evidence necessary for action < level of evidence necessary for scientific proof.

Second, the determination of the level of evidence required for scientific proof involves value judgements, based on epistemic and possibly also moral values. In summary, to the extent that the charge 'it is a value judgment' holds against the precautionary principle, it also holds against its rivals. Of course, different value systems may sometimes be in conflict. This is evident from the distinction between two types of errors in scientific practice. The first of these consists in concluding that there is a phenomenon or an effect when in fact there is none (type I error, false positive). The second consists in missing an existing phenomenon or effect. This is called an error of type II (false negative). According to the value structure system of pure science, errors of type I are in general regarded as much more problematic than those of type II. According to the value system of environmental decision-making, however, type II errors – such as believing a highly toxic substance to be harmless – may be the more serious ones (see also Hansson, 1999a and, e.g. Raffensperger and de Fur, 1999: 937).

3.5 Is the precautionary principle unscientific?

Finally, we can turn to what is one of the most worrying arguments against the precautionary principle, namely that it is 'unscientific' and 'marginalises the role of science' (Gray and Bewers, 1996).

According to the adherents of this view, the precautionary principle fails to pay enough respect to science, since it requires that precautionary measures be taken also against threats for which full scientific evidence has not been established. For instance, the use of a chemical substance can be prohibited by the precautionary principle even if we do not know whether it is (say) carcinogenic or not. And since many of the achievements of Western civilization during the past two millennia result from its

success in applying scientific results, there are strong reasons to believe that decisions taken in this manner (i.e. without full scientific evidence) will be of a lower quality, leading to worse outcomes, than decisions based on full scientific knowledge.

In spite of its convincing first appearances, this argument breaks down as soon as sufficient attention is paid to its key term 'unscientific'. There are two meanings of this word. A statement is unscientific in what we may call the weak sense if it is *not based on* science. It is unscientific in what we may call the strong sense if it *contradicts* science. Creationism is unscientific in the strong sense. Your aesthetic judgements are unscientific in the weak but presumably not in the strong sense.

The precautionary principle is certainly unscientific in the weak sense, but then so are all decision rules – including the rule that equates the evidence required for practical measures against a possible hazard with the evidence required for scientific proof that the hazard exists.

On the other hand, the precautionary principle is *not* unscientific in the strong sense. A rational decision-maker who applies the precautionary principle will use the same type of scientific evidence (hopefully, the best evidence available), and assign the same relative weights to different kinds of evidence, as a decision-maker who requires full scientific evidence before actions are taken. The difference lies in the amount of such evidence that they require for a decision to act against a possible hazard. The scientific part of the process, i.e. the production and interpretation of scientific evidence, does not differ between the two decision-makers. This shows that the precautionary principle does not contradict science, and also that it does not marginalize science as a tool in decision-making (cf. e.g. Santillo and Johnston, 1999; Hansson, 1999a).

4 Conclusions

In this article, we have countered five common charges against the precautionary principle.

First, the precautionary principle is, in principle, no more vague or ill-defined than other decision principles. Like them, it can be made precise through elaboration and practice. This is an area where more work is urgently needed for proponents of the precautionary principle.

Second, the precautionary principle need not be absolutist in the way that has been claimed. A way to avoid this is through combining the precautionary principle with a specification of the degree of scientific evidence required to trigger precaution, and/or with some version of the *de minimis* rule.

Third, the precautionary principle does not lead to increased risk-taking, unless the framing is too narrow, and then the same problem applies to other decision rules as well. Nevertheless, as the precautionary principle might deceive decision-makers into simplistically focusing upon a single, conspicuous threat while ignoring countervailing risks, it is important that the precautionary principle is applied in a reasonable manner. In particular, the precautionary principle should be applied also to the precautionary measures prescribed by the principle itself.

Fourth, the precautionary principle is indeed value-based. But so are all decision rules, and the precautionary principle is only value-based to the same extent as other decision rules.

Fifth and last, the precautionary principle is not unscientific other than in the weak sense of not being exclusively based on science (while, of course, it may use scientific information as an input). In that sense all decision rules are unscientific, including the rivals of the precautionary principle.

To sum up: these five common charges against the precautionary principle may, in various degrees, also be raised against competing decision rules. This is of course not necessarily an argument in favour of applying the precautionary principle. But it indicates that the critics of the precautionary principle might have to rethink their strategies. While the objections might sometimes be valid against some unreflected interpretations of the precautionary principle, they do not seem valid in principle. This should be encouraging for proponents of the precautionary principle. But it should also urge them to refine their principle. We hope that they will not find the strategies outlined in this article completely unhelpful in this process.

Notes

1 In fact, lack of full scientific evidence is a prerequisite for applying the principle. If scientific evidence *is* conclusive, we may of course demand measures to preclude harm, but that would be a case of prevention rather than precaution.

2 The distinction between prescriptive and argumentative versions of the precautionary principle is noted by Morris (2000: 1), who, however, terms them 'weak' and 'strong' versions. This terminology is, we believe, a less happy one, as the difference is one of kind rather than degree.

3 This is briefly noted in Sandin (1999: 895). It might be pointed out that in the English version of the Rio Declaration, the term 'the precautionary principle' is not used. Instead, the phrase 'the precautionary approach' is used. However, the Swedish version, for instance, uses 'försiktighetsprincipen', i.e. the precautionary principle.

4 It is interesting to note that the European Chemical Industry Council has moved from an action prescribing version of the precautionary principle (CEFIC, 1995) to an argumentative one (CEFIC, 1999).

5 This argument should not be confused with the claim that the precautionary principle pays too little attention to scientific findings in general, a claim to which we will return in section 3.5. Of course, we would never have known that R&D Orange No. 17

might, in very unlikely cases, cause cancer, if scientific research had not been carried out.

6   There are rare exceptions, however. For instance, the Swedish Chemicals Legislation from 1985 required a 'reasonable scientific foundation' in order to trigger precautionary measures (Hansson, 1991). Cf. Hickey and Walker (1995: 448).

7   By *de minimis non curat lex* is meant the legal principle that courts of law should not concern themselves with trifles. The concept of '*de minimis* risk' was derived from this legal principle in the early 1980s. Originally, it was used by the US Food and Drug Administration (FDA) as a motive for not obeying the so-called Delaney Amendment, which prohibits the use of *all* carcinogenic food additives, including substances that only increase the cancer risk by a very tiny fraction (FDA, 1999; Blumenthal, 1990; Weinberg, 1985).

8   In practice, *de minimis* is usually applied to the probability of certain outcomes, given that a certain alternative is implemented, such as the lifetime probability *p* of developing cancer, given that one has been exposed to a certain food additive. For analytical completeness, however, we might regard *de minimis* as applied to states of nature (the probability that nature is so constituted that the lifetime probability of developing cancer, given that one has been exposed to a certain food additive, is *p*).

9   Cf. Cross (1996: 925): 'Rhetorically, the precautionary principle may prove quite useful to advocates of one particular policy or another. Pragmatically, the principle is destructive, even self-destructive.'

10   See Gray and Hammitt (2000) for a discussion of risk tradeoffs in pesticide substitution. On risk tradeoff analysis in general, see Graham and Wiener (1995).

References

Blumenthal, D. (1990) Red No. 3 and Other Colorful Controversies, *FDA Consumer* May 1990.

Bodansky, D. (1991) Scientific uncertainty and the precautionary principle, *Environment* **33**, 4–5, 43–44.

Bodansky, D. (1992) Commentary: the precautionary principle, *Environment* **34**, 4–5.

Cameron, J. and Abouchar, J. (1991) The precautionary principle: a fundamental principle of law and policy for the protection of the global environment, *Boston College International and Comparative Law Review* **14**, 12–7.

CEC [Commission of the European Communities] (2000) *Communication from the Commission on the precautionary principle*, Brussels, 02.02.2000. COM 2000:1.

CEFIC (1995) The precautionary principle, industry and law-making, Position Paper 15/02/95.

CEFIC (1999) CEFIC position paper on the precautionary principle, Position Paper 08/12/99.

Charnley, G. (1999) President's message, *RISK Newsletter*, **19**, 2.

Charnley, G. (2000) 1999 Annual meeting: Past President's message: risk analysis under fire, *RISK Newsletter* **20**, 3.

Churchman, C.W. (1956) Science and decision making, *Philosophy of Science* **23**, 248–9.

Cross, F.B. (1995) When environmental regulations kill: the role of health/health analysis, *Ecology Law Quarterly* **22**, 229–84.

Cross, F.B. (1996) Paradoxical perils of the precautionary principle, *Washington and Lee Law Review* **53**, 851–925.

European Council (2000) Presidency Conclusions. Nice European Council Meeting 7, 8 and 9 December 2000. SN 400/00.

FDA (1995) Food additives: threshold of regulation for substances used in food contact articles; final rule, *US Federal Register* **60**, 36581–96.

FDA (1999) Milestones in US Food and Drug Law History, *FDA Backgrounder* 1999 (4).

Fiksel, J. (1987) De minimis risk: from concept to practice, in C. Whipple (ed.) *De Minimis Risk*, pp. 1–7. New York and London: Plenum Press.

Flamm, W.G., Lake, R.L., Lorentzen, R.J., Rulis, A.M., Schwartz, P.S. and Troxell, T.C. (1987) Carcinogenic potencies and establishment of a threshold of regulation for food contact substances, in C. Whipple (ed.) *De Minimis Risk*, pp. 87–92. New York and London: Plenum Press.

Graham J.D. and Wiener J.B. (eds) (1995) *Risk versus Risk: Tradeoffs in Protecting Health and the Environment.* Cambridge, MA: Harvard University Press.

Gray, G.M. and Hammitt, J.K. (2000) Risk/risk trade-offs in pesticide regulation: an exploratory analysis of the public health effects of a ban on organophosphate and carbamate pesticides, *Risk Analysis* **20**, 665–80.

Gray, J.S. and Bewers, M. (1996) Towards a scientific definition of the precautionary principle, *Marine Pollution Bulletin* **32**, 768–71.

Hansson, S.O. (1991) *The Burden of Proof in Toxicology*, KemI Report No. 9/91. Stockholm: The Swedish National Chemicals Inspectorate.

Hansson, S.O. (1999a) Adjusting scientific practices to the precautionary principle, *Human and Ecological Risk Assessment* **5**, 909–21.

Hansson, S.O. (1999b) The moral significance of indetectable effects, *Risk* **10**, 101–8.

Hickey, J.E. and Walker, V.R. (1995) Refining the precautionary principle in international environmental law, *Virginia Environmental Law Journal* **14**, 423–54.

Keeney, R.L. (1990) Mortality risks induced by economic expenditures, *Risk Analysis* **10**, 147–59.

Keeney, R.L. (1997) Estimating fatalities induced by the economic costs of regulation, *Journal of Risk and Uncertainty* **14**, 5–23.

Krämer, L. (2000) *EC Environmental Law*, 4th ed. London: Sweet & Maxwell.

Leach, J. (1968) Explanation and value neutrality, *British Journal for the Philosophy of Science* **19**, 93–108.

Manson, N. (1999) The precautionary principle, the catastrophe argument, and Pascal's wager, *Ends and Means* **4**, 12–16.

Martin, M. (1973) Value judgement and the acceptance of hypotheses in science and science education, *Philosophic Exchange* **1**, 83–100.

McKinney, W.J. (1996) Prediction and Rolston's environmental ethics: lessons from the philosophy of science, *Science and Engineering Ethics* **2**, 429–40.

Morris, J. (2000) Defining the precautionary principle, in J. Morris (ed.) *Rethinking Risk and the Precautionary Principle*, pp. 1–21. Oxford: Butterworth-Heinemann.

Nollkaemper, A. (1996) 'What you risk reveals what you value', and other dilemmas encountered in the legal assaults on risks, in D. Freestone and E. Hey (eds) *The Precautionary Principle and International Law*, pp. 73–94. Dordrecht: Kluwer Law International.

Pease, W.S. (1992) The role of cancer risk in the regulation of industrial pollution, *Risk Analysis* **12**, 253–65.

Raffensperger, C. and de Fur, P.L. (1999) Implementing the precautionary principle: rigorous science and solid ethics, *Human and Ecological Risk Assessment* **5**, 933–41.

Raffensperger, C. and Tickner, J. (eds) (1999) *Protecting Public Health and the Environment: Implementing the Precautionary Principle*. Washington, DC: Island Press.

Robinson, R. (1971) Arguing from ignorance, *Philosophical Quarterly* **21**, 97–108.

Rooney, P. (1992) On values in science: is the epistemic/non-epistemic distinction useful?, *Proceedings of the Philosophy of Science Association* **1**, 13–22.

Rudner, R. (1953) The scientist *qua* scientist makes value judgements, *Philosophy of Science* **20**, 1–6.

Sandin, P. (1999) Dimensions of the precautionary principle, *Human and Ecological Risk Assessment* **5**, 889–907.

Santillo, D. and Johnston, P. (1999) Is there a role for risk assessment in precautionary legislation?, *Human and Ecological Risk Assessment* **5**, 923–32.

Santillo, D., Johnston, P. and Stringer, R. (1999) The precautionary principle in practice: a mandate for anticipatory preventative action, in C. Raffensperger and J. Tickner (eds) *Protecting Public Health and the Environment: Implementing the Precautionary Principle*, pp. 36–50. Washington, DC: Island Press.

Stern, A.H. (1996) Derivation of a target concentration of Pb in soil based on deviation of adult blood pressure, *Risk Analysis* **16**, 201–10.

Suter, G.W. II, Cornaby, B.W., Hadden, C.T., Hull, R.N., Stack, M. and Zafran, F.A. (1995) An approach for balancing health and ecological risks at hazardous waste sites, *Risk Analysis* **15**, 221–31.

Swedish Government (1997) [Government Bill] Prop. 1997–98:45 [in Swedish].

Toda, M. (1976) The decision process: a perspective, *International Journal of General Systems* **3**, 79–88.

UNCED (1993) *The Earth Summit: The United Nations Conference on Environment and Development* (1992: Rio De Janeiro). Introduction and commentary by Stanley P. Johnson. London: Graham & Trotman.

Valerioano, I. (1995) Epistemic values in science, *Sorites* **1**, 81–95.

Walton, D. (1992) Nonfallacious arguments from ignorance, *American Philosophical Quarterly* **29**, 381–7.

Weinberg, A.M. (1985) Science and its limits: the regulator's dilemma, *Issues in Science and Technology* **2**, 59–72.

Whelan, E.M. (2000) Can too much safety be hazardous? A critical look at the 'precautionary principle', http://www.acsh.org/press/editorials/safety052300.html (accessed 30 May 2001).

Wildavsky, A. (1980) Richer is safer, *Public Interest* No. 60.

Wreen, M. (1984) Light from darkness, from ignorance knowledge, *Dialectica* **43**, 299–314.

Chapter 33

Sven Ove Hansson

RISK AND ETHICS: THREE APPROACHES

How is risk related to value, and in particular to moral value? This has often been discussed by risk researchers and by activists in various risk-related political issues. Increasingly it has also become the subject of philosophical analysis, but admittedly, the philosophical discussion on risk is still at a rather early stage. In this contribution, I will try to show how rich this subject is from the viewpoint of moral philosophy by briefly introducing three major approaches to it.[1] The three approaches are:

1   clarifying the value dependence of risk assessments;
2   analysing risks and risk decisions from an ethical point of view;
3   developing moral theory so that it can deal with issues of risk.

The first of these approaches is the most common one (MacLean 1985; Thomson 1985; Shrader-Frechette 1991; Cranor 1997; Hansson 1998). It has been the starting-point for most researchers who have worked in the area of risk and values, including myself.

Clarifying the value dependence of risk assessments

In order to understand why values are controversial in risk assessment, we need to have a look at the role of risk assessments in the risk decision process. The standard view on this process has developed out of attempts to systematize the work carried out by various national authorities, in particular those responsible for environmental protection and occupational safety. It was codified in an influential 1983 report by the American National Academy of Sciences (NAS) (National Research Council 1983). The report puts a strong emphasis on the division of the decision procedure into two distinct parts to be performed consecutively. The first of these, commonly called risk assessment, is a scientific undertaking. It consists of collecting and assessing the relevant information and, based on this, characterizing the nature and magnitude of the risk. The second procedure is called risk management. Contrary to risk assessment, this is not a scientific undertaking. Its starting-point is the outcome of risk assessment, which it combines with economic and technological information pertaining to various ways of reducing or eliminating the risk in question, and also with political and social information. Based on this, a decision is made on what measures – if any – should be taken to reduce the risk.

An essential difference between risk assessment and risk management, according to this view, is that values only appear in risk management.

Ideally, risk assessment is a value-free process. But is it value-free in practice, and can it at all be made value-free?

In order to answer this question we must draw the distinction between being value-free and being free of controversial values. There are many values that are shared by virtually everyone or by everyone who takes part in a particular discourse. Medical science provides good examples of this. When discussing analgesics, we take for granted that it is better if patients have less rather than more pain. There is no need to interrupt a discussion on this topic in order to point out that a statement that one analgesic is better than another depends on this value assumption. Similarly, in economics, it is usually taken for granted that it is better if we all become richer. Economists sometimes lose sight of the fact that this is a value judgment.

An important class of uncontroversial values are the epistemic (scientific) values that rule the conduct of science. In science, and in risk assessment, we value truth, avoidance of error, simplicity and explanatory power. There are good reasons why these can be called values, although they are of course not moral values. Carl Hempel called them epistemic utilities and delineated them as follows:

> [T]he utilities should reflect the value or disvalue which the different outcomes have from the point of view of pure scientific research rather than the practical advantages or disadvantages that might result from the application of an accepted hypothesis, according as the latter is true or false. Let me refer to the kind of utilities thus vaguely characterized as *purely scientific*, or *epistemic*, utilities. (Hempel 1960: 465)[2]

In a discussion of values in science we have use for the two distinctions just made, namely between epistemic and non-epistemic values,

and also between controversial and non-controversial values. The presence of *epistemic values* is obvious, but they are not very relevant for a discussion of what we usually mean by science being value-free or value-laden. Next, we have the *non-controversial non-epistemic* values. In the context of risk they may include values such as those expressed by the statement that it is good if the prevalence of cancer decreases. These values are relevant for the issue of value-ladenness, but they are often overlooked. Finally, we have the *controversial non-epistemic values*. These are the values that are at focus in a discussion of the value-ladenness of science and science-based practices such as risk assessment.

Are there any values from this third group in risk assessment, i.e. are there any controversial non-epistemic values? I will try to show that there are. More precisely, my claim is that non-epistemic values are unavoidable in risk assessment because of a rather interesting process: when scientific information is transferred to risk assessment, those of the epistemic values in science that concern error-avoidance are transformed into non-epistemic and often quite controversial values. Let us have a closer look at values of error-avoidance.

There are two major types of error that you can make in a scientific statement. Either you conclude that there is a phenomenon or an effect that is in fact not there. This is called an error of type I (false positive). Or you miss an existing phenomenon or effect. This is called an error of type II (false negative). In scientific practice, these types of error are treated very differently. Errors of type I are the serious errors in science. To make such an error means to draw an unwarranted conclusion, to believe something that should not be believed. Such errors lead us astray, and if too many of them are committed then scientific progress will be blocked by the pursuit of all sorts of blind alleys.

Errors of type II, on the other hand, are much less serious from a (purely) scientific point of view. To make such an error means that you keep an issue open instead of adopting a correct hypothesis. Of course, not everything can be kept open, and science must progress when there are reasonable grounds for (provisionally) closing an issue. Nevertheless, failing to proceed is in this context a much less serious error than walking in the wrong direction.

This difference in severity between the two types of error can also be expressed in terms of burdens of proof. When determining whether or not a scientific hypothesis should be accepted for the time being, the onus of proof falls to its adherents. Those who claim the existence of an as yet unproven effect – such as a toxic effect of a chemical substance – have the burden of proof.

As long as we stay in the realm of pure science, the values that determine where we put the burden of proof are epistemic values. However, risk assessment does not belong to the realm of pure science. Risk assessment has a decision-guiding purpose. Often the same or very similar questions are asked in a (purely) scientific context and in a risk assessment context. We can for instance ask the question 'Is the fruit of the bog bilberry poisonous?' as a purely scientific question. Then the intra-scientific burden of proof applies in the way that I just described. If the same question is followed by 'My four-year-old has picked a lot of them and wants to eat them now', then the burden of proof is, presumably, distributed differently.

It is important to note that this does not mean that we operate with different criteria of truth in the two contexts. Instead, the difference concerns our criteria for reasonable recommendations for action (Hansson 1997b). There are inconclusive indications that the bog bilberry may be poisonous. Therefore, when acting in intra-scientific contexts, the scientist in our example should not act *as if* it is known to be true that

these fruits are poisonous. She should not, for instance, write in a textbook that they are poisonous, or refrain from investigating whether they are toxic with the motivation that they are already known to be so. In contrast, the parent should *act as if* it is known to be true that these fruits are poisonous. She should make sure that the child does not eat them.

This example illustrates a general pattern. It would not seem rational – let alone morally defensible – for a decision-maker to ignore all preliminary indications of a possible danger that do not amount to full scientific proof. We typically wish to protect ourselves against suspected health hazards even if the evidence is much weaker than what is required for scientific proof (Rudner 1953). Therefore, in order to guide the type of decisions that we want to make, risk assessments have to be based on criteria for the burden of proof that differ in many cases from those used for intra-scientific purposes.

However, risk assessors are normally not trained in these philosophical distinctions. We sometimes find scientists unreflectingly applying intra-scientific standards of proof in risk assessment contexts. Since these criteria give absolute priority to the avoidance of type I errors over that of type II errors, the outcome may very well be that risks are taken that few of us would accept when the lives and health of our own families are at stake. By just applying standard scientific criteria out of context, a scientist can transform uncontroversial epistemic values into highly controversial non-epistemic values.[3]

The practice of using the same burden of proof in risk assessments that we use for intra-scientific purposes is often mistaken for a 'value-free' or 'more scientific' risk assessment. Risk assessors who use other criteria are often accused of unnecessarily importing values into risk assessment (Charnley 1999, 2000).[4] But of course there is no value-free burden-of-proof standard for risk assessment. Instead of pretending that there is, we should base the standards of evidence in risk assessment on explicitly chosen, transparent criteria that ensure that the risk assessment provides the information needed in the risk-management decisions that it is intended to support (Hansson 1997a, 1999b; Sandin and Hansson 2002).

The use of statistics in science is tailored to the intra-scientific burden of proof. This is perhaps most clearly seen in the use of significance testing. Statistical significance can be seen as a mathematical expression of the usual burden-of-proof standard in science (Cranor and Nutting 1990). Statistical significance is a statistical measure that tells us essentially how probable the observed phenomena are (or would be) under the assumption that the effect we are looking for does in fact not exist. Observations are regarded as statistically significant if this probability is less than .05. For scientific purposes, the standard procedure is to regard statistically insignificant results as inconclusive. This is a necessary (but not sufficient) precaution against type I errors. However, significance testing does not take type II errors into account. Therefore, it is unfortunate that the dividing line between statistical significance and non-significance has in practice often been unreflectingly used as a dividing line between 'the regulatory body should act as if an effect has been demonstrated' and 'the regulatory body should act as if no effect has been demonstrated' (Krewski *et al.* 1989: 9; cf. Leisenring and Ryan 1992). For this purpose, of course, tests of significance are not suited. Other statistical tests must be added that can help us avoid type II errors (Hansson 1995, 2002).

In addition to values related to the burden of proof, risk assessments may also contain other controversial values. I will give three examples of this.

The first example is comparisons with natural conditions. Radiation levels are frequently compared to the natural background, often with the tacit assumption that exposures lower than the natural background are unproblematic. Conversely, in public debates (but more seldom in risk assessments), risks associated with GMOs or synthetic chemicals are denounced as 'unnatural'. These references to (un)naturalness may be efficient rhetoric, but they are bad argumentation. That something is natural or unnatural does not show it to be good or bad. Furthermore, the words 'natural' and 'unnatural' are highly value-laden and very difficult to define in a consistent way (Hansson 2003a). Appeals to naturalness or unnaturalness are often a way to sneak values unnoticed into risk assessments.

Another common implicit value assumption in risk assessments concerns detectability. It is often assumed that effects that we cannot detect directly are a matter of no concern. Hence, if a risk assessment shows that there are no detectable effects, then it is assumed that nothing more needs to be known about the risk. The Health Physics Society wrote in a position statement:

> [The] estimate of risk should be limited to individuals receiving a dose of 5 rem in one year or a lifetime dose of 10 rem in addition to natural background. Below these doses, risk estimates should not be used; expressions of risk should only be qualitative emphasizing the inability to detect any increased health detriment (i.e., zero health effects is the most likely outcome). (Health Physics Society 1996)

It is difficult to find a morally acceptable argument why risks that we cannot detect should, for that reason, be acceptable or negligible.[5] Unfortunately, quite large effects can be indetectable. As a rough rule of thumb, epidemiological studies cannot reliably detect excess relative risks that are about 10 per cent or smaller. For the more common types of lethal disease, such as coronary disease and lung cancer, lifetime risks are of the order of magnitude of about 10 per cent. Therefore, even in the most sensitive studies, an increase in lifetime risk of the size 10^{-2} (10 per cent of 10 per cent) may be indetectable (i.e. indistinguishable from random variations). Very few people would regard a fatal risk of 1 in 100 as negligible. Therefore, we cannot in general consider a risk assessment to be finished when we know that there is no detectable risk. Even if a risk cannot be directly detected, there may be other means – such as extrapolations from animal experiments – that can be used to estimate its magnitude (Hansson 1999a).

My third example is relevant in almost all risk assessments. It concerns the way in which values relating to different persons are combined in a risk assessment. In this respect, standard risk analysis follows the principles of classical utilitarianism. All risks are taken to be fully comparable and additively aggregable. The 'total' risk is obtained by adding up all the individual risks. In risk-benefit analysis, benefits are added in the same way, and finally the sum of benefits is compared to the sum of risks in order to determine whether the total effect is positive or negative.

The practice of adding all risks and all benefits has immediate intuitive appeal. However, it is not as unproblematic as it might seem at first sight. Even if we assume that risks and benefits accruing to different persons are fully comparable in the sense that they can be measured in the same (additive) units, it does not follow that they should just be added up in the moral appraisal. The crucial issue here is not whether or not a benefit (or harm) for one person can be said to be greater than a benefit (or harm) for another person. The crucial issue is whether or not an action that brings about the greater benefit (smaller harm) compensates for the non-realization of a smaller benefit (greater harm) accruing to another person. Such interpersonal compensability does not follow automatically from interpersonal comparability (Hansson 2004a). The fact that it is worse for you to lose your thumb than for me to lose my little finger does not in itself make it allowable for you or anybody else to chop off my little finger in order to save your thumb.

An obvious alternative to this utilitarian approach is to treat each individual as a separate moral unit. Then risks and benefits pertaining to one and the same person can be weighed against each other, whereas risks and benefits for different persons are treated as incomparable. This type of risk-weighing can be found in certain applications that are strongly influenced by the individual-centred traditions of clinical medicine. Dietary advice is one example. Due to environmental contamination, health authorities recommend limits in the consumption of fish caught in certain waters. Such recommendations are based on endeavours to balance the negative health effects of the contaminants against the positive effects of fish as a constituent of the diet (Knuth et al. 2003). This balance is struck separately for each individual; thus positive effects for others (such as the fishing industry) are excluded from consideration.

Hence, we have different practices in different areas of risk assessment.[6] In most applications, all risks and benefits are added up; but in some applications individuals are instead treated as separate units of evaluation. The choice between these approaches is another hidden value assumption in risk assessments that needs to be made explicit.

It is important for the quality of risk-decision processes that the hidden value assumptions in risk assessments are uncovered. Major parts of this work require philosophical competence. Implicit value components of complex arguments have to be discovered, and conceptual distinctions relating to values have to be made. In addition, systematic evaluations of risk assessments should be performed in order to determine what factors influence their outcomes (Hansson and Rudén 2006). This is an area in which co-operations between philosophy and other disciplines can be very fruitful.

Analysing risks and risk decisions from an ethical point of view

I will now turn to the second approach to risk and ethics, namely the ethical analysis of risk problems. Let me begin by explaining why ethical analysis is needed here. Is not standard risk analysis, with its calculations of probabilities and probability-weighted outcomes, sufficient for decision-guidance?

Modern risk analysis is largely based on a quantitative methodology that is, from a decision-theoretical point of view, essentially an application of expected utility maximization (or expected disutility minimization). The severity of a risk is measured as the probability-weighted severity of the negative outcome that the risk refers to. Hence, a risk characterized by a probability p of a negative event with severity u has the same impact in the calculations as a negative event whose severity equals $p \times u$ and about which we are certain that it will occur. Beginning with the influential *Reactor Safety Study* (WASH-1400, the Rasmussen report) from 1975, risk has often not only been *measured by* but also *identified with* expected disutility (Rechard 1999). In other words, risk is defined as the product of probability and severity.[7]

Probabilistic risk analysis is a highly useful tool that provides risk managers with important information. However, it does not provide decision-makers with all the information that they need in order to make risk-management decisions. In particular, important ethical aspects

are not covered in these forms of risk analysis. Risks do not just 'exist' as free-floating entities; they are taken, run or imposed. Risk-taking and risk imposition involve problems of agency and interpersonal relationships that cannot be adequately expressed in a framework that operates exclusively with the probabilities and severities of outcomes. In order to appraise an action of risk-taking or risk imposition from a moral point of view, we also need to know who performs the action and with what intentions. For instance, it makes a moral difference if someone risks her own life or that of somebody else in order to earn a fortune for herself. It also makes a difference if risk-taking is freely chosen by the affected person or imposed against her will. Therefore, traditional quantitative analysis of risk needs to be supplemented with a systematic characterization of the ethical aspects of risk, including issues such as voluntariness, consent, intent and justice.

In recent joint work with Hélène Hermansson, we have proposed a framework for such a systematic analysis of risk (Hermansson and Hansson 2007). Our starting-point is that there are three central roles in social interactions on risks. In every risk-management problem there are people who are potentially exposed to a risk, and there are people who make decisions that affect the risk. Furthermore, since non-trivial risks are seldom taken unless they are associated with some benefit, in practically every risk-management problem there are people who gain from the risk being taken. We propose that relationships between those who have these three roles, namely the risk-exposed, the decision-maker and the beneficiary, are essential for identifying the ethical aspects of a risk management problem. It is important for the ethical analysis to know if two, or perhaps all three of these roles are filled by the same persons, if for instance the risk-exposed are the same (or partly the same) people as the beneficiaries, if the risk-exposed are themselves decision-makers, etc. It is also important to know if one of these groups is in some way dependent on one of the others, for instance if the risk-exposed are economically dependent on the decision-makers. For an example, suppose that a city's traffic planning exposes residents in a particular area to a much higher risk of accidents than those who live in other parts of the city. Our ethical appraisal of this risk-exposure can legitimately be influenced by whether these residents had a say in the decision (i.e. whether they are also decision-makers) and whether the risk-inducing arrangement of traffic has advantages for them or only for people living in other districts (i.e. whether they are also beneficiaries).

We propose that for practical purposes, analysis of the ethics of risk can be based on seven questions concerning the (pairwise) relationships between the three risk roles. In my experience these questions cover most of the salient ethical issues in common types of risk-management problem.

1 To what extent do the risk-exposed benefit from the risk exposure?
2 Is the distribution of risks and benefits fair?
3 Can the distribution of risks and benefits be made less unfair by redistribution or by compensation?
4 To what extent is the risk exposure decided by those who run the risk?
5 Do the risk-exposed have access to all relevant information about the risk?
6 Are there risk-exposed persons who cannot be informed or included in the decision process?
7 Does the decision-maker benefit from other people's risk exposure?

There are many types of moral problem that arise in issues of risk. I will mention one such problem that is both theoretically interesting

and practically relevant: the definition and implications of voluntariness. It should be obvious that voluntariness is a vague notion. Any sharp line between voluntary and involuntary exposure to risk has to be drawn rather arbitrarily. Nevertheless, for many practical purposes – not least legal purposes – a division between voluntary and involuntary has to be made. There is nothing unusual in this; the same type of vagueness applies to many other concepts that we use.

However, in addition to this, the distinction between voluntary and involuntary risk-exposure depends largely on social conventions that are taken for granted without much reflection. As one example of this, the risks associated with smoking are regarded as voluntary whereas risks associated with emissions from a nearby factory are regarded as involuntary. For many smokers, to quit smoking is much more difficult than to move to a safer neighbourhood. This example shows that our ascriptions of (in)voluntariness tend to be strongly influenced by what we consider to be a reasonable demand on a person. This is problematic since it means that our appraisals of voluntariness cannot be seen as inferentially prior to our ethical appraisals.

Many risks have complex causal backgrounds that make the appraisal of voluntariness particularly difficult. Often, self-harming actions by the affected person combine with actions by others to produce the harmful outcome. Hence, the drug addict cannot use drugs unless someone provides her with them. Similarly, the smoker can smoke only if companies continue to sell products that kill half of their customers. The case of smoking is particularly instructive. In order to defend their own activities, tobacco companies have sponsored antipaternalist campaigns that focus on the voluntariness of smoking and the right of smokers to smoke (Taylor 1984). The (implicit) inference seems to be that if the smoker has a moral right to harm herself, then the tobacco company has a moral right to provide her with the means to do so. The term *extended antipaternalism* can be used to denote the use of antipaternalist arguments for accepting actions and activities that harm or contribute to harming (consenting) others (Hansson 2005a).

This distribution of responsibilities in cases with mixed causality seems to be a result of social conventions rather than of consistently applied moral principles. Other cases with a similar structure to the tobacco case are treated differently. In particular, the manufacturers and distributors of heroin are commonly held responsible for the effects that their products have on the health and well-being of those who choose to buy and use them. (Cigarettes are legal and heroin illegal, but that does not settle the *moral* issue.)

A hypothetical example can clarify how our appraisals of responsibilities for risk exposure depend on social conventions. Suppose that a major soft-drinks company comes up with a new product that customers will become addicted to. The new soft drink has no serious immediate health effects, but in the long run it will give rise to pulmonary, cardiovascular and various malignant diseases, thereby ultimately shortening the lives of about half of the consumers who become addicted to it. Few would claim that the sale of such a product should be allowed. Yet, its properties are analogous to those of cigarettes. The difference is of course that cigarettes are socially accepted and that it is considered politically impossible to prohibit them.

In summary, our appraisals of the (in)voluntariness of risk exposures are difficult to reconstruct in terms of consistent moral principles. This is a good reason to subject them to careful moral analysis and criticism. Our dealings with risk depend largely on social conventions, some of which we may wish to change after carefully considering alternative approaches.

Developing moral theory so that it can deal with issues of risk

The third approach to ethics and risk is a much more ambitious project than the other two from the point of view of moral theory. Studies of risk put focus on problems that should be central in moral theorizing but have been sadly neglected. They can therefore be instrumental in improving moral theory.

Throughout the history of moral philosophy, moral theorizing has for the most part referred to a deterministic world in which the morally relevant properties of human actions are both well-determined and knowable. A possible defence of this could be that issues of risk (and uncertainty) belong to decision theory rather than to moral theory. In fact, it seems to often be assumed that decision theory takes value assignments for deterministic cases as given, and derives from them instructions for rational behaviour in an uncertain, unpredictable and indeterministic world. Suppose, for instance, that moral considerations have led us to attach well-determined values to two outcomes X and Y. Then decision theory provides us with a value to be attached to mixed options such as 50-per-cent-chance-of-X-and-50-per-cent-chance-of-Y. The crucial assumption is that, given well-determined probabilities and well-determined values of the basic non-probabilistic alternatives X and Y, the values of mixed options can be *derived*. Probabilities and the values of non-probabilistic alternatives are assumed to completely determine the value of probabilistic alternatives. This is the conventional wisdom, so conventional that it is seldom stated explicitly. I believe it to be grossly misleading, and this for two major reasons (Hansson 2001).

First, as I have already indicated, risk-taking and risk exposure often have moral implications in themselves that are not mediated by the consequences of the possible outcomes. Compare, for instance, the act of throwing down a brick on a person from a high building to the act of throwing down a brick from a high building without first making sure that there is nobody beneath who can be hit by the brick. The moral difference between these two acts is not obviously expressible in a probability calculus. We can easily develop the example so that both the probability and the severity of personal injury are the same for the two acts, but in spite of this very few people would regard them as morally equivalent.

Second, we need our intuitions about indeterministic events in the development of our system of morality. I have found no good reason to believe that our intuitions on deterministic objects are always more reliable than our intuitions on indeterministic objects. To the contrary, in many contexts we have more experience from uncertain than from certain objects of value. It does not then seem reasonable to disregard all our intuitions on the former category from our deliberations, and reconstruct value assignments to them that are based only on our intuitions on the latter type of objects.

As I have tried to show elsewhere, our current moral theories are not suitable to deal with issues of risk (Hansson 2003b). One of the problems with utilitarianism in this respect is that, as we have already seen, it tends to support full interpersonal compensatability of risks and benefits. This means that a risk for one person can always be outweighed by a slightly larger benefit for another. Deontological and rights-based theories have the opposite problem: they have difficulties in showing how a person can at all be justified in taking actions that expose others to risks. This was well expressed by Robert Nozick: 'Imposing how slight a probability of a harm that violates someone's rights also violates his rights?' (Nozick 1974: 7; cf. McKerlie 1986).

Making a long story short, I believe that a reasonable solution to the distributive issues raised in contexts of risk can be developed in a system of prima facie rights. Such a system should contain a prima facie right not to be put at risk by others: *everyone has a prima facie moral right not to be exposed to risk of negative impact, such as damage to her health or her property, through the actions of others.*

Since this is a prima facie right, it can be overridden. In practice, it has to be overridden in quite a few cases, since social life would be impossible if we were not allowed to expose each other to certain risks. To mention just one example: as car-drivers we put each other's lives at risk. In order to make the prima facie right to risk avoidance workable, we need a normatively reasonable account of the overriding considerations in view of which these and similar risk impositions can be accepted. This can be obtained by appeal to reciprocal exchanges of risks and benefits. Each of us takes risks in order to obtain benefits for ourselves. It is beneficial for all of us to extend this practice to mutual exchanges of risks and benefits. Hence, if others are allowed to drive a car, exposing me to certain risks, then in exchange I am allowed to drive a car and expose them to the corresponding risks. This (we may suppose) is to the benefit of all of us. In order to deal with the complexities of modern society, we also need to apply this principle to exchanges of different types of risk and benefit. We can then regard exposure of a person to a risk as acceptable if it is part of a social system of risk-taking that works to her advantage (Hansson 2003b).

This approach has the important advantage of recognizing each person's individual rights, but still making mutually beneficial adjustments possible. It does not allow you to stop your neighbour from driving a car, thereby creating a (small) risk for you as a pedestrian, since this method of transportation is also accessible to you, and beneficial to you as well. On the other hand it will prohibit exploitative arrangements in which someone is exposed to risks in order to achieve benefits only for others.

With further developments, this approach can help us to deal with the distributive issues in risk. However, it does not help us to deal with the equally fundamental issue of which risks we should accept. This is a matter that transcends the limit between morality and rational self-interest.

Of course the standard answer to this question is that we should apply expected utility theory. However, that theory has many weaknesses that come out very clearly in issues of risk (Hansson 1993). One important weakness that has been overlooked in the literature is its instability against the actual occurrence of a serious negative event that was included in the calculation. This can be seen by studying the post-accident argumentation after almost any accident. If the expected utility argumentation were followed to the end, then many accidents would be defended as consequences of a maximization of expected utility that is, in toto, beneficial. However, this type of reasoning is very rarely heard in practice. Seldom do we hear a company that was responsible for a deadly accident justify the loss of lives by saying that it was the result of a decision which, in terms of its total effects, produced far more good than harm. Instead, two other types of reactions are common. One of these is to regret one's shortcomings and agree that one should have done more to prevent the accident. The other is to claim that somebody else was responsible for the accident.

It should also be noted that accident investigation boards are instructed to answer the questions 'What happened? Why did it happen? How can a similar event be avoided?', not the question 'Was the accident defensible in an expected utility calculation?' Once a serious accident has happened, the application of expected utility maximization appears much less satisfactory than it did before the accident. In

this pragmatic sense, expected utility maximization is not a stable strategy.

We need a framework for argumentation that increases our ability to come up with risk decisions that we are capable of defending even if things do not go our way. Such a framework can be obtained by systematizing a common type of argument in everyday discussions about future possibilities, namely arguments that refer to how one might in the future come to evaluate the possible actions under consideration. These arguments are often stated in terms of predicted regret: 'Do not do that. You may come to regret it.' This is basically a sound type of argument. For our purposes, it has to be developed into a procedure in which future developments are systematically identified, and decision alternatives are evaluated under each of these possible developments. Such *hypothetical retrospection* can be used as a means to achieve more well-considered social decisions in issues of risk. However, it cannot be adequately accounted for in terms of regret-avoidance. Psychologically, regret is often unavoidable for the simple reason that it may arise in response to information that was not available at the time of decision. Therefore, regret-avoidance has to be replaced by more carefully carved-out methods and criteria for hypothetical retrospection (Hansson 2007).

For a simple example, consider a factory owner who has decided to install an expensive fire-alarm system in a building that is used only temporarily. When the building is taken out of use, the fire alarm has yet to be activated. The owner may nevertheless consider the decision to install it to have been right, since at the time of the decision other possible developments had to be considered in which the alarm would have been life-saving. This argument can be used, not only in actual retrospection, but also, in essentially the same way, in hypothetical retrospection before the decision. Alternatively, suppose that there is a fire in the building. The owner may then regret that he did not install a much more expensive but highly efficient sprinkler system. In spite of this regret he may consider the decision to have been correct since when he made it, he had to consider the alternative, much more probable development in which there was no fire, but the cost of the sprinklers would have made other investments impossible. Of course, this argument can be used in hypothetical retrospection just like the previous one. In this way, when we perform hypothetical retrospection from the perspective of a particular branch of future development, we can refer to each of the alternative branches and use it to develop either counter-arguments or supportive arguments. In short, in each branch we can refer to all the others.

Hypothetical retrospection aims at ensuring that, whatever happens, the decision one makes will be morally acceptable (permissible) from the perspective of actual retrospection. Just as we can improve our moral decisions by considering them from the perspective of other concerned individuals, we can also improve them by considering alternative future perspectives.

This proposal goes in the opposite direction to much current moral theory: it adds concreteness instead of abstracting from concrete detail. In my view, our moral intuitions are in the end all that we have to base our moral judgments on, and these intuitions are best suited to deal with concrete realistic situations. The concreteness gained through hypothetical retrospection has the advantage that our moral deliberations will be based on 'the full story' rather than on curtailed versions of it. More specifically, this procedure brings to our attention interpersonal relations that should be essential in a moral appraisal of risk and uncertainty, such as who exposes whom to a risk, who receives the benefits from whose exposure to risk, etc. It is only by staying away from such concreteness that standard utility-maximizing risk analysis can remain on the detached and depersonalized level of statistical lives and free-floating risks and benefits.

And this, by the way, is the major service that studies of risk can do to moral philosophy. Introducing problems of risk is an unusually efficient way to expose moral theory to some of the complexities of real life. There are good reasons to believe that the impact on moral philosophy can be thoroughgoing.

Notes

1   For a more general overview of the philosophy of risk, see Hansson (2004c).
2   On epistemic value, see also Levi (1962), Feleppa (1981) and Harsanyi (1983).
3   In many cases this transformation takes place partly in risk assessment and partly in the risk-management decisions on which it is based.
4   For a rebuttal, see Sandin *et al.* (2002).
5   In Hansson (2004b) I called this the 'ostrich's fallacy', honouring the biological folklore that the ostrich buries its head in the sand, believing that what it cannot see is no problem.
6   This difference corresponds to the dividing-line between the 'old' and 'new' schools of welfare economics. See Hansson (2006).
7   See Hansson (1993 and 2005b) for critical appraisals of this definition.

References

Charnley, G. (1999) 'President's message', *RISK Newsletter*, **19**(2): 2.
—— (2000) '1999 annual meeting: past president's message: risk analysis under fire', RISK newsletter, **20**(1): 3.
Cranor, C.F. (1997) 'The normative nature of risk assessment: features and possibilities', *Risk: Health, Safety and Environment*, **8**: 123–36.
—— and Nutting, K. (1990) 'Scientific and legal standards of statistical evidence in toxic tort and discrimination suits', *Law and Philosophy*, **9**: 115–56.
Feleppa, R. (1981) 'Epistemic utility and theory acceptance: comments on Hempel', *Synthese*, **46**: 413–20.

Hansson, S.O. (1993) 'The false promises of risk analysis', *Ratio*, 6: 16–26.

—— (1995) 'The detection level', *Regulatory Toxicology and Pharmacology*, 22: 103–9.

—— (1997a) 'The limits of precaution', *Foundations of Science*, 2: 293–306.

—— (1997b) 'Can we reverse the burden of proof?', *Toxicology Letters*, 90: 223–8.

—— (1998) *Setting the Limit: Occupational Health Standards and the Limits of Science*, Oxford: Oxford University Press.

—— (1999a) 'The moral significance of indetectable effects', *Risk*, 10: 101–8.

—— (1999b) 'Adjusting scientific practices to the precautionary principle', *Human and Ecological Risk Assessment*, 5: 909–21.

—— (2001) 'The modes of value', *Philosophical Studies*, 104: 33–46.

—— (2002) 'Replacing the no effect level (NOEL) with bounded effect levels (OBEL and LEBEL)', *Statistics in Medicine*, 21: 3071–8.

—— (2003a) 'Are natural risks less dangerous than technological risks?', *Philosophia Naturalis*, 40: 43–54.

—— (2003b) 'Ethical criteria of risk acceptance', *Erkenntnis*, 59: 291–309.

—— (2004a) 'Weighing risks and benefits', *Topoi*, 23: 145–52.

—— (2004b) 'Fallacies of risk', *Journal of Risk Research*, 7: 353–60.

—— (2004c) 'Philosophical perspectives on risk', *Techne*, 8(1): 10–35.

—— (2005a) 'Extended antipaternalism', *Journal of Medical Ethics*, 31: 97–100.

—— (2005b) 'Seven myths of risk', *Risk Management*, 7(2): 7–17.

—— (2006) 'Economic (ir)rationality in risk analysis', *Economics and Philosophy*, 22: 231–41.

—— (2007) 'Hypothetical retrospection', *Ethical Theory and Moral Practice*, 10: 145–157.

—— and Rudén, C. (2006) 'Evaluating the risk decision process', *Toxicology*, 218: 100–11.

Harsanyi, J.C. (1983) 'Bayesian decision theory, subjective and objective probabilities, and acceptance of empirical hypotheses', *Synthese*, 57: 341–65.

Health Physics Society (1996) *Radiation Risk in Perspective*, position statement or the Health Physics Society, adopted January 1996.

Hempel, C.G. (1960) 'Inductive inconsistencies', *Synthese*, 12: 439–69.

Hermansson, H. and Hansson, S.O. (2007) 'A three party model tool for ethical risk analysis', *Risk Management*, 9(3): 129–44.

Knuth, B.A., Connelly, N.A., Sheeshka, J., and J. (2003) 'Weighing health benefit and health risk information when consuming sport-caught fish', *Risk Analysis*, 23: 1185–97.

Krewski, D., Goddard, M.J. and Murdoch, D. (1989) 'Statistical considerations in the interpretation of negative carcinogenicity data', *Regulatory Toxicology and Pharmacology*, 9: 5–22.

Leisenring, W. and Ryan, L. (1992) 'Statistical properties of the NOAEL', *Regulatory Toxicology and Pharmacology*, 15: 161–71.

Levi, I. (1962) 'On the seriousness of mistakes', *Philosophy of Science*, 29: 47–65.

McKerlie, D. (1986) 'Rights and risk', *Canadian Journal of Philosophy*, 16: 239–51.

MacLean, D. (ed.) (1985) *Values at Risk*, Totowa, NJ: Rowman and Allanheld.

National Research Council (1983) *Risk Assessment in the Federal Government: Managing the Process*, Washington, DC: National Academy Press.

Nozick, R. (1974) *Anarchy, State, and Utopia*, New York: Basic Books.

Rechard, R.P. (1999) 'Historical relationship between performance assessment for radioactive waste disposal and other types of risk assessment', *Risk Analysis*, 19: 763–807.

Rudner, R. (1953) 'The scientist qua scientist makes value judgments', *Philosophy of Science*, 20: 1–6.

Sandin, P. and Hansson, S.O. (2002) 'The default value approach to the precautionary principle', *Journal of Human and Ecological Risk Assessment (HERA)*, 8: 463–71.

Sandin, P., Peterson, M., Hansson, S.O., Rudén, C. and Juthe, A. (2002) 'Five charges against the precautionary principle', *Journal of Risk Research*, 5: 287–99.

Shrader-Frechette, K. (1991) *Risk and Rationality: philosophical foundations for populist reforms*, Berkeley: University of California Press.

Taylor, P. (1984) *Smoke Ring: The Politics of Tobacco*, London: Bodley Head.

Thomson, P.B. (1985) 'Risking or being willing: Hamlet and the DC-10', *Journal of Value Inquiry*, 19: 301–10.

PART NINE

Scientific realism and antirealism

INTRODUCTION TO PART NINE

PEOPLE OFTEN LOOK to science for answers to the big questions of existence, such as those concerning the origin of the universe, the nature of matter, the nature of space and time and so on. These questions are central in metaphysics since it is concerned with the most general and fundamental questions about the nature of reality. However, there have always been philosophers and scientists who have denied that scientific theories give us knowledge of such matters. When Copernicus's book describing his heliocentric model of the solar system was published, it was with a preface, written by his friend Osiander who published the work after Copernicus's death, which stated that the job of the astronomer is not to describe the hidden workings of the world, but only to make accurate predictions about the positions of the heavenly bodies in the night sky. While that may have been said to avoid the perception of a challenge to Catholic orthodoxy, many others have expressed similar sentiments about the limitations of scientific theories. For example, the celebrated contemporary physicist Hawking (1996, 121) has said, 'I don't demand that a theory should correspond to reality ... All I'm concerned with is that the theory should predict the results of measurements'. This is an expression of a form of antirealism about science sometimes called 'instrumentalism'.

On the other hand, scientific realists think that scientific theories tell us about much more than the results of measurements. In the Scientific Revolution it was common to use the analogy of a clockwork device. It could be a clock face, or some more elaborate display of co-ordinated motions, perhaps a model of the solar system or a theatrical device. The phenomena of the world we experience are analogous to the face of the clock or the visible objects in motion. The job of the natural philosopher/scientist is then to tell us about the mechanism driving the phenomena that are hidden from the senses inside the machine. Realists argue that science can and has given us knowledge of the unobservable mechanisms that produce the phenomena. Antirealists argue that the job of science is only to enable us to predict and perhaps manipulate the world's behaviour, not to explain it in terms of hidden causes.

The debate between scientific realism and antirealism has many important connections to very general issues in epistemology and metaphysics. For example, many of the same arguments are discussed in connection to scepticism about the external world. We will find that many of the issues we have discussed already arise again in this section. For example, questions about explanation and laws and causation are fundamental to the scientific realism debate. Our reading from Pierre Duhem, the great physicist and historian of science, begins with the distinction between the idea that theories should explain experimental laws, and the idea that they should merely classify and systematize them. He defines explanation as 'to strip reality of a veil'. This is very similar to the idea we just encountered of opening up the case of a watch to see how the hidden workings make the hands go round on the face.

Duhem brings to bear a wealth of scientific, philosophical and historical knowledge in his discussion, and introduces many of the most fundamental issues and arguments in the scientific realism debate. One of his first examples concerns the ether theory of light. This is a case that many of the other readings also consider. The ether theory of light was a triumph of mathematical physics in the early nineteenth century and seemed to resolve the debate about whether light was a wave-like phenomenon or was in fact made up of special particles. Newton held the latter view and he was able to use it to explain certain phenomena; for example, the angle of incidence of a beam of light onto a mirror is equal to the angle of reflectance, just as a pool ball bounces off the cushion at the same angle to that at which it struck the cushion. So Newtonian mechanics can be applied to the particles supposed to compose light to explain that optical phenomenon. However, the particle theory did not so neatly explain diffraction phenomena, and when interference phenomena in the famous 'two-slit' experiment were observed the particle theory of light lacked an account of what is happening to the particles to produce the patterns of light observed. On the other hand, waves can be observed to interfere with each other: for example, sound waves in a noisy room. The ether theory of light developed by August Fresnel in the early nineteenth century described light as a transverse wave, meaning a wave whose direction of oscillation is perpendicular to its direction of propagation, like a water wave as opposed to a (longitudinal) sound wave. His mathematical description of such waves enabled the precise prediction of optical phenomena, including some that were previously unknown, to which we shall return below.

Duhem's point about the ether theory of light is that it posits a material with certain properties in order to provide an explanation, but that the explanatory component is metaphysical in nature and never proven to exist by the empirical success of the theory. In order to argue for this Duhem discusses the different styles of explanation and metaphysical pictures that had been current since the Scientific Revolution. He considers atomism, Aristotelianism, Cartesianism and a version of Newtonian metaphysics associated with Boscovitch. Atomism and Cartesianism were both popular in the seventeenth century; they have in common materialism and were opposed to Aristotelian styles of explanation in terms of forms and essences. The difference between atomism and Cartesianism is that the former requires the actuality of the vacuum and a limit to the divisibility of matter, while the latter supposes all of space to be filled with matter of some kind (a plenum). Both philosophies fit with the idea that nature is a gigantic machine whose fundamental mechanism is material bodies in motion, and this research programme was extremely successful. Descartes, for example, has a vortex theory of the formation of the solar system that is roughly preserved in contemporary physics.

However, the clockwork idea of the universe always had a serious difficulty with forces. When Newton introduced the idea of a universal gravitational force acting between all bodies instantaneously, many people criticized this proposal as being akin to magic rather than the kind of explanations in terms of matter contacting matter that had been replacing Aristotelian ideas. Descartes, in particular, advocated a form of materialist physics in which all action was local in the sense that distant material facts could only affect some thing via a chain of things nearby. This is how a clock works, of course. The parts such as hands and a chime may move without being visibly linked, but there are in fact hidden parts connecting them. The complaints of Newton's critics were of no avail because, unlike those who had previously invoked mysterious forces in physics, Newton had a precise mathematical theory and it made predictions, for example that the Earth would be fattened around the equator, a fact that was not previously known. There were other phenomena that would also seem to defy materialist styles of explanation. Magnetism and electricity were analogous to gravity in that respect, and Coulomb discovered a law of electrostatics that was an inverse square law, like Newton's.

One way of restoring the idea of local action to a theory of forces is to posit a medium, which we now call a 'field' (Boscovich). The gravitational or other force can be thought of as being transmitted from point to point of apparently empty space by the field that exists everywhere. Similarly, various phenomena including magnetism, but also those associated with heat flow and light, had convinced some physicists of the need to consider all kinds of 'subtle and imponderable' kinds of matter. The ether was one such thing. It was thought to pervade all of space and that we move through it without noticing constantly. In order that it could support transverse wave motions, Fresnel's ether had to be considered as a very elastic solid. Light waves were understood as propagating disturbances in this medium. That there was such a medium was later seen as required by the conservation of energy.

Recall that Duhem argues that when scientific theories are taken to be explanations this can only be because they are interpreted in terms of some metaphysics. The above tour through various conceptions of matter and accounts of what there is in physics illustrates that there are different styles of explanation and different things that can be taken as primitive, and that each of them has enjoyed some success. Duhem argues that metaphysics never determines the development of physics and so cannot ever be confirmed by the success of physics. The ether theory of light is at a certain point in history a mature, mathematical scientific theory that is very empirically successful. One could choose to believe that it also explains the behaviour of light in terms of some exotic substance, but that hypothesis looks from the perspective of later developments to be as flawed as Descartes's or Newton's conception of the nature of light. By the time Duhem is writing, light has been described by James Clerk Maxwell's theory of the electromagnetic field and the idea that light was vibrations in an elastic solid material was completely abandoned. Duhem goes on to explain and defend his own preferred view that scientific theories should not be understood as explaining the world but rather as producing a natural classification that provides a simple and unified description of it.

Duhem's idea of a natural classification makes the role of scientific theories to systematize and simplify the various different phenomena we observe. The best evidence we can have that we have found a natural classification is that it makes a novel prediction. This is a completely new type of phenomenon that the theory entails will be observed in certain circumstances. The ether theory of light famously made a novel prediction that a point source of light will cast a shadow from an opaque disk that will have a bright white spot in the centre of it. This surprising effect was observed just as Fresnel's theory predicted it. Hence, Duhem argues that the ether theory of light is a natural classification even though we ought not to take it to be explanatory.

Duhem's ideas are profound and have been very influential. He goes on to introduce what has become known as the Duhem thesis, namely that experiments can only ever test a theory in conjunction with auxiliary assumptions and background theories and not on its own. Once this is pointed out it is obvious, but its significance is profound. For example, it is not possible to test Newton's law of gravitation alone for one can only derive predictions from it if one makes assumptions about the masses of the planets and the Sun, the value of the universal gravitational constant, the initial positions and velocities of the planets relative to the Sun and so on. Hence, when a discrepancy with observation is found it is always possible to take it as refuting not the law of gravitation but one of the other components of the derivation. In fact, this is what happened when predictions of the orbit of Uranus were found to be erroneous and this was attributed not to a fault in the theory but to the existence of a hitherto unknown planet (Neptune) that was subsequently observed. This is a great example of how Popper's naive falsificationism has to be modified because it is not true that scientists are equally attached to all hypotheses. The law of gravity was very successful so it was obviously wise to consider a mistake elsewhere. Of course, the diagnosis of the problem did lead to a new prediction and so was not an *ad hoc* hypothesis in accordance with Popper's methodology in that respect.

The Duhem thesis is associated with a very important argument against scientific realism called 'underdetermination'. The argument is that the evidence we have, or even all the evidence we could ever have, is logically compatible with more than one theory, and since only empirical evidence should determine our theory choice, it is indeed underdetermined no matter how much evidence we have. The Duhem thesis lends support to this because if we take any given hypothesis, say, that light is a wave, then this only has empirical consequences if we make other assumptions, for example, that it is longitudinal or transverse, that it has a certain wavelength and velocity of propagation and so on. If our experiments contradict the theory's predictions we can change parts of it other than the fundamental thesis that light is wave-like. We have already seen other examples. When the orbit of Mercury was found to be at odds with Newtonian gravitation, astronomers resorted to the tactic that had worked before of positing another planet. However, in this case it was subsequently decided that there is no such planet, that Newtonian gravitation is not correct and that Einstein's general theory of relativity must be used to give the correct result. Duhem argues that there are no rules that govern how we should react to evidence in modifying our theories. Sometimes a peripheral assumption must be changed; on other occasions the central theory must be abandoned. According to Duhem the only way for scientists to solve the problem is by using 'good sense'.

Many philosophers have been unsatisfied with Duhem's appeal to a primitive and unexplained idea of good sense and they have tried to say more about the criteria that should be used to solve the underdetermination problem. We have already seen how Duhem makes a great deal of simplicity as a virtue of theories. Our next author's positive views bear some resemblance to Duhem's since he also has a lot to say about simplicity. However, the reading from the great mathematician and physicist Henri Poincaré begins with the second very important line of argument against scientific realism. This is based on the history of theory change and of theories that were predictively successful but whose central metaphysical commitments are now regarded as false. For example, the ether theory of light discussed above was replaced by Maxwellian electrodynamics, as we noted.

Like Duhem, Poincaré is not a scientific realist. Both of them were sceptics about the atom who were writing before the revolution in physics brought about by the discovery of atomic structure and quantum mechanics. However, Poincaré points out that, while theory change makes the metaphysical framework associated with the ether theory of light redundant, the equations of the theory retain their power. Indeed, Poincaré argues that the 'relations' expressed by the equations are true even though the objects related by them are unknown. This is a statement of 'structural realism' to which we return below. According to Poincaré, theories express relational or structural facts about the world that can be understood in terms of different pictures of unobservable entities as theories evolve. So, he argues, that light is an oscillatory phenomenon is preserved from the ether theory to Maxwell's theory even though our conception of the nature of the oscillations changes dramatically. Another example is provided by the theory of the Carnot cycle in thermodynamics, which was originally developed in terms of the flow of an indestructible material fluid called 'caloric', which was thought to be heat. There is no such stuff according to later physics but the Carnot cycle remains valid. Poincaré offers various other examples and argues, like Duhem, that many different explanations in terms of different metaphysics are possible but that none should be taken to be true.

In the next section of the reading from Poincaré he sets out his positive view of the objectivity of science. Like Duhem, he thinks that the job of theories is to unify the phenomena, and Poincaré emphasizes again his idea that science describes a 'system of relations'. When theories try to tell us what heat or light really are they are doomed to failure for they can only provide us with crude images. What they can do is tell us how heat phenomena are related, and the relational structure upon which he argues science is converging is objective because any rational beings would arrive at the same system because of its overall simplicity and convenience.

A much more recent attack on scientific realism is that of Bas van Fraassen. His positive view, 'constructive empiricism', is the most influential form of antirealism in contemporary philosophy of science. In the reading he begins with the definition of scientific realism, and like Duhem his definition is in terms of its aim. Realists think that science aims at truth; constructivism empiricists think it aims at empirical adequacy. The latter is understood as truth with respect to what is observable. Van Fraassen therefore relies upon the distinction between what is observable and what is unobservable, and this has been the subject of much discussion. He argues that while the boundary between the observable and the unobservable may be vague there are nonetheless clear cases sufficient for the distinction to be meaningful and epistemologically important. It is important to note though that this is not the same distinction as that between theoretical and observational language. In general, van Fraassen eschews the emphasis other antirealists have put on language. While he denies that science aims at truth, as van Fraassen explains, he does think that scientific theories must be taken literally, and hence that talk of atoms and so on cannot be translated in talk about observable phenomena as the logical positivists thought.

Van Fraassen explains and then criticizes the most important form of argument for scientific realism, namely 'inference to the best explanation'. In the first instance, the scientific realist argues that we ought to believe in, say, electrons, because they are part of the best explanation for the phenomena we observe. They also deploy what is called the 'no-miracles argument' for scientific realism, according to which the best explanation of the success of science as a whole, and especially of its novel predictive success, is that our theories are describing the deep structure of reality behind the phenomena. Inference to the best explanation is vital to ordinary reasoning and it is argued that it is the only way to defeat Cartesian scepticism about the external world. However, van Fraassen denies it is rationally compelling and that, in any case, other explanations of the success of science are available.

The reading from Larry Laudan presents in some detail the argument against scientific realism that has become known as the pessimistic meta-induction. This is the argument with which the reading from Poincaré began. Laudan is concerned to directly respond to the no-miracles argument for scientific realism. He argues against a number of realist theses that posit logical and explanatory connections between approximate truth, empirical success and the successful reference of theoretical terms. For example, Laudan argues that successful reference of theoretical terms is not a necessary condition for empirical success, and that successful reference does not explain empirical success. Consider, for example, the ether theory of light. We have seen that it is empirically successful but, Laudan argues, there is no ether and therefore the theory cannot be said to be approximately true. Other examples include the caloric theory of heat, the phlogiston theory of combustion and Newtonian mechanics, all of which were empirically successful and all of which say false things about the hidden nature of reality.

Realists have responded to Laudan in various ways. Most have tried to argue in some way that the theories he discusses are approximately true after all. Some have argued that terms like 'ether' and 'caloric' can be understood as referring after all. These responses to Laudan are discussed in John Worrall's paper. He weighs up the issues we have been discussing, and reviews both the no-miracles argument, and the pessimistic meta-induction, and the underdetermination argument and van Fraassen's critique of scientific realism. Worrall would like to have the best of all worlds. He is inspired by Duhem and especially Poincaré. Both of these philosophers of science were well aware of both the positive arguments for scientific realism, the strongest being from novel predictive success, but also of the problem realism faces with the historical record to which Laudan draws attention. Worrall elaborates a position called structural realism that has been very influential. A very different response to the arguments about scientific realism we have discussed is due to Ian Hacking. He argues that philosophers of science have been too focused on theories and have paid insufficient attention to experimentation and scientific practice. On Hacking's view it is our practical interventions using the entities of science that provide the best grounds for believing them. After spending time with physicists at work in a laboratory and watching them manipulate electrons for various purposes, he famously pronounced 'if you can spray them then they are real'. In the end, we may agree with Hacking that whatever the theoretical debates about science, the reason we bother with them and the reason science is accorded such a privileged status is because it is so successful when it comes to practical matters.

Further reading

Boyd, R. 1983 'On the current status of scientific realism', *Erkenntnis* 19:45–90. Reprinted in Richard Boyd, Philip Gasper and J.D. Trout (eds) *The Philosophy of Science*. MIT Press, 1999

Churchland, P.M. and Hooker, C.A. 1985 *Images of Science: Essays on Realism and Empiricism, With a Reply from Bas C. van Fraassen*. University of Chicago Press.

Fine, A. 1984 'The natural ontological attitude', chapter 4 of Leplin, J. (ed.) *Scientific Realism*. University of California Press.

Ladyman, J. 2002 *Understanding Philosophy of Science*. Routledge.

Ladyman, J. and Ross, D. with Spurrett, D. and Collier, J.G. 2007 *Every Thing Must Go: Metaphysics Naturalized*. Oxford University Press.

Leplin, J. 1997 *A Novel Defense of Scientific Realism*. Oxford University Press.

Psillos, S. 1999 *Scientific Realism: How Science Tracks Truth*. Taylor and Francis.

Reference

Hawking, S. and Penrose, R. 1996 *The Nature of Time*. Princeton University Press.

Chapter 34

Pierre Duhem

THE AIM AND STRUCTURE OF PHYSICAL THEORY

1 Physical theory and metaphysical explanation

1.1 Physical theory considered as explanation

The first question we should face is: What is the aim of a physical theory? To this question diverse answers have been made, but all of them may be reduced to two main principles:

"A physical theory," certain logicians have replied, "has for its object the *explanation* of a group of laws experimentally established."

"A physical theory," other thinkers have said, "is an abstract system whose aim is to *summarize* and *classify logically* a group of experimental laws without claiming to explain these laws."

We are going to examine these two answers one after the other, and weigh the reasons for accepting or rejecting each of them. We begin with the first, which regards a physical theory as an explanation.

But, first, what is an explanation?

To explain (explicate, *explicare*) is to strip reality of the appearances covering it like a veil, in order to see the bare reality itself.

The observation of physical phenomena does not put us into relation with the reality hidden under the sensible appearances, but enables us to apprehend the sensible appearances themselves in a particular and concrete form. Besides, experimental laws do not have material reality

for their object, but deal with these sensible appearances, taken, it is true, in an abstract and general form. Removing or tearing away the veil from these sensible appearances, theory proceeds into and underneath them, and seeks what is really in bodies.

For example, string or wind instruments have produced sounds to which we have listened closely and which we have heard become stronger or weaker, higher or lower, in a thousand nuances productive in us of auditory sensations and musical emotions; such are the acoustic facts.

These particular and concrete sensations have been elaborated by our intelligence, following the laws by which it functions, and have provided us with such general and abstract notions as intensity, pitch, octave, perfect major or minor chord, timbre, etc. The experimental laws of acoustics aim at the enunciation of fixed relations among these and other equally abstract and general notions. A law, for example, teaches us what relation exists between the dimensions of two strings of the same metal which yield two sounds of the same pitch or two sounds an octave apart.

But these abstract notions – sound intensity, pitch, timbre, etc. – depict to our reason no more than the general characteristics of our sound perceptions; these notions get us to know sound as it is in relation to us, not as it is by itself in sounding bodies. This reality whose external veil alone appears in our sensations is made known to us by theories of acoustics. The latter are to teach us that where our perceptions grasp only that appearance we call sound, there is in reality a very small and very rapid periodic motion; that intensity and pitch are only external aspects of the amplitude and frequency of this motion; and that timbre is the apparent manifestation of the real structure of this motion, the complex sensation which results from the diverse vibratory motions into which we can analyze it. Acoustic theories are therefore explanations.

The explanation which acoustic theories give of experimental laws governing sound claims to give us certainty; it can in a great many cases make us see with our own eyes the motions to which it attributes these phenomena, and feel them with our fingers.

Most often we find that physical theory cannot attain that degree of perfection; it cannot offer itself as a *certain* explanation of sensible appearances, for it cannot render accessible to the senses the reality it proclaims as residing underneath those appearances. It is then content with proving that all our perceptions are produced *as if* the reality were what it asserts; such a theory is a hypothetical explanation.

Let us, for example, take the set of phenomena observed with the sense of sight. The rational analysis of these phenomena leads us to conceive certain abstract and general notions expressing the properties we come across in every perception of light: a simple or complex color, brightness, etc. Experimental laws of optics make us acquainted with fixed relations among these abstract and general notions as well as among other analogous notions. One law, for instance, connects the intensity of yellow light reflected by a thin plate with the thickness of the plate and the angle of incidence of the rays which illuminate it.

Of these experimental laws the vibratory theory of light gives a hypothetical explanation. It supposes that all the bodies we see, feel, or weigh are immersed in an imponderable, unobservable medium called the ether. To this ether certain mechanical properties are attributed; the theory states that all simple light is a transverse vibration, very small and very rapid, of this ether, and that the frequency and amplitude of this vibration characterize the color of this light and its brightness; and, without enabling us to perceive the ether, without putting us in a position to observe directly the back-and-forth motion of light vibration, the theory tries to prove that its postulates entail consequences agreeing at every point with the laws furnished by experimental optics.

1.2 *According to the foregoing opinion, theoretical physics is subordinate to metaphysics*

When a physical theory is taken as an explanation, its goal is not reached until every sensible appearance has been removed in order to grasp the physical reality. For example, Newton's research on the dispersion of light has taught us to decompose the sensation we experience of light emanating from the sun; his experiments have shown us that this light is complex and resolvable into a certain number of simpler light phenomena, each associated with a determinate and invariable color. But these simple or monochromatic light data are abstract and general representations of certain sensations; they are sensible appearances, and we have only dissociated a complicated appearance into other simpler appearances. But we have not reached the real thing, we have not given an explanation of the color effects, we have not constructed an optical theory.

Thus, it follows that in order to judge whether a set of propositions constitutes a physical theory or not, we must inquire whether the notions connecting these propositions express, in an abstract and general form, the elements which really go to make up material things, or merely represent the universal properties perceived.

For such an inquiry to make sense or to be at all possible, we must first of all regard as certain the following affirmation: Under the sensible appearances, which are revealed in our perceptions, there is a reality distinct from these appearances.

This point granted, and without it the search for a physical explanation could not be conceived, it is impossible to recognize having reached such an explanation until we have answered this next question: What is the nature of the elements which constitute material reality?

Now these two questions – Does there exist a material reality distinct from sensible appearances? and What is the nature of this reality? – do not have their source in experimental method, which is acquainted only with sensible appearances and can discover nothing beyond them. The resolution of these questions transcends the methods used by physics; it is the object of metaphysics.

Therefore, *if the aim of physical theories is to explain experimental laws, theoretical physics is not an autonomous science; it is subordinate to metaphysics.*

1.3 According to the foregoing opinion, the value of a physical theory depends on the metaphysical system one adopts

The propositions which make up purely mathematical sciences are, to the highest degree, universally accepted truths. The precision of language and the rigor of the methods of demonstration leave no room for any permanent divergences among the views of different mathematicians; over the centuries doctrines are developed by continuous progress without new conquests causing the loss of any previously acquired domains.

There is no thinker who does not wish for the science he cultivates a growth as calm and as regular as that of mathematics. But if there is a science for which this wish seems particularly legitimate, it is indeed theoretical physics, for of all the well-established branches of knowledge it surely is the one which least departs from algebra and geometry.

Now, to make physical theories depend on metaphysics is surely not the way to let them enjoy the privilege of universal consent. In fact, no philosopher, no matter how confident he may be in the value of the methods used in dealing with metaphysical problems, can dispute the following empirical truth: Consider in review all the domains of man's intellectual activity; none of the systems of thought arising in different eras or the contemporary systems born of different schools will appear more profoundly distinct, more sharply separated, more violently opposed to one another, than those in the field of metaphysics.

If theoretical physics is subordinated to metaphysics, the divisions separating the diverse metaphysical systems will extend into the domain of physics. A physical theory reputed to be satisfactory by the sectarians of one metaphysical school will be rejected by the partisans of another school.

Consider, for example, the theory of the action exerted by a magnet on iron, and suppose for a moment that we are Aristotelians.

What does the metaphysics of Aristotle teach us concerning the real nature of bodies? Every substance – in particular, every material substance – results from the union of two elements: one permanent (matter) and one variable (form). Through its permanence, the piece of matter before me remains always and in all circumstances the same piece of iron. Through the variations which its form undergoes, through the *alterations* that it experiences, the properties of this same piece of iron may change according to circumstances; it may be solid or liquid, hot or cold, and assume such and such a shape.

Placed in the presence of a magnet, this piece of iron undergoes a special alteration in its form, becoming more intense with the proximity of the magnet. This alteration corresponds to the appearance of two poles and gives the piece of iron a principle of movement such that one pole tends to draw near the pole opposite to it on the magnet and the other to be repelled by the one designated as the like pole on the magnet.

Such for the Aristotelian philosopher is the reality hidden under the magnetic phenomena; when we have analyzed all these phenomena by reducing them to the properties of the magnetic quality of the two poles, we have given a complete explanation and formulated a theory altogether satisfactory. It was such a theory that Niccolo Cabeo constructed in 1629 in his remarkable work on magnetic philosophy.[1]

If an Aristotelian declares he is satisfied with the theory of magnetism as Father Cabeo conceives it, the same will not be true of a Newtonian philosopher faithful to the cosmology of Father Boscovich.

According to the natural philosophy which Boscovich has drawn from the principles of Newton and his disciples,[2] to explain the laws of the action which the magnet exerts on the iron by a magnetic alteration of the substantial form of the iron is to explain nothing at all; we are really concealing our ignorance of reality under words that sound deep but are hollow.

Material substance is not composed of matter and form; it can be resolved into an immense number of points, deprived of extension and shape but having mass; between any two of these points is exerted a mutual attraction or repulsion proportional to the product of the masses and to a certain function of the distance separating them. Among these points there are some which form the bodies themselves. A mutual action takes place among the latter points, and as soon as the distances separating them exceed a certain limit, this action becomes the universal gravitation studied by Newton. Other points, deprived of this action of gravity, compose weightless fluids such as electric fluids and calorific fluid. Suitable assumptions about the masses of all these material points, about their distribution, and about the form of the functions of the distance on which their mutual actions depend are to account for all physical phenomena.

For example, in order to explain magnetic effects, we imagine that each molecule of iron carries equal masses of south magnetic fluid and north magnetic fluid; that the distribution of the fluids about this molecule is governed by the laws of mechanics; that two magnetic masses exert on one another an action proportional to the product of those masses and to the inverse square of the distance between them; finally, that this action is a repulsion or an attraction according to whether the masses are of the same or of different kinds. Thus was developed the theory of magnetism which, inaugurated by Franklin, Oepinus, Tobias Mayer, and Coulomb, came to full flower in the classical memoirs of Poisson.

Does this theory give an explanation of magnetic phenomena capable of satisfying an atomist? Surely not. Among some portions of magnetic fluid distant from one another, the theory admits the existence of actions of attraction or repulsion; for an atomist such actions at a distance amount to appearances which cannot be taken for realities.

According to the atomistic teachings, matter is composed of very small, hard, and rigid bodies of diverse shapes, scattered profusely in the void. Separated from each other, two such corpuscles cannot in any way influence each other; it is only when they come in contact with one another that their impenetrable natures clash and that their motions are modified according to fixed laws. The magnitudes, shapes, and masses of the atoms, and the rules governing their impact alone provide the sole satisfactory explanation which physical laws can admit.

In order to explain in an intelligible manner the various motions which a piece of iron undergoes in the presence of a magnet, we have to imagine that floods of magnetic corpuscles escape from the magnet in compressed, though invisible and intangible, streams, or else are precipitated toward the magnet. In their rapid course these corpuscles collide in various ways with the molecules of the iron, and from these collisions arise the forces which a superficial philosophy attributed to magnetic attraction and repulsion. Such is the principle of a theory of the magnet's properties already outlined by Lucretius, developed by Gassendi in the seventeenth century, and often taken up again since that time.

Shall we not find more minds, difficult to satisfy, who condemn this theory for not explaining anything at all and for taking appearances for reality? Here is where the Cartesians appear.

According to Descartes, matter is essentially identical with the extended in length, breadth, and depth, as the language of geometry goes; we have to consider only its various shapes and motions. Matter for the Cartesians is, if you please, a kind of vast fluid, incompressible and absolutely homogeneous. Hard, unbreakable atoms and the empty spaces separating them are merely so many appearances, so many illusions. Certain parts of the universal fluid may be animated by constant whirling or vortical motions; to the coarse eyes of the atomist these whirlpools or vortices will look like individual corpuscles. The intermediary fluid transmits

from one vortex to the other forces which Newtonians, through insufficient analysis, will take for actions at a distance. Such are the principles of the physics first sketched by Descartes, which Malebranche investigated further, and to which W. Thomson, aided by the hydrodynamic researches of Cauchy and Helmholtz, has given the elaboration and precision characteristic of present-day mathematical doctrines.

This Cartesian physics cannot dispense with a theory of magnetism; Descartes had already tried to construct such a theory. The corkscrews of "subtle matter" with which Descartes, not without some naïveté, in his theory replaced the magnetic corpuscles of Gassendi were succeeded, among the Cartesians of the nineteenth century, by the vortices conceived more scientifically by Maxwell.

Thus we see each philosophical school glorifying a theory which reduces magnetic phenomena to the elements with which it composes the essence of matter, but the other schools rejecting this theory, in which their principles do not let them recognize a satisfactory explanation of magnetism.

1.4 The quarrel over occult causes

There is one form of criticism which very often occurs when one cosmological school attacks another school: the first accuses the second of appealing to "occult causes."

The great cosmological schools, the Aristotelian, the Newtonian, the atomistic, and the Cartesian, may be arranged in an order such that each admits the existence in matter of a smaller number of essential properties than the preceding schools are willing to admit.

The Aristotelian school composes the substance of bodies out of only two elements, matter and form; but this form may be affected by qualities whose number is not limited. Each physical property can thus be attributed to a special quality: a *sensible* quality, directly accessible to our perception, like weight, solidity, fluidity, heat, or brightness; or else an *occult* quality whose effects alone will appear in an indirect manner, as with magnetism or electricity.

The Newtonians reject this endless multiplication of qualities in order to simplify, to a high degree, the notion of material substance: in the elements of matter they leave only masses, mutual actions, and shapes, when they do not go as far as Boscovich and several of his successors, who reduce the elements to unextended points.

The atomistic school goes further: its material elements preserve mass, shape, and hardness. But the forces through which the elements act on one another, according to the Newtonian school, disappear from the domain of realities; they are regarded merely as appearances and fictions.

Finally, the Cartesians push to the limit this tendency to strip material substances of various properties: they reject the hardness of atoms and even the distinction between plenum and void, in order to identify matter, as Leibniz said, with "completely naked extension and its modification."[3]

Thus each cosmological school admits in its explanations certain properties of matter which the next school refuses to take as real, for the latter regards them as mere words designating more deeply hidden realities without revealing them; it groups them, in short, with the occult qualities created in so much profusion by scholasticism.

It is hardly necessary to recall that all the cosmological schools other than the Aristotelian have agreed in attacking the latter for the arsenal of qualities which it stored in substantial form, an arsenal which added a new quality each time a new phenomenon had to be explained. But Aristotelian physics has not been the only one obliged to meet such criticisms.

The Newtonians who endow material elements with attractions and repulsions acting at a distance seem to the atomists and Cartesians to be adopting one of those purely verbal explanations usual with the old Scholasticism. Newton's *Principia* had hardly been published when his work excited the sarcasm of the atomistic clan grouped around Huygens. "So far as concerns the cause of the tides given by Mr. Newton," Huygens wrote Leibniz, "I am far from satisfied, nor do I feel happy about any of his other theories built on his principle of attraction, which to me appears absurd."[4]

If Descartes had been alive at that time, he would have used a language similar to that of Huygens. In fact, Father Mersenne had submitted to Descartes a work by Roberval[5] in which the author adopted a form of universal gravitation long before Newton. On April 20, 1646 Descartes expressed his opinion as follows:

Nothing is more absurd than the assumption added to the foregoing; the author assumes that a certain property is inherent in each of the parts of the world's matter and that, by the force of this property, the parts are carried toward one another and attract each other. He also assumes that a like property inheres in each part of the earth considered in relation with the other parts of the earth, and that this property does not in any way disturb the preceding one. In order to understand this, we must not only assume that each material particle is animated, and even animated by a large number of diverse souls that do not disturb each other, but also that these souls of material particles are endowed with knowledge of a truly divine sort, so that they may know without any medium what takes place at very great distances and act accordingly.[6]

The Cartesians agree, then, with the atomists when it comes to condemning as an occult quality the action at a distance which Newtonians invoke in their theories; but turning next against the atomists, the Cartesians deal just as harshly with the hardness and indivisibility attributed to corpuscles by the atomists. The Cartesian Denis Papin wrote to the atomist Huygens: "Another thing that bothers me is . . . that you believe that perfect hardness is of the essence of bodies; it seems to me that you are there assuming an inherent quality which takes us beyond mathematical or mechanical principles."[7] The atomist Huygens, it is true, did not deal less harshly with Cartesian opinion: "Your other difficulty," he replied to Papin, "is that I assume hardness to be of the essence of bodies whereas you and Descartes admit only their extension. By which I see that you have not yet rid yourself of that opinion which I have for a long time judged very absurd."[8]

1.5 No metaphysical system suffices in constructing a physical theory

Each of the metaphysical schools scolds its rivals for appealing in its explanations to notions which are themselves unexplained and are really occult qualities. Could not this criticism be nearly always applied to the scolding school itself?

In order for the philosophers belonging to a certain school to declare themselves completely satisfied with a theory constructed by the physicists of the same school, all the principles used in this theory would have to be deduced from the metaphysics professed by that school. If an appeal is made, in the course of the explanation of a physical phenomenon, to some law which that metaphysics is powerless to justify, then no explanation will be forthcoming and physical theory will have failed in its aim.

Now, no metaphysics gives instruction exact enough or detailed enough to make it possible to derive all the elements of a physical theory from it.

In fact, the instruction furnished by a metaphysical doctrine concerning the real nature of bodies consists most often of negations. The Aristotelians, like the Cartesians, deny the possibility of empty space; the Newtonians reject any quality which is not reducible to a force acting among material points; the atomists and Cartesians deny any action at a distance; the Cartesians do not recognize among the diverse parts of matter any distinctions other than shape and motion.

All these negations are appropriately argued when it is a matter of condemning a theory proposed by an adverse school; but they appear singularly sterile when we wish to derive the principles of a physical theory.

Descartes, for example, denied that there is anything else in matter than extension in length, breadth, depth, and its diverse modes – that is to say, shapes and motions; but with these data alone, he could not even begin to sketch the explanation of a physical law.

At the very least, before attempting the construction of any theory, he would have had to know the general laws governing diverse motions. Hence, he proceeded from his metaphysical principles to attempt first of all to deduce a dynamics.

The perfection of God requires him to be immutable in his plans; from this immutability the following consequence is drawn: God preserves as constant the quantity of motion that he gave the world in the beginning.

But this constancy of the quantity of motion in the world is still not a sufficiently precise or definite principle to make it possible for us to write any equation of dynamics. We must state it in a quantitative form, and that means translating the hitherto very vague notion of "quantity of motion" into a completely determined algebraic expression.

What, then, will be the mathematical meaning to be attached by the physicist to the words "quantity of motion"?

According to Descartes, the quantity of motion of each material particle will be the product of its mass – or of its volume, which in Cartesian physics is identical with its mass – times the velocity with which it is animated, and the quantity of motion of all matter in its entirety will be the sum of the quantities of motion of its diverse parts. This sum should in any physical change retain a constant value.

Certainly the combination of algebraic magnitudes through which Descartes proposed to translate the notion of "quantity of motion" satisfies the requirements imposed in advance by our instinctive knowledge of such a translation. It is zero for a whole at rest, and always positive for a group of bodies agitated by a certain motion; its value increases when a determined mass increases the velocity of its movement; it increases again when a given velocity affects a larger mass. But an infinity of other expressions might just as well have satisfied these requirements: for the velocity we might notably have substituted the square of the velocity. The algebraic expression obtained would then have coincided with what Leibniz was to call "living force"; instead of drawing from divine immutability the constancy of the Cartesian quantity of motion in the world, we should have deduced the constancy of the Leibnizian living force.

Thus, the law which Descartes proposed to place at the base of dynamics undoubtedly agrees with the Cartesian metaphysics; but this agreement is not necessary. When Descartes reduced certain physical effects to mere consequences of such a law, he proved, it is true, that these effects do not contradict his principles of philosophy, but he did not give an explanation of the law by means of these principles.

What we have just said about Cartesianism can be repeated about any metaphysical doctrine which claims to terminate in a physical theory; in this theory there are always posited certain hypotheses which do not have as their grounds the principles of the metaphysical doctrine. Those who follow the thought of Boscovich admit that all the attractions or repulsions which are observable at a perceptible distance vary inversely with the square of the distance. It is this hypothesis which permits them to construct three systems of mechanics: celestial, electrical, and magnetic; but this form of law is dictated to them by the desire to have their explanations agree with the facts and not by the requirements of their philosophy. The atomists admit that a certain law governs the collisions of corpuscles; but this law is a singularly bold extension to the atomic world of another law which is permissible only when masses big enough to be observed are considered; it is not deduced from the Epicurean philosophy.

We cannot therefore derive from a metaphysical system all the elements necessary for the construction of a physical theory. The latter always appeals to propositions which the metaphysical system has not furnished and which consequently remain mysteries for the partisans of that system. At the root of the explanations it claims to give there always lies the unexplained.

2 Physical theory and natural classification

2.1 *What is the true nature of a physical theory and the operations constituting it?*

While we regard a physical theory as a hypothetical explanation of material reality, we make it dependent on metaphysics. In that way, far from giving it a form to which the greatest number of minds can give their assent, we limit its acceptance to those who acknowledge the philosophy it insists on. But even they cannot be entirely satisfied with this theory since it does not draw all its principles from the metaphysical doctrine from which it is claimed to be derived.

These thoughts, discussed in the preceding chapter [sic], lead us quite naturally to ask the following two questions:

Could we not assign an aim to physical theory that would render it *autonomous?* Based on principles which do not arise from any metaphysical doctrine, physical theory might be judged in its own terms without including the opinions of physicists who depend on the philosophical schools to which they may belong.

Could we not conceive a method which might be *sufficient* for the construction of a physical theory? Consistent with its own definition the theory would employ no principle and have no recourse to any procedure which it could not legitimately use.

We intend to concentrate on this aim and this method, and to study both.

Let us posit right now a definition of physical theory; the sequel of this book will clarify it and will develop its complete content: A physical theory is not an explanation. It is a system of mathematical proposition, deduced from a small number of principles, which aim to represent as simply, as completely, and as exactly as possible a set of experimental laws.

In order to start making this definition somewhat more precise, let us characterize the four successive operations through which a physical theory is formed:

1. Among the physical properties which we set ourselves to represent we select those we regard as simple properties so that the others will supposedly be groupings or combinations of them. We make them correspond to a certain group of mathematical symbols, numbers, and magnitudes, through appropriate methods of measurement. These mathematical symbols have no connection of an intrinsic nature with the properties they represent; they bear to the latter only the relation of sign to thing signified. Through methods of measurement we can make each state of a physical property correspond to a

value of the representative symbol, and vice versa.

2. We connect the different sorts of magnitudes, thus introduced, by means of a small number of propositions which will serve as principles in our deductions. These principles may be called "hypotheses" in the etymological sense of the word for they are truly the grounds on which the theory will be built; but they do not claim in any manner to state real relations among the real properties of bodies. These hypotheses may then be formulated in an arbitrary way. The only absolutely impassable barrier which limits this arbitrariness is logical contradiction either among the terms of the same hypothesis or among the various hypotheses of the same theory.

3. The diverse principles or hypotheses of a theory are combined together according to the rules of mathematical analysis. The requirements of algebraic logic are the only ones which the theorist has to satisfy in the course of this development. The magnitudes on which his calculations bear are not claimed to be physical realities, and the principles he employs in his deductions are not given as stating real relations among those realities; therefore it matters little whether the operations he performs do or do not correspond to real or conceivable physical transformations. All that one has the right to demand of him is that his syllogisms be valid and his calculations accurate.

4. The various consequences thus drawn from the hypotheses may he [sic] translated into as many judgments bearing on the physical properties of the bodies. The methods appropriate for defining and measuring these physical properties are like the vocabulary and key permitting one to make this translation. These judgments are compared with the experimental laws which the theory is intended to represent. If they agree with these laws to the degree of approximation corresponding to the measuring procedures

employed, the theory has attained its goal, and is said to be a good theory; if not, it is a bad theory, and it must be modified or rejected.

Thus a true theory is not a theory which gives an explanation of physical appearances in conformity with reality; it is a theory which represents in a satisfactory manner a group of experimental laws. A false theory is not an attempt at an explanation based on assumptions contrary to reality; it is a group of propositions which do not agree with the experimental laws. *Agreement with experiment is the sole criterion of truth for a physical theory.*

The definition we have just outlined distinguishes four fundamental operations in a physical theory: (1) the definition and measurement of physical magnitudes; (2) the selection of hypotheses; (3) the mathematical development of the theory; (4) the comparison of the theory with experiment.

Each one of these operations will occupy us in detail as we proceed with this book [sic], for each of them presents difficulties calling for minute analysis. But right now it is possible for us to answer a few questions and to refute a few objections raised by the present definition of physical theory.

2.2 *What is the utility of a physical theory? Theory considered as an economy of thought*

And first, of what use is such a theory?

Concerning the very nature of things, or the realities hidden under the phenomena we are studying, a theory conceived on the plan we have just drawn teaches us absolutely nothing, and does not claim to teach us anything. Of what use is it, then? What do physicists gain by replacing the laws which experimental method furnishes directly with a system of mathematical propositions representing those laws?

First of all, instead of a great number of laws offering themselves as independent of one another, each having to be learnt and remembered on its own account, physical theory substitutes a very small number of propositions, viz., fundamental hypotheses. The hypotheses once known, mathematical deduction permits us with complete confidence to call to mind all the physical laws without omission or repetition. Such condensing of a multitude of laws into a small number of principles affords enormous relief to the human mind, which might not be able without such an artifice to store up the new wealth it acquires daily.

The reduction of physical laws to theories thus contributes to that "intellectual economy" in which Ernst Mach sees the goal and directing principle of science.[9]

The experimental law itself already represented a first intellectual economy. The human mind had been facing an enormous number of concrete facts, each complicated by a multitude of details of all sorts; no man could have embraced and retained a knowledge of all these facts; none could have communicated this knowledge to his fellows. Abstraction entered the scene. It brought about the removal of everything private or individual from these facts, extracting from their total only what was general in them or common to them, and in place of this cumbersome mass of facts it has substituted a single proposition, occupying little of one's memory and easy to convey through instruction: it has formulated a physical law.

Thus, instead of noting individual cases of light-refraction, we can mentally reconstruct all present and future cases, if we know that the incident ray, the refracted ray, and the perpendicular lie in the same plane and that $\sin i / \sin r = n$. Here, instead of the numberless cases of refraction in different combinations of matter and under all different angles of incidence, we have simply to note the rule above stated and the values of n — which is much easier. The economical purpose here is unmistakable.[10]

The economy achieved by the substitution of the law for the concrete facts is redoubled by the mind when it condenses experimental laws into theories. What the law of refraction is to the innumerable facts of refraction, optical theory is to the infinitely varied laws of light phenomena.

Among the effects of light only a very small number had been reduced to laws by the ancients; the only laws of optics they knew were the law of the rectilinear propagation of light and the laws of reflection. This meager contingent was reinforced in Descartes' time by the law of refraction. An optics so slim could do without theory; it was easy to study and teach each law by itself.

Today, on the contrary, how can a physicist who wishes to study optics, as we know it, acquire even a superficial knowledge of this enormous domain without the aid of a theory? Consider the effects of simple refraction, of double refraction by uniaxial or biaxial crystals, of reflection on isotropic or crystalline media, of interference, of diffraction, of polarization by reflection and by simple or double refraction, of chromatic polarization, of rotary polarization, etc. Each one of these large categories of phenomena may occasion the statement of a large number of experimental laws whose number and complication would frighten the most capable and retentive memory.

Optical theory supervenes, takes possession of these laws, and condenses them into a small number of principles. From these principles we can always, through regular and sure calculation, extract the law we wish to use. It is no longer necessary, therefore, to keep watch over the knowledge of all these laws; the knowledge of the principles on which they rest is sufficient.

This example enables us to take firm hold of the way the physical sciences progress. The experimenter constantly brings to light facts hitherto unsuspected and formulates new laws, and the theorist constantly makes it possible to store up these acquisitions by imagining more condensed representations, more economical systems. The development of physics incites a continual struggle between "nature that does not tire of providing" and reason that does not wish "to tire of conceiving."

2.3 Theory considered as classification

Theory is not solely an economical representation of experimental laws; it is also a *classification* of these laws.

Experimental physics supplies us with laws all lumped together and, so to speak, on the same plane, without partitioning them into groups of laws united by a kind of family tie. Very often quite accidental causes or rather superficial analogies have led observers in their research to bring together different laws. Newton put into the same work the laws of the dispersion of light crossing a prism and the laws of the colors adorning a soap bubble, simply because of the colors that strike the eye in these two sorts of phenomena.

On the other hand, theory, by developing the numerous ramifications of the deductive reasoning which connects principles to experimental laws, establishes an order and a classification among these laws. It brings some laws together, closely arranged in the same group; it separates some of the others by placing them in two groups very far apart. Theory gives, so to speak, the table of contents and the chapter headings under which the science to be studied will be methodically divided, and it indicates the laws which are to be arranged under each of these chapters.

Thus, alongside the laws which govern the spectrum formed by a prism it arranges the laws governing the colors of the rainbow; but the laws according to which the colors of Newton's rings are ordered go elsewhere to join the laws of fringes discovered by Young and Fresnel; still in another category, the elegant coloration analyzed by Grimaldi is considered related to the diffraction spectra produced by Fraunhofer. The laws of all these phenomena, whose striking colors lead to their confusion in the eyes of the simple observer, are, thanks to the efforts of the theorist, classified and ordered.

These classifications make knowledge convenient to use and safe to apply. Consider those utility cabinets where tools for the same purpose lie side by side, and where partitions logically separate instruments not designed for the same task: the worker's hand quickly grasps, without fumbling or mistake, the tool needed. Thanks to theory, the physicist finds with certitude, and without omitting anything useful or using anything superfluous, the laws which may help him solve a given problem.

Order, wherever it reigns, brings beauty with it. Theory not only renders the group of physical laws it represents easier to handle, more convenient, and more useful, but also more beautiful.

It is impossible to follow the march of one of the great theories of physics, to see it unroll majestically its regular deductions starting from initial hypotheses, to see its consequences represent a multitude of experimental laws down to the smallest detail, without being charmed by the beauty of such a construction, without feeling keenly that such a creation of the human mind is truly a work of art.

2.4 A theory tends to be transformed into a natural classification[11]

This esthetic emotion is not the only reaction that is produced by a theory arriving at a high degree of perfection. It persuades us also to see a natural classification in a theory.

Now first, what is a natural classification? For example, what does a naturalist mean in proposing a natural classification of vertebrates?

The classification he has imagined is a group of intellectual operations not referring to concrete individuals but to abstractions, species; these species are arranged, in groups, the more particular under the more general. In order to form these groups the naturalist considers the diverse organs — vertebral column, cranium, heart, digestive tube, lungs, swim-bladder — not in the particular and concrete forms they assume in each individual, but in the abstract, general, schematic forms which fit all the species of the same group. Among these organs thus transfigured by abstraction he establishes comparisons, and notes analogies and differences; for example, he declares the swim-bladder of fish analogous to the lung of vertebrates. These homologies are purely ideal connections, not referring to real organs but to generalized and simplified conceptions formed in the mind of the naturalist; the classification is only a synoptic table which summarizes all these comparisons.

When the zoologist asserts that such a classification is natural, he means that those ideal connections established by his reason among abstract conceptions correspond to real relations among the associated creatures brought together and embodied in his abstractions. For example, he means that the more or less striking resemblances which he has noted among various species are the index of a more or less close blood-relationship, properly speaking, among the individuals composing these species; that the cascades through which he translates the subordination of classes, of orders, of families, and of genera reproduce the genealogical tree in which the various vertebrates are branched out from the same trunk and root. These relations of real family affiliation can be established only by comparative anatomy; to grasp them in themselves and put them in evidence is the business of physiology and of paleontology. However, when he contemplates the order which his methods of comparison introduce into the confused multitude of animals, the anatomist cannot assert these relations, the proof of which transcends his methods. And if physiology and paleontology should someday demonstrate to him that the relationship imagined by him cannot be, that the evolutionist hypothesis is controverted, he would continue to believe that the plan drawn by his classification depicts real relations among animals; he would admit being deceived about the nature of these relations but not about their existence.

The neat way in which each experimental law finds its place in the classification created by the physicist and the brilliant clarity imparted to this group of laws so perfectly ordered persuade us in an overwhelming manner that such a classification is not purely artificial, that such an order does not result from a purely arbitrary grouping imposed on laws by an ingenious organizer. Without being able to explain our conviction, but also without being able to get rid of it, we see in the exact ordering of this system the mark by which a natural classification is recognized. Without claiming to explain the reality hiding under the phenomena whose laws we group, we feel that the groupings established by our theory correspond to real affinities among the things themselves.

The physicist who sees in every theory an explanation is convinced that he has grasped in light vibration the proper and intimate basis of the quality which our senses reveal in the form of light and color; he believes in an ether, a body whose parts are excited by this vibration into a rapid to-and-fro motion.

Of course, we do not share these illusions. When, in the course of an optical theory, we talk about luminous vibration, we no longer think of a real to-and-fro motion of a real body; we imagine only an abstract magnitude, i.e., a pure,

geometrical expression. It is a periodically variable length which helps us state the hypotheses of optics, and to regain by regular calculations the experimental laws governing light. This vibration is to our mind a *representation*, and not an *explanation*.

But when, after much groping, we succeed in formulating with the aid of this vibration a body of fundamental hypotheses, when we see in the plan drawn by these hypotheses a vast domain of optics, hitherto encumbered by so many details in so confused a way, become ordered and organized, it is impossible for us to believe that this order and this organization are not the reflected image of a real order and organization; that the phenomena which are brought together by the theory, e.g., interference bands and colorations of thin layers, are not in truth slightly different manifestations of the same property of light; and that phenomena separated by the theory, e.g., the spectra of diffraction and of dispersion, do not have good reasons for being in fact essentially different.

Thus, physical theory never gives us the explanation of experimental laws; it never reveals realities hiding under the sensible appearances; but the more complete it becomes, the more we apprehend that the logical order in which theory orders experimental laws is the reflection of an ontological order, the more we suspect that the relations it establishes among the data of observation correspond to real relations among things,[12] and the more we feel that theory tends to be a natural classification.

The physicist cannot take account of this conviction. The method at his disposal is limited to the data of observation. It therefore cannot prove that the order established among experimental laws reflects an order transcending experience; which is all the more reason why his method cannot suspect the nature of the real relations corresponding to the relations established by theory.

But while the physicist is powerless to justify this conviction, he is nonetheless powerless to rid his reason of it. In vain is he filled with the idea that his theories have no power to grasp reality, and that they serve only to give experimental laws a summary and classificatory representation. He cannot compel himself to believe that a system capable of ordering so simply and so easily a vast number of laws, so disparate at first encounter, should be a purely artificial system. Yielding to an intuition which Pascal would have recognized as one of those reasons of the heart "that reason does not know," he asserts his faith in a real order reflected in his theories more clearly and more faithfully as time goes on.

Thus the analysis of the methods by which physical theories are constructed proves to us with complete evidence that these theories cannot be offered as explanations of experimental laws; and, on the other hand, an act of faith, as incapable of being justified by this analysis as of being frustrated by it, assures us that these theories are not a purely artificial system, but a natural classification. And so, we may here apply that profound thought of Pascal: "We have an impotence to prove, which cannot be conquered by any dogmatism; we have an idea of truth which cannot be conquered by any Pyrrhonian skepticism."

2.5 Theory anticipating experiment

There is one circumstance which shows with particular clarity our belief in the natural character of a theoretical classification; this circumstance is present when we ask of a theory that it tell us the result of an experiment before it has occurred, when we give it the bold injunction: "Be a prophet for us."

A considerable group of experimental laws had been established by investigators; the theorist has proposed to condense the laws into a very small number of hypotheses, and has succeeded in doing so; each one of the experimental laws is correctly represented by a consequence of these hypotheses.

But the consequences that can be drawn from these hypotheses are unlimited in number; we can, then, draw some consequences which do not correspond to any of the experimental laws previously known, and which simply represent possible experimental laws.

Among these consequences, some refer to circumstances realizable in practice, and these are particularly interesting, for they can be submitted to test by facts. If they represent exactly the experimental laws governing these facts, the value of the theory will be augmented, and the domain governed by the theory will annex new laws. If, on the contrary, there is among these consequences one which is sharply in disagreement with the facts whose law was to be represented by the theory, the latter will have to be more or less modified, or perhaps completely rejected.

Now, on the occasion when we confront the predictions of the theory with reality, suppose we have to bet for or against the theory; on which side shall we lay our wager?

If the theory is a purely artificial system, if we see in the hypotheses on which it rests statements skillfully worked out so that they represent the experimental laws already known, but if the theory fails to hint at any reflection of the real relations among the invisible realities, we shall think that such a theory will fail to confirm a new law. That, in the space left free among the drawers adjusted for other laws, the hitherto unknown law should find a drawer already made into which it may be fitted exactly would be a marvelous feat of chance. It would be folly for us to risk a bet on this sort of expectation.

If, on the contrary, we recognize in the theory a natural classification, if we feel that its principles express profound and real relations among things, we shall not be surprised to see its consequences anticipating experience and stimulating the discovery of new laws; we shall bet fearlessly in its favor.

The highest test, therefore, of our holding a classification as a natural one is to ask it to indicate in advance things which the future alone will reveal. And when the experiment is made and confirms the predictions obtained from our theory, we feel strengthened in our conviction that the relations established by our reason among abstract notions truly correspond to relations among things.

Thus, modern chemical symbolism, by making use of developed formulas, establishes a classification in which diverse compounds are ordered. The wonderful order this classification brings about in the tremendous arsenal of chemistry already assures us that the classification is not a purely artificial system. The relations of analogy and derivation by substitution it establishes among diverse compounds have meaning only in our mind; yet, we are convinced that they correspond to kindred relations among substances themselves, whose nature remains deeply hidden but whose reality does not seem doubtful. Nevertheless, for this conviction to change into overwhelming certainty, we must see the theory write in advance the formulas of a multitude of bodies and, yielding to these indications, synthesis must bring to light a large number of substances whose composition and several properties we should know even before they exist.

Just as the syntheses announced in advance sanction chemical notation as a natural classification, so physical theory will prove that it is the reflection of a real order by anticipating observation.

Now the history of physics provides us with many examples of this clairvoyant guesswork; many a time has a theory forecast laws not yet observed, even laws which appear improbable, stimulating the experimenter to discover them and guiding him toward that discovery.

The Académie des Sciences had set, as the subject for the physics prize that was to be awarded in the public meeting of March 1819, the general examination of the phenomena of the diffraction of light. Two memoirs were presented, and one by Fresnel was awarded the prize, the commission of judges consisting of Biot, Arago, Laplace, Gay-Lussac, and Poisson.

From the principles put forward by Fresnel, Poisson deduced through an elegant analysis the following strange consequence: If a small, opaque, and circular screen intercepts the rays emitted by a point source of light, there should exist behind the screen, on the very axis of this screen, points which are not only bright, but which shine exactly as though the screen were not interposed between them and the source of light.

Such a corollary, so contrary, it seems, to the most obvious experimental certainties, appeared to be a very good ground for rejecting the theory of diffraction proposed by Fresnel. Arago had confidence in the natural character arising from the clairvoyance of this theory. He tested it, and observation gave results which agreed absolutely with the improbable predictions from calculation.[13]

Thus physical theory, as we have defined it, gives to a vast group of experimental laws a condensed representation, favorable to intellectual economy.

It classifies these laws and, by classifying, renders them more easily and safely utilizable. At the same time, putting order into the whole, it adds to their beauty.

It assumes, while being completed, the characteristics of a natural classification. The groups it establishes permit hints as to the real affinities of things.

This characteristic of natural classification is marked, above all, by the fruitfulness of the theory which anticipates experimental laws not yet observed, and promotes their discovery.

That sufficiently justifies the search for physical theories, which cannot be called a vain and idle task even though it does not pursue the explanation of phenomena . . .

[. . .]

4 Physical theory and experiment

4.1 The experimental testing of a theory does not have the same logical simplicity in physics as in physiology

The sole purpose of physical theory is to provide a representation and classification of experimental laws; the only test permitting us to judge a physical theory and pronounce it good or bad is the comparison between the consequences of this theory and the experimental laws it has to represent, and classify. Now that we have minutely analyzed the characteristics of a physical experiment and of a physical law, we can establish the principles that should govern the comparison between experiment and theory; we can tell how we shall recognize whether a theory is confirmed or weakened by facts.

When many philosophers talk about experimental sciences, they think only of sciences still close to their origins, e.g., physiology or certain branches of chemistry where the experimenter reasons directly on the facts by a method which is only common sense brought to greater attentiveness but where mathematical theory has not yet introduced its symbolic representations. In such sciences the comparison between the deductions of a theory and the facts of experiment is subject to very simple rules. These rules were formulated in a particularly forceful manner by Claude Bernard, who would condense them into a single principle, as follows:

The experimenter should suspect and stay away from fixed ideas, and always preserve his freedom of mind.

The first condition that has to be fulfilled by a scientist who is devoted to the investigation of natural phenomena is to preserve a complete freedom of mind based on philosophical doubt.[14]

If a theory suggests experiments to be done, so much the better: ". . . we can follow our judgment and our thought, give free rein to our imagination provided that all our ideas are only pretexts for instituting new experiments that may furnish us probative facts or unexpected and fruitful ones."[15] Once the experiment is done and the results clearly established, if a theory takes them over in order to generalize them, coordinate them, and draw from them new subjects for experiment, still so much the better: ". . . if one is imbued with the principles of experimental method, there is nothing to fear; for so long as the idea is a right one, it will go on being developed; when it is an erroneous idea, experiment is there to correct it."[16] But so long as the experiment lasts, the theory should remain waiting, under strict orders to stay outside the door of the laboratory; it should keep silent and leave the scientist without disturbing him while he faces the facts directly; the facts must be observed without a preconceived idea and gathered with the same scrupulous impartiality, whether they confirm or contradict the predictions of the theory. The report that the observer will give us of his experiment should be a faithful and scrupulously exact reproduction of the phenomena, and should not let us even guess what system the scientist places his confidence in or distrusts.

Men who have an excessive faith in their theories or in their ideas are not only poorly disposed to make discoveries but they also make very poor observations. They necessarily observe with a preconceived idea and, when they have begun an experiment, they want to see in its results only a confirmation of their theory. Thus they distort observation and often neglect very important facts because they go counter to their goal. That is what made us say elsewhere that we must never do experiments in order to confirm our ideas but merely to check them. . . . But it quite naturally happens that those who believe too much in their own theories do not sufficiently believe in the theories of others. Then the dominant idea of these condemners of others is to find fault with the theories of the latter and to seek to contradict them. The setback for science remains the same. They are doing experiments only in order to destroy a theory instead of doing them in order to look for the truth. They also make poor observations because they take into the results of their experiments only what fits their purpose, by neglecting what is unrelated to it, and by very carefully avoiding whatever might go in the direction of the idea they wish to combat. Thus one is led by two parallel paths to the same result, that is to say, to falsifying science and the facts.

The conclusion of all this is that it is necessary to obliterate one's opinion as well as that of others when faced with the decisions of the experiment; . . . we must accept the results of experiment just as they present themselves with all that is unforeseen and accidental in them.[17]

Here, for example, is a physiologist who admits that the anterior roots of the spinal nerve contain the motor nerve-fibers and the posterior roots the sensory fibers. The theory he accepts leads him to imagine an experiment: if he cuts a certain anterior root, he ought to be suppressing the mobility of a certain part of the body without

destroying its sensibility; after making the section of this root, when he observes the consequences of his operation and when he makes a report of it, he must put aside all his ideas concerning the physiology of the spinal nerve; his report must be a raw description of the facts; he is not permitted to overlook or fail to mention any movement or quiver contrary to his predictions or to attribute it to some secondary cause unless some special experiment has given evidence of this cause; he must, if he does not wish to be accused of scientific bad faith, establish an absolute separation or watertight compartment between the consequences of his theoretical deductions and the establishing of the facts shown by his experiments.

Such a rule is not by any means easily followed; it requires of the scientist an absolute detachment from his own thought and a complete absence of animosity when confronted with the opinion of another person; neither vanity nor envy ought to be countenanced by him. As Bacon put it, he should never show eyes lustrous with human passions. Freedom of mind, which constitutes the sole principle of experimental method, according to Claude Bernard, does not depend merely on intellectual conditions, but also on moral conditions, making its practice rarer and more meritorious.

But if experimental method as just described is difficult to practice, the logical analysis of it is very simple. This is no longer the case when the theory to be subjected to test by the facts is not a theory of physiology but a theory of physics. In the latter case, in fact, it is impossible to leave outside the laboratory door the theory that we wish to test, for without theory it is impossible to regulate a single instrument or to interpret a single reading. We have seen that in the mind of the physicist there are constantly present two sorts of apparatus: one is the concrete apparatus in glass and metal, manipulated by him, the other is the schematic and abstract apparatus which theory substitutes for the concrete apparatus and on which the physicist does his reasoning. For these two ideas are indissolubly connected in his intelligence, and each necessarily calls on the other; the physicist can no sooner conceive the concrete apparatus without associating with it the idea of the schematic apparatus than a Frenchman can conceive an idea without associating it with the French word expressing it. This radical impossibility, preventing one from dissociating physical theories from the experimental procedures appropriate for testing these theories, complicates this test in a singular way, and obliges us to examine the logical meaning of it carefully.

Of course, the physicist is not the only one who appeals to theories at the very time he is experimenting or reporting the results of his experiments. The chemist and the physiologist when they make use of physical instruments, e.g., the thermometer, the manometer, the calorimeter, the galvanometer, and the saccharimeter, implicitly admit the accuracy of the theories justifying the use of these pieces of apparatus as well as of the theories giving meaning to the abstract ideas of temperature, pressure, quantity of heat, intensity of current, and polarized light, by means of which the concrete indications of these instruments are translated. But the theories used, as well as the instruments employed, belong to the domain of physics; by accepting with these instruments the theories without which their readings would be devoid of meaning, the chemist and the physiologist show their confidence in the physicist, whom they suppose to be infallible. The physicist, on the other hand, is obliged to trust his own theoretical ideas or those of his fellow-physicists. From the standpoint of logic, the difference is of little importance; for the physiologist and chemist as well as for the physicist, the statement of the result of an experiment implies, in general, an act of faith in a whole group of theories.

4.2 An experiment in physics can never condemn an isolated hypothesis but only a whole theoretical group

The physicist who carries out an experiment, or gives a report of one, implicitly recognizes the accuracy of a whole group of theories. Let us accept this principle and see what consequences we may deduce from it when we seek to estimate the role and logical import of a physical experiment.

In order to avoid any confusion we shall distinguish two sorts of experiments: experiments of *application*, which we shall first just mention, and experiments of *testing*, which will be our chief concern.

You are confronted with a problem in physics to be solved practically; in order to produce a certain effect you wish to make use of knowledge acquired by physicists; you wish to light an incandescent bulb; accepted theories indicate to you the means for solving the problem; but to make use of these means you have to secure certain information; you ought, I suppose, to determine the electromotive force of the battery of generators at your disposal; you measure this electromotive force: that is what I call an experiment of application. This experiment does not aim at discovering whether accepted theories are accurate or not; it merely intends to draw on these theories. In order to carry it out, you make use of instruments that these same theories legitimize; there is nothing to shock logic in this procedure.

But experiments of application are not the only ones the physicist has to perform; only with their aid can science aid practice, but it is not through them that science creates and develops itself; besides experiments of application, we have experiments of testing.

A physicist disputes a certain law; he calls into doubt a certain theoretical point. How will he justify these doubts? How will he demonstrate the inaccuracy of the law? From the proposition under indictment he will derive the prediction of an experimental fact; he will bring into existence the conditions under which this fact should be produced; if the predicted fact is not produced, the proposition which served as the basis of the prediction will be irremediably condemned.

F. E. Neumann assumed that in a ray of polarized light the vibration is parallel to the plane of polarization, and many physicists have doubted this proposition. How did O. Wiener undertake to transform this doubt into a certainty in order to condemn Neumann's proposition? He deduced from this proposition the following consequence: If we cause a light beam reflected at 45° from a plate of glass to interfere with the incident beam polarized perpendicularly to the plane of incidence, there ought to appear alternately dark and light interference bands parallel to the reflecting surface; he brought about the conditions under which these bands should have been produced and showed that the predicted phenomenon did not appear, from which he concluded that Neumann's proposition is false, viz., that in a polarized ray of light the vibration is not parallel to the plane of polarization.

Such a mode of demonstration seems as convincing and as irrefutable as the proof by reduction to absurdity customary among mathematicians; moreover, this demonstration is copied from the reduction to absurdity, experimental contradiction playing the same role in one as logical contradiction plays in the other.

Indeed, the demonstrative value of experimental method is far from being so rigorous or absolute: the conditions under which it functions are much more complicated than is supposed in what we have just said; the evaluation of results is much more delicate and subject to caution.

A physicist decides to demonstrate the inaccuracy of a proposition; in order to deduce from this proposition the prediction of a phenomenon and institute the experiment which is to show whether this phenomenon is or is not produced, in order to interpret the results of this experiment and establish that the predicted phenomenon is not produced, he does not confine himself to making use of the proposition in question; he makes use also of a whole group of theories accepted by him as beyond dispute. The prediction of the phenomenon, whose nonproduction is to cut off debate, does not derive from the proposition challenged if taken by itself, but from the proposition at issue joined to that whole group of theories; if the predicted phenomenon is not produced, not only is the proposition questioned at fault, but so is the whole theoretical scaffolding used by the physicist. The only thing the experiment teaches us is that among the propositions used to predict the phenomenon and to establish whether it would be produced, there is at least one error; but where this error lies is just what it does no tell us. The physicist may declare that this error is contained in exactly the proposition he wishes to refute, but is he sure it is not in another proposition? If he is, he accepts implicitly the accuracy of all the other propositions he has used, and the validity of his conclusion is as great as the validity of his confidence.

Let us take as an example the experiment imagined by Zenker and carried out by O. Wiener. In order to predict the formation of bands in certain circumstances and to show that these did not appear, Wiener did not make use merely of the famous proposition of F. E. Neumann, the proposition which he wished to refute; he did not merely admit that in a polarized ray vibrations are parallel to the plane of polarization; but he used, besides this, propositions, laws, and hypotheses constituting the optics commonly accepted: he admitted that light consists in simple periodic vibrations, that these vibrations are normal to the light ray, that at each point the mean kinetic energy of the vibratory motion is a measure of the intensity of light, that the more or less complete attack of the gelatine coating on a photographic plate indicates the various degrees of this intensity. By joining these propositions, and many others that would take too long to enumerate, to Neumann's proposition, Wiener was able to formulate a forecast and establish that the experiment belied it. If he attributed this solely to Neumann's proposition, if it alone bears the responsibility for the error this negative result has put in evidence, then Wiener was taking all the other propositions he invoked as beyond doubt. But this assurance is not imposed as a matter of logical necessity; nothing stops us from taking Neumann's proposition as accurate and shifting the weight of the experimental contradiction to some other proposition of the commonly accepted optics; as H. Poincaré has shown, we can very easily rescue Neumann's hypothesis from the grip of Wiener's experiment on the condition that we abandon in exchange the hypothesis which takes the mean kinetic energy as the measure of the light intensity; we may, without being contradicted by the experiment, let the vibration be parallel to the plane of polarization, provided that we measure the light intensity by the mean potential energy of the medium deforming; the vibratory motion.

These principles are so important that it will be useful to apply them to another example; again we choose an experiment regarded as one of the most decisive ones in optics.

We know that Newton conceived the emission theory for optical phenomena. The emission theory supposes light to be formed of extremely thin projectiles, thrown out with very great speed by the sun and other sources of light; these projectiles penetrate all transparent bodies; on account of the various parts of the media through which they move, they undergo attractions and repulsions; when the distance separating the acting particles is very small these actions are very powerful, and they vanish when the masses between which they act are appreciably far from each other. These essential hypotheses joined to several others, which we pass over without mention, lead to the formulation of a complete theory of reflection and refraction of light; in particular, they imply the following proposition: The index of refraction of light passing from one medium into another is equal to the velocity of the light projectile within the medium it penetrates, divided by the velocity of the same projectile in the medium it leaves behind.

This is the proposition that Arago chose in order to show that the theory of emission is in contradiction with the facts. From this proposition a second follows: Light travels faster in water than in air. Now Arago had indicated an appropriate procedure for comparing the velocity of light in air with the velocity of light in water; the procedure, it is true, was inapplicable, but Foucault modified the experiment in such a way that it could be carried out; he found that the light was propagated less rapidly in water than in air. We may conclude from this, with Foucault, that the system of emission is incompatible with the facts.

I say the *system* of emission and not the *hypothesis* of emission; in fact, what the experiment declares stained with error is the whole group of propositions accepted by Newton, and after him by Laplace and Biot, that is, the whole theory from which we deduce the relation between the index of refraction and the velocity of light in various media. But in condemning this system as a whole by declaring it stained with error, the experiment does not tell us where the error lies. Is it in the fundamental hypothesis that light consists in projectiles thrown out with great speed by luminous bodies? Is it in some other assumption concerning the actions experienced by light corpuscles due to the media through which they move? We know nothing about that. It would be rash to believe, as Arago seems to have thought, that Foucault's experiment condemns once and for all the very hypothesis of emission, i.e., the assimilation of a ray of light to a swarm of projectiles. If physicists had attached some value to this task, they would undoubtedly have succeeded in founding on this assumption a system of optics that would agree with Foucault's experiment.

In sum, the physicist can never subject an isolated hypothesis to experimental test but only a whole group of hypotheses; when the experiment is in disagreement with his predictions, what he learns is that at least one of the hypotheses constituting this group is unacceptable and ought to be modified; but the experiment does not designate which one should be changed.

We have gone a long way from the conception of the experimental method arbitrarily held by persons unfamiliar with its actual functioning. People generally think that each one of the hypotheses employed in physics can be taken in isolation, checked by experiment, and then, when many varied tests have established its validity, given a definitive place in the system of physics. In reality, this is not the case. Physics is not a machine which lets itself be taken apart; we cannot try each piece in isolation and, in order to adjust it wait until its solidity has been carefully checked. Physical science is a system that must be taken as a whole; it is an organism in which one part cannot be made to function except when the parts that are most remote from it are called into play, some more so than others, but all to some degree. If something goes wrong, if some discomfort is felt in the functioning of the organism, the physicist will have to ferret out through its effect on the entire system which organ needs to be remedied or modified without the possibility of isolating this

organ and examining it apart. The watchmaker to whom you give a watch that has stopped separates all the wheel-works and examines them one by one until he finds the part that is defective or broken. The doctor to whom a patient appears cannot dissect him in order to establish his diagnosis; he has to guess the seat and cause of the ailment solely by inspecting disorders affecting the whole body. Now, the physicist concerned with remedying a limping theory resembles the doctor and not the watchmaker.

4.3 A "crucial experiment" is impossible in physics

Let us press this point further, for we are touching on one of the essential features of experimental method, as it is employed in physics.

Reduction to absurdity seems to be merely a means of refutation, but it may become a method of demonstration: in order to demonstrate the truth of a proposition it suffices to corner anyone who would admit the contradictory of the given proposition into admitting an absurd consequence. We know to what extent the Greek geometers drew heavily on this mode of demonstration.

Those who assimilate experimental contradiction to reduction to absurdity imagine that in physics we may use a line of argument similar to the one Euclid employed so frequently in geometry. Do you wish to obtain from a group of phenomena a theoretically certain and indisputable explanation? Enumerate all the hypotheses that can be made to account for this group of phenomena; then, by experimental contradiction eliminate all except one; the latter will no longer be a hypothesis, but will become a certainty.

Suppose, for instance, we are confronted with only two hypotheses. Seek experimental conditions such that one of the hypotheses forecasts the production of one phenomenon and the other the production of quite a different effect; bring these conditions into existence and observe what happens; depending on whether you observe the first or the second of the predicted phenomena, you will condemn the second or the first hypothesis; the hypothesis not condemned will be henceforth indisputable; debate will be cut off, and a new truth will be acquired by science. Such is the experimental test that the author of the *Novum Organum* called the "*fact of the cross*, borrowing this expression from the crosses which at an intersection indicate the various roads."

We are confronted with two hypotheses concerning the nature of light; for Newton, Laplace, or Biot light consisted of projectiles hurled with extreme speed, but for Huygens, Young, or Fresnel light consisted of vibrations whose waves are propagated within an ether. These are the only two possible hypotheses as far as one can see: either the motion is carried away by the body it excites and remains attached to it, or else it passes from one body to another. Let us pursue the first hypothesis; it declares that light travels more quickly in water than in air; but if we follow the second, it declares that light travels more quickly in air than in water. Let us set up Foucault's apparatus; we set into motion the turning mirror; we see two luminous spots formed before us, one colorless, the other greenish. If the greenish band is to the left of the colorless one, it means that light travels faster in water than in air, and that the hypothesis of vibrating waves is false. If, on the contrary, the greenish band is to the right of the colorless one, that means that light travels faster in air than in water, and that the hypothesis of emissions is condemned. We look through the magnifying glass used to examine the two luminous spots, and we notice that the greenish spot is to the right of the colorless one; the debate is over; light is not a body, but a vibratory wave

motion propagated by the ether; the emission hypothesis has had its day; the wave hypothesis has been put beyond doubt, and the crucial experiment has made it a new article of the scientific credo.

What we have said in the foregoing paragraph shows how mistaken we should be to attribute to Foucault's experiment so simple a meaning and so decisive an importance; for it is not between two hypotheses, the emission and wave hypotheses, that Foucault's experiment judges trenchantly; it decides rather between two sets of theories each of which has to be taken as a whole, i.e., between two entire systems, Newton's optics and Huygens' optics.

But let us admit for a moment that in each of these systems everything is compelled to be necessary by strict logic, except a single hypothesis; consequently, let us admit that the facts, in condemning one of the two systems, condemn once and for all the single doubtful assumption it contains. Does it follow that we can find in the "crucial experiment" an irrefutable procedure for transforming one of the two hypotheses before us into a demonstrated truth? Between two contradictory theorems of geometry there is no room for a third judgment; if one is false, the other is necessarily true. Do two hypotheses in physics ever constitute such a strict dilemma? Shall we ever dare to assert that no other hypothesis is imaginable? Light may be a swarm of projectiles, or it may be a vibratory motion whose waves are propagated in a medium; is it forbidden to be anything else at all? Arago undoubtedly thought so when he formulated this incisive alternative: Does light move more quickly in water than in air? "Light is a body. If the contrary is the case, then light is a wave." But it would be difficult for us to take such a decisive stand; Maxwell, in fact, showed that we might just as well attribute light to a periodical electrical disturbance that is propagated within a dielectric medium.

Unlike the reduction to absurdity employed by geometers, experimental contradiction does not have the power to transform a physical hypothesis into an indisputable truth; in order to confer this power on it, it would be necessary to enumerate completely the various hypotheses which may cover a determinate group of phenomena; but the physicist is never sure he has exhausted all the imaginable assumptions. The truth of a physical theory is not decided by heads or tails.

4.4 Criticism of the Newtonian method. First example: celestial mechanics

It is illusory to seek to construct by means of experimental contradiction a line of argument in imitation of the reduction to absurdity; but the geometer is acquainted with other methods for attaining certainty than the method of reducing to an absurdity; the direct demonstration in which the truth of a proposition is established by itself and not by the refutation of the contradictory proposition seems to him the most perfect of arguments. Perhaps physical theory would be more fortunate in its attempts if it sought to imitate direct demonstration. The hypotheses from which it starts and develops its conclusions would then be tested one by one; none would have to be accepted until it presented all the certainty that experimental method can confer on an abstract and general proposition; that is to say, each would necessarily be either a law drawn from observation by the sole use of those two intellectual operations called induction and generalization, or else a corollary mathematically deduced from such laws. A theory based on such hypotheses would then not present anything arbitrary or doubtful; it would deserve all the confidence merited by the faculties which serve us in formulating natural laws.

It was this sort of physical theory that Newton had in mind when, in the "General Scholium"

which crowns his *Principia*, he rejected so vigorously as outside of natural philosophy any hypothesis that induction did not extract from experiment; when he asserted that in a sound physics every proposition should be drawn from phenomena and generalized by induction.

The ideal method we have just described therefore deserves to be named the Newtonian method. Besides, did not Newton follow this method when he established the system of universal attraction, thus adding to his precepts the most magnificent of examples? Is not his theory of gravitation derived entirely from the laws which were revealed to Kepler by observation, laws which problematic reasoning transforms and whose consequences induction generalizes?

This first law of Kepler's, "The radial vector from the sun to a planet sweeps out an area proportional to the time during which the planet's motion is observed," did, in fact, teach Newton that each planet is constantly subjected to a force directed toward the sun.

The second law of Kepler's, "The orbit of each planet is an ellipse having the sun at one focus," taught him that the force attracting a given planet varies with the distance of this planet from the sun, and that it is in an inverse ratio to the square of this distance.

The third law of Kepler's, "The squares of the periods of revolution of the various planets are proportional to the cubes of the major axes of their orbits," showed him that different planets would, if they were brought to the same distance from the sun, undergo in relation to it attractions proportional to their respective masses.

The experimental laws established by Kepler and transformed by geometric reasoning yield all the characteristics present in the action exerted by the sun on a planet; by induction Newton generalized the result obtained; he allowed this result to express the law according to which any portion of matter acts on any other portion whatsoever, and he formulated this great principle: "Any two bodies whatsoever attract each other with a force which is proportional to the product of their masses and in inverse ratio to the square of the distance between them." The principle of universal gravitation was found, and it was obtained, without any use having been made of any fictive hypothesis, by the inductive method the plan of which Newton outlined.

Let us again examine this application of the Newtonian method, this time more closely; let us see if a somewhat strict logical analysis will leave intact the appearance of rigor and simplicity that this very summary exposition attributes to it.

In order to assure this discussion of all the clarity it needs, let us begin by recalling the following principle, familiar to all those who deal with mechanics: We cannot speak of the force which attracts a body in given circumstances before we have designated the supposedly fixed term of reference to which we relate the motion of all bodies; when we change this point of reference or term of comparison, the force representing the effect produced on the observed body by the other bodies surrounding it changes in direction and magnitude according to the rules stated by mechanics with precision.

That posited, let us follow Newton's reasoning.

Newton first took the sun as the fixed point of reference; he considered the motions affecting the different planets by reference to the sun; he admitted Kepler's laws as governing these motions, and derived the following proposition: If the sun is the point of reference in relation to which all forces are compared, each planet is subjected to a force directed toward the sun, a force proportional to the mass of the planet and to the inverse square of its distance from the sun. Since the latter is taken as the reference point, it is not subject to any force.

In an analogous manner Newton studied the motion of the satellites and for each of these he chose as a fixed reference point the planet which the satellite accompanies, the earth in the case of the moon, Jupiter in the case of the masses moving around Jupiter. Laws just like Kepler's were taken as governing these motions, from which it follows that we can formulate the following proposition: If we take as a fixed reference point the planet accompanied by a satellite, this satellite is subject to a force directed toward the planet varying inversely with the square of the distance. If, as happens with Jupiter, the same planet possesses several satellites, these, satellites, were they at the same distance from the planet, would be acted on by the latter with forces proportional to their respective masses. The planet is itself not acted on by the satellite.

Such, in very precise form, are the propositions which Kepler's laws of planetary motion and the extension of these laws to the motions of satellites authorize us to formulate. For these propositions Newton substituted another which may be stated as follows: Any two celestial bodies whatsoever exert on each other a force of attraction in the direction of the straight line joining them, a force proportional to the product of their masses and to the inverse square of the distance between them. This statement presupposes all motions and forces to be related to the same reference point; the latter is an ideal standard of reference which may well be conceived by the geometer but which does not characterize in an exact and concrete manner the position in the sky of any body.

Is this principle of universal gravitation merely a generalization of the two statements provided by Kepler's laws and their extension to the motion of satellites? Can induction derive it from these two statements? Not at all. In fact, not only is it more general than these two statements and unlike them, but it contradicts them. The student of mechanics who accepts the principle of universal attraction can calculate the magnitude and direction of the forces between the various planets and the sun when the latter is taken as the reference point, and if he does he finds that these forces are not what our first statement would require. He can determine the magnitude and direction of each of the forces between Jupiter and its satellites when we refer all the motions to the planet, assumed to be fixed, and if he does he notices that these forces are not what our second statement would require.

The principle of universal gravity, very far from being derivable by generalization and induction from the observational laws of Kepler, formally contradicts these laws. If Newton's theory is correct, Kepler's laws are necessarily false.

Kepler's laws based on the observation of celestial motions do not transfer their immediate experimental certainty to the principle of universal weight, since if, on the contrary, we admit the absolute exactness of Kepler's laws, we are compelled to reject the proposition on which Newton based his celestial mechanics. Far from adhering to Kepler's laws, the physicist who claims to justify the theory of universal gravitation finds that he has, first of all, to resolve a difficulty in these laws: he has to prove that his theory, incompatible with the exactness of Kepler's laws, subjects the motions of the planets and satellites to other laws scarcely different enough from the first laws for Tycho Brahé, Kepler, and their contemporaries to have been able to discern the deviations between the Keplerian and Newtonian orbits. This proof derives from the circumstances that the sun's mass is very large in relation to the masses of the various planets and the mass of a planet is very large in relation to the masses of its satellites.

Therefore, if the certainty of Newton's theory does not emanate from the certainty of Kepler's laws, how will this theory prove its validity? It will calculate, with all the high degree of approximation that the constantly perfected methods of algebra involve, the perturbations which at each instant remove every heavenly body from the orbit assigned to it by Kepler's laws; then it will compare the calculated perturbations with the perturbations observed by means of the most precise instruments and the most scrupulous methods. Such a comparison will not only bear on this or that part of the Newtonian principle, but will involve all its parts at the same time; with those it will also involve all the principles of dynamics; besides, it will call in the aid of all the propositions of optics, the statics of gases, and the theory of heat, which are necessary to justify the properties of telescopes in their construction, regulation, and correction, and in the elimination of the errors caused by diurnal or annual aberration and by atmospheric refraction. It is no longer a matter of taking, one by one, laws justified by observation, and raising each of them by induction and generalization to the rank of a principle; it is a matter of comparing the corollaries of a whole group of hypotheses to a whole group of facts.

Now, if we seek out the causes which have made the Newtonian method fail in this case for which it was imagined and which seemed to be the most perfect application for it, we shall find them in that double character of any law made use of by theoretical physics: This law is symbolic and approximate.

Undoubtedly, Kepler's laws bear quite directly on the very objects of astronomical observation; they are as little symbolic as possible. But in this purely experimental form they remain inappropriate for suggesting the principle of universal gravitation; in order to acquire this fecundity they must be transformed and must yield the characters of the forces by which the sun attracts the various planets.

Now this new form of Kepler's laws is a symbolic form; only dynamics gives meanings to the words "force" and "mass," which serve to state it, and only dynamics permits us to substitute the new symbolic formulas for the old realistic formulas, to substitute statements relative to "forces" and "masses" for laws relative to orbits. The legitimacy of such a substitution implies full confidence in the laws of dynamics.

And in order to justify this confidence let us not proceed to claim that the laws of dynamics were beyond doubt at the time Newton made use of them in symbolically translating Kepler's laws; that they had received enough empirical confirmation to warrant the support of reason. In fact, the laws of dynamics had been subjected up to that time to only very limited and very crude tests. Even their enunciations had remained very vague and involved; only in Newton's *Principia* had they been for the first time formulated in a precise manner. It was in the agreement of the facts with the celestial mechanics which Newton's labors gave birth to that they received their first convincing verification.

Thus the translation of Kepler's laws into symbolic laws, the only kind useful for a theory, presupposed the prior adherence of the physicist to a whole group of hypotheses. But, in addition, Kepler's laws being only approximate laws, dynamics permitted giving them an infinity of different symbolic translations. Among these various forms, infinite in number, there is one and only one which agrees with Newton's principle. The observations of Tycho Brahé, so felicitously reduced to laws by Kepler, permit the theorist to choose this form, but they do not constrain him to do so, for there is an infinity of others they permit him to choose.

The theorist cannot, therefore, be content to invoke Kepler's laws in order to justify his choice. If he wishes to prove that the principle he has adopted is truly a principle of natural classification for celestial motions, he must show that the observed perturbations are in agreement with those which had been calculated in advance; he has to show how from the course of

Uranus he can deduce the existence and position of a new planet, and find Neptune in an assigned direction at the end of his telescope.

4.5 Criticism of the Newtonian method (continued). Second example: electrodynamics

Nobody after Newton except Ampère has more clearly declared that all physical theory should be derived from experience by induction only; no work has been more closely modelled after Newton's *Philosophiae naturalis Principia mathematica* than Ampère's *Théorie mathématique des phénomènes èlectrodynamiques uniquement déduite de l'expérience*.

> The epoch marked by the works of Newton in the history of the sciences is not only one of the most important discoveries that man has made concerning the causes of the great phenomena of nature, but it is also the epoch in which the human mind opened a new route in the sciences whose object is the study of these phenomena.

These are the lines with which Ampère began the exposition of his *Théorie mathématique*; he continued in the following terms:

> "Newton was far from thinking" that the law of universal weight "could be discovered by starting from more or less plausible abstract considerations. He established the fact that it had to be deduced from observed facts, or rather from those empirical laws which, like those of Kepler, are but results generalized from a great number of facts.
>
> "To observe the facts first, to vary their circumstances as far as possible, to make precise measurements along with this first task in order to deduce from them general laws based only on experience, and to deduce from these laws, independently of any hypothesis about the nature of the forces

producing the phenomena, the mathematical value of these forces, i.e., the formula representing them – that is the course Newton followed. It has been generally adopted in France by the scientists to whom physics owes the enormous progress it has made in recent times, and it has served me as a guide in all my research on electrodynamic phenomena. I have consulted only experience in order to establish the laws of these phenomena, and I have deduced from them the formula which can only represent the forces to which they are due; I have made no investigation about the cause itself assignable to these forces, well convinced that any investigation of this kind should be preceded simply by experimental knowledge of the laws and of the determination, deduced solely from these laws, of the value of the elementary force."

Neither very close scrutiny nor great perspicacity is needed in order to recognize that the *Théorie mathématique des phénomènes électrodynamiques* does not in any way proceed according to the method prescribed by Ampère, and to see that it is not "deduced only from experience" (*uniquement déduite de l'expérience*). The facts of experience taken in their primitive rawness cannot serve mathematical reasoning; in order to feed this reasoning they have to be transformed and put into a symbolic form. This transformation Ampère did make them undergo. He was not content merely with reducing the metal apparatus in which currents flow to simple geometric figures; such an assimilation imposes itself too naturally to give way to any serious doubt. Neither was he content merely to use the notion of force, borrowed from mechanics, and various theorems constituting this science; at the time he wrote, these theorems might be considered as beyond dispute. Besides all this, he appealed to a whole set of entirely new

hypotheses which are entirely gratuitous and sometimes even rather surprising. Foremost among these hypotheses it is appropriate to mention the intellectual operation by which he decomposed into infinitely small elements the electric current, which, in reality, cannot be broken without ceasing to exist; then the supposition that all real electrodynamic actions are resolved into fictive actions involving the pairs that the elements of current form, one pair at a time; then the postulate that the mutual actions of two elements are reduced to two forces applied to the elements in the direction of the straight line joining them, forces equal and opposite in direction; then the postulate that the distance between two elements enters simply into the formula of their mutual action by the inverse of a certain power.

These diverse assumptions are so little self-evident and so little necessary that several of them have been criticized or rejected by Ampère's successors; other hypotheses equally capable of translating symbolically the fundamental experiments of electrodynamics have been proposed by other physicists, but none of them has succeeded in giving this translation without formulating some new postulate, and it would be absurd to claim to do so.

The necessity which leads the physicist to translate experimental facts symbolically before introducing them into his reasoning, renders the purely inductive path Ampère drew impracticable; this path is also forbidden to him because each of the observed laws is not exact but merely approximate.

Ampère's experiments have the grossest degree of approximation. He gave a symbolic translation of the facts observed in a form appropriate for the success of his theory, but how easily he might have taken advantage of the uncertainty of the observations in order to give quite a different translation! Let us listen to Wilhelm Weber:

"Ampère made a point of expressly indicating in the title of his memoir that his mathematical theory of electrodynamic phenomena is *deduced only from experiment*, and indeed in his book we find expounded in detail the simple as well as ingenious method which led him to his goal. There we find, presented with all the precision and scope desirable, the exposition of his experiments, the deductions that he draws from them for theory, and the description of the instruments he employs. But in fundamental experiments, such as we have here, it is not enough to indicate the general meaning of an experiment, to describe the instruments used in performing it, and to tell in a general way that it has yielded the result expected; it is indispensable to go into the details of the experiment itself, to say how often it has been repeated, how the conditions were modified, and what the effect of these modifications has been; in a word, to compose a sort of brief of all the circumstances permitting the reader to sit in judgment on the degree of reliability and certainty of the result. Ampère does *not* give these precise details concerning his experiments, and the demonstration of the fundamental law of electrodynamics still awaits this indispensable supplementation. The fact of the mutual attraction of two conducting wires has been verified over and over again and is beyond all dispute; but these verifications have always been made under conditions and by such means that no *quantitative* measurement was possible and these measurements are far from having reached the degree of precision required for considering the law of these phenomena demonstrated.

"More than once, Ampère has drawn from the *absence* of any electrodynamic action the same consequences as from a measurement that would have given him a result equal to zero, and by this artifice, with great sagacity and with even greater skill, he has succeeded in bringing together the data necessary for the establishment and demonstration of his theory; but these *negative* experiments with which we must be content in the absence of direct *positive* measurements," those experiments in which all passive resistances, all friction, all causes of error tend precisely to produce the effect we wish to observe, "cannot have all the value or demonstrative force of those positive measurements, especially when they are not obtained with the procedures and under the conditions of true measurement, which are moreover impossible to obtain with the instruments Ampère has employed."[18]

Experiments with so little precision leave the physicist with the problem of choosing between an infinity of equally possible symbolic translations, and confer no certainty on a choice they do not impose; only intuition, guessing the form of theory to be established, directs this choice. This role of intuition is particularly important in the work of Ampère; it suffices to run through the writings of this great geometer in order to recognize that his fundamental formula of electrodynamics was found quite completely by a sort of divination, that his experiments were thought up by him as afterthoughts and quite purposefully combined so that he might be able to expound according to the Newtonian method a theory that he had constructed by a series of postulates.

Besides, Ampère had too much candor to dissimulate very learnedly that what was artificial in his exposition was *entirely deduced from experiment*; at the end of his *Théorie mathématique des phénomènes électrodynamiques* he wrote the following lines: "I think I ought to remark in finishing this memoir that I have not yet had the time to construct the instruments represented in

Diagram 4 of the first plate and in Diagram 20 of the second plate. The experiments for which they were intended have not yet been done." Now the first of the two sets of apparatus in question aimed to bring into existence the last of the four fundamental cases of equilibrium which are like columns in the edifice constructed by Ampère: it is with the aid of the experiment for which this apparatus was intended that we were to determine the power of the distance according to which electrodynamic actions proceed. Very far from its being the case that Ampère's electrodynamic theory was *entirely deduced from experiment*, experiment played a very feeble role in its formation: it was merely the occasion which awakened the intuition of this physicist of genius, and his intuition did the rest.

It was through the research of Wilhelm Weber that the very intuitive theory of Ampère was first subjected to a detailed comparison with the facts; but this comparison was not guided by the Newtonian method. Weber deduced from Ampère's theory, taken as a whole, certain effects capable of being calculated; the theorems of statics and of dynamics, and also even certain propositions of optics, permitted him to conceive an apparatus, the electrodynamometer, by means of which these same effects may be subjected to precise measurements; the agreement of the calculated predictions with the results of the measurements no longer, then, confirms this or that isolated proposition of Ampère's theory, but the whole set of electrodynamical, mechanical, and optical hypotheses that must be invoked in order to interpret each of Weber's experiments.

Hence, where Newton had failed, Ampère in his turn just stumbled. That is because two inevitable rocky reefs make the purely inductive course impracticable for the physicist. In the first place, no experimental law can serve the theorist before it has undergone an interpretation transforming it into a symbolic law; and this interpretation implies adherence to a whole set of theories. In the second place, no experimental law is exact but only approximate, and is therefore susceptible to an infinity of distinct symbolic translations; and among all these translations the physicist has to choose one which will provide him with a fruitful hypothesis without his choice being guided by experiment at all.

This criticism of the Newtonian method brings us back to the conclusions to which we have already been led by the criticism of experimental contradiction and of the crucial experiment. These conclusions merit our formulating them with the utmost clarity. Here they are:

To seek to separate each of the hypotheses of theoretical physics from the other assumptions on which this science rests in order to subject it in isolation to observational test is to pursue a chimera; for the realization and interpretation of no matter what experiment in physics imply adherence to a whole set of theoretical propositions.

The only experimental check on a physical theory which is not illogical consists in comparing the *entire system of the physical theory with the whole group of experimental laws*, and in judging whether the latter is represented by the former in a satisfactory manner.

4.6 Consequences relative to the teaching of physics

Contrary to what we have made every effort to establish, it is generally accepted that each hypothesis of physics may be separated from the group and subjected in isolation to experimental test. Of course, from this erroneous principle false consequences are deduced concerning the method by which physics should be taught. People would like the professor to arrange all the hypotheses of physics in a certain order, to take the first one, enounce it, expound its experimental verifications, and then when the latter have been recognized as sufficient, declare the hypothesis accepted. Better still, people would like him to formulate this first hypothesis by inductive generalization of a purely experimental law; he would begin this operation again on the second hypothesis, on the third, and so on until all of physics was constituted. Physics would be taught as geometry is: hypotheses would follow one another as theorems follow one another; the experimental test of each assumption would replace the demonstration of each proposition; nothing which is not drawn from facts or immediately justified by facts would be promulgated.

Such is the ideal which has been proposed by many teachers, and which several perhaps think they have attained. There is no lack of authoritative voices inviting them to the pursuit of this ideal. M. Poincaré says: "It is important not to multiply hypotheses excessively, but to make them only one after the other. If we construct a theory based on multiple hypotheses, and experiment condemns the theory, which one among our premises is it necessary to change? It will be impossible to know. And if, on the other hand, the experiment succeeds, shall we think we have verified all these hypotheses at the same time? Shall we think we have determined several unknowns with a single equation?"[19]

In particular, the purely inductive method whose laws Newton formulated is given by many physicists as the only method permitting one to expound rationally the science of nature. Gustave Robin says: "The science we shall make will be only a combination of simple inductions suggested by experience. As to these inductions, we shall formulate them always in propositions easy to retain and *susceptible of direct verification*, never losing sight of the fact that *a hypothesis cannot be verified by its consequences.*"[20] This is the Newtonian

method recommended if not prescribed for those who plan to teach physics in the secondary schools. They are told: "The procedures of mathematical physics are not adequate for secondary-school instruction, for they consist in starting from hypotheses or from definitions posited a priori in order to deduce from them conclusions which will be subjected to experimental check. This method may be suitable for specialized classes in mathematics, but it is wrong to apply it at present in our elementary courses in mechanics, hydrostatics, and optics. Let us replace it by the inductive method."[21]

The arguments we have developed have established more than sufficiently the following truth: It is as impracticable for the physicist to follow the inductive method whose practice is recommended to him as it is for the mathematician to follow that perfect deductive method which would consist in defining and demonstrating everything, a method of inquiry to which certain geometers seem passionately attached, although Pascal properly and rigorously disposed of it a long time ago. Therefore, it is clear that those who claim to unfold the series of physical principles by means of this method are naturally giving an exposition of it that is faulty at some point.

Among the vulnerable points noticeable in such an exposition, the most frequent and, at the same time, the most serious, because of the false ideas it deposits in the minds of students, is the "fictitious experiment." Obliged to invoke a principle which has not really been drawn from facts or obtained by induction, and averse, moreover, to offering this principle for what it is, namely, a postulate, the physicist invents an imaginary experiment which, were it carried out with success, would possibly lead to the principle whose justification is desired.

To invoke such a fictitious experiment is to offer an experiment to be done for an experiment done; this is justifying a principle not by means of facts observed but by means of facts whose existence is predicted, and this prediction has no other foundation than the belief in the principle supported by the alleged experiment. Such a method of demonstration implicates him who trusts it in a vicious circle; and he who teaches it without making it exactly clear that the experiment cited has not been done commits an act of bad faith.

At times the fictitious experiment described by the physicist could not, if we attempted to bring it about, yield a result of any precision; the very indecisive and rough results it would produce could undoubtedly be put into agreement with the proposition claimed to be warranted; but they would agree just as well with certain very different propositions; the demonstrative value of such an experiment would therefore be very weak and subject to caution. The experiment that Ampère imagined in order to prove that electrodynamic actions proceed according to the inverse square of the distance, but which he did not perform, gives us a striking example of such a fictitious experiment.

But there are worse things. Very often the fictitious experiment invoked is not only not realized but incapable of being realized; it presupposes the existence of bodies not encountered in nature and of physical properties which have never been observed. Thus Gustave Robin, in order to give the principles of chemical mechanics the purely inductive exposition that he wishes, creates at will what he calls witnessing bodies (*corps témoins*), bodies which by their presence alone are capable of agitating or stopping a chemical reaction.[22] Observation has never revealed such bodies to chemists.

The unperformed experiment, the experiment which would not be performed with precision, and the absolutely unperformable experiment do not exhaust the diverse forms assumed by the fictitious experiment in the

writings of physicists who claim to be following the experimental method; there remains to be pointed out a form more illogical than all the others, namely, the absurd experiment. The latter claims to prove a proposition which is contradictory if regarded as the statement of an experimental fact.

The most subtle physicists have not always known how to guard against the intervention of the absurd experiment in their expositions. Let us quote, for instance, some lines taken from J. Bertrand: "If we accept it as an experimental fact that electricity is carried to the surface of bodies, and as a necessary principle that the action of free electricity on the points of conductors should be null, we can deduce from these two conditions, supposing they are strictly satisfied, that electrical attractions and repulsions are inversely proportional to the square of the distance."[23]

Let us take the proposition "There is no electricity in the interior of a conducting body when electrical equilibrium is established in it," and let us inquire whether it is possible to regard it as the statement of an experimental fact. Let us weigh the exact sense of the words figuring in the statement, and particularly, of the word interior. In the sense we must give this word in this proposition, a point interior to a piece of electrified copper is a point taken within the mass of copper. Consequently, how can we go about establishing whether there is or is not any electricity at this point? It would be necessary to place a testing body there, and to do that it would be necessary to take away beforehand the copper that is there, but then this point would no longer be within the mass of copper; it would be outside that mass. We cannot without falling into a logical contradiction take our proposition as a result of observation.

What, therefore, is the meaning of the experiments by which we claim to prove this proposition? Certainly, something quite different from what we make them say. We hollow out a cavity in a conducting mass and note that the walls of this cavity are not charged. This observation proves nothing concerning the presence or absence of electricity at points deep within the conducting mass. In order to pass from the experimental law noted to the law stated we play on the word interior. Afraid to base electrostatics on a postulate, we base it on a pun.

If we simply turn the pages of the treatises and manuals of physics we can collect any number of fictitious experiments; we should find there abundant illustrations of the various forms that such an experiment can assume, from the merely unperformed experiment to the absurd experiment. Let us not waste time on such a fastidious task. What we have said suffices to warrant the following conclusion: The teaching of physics by the purely inductive method such as Newton defined it is a chimera. Whoever claims to grasp this mirage is deluding himself and deluding his pupils. He is giving them, as facts seen, facts merely foreseen; as precise observations, rough reports; as performable procedures, merely ideal experiments; as experimental laws, propositions whose terms cannot be taken as real without contradiction. The physics he expounds is false and falsified.

Let the teacher of physics give up this ideal inductive method which proceeds from a false idea, and reject this way of conceiving the teaching of experimental science, a way which dissimulates and twists its essential character. If the interpretation of the slightest experiment in physics presupposes the use of a whole set of theories, and if the very description of this experiment requires a great many abstract symbolic expressions whose meaning and correspondence with the facts are indicated only by theories, it will indeed be necessary for the physicist to decide to develop a long chain of hypotheses and deductions before trying the slightest comparison between the

theoretical structure and the concrete reality; also, in describing experiments verifying theories already developed, he will very often have to anticipate theories to come. For example, he will not be able to attempt the slightest experimental verification of the principles of dynamics before he has not only developed the chain of propositions of general mechanics but also laid the foundations of celestial mechanics; and he will also have to suppose as known, in reporting the observations verifying this set of theories, the laws of optics which alone warrant the use of astronomical instruments.

Let the teacher therefore develop, in the first place, the essential theories of the science; without doubt, by presenting the hypotheses on which these theories rest, it is necessary for him to prepare their acceptance; it is good for him to point out the data of common sense, the facts gathered by ordinary observation or simple experiments or those scarcely analyzed which have led to formulating these hypotheses. To this point, moreover, we shall insist on returning in the next chapter [sic]; but we must proclaim loudly that these facts sufficient for suggesting hypotheses are not sufficient to verify them; it is only after he has constituted an extensive body of doctrine and constructed a complete theory that he will be able to compare the consequences of this theory with experiment.

Instruction ought to get the student to grasp this primary truth: Experimental verifications are not the base of theory but its crown. Physics does not make progress in the way geometry does: the latter grows by the continual contribution of a new theorem demonstrated once and for all and added to theorems already demonstrated; the former is a symbolic painting in which continual re-touching gives greater comprehensiveness and unity, and the *whole* of which gives a picture resembling more and more the *whole* of the experimental facts, whereas each detail of this picture cut off and isolated from the whole loses all meaning and no longer represents anything.

To the student who will not have perceived this truth, physics will appear as a monstrous confusion of fallacies of reasoning in circles and begging the question; if he is endowed with a mind of high accuracy, he will repel with disgust these perpetual defiances of logic; if he has a less accurate mind, he will learn by heart here words with inexact meaning, these descriptions of unperformed and unperformable experiments, and lines of reasoning which are sleight-of-hand passes, thus losing in such unreasoned memory work the little correct sense and critical mind he used to possess.

The student who, on the other hand, will have seen clearly the ideas we have just formulated will have done more than learned a certain number of propositions of physics; he will have understood the nature and true method of experimental science.[24]

4.7 Consequences relative to the mathematical development of physical theory

Through the preceding discussions the exact nature of physical theory and of its relations with experiment emerge more and more clearly and precisely.

The materials with which this theory is constructed are, on the one hand, the mathematical symbols serving to represent the various quantities and qualities of the physical world, and, on the other hand, the general postulates serving as principles. With these materials theory builds a logical structure; in drawing the plan of this structure it is hence bound to respect scrupulously the laws that logic imposes on all deductive reasoning and the rules that algebra prescribes for any mathematical operation.

The mathematical symbols used in theory have meaning only under very definite conditions; to define these symbols is to enumerate these conditions. Theory is forbidden to make use of these signs outside these conditions. Thus, an absolute temperature by definition can be positive only, and by definition the mass of a body is invariable; never will theory in its formulas give a zero or negative value to absolute temperature, and never in its calculations will it make the mass of a given body vary.

Theory is in principle grounded on postulates, that is to say, on propositions that it is at leisure to state as it pleases, provided that no contradiction exists among the terms of the same postulate or between two distinct postulates. But once these postulates are set down it is bound to guard them with jealous rigor. For instance, if it has placed at the base of its system the principle of the conservation of energy, it must forbid any assertion in disagreement with this principle.

These rules bring all their weight to bear on a physical theory that is being constructed; a single default would make the system illogical and would oblige us to upset it in order to reconstruct another; but they are the only limitations imposed. In the course of its development, *a physical theory is free to choose any path it pleases provided that it avoids any logical contradiction; in particular, it is free not to take account of experimental facts.*

This is no longer the case when the theory has reached its complete development. When the logical structure has reached its highest point it becomes necessary to compare the set of mathematical propositions obtained as conclusions from these long deductions with the set of experimental facts; by employing the adopted procedures of measurement we must be sure that the second set finds in the first a sufficiently similar image, a sufficiently precise and complete symbol. If this agreement between the conclusions of theory and the facts of experiment were not to manifest a satisfactory approximation, the theory might well be logically constructed, but it should nonetheless be rejected because it would be contradicted by observation, because it would be *physically* false.

This comparison between the conclusions of theory and the truths of experiment is therefore indispensable, since only the test of facts can give physical validity to a theory. But this test by facts should bear exclusively on the conclusions of a theory, for only the latter are offered as an image of reality; the postulates serving as points of departure for the theory and the intermediary steps by which we go from the postulates to the conclusions do not have to be subject to this test.

We have in the foregoing pages very thoroughly analyzed the error of those who claim to subject one of the fundamental postulates of physics directly to the test of facts through a procedure such as a crucial experiment; and especially the error of those who accept as principles only "inductions consisting exclusively in erecting into general laws not the interpretation but *the very result of a very large number of experiments.*"[25]

There is another error lying very close to this one; it consists in requiring that all the operations performed by the mathematician connecting postulates with conclusions should have *a physical meaning*, in wishing "to reason only about *performable operations*," and in "introducing only magnitudes accessible to experiment."[26]

According to this requirement any magnitude introduced by the physicist in his formulas should be connected through a process of measurement to a property of a body; any algebraic operation performed on these magnitudes should be translated into concrete language by the employment of these processes of measurement; thus translated, it should express a real or possible fact.

Such a requirement, legitimate when it comes to the final formulas at the end of a theory, has no justification if applied to the intermediary formulas and operations establishing the transition from postulates to conclusions.

Let us take an example.

J. Willard Gibbs studied the theory of the dissociation of a perfect composite gas into its elements, also regarded as perfect gases. A formula was obtained expressing the law of chemical equilibrium internal to such a system. I propose to discuss this formula. For this purpose, keeping constant the pressure supporting the gaseous mixture, I consider the absolute temperature appearing in the formula and I make it vary from 0 to $+ \infty$.

If we wish to attribute a physical meaning to this mathematical operation, we shall be confronted with a host of objections and difficulties. No thermometer can reveal temperatures below a certain limit, and none can determine temperatures high enough; this symbol which we call "absolute temperature" cannot be translated through the means of measurement at our disposal into something having a concrete meaning unless its numerical value remains between a certain minimum and a certain maximum. Moreover, at temperatures sufficiently low this other symbol which thermodynamics calls "a perfect gas" is no longer even an approximate image of any real gas.

These difficulties and many others, which it would take too long to enumerate, disappear if we heed the remarks we have formulated. In the construction of the theory, the discussion we have just given is only an intermediary step, and there is no justification for seeking a physical meaning in it. Only when this discussion shall have led us to a series of propositions, shall we have to submit these propositions to the test of facts; then we shall inquire whether, within the limits in which the absolute temperature may be translated into concrete readings of a thermometer and the idea of a perfect gas is approximately embodied in the fluids we observe, the conclusions of our discussion agree with the results of experiment.

By requiring that mathematical operations by which postulates produce their consequences shall always have a physical meaning, we set unjustifiable obstacles before the mathematician and cripple his progress. G. Robin goes so far as to question the use of the differential calculus; if Professor Robin is intent on constantly and scrupulously satisfying this requirement, he would practically be unable to develop any calculation; theoretical deduction would be stopped in its tracks from the start. A more accurate idea of the method of physics and a more exact line of demarcation between the propositions which have to submit to factual test and those which are free to dispense with it would give back to the mathematician all his freedom and permit him to use all the resources of algebra for the greatest development of physical theories.

4.8 Are certain postulates of physical theory incapable of being refuted by experiment?

We recognize a correct principle by the facility with which it straightens out the complicated difficulties into which the use of erroneous principles brought us.

If, therefore, the idea we have put forth is correct, namely, that comparison is established necessarily between the *whole* of theory and the *whole* of experimental facts, we ought in the light of this principle to see the disappearance of the obscurities in which we should be lost by thinking that we are subjecting each isolated theoretical hypothesis to the test of facts.

Foremost among the assertions in which we shall aim at eliminating the appearance of paradox, we shall place one that has recently been often formulated and discussed. Stated first by G. Milhaud in connection with the *"pure bodies"* of chemistry,[27] it has been developed at length and forcefully by H. Poincaré with regard to principles of mechanics;[28] Edouard Le Roy has also formulated it with great clarity.[29]

That assertion is as follows: Certain fundamental hypotheses of physical theory cannot be contradicted by any experiment, because they constitute in reality *definitions*, and because certain expressions in the physicist's usage take their meaning only through them.

Let us take one of the examples cited by Le Roy:

When a heavy body falls freely, the acceleration of its fall is constant. Can such a law be contradicted by experiment? No, for it constitutes the very definition of what is meant by "falling freely." If while studying the fall of a heavy body we found that this body does not fall with uniform acceleration, we should conclude not that the stated law is false, but that the body does not fall freely, that some cause obstructs its motion, and that the deviations of the observed facts from the law as stated would serve to discover this cause and to analyze its effects.

Thus, M. Le Roy concludes, "laws are verifiable, taking things strictly . . ., because they constitute the very criterion by which we judge appearances as well as the methods that it would be necessary to utilize in order to submit them to an inquiry whose precision is capable of exceeding any assignable limit."

Let us study again in greater detail, in the light of the principles previously set down, what this comparison is between the law of falling bodies and experiment.

Our daily observations have made us acquainted with a whole category of motions which we have brought together under the name of motions of heavy bodies; among these motions is the falling of a heavy body when it is not hindered by any obstacle. The result of this is that the words "free fall of a heavy body" have a meaning for the man who appeals only to the knowledge of common sense and who has no notion of physical theories.

On the other hand, in order to classify the laws of motion in question the physicist has created a theory, the theory of weight, an important application of rational mechanics. In that theory, intended to furnish a symbolic representation of reality, there is also the question of "free fall of a heavy body," and as a consequence of the hypotheses supporting this whole scheme free fall must necessarily be a uniformly accelerated motion.

The words "free fall of a heavy body" now have two distinct meanings. For the man ignorant of physical theories, they have their *real* meaning, and they mean what common sense means in pronouncing them; for the physicist they have a *symbolic* meaning, and mean "uniformly accelerated motion." Theory would not have realized its aim if the second meaning were not the sign of the first, if a fall regarded as free by common sense were not also regarded as uniformly accelerated, or *nearly* uniformly accelerated, since common-sense observations are essentially devoid of precision, according to what we have already said.

This agreement, without which the theory would have been rejected without further examination, is finally arrived at: a fall declared by common sense to be nearly free is also a fall whose acceleration is nearly constant. But noticing this crudely approximate agreement does not satisfy us; we wish to push on and surpass the degree of precision which common sense can claim. With the aid of the theory that we have imagined, we put together apparatus enabling us to recognize with sensitive accuracy whether the fall of a body is or is not uniformly accelerated; this apparatus shows us that a certain fall regarded by common sense as a free fall has a slightly variable acceleration. The proposition which in our theory gives its symbolic meaning to the words "free fall" does not represent with sufficient accuracy the properties of the real and concrete fall that we have observed.

Two alternatives are then open to us.

In the first place, we can declare that we were right in regarding the fall studied as a free fall and in requiring that the theoretical definition

of these words agree with our observations. In this case, since our theoretical definition does not satisfy this requirement, it must be rejected; we must construct another mechanics on new hypotheses, a mechanics in which the words "free fall" no longer signify "uniformly accelerated motion," but "fall whose acceleration varies according to a certain law."

In the second alternative, we may declare that we were wrong in establishing a connection between the concrete fall we have observed and the symbolic free fall defined by our theory, that the latter was too simplified a scheme of the former, that in order to represent suitably the fall as our experiments have reported it the theorist should give up imagining a weight falling freely and think in terms of a weight hindered by certain obstacles like the resistance of the air, that in picturing the action of these obstacles by means of appropriate hypotheses he will compose a more complicated scheme than a free weight but one more apt to reproduce the details of the experiment; in short, in accord with the language we have previously established (Ch. IV, Sec. 3 [sic]), we may seek to eliminate by means of suitable "corrections" the "causes of error," such as air resistance, which influenced our experiment.

M. Le Roy asserts that we shall prefer the second to the first alternative, and he is surely right in this. The reasons dictating this choice are easy to perceive. By taking the first alternative we should be obliged to destroy from top to bottom a very vast theoretical system which represents in a most satisfactory manner a very extensive and complex set of experimental laws. The second alternative, on the other hand, does not make us lose anything of the terrain already conquered by physical theory; in addition, it has succeeded in so large a number of cases that we can bank with interest on a new success. But in this confidence accorded the law of fall of weights, we see nothing analogous to the certainty that a mathematical definition draws from its very essence, that is, to the kind of certainty we have when it would be foolish to doubt that the various points on a circumference are all equidistant from the center.

We have here nothing more than a particular application of the principle set down in Section 2 of this chapter. A disagreement between the concrete facts constituting an experiment and the symbolic representation which theory substitutes for this experiment proves that some part of this symbol is to be rejected. But which part? This the experiment does not tell us; it leaves to our sagacity the burden of guessing. Now among the theoretical elements entering into the composition of this symbol there is always a certain number which the physicists of a certain epoch agree in accepting without test and which they regard as beyond dispute. Hence, the physicist who wishes to modify this symbol will surely bring his modification to bear on elements other than those just mentioned.

But what impels the physicist to act thus is not logical necessity. It would be awkward and ill inspired for him to do otherwise, but it would not be doing something logically absurd; he would not for all that be walking in the footsteps of the mathematician mad enough to contradict his own definitions. More than this, perhaps some day by acting differently, by refusing to invoke causes of error and take recourse to corrections in order to reestablish agreement between the theoretical scheme and the fact, and by resolutely carrying out a reform among the propositions declared untouchable by common consent, he will accomplish the work of a genius who opens a new career for a theory.

Indeed, we must really guard ourselves against believing forever warranted those hypotheses which have become universally adopted conventions, and whose certainty seems to break through experimental contradiction by throwing the latter back on more doubtful

assumptions. The history of physics shows us that very often the human mind has been led to overthrow such principles completely, though they have been regarded by common consent for centuries as inviolable axioms, and to rebuild its physical theories on new hypotheses.

Was there, for instance, a clearer or more certain principle for thousands of years than this one: In a homogeneous medium, light is propagated in a straight line? Not only did this hypothesis carry all former optics, catoptrics, and dioptrics, whose elegant geometric deductions represented at will an enormous number of facts, but it had become, so to speak, the physical definition of a straight line. It is to this hypothesis that any man wishing to make a straight line appeals, the carpenter who verifies the straightness of a piece of wood, the surveyor who lines up his sights, the geodetic surveyor who obtains a direction with the help of the pinholes of his alidade, the astronomer who defines the position of stars by the optical axis of his telescope. However, the day came when physicists tired of attributing to some cause of error the diffraction effects observed by Grimaldi, when they resolved to reject the law of the rectilinear propagation of light and to give optics entirely new foundations; and this bold resolution was the signal of remarkable progress for physical theory.

4.9 On hypotheses whose statement has no experimental meaning

This example, as well as others we could add from the history of science, should show that it would be very imprudent for us to say concerning a hypothesis commonly accepted today: "We are certain that we shall never be led to abandon it because of a new experiment, no matter how precise it is." Yet M. Poincaré does not hesitate to enunciate it concerning the principles of mechanics.[30]

To the reasons already given to prove that these principles cannot be reached by experimental refutation, M. Poincaré adds one which seems even more convincing: Not only can these principles not be refuted by experiment because they are the universally accepted rules serving to discover in our theories the weak spots indicated by these refutations, but also, they cannot be refuted by experiment because *the operation which would claim to compare them with the facts would have no meaning.*

Let us explain that by an illustration.

The principle of inertia teaches us that a material point removed from the action of any other body moves in a straight line with uniform motion. Now, we can observe only relative motions; we cannot, therefore, give an experimental meaning to this principle unless we assume a certain point chosen or a certain geometric solid taken as a fixed reference point to which the motion of the material point is related. The fixation of this reference frame constitutes an integral part of the statement of the law, for if we omitted it, this statement would be devoid of meaning. There are as many different laws as there are distinct frames of reference. We shall be stating one law of inertia when we say that the motion of an isolated point assumed to be seen from the earth is rectilinear and uniform, and another when we repeat the same sentence in referring the motion to the sun, and still another if the frame of reference chosen is the totality of fixed stars. But then, one thing is indeed certain, namely, that whatever the motion of a material point is, when seen from a first frame of reference, we can always and in infinite ways choose a second frame of reference such that seen from the latter our material point appears to move in a straight line with uniform motion. We cannot, therefore, attempt an experimental verification of the principle of inertia; false when we refer the motions to one frame of reference, it will become true when selection is made of another term of comparison, and we shall always be free to choose the latter. If the law of inertia stated by taking the earth as a frame of reference is contradicted by an observation, we shall substitute for it the law of inertia whose statement refers the motion to the sun; if the latter in its turn is contraverted, we shall replace the sun in the statement of the law by the system of fixed stars, and so forth. It is impossible to stop this loophole.

The principle of the equality of action and reaction, analyzed at length by M. Poincaré,[31] provides room for analogous remarks. This principle may be stated thus: "The center of gravity of an isolated system can have only a uniform rectilinear motion."

This is the principle that we propose to verify by experiment. "Can we make this verification? For that it would be necessary for isolated systems to exist. Now, these systems do not exist; the only isolated system is the whole universe.

"But we can observe only relative motions; the absolute motion of the center of the universe will therefore be forever unknown. We shall never be able to know if it is rectilinear and uniform or, better still, the question has no meaning. Whatever facts we may observe, we shall hence always be free to assume our principle is true."

Thus many a principle of mechanics has a form such that it is absurd to ask one's self: "Is this principle in agreement with experiment or not?" This strange character is not peculiar to the principles of mechanics; it also marks certain fundamental hypotheses of our physical or chemical theories.[32]

For example, chemical theory rests entirely on the "law of multiple proportions"; here is the exact statement of this law:

Simple bodies A, B, and C may by uniting in various proportions form various compounds M, M′, ... The masses of the bodies A, B, and C combining to form the compound M are to one another as the three numbers a, b, and c. Then the masses of the elements A, B, and C combining to form the compound M′ will be to one another as the numbers xa, yb, and zc (x, y, and z being three whole numbers).

Is this law perhaps subject to experimental test? Chemical analysis will make us acquainted with the chemical composition of the body M′ not exactly but with a certain approximation. The uncertainty of the results obtained can be extremely small; it will never be strictly zero. Now, in whatever relations the elements A, B, and C are combined within the compound M′, we can always represent these relations, with as close an approximation as you please, by the mutual relations of three products xa, yb, and zc, where x, y, and z are whole numbers; in other words, whatever the results given by the chemical analysis of the compound M′, we are always sure to find three integers x, y, and z thanks to which the law of multiple proportions will be verified with a precision greater than that of the experiment. Therefore, no chemical analysis, no matter how refined, will ever be able to show the law of multiple proportions to be wrong.

In like manner, all crystallography rests entirely on the "law of rational indices" which is formulated in the following way:

A trihedral being formed by three faces of a crystal, a fourth face cuts the three edges of this trihedral at distances from the summit which are proportional to one another as three given numbers, the parameters of the crystal. Any other face whatsoever should cut these same edges at distances from the summit which are to one another as xa, yb, and zc, where x, y, and z are three integers, the indices of the new face of the crystal.

The most perfect protractor determines the direction of a crystal's face only with a certain degree of approximation; the relations among the three segments that such a face makes on the edges of the fundamental trihedral are always able to get by with a certain error; now, however small this error is, we can always choose three numbers x, y, and z such that the mutual relations of these segments are represented with the least amount of error by the mutual relations of the three numbers xa, yb, and zc; the crystallographer who would claim that the law of rational indices is made justifiable by his protractor would surely not have understood the very meaning of the words he is employing.

The law of multiple proportions and the law of rational indices are mathematical statements deprived of all physical meaning. A mathematical statement has physical meaning only if it retains a meaning when we introduce the word "nearly" or "approximately." This is not the case with the statements we have just alluded to. Their object really is to assert that certain relations are *commensurable* numbers. They would degenerate into mere truisms if they were made to declare that these relations are approximately commensurable, for any incommensurable relation whatever is always approximately commensurable; it is even as near as you please to being commensurable.

Therefore, it would be absurd to wish to subject certain principles of mechanics to *direct* experimental test; it would be absurd to subject the law of multiple proportions or the law of rational indices to this *direct* test.

Does it follow that these hypotheses placed beyond the reach of direct experimental refutation have nothing more to fear from experiment? That they are guaranteed to remain immutable no matter what discoveries observation has in store for us? To pretend so would be a serious error.

Taken in isolation these different hypotheses have no experimental meaning; there can be no question of either confirming or contradicting them by experiment. But these hypotheses enter as essential foundations into the construction of certain theories of rational mechanics, of chemical theory, of crystallography. The object of these theories is to represent experimental laws; they are schematisms intended essentially to be compared with facts.

Now this comparison might some day very well show us that one of our representations is ill adjusted to the realities it should picture, that the corrections which come and complicate our schematism do not produce sufficient concordance between this schematism and the facts, that the theory accepted for a long time without dispute should be rejected, and that an entirely different theory should be constructed on entirely different or new hypotheses. On that day some one of our hypotheses, which taken in isolation defied direct experimental refutation, will crumble with the system it supported under the weight of the contradictions inflicted by reality on the consequences of this system taken as a whole.[33]

In truth, hypotheses which by themselves have no physical meaning undergo experimental testing in exactly the same manner as other hypotheses. Whatever the nature of the hypothesis is, we have seen at the beginning of this chapter that it is never in isolation contradicted by experiment; experimental contradiction always bears as a whole on the entire group constituting a theory without any possibility of designating which proposition in this group should be rejected.

There thus disappears what might have seemed paradoxical in the following assertion: Certain physical theories rest on hypotheses which do not by themselves have any physical meaning.

4.10 Good sense is the judge of hypotheses which ought to be abandoned

When certain consequences of a theory are struck by experimental contradiction, we learn that this theory should be modified but we are not told by the experiment what must be changed. It leaves to the physicist the task of finding out the weak spot that impairs the whole system. No absolute principle directs this inquiry, which different physicists may conduct in very different ways without having the right to accuse one another of illogicality. For instance, one may be obliged to safeguard certain fundamental hypotheses while he tries to reestablish harmony between the consequences of the theory and the facts by complicating the schematism in which these hypotheses are applied, by invoking various causes of error, and by multiplying corrections. The next physicist, disdainful of these complicated artificial procedures, may decide to change some one [sic] of the essential assumptions supporting the entire system. The first physicist does not have the right to condemn in advance the boldness of the second one, nor does the latter have the right to treat the timidity of the first physicist as absurd. The methods they follow are justifiable only by experiment, and if they both succeed in satisfying the requirements of experiment each is logically permitted to declare himself content with the work that he has accomplished.

That does not mean that we cannot very properly prefer the work of one of the two to that of the other. Pure logic is not the only rule for our judgments; certain opinions which do not fall under the hammer of the principle of contradiction are in any case perfectly unreasonable. These motives which do not proceed from logic and yet direct our choices, these "reasons which reason does not know" and which speak to the ample "mind of finesse" but not to the "geometric mind," constitute what is appropriately called good sense.

Now, it may be good sense that permits us to decide between two physicists. It may be that we do not approve of the haste with which the second one upsets the principles of a vast and harmoniously constructed theory whereas a modification of detail, a slight correction, would have sufficed to put these theories in accord with the facts. On the other hand, it may be that we may find it childish and unreasonable for the first physicist to maintain obstinately at any cost, at the price of continual repairs and many tangled-up stays, the worm-eaten columns of a building tottering in every part, when by razing these columns it would be possible to construct a simple, elegant, and solid system.

But these reasons of good sense do not impose themselves with the same implacable rigor that the prescriptions of logic do. There is something vague and uncertain about them; they do not reveal themselves at the same time with the same degree of clarity to all minds. Hence, the possibility of lengthy quarrels between the adherents of an old system and the partisans of a new doctrine, each camp claiming to have good sense on its side, each party finding the reasons of the adversary inadequate. The history of physics would furnish us with innumerable illustrations of these quarrels at all times and in all domains. Let us confine ourselves to the tenacity and ingenuity with which Biot by a continual bestowal of corrections and accessory hypotheses maintained the emissionist doctrine in optics, while Fresnel opposed this doctrine constantly with new experiments favoring the wave theory.

In any event this state of indecision does not last forever. The day arrives when good sense comes out so clearly in favor of one of the two sides that the other side gives up the struggle even though pure logic would not forbid its continuation. After Foucault's experiment had shown that light traveled faster in air than in water, Biot gave up supporting the emission hypothesis; strictly, pure logic would not have compelled him to give it up, for Foucault's experiment was not the crucial experiment that Arago thought he saw in it, but by resisting wave optics for a longer time Biot would have been lacking in good sense.

Since logic does not determine with strict precision the time when an inadequate hypothesis should give way to a more fruitful assumption, and since recognizing this moment belongs to good sense, physicists may hasten this judgment and increase the rapidity of scientific progress by trying consciously to make good sense within themselves more lucid and more vigilant. Now nothing contributes more to entangle good sense and to disturb its insight than passions and interests. Therefore, nothing will delay the decision which should determine a fortunate reform in a physical theory more than the vanity which makes a physicist too indulgent towards his own system and too severe towards the system of another. We are thus led to the conclusion so clearly expressed by Claude Bernard: The sound experimental criticism of a hypothesis is subordinated to certain moral conditions; in order to estimate correctly the agreement of a physical theory with the facts, it is not enough to be a good mathematician and skillful experimenter; one must also be an impartial and faithful judge.

Notes

1   Nicolaus Cabeus, S. J., *Philosophia magnetica, in qua magnetis natura penitus explicatur et omnium quae hoc lapide cernuntur causae propriae afferuntur, multa quoque dicuntur de electricis et aliis attractionibus, et eorum causis* (Cologne: Joannem Kinckium, 1629).

2   P. Rogerio Josepho Boscovich, S. J., *Theoria philosophiae naturalis redacta ad unicam legem virium in natura existentium* (Vienna, 1758).

3   G.W. Leibniz, *Oeuvres*, ed. Gerhardt, IV, 464. (Translator's note: See Leibniz, Selections [Charles Scribner's Sons, 1951], pp. 100ff.).

4   Christian Huygens to G.W. Leibniz, Nov. 18, 1690, *Oeuvres complètes de Huygens, Correspondance*, 10 vols. (The Hague, 1638–1695), ix, 52. (Translator's note: The complete edition of Huygens' Collected

Works was published in twenty-two volumes by the Holland Society of Sciences [Haarlem, 1950].)

5 Aristarchi Samii "De mundi systemate, partibus et motibus ejusdem, liber singularis" (Paris, 1643). This work was reproduced in 1647 in Volume iii of the Cogitata physico-mathematica by Marin Mersenne. [See pp. 242–243, below. [sic]]

6 R. Descartes, Correspondance, ed. P. Tannery and C. Adam, Vol. iv (Paris, 1893), Letter clxxx, p. 396.

7 Denis Papin to Christian Huygens, June 18, 1690, Oeuvres complètes de Huygens . . ., ix, 429.

8 Christian Huygens to Denis Papin, Sept. 2, 1690, ibid., ix, 484.

9 E. Mach, "Die ökonomische Natur der physika-lischen Forschung," Populär-wissenschaftliche Vorlesungen (3rd ed.; Leipzig, 1903), Ch. xiii, p. 215. (Translator's note: Translated by T.J. McCormack, "The Economical Nature of Physical Research," Mach's Popular Scientific Lectures [3rd ed.; La Salle, Ill.: Open Court, 1907], Ch. xiii.)
   See also E. Mach, La Mécanique; exposé historique et critique de son développement (Paris, 1904), Ch. iv, Sec. 4: "La Science comme économie de la pensée," p. 449. (Translator's note: Translated from the German 2nd ed. by T.J. McCormack, The Science of Mechanics: a Critical and Historical Account of Its Development [Open Court, 1902], Ch. iv, Sec. iv: "The Economy of Science," pp. 481–494.)

10 E. Mach, La Mécanique . . ., p. 453. (Translator's note: Translated in The Science of Mechanics . . ., p. 485.)

11 We have already noted natural classification as the ideal form toward which physical theory tends in "L'Ecole anglaise et les théories physiques," Art. 6, Revue des questions scientifiques, October 1893.

12 Cf. II. Poincaré, La Science et l'Hypothèse (Paris, 1903), p. 190. (Translator's note: Translated by Bruce Halsted, "Science and Hypothesis" in Foundations of Science [Lancaster, Pa.: Science Press, 1905].)

13 Oeuvres complètes d'Augustin Fresnel, 3 vols. (Paris, 1866–1870), i, 236, 365, 368.

14 Claude Bernard, Introduction à la Médecine expérimentale (Paris, 1865), p. 63. (Translator's note: Translated into English by H.C. Greene, An Introduction to Experimental Medicine [New York: Henry Schuman, 1949].)

15 Claude Bernard, Introduction à la Médecine expérimentale (Paris, 1865), p. 64.

16 Ibid., p. 70.

17 Ibid., p. 67.

18 Wilhelm Weber, Electrodynamische Maassbestimmungen (Leipzig, 1846). Translated into French in Collection de Mémoires relatifs à la Physique (Société française de Physique), Vol. iii: Mémoires sur l'Electrodynamique.

19 H. Poincaré, Science et Hypothèse, p. 179.

20 G. Robin, Oeuvres scientifiques. Thermodynamique générale (Paris, 1901), Introduction, p. xii.

21 Note on a lecture of M. Joubert, inspector-general of secondary-school instruction, L'Enseignement secondaire, April 15, 1903.

22 G. Robin, op. cit., p. ii.

23 J. Bertrand, Leçons sur la Théorie mathématique de l'Electricité (Paris, 1890), p. 71.

24 It will be objected undoubtedly that such teaching of physics would be hardly accessible to young minds; the answer is simple: Do not teach physics to minds not yet ready to assimilate it. Mme. de Sévigné used to say, speaking of young children: "Before you give them the food of a truckdriver, find out if they have the stomach of a truckdriver."

25 G. Robin, op. cit., p. xiv.

26 Op. cit.

27 G. Milhaud, "La Science rationnelle," Revue de Métaphysique et de Morale, iv (1896), 280. Reprinted in Le Rationnel (Paris, 1898), p. 45.

28 H. Poincaré, "Sur les Principes de la Mécanique," Bibliothèque du Congrès International de Philosophie iii: Logique et Histoire des Sciences (Paris, 1901), p. 457; "Sur la valeur objective des théories physiques," Revue de Métaphysique et de Morale, x (1902), 263; La Science et l'Hypothèse, p. 110.

29 E. Le Roy, "Un positivisme nouveau," Revue de Métaphysique et de Morale, ix (1901), 143–144.

30 H. Poincaré, "Sur les Principes de la Mécanique," Bibliothèque du Congrès international de Philosophie, Sec. iii: "Logique et Histoire des Sciences" (Paris, 1901), pp. 475, 491.

31 Ibid., pp. 472ff.

32 P. Duhem, Le Mixte et la Combinaison chimique: Essai sur l'évolution d'une idée (Paris, 1902), pp. 159–161.

33 At the International Congress of Philosophy held in Paris in 1900, M. Poincaré developed this

conclusion: "Thus is explained how experiment may have been able to edify (or suggest) the principles of mechanics, but will never be able to overthrow them." Against this conclusion, M. Hadamard offered various remarks, among them the following:

"Moreover, in conformity with a remark of M. Duhem, it is not an isolated hypothesis but the whole group of the hypotheses of mechanics that we can try to verify experimentally." *Revue de Métaphysique et de Morale*, viii (1900), 559.

Chapter 35

Henri Poincaré

THE THEORIES OF MODERN PHYSICS

Significance of physical theories

The ephemeral nature of scientific theories takes by surprise the man of the world. Their brief period of prosperity ended, he sees them abandoned one after another, he sees ruins piled upon ruins; he predicts that the theories in fashion today will in a short time succumb in their turn, and he concludes that they are absolutely in vain. This is what he calls the *bankruptcy of science*.

His skepticism is superficial; he does not take into account the object of scientific theories and the part they play, or he would understand that the ruins may be still good for something. No theory seemed established on firmer ground than Fresnel's, which attributed light to the movements of the ether. Then if Maxwell's theory is today preferred, does that mean that Fresnel's work was in vain? No; for Fresnel's object was not to know whether there really is an ether, if it is or is not formed of atoms, if these atoms really move in this way or that; his object was to predict optical phenomena.

This Fresnel's theory enables us to do today as well as it did before Maxwell's time. The differential equations are always true, they may be always integrated by the same methods and the results of this integration still preserve their value. It cannot be said that this is reducing physical theories to simple practical recipes; these equations express relations, and if the equations remain true, it is because the relations preserve their reality. They teach us now, as they did then, that there is such and such a relation between this thing and that; only, the something which we then called *motion*, we now call *electric current*. But these are merely names of the images we substituted for the real objects which Nature will hide forever from our eyes. The true relations between these real objects are the only reality we can attain, and the sole condition is that the same relations shall exist between these objects as between the images we are forced to put in their place. If the relations are known to us, what does it matter if we think it convenient to replace one image by another?

That a given periodic phenomenon (an electric oscillation, for instance) is really due to the vibration of a given atom, which, behaving like a pendulum, is really displaced in this manner or that, all this is neither certain nor essential. But that there is between the electric oscillation, the movement of the pendulum, and all periodic phenomena an intimate relationship which corresponds to a profound reality; that this relationship, this similarity, or rather this parallelism, is continued in the details; that it is a consequence of more general principles such as that of the conservation of energy, and that of least action; this we may affirm; this is the truth which will ever remain the same in whatever garb we may see fit to clothe it.

Many theories of dispersion have been proposed. The first were imperfect, and contained but little truth. Then came that of Helmholtz, and this in its turn was modified in different ways; its author himself conceived another theory, founded on Maxwell's principles. But the remarkable thing is, that all the scientists who followed Helmholtz obtain the same equations, although their starting-points were to all appearance widely separated. I venture to say that these theories are all simultaneously true; not merely because they express a true relation – that between absorption and abnormal dispersion. In the premisses of these theories the part that is true is the part common to all: it is the affirmation of this or that relation between certain things, which some call by one name and some by another.

The kinetic theory of gases has given rise to many objections, to which it would be difficult to find an answer were it claimed that the theory is absolutely true. But all these objections do not alter the fact that it has been useful, particularly in revealing to us one true relation which would otherwise have remained profoundly hidden – the relation between gaseous and osmotic pressures. In this sense, then, it may be said to be true.

When a physicist finds a contradiction between two theories which are equally dear to him, he sometimes says: 'Let us not be troubled, but let us hold fast to the two ends of the chain, lest we lose the intermediate links.' This argument of the embarrassed theologian would be ridiculous if we were to attribute to physical theories the interpretation given them by the man of the world. In case of contradiction one of them at least should be considered false. But this is no longer the case if we only seek in them what should be sought. It is quite possible that

they both express true relations, and that the contradictions only exist in the images we have formed to ourselves of reality. To those who feel that we are going too far in our limitations of the domain accessible to the scientist, I reply: these questions which we forbid you to investigate, and which you so regret, are not only insoluble, they are illusory and devoid of meaning.

Such a philosopher claims that all physics can be explained by the mutual impact of atoms. If he simply means that the same relations obtain between physical phenomena as between the mutual impact of a large number of billiard balls – well and good! this is verifiable, and perhaps is true. But he means something more, and we think we understand him, because we think we know what an impact is. Why? Simply because we have often watched a game of billiards. Are we to understand that God experiences the same sensations in the contemplation of His work that we do in watching a game of billiards? If it is not our intention to give his assertion this fantastic meaning, and if we do not wish to give it the more restricted meaning I have already mentioned, which is the sound meaning, then it has no meaning at all. Hypotheses of this kind have therefore only a metaphorical sense. The scientist should no more banish them than a poet banishes metaphor; but he ought to know what they are worth. They may be useful to give satisfaction to the mind, and they will do no harm as long as they are only indifferent hypotheses.

These considerations explain to us why certain theories, that were thought to be abandoned and definitively condemned by experiment, are suddenly revived from their ashes and begin a new life. It is because they expressed true relations, and had not ceased to do so when for some reason or other we felt it necessary to enunciate the same relations in another language. Their life had been latent, as it were.

Barely fifteen years ago, was there anything more ridiculous, more quaintly old-fashioned, than the fluids of Coulomb? And yet, here they are re-appearing under the name of *electrons*. In what do these permanently electrified molecules differ from the electric molecules of Coulomb? It is true that in the electrons the electricity is supported by a little, a very little matter; in other words, they have mass. Yet Coulomb did not deny mass to his fluids, or if he did, it was with reluctance. It would be rash to affirm that the belief in electrons will not also undergo an eclipse, but it was nonetheless curious to note this unexpected renaissance.

But the most striking example is Carnot's principle. Carnot established it, starting from false hypotheses. When it was found that heat was indestructible, and may be converted into work, his ideas were completely abandoned; later, Clausius returned to them, and to him is due their definitive triumph. In its primitive form, Carnot's theory expressed in addition to true relations, other inexact relations, the *débris* of old ideas; but the presence of the latter did not alter the reality of the others. Clausius had only to separate them, just as one lops off dead branches.

The result was the second fundamental law of thermodynamics. The relations were always the same, although they did not hold, at least to all appearance, between the same objects. This was sufficient for the principle to retain its value. Nor have the reasonings of Carnot perished on this account; they were applied to an imperfect conception of matter, but their form — i.e., the essential part of them, remained correct. What I have just said throws some light at the same time on the role of general principles, such as those of the principle of least action or of the conservation of energy. These principles are of very great value. They were obtained in the search for what there was in common in the enunciation of numerous physical laws; they thus represent the quintessence of innumerable observations. However, from their very generality results a consequence to which I have called attention in Chapter VIII [sic] — namely, that they are no longer capable of verification. As we cannot give a general definition of energy, the principle of the conservation of energy simply signifies that there is a *something* which remains constant. Whatever fresh notions of the world may be given us by future experiments, we are certain beforehand that there is something which remains constant, and which may be called *energy*. Does this mean that the principle has no meaning and vanishes into a tautology? Not at all. It means that the different things to which we give the name of *energy* are connected by a true relationship; it affirms between them a real relation. But then, if this principle has a meaning, it may be false; it may be that we have no right to extend indefinitely its applications, and yet it is certain beforehand to be verified in the strict sense of the word. How, then, shall we know when it has been extended as far as is legitimate? Simply when it ceases to be useful to us — i.e., when we can no longer use it to predict correctly new phenomena. We shall be certain in such a case that the relation affirmed is no longer real, for otherwise it would be fruitful; experiment without directly contradicting a new extension of the principle will nevertheless have condemned it.

Physics and mechanism

Most theorists have a constant predilection for explanations borrowed from physics, mechanics, or dynamics. Some would be satisfied if they could account for all phenomena by the motion of molecules attracting one another according to certain laws. Others are more exact; they would suppress attractions acting at a distance; their molecules would follow rectilinear paths, from which they would only be deviated by impacts.

Others again, such as Hertz, suppress the forces as well, but suppose their molecules subjected to geometrical connections analogous, for instance, to those of articulated systems; thus, they wish to reduce dynamics to a kind of kinematics. In a word, they all wish to bend nature into a certain form, and unless they can do this they cannot be satisfied. Is Nature flexible enough for this?

We shall examine this question in Chapter XII [sic], *àpropos* of Maxwell's theory. Every time that the principles of least action and energy are satisfied, we shall see that not only is there always a mechanical explanation possible, but that there is an unlimited number of such explanations. By means of a well-known theorem due to Königs, it may be shown that we can explain everything in an unlimited number of ways, by connections after the manner of Hertz, or, again, by central forces. No doubt it may be just as easily demonstrated that everything may be explained by simple impacts. For this, let us bear in mind that it is not enough to be content with the ordinary matter of which we are aware by means of our senses, and the movements of which we observe directly. We may conceive of ordinary matter as either composed of atoms, whose internal movements escape us, our senses being able to estimate only the displacement of the whole; or we may imagine one of those subtle fluids, which under the name of *ether* or other names, have from all time played so important a role in physical theories. Often we go further, and regard the ether as the only primitive, or even as the only true matter. The more moderate consider ordinary matter to be condensed ether, and there is nothing startling in this conception; but others only reduce its importance still further, and see in matter nothing more than the geometrical locus of singularities in the ether. Lord Kelvin, for instance, holds what we call matter to be only the locus of those points at which the ether is animated by vortex motions. Riemann believes it to be locus of those points at which ether is

constantly destroyed; to Wiechert or Larmor, it is the locus of the points at which the ether has undergone a kind of torsion of a very particular kind. Taking any one of these points of view, I ask by what right do we apply to the ether the mechanical properties observed in ordinary matter, which is but false matter? The ancient fluids, caloric, electricity, etc., were abandoned when it was seen that heat is not indestructible. But they were also laid aside for another reason. In materialising them, their individuality was, so to speak, emphasised – gaps were opened between them; and these gaps had to be filled in when the sentiment of the unity of Nature became stronger, and when the intimate relations which connect all the parts were perceived. In multiplying the fluids, not only did the ancient physicists create unnecessary entities, but they destroyed real ties. It is not enough for a theory not to affirm false relations; it must not conceal true relations.

Does our ether actually exist? We know the origin of our belief in the ether. If light takes several years to reach us from a distant star, it is no longer on the star, nor is it on the earth. It must be somewhere, and supported, so to speak, by some material agency.

The same idea may be expressed in a more mathematical and more abstract form. What we note are the changes undergone by the material molecules. We see, for instance, that the photographic plate experiences the consequences of a phenomenon of which the incandescent mass of a star was the scene several years before. Now, in ordinary mechanics, the state of the system under consideration depends only on its state at the moment immediately preceding; the system therefore satisfies certain differential equations. On the other hand, if we did not believe in the ether, the state of the material universe would depend not only on the state immediately preceding, but also on much older states; the system would satisfy equations of finite differences. The ether was invented to escape this breaking down of the laws of general mechanics.

Still, this would only compel us to fill the interplanetary space with ether, but not to make it penetrate into the midst of the material media. Fizeau's experiment goes further. By the interference of rays which have passed through the air or water in motion, it seems to show us two different media penetrating each other, and yet being displaced with respect to each other. The ether is all but in our grasp. Experiments can be conceived in which we come closer still to it. Assume that Newton's principle of the equality of action and re-action is not true if applied to matter *alone*, and that this can be proved. The geometrical sum of all the forces applied to all the molecules would no longer be zero. If we did not wish to change the whole of the science of mechanics, we should have to introduce the ether, in order that the action which matter apparently undergoes should be counterbalanced by the re-action of matter on something.

Or again, suppose we discover that optical and electrical phenomena are influenced by the motion of the earth. It would follow that those phenomena might reveal to us not only the relative motion of material bodies, but also what would seem to be their absolute motion. Again, it would be necessary to have an ether in order that these so-called absolute movements should not be their displacements with respect to empty space, but with respect to something concrete.

Will this ever be accomplished? I do not think so, and I shall explain why; and yet, it is not absurd, for others have entertained this view. For instance, if the theory of Lorentz, of which I shall speak in more detail in Chapter XIII [sic], were true, Newton's principle would not apply to matter *alone*, and the difference would not be very far from being within reach of experiment. On the other hand, many experiments have been made on the influence of the motion of the earth. The results have always been negative. But if these experiments have been undertaken, it is because we have not been certain beforehand; and indeed, according to current theories, the compensation would be only approximate, and we might expect to find accurate methods giving positive results. I think that such a hope is illusory; it was nonetheless interesting to show that a success of this kind would, in a certain sense, open to us a new world.

And now allow me to make a digression; I must explain why I do not believe, in spite of Lorentz, that more exact observations will ever make evident anything else but the relative displacements of material bodies. Experiments have been made that should have disclosed the terms of the first order; the results were nugatory. Could that have been by chance? No one has admitted this; a general explanation was sought, and Lorentz found it. He showed that the terms of the first order should cancel each other, but not the terms of the second order. Then more exact experiments were made, which were also negative; neither could this be the result of chance. An explanation was necessary, and was forthcoming; they always are; hypotheses are what we lack the least. But this is not enough. Who is there who does not think that this leaves to chance far too important a role? Would it not also be a chance that this singular concurrence should cause a certain circumstance to destroy the terms of the first order, and that a totally different but very opportune circumstance should cause those of the second order to vanish? No; the same explanation must be found for the two cases, and everything tends to show that this explanation would serve equally well for the terms of the higher order, and that the mutual destruction of these terms will be rigorous and absolute.

The present state of physics

Two opposite tendencies may be distinguished in the history of the development of physics. On the one hand, new relations are continually being discovered between objects which seemed destined to remain forever unconnected; scattered facts cease to be strangers to each other and tend to be marshalled into an imposing synthesis. The march of science is towards unity and simplicity.

On the other hand, new phenomena are continually being revealed; it will be long before they can be assigned their place – sometimes it may happen that to find them a place a corner of the edifice must be demolished. In the same way, we are continually perceiving details ever more varied in the phenomena we know, where our crude senses used to be unable to detect any lack of unity. What we thought to be simple becomes complex, and the march of science seems to be towards diversity and complication.

Here, then, are two opposing tendencies, each of which seems to triumph in turn. Which will win? If the first wins, science is possible; but nothing proves this *à priori*, and it may be that after unsuccessful efforts to bend Nature to our ideal of unity in spite of herself, we shall be submerged by the ever-rising flood of our new riches and compelled to renounce all idea of classification – to abandon our ideal, and to reduce science to the mere recording of innumerable recipes.

In fact, we can give this question no answer. All that we can do is to observe the science of today, and compare it with that of yesterday. No doubt after this examination we shall be in a position to offer a few conjectures.

Half a century ago hopes ran high indeed. The unity of force had just been revealed to us by the discovery of the conservation of energy and of its transformation. This discovery also showed that the phenomena of heat could be explained by molecular movements. Although the nature of these movements was not exactly known, no one doubted but that they would be ascertained before long. As for light, the work

seemed entirely completed. So far as electricity was concerned, there was not so great an advance. Electricity had just annexed magnetism. This was a considerable and a definitive step towards unity. But how was electricity in its turn to be brought into the general unity, and how was it to be included in the general universal mechanism? No one had the slightest idea. As to the possibility of the inclusion, all were agreed; they had faith. Finally, as far as the molecular properties of material bodies are concerned, the inclusion seemed easier, but the details were very hazy. In a word, hopes were vast and strong, but vague.

Today, what do we see? In the first place, a step in advance – immense progress. The relations between light and electricity are now known; the three domains of light, electricity, and magnetism, formerly separated, are now one; and this annexation seems definitive.

Nevertheless the conquest has caused us some sacrifices. Optical phenomena become particular cases in electric phenomena; as long as the former remained isolated, it was easy to explain them by movements which were thought to be known in all their details. That was easy enough; but any explanation to be accepted must now cover the whole domain of electricity. This cannot be done without difficulty.

The most satisfactory theory is that of Lorentz; it is unquestionably the theory that best explains the known facts, the one that throws into relief the greatest number of known relations, the one in which we find most traces of definitive construction. That it still possesses a serious fault I have shown above. It is in contradiction with Newton's law that action and re-action are equal and opposite – or rather, this principle according to Lorentz cannot be applicable to matter alone; if it be true, it must take into account the action of the ether on matter, and the re-action of the matter on the ether. Now, in the new order, it is very likely that things do not happen in this way.

However this may be, it is due to Lorentz that the results of Fizeau on the optics of moving bodies, the laws of normal and abnormal dispersion and of absorption are connected with each other and with the other properties of the ether, by bonds which no doubt will not be readily severed. Look at the ease with which the new Zeeman phenomenon found its place, and even aided the classification of Faraday's magnetic rotation, which had defied all Maxwell's efforts. This facility proves that Lorentz's theory is not a mere artificial combination which must eventually find its solvent. It will probably have to be modified, but not destroyed.

The only object of Lorentz was to include in a single whole all the optics and electro-dynamics of moving bodies; he did not claim to give a mechanical explanation. Larmor goes further; keeping the essential part of Lorentz's theory, he grafts upon it, so to speak, MacCullagh's ideas on the direction of the movement of the ether. MacCullagh held that the velocity of the ether is the same in magnitude and direction as the magnetic force. Ingenious as is this attempt, the fault in Lorentz's theory remains, and is even aggravated. According to Lorentz, we do not know what the movements of the ether are; and because we do not know this, we may suppose them to be movements compensating those of matter, and re-affirming that action and re-action are equal and opposite. According to Larmor we know the movements of the ether, and we can prove that the compensation does not take place.

If Larmor has failed, as in my opinion he has, does it necessarily follow that a mechanical explanation is impossible? Far from it. I said above that as long as a phenomenon obeys the two principles of energy and least action, so long it allows of an unlimited number of mechanical explanations. And so with the phenomena of optics and electricity.

But this is not enough. For a mechanical explanation to be good it must be simple; to choose it from among all the explanations that are possible there must be other reasons than the necessity of making a choice. Well, we have no theory as yet which will satisfy this condition and consequently be of any use. Are we then to complain? That would be to forget the end we seek, which is not the mechanism; the true and only aim is unity.

We ought therefore to set some limits to our ambition. Let us not seek to formulate a mechanical explanation; let us be content to show that we can always find one if we wish. In this we have succeeded. The principle of the conservation of energy has always been confirmed, and now it has a fellow in the principle of least action, stated in the form appropriate to physics. This has also been verified, at least as far as concerns the reversible phenomena which obey Lagrange's equations – in other words, which obey the most general laws of physics. The irreversible phenomena are much more difficult to bring into line; but they, too, are being co-ordinated and tend to come into the unity. The light which illuminates them comes from Carnot's principle. For a long time thermo-dynamics was confined to the study of the dilatations of bodies and of their change of state. For some time past it has been growing bolder, and has considerably extended its domain. We owe to it the theories of the voltaic cell and of their thermo-electric phenomena; there is not a corner in physics which it has not explored, and it has even attacked chemistry itself. The same laws hold good; everywhere, disguised in some form or other, we find Carnot's principle; everywhere also appears that eminently abstract concept of entropy which is as universal as the concept of energy, and like it, seems to conceal a reality. It seemed that radiant heat must escape, but recently that, too, has been brought under the same laws.

In this way fresh analogies are revealed which may be often pursued in detail; electric resistance resembles the viscosity of fluids; hysteresis would rather be like the friction of solids. In all cases friction appears to be the type most imitated by the most diverse irreversible phenomena, and this relationship is real and profound.

A strictly mechanical explanation of these phenomena has also been sought, but, owing to their nature, it is hardly likely that it will be found. To find it, it has been necessary to suppose that the irreversibility is but apparent, that the elementary phenomena are reversible and obey the known laws of dynamics. But the elements are extremely numerous, and become blended more and more, so that to our crude sight all appears to tend towards uniformity − i.e., all seems to progress in the same direction, and that without hope of return. The apparent irreversibility is therefore but an effect of the law of great numbers. Only a being of infinitely subtle senses, such as Maxwell's demon, could unravel this tangled skein and turn back the course of the universe.

This conception, which is connected with the kinetic theory of gases, has cost great effort and has not, on the whole, been fruitful; it may become so. This is not the place to examine if it leads to contradictions, and if it is in conformity with the true nature of things.

Let us notice, however, the original ideas of M. Gouy on the Brownian movement. According to this scientist, this singular movement does not obey Carnot's principle. The particles which it sets moving would be smaller than the meshes of that tightly drawn net; they would thus be ready to separate them, and thereby to set back the course of the universe. One can almost see Maxwell's demon at work.[1]

To resume, phenomena long known are gradually being better classified, but new phenomena come to claim their place, and most of them, like the Zeeman effect, find it at once. Then we have the cathode rays, the X-rays, uranium and radium rays; in fact, a whole world of which none had suspected the existence. How many unexpected guests to find a place for! No one can yet predict the place they will occupy, but I do not believe they will destroy the general unity; I think that they will rather complete it. On the one hand, indeed, the new radiations seem to be connected with the phenomena of luminosity; not only do they excite fluorescence, but they sometimes come into existence under the same conditions as that property; neither are they unrelated to the cause which produces the electric spark under the action of ultra-violet light. Finally, and most important of all, it is believed that in all these phenomena there exist ions, animated, it is true, with velocities far greater than those of electrolytes. All this is very vague, but it will all become clearer.

Phosphorescence and the action of light on the spark were regions rather isolated, and consequently somewhat neglected by investigators. It is to be hoped that a new path will now be made which will facilitate their communications with the rest of science. Not only do we discover new phenomena, but those we think we know are revealed in unlooked-for aspects. In the free ether the laws preserve their majestic simplicity, but matter properly so called seems more and more complex; all we can say of it is but approximate, and our formulæ are constantly requiring new terms.

But the ranks are unbroken, the relations that we have discovered between objects we thought simple still hold good between the same objects when their complexity is recognised, and that alone is the important thing. Our equations become, it is true, more and more complicated, so as to embrace more closely the complexity of nature; but nothing is changed in the relations which enable these equations to be derived from each other. In a word, the form of these equations persists. Take for instance the laws of reflection. Fresnel established them by a simple and attractive theory which experiment seemed to confirm. Subsequently, more accurate researches have shown that this verification was but approximate; traces of elliptic polarisation were detected everywhere. But it is owing to the first approximation that the cause of these anomalies was found in the existence of a transition layer, and all the essentials of Fresnel's theory have remained. We cannot help reflecting that all these relations would never have been noted if there had been doubt in the first place as to the complexity of the objects they connect. Long ago it was said: If Tycho had had instruments ten times as precise, we would never have had a Kepler, or a Newton, or Astronomy. It is a misfortune for a science to be born too late, when the means of observation have become too perfect. That is what is happening at this moment with respect to physical chemistry; the founders are hampered in their general grasp by third and fourth decimal places; happily they are men of robust faith. As we get to know the properties of matter better we see that continuity reigns. From the work of Andrews and Van der Waals, we see how the transition from the liquid to the gaseous state is made, and that it is not abrupt. Similarly, there is no gap between the liquid and solid states, and in the proceedings of a recent Congress we see memoirs on the rigidity of liquids side by side with papers on the flow of solids.

With this tendency there is no doubt a loss of simplicity. Such and such an effect was represented by straight lines; it is now necessary to connect these lines by more or less complicated curves. On the other hand, unity is gained. Separate categories quieted but did not satisfy the mind.

Finally, a new domain, that of chemistry, has been invaded by the method of physics, and we see the birth of physical chemistry. It is still quite young, but already it has enabled us to connect such phenomena as electrolysis, osmosis, and the movements of ions.

From this cursory exposition what can we conclude? Taking all things into account, we have approached the realisation of unity. This has not been done as quickly as was hoped fifty years ago, and the path predicted has not always been followed; but, on the whole, much ground has been gained.

Note

1 Clerk-Maxwell imagined some supernatural agency at work, sorting molecules in a gas of uniform temperature into (a) those possessing kinetic energy above the average, (b) those possessing kinetic energy below the average. – [Tr]

Chapter 36

Larry Laudan

A CONFUTATION OF CONVERGENT REALISM

The positive argument for realism is that it is the only philosophy that doesn't make the success of science a miracle.

H. Putnam (1975)

1 The problem

It is becoming increasingly common to suggest that epistemological realism is an empirical hypothesis, grounded in, and to be authenticated by its ability to explain the workings of science. A growing number of philosophers (including Boyd, Newton-Smith, Shimony, Putnam, Friedman and Niiniluoto) have argued that the theses of epistemic realism are open to empirical test. The suggestion that epistemological doctrines have much the same empirical status as the sciences is a welcome one: for, whether it stands up to detailed scrutiny or not, it marks a significant facing-up by the philosophical community to one of the most neglected (and most notorious) problems of philosophy: the status of epistemological claims.

But there are potential hazards as well as advantages associated with the 'scientizing' of epistemology. Specifically, once one concedes that epistemic doctrines are to be tested in the court of experience, it is possible that one's favorite epistemic theories may be refuted rather than confirmed. It is the thesis of this paper that precisely such a fate afflicts a form of realism advocated by those who have been in the vanguard of the move to show that realism is supported by an empirical study of the development of science. Specifically, I shall show that epistemic realism, at least in certain of its extant forms, is neither supported by, nor has it made sense of, much of the available historical evidence.

2 Convergent realism

Like other philosophical -isms, the term 'realism' covers a variety of sins. Many of these will not be at issue here. For instance, 'semantic realism' (in brief, the claim that all theories have truth values and that some theories – we know not which – are true) is not in dispute. Nor shall I discuss what one might call 'intentional realism' (i.e., the view that theories are generally intended by their proponents to assert the existence of entities corresponding to the terms in those theories). What I shall focus on instead are certain forms of *epistemological* realism. As Hilary Putnam has pointed out, although such realism has become increasingly fashionable, "very little is said about what realism *is*" (1978). The lack of specificity about what realism asserts makes it difficult to evaluate its claims, since many formulations are too vague and sketchy to get a grip on. At the same time, any efforts to formulate the realist position with greater precision lay the critic open to charges of attacking a straw man. In the course of this paper, I shall attribute several theses to the realists. Although there is probably no realist who subscribes to all of them, most of them have been defended by some self-avowed realist or other; taken together, they are perhaps closest to that version of realism advocated by Putnam, Boyd and Newton-Smith. Although I believe the views I shall be discussing can be legitimately attributed to certain contemporary philosophers (and will frequently cite the textual evidence for such attributions), it is not crucial to my case that such attributions can be made. Nor will I claim to do justice to the complex epistemologies of those whose work I will criticize. My aim, rather, is to explore certain epistemic claims which those who are realists might be tempted (and in some cases have been tempted) to embrace. If my arguments are sound, we will discover that some of the most intuitively tempting versions of realism prove to be chimeras.

The form of realism I shall discuss involves variants of the following claims:

R1) Scientific theories (at least in the 'mature' sciences) are typically approximately true and more recent theories are closer to the truth than older theories in the same domain;

R2) The observational and theoretical terms within the theories of a mature science genuinely refer (roughly, there are substances in the world that correspond to the ontologies presumed by our best theories);

R3) Successive theories in any mature science will be such that they 'preserve' the theoretical relations and the apparent referents of earlier theories (i.e., earlier theories will be 'limiting cases' of later theories).[1]

R4) Acceptable new theories do and should explain why their predecessors were successful insofar as they were successful.

To these semantic, methodological and epistemic theses is conjoined an important meta-philosophical claim about how realism is to be evaluated and assessed. Specifically, it is maintained that:

R5) Theses (R1)–(R4) entail that ('mature') scientific theories should be successful; indeed, these theses constitute the best, if not the only, explanation for the success of science. The empirical success of science (in the sense of giving detailed explanations and accurate predictions) accordingly provides striking empirical confirmation for realism.

I shall call the position delineated by (R1) to (R5) *convergent epistemological realism*, or CER for short. Many recent proponents of CER maintain that (R1), (R2), (R3), and (R4) are empirical hypotheses which, via the linkages postulated in (R5), can be tested by an investigation of science itself. They propose two elaborate abductive arguments. The structure of the first, which is germane to (R1) and (R2), is something like this:

1. If scientific theories are approximately true, they will typically be empirically successful;

2. If the central terms in scientific theories genuinely refer, those theories will generally be empirically successful;

3. Scientific theories are empirically successful.

---

4. (Probably) Theories are approximately true and their terms genuinely refer.

The argument relevant to (R3) is of slightly different form, specifically:

II 1.   If the earlier theories in a 'mature' science are approximately true and if the central terms of those theories genuinely refer, then later more successful theories in the same science will preserve the earlier theories as limiting cases;

2.   Scientists seek to preserve earlier theories as limiting cases and generally succeed.

3.   (Probably) Earlier theories in a 'mature' science are approximately true and genuinely referential.

Taking the success of present and past theories as givens, proponents of CER claim that if CER were true, it would follow that the success and the progressive success of science would be a matter of course. Equally, they allege that if CER were false, the success of science would be 'miraculous' and without explanation.[2] Because (on their view) CER explains the fact that science is successful, the theses of CER are thereby confirmed by the success of science and non-realist epistemologies are discredited by the latter's alleged inability to explain both the success of current theories and the progress which science historically exhibits.

As Putnam and certain others (e.g., Newton-Smith) see it, the fact that statements about reference (R2, R3) or about approximate truth (R1, R3) function in the explanation of a contingent state of affairs, establishes that "the notions of 'truth' and 'reference' have a causal explanatory role in epistemology" (Putnam 1978, p. 21).[3] In one fell swoop, both epistemology and semantics are 'naturalized' and, to top it all off, we get an explanation of the success of science into the bargain!

The central question before us is whether the realist's assertions about the interrelations between truth, reference and success are sound. It will be the burden of this paper to raise doubts about both I and II. Specifically, I shall argue that four of the five premises of those abductions are either false or too ambiguous to be acceptable. I shall also seek to show that, even if the premises were true, they would not warrant the conclusions which realists draw from them. Sections 3 through 5 of this essay deal with the first abductive argument; section 6 deals with the second.

3   Reference and success

The specifically referential side of the 'empirical' argument for realism has been developed chiefly by Putnam, who talks explicitly of reference rather more than most realists. On the other hand, reference is usually implicitly smuggled in, since most realists subscribe to the (ultimately referential) thesis that "the world probably contains entities very like those postulated by our most successful theories."

If R2 is to fulfill Putnam's ambition that reference can explain the success of science, and that the success of science establishes the presumptive truth of R2, it seems he must subscribe to claims similar to these:

S1)   The theories in the advanced or mature sciences are successful;

S2)   A theory whose central terms genuinely refer will be a successful theory;

S3)   If a theory is successful, we can reasonably infer that its central terms genuinely refer;

S4)   All the central terms in theories in the mature sciences do refer.

There are complex interconnections here. (S2) and (S4) explain (S1), while (S1) and (S3) provide the warrant for (S4). Reference explains success and success warrants a presumption of reference. The arguments are plausible, given the premises. But there is the rub, for with the possible exception of (S1), none of the premises is acceptable.

The first and toughest nut to crack involves getting clearer about the nature of that 'success' which realists are concerned to explain. Although Putnam, Sellars and Boyd all take the success of certain sciences as a given, they say little about what this success amounts to. So far as I can see, they are working with a largely *pragmatic* notion to be cashed out in terms of a theory's workability or applicability. On this account, we would say that a theory is success-ful if it makes substantially correct predictions, if it leads to efficacious interventions in the natural order, if it passes a battery of standard tests. One would like to be able to be more specific about what success amounts to, but the lack of a coherent theory of confirmation makes further specificity very difficult.

Moreover, the realist must be wary – at least for these purposes – of adopting too strict a notion of success, for a highly robust and stringent construal of 'success' would defeat the realist's purposes. What he wants to explain, after all, is why science in general has worked so well. If he were to adopt a very demanding characterization of success (such as those advocated by inductive logicians or Popperians) then it would probably turn out that science has been largely 'unsuccessful' (because it does not have high confirmation) and the realist's avowed explanandum would thus be a non-problem. Accordingly, I shall assume that a theory is 'successful' so long as it has worked well, i.e., so long as it has functioned in a variety of explanatory contexts, has led to confirmed predictions and has been of broad explanatory scope. As I understand the realist's position, his concern is

to explain why certain theories have enjoyed this kind of success.

If we construe 'success' in this way, (S1) can be conceded. Whether one's criterion of success is broad explanatory scope, possession of a large number of confirming instances, or conferring manipulative or predictive control, it is clear that science is, by and large, a successful activity.

What about (S2)? I am not certain that any realist would or should endorse it, although it is a perfectly natural construal of the realist's claim that 'reference explains success'. The notion of reference that is involved here is highly complex and unsatisfactory in significant respects. Without endorsing it, I shall use it frequently in the ensuing discussion. The realist sense of reference is a rather liberal one, according to which the terms in a theory may be genuinely referring even if many of the claims the theory makes about the entities to which it refers are false. Provided that there are entities which "approximately fit" a theory's description of them, Putnam's charitable account of reference allows us to say that the terms of a theory genuinely refer.[4] On this account (and these are Putnam's examples), Bohr's 'electron', Newton's 'mass', Mendel's 'gene', and Dalton's 'atom' are all referring terms, while 'phlogiston' and 'aether' are not (Putnam 1978, pp. 20–22).

Are genuinely referential theories (i.e., theories whose central terms genuinely refer) invariably or even generally successful at the empirical level, as (S2) states? There is ample evidence that they are not. The chemical atomic theory in the eighteenth century was so remarkably unsuccessful that most chemists abandoned it in favor of a more phenomenological, elective affinity chemistry. The Proutian theory that the atoms of heavy elements are composed of hydrogen atoms had, through most of the nineteenth century, a strikingly unsuccessful career, confronted by a long string of apparent refutations. The Wegenerian theory that the continents are carried by large subterranean objects moving laterally across the earth's surface was, for some thirty years in the recent history of geology, a strikingly unsuccessful theory until, after major modifications, it became the geological orthodoxy of the 1960s and 1970s. Yet all of these theories postulated basic entities which (according to Putnam's 'principle of charity') genuinely exist.

The realist's claim that we should expect referring theories to be empirically successful is simply false. And, with a little reflection, we can see good reasons why it should be. To have a genuinely referring theory is to have a theory which "cuts the world at its joints", a theory which postulates entities of a kind that really exist. But a genuinely referring theory need not be such that all – or even most – of the specific claims it makes about the properties of those entities and their modes of interaction are true. Thus, Dalton's theory makes many claims about atoms which are false; Bohr's early theory of the electron was similarly flawed in important respects. Contra-(S2), genuinely referential theories need not be strikingly successful, since such theories may be 'massively false' (i.e., have far greater falsity content than truth content).

(S2) is so patently false that it is difficult to imagine that the realist need be committed to it. But what else will do? The (Putnamian) realist wants attributions of reference to a theory's terms to function in an explanation of that theory's success. The simplest and crudest way of doing that involves a claim like (S2). A less outrageous way of achieving the same end would involve the weaker,

(S2′) A theory whose terms refer will usually (but not always) be successful.

Isolated instances of referring but unsuccessful theories, sufficient to refute (S2), leave (S2′)

unscathed. But, if we were to find a broad range of referring but unsuccessful theories, that would be evidence against (S2′). Such theories can be generated at will. For instance, take any set of terms which one believes to be genuinely referring. In any language rich enough to contain negation, it will be possible to construct indefinitely many unsuccessful theories, all of whose substantive terms are genuinely referring. Now, it is always open to the realist to claim that such 'theories' are not really theories at all, but mere conjunctions of isolated statements – lacking that sort of conceptual integration we associate with 'real' theories. Sadly a parallel argument can be made for genuine theories. Consider, for instance, how many inadequate versions of the atomic theory there were in the 2000 years of atomic 'speculating', before a genuinely successful theory emerged. Consider how many unsuccessful versions there were of the wave theory of light before the 1820s, when a successful wave theory first emerged. Kinetic theories of heat in the seventeenth and eighteenth century, developmental theories of embryology before the late nineteenth century sustain a similar story. (S2′), every bit as much as (S2), seems hard to reconcile with the historical record.

As Richard Burian has pointed out to me (in personal communication), a realist might attempt to dispense with both of those theses and simply rest content with (S3) alone. Unlike (S2) and (S2′), (S3) is not open to the objection that referring theories are often unsuccessful, for it makes no claim that referring theories are always or generally successful. But (S3) has difficulties of its own. In the first place, it seems hard to square with the fact that the central terms of many relatively successful theories (e.g., aether theories, phlogistic theories) are evidently non-referring. I shall discuss this tension in detail below. More crucial for our purposes here is that (S3) is *not strong enough* to permit the realist to utilize reference to explain success. Unless genuineness of reference entails that all or most referring theories will be successful, then the fact that a theory's terms refer scarcely provides a convincing explanation of that theory's success. If, as (S3) allows, many (or even most) referring theories can be unsuccessful, how can the fact that a successful theory's terms refer be taken to explain why it is successful? (S3) may or may not be true; but in either case it arguably gives the realist no explanatory access to scientific success.

A more plausible construal of Putnam's claim that reference plays a role in explaining the success of science involves a rather more indirect argument. It might be said (and Putnam does say this much) that we can explain why a theory is successful by assuming that the theory is true or approximately true. Since a theory can only be true or nearly true (in any sense of those terms open to the realist) if its terms genuinely refer, it might be argued that reference gets into the act willy-nilly when we explain a theory's success in terms of its truth(like) status. On this account, reference is piggy-backed on approximate truth. The viability of this indirect approach is treated at length in section 4 below so I shall not discuss it here except to observe that if the only contact point between reference and success is provided through the medium of approximate truth, then the link between reference and success is extremely tenuous.

What about (S3), the realist's claim that success creates a rational presumption of reference? We have already seen that (S3) provides no explanation of the success of science, but does it have independent merits? The question specifically is whether the success of a theory provides a warrant for concluding that its central terms refer. Insofar as this is – as certain realists suggest – an empirical question, it requires us to inquire whether past theories which have been successful are ones whose central terms genuinely referred (according to the realist's own account of reference).

A proper empirical test of this hypothesis would require extensive sifting of the historical record of a kind that is not possible to perform here. What I can do is to mention a range of once successful, but (by present lights) non-referring, theories. A fuller list will come later (see section 5), but for now we shall focus on a whole family of related theories, namely, the subtle fluids and aethers of eighteenth and nineteenth century physics and chemistry.

Consider specifically the state of aetherial theories in the 1830s and 1840s. The electrical fluid, a substance which was generally assumed to accumulate on the surface rather than permeate the interstices of bodies, had been utilized to explain *inter alia* the attraction of oppositely charged bodies, the behavior of the Leyden jar, the similarities between atmospheric and static electricity and many phenomena of current electricity. Within chemistry and heat theory, the caloric aether had been widely utilized since Boerhaave (by, among others, Lavoisier, Laplace, Black, Rumford, Hutton, and Cavendish) to explain everything from the role of heat in chemical reactions to the conduction and radiation of heat and several standard problems of thermometry. Within the theory of light, the optical aether functioned centrally in explanations of reflection, refraction, interference, double refraction, diffraction and polarization. (Of more than passing interest, optical aether theories had also made some very startling predictions, e.g., Fresnel's prediction of a bright spot at the center of the shadow of a circular disc; a surprising prediction which, when tested, proved correct. If that does not count as empirical success, nothing does!) There were also gravitational (e.g., LeSage's) and physiological (e.g., Hartley's) aethers which enjoyed some measure of empirical success. It would be difficult to find a family of theories in this period which were as successful as aether theories; compared to them, nineteenth century atomism (for instance), a genuinely referring theory (on realist accounts), was a dismal failure. Indeed, on any account of empirical success which I can conceive of, non-referring nineteenth-century aether theories were more successful than contemporary, referring atomic theories. In this connection, it is worth recalling the remark of the great theoretical physicist, J.C. Maxwell, to the effect that the aether was better confirmed than any other theoretical entity in natural philosophy!

What we are confronted by in nineteenth-century aether theories, then, is a wide variety of once successful theories, whose central explanatory concept Putnam singles out as a prime example of a non-referring one (Putnam 1978, p. 22). What are (referential) realists to make of this historical case? On the face of it, this case poses two rather different kinds of challenges to realism: (1) it suggests that (S3) is a dubious piece of advice in that *there can be* (and have been) *highly successful theories some central terms of which are non-referring;* and (2) it suggests that *the realist's claim that he can explain why science is successful is false at least insofar as a part of the historical success of science has been success exhibited by theories whose central terms did not refer.*

But perhaps I am being less than fair when I suggest that the realist is committed to the claim that *all* the central terms in a successful theory refer. It is possible that when Putnam, for instance, says that "terms in a mature [or successful] science typically refer" (Putnam 1978, p. 20), he only means to suggest that *some* terms in a successful theory or science genuinely refer. Such a claim is fully consistent with the fact that certain other terms (e.g., 'aether') in certain successful, mature sciences (e.g., nineteenth-century physics) are nonetheless non-referring. Put differently, the realist might argue that the success of a theory warrants the claim that at least some (but not necessarily all) of its central concepts refer.

Unfortunately, such a weakening of (S3) entails a theory of evidential support which can scarcely give comfort to the realist. After all, part of what separates the realist from the positivist is the former's belief that the evidence for a theory is evidence for *everything* which the theory asserts. Where the stereotypical positivist argues that the evidence selectively confirms only the more 'observable' parts of a theory, the realist generally asserts (in the language of Boyd) that:

> the sort of evidence which ordinarily counts in favor of the acceptance of a scientific law or theory is, ordinarily, evidence for the (at least approximate) truth of the law or theory as an account of the causal relations obtaining between the entities ["observation or theoretical"] quantified over in the law or theory in question.
>
> (Boyd 1973, p. 1)[5]

For realists such as Boyd, either all parts of a theory (both observational and non-observational) are confirmed by successful tests or none are. In general, realists have been able to utilize various holistic arguments to insist that it is not merely the lower-level claims of a well-tested theory which are confirmed but its deep-structural assumptions as well. This tactic has been used to good effect by realists in establishing that inductive support 'flows upward' so as to authenticate the most 'theoretical' parts of our theories. Certain latter-day realists (e.g., Glymour) want to break out of this holist web and argue that certain components of theories can be 'directly' tested. This approach runs the very grave risk of undercutting what the realist desires most: a rationale for taking our deepest-structure theories seriously, and a justification for linking reference and success. After all, if the tests to which we subject our theories only test *portions* of those theories, then even highly successful theories may well have central terms which are non-referring and central tenets which, because untested, we have no grounds for believing to be approximately true. Under those circumstances, a theory might be highly successful and yet contain important constituents which were patently false. Such a state of affairs would wreak havoc with the realist's presumption (R1) that success betokens approximate truth. In short, to be less than a holist about theory testing is to put at risk precisely that predilection for deep-structure claims which motivates much of the realist enterprise.

There is, however, a rather more serious obstacle to this weakening of referential realism. It is true that by weakening (S3) to only certain terms in a theory, one would immunize it from certain obvious counter-examples. But such a maneuver has debilitating consequences for other central realist theses. Consider the realist's thesis (R3) about the retentive character of inter-theory relations (discussed below in detail). The realist both recommends as a matter of policy and claims as a matter of fact that successful theories are (and should be) rationally replaced only by theories which preserve reference for the central terms of their successful predecessors. The rationale for the normative version of this retentionist doctrine is that the terms in the earlier theory, *because it was successful,* must have been referential and thus a constraint on any successor to that theory is that reference should be retained for such terms. This makes sense just in case success provides a blanket warrant for presumption of reference. But if (S3) were weakened so as to say merely that it is reasonable to assume that *some* of the terms in a successful theory genuinely refer, then the realist would have no rationale for his retentive theses (variants of R3), which have been a central pillar of realism for several decades.[6]

Something apparently has to give. A version of (S3) strong enough to license (R3) seems incompatible with the fact that many successful theories contain non-referring central terms. But

any weakening of (S3) dilutes the force of, and removes the rationale for, the realist's claims about convergence, retention and correspondence in inter-theory relations.[7] If the realist once concedes that some unspecified set of the terms of a successful theory may well not refer, then his proposals for restricting "the class of candidate theories" to those which retain reference for the *prima facie* referring terms in earlier theories is without foundation. (Putnam 1975, p. 22)

More generally, we seem forced to say that such linkages as there are between reference and success are rather murkier than Putnam's and Boyd's discussions would lead us to believe. If the realist is going to make his case for CER, it seems that it will have to hinge on approximate truth, (R1), rather than reference, (R2).

4 Approximate truth and success: the 'downward path'

Ignoring the referential turn among certain recent realists, most realists continue to argue that, at bottom, epistemic realism is committed to the view that successful scientific theories, even if strictly false, are nonetheless 'approximately true' or 'close to the truth' or 'verisimilar'.[8] The claim generally amounts to this pair:

(T1) if a theory is approximately true, then it will be explanatorily successful; and

(T2) if a theory is explanatorily successful, then it is probably approximately true.

What the realist would *like* to be able to say, of course, is:

(T1′) if a theory is true, then it will be successful.

(T1′) is attractive because self-evident. But most realists balk at invoking (T1′) because they are (rightly) reluctant to believe that we can reasonably presume of any given scientific theory that it is true. If all the realist could explain was the success of theories which were true *simpliciter*, his explanatory repertoire would be acutely limited. As an attractive move in the direction of broader explanatory scope, (T1) is rather more appealing. After all, presumably many theories which we believe to be false (e.g., Newtonian mechanics, thermodynamics, wave optics) were − and still are − highly successful across a broad range of applications.

Perhaps, the realist evidently conjectures, we can find an *epistemic* account of that pragmatic success by assuming such theories to be 'approximately true'. But we must be wary of this potential sleight of hand. It may be that there is a connection between success and approximate truth; but *if there is such a connection it must be independently argued for*. The acknowledgedly uncontroversial character of (T1′) must not be surreptitiously invoked − as it sometimes seems to be − in order to establish (T1). When (T1′)'s antecedent is appropriately weakened by speaking of approximate truth, it is by no means clear that (T1) is sound.

Virtually all the proponents of epistemic realism take it as unproblematic that if a theory were approximately true, it would deductively follow that the theory would be a relatively successful predictor and explainer of observable phenomena. Unfortunately, few of the writers of whom I am aware have defined what it means for a statement or theory to be 'approximately true'. Accordingly, it is impossible to say whether the alleged entailment is genuine. This reservation is more than perfunctory. Indeed, on the best known account of what it means for a theory to be approximately true, it does *not* follow that an approximately true theory will be explanatorily successful.

Suppose, for instance, that we were to say in a Popperian vein that a theory, T_1, is approximately true if its truth content is greater than its falsity content, i.e.,

$$Ct_T(T_1) \gg Ct_F(T_1).[9]$$

(Where $Ct_T(T_1)$ is the cardinality of the set of true sentences entailed by T_1, and $Ct_F(T_1)$ is the cardinality of the set of false sentences entailed by T_1.) When approximate truth is so construed, it does *not* logically follow that an arbitrarily selected class of a theory's entailments (namely, some of its observable consequences) will be true. Indeed, it is entirely conceivable that a theory might be approximately true in the indicated sense and yet be such that *all* of its thus far tested consequences *are false*.[10]

Some realists concede their failure to articulate a coherent notion of approximate truth or verisimilitude, but insist that this failure in no way compromises the viability of (T1). Newton-Smith, for instance, grants that "no one has given a satisfactory analysis of the notion of verisimilitude" (forthcoming, p. 16), but insists that the concept can be legitimately invoked "even if one cannot at the time give a philosophically satisfactory analysis of it." He quite rightly points out that many scientific concepts were explanatorily useful long before a philosophically coherent analysis was given for them. But the analogy is unseemly, for what is being challenged is not whether the concept of approximate truth is philosophically rigorous but rather whether it is even clear enough for us to ascertain whether it entails what it purportedly explains. Until someone provides a clearer analysis of approximate truth than is now available, it is not even clear whether truth-likeness would explain success, let alone whether, as Newton-Smith insists, "the concept of verisimilitude is *required* in order to give a satisfactory theoretical explanation of an aspect of the scientific enterprise." If the realist would de-mystify the 'miraculousness' (Putnam) or the 'mysteriousness' (Newton-Smith[11]) of the success of science, he needs more than a promissory note that somehow, someday, someone will show that approximately true theories must be successful theories.[12]

Whether there is some definition of approximate truth which does indeed entail that approximately true theories will be predictively successful (and yet still probably false) is not clear.[13] What can be said is that, promises to the contrary notwithstanding, *none* of the proponents of realism has yet articulated a coherent account of approximate truth which entails that approximately true theories will, across the range where we can test them, be successful predictors. Further difficulties abound. Even if the realist had a semantically adequate characterization of approximate or partial truth, and even if that semantics entailed that most of the consequences of an approximately true theory would be true, he would still be without any criterion that would *epistemically* warrant the ascription of approximate truth to a theory. As it is, the realist seems to be long on intuitions and short on either a semantics or an epistemology of approximate truth.

These should be urgent items on the realists' agenda since, until we have a coherent account of what approximate truth is, central realist theses like (R1), (T1) and (T2) are just so much mumbo-jumbo.

5 Approximate truth and success: the 'upward path'

Despite the doubts voiced in section 4, let us grant for the sake of argument that if a theory is approximately true, then it will be successful. Even granting (T1), is there any plausibility to the suggestion of (T2) that explanatory success can be taken as a rational warrant for a judgment of approximate truth? The answer seems to be "no".

To see why, we need to explore briefly one of the connections between 'genuinely referring' and being 'approximately true'. However the latter is understood, I take it that *a realist would never want to say that a theory was approximately true if its central theoretical terms failed to refer.* If there were nothing like genes, then a genetic theory, no matter how well confirmed it was, would not be approximately true. If there were no entities similar to atoms, no atomic theory could be approximately true; if there were no sub-atomic particles, then no quantum theory of chemistry could be approximately true. In short, a necessary condition – especially for a scientific realist – for a theory being close to the truth is that its central explanatory terms genuinely refer. (An *instrumentalist*, of course, could countenance the weaker claim that a theory was approximately true so *long* as its directly testable consequences were close to the observable values. But as I argued above, the realist must take claims about approximate truth to refer alike to the observable and the deep-structural dimensions of a theory.)

Now, what the history of science offers us is a plethora of theories which were both successful and (so far as we can judge) non-referential with respect to many of their central explanatory concepts. I discussed earlier one specific family of theories which fits this description. Let me add a few more prominent examples to the list:

- the crystalline spheres of ancient and medieval astronomy;
- the humoral theory of medicine;
- the effluvial theory of static electricity;
- 'catastrophist' geology, with its commitment to a universal (Noachian) deluge;
- the phlogiston theory of chemistry;
- the caloric theory of heat;
- the vibratory theory of heat;
- the vital force theories of physiology;
- the electromagnetic aether;
- the optical aether;
- the theory of circular inertia;
- theories of spontaneous generation.

This list, which could be extended *ad nauseam*, involves in every case a theory which was once successful and well confirmed, but which contained central terms which (we now believe) were non-referring. Anyone who imagines that the theories which have been successful in the history of science have also been, with respect to their central concepts, genuinely referring theories has studied only the more 'whiggish' versions of the history of science (i.e., the ones which recount only those past theories which are referentially similar to currently prevailing ones).

It is true that proponents of CER sometimes hedge their bets by suggesting that their analysis applies exclusively to 'the mature sciences' (e.g., Putnam and Krajewski). This distinction between mature and immature sciences proves convenient to the realist since he can use it to dismiss any *prima facie* counter-example to the empirical claims of CER on the grounds that the example is drawn from an 'immature' science. But this insulating maneuver is unsatisfactory in two respects. In the first place, it runs the risk of making CER vacuous since these authors generally define a mature science as one in which correspondence or limiting case relations obtain invariably between any successive theories in the science once it has passed 'the threshold of maturity'. Krajewski grants the tautological character of this view when he notes that "the thesis that there is [correspondence] among successive theories becomes, indeed, analytical" (1977, p. 91). Nonetheless, he believes that there is a version of the maturity thesis which "may be and must be tested by the history of science". That version is that "every branch of science crosses at some period the threshold of maturity". But the testability of this hypothesis is dubious at best. There is no historical observation which could conceivably *refute* it since, even if we discovered that no sciences yet possessed 'corresponding' theories, it could be maintained that eventually every science will

become corresponding. It is equally difficult to *confirm* it since, even if we found a science in which corresponding relations existed between the latest theory and its predecessor, we would have no way of knowing whether that relation will continue to apply to subsequent changes of theory in that science. In other words, the much-vaunted empirical testability of realism is seriously compromised by limiting it to the mature sciences.

But there is a second unsavory dimension to the restriction of CER to the 'mature' sciences. The realists' avowed aim, after all, is to explain why science is successful: that is the 'miracle' which they allege the non-realists leave unaccounted for. The fact of the matter is that parts of science, including many 'immature' sciences, have been successful for a very long time; indeed, many of the theories I alluded to above were empirically successful by any criterion I can conceive of (including fertility, intuitively high confirmation, successful prediction, etc.). If the realist restricts himself to explaining only how the 'mature' sciences work (and recall that very few sciences indeed are yet 'mature' as the realist sees it), then he will have completely failed in his ambition to explain why science in general is successful. Moreover, several of the examples I have cited above come from the history of mathematical physics in the last century (e.g., the electromagnetic and optical aethers) and, as Putnam himself concedes, "*physics* surely counts as a 'mature' science if any science does" (1978, p. 21). Since realists would presumably insist that many of the central terms of the theories enumerated above do not genuinely refer, it follows that none of those theories could be approximately true (recalling that the former is a necessary condition for the latter). Accordingly, cases of this kind cast very grave doubts on the plausibility of (T2), i.e., the claim that nothing succeeds like approximate truth.

I daresay that for every highly successful theory in the past of science which we now believe to be a genuinely referring theory, one could find half a dozen once successful theories which we now regard as substantially non-referring. If the proponents of CER are the empiricists they profess to be about matters epistemological, cases of this kind and this frequency should give them pause about the well-foundedness of (T2).

But we need not limit our counter-examples to non-referring theories. There were many theories in the past which (so far as we can tell) were both genuinely referring and empirically successful which we are nonetheless loath to regard as approximately true. Consider, for instance, virtually all those geological theories prior to the 1960s which denied any lateral motion to the continents. Such theories were, by any standard, highly successful (and apparently referential); but would anyone today be prepared to say that their constituent theoretical claims – committed as they were to laterally stable continents – are almost true? Is it not the fact of the matter that structural geology was a successful science between (say) 1920 and 1960, even though geologists were fundamentally mistaken about many – perhaps even most – of the basic mechanisms of tectonic construction? Or what about the chemical theories of the 1920s which assumed that the atomic nucleus was structurally homogenous? Or those chemical and physical theories of the late nineteenth century which explicitly assumed that matter was neither created nor destroyed? I am aware of no sense of approximate truth (available to the realist) according to which such highly successful, but evidently false, theoretical assumptions could be regarded as 'truthlike'.

More generally, the realist needs a riposte to the *prima facie* plausible claim that there is no necessary connection between increasing the accuracy of our deep-structural characterizations

of nature and improvements at the level of phenomenological explanations, predictions and manipulations. It *seems* entirely conceivable intuitively that the theoretical mechanisms of a new theory, T_2, might be closer to the mark than those of a rival T_1 and yet T_1 might be more accurate at the level of testable predictions. In the absence of an argument that greater correspondence at the level of unobservable claims is more likely than not to reveal itself in greater accuracy at the experimental level, one is obliged to say that the realist's hunch that increasing deep-structural fidelity must manifest itself pragmatically in the form of heightened experimental accuracy has yet to be made cogent. (Equally problematic, of course, is the inverse argument to the effect that increasing experimental accuracy betokens greater truthlikeness at the level of theoretical, i.e., deep-structural, commitments.)

6 Confusions about convergence and retention

Thus far, I have discussed only the static or synchronic versions of CER, versions which make absolute rather than relative judgments about truthlikeness. Of equal appeal have been those variants of CER which invoke a notion of what is variously called convergence, correspondence or cumulation. Proponents of the diachronic version of CER supplement the arguments discussed above ((S1)–(S4) and (T1)–(T2)) with an additional set. They tend to be of this form:

C1)   If earlier theories in a scientific domain are successful and thereby, according to realist principles (e.g., (S3) above), approximately true, then scientists should only accept later theories which retain appropriate portions of earlier theories;

C2)   As a matter of fact, scientists do adopt the strategy of (C1) and manage to produce new, more successful theories in the process;

C3) The 'fact' that scientists succeed at retaining appropriate parts of earlier theories in more successful successors shows that the earlier theories did genuinely refer and that they were approximately true. And thus, the strategy propounded in (C1) is sound.[14]

Perhaps the prevailing view here is Putnam's and (implicitly) Popper's, according to which rationally-warranted successor theories in a 'mature' science must (a) contain reference to the entities apparently referred to in the predecessor theory (since, by hypothesis, the terms in the earlier theory refer), and (b) contain the 'theoretical laws' and 'mechanisms' of the predecessor theory as limiting cases. As Putnam tells us, a 'realist' should insist that *any* viable successor to an old theory T_1 must "contain the laws of T_1 as a limiting case" (1978, p. 21). John Watkins, a like-minded convergentist, puts the point this way:

It typically happens in the history of science that when some hitherto dominant theory T is superceded by T^1, T^1 is in the relation of correspondence to T [i.e., T is a 'limiting case' of T^1] (1978, pp. 376–377).

Numerous recent philosophers of science have subscribed to a similar view, including Popper, Post, Krajewski, and Koertge.[15]

This form of retention is not the only one to have been widely discussed. Indeed, realists have espoused a wide variety of claims about what is or should be retained in the transition from a once successful predecessor (T_1) to a successor (T_2) theory. Among the more important forms of realist retention are the following cases: (1) T_2 entails T_1 (Whewell); (2) T_2 retains the true consequences or truth content of T_1 (Popper);

(3) T_2 retains the 'confirmed' portions of T_1 (Post, Koertge); (4) T_2 preserves the theoretical laws and mechanisms of T_1 (Boyd, McMullin, Putnam); (5) T_2 preserves T_1 as a limiting case (Watkins, Putnam, Krajewski); (6) T_2 explains why T_1 succeeded insofar as T_1 succeeded (Sellars); (7) T_2 retains reference for the central terms of T_1 (Putnam, Boyd).

The question before us is whether, when retention is understood in *any* of these senses, the realist's theses about convergence and retention are correct.

6.1 Do scientists adopt the 'retentionist' strategy of CER?

One part of the convergent realist's argument is a claim to the effect that scientists generally adopt the strategy of seeking to preserve earlier theories in later ones. As Putnam puts it:

preserving the *mechanisms* of the earlier theory as often as possible, which is what scientists try to do . . . That scientists try to do this . . . is a fact, and that this strategy has led to important discoveries . . . is also a fact (1978, p. 20).[16]

In a similar vein, Szumilewicz (although not stressing realism) insists that many eminent scientists made it a main heuristic requirement of their research programs that a new theory stand in a relation of 'correspondence' with the theory it supersedes (1977, p. 348). If Putnam and the other retentionists are right about the strategy which most scientists have adopted, we should expect to find the historical literature of science abundantly provided with (a) proofs that later theories do indeed contain earlier theories as limiting cases, or (b) outright rejections of later theories which fail to contain earlier theories. Except on rare occasions (coming primarily from the history of mechanics), one finds neither of these concerns prominent in the literature of science. For instance, to the best of my knowledge, literally no one criticized the wave theory of light because it did not preserve the theoretical mechanisms of the earlier corpuscular theory; no one faulted Lyell's uniformitarian geology on the grounds that it dispensed with several causal processes prominent in catastrophist geology; Darwin's theory was not criticized by most geologists for its failure to retain many of the mechanisms of Lamarckian 'evolutionary theory'.

For all the realist's confident claims about the prevalence of a retentionist strategy in the sciences, I am aware of *no* historical studies which would sustain as a *general* claim his hypothesis about the evaluative strategies utilized in science. Moreover, insofar as Putnam and Boyd claim to be offering "an explanation of the [retentionist] behavior of scientists" (Putnam 1978, p. 21), they have the wrong explanandum, for if there is any widespread strategy in science, it is one which says, "accept an empirically successful theory, regardless of whether it contains the theoretical laws and mechanisms of its predecessors".[17] Indeed, one could take a leaf from the realist's (C2) and claim that the success of the strategy of assuming that earlier theories do not generally refer shows that it is true that earlier theories generally do not!

(One might note in passing how often, and on what evidence, realists imagine that they are speaking for the scientific majority. Putnam, for instance, claims that "realism is, so to speak, 'science's philosophy of science'" and that "science taken at 'face value' *implies* realism" (1978, p. 37).[18] Hooker insists that to be a realist is to take science "seriously" (1974, pp. 467–472), as if to suggest that conventionalists, instrumentalists and positivists such as Duhem, Poincaré, and Mach did not take science seriously. The willingness of some realists to attribute realist strategies to working scientists

– on the strength of virtually no empirical research into the principles which in *fact* have governed scientific practice – raises doubts about the seriousness of their avowed commitment to the empirical character of epistemic claims.)

6.2 Do later theories preserve the mechanisms, models, and laws of earlier theories?

Regardless of the explicit strategies to which scientists have subscribed, are Putnam and several other retentionists right that later theories "typically" entail earlier theories, and that "earlier theories are, very often, limiting cases of later theories"?[19] Unfortunately, answering this question is difficult, since "typically" is one of those weasel words which allows for much hedging. I shall assume that Putnam and Watkins mean that "most of the time (or perhaps in most of the important cases) successor theories contain predecessor theories as limiting cases". So construed, the claim is patently false. Copernican astronomy did not retain all the key mechanisms of Ptolemaic astronomy (e.g., motion along an equant); Newton's physics did not retain all (or even most of) the 'theoretical laws' of Cartesian mechanics, astronomy and optics; Franklin's electrical theory did not contain its predecessor (Nollet's) as a limiting case. Relativistic physics did not retain the aether, nor the mechanisms associated with it; statistical mechanics does not incorporate all the mechanisms of thermodynamics; modern genetics does not have Darwinian pangenesis as a limiting case; the wave theory of light did not appropriate the mechanisms of corpuscular optics; modern embryology incorporates few of the mechanisms prominent in classical embryological theory. As I have shown elsewhere,[20] loss occurs at virtually every level: the confirmed predictions of earlier theories are sometimes not explained by later ones; even the 'observable' laws explained by earlier theories are not always retained, not even as limiting cases; theoretical processes and mechanisms of earlier theories are, as frequently as not, treated as flotsam.

The point is that some of the most important theoretical innovations have been due to a willingness of scientists to violate the cumulationist or retentionist constraint which realists enjoin 'mature' scientists to follow.

There is a deep reason why the convergent realist is wrong about these matters. It has to do, in part, with the role of ontological frameworks in science and with the nature of limiting case relations. As scientists use the term 'limiting case', T_1 can be a limiting case of T_2 only if (a) *all* the variables (observable and theoretical) assigned a value in T_1 are assigned a value by T_2 and (b) the values assigned to every variable of T_1 are the same as, or very close to, the values T_2 assigns to the corresponding variable when certain initial and boundary conditions – consistent with T_2[21] – are specified. This seems to require that T_1 can be a limiting case of T_2 only if *all* the entities postulated by T_1 occur in the ontology of T_2. Whenever there is a change of ontology accompanying a theory transition such that T_2 (when conjoined with suitable initial and boundary conditions) fails to capture T_1's ontology, then T_1 *cannot* be a limiting case of T_2. Even where the ontologies of T_1 and T_2 overlap appropriately (i.e., where T_2's ontology embraces all of T_1's), T_1 is a limiting case of T_2 only if *all* the laws of T_1 can be derived from T_2, given appropriate limiting conditions. It is important to stress that *both* these conditions (among others) must be satisfied before one theory can be a limiting case of another. Where 'closet positivists' might be content with capturing only the formal mathematical relations or only the observable consequences of T_1 within a successor, T_2, any genuine realist must insist that T_1's underlying ontology is preserved

in T_2's *for it is that ontology above all which he alleges to be approximately true.*

Too often, philosophers (and physicists) infer the existence of a limiting case relation between T_1 and T_2 on substantially less than this. For instance, many writers have claimed one theory to be a limiting case of another when only some, but not all, of the laws of the former are 'derivable' from the latter. In other cases, one theory has been said to be a limiting case of a successor when the mathematical laws of the former find homologies in the latter but where the former's ontology is not fully extractable from the latter's.

Consider one prominent example which has often been misdescribed, namely, the transition from the classical aether theory to relativistic and quantum mechanics. It can, of course, be shown that *some* 'laws' of classical mechanics are limiting cases of relativistic mechanics. But there are other laws and general assertions made by the classical theory (e.g., claims about the density and fine structure of the aether, general laws about the character of the interaction between aether and matter, models and mechanisms detailing the compressibility of the aether) which could not conceivably be limiting cases of modern mechanics. The reason is a simple one: a theory cannot assign values to a variable which does not occur in that theory's language (or, more colloquially, it cannot assign properties to entities whose existence it does not countenance). Classical aether physics contained a number of postulated mechanisms for dealing *inter alia* with the transmission of light through the aether. Such mechanisms could not possibly appear in a successor theory like the special theory of relativity which denies the very existence of an aetherial medium and which accomplishes the explanatory tasks performed by the aether via very different mechanisms.

Nineteenth-century mathematical physics is replete with similar examples of evidently successful mathematical theories which, because some of their variables refer to entities whose existence we now deny, cannot be shown to be limiting cases of our physics. As Adolf Grünbaum has cogently argued, when we are confronted with two incompatible theories, T_1 and T_2, such that T_2 does not 'contain' all of T_1's ontology, then not all the mechanisms and theoretical laws of T_1, which involve those entities of T_1 not postulated by T_2 can possibly be retained – not even as limiting cases – in T_2 (1976, pp. 1–23). This result is of some significance. What little plausibility convergent or retentive realism has enjoyed derives from the presumption that it correctly describes the relationship between classical and post-classical mechanics and gravitational theory. Once we see that even in this *prima facie* most favorable case for the realist (where *some* of the laws of the predecessor theory are genuinely limiting cases of the successor), changing ontologies or conceptual frameworks make it impossible to capture many of the central theoretical laws and mechanisms postulated by the earlier theory, then we can see how misleading is Putnam's claim that "what scientists try to do" is to preserve

> the *mechanisms* of the earlier theory as often as possible – or to show that they are 'limiting cases' of new mechanisms . . . (1978, p. 20).

Where the mechanisms of the earlier theory involve entities whose existence the later theory denies, no scientist does (or should) feel any compunction about wholesale repudiation of the earlier mechanisms.

But even where there is no change in basic ontology, many theories (even in 'mature sciences' like physics) fail to retain all the explanatory successes of their predecessors. It is well known that statistical mechanics has yet to capture the irreversibility of macro-thermodynamics as a genuine limiting case.

Classical continuum mechanics has not yet been reduced to quantum mechanics or relativity. Contemporary field theory has yet to replicate the classical thesis that physical laws are invariant under reflection in space. If scientists had accepted the realist's constraint (namely, that new theories must have old theories as limiting cases), neither relativity nor statistical mechanics would have been viewed as viable theories. It has been said before, but it needs to be reiterated over and again: *a proof of the existence of limiting relations between selected components of two theories is a far cry from a systematic proof that one theory is a limiting case of the other.* Even if classical and modern physics stood to one another in the manner in which the convergent realist erroneously imagines they do, his hasty generalization that theory successions in all the advanced sciences show limiting case relations is patently false.[22] But, as this discussion shows, not even the realist's paradigm case will sustain the claims he is apt to make about it.

What this analysis underscores is just how reactionary many forms of convergent epistemological realism are. If one took seriously CER's advice to reject any new theory which did not capture existing mature theories as referential and existing laws and mechanisms as approximately authentic, then any prospect for deep-structure, ontological changes in our theories would be foreclosed. Equally outlawed would be any significant repudiation of our theoretical models. In spite of his commitment to the growth of knowledge, the realist would unwittingly freeze science in its present state by forcing all future theories to accommodate the ontology of contemporary ('mature') science and by foreclosing the possibility that some future generation may come to the conclusion that some (or even most) of the central terms in our best theories are no more referential than was 'natural place', 'phlogiston', 'aether', or 'caloric'.

6.3 Could theories converge in ways required by the realist?

These instances of violations in genuine science of the sorts of continuity usually required by realists are by themselves sufficient to show that the form of scientific growth which the convergent realist takes as his explicandum is often absent, even in the 'mature' sciences. But we can move beyond these specific cases to show in principle that the kind of cumulation demanded by the realist is unattainable. Specifically, by drawing on some results established by David Miller and others, the following can be shown:

a)   the familiar requirement that a successor theory, T_2, must both preserve as true the true consequences of its predecessor, T_1, and explain T_1's anomalies is contradictory;

b)   that if a new theory, T_2, involves a change in the ontology or conceptual framework of a predecessor, T_1, then T_1 will have true and determinate consequences not possessed by T_2;

c)   that if two theories, T_1 and T_2, disagree, then each will have true and determinate consequences not exhibited by the other.

In order to establish these conclusions, one needs to utilize a 'syntactic' view of theories according to which a theory is a conjunction of statements and its consequences are defined à la Tarski in terms of content classes. Needless to say, this is neither the only, nor necessarily the best, way of thinking about theories; but it happens to be the way in which most philosophers who argue for convergence and retention (e.g., Popper, Watkins, Post, Krajewski, and Niiniluoto) tend to conceive of theories. What can be said is that if one utilizes the Tarskian conception of a theory's content and its consequences as they do, then the familiar convergentist theses alluded to in (a) through (c) make no sense.

The elementary but devastating consequences of Miller's analysis establish that virtually any effort to link scientific progress or growth to the wholesale retention of a predecessor theory's Tarskian content or logical consequences or true consequences or observed consequences or confirmed consequences is evidently doomed. Realists have not only got their history wrong insofar as they imagine that cumulative retention has prevailed in science, but we can see that – given their views on what should be retained through theory change – history could not possibly have been the way their models require it to be. The realists' strictures on cumulativity are as ill-advised normatively as they are false historically.

Along with many other realists, Putnam has claimed that "the mature sciences do converge . . . and that that convergence has great explanatory value for the theory of science" (1978, p. 37). As this section should show, Putnam and his fellow realists are arguably wrong on *both* counts. Popper once remarked that "no theory of knowledge should attempt to explain why we are successful in our attempts to explain things" (1972, p. 23). Such a dogma is too strong. But what the foregoing analysis shows is that an occupational hazard of recent epistemology is imagining that convincing explanations of our success come easily or cheaply.

6.4 Should new theories explain why their predecessors were successful?

An apparently more modest realism than that outlined above is familiar in the form of the requirement (R4) often attributed to Sellars – that every satisfactory new theory must be able to explain why its predecessor was successful insofar as it was successful. On this view, viable new theories need not preserve all the content of their predecessors, nor capture those predecessors as limiting cases. Rather, it is simply insisted that a viable new theory, T_N, must explain why, when we conceive of the world according to the old theory T_0, there is a range of cases where our T_0-guided expectations were correct or approximately correct.

What are we to make of this requirement? In the first place, it is clearly *gratuitous*. If T_N has more confirmed consequences (and greater conceptual simplicity) than T_0, then T_N is preferable to T_0 even if T_N cannot explain why T_0 is successful. Contrariwise, if T_N has fewer confirmed consequences than T_0, then T_N cannot be rationally preferred to T_0 even if T_N explains why T_0 is successful. In short, a theory's ability to explain why a rival is successful is neither a necessary nor a sufficient condition for saying that it is better than its rival.

Other difficulties likewise confront the claim that new theories should explain why their predecessors were successful. Chief among them is the ambiguity of the notion itself. One way to show that an older theory, T_0 was successful is to show that it shares many confirmed consequences with a newer theory, T_N, which is highly successful. But this is not an 'explanation' that a scientific realist could accept, since it makes no reference to, and thus does not depend upon, an epistemic assessment of either T_0 or T_N. (After all, an instrumentalist could quite happily grant that if T_N 'saves the phenomena' then T_0 – insofar as some of its observable consequences overlap with or are experimentally indistinguishable from those of T_N – should also succeed at saving the phenomena.)

The intuition being traded on in this persuasive account is that the pragmatic success of a new theory, combined with a partial comparison of the respective consequences of the new theory and its predecessor, will sometimes put us in a position to say when the older theory

worked and when it failed. But such comparisons as can be made in this manner do not involve *epistemic* appraisals of either the new or the old theory *qua* theories. Accordingly, the possibility of such comparisons provides no argument for epistemic realism.

What the realist apparently needs is an *epistemically* robust sense of 'explaining the success of a predecessor'. Such an epistemic characterization would presumably begin with the claim that T_N, the new theory, was approximately true and would proceed to show that the 'observable' claims of its predecessor, T_0, deviated only slightly from (some of) the 'observable' consequences of T_N. It would then be alleged that the (presumed) approximate truth of T_N and the partially overlapping consequences of T_0 and T_N jointly explained why T_0 was successful in so far as it was successful. But this is a *non-sequitur*. As I have shown above, the fact that a T_N is approximately true does not even explain why it is successful; how, under those circumstances, can the approximate truth of T_N explain why some theory different from T_N is successful? Whatever the nature of the relations between T_N and T_0 (entailment, limiting case, etc.), the epistemic ascription of approximate truth to either T_0 or T_N (or both) apparently leaves untouched questions of how successful T_0 or T_N are.

The idea that new theories should explain why older theories were successful (insofar as they were) originally arose as a rival to the 'levels' picture of explanation according to which new theories fully explained – because they entailed – their predecessors. It is clearly an improvement over the levels picture (for it does recognize that later theories generally do not entail their predecessors). But when it is formulated as a general thesis about inter-theory relations, designed to buttress a realist epistemology, it is difficult to see how this position avoids difficulties similar to those discussed in earlier sections.

7 The realists' ultimate 'petitio principii'

It is time to step back a moment from the details of the realists' argument to look at its general strategy. Fundamentally, the realist is utilizing, as we have seen, an abductive inference which proceeds from the success of science to the conclusion that science is approximately true, verisimilar, or referential (or any combination of these). This argument is meant to show the sceptic that theories are not ill-gotten, the positivist that theories are not reducible to their observational consequences, and the pragmatist that classical epistemic categories (e.g., 'truth', 'falsehood') are a relevant part of meta-scientific discourse.

It is little short of remarkable that realists would imagine that their critics would find the argument compelling. As I have shown elsewhere (1978), ever since antiquity critics of epistemic realism have based their scepticism upon a deep-rooted conviction that the fallacy of affirming the consequent is indeed fallacious. When Sextus or Bellarmine or Hume doubted that certain theories which saved the phenomena were warrantable as true, their doubts were based on a belief that the exhibition that a theory had some true consequences left entirely open the truth-status of the theory. Indeed, many non-realists have been non-realists precisely because they believed that false theories, as well as true ones, could have true consequences.

Now enters the new breed of realist (e.g., Putnam, Boyd and Newton-Smith) who wants to argue that epistemic realism can reasonably be presumed to be true by virtue of the fact that it has true consequences. But this is a monumental case of begging the question. The non-realist refuses to admit that a *scientific* theory can be warrantedly judged to be true simply because it has some true consequences. Such non-realists are not likely to be impressed by the claim that a *philosophical* theory like realism can be warranted

as true because it arguably has some true consequences. If non-realists are chary about first-order abductions to avowedly true conclusions, they are not likely to be impressed by second-order abductions, particularly when, as I have tried to show above, the premises and conclusions are so indeterminate.

But, it might be argued, the realist is not out to convert the intransigent sceptic or the determined instrumentalist.[23] He is perhaps seeking, rather, to show that realism can be tested like any other scientific hypothesis, and that realism is at least as well confirmed as some of our best scientific theories. Such an analysis, however plausible initially, will not stand up to scrutiny. I am aware of no realist who is willing to say that a *scientific* theory can be reasonably presumed to be true or even regarded as well confirmed just on the strength of the fact that its thus far tested consequences are true. Realists have long been in the forefront of those opposed to *ad hoc* and *post hoc* theories. Before a realist accepts a scientific hypothesis, he generally wants to know whether it has explained or predicted more than it was devised to explain; he wants to know whether it has been subjected to a battery of controlled tests; whether it has successfully made novel predictions; whether there is independent evidence for it.

What, then, of realism itself as a 'scientific' hypothesis?[24] Even if we grant (contrary to what I argued in section 4) that realism entails and thus explains the success of science, ought that (hypothetical) success warrant, by the realist's own construal of scientific acceptability, the acceptance of realism? Since realism was devised in order to explain the success of science, it remains purely *ad hoc* with respect to that success. If realism has made some novel predictions or been subjected to carefully controlled tests, one does not learn about it from the literature of contemporary realism. At the risk of apparent inconsistency, the realist repudiates the instrumentalist's view that saving the phenomena is a significant form of evidential support while endorsing realism itself on the transparently instrumentalist grounds that it is confirmed by those very facts it was invented to explain. No proponent of realism has sought to show that realism satisfies those stringent empirical demands which the realist himself minimally insists on when appraising scientific theories. The latter-day realist often calls realism a 'scientific' or 'well-tested' hypothesis, but seems curiously reluctant to subject it to those controls which he otherwise takes to be a *sine qua non* for empirical well-foundedness.

8 Conclusion

The arguments and cases discussed above seem to warrant the following conclusions:

1. The fact that a theory's central terms refer does not entail that it will be successful; and a theory's success is no warrant for the claim that all or most of its central terms refer.

2. The notion of approximate truth is presently too vague to permit one to judge whether a theory consisting entirely of approximately true laws would be empirically successful; what is clear is that a theory may be empirically successful even if it is not approximately true.

3. Realists have no explanation whatever for the fact that many theories which are not approximately true and whose 'theoretical' terms seemingly do not refer are nonetheless often successful.

4. The convergentist's assertion that scientists in a 'mature' discipline usually preserve, or seek to preserve, the laws and mechanisms of earlier theories in later ones is probably false; his assertion that when such laws are preserved in a successful successor, we can explain the success of the latter by virtue of the truthlikeness of the preserved laws and mechanisms, suffers from all the defects noted above confronting approximate truth.

5. Even if it could be shown that referring theories and approximately true theories would be successful, the realists' argument that successful theories are approximately true and genuinely referential takes for granted precisely what the non-realist denies (namely, that explanatory success betokens truth).

6. It is not clear that acceptable theories either *do* or *should* explain why their predecessors succeeded or failed. If a theory is better supported than its rivals and predecessors, then it is not epistemically decisive whether it explains why its rivals worked.

7. If a theory has once been falsified, it is unreasonable to expect that a successor should retain either all of its content or its confirmed consequences or its theoretical mechanisms.

8. Nowhere has the realist established – except by fiat – that non-realist epistemologists lack the resources to explain the success of science.

With these specific conclusions in mind, we can proceed to a more global one: it is not yet established – Putnam, Newton-Smith and Boyd notwithstanding – that realism can explain *any* part of the success of science. What is very clear is that realism *cannot*, even by its own lights, explain the success of those many theories whose central terms have evidently not referred and whose theoretical laws and mechanisms were not approximately true. The inescapable conclusion is that insofar as many realists are concerned with explaining how science works and with assessing the adequacy of their epistemology by that standard, they have thus far failed to explain very much. Their epistemology is confronted by anomalies which seem beyond its resources to grapple with.

It is important to guard against a possible misinterpretation of this essay. *Nothing* I have said here refutes the possibility in principle of a realistic epistemology of science. To conclude as much would be to fall prey to the same inferential prematurity with which many realists have rejected in principle the possibility of explaining science in a non-realist way. My task here is, rather, that of reminding ourselves that there is a difference between wanting to believe something and having good reasons for believing it. All of us would like realism to be true; we would like to think that science works because it has got a grip on how things really are. But such claims have yet to be made out. Given the *present* state of the art, it can only be wish fulfillment that gives rise to the claim that realism, and realism alone, explains why science works.

Notes

I am indebted to all of the following for clarifying my ideas on these issues and for saving me from some serious errors: Peter Achinstein, Richard Burian, Clark Glymour, Adolf Grünbaum, Gary Gutting, Allen Janis, Lorenz Krüger, James Lennox, Andrew Lugg, Peter Machamer, Nancy Maull, Ernan McMullin, Ilkka Niiniluoto, Nicholas Rescher, Ken Schaffner, John Worrall, Steven Wykstra.

1 Putnam, evidently following Boyd, sums up (Rl) to (R3) in these words:
   "1) Terms in a mature science typically *refer*.
   2) The laws of a theory belonging to a mature science are typically approximately true ... I will only consider [new] theories ... which have this property – [they] contain the [theoretical] laws of [their predecessors] as a limiting case" (1978, pp. 20–21).

2 Putnam insists, for instance, that if the realist is wrong about theories being referential, then "the success of science is a miracle". (Putnam 1975, p. 69).

3 Boyd remarks: "scientific realism offers an *explanation* for the legitimacy of ontological commitment to theoretical entities" (Putnam 1978, Note 10, p. 2). It allegedly does so by explaining why theories containing theoretical entities work so well: because such entities genuinely exist.

4 Whether one utilizes Putnam's earlier or later versions of realism is irrelevant for the central arguments of this essay.

5 See also p. 3: "experimental evidence for a theory is evidence for the truth of even its non-observational laws". See also (Sellars 1963, p. 97).

6 A caveat is in order here. *Even if* all the central terms in some theory refer, it is not obvious that every rational successor to that theory must preserve all the referring terms of its predecessor. One can easily imagine circumstances when the new theory is preferable to the old one even though the range of application of the new theory is less broad than the old. When the range is so restricted, it may well be entirely appropriate to drop reference to some of the entities which figured in the earlier theory.

7 For Putnam and Boyd both "it will be a constraint on T_2 [i.e., any new theory in a domain] ... that T_2 must have this property, the property that *from its standpoint* one can assign referents to the terms of T_1 [i.e., an earlier theory in the same domain]" (Putnam 1978, p. 22). For Boyd, see (1973, p. 8): "new theories should, *prima facie*, resemble current theories with respect to their accounts of causal relations among theoretical entities".

8 For just a small sampling of this view, consider the following: "The claim of a realist ontology of science is that the only way of explaining why the models of science function so successfully ... is that they approximate in some way the structure of the object" (McMullin 1970, pp. 63–64); "the continued success [of confirmed theories] can be *explained* by the hypothesis that they are in fact close to the truth ..." (Niiniluoto 1980, p. 448); the claim that "the laws of a theory belonging to a mature science are typically approximately *true* ... [provides] an *explanation* of the behavior of scientists and the success of science" (Putnam 1978, pp. 20–21). Smart, Sellars, and Newton-Smith, among others, share a similar view.

9 Although Popper is generally careful not to assert that actual historical theories exhibit ever increasing truth content (for an exception, see his (1963, p. 220)), other writers have been more bold. Thus, Newton-Smith writes that "the historically generated sequence of theories of a mature

science" is a sequence in which succeeding theories are increasing in truth content without increasing in falsity content" (forthcoming, p. 2).

10 On the more technical side, Niiniluoto has shown that a theory's degree of corroboration co-varies with its "estimated verisimilitude" (1977, pp. 121–147 and 1980). Roughly speaking, 'estimated truthlikeness' is a measure of how closely (the content of) a theory corresponds to *what we take to be* the best conceptual systems that we so far have been able to find (1980, pp. 443ff.). If Niiniluoto's measures work it follows from the above-mentioned co-variance that an empirically successful theory will have a high degree of estimated truthlikeness. But because estimated truthlikeness and genuine verisimilitude are not necessarily related (the former being parasitic on existing evidence and available conceptual systems), it is an open question whether – as Niiniluoto asserts – the continued success of highly confirmed theories can be *explained* by the hypothesis that they in fact are close to the truth at least in the relevant respects. Unless I am mistaken, this remark of his betrays a confusion between 'true verisimilitude' (to which we have no epistemic access) and 'estimated verisimilitude' (which is accessible but non-epistemic).

11 Newton-Smith claims that the increasing predictive success of science through time "would be totally mystifying . . . if it were not for the fact that theories are capturing more and more truth about the world" (forthcoming, p. 15).

12 I must stress again that I am *not* denying that there *may* be a connection between approximate truth and predictive success. I am only observing that until the realists show us what that connection is, they should be more reticent than they are about claiming that realism can explain the success of science.

13 A *non-realist* might argue that a theory is approximately true just in case all its *observable* consequences are true or within a specified interval from the true value. Theories that were "approximately true" in this sense would indeed be demonstrably successful. But, the realist's (otherwise commendable) commitment to taking

seriously the theoretical claims of a theory precludes him from utilizing any such construal of approximate truth, since he wants to say that the theoretical as well as the observational consequences are approximately true.

14 If this argument, which I attribute to the realists, seems a bit murky, I challenge any reader to find a more clear-cut one in the literature! Overt formulations of this position can be found in Putnam, Boyd and Newton-Smith.

15 Popper: "a theory which has been well corroborated can only be superseded by one . . . [which] *contains* the old well-corroborated theory – or at least a good approximation to it" (1959, p. 276).

Post: "I shall even claim that, as a matter of empirical historical fact, [successor] theories [have] always explained the *whole* of [the well-confirmed part of their predecessors]" (1971, p. 229).

Koertge: "nearly all pairs of successive theories in the history of science stand in a correspondence relation and . . . where there is no correspondence to begin with, the new theory will be developed in such a way that it comes more nearly into correspondence with the old" (1973, p. 176–177). Among other authors who have defended a similar view, one should mention (Fine 1967, p. 231 ff.), (Kordig 1971, pp. 119–125), (Margenau 1950) and (Sklar 1967, pp. 190–224).

16 Putnam fails to point out that it is also a fact that many scientists do *not* seek to preserve earlier mechanisms and that theories which have not preserved earlier theoretical mechanisms (whether the germ theory of disease, plate tectonics, or wave optics) have led to important discoveries is also a fact.

17 I have written a book about this strategy, (Laudan 1977).

18 After the epistemological and methodological battles about science during the last three hundred years, it should be fairly clear that science, taken at its face value, *implies* no particular epistemology.

19 (Putnam 1978, pp. 20, 123).

20 (Laudan 1976, pp. 467–472).

21 This matter of limiting conditions consistent with the 'reducing' theory is curious. Some of the best-known expositions of limiting case relations

depend (as Krajewski has observed) upon showing an earlier theory to be a limiting case of a later theory only by adopting limiting assumptions *explicitly denied by the later theory.* For instance, several standard textbook discussions present (a portion of) classical mechanics as a limiting case of special relativity, provided c approaches infinity. But special relativity is committed to the claim that c is a constant. Is there not something suspicious about a 'derivation' of T_1 from a T_2 which essentially involves an assumption inconsistent with T_2? If T_2 is correct, then it forbids the adoption of a premise commonly used to derive T_1, as a limiting case. (It should be noted that most such proofs can be re-formulated unobjectionably, e.g., in the relativity case, by letting $v \rightarrow 0$ rather than $c \rightarrow \infty$.)

22 As Mario Bunge has cogently put it: "The popular view on inter-theory relations ... that every new theory includes (as regards its extension) its predecessors ... is philosophically superficial, ... and it is false as a historical hypothesis concerning the advancement of science" (1970, pp. 309–310).

23 I owe the suggestion of this realist response to Andrew Lugg.

24 I find Putnam's views on the 'empirical' or 'scientific' character of realism rather perplexing. At some points, he seems to suggest that realism is both empirical and scientific. Thus, he writes: "If realism is an explanation of this fact [namely, that science is successful], realism must itself be an over-arching scientific *hypothesis*" (1978, p. 19). Since Putnam clearly maintains the antecedent, he seems committed to the consequent. Elsewhere he refers to certain realist tenets as being "our highest level empirical generalizations about knowledge" (p. 37). He says moreover that realism "could be false", and that "facts are relevant to its support (or to criticize it)" (pp. 78–79). Nonetheless, for reasons he has not made clear, Putnam wants to deny that realism is either scientific or a hypothesis (p. 79). How realism can consist of doctrines which 1) explain facts about the world, 2) are empirical generalizations about knowledge, and 3) can be confirmed or falsified by evidence and yet be neither scientific nor hypothetical is left opaque.

References

Boyd, R. (1973), "Realism, Underdetermination, and a Causal Theory of Evidence", *Noûs* **7**: 1–12.

Bunge, M. (1970), "Problems Concerning Intertheory Relations", Weingartner, P. and Zecha, G. (eds), *Induction, Physics and Ethics*: 285–315. Dordrecht: Reidel.

Fine, A. (1967), "Consistency, Derivability and Scientific Change", *Journal of Philosophy* **64**: 231ff.

Grünbaum, Adolf (1976), "Can a Theory Answer More Questions than One of its Rivals?", *British Journal for Philosophy of Science* **27**: 1–23.

Hooker, Clifford (1974), "Systematic Realism", *Synthese* **26**: 409–497.

Koertge, N. (1973), "Theory Change in Science", Pearce, G. and Maynard, P. (eds), *Conceptual Change*: 167–198. Dordrecht: Reidel.

Kordig, C. (1971), "Scientific Transitions, Meaning Invariance, and Derivability", *Southern Journal of Philosophy*: **9**(2): 119–125.

Krajewski, W. (1977), *Correspondence Principle and Growth of Science*. Dordrecht: Reidel.

Laudan, L. (1976), "Two Dogmas of Methodology", *Philosophy of Science* **43**: 467–472.

Laudan, L. (1977), *Progress and its Problems*. California: University of California Press.

Laudan, L. (1978), "Ex-Huming Hacking", *Erkenntnis* **13**: 417–435.

Margenau, H. (1950), *The Nature of Physical Reality*. New York: McGraw-Hill.

McMullin, Ernan (1970), "The History and Philosophy of Science: A Taxonomy", Stuewer, R. (ed.), *Minnesota Studies in the Philosophy of Science V*: 12–67. Minneapolis: University of Minnesota Press.

Newton-Smith, W. (1978), "The Underdetermination of Theories by Data", *Proceedings of the Aristotelian Society*: 71–91.

Newton-Smith, W. (forthcoming), "In Defense of Truth".

Niiniluoto, Ilkka (1977), "On the Truthlikeness of Generalizations", Butts, R. and Hintikka, J. (eds), *Basic Problems in Methodology and Linguistics*: 121–147. Dordrecht: Reidel.

Niiniluoto, Ilkka (1980), "Scientific Progress", *Synthese* **45**: 427–62.

Popper, K. (1959), *Logic of Scientific Discovery*. New York: Basic Books.

Popper, K. (1963), *Conjectures and Refutations*. London: Routledge & Kegan Paul.

Popper, K. (1972), *Objective Knowledge*. Oxford: Oxford University Press.

Post, H.R. (1971), "Correspondence, Invariance and Heuristics: In Praise of Conservative Induction", *Studies in the History and Philosophy of Science* **2**: 213–255.

Putnam, H. (1975), *Mathematics, Matter and Method*, Vol. 1. Cambridge: Cambridge University Press.

Putnam, H. (1978), *Meaning and the Moral Sciences*. London: Routledge & Kegan Paul.

Sellars W. (1963), *Science, Perception and Reality*. New York: The Humanities Press.

Sklar, L. (1967), "Types of Inter-Theoretic Reductions", *British Journal for Philosophy of Science* **18**: 190–224.

Szumilewicz, I. (1977), "Incommensurability and the Rationality of the Development of Science", *British Journal for Philosophy of Science* **28**: 348.

Watkins, John (1978), "Corroboration and the Problem of Content-Comparison", Radnitzky and Andersson (eds), *Progress and Rationality in Science*: 339–378. Dordrecht: Reidel.

Chapter 37

Bas van Fraassen

ARGUMENTS CONCERNING SCIENTIFIC REALISM

The rigour of science requires that we distinguish well the undraped figure of nature itself from the gay-coloured vesture with which we clothe it at our pleasure.

Heinrich Hertz, quoted
by Ludwig Boltzmann, letter to
Nature, 28 February 1895

In our century, the first dominant philosophy of science was developed as part of logical positivism. Even today, such an expression as 'the received view of theories' refers to the views developed by the logical positivists, although their heyday preceded the Second World War.

In this chapter I shall examine, and criticize, the main arguments that have been offered for scientific realism. These arguments occurred frequently as part of a critique of logical positivism. But it is surely fair to discuss them in isolation, for even if scientific realism is most easily understood as a reaction against positivism, it should be able to stand alone. The alternative view which I advocate – for lack of a traditional name I shall call it *constructive empiricism* – is equally at odds with positivist doctrine.

1 Scientific realism and constructive empiricism

In philosophy of science, the term 'scientific realism' denotes a precise position on the question of how a scientific theory is to be understood, and what scientific activity really is. I shall attempt to define this position, and to canvass its possible alternatives. Then I shall indicate, roughly and briefly, the specific alternative which I shall advocate and develop in later chapters [*sic*].

1.1 Statement of scientific realism

What exactly is scientific realism? A naïve statement of the position would be this: the picture which science gives us of the world is a true one, faithful in its details, and the entities postulated in science really exist: the advances of science are discoveries, not inventions. That statement is too naïve; it attributes to the scientific realist the belief that today's theories are correct. It would mean that the philosophical position of an earlier scientific realist such as C. S. Peirce had been refuted by empirical findings. I do not suppose that scientific realists wish to be committed, as such, even to the claim that science will arrive in due time at theories true in all respects — for the growth of science might be an endless self-correction; or worse, Armageddon might occur too soon.

But the naïve statement has the right flavour. It answers two main questions: it characterizes a scientific theory as a story about what there really is, and scientific activity as an enterprise of discovery, as opposed to invention. The two questions of what a scientific theory is, and what a scientific theory does, must be answered by any philosophy of science. The task we have at this point is to find a statement of scientific realism that shares these features with the naïve statement, but does not saddle the realists with unacceptably strong consequences. It is especially important to make the statement as weak as possible if we wish to argue against it, so as not to charge at windmills.

As clues I shall cite some passages most of which will also be examined below in the contexts of the authors' arguments. A statement of Wilfrid Sellars is this:

> to have good reason for holding a theory is *ipso facto* to have good reason for holding that the entities postulated by the theory exist.

This addresses a question of epistemology, but also throws some indirect light on what it is in Sellars's opinion, to hold a theory. Brian Ellis, who calls himself a scientific entity realist rather than a scientific realist, appears to agree with that statement of Sellars, but gives the following formulation of a stronger view:

> I understand scientific realism to be the view that the theoretical statements of science are, or purport to be, true generalized descriptions of reality.[1]

This formulation has two advantages: It focuses on the understanding of the theories without reference to reasons for belief, and it avoids the suggestion that to be a realist you must believe current scientific theories to be true. But it gains the latter advantage by use of the word 'purport', which may generate its own puzzles.

Hilary Putnam, in a passage which I shall cite again in Section 7, gives a formulation which he says he learned from Michael Dummett:

> A realist (with respect to a given theory or discourse) holds that (1) the sentences of that theory are true or false; and (2) that what makes them true or false is something external — that is to say, it is not (in general) our sense data, actual or potential, or the structure of our minds, or our language, etc.[29]

He follows this soon afterwards with a further formulation which he credits to Richard Boyd:

> That terms in mature scientific theories typically refer (this formulation is due to Richard Boyd), that the theories accepted in a mature science are typically approximately true, that the same term can refer to the same thing even when it occurs in different theories — these statements are viewed by the scientific realist . . . as part of any adequate scientific description of science and its relations to its objects.[33]

None of these were intended as definitions. But they show I think that truth must play an important role in the formulation of the basic realist position. They also show that the formulation must incorporate an answer to the question what it is to *accept* or *hold* a theory. I shall now propose such a formulation, which seems to me to make sense of the above remarks, and also renders intelligible the reasoning by realists which I shall examine below — without burdening them with more than the minimum required for this.

Science aims to give us, in its theories, a literally true story of what the world is like: and acceptance of a scientific theory involves the belief that it is true. This is the correct statement of scientific realism.

Let me defend this formulation by showing that it is quite minimal, and can be agreed to by anyone who considers himself a scientific realist. The naïve statement said that science tells a true story; the correct statement says only that it is the aim of science to do so. The aim of science is of course not to be identified with individual scientists' motives. The aim of the game of chess is to checkmate your opponent; but the motive for playing may be fame, gold, and glory. What the aim is determines what counts as success in the enterprise as such; and this aim may be pursued for any number of reasons. Also, in calling something *the* aim, I do not deny that there are other subsidiary aims which may or may not be means to that end: everyone will readily agree that simplicity, informativeness, predictive power, explanation are (also) virtues. Perhaps my formulation can even be accepted by any philosopher who considers the most important aim of science to be something which only *requires* the finding of true theories – given that I wish to give the weakest formulation of the doctrine that is generally acceptable.

I have added 'literally' to rule out as realist such positions as imply that science is true if 'properly understood' but literally false or meaningless. For that would be consistent with conventionalism, logical positivism, and instrumentalism. I will say more about this below; and also in Section 7 where I shall consider Dummett's views further.

The second part of the statement touches on epistemology. But it only equates acceptance of a theory with belief in its truth.[2] It does not imply that anyone is ever rationally warranted in forming such a belief. We have to make room for the epistemological position, today the subject of considerable debate, that a rational person never assigns personal probability 1 to any proposition except a tautology. It would, I think, be rare for a scientific realist to take this stand in epistemology, but it is certainly possible.[3]

To understand qualified acceptance we must first understand acceptance *tout court*. If acceptance of a theory involves the belief that it is true, then tentative acceptance involves the tentative adoption of the belief that it is true. If belief comes in degrees, so does acceptance, and we may then speak of a degree of acceptance involving a certain degree of belief that the theory is true. This must of course be distinguished from belief that the theory is approximately true, which seems to mean belief that some member of a class centring on the mentioned theory is (exactly) true. In this way the proposed formulation of realism can be used regardless of one's epistemological persuasion.

1.2 *Alternatives to realism*

Scientific realism is the position that scientific theory construction aims to give us a literally true story of what the world is like, and that acceptance of a scientific theory involves the belief that it is true. Accordingly, anti-realism is a position according to which the aim of science can well be served without giving such a literally true story, and acceptance of a theory may properly involve something less (or other) than belief that it is true.

What does a scientist do then, according to these different positions? According to the realist, when someone proposes a theory, he is asserting it to be true. But according to the anti-realist, the proposer does not assert the theory to be true; *he displays* it, and claims certain virtues for it. These virtues may fall short of truth: empirical adequacy, perhaps; comprehensiveness, acceptability for various purposes. This will have to be spelt out, for the details here are not determined by the denial of realism. For now we must concentrate on the key notions that allow the generic division.

The idea of a literally true account has two aspects: the language is to be literally construed; and so construed, the account is true. This divides the anti-realists into two sorts. The first sort holds that science is or aims to be true, properly (but not literally) construed. The second holds that the language of science should be literally construed, but its theories need not be true to be good. The anti-realism I shall advocate belongs to the second sort.

It is not so easy to say what is meant by a literal construal. The idea comes perhaps from theology, where fundamentalists construe the Bible literally, and liberals have a variety of allegorical, metaphorical, and analogical interpretations, which 'demythologize'. The problem of explicating 'literal construal' belongs to the philosophy of language. In Section 7 below, where I briefly examine some of Michael Dummett's views, I shall emphasize that 'literal' does not mean 'truth-valued'. The term 'literal' is well enough understood for general philosophical use, but if we try to explicate it we find ourselves in the midst of the problem of giving an adequate account of natural language. It would be bad tactics to link an inquiry into science to a commitment to some solution to that problem. The following remarks, and those in Section 7, should fix the usage of 'literal' sufficiently for present purposes.

The decision to rule out all but literal construals of the language of science, rules out those forms of anti-realism known as *positivism* and *instrumentalism*. First, on a literal construal, the apparent statements of science really are statements, *capable of* being true or false. Secondly, although a literal construal can elaborate, it cannot change logical relationships. (It is possible to elaborate, for instance, by identifying what the terms designate. The 'reduction' of the language of phenomenological thermodynamics to that of

statistical mechanics is like that: bodies of gas are identified as aggregates of molecules, temperature as mean kinetic energy, and so on.) On the positivists' interpretation of science, theoretical terms have meaning only through their connection with the observable. Hence they hold that two theories may in fact *say the same thing* although in form they contradict each other. (Perhaps the one says that all matter consists of atoms, while the other postulates instead a universal continuous medium; they will say the same thing nevertheless if they agree in their observable consequences, according to the positivists.) But two theories which contradict each other in such a way can 'really' be saying the same thing only if they are not literally construed. Most specifically, if a theory says that something exists, then a literal construal may elaborate on what that something is, but will not remove the implication of existence.

There have been many critiques of positivist interpretations of science, and there is no need to repeat them. I shall add some specific criticisms of the positivist approach in the next chapter [*sic*].

1.3 Constructive empiricism

To insist on a literal construal of the language of science is to rule out the construal of a theory as a metaphor or simile, or as intelligible only after it is 'demythologized' or subjected to some other sort of 'translation' that does not preserve logical form. If the theory's statements include 'There are electrons', then the theory says that there are electrons. If in addition they include 'Electrons are not planets', then the theory says, in part, that there are entities other than planets.

But this does not settle very much. It is often not at all obvious whether a theoretical term refers to a concrete entity or a mathematical entity. Perhaps one tenable interpretation of classical physics is that there are no concrete entities which are forces – that 'there are forces such that . . .' can always be understood as a mathematical statement asserting the existence of certain functions. That is debatable.

Not every philosophical position concerning science which insists on a literal construal of the language of science is a realist position. For this insistence relates not at all to our epistemic attitudes toward theories, nor to the aim we pursue in constructing theories, but only to the correct understanding of *what a theory says*. (The fundamentalist theist, the agnostic, and the atheist presumably agree with each other (though not with liberal theologians) in their understanding of the statement that God, or gods, or angels exist.) After deciding that the language of science must be literally understood, we can still say that there is no need to believe good theories to be true, nor to believe *ipso facto* that the entities they postulate are real.

Science aims to give us theories which are empirically adequate; and acceptance of a theory involves as belief only that it is empirically adequate. This is the statement of the anti-realist position I advocate; I shall call it *constructive empiricism*.

This formulation is subject to the same qualifying remarks as that of scientific realism in Section 1.1 above. In addition it requires an explication of 'empirically adequate'. For now, I shall leave that with the preliminary explication that a theory is empirically adequate exactly if what it says about the observable things and events in this world, is true – exactly if it 'saves the phenomena'. A little more precisely: such a theory has at least one model that all the actual phenomena fit inside. I must emphasize that this refers to *all* the phenomena: these are not exhausted by those actually observed, nor even by those observed at some time, whether past, present, or future. The whole of the next chapter will be devoted to the explication of this term [*sic*], which is intimately bound up with our conception of the structure of a scientific theory.

The distinction I have drawn between realism and anti-realism, in so far as it pertains to acceptance, concerns only how much belief is involved therein. Acceptance of theories (whether full, tentative, to a degree, etc.) is a phenomenon of scientific activity which clearly involves more than belief. One main reason for this is that we are never confronted with a complete theory. So if a scientist accepts a theory, he thereby involves himself in a certain sort of research programme. That programme could well be different from the one acceptance of another theory would have given him, even if those two (very incomplete) theories are equivalent to each other with respect to everything that is observable – in so far as they go.

Thus acceptance involves not only belief but a certain commitment. Even for those of us who are not working scientists, the acceptance involves a commitment to confront any future phenomena by means of the conceptual resources of this theory. It determines the terms in which we shall seek explanations. If the acceptance is at all strong, it is exhibited in the person's assumption of the role of explainer, in his willingness to answer questions *ex cathedra*. Even if you do not accept a theory, you can engage in discourse in a context in which language use is guided by that theory – but acceptance produces such contexts. There are similarities in all of this to ideological commitment. A commitment is of course not true or false: The confidence exhibited is that it will be *vindicated*.

This is a preliminary sketch of the *pragmatic* dimension of theory acceptance. Unlike the epistemic dimension, it does not figure overtly in the disagreement between realist and anti-realist. But because the amount of belief involved in acceptance is typically less according to anti-realists, they will tend to make more of the pragmatic aspects. It is as well to note here the important difference. Belief that a theory is true, or that it is empirically adequate, does not imply, and is not implied by, belief that full acceptance of the theory will be vindicated. To see this, you need only consider here a person who has quite definite beliefs about the future of the human race, or about the scientific community and the influences thereon and practical limitations we have. It might well be, for instance, that a theory which is empirically adequate will not combine easily with some other theories which we have accepted in fact, or that Armageddon will occur before we succeed. Whether belief that a theory is true, or that it is empirically adequate, can be equated with belief that acceptance of it would, under ideal research conditions, be vindicated in the long run, is another question. It seems to me an irrelevant question within philosophy of science, because an affirmative answer would not obliterate the distinction we have already established by the preceding remarks. (The question may also assume that counterfactual statements are objectively true or false, which I would deny.)

Although it seems to me that realists and anti-realists need not disagree about the pragmatic aspects of theory acceptance, I have mentioned it here because I think that typically they do. We shall find ourselves returning time and again, for example, to requests for explanation to which realists typically attach an objective validity which anti-realists cannot grant.

2 The theory/observation 'dichotomy'

For good reasons, logical positivism dominated the philosophy of science for thirty years. In 1960, the first volume of *Minnesota Studies in the Philosophy of Science* published Rudolf Carnap's 'The Methodological Status of Theoretical Concepts', which is, in many ways, the culmination of the positivist programme. It interprets science by relating it to an observation language

(a postulated part of natural language which is devoid of theoretical terms). Two years later this article was followed in the same series by Grover Maxwell's 'The Ontological Status of Theoretical Entities', in title and theme a direct counter to Carnap's. This is the *locus classicus* for the new realists' contention that the theory/observation distinction cannot be drawn.

I shall examine some of Maxwell's points directly, but first a general remark about the issue. Such expressions as 'theoretical entity' and 'observable–theoretical dichotomy' are, on the face of it, examples of category mistakes. Terms or concepts are theoretical (introduced or adapted for the purposes of theory construction); entities are observable or unobservable. This may seem a little point, but it separates the discussion into two issues. Can we divide our language into a theoretical and non-theoretical part? On the other hand, can we classify objects and events into observable and unobservable ones?

Maxwell answers both questions in the negative, while not distinguishing them too carefully. On the first, where he can draw on well-known supportive essays by Wilfrid Sellars and Paul Feyerabend, I am in total agreement. All our language is thoroughly theory-infected. If we could cleanse our language of theory-laden terms, beginning with the recently introduced ones like 'VHF receiver', continuing through 'mass' and 'impulse' to 'element' and so on into the prehistory of language formation, we would end up with nothing useful. The way we talk, and scientists talk, is guided by the pictures provided by previously accepted theories. This is true also, as Duhem already emphasized, of experimental reports. Hygienic reconstructions of language such as the positivists envisaged are simply not on. I shall return to this criticism of positivism in the next chapter [*sic*].

But does this mean that we must be scientific realists? We surely have more tolerance of ambiguity than that. The fact that we let our language be guided by a given picture, at some point, does not show how much we believe about that picture. When we speak of the sun coming up in the morning and setting at night, we are guided by a picture now explicitly disavowed. When Milton wrote *Paradise Lost* he deliberately let the old geocentric astronomy guide his poem, although various remarks in passing clearly reveal his interest in the new astronomical discoveries and speculations of his time. These are extreme examples, but show that no immediate conclusions can be drawn from the theory-ladenness of our language.

However, Maxwell's main arguments are directed against the observable–unobservable distinction. Let us first be clear on what this distinction was supposed to be. The term 'observable' classifies putative entities (entities which may or may not exist). A flying horse is observable – that is why we are so sure that there aren't any – and the number seventeen is not. There is supposed to be a correlate classification of human acts: an unaided act of perception, for instance, is an observation. A calculation of the mass of a particle from the deflection of its trajectory in a known force field, is not an observation of that mass.

It is also important here not to confuse *observing* (an entity, such as a thing, event, or process) and *observing that* (something or other is the case). Suppose one of the Stone Age people recently found in the Philippines is shown a tennis ball or a car crash. From his behaviour, we see that he has noticed them; for example, he picks up the ball and throws it. But he has not seen *that* it is a tennis ball, or *that* some event is a car crash, for he does not even have those concepts. He cannot get that information through perception; he would first have to learn a great deal. To say that he does not see the same things and events as we do, however, is just silly; it is a pun which trades on the ambiguity between seeing and seeing that. (The truth-conditions for our statement '*x* observes *that A*' must be such that what concepts

x has, presumably related to the language *x* speaks if he is human, enter as a variable into the correct truth definition, in some way. To say that *x* observed the tennis ball, therefore, does not imply at all that *x* observed that it was a tennis ball; that would require some conceptual awareness of the game of tennis.)

The arguments Maxwell gives about observability are of two sorts: one directed against the possibility of drawing such distinctions, the other against the importance that could attach to distinctions that can be drawn.

The first argument is from the continuum of cases that lie between direct observation and inference:

> there is, in principle, a continuous series beginning with looking through a vacuum and containing these as members: looking through a windowpane, looking through glasses, looking through binoculars, looking through a low-power microscope, looking through a high-power microscope, etc., in the order given. The important consequence is that, so far, we are left without criteria which would enable us to draw a non-arbitrary line between 'observation' and 'theory'.[4]

This continuous series of supposed acts of observation does not correspond directly to a continuum in what is supposed observable. For if something can be seen through a window, it can also be seen with the window raised. Similarly, the moons of Jupiter can be seen through a telescope; but they can also be seen without a telescope if you are close enough. That something is observable does not automatically imply that the conditions are right for observing it now. The principle is:

> X is observable if there are circumstances which are such that, if X is present to us under those circumstances, then we observe it.

This is not meant as a definition, but only as a rough guide to the avoidance of fallacies.

We may still be able to find a continuum in what is supposed detectable: perhaps some things can only be detected with the aid of an optical microscope, at least; perhaps some require an electron microscope, and so on. Maxwell's problem is: where shall we draw the line between what is observable and what is only detectable in some more roundabout way?

Granted that we cannot answer this question without arbitrariness, what follows? That 'observable' is a *vague predicate*. There are many puzzles about vague predicates, and many sophisms designed to show that, in the presence of vagueness, no distinction can be drawn at all. In Sextus Empiricus, we find the argument that incest is not immoral, for touching your mother's big toe with your little finger is not immoral, and all the rest differs only by degree. But predicates in natural language are almost all vague, and there is no problem in their use; only in formulating the logic that governs them.[5] A vague predicate is usable provided it has clear cases and clear counter-cases. Seeing with the unaided eye is a clear case of observation. Is Maxwell then perhaps challenging us to present a clear counter-case? Perhaps so, for he says 'I have been trying to support the thesis that any (non-logical) term is a *possible* candidate for an observation term.'

A look through a telescope at the moons of Jupiter seems to me a clear case of observation, since astronauts will no doubt be able to see them as well from close up. But the purported observation of micro-particles in a cloud chamber seems to me a clearly different case – if our theory about what happens there is right. The theory says that if a charged particle traverses a chamber filled with saturated vapour, some atoms in the neighbourhood of its path are ionized. If this vapour is decompressed, and hence becomes super-saturated, it condenses in droplets on the ions, thus marking the path of the particle. The resulting silver-grey line is similar (physically as well as in appearance) to the vapour trail left in the sky when a jet passes. Suppose I point to such a trail and say: 'Look, there is a jet!'; might you not say: 'I see the vapour trail, but where is the jet?' Then I would answer: 'Look just a bit ahead of the trail . . . there! Do you see it?' Now, in the case of the cloud chamber this response is not possible. So while the particle is detected by means of the cloud chamber, and the detection is based on observation, it is clearly not a case of the particle's being observed.

As a second argument, Maxwell directs our attention to the 'can' in 'what is observable is what can be observed.' An object might of course be temporarily unobservable – in a rather different sense: it cannot be observed in the circumstances in which it actually is at the moment, but could be observed if the circumstances were more favourable. In just the same way, I might be temporarily invulnerable or invisible. So we should concentrate on 'observable' *tout court*, or on (as he prefers to say) 'unobservable in principle'. This Maxwell explains as meaning that the relevant scientific theory *entails* that the entities cannot be observed in any circumstances. But this never happens, he says, because the different circumstances could be ones in which we have different sense organs – electron-microscope eyes, for instance.

This strikes me as a trick, a change in the subject of discussion. I have a mortar and pestle made of copper and weighing about a kilo. Should I call it breakable because a giant could break it? Should I call the Empire State Building portable? Is there no distinction between a portable and a console record player? The human organism is, from the point of view of physics, a certain kind of measuring apparatus. As such it has certain inherent limitations – which will be described in detail in the final physics and biology. It is these limitations to which the 'able' in 'observable' refers – our limitations, *qua* human beings.

As I mentioned, however, Maxwell's article also contains a different sort of argument: even if there is a feasible observable/unobservable distinction, this distinction has no importance. The point at issue for the realist is, after all, the reality of the entities postulated in science. Suppose that these entities could be classified into observables and others; what relevance should that have to the question of their existence?

Logically, none. For the term 'observable' classifies putative entities, and has logically nothing to do with existence. But Maxwell must have more in mind when he says: 'I conclude that the drawing of the observational–theoretical line at any given point is an accident and a function of our physiological make-up, . . . and, therefore, that it has no ontological significance whatever.'[6] No ontological significance if the question is only whether 'observable' and 'exists' imply each other – for they do not; but significance for the question of scientific realism?

Recall that I defined scientific realism in terms of the aim of science, and epistemic attitudes. The question is what aim scientific activity has, and how much we shall believe when we accept a scientific theory. What is the proper form of acceptance: belief that the theory, as a whole, is true; or something else? To this question, what is observable by us seems eminently relevant. Indeed, we may attempt an answer at this point: to accept a theory is (for us) to believe that it is empirically adequate – that what the theory says *about what is observable* (by us) is true.

It will be objected at once that, on this proposal, what the anti-realist decides to believe about the world will depend in part on what he believes to be his, or rather the epistemic community's, accessible range of evidence. At present, we count the human race as the epistemic community to which we belong; but this race may mutate, or that community may be increased by adding other animals (terrestrial or extra-terrestrial) through relevant ideological or moral decisions ('to count them as persons'). Hence the anti-realist would, on my proposal, have to accept conditions of the form

> If the epistemic community changes in fashion Y, then my beliefs about the world will change in manner Z.

To see this as an objection to anti-realism is to voice the requirement that our epistemic policies should give the same results independent of our beliefs about the range of evidence accessible to us. That requirement seems to me in no way rationally compelling; it could be honoured, I should think, only through a thorough-going scepticism or through a commitment to wholesale leaps of faith. But we cannot settle the major questions of epistemology *en passant* in philosophy of science; so I shall just conclude that it is, on the face of it, not irrational to commit oneself only to a search for theories that are empirically adequate, ones whose models fit the observable phenomena, while recognizing that what counts as an observable phenomenon is a function of what the epistemic community is (that *observable* is *observable-to-us*).

The notion of empirical adequacy in this answer will have to be spelt out very carefully if it is not to bite the dust among hackneyed objections. I shall try to do so in the next chapter [*sic*]. But the point stands: even if observability has nothing to do with existence (is, indeed, too anthropocentric for that), it may still have much to do with the proper epistemic attitude to science.

3 Inference to the best explanation

A view advanced in different ways by Wilfrid Sellars, J. J. C. Smart, and Gilbert Harman is that the canons of rational inference require scientific realism. If we are to follow the same patterns of inference with respect to this issue as we do in science itself, we shall find ourselves irrational unless we assert the truth of the scientific theories we accept. Thus Sellars says: 'As I see it, to have good reason for holding a theory is *ipso facto* to have good reason for holding that the entities postulated by the theory exist.'[7]

The main rule of inference invoked in arguments of this sort is the rule of *inference to the best explanation*. The idea is perhaps to be credited to C. S. Peirce,[8] but the main recent attempts to explain this rule and its uses have been made by Gilbert Harman.[9] I shall only present a simplified version. Let us suppose that we have evidence E, and are considering several hypotheses, say H and H'. The rule then says that we should infer H rather than H' exactly if H is a better explanation of E than H' is. (Various qualifications are necessary to avoid inconsistency: we should always try to move to the best over-all explanation of all available evidence.)

It is argued that we follow this rule in all 'ordinary' cases; and that if we follow it consistently everywhere, we shall be led to scientific realism, in the way Sellars's dictum suggests. And surely there are many telling 'ordinary' cases: I hear scratching in the wall, the patter of little feet at midnight, my cheese disappears — and I infer that a mouse has come to live with me. Not merely that these apparent signs of mousely presence will continue, not merely that all the observable phenomena will be as if there is a mouse; but that there really is a mouse.

Will this pattern of inference also lead us to belief in unobservable entities? Is the scientific realist simply someone who consistently follows the rules of inference that we all follow in more mundane contexts? I have two objections to the idea that this is so.

First of all, what is meant by saying that we all follow a certain rule of inference? One meaning might be that we deliberately and consciously 'apply' the rule, like a student doing a logic exercise. That meaning is much too literalistic and restrictive; surely all of mankind follows the rules of logic much of the time, while only a fraction can even formulate them. A second meaning is that we act in accordance with the rules in a sense that does not require conscious deliberation. That is not so easy to make precise, since each logical rule is a rule of permission (*modus ponens* allows you to infer B from A and (if A then B), but does not forbid you to infer (B or A) instead). However, we might say that a person behaved in accordance with a set of rules in that sense if every conclusion he drew could be reached from his premisses via those rules. But this meaning is much too loose; in this sense we always behave in accordance with the rule that any conclusion may be inferred from any premiss. So it seems that to be following a rule, I must be willing to believe all conclusions it allows, while definitely unwilling to believe conclusions at variance with the ones it allows — or else, change my willingness to believe the premisses in question.

Therefore the statement that we all follow a certain rule in certain cases, is a *psychological hypothesis* about what we are willing and unwilling to do. It is an empirical hypothesis, to be confronted with data, and with rival hypotheses. Here is a rival hypothesis: we are always willing to believe that the theory which best explains the evidence, is empirically adequate (that all the observable phenomena are as the theory says they are).

In this way I can certainly account for the many instances in which a scientist appears to argue for the acceptance of a theory or hypothesis, on the basis of its explanatory success. (A number of such instances are related by Thagard.[8]) For, remember: I equate the acceptance of a scientific theory with the belief that it is empirically adequate. We have therefore two rival hypotheses concerning these instances of scientific inference, and the one is apt in a realist account, the other in an anti-realist account.

Cases like the mouse in the wainscoting cannot provide telling evidence between those rival hypotheses. For the mouse is an observable thing; therefore 'there is a mouse in the wainscoting' and 'All observable phenomena are as if there is a mouse in the wainscoting' are totally equivalent; each implies the other (given what we know about mice).

It will be countered that it is less interesting to know whether people do follow a rule of inference than whether they ought to follow it. Granted; but the premiss that we all follow the rule of inference to the best explanation when it comes to mice and other mundane matters – that premiss is shown wanting. It is not warranted by the evidence, because that evidence is not telling for the premiss *as against* the alternative hypothesis I proposed, which is a relevant one in this context.

My second objection is that even if we were to grant the correctness (or worthiness) of the rule of inference to the best explanation, the realist needs some further premiss for his argument. For this rule is only one that dictates a choice when given a set of rival hypotheses. In other words, we need to be committed to belief in one of a range of hypotheses before the rule can be applied. Then, under favourable circumstances, it will tell us which of the hypotheses in that range to choose. The realist asks us to choose between different hypotheses that explain the regularities in certain ways; but his opponent always wishes to choose among hypotheses of the form 'theory T_i is empirically adequate'. So the realist will need his special extra premiss that every universal regularity in nature needs an explanation, before the rule will make realists of us all. And that is just the premiss that distinguishes the realist from his opponents (and which I shall examine in more detail in Sections 4 and 5 below).

The logically minded may think that the extra premiss can be bypassed by logical *léger-de-main*. For suppose the data are that all facts observed so far accord with theory T; then T is one possible explanation of those data. A rival is not-T (that T is false). This rival is a very poor explanation of the data. So we *always* have a set of rival hypotheses, and the rule of inference to the best explanation leads us unerringly to the conclusion that T is true. Surely I am committed to the view that T is true or T is false?

This sort of epistemological rope-trick does not work of course. To begin, I am committed to the view that T is true or T is false, but not thereby committed to an inferential move to one of the two! The rule operates only if I have decided not to remain neutral between these two possibilities.

Secondly, it is not at all likely that the rule will be applicable to such logically concocted rivals. Harman lists various criteria to apply to the evaluation of hypotheses *qua* explanations.[10] Some are rather vague, like simplicity (but is simplicity not a reason to use a theory whether you believe it or not?). The precise ones come from statistical theory which has lately proved of wonderful use to epistemology:

H is a better explanation than H′ (*ceteris paribus*) of E, provided:
(a) $P(H) > P(H')$ – H has higher probability than H
(b) $P(E/H) > P(E/H')$ – H bestows higher probability on E than H′ does.

The use of 'initial' or a priori probabilities in (a) – the initial plausibility of the hypotheses themselves – is typical of the so-called *Bayesians*. More traditional statistical practice suggests only the use of (b). But even that supposes that H and H′ bestow definite probabilities on E. If H′ is simply the denial of H, that is not generally the case. (Imagine that H says that the probability of

E equals ¾. The very most that not-H will entail is that the probability of E is some number other than ¾; and usually it will not even entail that much, since H will have other implications as well.)

Bayesians tend to cut through this 'unavailability of probabilities' problem by hypothesizing that everyone has a specific subjective probability (degree of belief) for every proposition he can formulate. In that case, no matter what E, H, H' are, all these probabilities really are (in principle) available. But they obtain this availability by making the probabilities thoroughly subjective. I do not think that scientific realists wish their conclusions to hinge on the subjectively established initial plausibility of there being unobservable entities, so I doubt that this sort of Bayesian move would help here. (This point will come up again in a more concrete form in connection with an argument by Hilary Putnam.)

I have kept this discussion quite abstract; but more concrete arguments by Sellars, Smart, and Putnam will be examined below. It should at least be clear that there is no open-and-shut argument from common sense to the unobservable. Merely following the ordinary patterns of inference in science does not obviously and automatically make realists of us all.

4 Limits of the demand for explanation

In this section and the next two, I shall examine arguments for realism that point to explanatory power as a criterion for theory choice. That this is indeed a criterion I do not deny. But these arguments for realism succeed only if the demand for explanation is supreme – if the task of science is unfinished, *ipso facto*, as long as any pervasive regularity is left unexplained. I shall object to this line of argument, as found in the writings of Smart, Reichenbach, Salmon, and Sellars, by arguing that such an unlimited

demand for explanation leads to a demand for hidden variables, which runs contrary to at least one major school of thought in twentieth-century physics. I do not think that even these philosophers themselves wish to saddle realism with logical links to such consequences: but realist yearnings were born among the mistaken ideals of traditional metaphysics.

In his book *Between Science and Philosophy*, Smart gives two main arguments for realism. One is that only realism can respect the important distinction between *correct* and *merely useful* theories. He calls 'instrumentalist' any view that locates the importance of theories in their use, which requires only empirical adequacy, and not truth. But how can the instrumentalist explain the usefulness of his theories?

> Consider a man (in the sixteenth century) who is a realist about the Copernican hypothesis but instrumentalist about the Ptolemaic one. He can explain the instrumental usefulness of the Ptolemaic system of epicycles because he can prove that the Ptolemaic system can produce almost the same predictions about the apparent motions of the planets as does the Copernican hypothesis. Hence the assumption of the realist truth of the Copernican hypothesis explains the instrumental usefulness of the Ptolemaic one. Such an explanation of the instrumental usefulness of certain theories would not be possible if *all* theories were regarded as merely instrumental.[11]

What exactly is meant by 'such an explanation' in the last sentence? If no theory is assumed to be true, then no theory has its usefulness explained as following from the truth of another one – granted. But would we have less of an explanation of the usefulness of the Ptolemaic hypothesis if we began instead with the premiss that the Copernican gives implicitly a very accurate description of the motions of the planets as observed from earth? This would not assume the truth of Copernicus's heliocentric hypothesis, but would still entail that Ptolemy's simpler description was also a close approximation of those motions.

However, Smart would no doubt retort that such a response pushes the question only one step back: what explains the accuracy of predictions based on Copernicus's theory? If I say, the empirical adequacy of that theory, I have merely given a verbal explanation. For of course Smart does not mean to limit his question to actual predictions – it really concerns all actual and possible predictions and retrodictions. To put it quite concretely: what explains the fact that all observable planetary phenomena fit Copernicus's theory (if they do)? From the medieval debates, we recall the nominalist response that the basic regularities are merely brute regularities, and have no explanation. So here the anti-realist must similarly say: that the observable phenomena exhibit these regularities, because of which they fit the theory, is merely a brute fact, and may or may not have an explanation in terms of unobservable facts 'behind the phenomena' – it really does not matter to the goodness of the theory, nor to our understanding of the world.

Smart's main line of argument is addressed to exactly this point. In the same chapter he argues as follows. Suppose that we have a theory T which postulates micro-structure directly, and macro-structure indirectly. The statistical and approximate laws about macroscopic phenomena are only partially spelt out perhaps, and in any case derive from the precise (deterministic or statistical) laws about the basic entities. We now consider theory T', which is part of T, and says only what T says about the macroscopic phenomena. (How T' should be characterized I shall leave open, for that does not affect the argument here.) Then he continues:

I would suggest that the realist could (say) . . . that the success of T' is explained by the fact that the original theory T is true of the things that it is ostensibly about; in other words by the fact that there really are electrons or whatever is postulated by the theory T. If there were no such things, and if T were not true in a realist way, would not the success of T' be quite inexplicable? One would have to suppose that there were innumerable lucky accidents about the behaviour mentioned in the observational vocabulary, so that they behaved miraculously *as if* they were brought about by the nonexistent things ostensibly talked about in the theoretical vocabulary.[12]

In other passages, Smart speaks similarly of 'cosmic coincidences'. The regularities in the observable phenomena must be explained in terms of deeper structure, for otherwise we are left with a belief in lucky accidents and coincidences on a cosmic scale.

I submit that if the demand for explanation implicit in these passages were precisely formulated, it would at once lead to absurdity. For if the mere fact of postulating regularities, without explanation, makes T' a poor theory, T will do no better. If, on the other hand, there is some precise limitation on what sorts of regularities can be postulated as basic, the context of the argument provides no reason to think that T' must automatically fare worse than T.

In any case, it seems to me that it is illegitimate to equate being a lucky accident, or a coincidence, with having no explanation. It was by coincidence that I met my friend in the market – but I can explain why I was there, and he can explain why he came, so together we can explain how this meeting happened. We call it a coincidence, not because the occurrence was inexplicable, but because we did not severally go to the market in order to meet.[13] There cannot be a requirement upon science to provide a theoretical elimination of coincidences, or accidental correlations in general, for that does not even make sense. There is nothing here to motivate the demand for explanation, only a restatement in persuasive terms.

5 The principle of the common cause

Arguing against Smart, I said that if the demand for explanation implicit in his arguments were precisely formulated, it would lead to absurdity. I shall now look at a precise formulation of the demand for explanation: Reichenbach's principle of the common cause. As Salmon has recently pointed out, if this principle is imposed as a demand on our account of what there is in the world, then we are led to postulate the existence of unobservable events and processes.[14]

I will first state the argument, and Reichenbach's principle, in a rough, intuitive form, and then look at its precise formulation. Suppose that two sorts of events are found to have a correlation. A simple example would be that one occurs whenever the other does; but the correlation may only be statistical. There is apparently a significant correlation between heavy cigarette-smoking and cancer, though merely a statistical one. Explaining such a correlation requires finding what Reichenbach called a *common cause*. But, the argument runs, there are often among observable events no common causes of given observable correlations. Therefore, scientific explanation often requires that there be certain unobservable events.

Reichenbach held it to be a principle of scientific methodology that every statistical correlation (at least, every positive dependence) must be explained through common causes. This means then that the very project of science will necessarily lead to the introduction of unobservable structure behind the phenomena. Scientific explanation will be impossible unless there are unobservable entities; but the aim of science is to provide scientific explanation; therefore, the aim of science can only be served if it is true that there are unobservable entities.

To examine this argument, we must first see how Reichenbach arrived at his notion of common causes and how he made it precise. I will then argue that his principle cannot be a general principle of science at all, and secondly, that the postulation of common causes (when it does occur) is also quite intelligible without scientific realism.

Reichenbach was one of the first philosophers to recognize the radical 'probabilistic turn' of modern physics. The classical ideal of science had been to find a method of description of the world so fine that it could yield deterministic laws for all processes. This means that, if such a description be given of the state of the world (or, more concretely, of a single isolated system) at time t, then its state at later time t + d is uniquely determined. What Reichenbach argued very early on is that this ideal has a factual presupposition: it is not logically necessary that such a fine method of description exists, even in principle.[15] This view became generally accepted with the development of quantum mechanics.

So Reichenbach urged philosophers to abandon that classical ideal as the standard of completeness for a scientific theory. Yet it is clear that, if science does not seek for deterministic laws relating events to what happened before them, it does seek for *some* laws. And so Reichenbach proposed that the correct way to view science is as seeking for 'common causes' of a probabilistic or statistical sort.

We can make this precise using the language of probability theory. Let A and B be two events; we use P to designate their probability of occurrence. Thus $P(A)$ is the probability that A occurs and $P(A\&B)$ the probability that both A and B occur. In addition, we must consider the probability that A occurs *given that* B occurs. Clearly the probability of rain *given that* the sky is over-cast, is higher than the probability of rain in general. We say that B is statistically relevant to A if the probability of A given B – written $P(A/B)$ – is different from $P(A)$. If $P(A/B)$ is higher than $P(A)$, we say that there is a positive correlation. Provided A and B are events which have some positive likelihood of occurrence (i.e. $P(A)$, $P(B)$ are not zero), this is a symmetric relationship. The precise definitions are these:

(a) the probability of A given B is defined provided $P(B) \neq 0$, and is

$$P(A / B) = \frac{P(A \& B)}{P(B)}$$

(b) B is statistically relevant to A exactly if $P(A/B) \neq P(A)$

(c) there is a positive correlation between A and B exactly if $P(A\&B) > P(A).P(B)$

(d) from (a) and (c) it follows that, if $P(A) \neq 0$ and $P(B) \neq 0$, then there is a positive correlation between A and B exactly if

$$P(A/B) > P(A),$$

and also if and only if

$$P(B/A) > P(B)$$

To say therefore that there is a positive correla-tion between cancer and heavy cigarette-smoking, is to say that the incidence of cancer among heavy cigarette-smokers is greater than it is in the general population. But because of the symmetry of A and B in (d), this statement by itself gives no reason to think that the smoking produces the cancer rather than the cancer producing the smoking, or both being produced by some other factor, or by several other factors, if any.

We are speaking here of facts relating to the same time. The cause we seek in the past: heavy

smoking at one time is followed (with certain probabilities) by heavy smoking at a later time, and also by being cancerous at that later time. We have in this past event C really found the *common cause* of this present correlation if

$$P(A/B\&C) = P(A/C)$$

We may put this as follows: relative to the infor-mation that C has occurred, A and B are statisti-cally independent. We can define the probability of an event X, whether by itself or conditional on another event Y, *relative to* C as follows:

(e) the probability relative to C is defined as

$$P_c(X) = P(X/C)$$

$$P_c(X/Y) = P_c(X\&Y) \div P_c(Y)$$

$$= P(X/Y\&C)$$

provided $P_c(Y) \neq 0, P(C) \neq 0$

So to say that C is the common cause for the correlation between A and B is to say that, relative to C there is no such correlation. C explains the correlation, because we notice a correlation only as long as we do not take C into account.

Reichenbach's *Principle of the Common Cause* is that *every* relation of positive statistical relevance must be explained by statistical past common causes, in the above way.[16] To put it quite precisely and in Reichenbach's own terms:

If coincidences of two events A and B occur more frequently than would correspond to their independent occurrence, that is, if the events satisfy the relation

(1) $P(A\&B) > P(A).P(B),$

then there exists a common cause C for these events such that the fork ACB is *conjunctive*, that is, satisfies relations (2)–(5) below:

(2) $P(A\&B/C) = P(A/C).P(B/C)$
(3) $P(A\&B/\overline{C}) = P(A/\overline{C}).P(B/\overline{C})$
(4) $P(A/C) > P(A/\overline{C})$
(5) $P(B/C) > P(B/\overline{C})$

(1) follows logically from (2)–(5).

This principle of the common cause is at once precise and persuasive. It may be regarded as a formulation of the conviction that lies behind such arguments as that of Smart, requiring the elimination of 'cosmic coincidence' by science. But it is not a principle that guides twentieth-century science, because it is too close to the demand for deterministic theories of the world that Reichenbach wanted to reject. I shall show this by means of a schematic example; but this example will incorporate the sorts of non-classical correlations which distinguish quantum mechanics from classical physics. I refer here to the correlations exhibited by the thought experiment of Einstein, Podolski, and Rosen in their famous paper 'Can Quantum-Mechanical Description of Reality be Considered Complete?' These correlations are not merely theoretical: they are found in many actual experiments, such as Compton scattering and photon pair production. I maintain in addition that correlations sufficiently similar to refute the principle of the common cause must appear in almost any indeterministic theory of sufficient complexity.[17] Imagine that you have studied the behaviour of a system or object which, after being in state S, always goes into another state which may be characterized by various attributes $F_1, \ldots, F_n$ and $G_1, \ldots, G_n$. Suppose that you have come to the conclusion that this transition is genuinely indeterministic, but you can propose a theory about the transition probabilities:

(8)   (a) $P(F_i/S) = 1/n$
      (b) $P(G_i/S) = 1/n$
      (c) $P(F_i \equiv G_i/S) = 1$

where $\equiv$ means if *and only if* or *when and exactly when*. In other words, it is pure chance whether the state to which S transits is characterized by a given one of the F–attributes, and similarly for the G–attributes, but certain that it is characterized by F_1 if it is characterized by G_1, by F_2 if by G_2, and so on.

If we are convinced that this is an irreducible, indeterministic phenomenon, so that S is a complete description of the initial state, then we have a violation of the principle of the common cause. For from (8) we can deduce

(9) $P(F_i/S).P(G_i/S) = 1/n^2$
      $P(F_i\&G_i/S)=P(F_i/S) = 1/n$

which numbers are equal only if n is zero or one – the deterministic case. In all other cases, S does not qualify as the common cause of the new state's being F_i and G_i, and if S is complete, nothing else can qualify either.

The example I have given is schematic and simplified, and besides its indeterminism, it also exhibits a certain discontinuity, in that we discuss the transition of a system from one state S into a new state. In classical physics, if a physical quantity changed its value from i to j it would do so by taking on all the values between i and j in succession, that is, changing continuously. Would Reichenbach's principle be obeyed at least in some non-trivial, indeterministic theory in which all quantities have a continuous spectrum of values and all change is continuous? I think not, but I shall not argue this further. The question is really academic, for if the principle requires that, then it is also not acceptable to modern physical science.

Could one change a theory which violates Reichenbach's principle into one that obeys it, without upsetting its empirical adequacy? Possibly; one would have to deny that the attribution of state S gives complete information about the system at the time in question, and postulate *hidden parameters* that underlie these states. Attempts to do so for quantum mechanics are referred to as *hidden variable theories*, but it can be shown that if such a theory is empirically equivalent to orthodox quantum mechanics, then it still exhibits non-local correlations of a non-classical sort, which would still violate Reichenbach's principle. But again, the question is academic, since modern physics does not recognize the need for such hidden variables.

Could Reichenbach's principle be weakened so as to preserve its motivating spirit, while eliminating its present unacceptable consequences? As part of a larger theory of explanation (which I shall discuss later), Wesley Salmon has proposed to disjoin equation (2) above with

$$(2^*) \quad P(A\&B/C) > P(A/C) \cdot P(B/C)$$

in which case C would still qualify as common cause. Note that in the schematic example I gave, S would then qualify as a common cause for the events F_i and G_i.

But so formulated, the principle yields a regress. For suppose (2^*) is true. Then we note a positive correlation *relative to* C:

$$P_c(A\&B) > P_c(A) \cdot P_c(B)$$

to which the principle applies and for which it demands a common cause C'. This regress stops only if, at some point, the exhibited common cause satisfies the original equation (2), which brings us back to our original situation; or if some other principle is used to curtail the demand for explanation.

In any case, weakening the principle in various ways (and certainly it will have to be weakened if it is going to be acceptable in any sense) will remove the force of the realist arguments. For any weakening is an agreement to leave some sorts of 'cosmic coincidence' unexplained. But that is to admit the tenability of the nominalist/empiricist point of view, for the demand for explanation ceases then to be a scientific 'categorical imperative'.

Nevertheless, there is a problem here that should be faced. Without a doubt, many scientific enterprises can be characterized as searches for common causes to explain correlations. What is the anti-realist to make of this? Are they not searches for explanatory realities behind the phenomena?

I think that there are two senses in which a principle of common causes is operative in the scientific enterprise, and both are perfectly intelligible without realism.

To the anti-realist, all scientific activity is ultimately aimed at greater knowledge of what is observable. So he can make sense of a search for common causes only if that search aids the acquisition of that sort of knowledge. But surely it does! When past heavy smoking is postulated as a causal factor for cancer, this suggests a further correlation between cancer and either irritation of the lungs, or the presence of such chemicals as nicotine in the bloodstream, or both. The postulate will be vindicated if such suggested further correlations are indeed found, and will, if so, have aided in the search for larger scale correlations among observable events.[18] This view reduces the Principle of Common Cause from a regulative principle for all scientific activity to one of its tactical maxims.

There is a second sense in which the principle of the common cause may be operative: as advice for the construction of theories and models. One way to construct a model for a set of observable correlations is to exhibit hidden variables with which the observed ones are individually correlated. This is a theoretical enterprise, requiring mathematical embedding or existence proofs. But if the resulting theory is then claimed to be empirically adequate, there is no claim that all aspects of the model correspond to 'elements of reality'. As a theoretical directive, or as a practical maxim, the principle of the common cause may well be operative in science – but not as a demand for explanation which would produce the metaphysical baggage of hidden parameters that carry no new empirical import.

6 Limits to explanation: a thought experiment

Wilfrid Sellars was one of the leaders of the return to realism in philosophy of science and has, in his writings of the past three decades, developed a systematic and coherent scientific realism. I have discussed a number of his views and arguments elsewhere; but will here concentrate on some aspects that are closely related to the arguments of Smart, Reichenbach, and Salmon just examined.[19] Let me begin by setting the stage in the way Sellars does.

There is a certain over-simplified picture of science, the 'levels picture', which pervades positivist writings and which Sellars successfully demolished.[20] In that picture, singular observable facts ('this crow is black') are scientifically explained by general observable regularities ('all crows are black') which in turn are explained by highly theoretical hypotheses not restricted in what they say to the observable. The three levels are commonly called those of *fact*, of *empirical law*, and of *theory*. But, as Sellars points out, theories do not explain, or even entail such empirical laws – they only show why observable things obey these so-called laws to the extent they do.[21] Indeed, perhaps we have no such empirical laws at all: all crows are black – except albinos; water

boils at 100°C – provided atmospheric pressure is normal; a falling body accelerates – provided it is not intercepted, or attached to an aeroplane by a static line; and so forth. On the level of the observable we are liable to find only putative laws heavily subject to unwritten *ceteris paribus* qualifications.

This is, so far, only a methodological point. We do not really expect theories to 'save' our common everyday generalizations, for we ourselves have no confidence in their strict universality. But a theory which says that the micro-structure of things is subject to *some* exact, universal regularities, must imply the same for those things themselves. This, at least, is my reaction to the points so far. Sellars, however, sees an inherent inferiority in the description of the observable alone, an incompleteness which requires (*sub specie* the aims of science) an introduction of an unobservable reality behind the phenomena. This is brought out by an interesting 'thought-experiment'.

Imagine that at some early stage of chemistry it had been found that different samples of gold dissolve in *aqua regia* at different rates, although 'as far as can be observationally determined, the specimens and circumstances are identical'.[22] Imagine further that the response of chemistry to this problem was to postulate two distinct micro-structures for the different samples of gold. Observationally unpredictable variation in the rate of dissolution is explained by saying that the samples are mixtures (not compounds) of these two (observationally identical) substances, each of which has a fixed rate of dissolution.

In this case we have explanation through laws which have no observational counterparts that can play the same role. Indeed, no explanation seems possible unless we agree to find our physical variables outside the observable. But science aims to explain, must try to explain, and so must require a belief in this unobservable micro-structure. So Sellars contends.

There are at least three questions before us. Did this postulation of micro-structure really have no new consequences for the observable phenomena? Is there really such a demand upon science that it must explain – even if the means of explanation bring no gain in empirical predictions? And thirdly, could a *different* rationale exist for the use of a micro-structure picture in the development of a scientific theory in a case like this?

First, it seems to me that these hypothetical chemists did postulate new observable regularities as well. Suppose the two substances are *A* and B, with dissolving rates *x* and *x* + *y* and that every gold sample is a mixture of these substances. Then it follows that every gold sample dissolves at a rate no lower than *x* and no higher than *x* + *y*; *and* that between these two any value may be found – to within the limits of accuracy of gold mixing. None of this is implied by the data that different samples of gold have dissolved at various rates between *x* and *x* + *y*. So Sellar's first contention is false.

We may assume, for the sake of Sellars's example, that there is still no way of predicting dissolving rates any further. Is there then a categorical demand upon science to explain this variation which does not depend on other observable factors? We have seen that a precise version of such a demand (Reichenbach's principle of the common cause) could result automatically in a demand for hidden variables, providing a 'classical' underpinning for indeterministic theories. Sellars recognized very well that a demand for hidden variables would run counter to the main opinions current in quantum physics. Accordingly he mentions '. . . . the familiar point that the irreducibly and lawfully statistical ensembles of quantum-mechanical theory are mathematically inconsistent with the assumption of hidden variables.'[23] Thus, he restricts the demand for explanation, in effect, to just those cases where

it is *consistent* to add hidden variables to the theory. And consistency is surely a logical stopping-point.

This restriction unfortunately does not prevent the disaster. For while there are a number of proofs that hidden variables cannot be supplied so as to turn quantum mechanics into a classical sort of deterministic theory, those proofs are based on requirements much stronger than consistency. To give an example, one such assumption is that two distinct physical variables cannot have the same statistical distributions in measurement on all possible states.[24] Thus it is assumed that, if we cannot point to some possible difference in empirical predictions, then there is no real difference at all. If such requirements were lifted, and consistency alone were the criterion, hidden variables could indeed be introduced. I think we must conclude that science, in contrast to scientific realism, does not place an overriding value on explanation in the absence of any gain for empirical results.

Thirdly, then, let us consider how an anti-realist could make sense of those hypothetical chemists' procedure. After pointing to the new empirical implications which I mentioned three paragraphs ago, he would point to methodological reasons. By imagining a certain sort of micro-structure for gold and other metals, say, we might arrive at a theory governing many observationally disparate substances; and this might then have implications for new, wider empirical regularities when such substances interact. This would only be a hope, of course; no hypothesis is guaranteed to be fruitful – but the point is that the true demand on science is not for explanation *as such*, but for imaginative pictures which have a hope of suggesting new statements of observable regularities and of correcting old ones. This point is exactly the same as that for the principle of the common cause.

7 Demons and the ultimate argument

Hilary Putnam, in the course of his discussions of realism in logic and mathematics, advanced several arguments for scientific realism as well. In *Philosophy of Logic* he concentrates largely on indispensability arguments – concepts of mathematical entities are indispensable to non-elementary mathematics, theoretical concepts are indispensable to physics.[25] Then he confronts the philosophical position of Fictionalism, which he gleans from the writings of Vaihinger and Duhem:

> (T)he fictionalist says, in substance, 'Yes, certain concepts ... are indispensable, but no, that has no tendency to show that entities corresponding to those concepts actually exist. It only shows that those 'entities' are useful fictions'.[26]

Glossed in terms of theories: even if certain kinds of theories are indispensable for the advance of science, that does not show that those theories are true *in toto*, as well as empirically correct.

Putnam attacks this position in a roundabout way, first criticizing bad arguments against Fictionalism, and then garnering his reasons for rejecting Fictionalism from that discussion. The main bad reason he sees is that of Verificationism. The logical positivists adhered to the verificationist theory of meaning; which is roughly that the total cognitive content of an assertion, all that is meaningful in it, is a function of what empirical results would verify or refute it. Hence, they would say that there are no real differences between two hypotheses with the same empirical content. Consider two theories of what the world is like: Rutherford's atomic theory, and Vaihinger's hypothesis that, although perhaps there are no electrons and such, the observable world is nevertheless exactly as if

Rutherford's theory were true. The Verificationist would say: these two theories, although Vaihinger's appears to be consistent with the denial of Rutherford's, amount to exactly the same thing.

Well, they don't, because the one says that there are electrons, and the other allows that there may not be. Even if the observable phenomena are as Rutherford says, the unobservable may be different. However, the positivists would say, if you argue that way, then you will automatically become a prey to scepticism. You will have to admit that there are possibilities you cannot prove or disprove by experiment, and so you will have to say that we just cannot know what the world is like. Worse; you will have no reason to reject any number of outlandish possibilities; demons, witchcraft, hidden powers collaborating to fantastic ends.

Putnam considers this argument for Verificationism to be mistaken, and his answer to it, strangely enough, will also yield an answer to the Fictionalism rejected by the verificationist. To dispel the bogey of scepticism, Putnam gives us a capsule introduction to contemporary (Bayesian) epistemology: Rationality requires that if two hypotheses have all the same testable consequences (consequences for evidence that could be gathered), then we should not accept the one which is *a priori the less plausible*. Where do we get our a priori plausibility orderings? These we supply ourselves, either individually or as communities: to accept a plausibility ordering is neither

to make a judgment of empirical fact nor to state a theorem of deductive logic; it is to take a methodological stand. One can only say whether the demon hypothesis is 'crazy' or not if one has taken such a stand; I report the stand I have taken (and, speaking as one who has taken this stand, I add: and the stand all rational men take, implicitly or explicitly).[27]

On this view, the difference between Rutherford and Vaihinger, or between Putnam and Duhem, is that (although they presumably agree on the implausibility of demons) they disagree on the a priori plausibility of electrons. Does each simply report the stand he has taken, and add: this is, in my view, the stand of all rational men? How disappointing.

Actually, it does not quite go that way. Putnam has skilfully switched the discussion from electrons to demons, and asked us to consider how we could rule out their existence. As presented, however, Vaihinger's view differed from Rutherford's by being logically weaker – it only withheld assent to an existence assertion. It follows automatically that Vaihinger's view cannot be a priori less plausible than Rutherford's. Putnam's ideological manœuvre could at most be used to accuse an 'atheistic' anti-realist of irrationality (relative to Putnam's own stand, of course) – not one of the agnostic variety.

Putnam concludes this line of reasoning by asking what more could be wanted as evidence for the truth of a theory than what the realist considers sufficient: 'But then ... what *further* reasons could one want before one regarded it as rational to *believe* a theory?'[28] The answer is: *none* – at least if he equates reasons here either with empirical evidence or with compelling arguments. (Inclining reasons are perhaps another matter, especially because Putnam uses the phrase 'rational to believe' rather than 'irrational not to believe'.) Since Putnam has just done us the service of refuting Verificationism, this answer 'none' cannot convict us of irrationality. He has himself just argued forcefully that theories could agree in empirical content and differ in truth-value. Hence, a realist will have to make a leap of faith. The decision to leap is subject to rational scrutiny, but not *dictated* by reason and evidence.

In a further paper, 'What is Mathematical Truth?', Putnam continues the discussion of scientific realism, and gives what I shall call the *Ultimate Argument*. He begins with a formulation of realism which he says he learned from Michael Dummett:

> A realist (with respect to a given theory or discourse) holds that (1) the sentences of that theory are true or false; and (2) that what makes them true or false is something external – that is to say, it is not (in general) our sense data, actual or potential, or the structure of our minds, or our language, etc.[29]

This formulation is quite different from the one I have given even if we instantiate it to the case in which that theory or discourse is science or scientific discourse. Because the wide discussion of Dummett's views has given some currency to his usage of these terms, and because Putnam begins his discussion in this way, we need to look carefully at this formulation.

In my view, Dummett's usage is quite idiosyncratic. Putnam's statement, though very brief, is essentially accurate. In his 'Realism', Dummett begins by describing various sorts of realism in the traditional fashion, as disputes over whether there really exist entities of a particular type. But he says that in some cases he wishes to discuss, such as the reality of the past and intuitionism in mathematics, the central issues seem to him to be about other questions. For this reason he proposes a new usage: he will take such disputes

> as relating, not to a class of entities or a class of terms, but to a class of *statements* ... Realism I characterize as the belief that statements of the disputed class possess an objective truth-value, independently of our means of knowing it: they are true or false in virtue of a reality existing independently of us. The anti-realist opposes to this the view that statements of the disputed class are to be understood only by reference to the sort of thing

which we count as evidence for a statement of that class.[30]

Dummett himself notes at once that nominalists are realists in this sense.[31] If, for example, you say that abstract entities do not exist, and sets are abstract entities, hence sets do not exist, then you will certainly accord a truth-value to every statement of set theory. It might be objected that if you take this position then you have a decision procedure for determining the truth-values of these statements (*false* for existentially quantified ones, *true* for universal ones, apply truth tables for the rest). Does that not mean that, on your view, the truth-values are not independent of our knowledge? Not at all; for you clearly believe that if we had not existed, and *a fortiori* had had no knowledge, the state of affairs with respect to abstract entities would be the same.

Has Dummett perhaps only laid down a necessary condition for realism, in his definition, for the sake of generality? I do not think so. In discussions of quantum mechanics we come across the view that the particles of microphysics are real, and obey the principles of the theory, but at any time t when 'particle x has exact momentum p' is true then 'particle x has position q' is neither true nor false. In any traditional sense, this is a realist position with respect to quantum mechanics.

We note also that Dummett has, at least in this passage, taken no care to exclude non-literal construals of the theory, as long as they are truth-valued. The two are not the same; when Strawson construed 'The king of France in 1905 is bald' as neither true nor false, he was not giving a non-literal construal of our language. On the other hand, people tend to fall back on non-literal construals typically in order to be able to say, 'properly construed, the theory is true.'[32]

Perhaps Dummett is right in his assertion that what is really at stake, in realist disputes of various sorts, is questions about language – or, if not really at stake, at least the only serious philosophical problems in those neighborhoods. Certainly the arguments in which he engages are profound, serious, and worthy of our attention. But it seems to me that his terminology ill accords with the traditional one. Certainly I wish to define scientific realism so that it need not imply that all statements in the theoretical language are true or false (only that they are all capable of being true or false, that is, there are conditions for each under which it has a truth-value); to imply nevertheless that the aim is that the theories should be true. And the contrary position of constructive empiricism is not anti-realist in Dummett's sense, since it also assumes scientific statements to have truth-conditions entirely independent of human activity or knowledge. But then, I do not conceive the dispute as being about language at all.

In any case Putnam himself does not stick with this weak formulation of Dummett's. A little later in the paper he directs himself to scientific realism *per se*, and formulates it in terms borrowed, he says, from Richard Boyd. The new formulation comes in the course of a new argument for scientific realism, which I shall call the Ultimate Argument:

the positive argument for realism is that it is the only philosophy that doesn't make the success of science a miracle. That terms in mature scientific theories typically refer (this formulation is due to Richard Boyd), that the theories accepted in a mature science are typically approximately true, that the same term can refer to the same thing even when it occurs in different theories – these statements are viewed by the scientific realist not as necessary truths but as part of the only scientific explanation of the success of science, and hence as part of any adequate scientific description of science and its relations to its objects.[33]

Science, apparently, is required to explain its own success. There is this regularity in the world, that scientific predictions are regularly fulfilled; and this regularity, too, needs an explanation. Once *that* is supplied we may perhaps hope to have reached the *terminus de jure?*

The explanation provided is a very traditional one – *adequatio ad rem*, the 'adequacy' of the theory to its objects, a kind of mirroring of the structure of things by the structure of ideas – Aquinas would have felt quite at home with it.

Well, let us accept for now this demand for a scientific explanation of the success of science. Let us also resist construing it as merely a restatement of Smart's 'cosmic coincidence' argument, and view it instead as the question why we have successful scientific theories at all. Will this realist explanation with the Scholastic look be a scientifically acceptable answer? I would like to point out that science is a biological phenomenon, an activity by one kind of organism which facilitates its interaction with the environment. And this makes me think that a very different kind of scientific explanation is required.

I can best make the point by contrasting two accounts of the mouse who runs from its enemy, the cat. St. Augustine already remarked on this phenomenon, and provided an intentional explanation: the mouse *perceives that* the cat is its enemy, hence the mouse runs. What is postulated here is the 'adequacy' of the mouse's thought to the order of nature: the relation of enmity is correctly reflected in his mind. But the Darwinist says: Do not ask why the *mouse* runs from its enemy. Species which did not cope with their natural enemies no longer exist. That is why there are only ones who do.

In just the same way, I claim that the success of current scientific theories is no miracle. It is not even surprising to the scientific (Darwinist) mind. For any scientific theory is born into a life of fierce competition, a jungle red in tooth and claw. Only the successful theories survive – the ones which *in fact* latched on to actual regularities in nature.[34]

Notes

1 Brian Ellis, *Rational Belief Systems* (Oxford: Blackwell, 1979), p. 28.

2 Hartry Field has suggested that 'acceptance of a scientific theory involves the belief that it is true' be replaced by 'any reason to think that any part of a theory is not, or might not be, true, is reason not to accept it.' The drawback of this alternative is that it leaves open what epistemic attitude acceptance of a theory does involve. This question must also be answered, and as long as we are talking about full acceptance – as opposed to tentative or partial or otherwise qualified acceptance – I cannot see how a realist could do other than equate that attitude with full belief. (That theories believed to be false are used for practical problems, for example, classical mechanics for orbiting satellites, is of course a commonplace.) For if the aim is truth, and acceptance requires belief that the aim is served ... I should also mention the statement of realism at the beginning of Richard Boyd, 'Realism, Underdetermination, and a Causal Theory of Evidence', *Noûs*, **7** (1973), 1–12. Except for some doubts about his use of the terms 'explanation' and 'causal relation' I intend my statement of realism to be entirely in accordance with his. Finally, see C.A. Hooker, 'Systematic Realism', *Synthese*, **26** (1974), 409–97; esp. pp. 409 and 426.

3 More typical of realism, it seems to me, is the sort of epistemology found in Clark Glymour's forthcoming book, *Theory and Evidence* (Princeton: Princeton University Press, 1980), except of course that there it is fully and carefully developed in one specific fashion. (See esp. his chapter 'Why I am not a Bayesian' for the present issue.) But I see no reason why a realist, as such, could not be a Bayesian of the type of Richard Jeffrey, even if the Bayesian position has in the past been linked with anti-realist and even instrumentalist views in philosophy of science.

4 G. Maxwell, 'The Ontological Status of Theoretical Entities' Minnesota Studies in Philosophy of Science, III (1962), p. 7.

5 There is a great deal of recent work on the logic of vague predicates; especially important, to my mind, is that of Kit Fine ('Vagueness, Truth, and Logic', Synthese, **30** (1975), 265–300) and Hans Kamp. The latter is currently working on a new theory of vagueness that does justice to the 'vagueness of vagueness' and the context dependence of standards of applicability for predicates.

6 Op. cit., p. 15. In the next chapter I shall discuss further how observability should be understood. At this point, however, I may be suspected of relying on modal distinctions which I criticize elsewhere. After all, I am making a distinction between human limitations, and accidental factors. A certain apple was dropped into the sea in a bag of refuse, which sank; relative to that information it is necessary that no one ever observed the apple's core. That information, however, concerns an accident of history, and so it is not human limitations that rule out observation of the apple core. But unless I assert that some facts about humans are essential, or physically necessary, and others accidental, how can I make sense of this distinction? This question raises the difficulty of a philosophical retrenchment for modal language. This I believe to be possible through an ascent to pragmatics. In the present case, the answer would be, to speak very roughly, that the scientific theories we accept are a determining factor for the set of features of the human organism counted among the limitations to which we refer in using the term 'observable'. The issue of modality will occur explicitly again in the chapter on probability [sic].

7 Science, Perception and Reality (New York: Humanities Press, 1962); cf. the footnote on p. 97. See also my review of his Studies in Philosophy and its History, in Annals of Science, January 1977.

8 Cf. P. Thagard, doctoral dissertation, University of Toronto, 1977, and 'The Best Explanation: Criteria for Theory Choice', Journal of Philosophy, **75** (1978), 76–92.

9 'The Inference to the Best Explanation', Philosophical Review, **74** (1965), 88–95 and 'Knowledge,

Inference, and Explanation', American Philosophical Quarterly, **5** (1968), 164–73. Harman's views were further developed in subsequent publications (Noûs, 1967; Journal of Philosophy, 1968; in M. Swain (ed.), Induction, 1970; in H.-N. Castaneda (ed.), Action, Thought, and Reality, 1975: and in his book Thought, Ch. 10). I shall not consider these further developments here.

10 See esp. 'Knowledge, Inference, and Explanation', p. 169.

11 J.J.C. Smart, Between Science and Philosophy (New York: Random House, 1968), p. 150.

12 Ibid., pp. 150f.

13 This point is clearly made by Aristotle, Physics, II, Chs. 4–6 (see esp. 196^a 1-20: 196^b 20–197^a 12).

14 W. Salmon, 'Theoretical Explanation', pp. 118–45 in S. Korner (ed.), Explanation (Oxford: Blackwell, 1975). In a later paper, 'Why ask, "Why?"?' (Presidential Address, Proc. American Philosophical Association **51** (1978), 683–705), Salmon develops an argument for realism like that of Smart's about coincidences, and adds that the demand for a common cause to explain apparent coincidences formulates the basic principle behind this argument. However, he has weakened the common cause principle so as to escape the objections I bring in this section. It seems to me that his argument for realism is also correspondingly weaker. As long as there is no universal demand for a common cause for every pervasive regularity or correlation, there is no argument for realism here. There is only an explanation of why it is satisfying to the mind to postulate explanatory, if unobservable, mechanisms when we can. There is no argument in that the premises do not compel the realist conclusion. Salmon has suggested in conversation that we should perhaps impose the universal demand that only correlations among spatio-temporally (approximately) coincident events are allowed to remain without explanation. I do not see a rationale for this; but also, it is a demand not met by quantum mechanics in which there are nonlocal correlations (as in the Einstein-Podolski-Rosen 'paradox'); orthodox physics refuses to see these correlations as genuinely paradoxical. I shall discuss Salmon's more recent theory in Ch. 4 [sic].

These are skirmishes; on a more basic level I wish to maintain that there is sufficient satisfaction for the mind if we can construct theories in whose models correlations and apparent coincidences are traceable back to common causes – without adding that all features of these models correspond to elements of reality. See further my 'Rational Belief and the Common Cause Principle', in R. McLaughlin's forthcoming collection of essays on Salmon's philosophy of science.

15 H. Reichenbach, *Modern Philosophy of Science* (London: Routledge & Kegan Paul, 1959), Chs 3 and 5. From a purely logical point of view this is not so. Suppose we define the predicate P(−m) to apply to a thing at time t exactly if the predicate P applies to it at time t + m. In that case, description of its 'properties' at time t, using predicate P(−m), will certainly give the information whether the thing is P at time t + m. But such a defined predicate 'has no physical significance', its application cannot be determined by any observations made at or prior to time t. Thus Reichenbach was assuming certain criteria of adequacy on what counts as a description for empirical science; and surely he was right in this.

16 H. Reichenbach, *The Direction of Time* (Berkeley: University of California, 1963), Sect. 19, pp. 157–63; see also Sects. 22 and 23.

17 The paper by Einstein, Podolski, and Rosen appeared in the *Physical Review*, **47** (1935); 777–80: their thought experiment and Compton scattering are discussed in Part I of my 'The Einstein-Podolski-Rosen Paradox', *Synthese*, **29** (1974), 291–309. An elegant general result concerning the extent to which the statistical 'explanation' of a correlation by means of a third variable requires determinism is the basic lemma in P. Suppes and M. Zznotti, 'On the Determinism of Hidden Variable Theories with Strict Correlation and Conditional Statistical Independence of Observables', pp. 445–55 in P. Suppes (ed.), *Logic and Probability in Quantum Mechanics* (Dordrecht: Reidel Pub. Co., 1976). This book also contains a reprint of the preceding paper.

18 There is another way: if the correlation between A and B is known, but only within inexact limits, the postulation of the common cause C by a theory which specifies P(A/C) and P(B/C) will then entail an exact statistical relationship between A and B, which can be subjected to further experiment.

19 See my 'Wilfrid Sellars on Scientific Realism', *Dialogue*, **14** (1975), 606–16; W. Sellars, 'Is Scientific Realism Tenable?', pp. 307–34 in F. Suppe and P. Asquith (eds), *PSA 1976* (East Lansing, Mich.: Philosophy of Science Association, 1977), vol. II; and my 'On the Radical Incompleteness of the Manifest Image', ibid., 335–43; and see n. 7 above.

20 W. Sellars, 'The Language of Theories', in his *Science, Perception, and Reality* (London: Routledge & Kegan Paul, 1963).

21 Op. cit., p. 121.

22 Ibid., p. 121.

23 Ibid., p. 123.

24 See my 'Semantic Analysis of Quantum Logic', in C.A. Hooker (ed.), *Contemporary Research in the Foundations of Philosophy of Quantum Theory* (Dordrecht: Reidel; 1973), Part III, Sects. 5 and 6.

25 Hilary Putnam, *Philosophy of Logic* (New York: Harper and Row, 1971) see also my review of this in *Canadian Journal of Philosophy*, **4** (1975), 731–43. Since Putnam's metaphysical views have changed drastically during the last few years, my remarks apply only to his views as they then appeared in his writings.

26 Op. cit., p. 63.

27 Ibid., p. 67.

28 Ibid., p. 69.

29 Hilary Putnam, *Mathematics, Matter and Method* (Cambridge: Cambridge University Press, 1975), vol. I, pp. 69f.

30 Michael Dummett, *Truth and Other Enigmas* (Cambridge, Mass.: Harvard University Press, 1978), p. 146 (see also pp. 358–61).

31 Dummett adds to the cited passage that he realizes that his characterization does not include all the disputes he had mentioned, and specifically excepts nominalism about abstract entities. However, he includes scientific realism as an example (op. cit., pp. 146f.).

32 This is especially relevant here because the 'translation' that connects Putnam's two foundations of mathematics (existential and modal) as discussed

in this essay, is not a literal construal: it is a mapping presumably preserving statementhood and theoremhood, but it does not preserve logical form.

33 Putnam, op. cit., p. 73 (n. 29 above). The argument is reportedly developed at greater length in Boyd's forthcoming book *Realism and Scientific Epistemology* (Cambridge University Press).

34 Of course, we can ask specifically why the mouse is one of the surviving species, how it survives, and answer this, on the basis of whatever scientific theory we accept, in terms of its brain and environment. The analogous question for theories would be why, say, Balmer's formula for the line spectrum of hydrogen survives as a successful hypothesis. In that case too we explain, on the basis of the physics we accept now, why the spacing of those lines satisfies the formula. Both the question and the answer are very different from the global question of the success of science, and the global answer of realism. The realist may now make the further objection that the anti-realist cannot answer the question about the mouse specifically, nor the one about Balmer's formula, in this fashion, since the answer is in part an assertion that the scientific theory, used as basis of the explanation, is true. This is a quite different argument, which I shall take up in Ch.4, Sect.4, and Ch. 5 [sic].

In his most recent publications and lectures Hilary Putnam has drawn a distinction between two doctrines, metaphysical realism and internal realism. He denies the former, and identifies his preceding scientific realism as the latter. While I have at present no commitment to either side of the metaphysical dispute, I am very much in sympathy with the critique of Platonism in philosophy of mathematics which forms part of Putnam's arguments. Our disagreement about scientific (internal) realism would remain of course, whenever we came down to earth after deciding to agree or disagree about metaphysical realism, or even about whether this distinction makes sense at all.

Chapter 38

Ian Hacking

EXPERIMENTATION AND SCIENTIFIC REALISM

Experimental physics provides the strongest evidence for scientific realism. Entities that in principle cannot be observed are regularly manipulated to produce new phenomena and to investigate other aspects of nature. They are tools, instruments not for thinking but for doing.

The philosopher's standard "theoretical entity" is the electron. I shall illustrate how electrons have become experimental entities, or experimenters' entities. In the early stages of our discovery of an entity, we may test hypotheses about it. Then it is merely an hypothetical entity. Much later, if we come to understand some of its causal powers and to use it to build devices that achieve well understood effects in other parts of nature, then it assumes quite a different status.

Discussions about scientific realism or anti-realism usually talk about theories, explanation and prediction. Debates at that level are necessarily inconclusive. Only at the level of experimental practice is scientific realism unavoidable. But this realism is not about theories and truth. The experimentalist need only be a realist about the entities used as tools.

A plea for experiments

No field in the philosophy of science is more systematically neglected than experiment. Our grade school teachers may have told us that scientific method is experimental method, but histories of science have become histories of theory. Experiments, the philosophers say, are of value only when they test theory. Experimental work, they imply, has no life of its own. So we lack even a terminology to describe the many varied roles of experiment. Nor has this one-sidedness done theory any good, for radically different types of theory are used to think about the same physical phenomenon (e.g., the magneto-optical effect). The philosophers of theory have not noticed this and so misreport even theoretical inquiry.[1]

Different sciences at different times exhibit different relationships between "theory" and "experiment." One chief role of experiment is the creation of phenomena. Experimenters bring into being phenomena that do not naturally exist in a pure state. These phenomena are the touchstones of physics, the keys to nature and the source of much modern technology. Many are what physicists after the 1870s began to call "effects": the photo-electric effect, the Compton effect, and so forth. A recent high-energy extension of the creation of phenomena is the creation of "events," to use the jargon of the trade. Most of the phenomena, effects and events created by the experimenter are like plutonium: they do not exist in nature except possibly on vanishingly rare occasions.[2]

In this paper I leave aside questions of methodology, history, taxonomy and the purpose of experiment in natural science. I turn to the purely philosophical issue of scientific realism. Call it simply "realism" for short. There are two basic kinds: realism about entities and realism about theories. There is no agreement on the precise definition of either. Realism about theories says we try to form true theories about the world, about the inner constitution of matter and about the outer reaches of space. This realism gets its bite from optimism; we think we can do well in this project, and have already had partial success.

Realism about entities – and I include processes, states, waves, currents, interactions, fields, black holes and the like among entities – asserts the existence of at least some of the entities that are the stock in trade of physics.[3]

The two realisms may seem identical. If you believe a theory, do you not believe in the existence of the entities it speaks about? If you believe in some entities, must you not describe them in some theoretical way that you accept? This seeming identity is illusory. The vast majority of experimental physicists are realists about entities without a commitment to realism about theories. The experimenter is convinced of the existence of plenty of "inferred" and "unobservable" entities. But no one in the lab believes in the literal truth of present theories about those entities. Although various properties are confidently ascribed to electrons, most of these properties can be embedded in plenty of different inconsistent theories about which the experimenter is agnostic. Even people working on adjacent parts of the same large experiment will use different and mutually incompatible accounts of what an electron is. That is because different parts of the experiment will make different uses of electrons, and the models that are useful for making calculations about one use may be completely haywire for another use.

Do I describe a merely sociological fact about experimentalists? It is not surprising, it will be said, that these good practical people are realists. They need that for their own self-esteem. But the self-vindicating realism of experimenters shows nothing about what actually exists in the world. In reply I repeat the distinction between realism about entities and realism about theories and models. Anti-realism about models is perfectly coherent. Many research workers may in fact hope that their theories and even their mathematical models "aim at the truth," but they seldom suppose that any particular model is more than adequate for a purpose. By and large most experimenters seem to be instrumentalists about the models they use. The models are products of the intellect, tools for thinking and calculating. They are essential for writing up grant proposals to obtain further funding. They are rules of thumb used to get things done. Some experimenters are instrumentalists about theories and models, while some are not. That is a sociological fact. But experimenters are realists about the entities that they use in order to investigate other hypotheses or hypothetical entities. That is not a sociological fact. Their enterprise would be incoherent without it. But their enterprise is not incoherent. It persistently creates new phenomena that become regular technology. My task is to show that realism about entities is a necessary condition for the coherence of most experimentation in natural science.

Our debt to Hilary Putnam

It was once the accepted wisdom that a word like "electron" gets its meaning from its place in a network of sentences that state theoretical laws. Hence arose the infamous problems of incommensurability and theory change. For if a theory is modified, how could a word like "electron" retain its previous meaning? How could different theories about electrons be compared, since the very word "electron" would differ in meaning from theory to theory?

Putnam saves us from such questions by inventing a referential model of meaning. He says that meaning is a vector, refreshingly like a dictionary entry. First comes the syntactic marker (part of speech). Next the semantic marker (general category of thing signified by the word). Then the stereotype (clichés about the natural kind, standard examples of its use and present day associations. The stereotype is subject to change as opinions about the kind are modified). Finally there is the actual reference of the word, the very stuff, or thing, it denotes if it denotes anything. (Evidently dictionaries cannot include this in their entry, but pictorial dictionaries do their best by inserting illustrations whenever possible.)[4]

Putnam thought we can often guess at entities that we do not literally point to. Our initial guesses may be jejune or inept, and not every naming of an invisible thing or stuff pans out. But when it does, and we frame better and better ideas, then Putnam says that although the stereotype changes, we refer to the same kind of thing or stuff all along. We and Dalton alike spoke about the same stuff when we spoke of (inorganic) acids. J. J. Thomson, Lorentz, Bohr and Millikan were, with their different theories and observations, speculating about the same kind of thing, the electron.

There is plenty of unimportant vagueness about when an entity has been successfully "dubbed," as Putnam puts it. "Electron" is the name suggested by G. Johnstone Stoney in 1891 as the name for a natural unit of electricity. He had drawn attention to this unit in 1874. The name was then applied in 1897 by J. J. Thomson to the subatomic particles of negative charge of which cathode rays consist. Was Johnstone Stoney referring to the electron? Putnam's account does not require an unequivocal answer.

Standard physics books say that Thomson discovered the electron. For once I might back theory and say Lorentz beat him to it. What Thomson did was to measure the electron. He showed its mass is 1/1800 that of hydrogen. Hence it is natural to say that Lorenz merely postulated the particle of negative charge, while Thomson, determining its mass, showed that there is some such real stuff beaming off a hot cathode.

The stereotype of the electron has regularly changed, and we have at least two largely incompatible stereotypes, the electron as cloud and the electron as particle. One fundamental enrichment of the idea came in the 1920s. Electrons, it was found, have angular momentum, or "spin." Experimental work by Stern and Gerlach first indicated this, and then Goudsmit and Uhlenbeck provided the theoretical understanding of it in 1925. Whatever we think about Johnstone Stoney, others − Lorentz, Bohr, Thomson and Goudsmit − were all finding out more about the same kind of thing, the electron.

We need not accept the fine points of Putnam's account of reference in order to thank him for providing a new way to talk about meaning. Serious discussions of inferred entities need no longer lock us into pseudo-problems of incommensurability and theory change. Twenty-five years ago the experimenter who believed that electrons exist, without giving much credence to any set of laws about electrons, would have been dismissed as philosophically incoherent. We now realize it was the philosophy that was wrong, not the experimenter. My own relationship to Putnam's account of meaning is like the experimenter's relationship to a theory. I don't literally believe Putnam, but I am happy to employ his account as an alternative to the unpalatable account in fashion some time ago.

Putnam's philosophy is always in flux. At the time of this writing, July 1981, he rejects any "metaphysical realism" but allows "internal realism."[5] The internal realist acts, in practical affairs, as if the entities occurring in his working theories did in fact exist. However, the direction of Putnam's metaphysical anti-realism is no longer scientific. It is not peculiarly about natural science. It is about chairs and livers too. He thinks that the world does not naturally break up into our classifications. He calls himself a transcendental idealist. I call him a transcendental nominalist. I use the word "nominalist" in the old fashioned way, not meaning opposition to "abstract entities" like sets, but meaning the doctrine that there is no nonmental classification in nature that exists over and above our own human system of naming.

There might be two kinds of Putnamian internal realist – the instrumentalist and the scientific realist. The former is, in practical affairs where he uses his present scheme of concepts, a realist about livers and chairs, but he thinks that electrons are mental constructs only. The latter thinks that livers, chairs, and electrons are probably all in the same boat, that is, real at least within the present system of classification. I take Putnam to be an internal scientific realist rather than an internal instrumentalist. The fact that either doctrine is compatible with transcendental nominalism and internal realism shows that our question of scientific realism is almost entirely independent of Putnam's present philosophy.

Interfering

Francis Bacon, the first and almost last philosopher of experiments, knew it well: the experimenter sets out "to twist the lion's tail." Experimentation is interference in the course of nature; "nature under constraint and vexed; that is to say, when by art and the hand of man she is forced out of her natural state, and squeezed and moulded."[6] The experimenter is convinced of the reality of entities some of whose causal

properties are sufficiently well understood that they can be used to interfere *elsewhere* in nature. One is impressed by entities that one can use to test conjectures about other more hypothetical entities. In my example, one is sure of the electrons that are used to investigate weak neutral currents and neutral bosons. This should not be news, for why else are we (non-sceptics) sure of the reality of even macroscopic objects, but because of what we do with them, what we do to them, and what they do to us?

Interference and intervention are the stuff of reality. This is true, for example, at the borderline of observability. Too often philosophers imagine that microscopes carry conviction because they help us see better. But that is only part of the story. On the contrary, what counts is what we can do to a specimen under a microscope, and what we can see ourselves doing. We stain the specimen, slice it, inject it, irradiate it, fix it. We examine it using different kinds of microscopes that employ optical systems that rely on almost totally unrelated facts about light. Microscopes carry conviction because of the great array of interactions and interferences that are possible. When we see something that turns out not to be stable under such play, we call it an artefact and say it is not real.[7]

Likewise, as we move down in scale to the truly un-seeable, it is our power to use unobservable entities that make us believe they are there. Yet I blush over these words "see" and "observe." John Dewey would have said that a fascination with seeing-with-the-naked-eye is part of the Spectator Theory of Knowledge that has bedeviled philosophy from earliest times. But I don't think Plato or Locke or anyone before the nineteenth century was as obsessed with the sheer opacity of objects as we have been since. My own obsession with a technology that manipulates objects is, of course a twentieth-century counterpart to positivism and phenomenology. Their proper rebuttal is not a restriction to a narrower domain of reality, namely to what can be positivistically "seen" (with the eye), but an extension to other modes by which people can extend their consciousness.

Making

Even if experimenters are realists about entities, it does not follow that they are right. Perhaps it is a matter of psychology: the very skills that make for a great experimenter go with a certain cast of mind that objectifies whatever it thinks about. Yet this will not do. The experimenter cheerfully regards neutral bosons as merely hypothetical entities, while electrons are real. What is the difference?

There are an enormous number of ways to make instruments that rely on the causal properties of electrons in order to produce desired effects of unsurpassed precision. I shall illustrate this. The argument – it could be called the experimental argument for realism – is not that we infer the reality of electrons from our success. We do not make the instruments and then infer the reality of the electrons, as when we test a hypothesis, and then believe it because it passed the test. That gets the time-order wrong. By now we design apparatus relying on a modest number of home truths about electrons to produce some other phenomenon that we wish to investigate.

That may sound as if we believe in the electrons because we predict how our apparatus will behave. That too is misleading. We have a number of general ideas about how to prepare polarized electrons, say. We spend a lot of time building prototypes that don't work. We get rid of innumerable bugs. Often we have to give up and try another approach. Debugging is not a matter of theoretically explaining or predicting what is going wrong. It is partly a matter of getting rid of "noise" in the apparatus. "Noise" often means all the events that are not understood by any theory. The instrument must be able to isolate,

physically, the properties of the entities that we wish to use, and damp down all the other effects that might get in our way. *We are completely convinced of the reality of electrons when we regularly set out to build – and often enough succeed in building – new kinds of devices that use various well understood causal properties of electrons to interfere in other more hypothetical parts of nature.*

It is not possible to grasp this without an example. Familiar historical examples have usually become encrusted by false theory-oriented philosophy or history. So I shall take something new. This is a polarizing electron gun whose acronym is PEGGY II. In 1978 it was used in a fundamental experiment that attracted attention even in *The New York Times*. In the next section I describe the point of making PEGGY II. So I have to tell some new physics. You can omit this and read only the engineering section that follows. Yet it must be of interest to know the rather easy-to-understand significance of the main experimental results, namely, (1) parity is not conserved in scattering of polarized electrons from deuterium, and (2) more generally, parity is violated in weak neutral current interactions.[8]

Methodological remark

In the following section I retail a little current physics; in the section after that I describe how a machine has been made. It is the latter that matters to my case, not the former. Importantly, even if present quantum electrodynamics turns out to need radical revision, the machine, called PEGGY II, will still work. I am concerned with how it was made to work, and why. I shall sketch far more sheer engineering than is seen in philosophy papers. My reason is that the engineering is incoherent unless electrons are taken for granted. One cannot say this by merely reporting, "Oh, they made an electron gun for shooting polarized electrons." An immense practical knowledge of how to manipulate electrons, of what sorts of things they will do reliably and how they tend to misbehave – that is the kind of knowledge which grounds the experimenter's realism about electrons. You cannot grasp this kind of knowledge in the abstract, for it is practical knowledge. So I must painfully introduce the reader to some laboratory physics. Luckily it is a lot of fun.

Parity and weak neutral currents

There are four fundamental forces in nature, not necessarily distinct. Gravity and electromagnetism are familiar. Then there are the strong and weak forces, the fulfillment of Newton's program, in the *Optics*, which taught that all nature would be understood by the interaction of particles with various forces that were effective in attraction or repulsion over various different distances (i.e., with different rates of extinction).

Strong forces are 100 times stronger than electromagnetism but act only for a miniscule distance, at most the diameter of a proton. Strong forces act on "hadrons," which include protons, neutrons, and more recent particles, but not electrons or any other members of the class of particles called "leptons."

The weak forces are only 1/10,000 times as strong as electromagnetism, and act over a distance 1/100 times smaller than strong forces. But they act on both hadrons and leptons, including electrons. The most familiar example of a weak force may be radioactivity.

The theory that motivates such speculation is quantum electrodynamics. It is incredibly successful, yielding many predictions better than one part in a million, a miracle in experimental physics. It applies over distances ranging from diameters of the earth to 1/100 the diameter of the proton. This theory supposes that all the forces are "carried" by some sort of particle. Photons do the job in electromagnetism. We hypothesize "gravitons" for gravity.

In the case of interactions involving weak forces, there are charged currents. We postulate that particles called bosons carry these weak forces.[9] For charged currents, the bosons may be positive or negative. In the 1970s there arose the possibility that there could be weak "neutral" currents in which no charge is carried or exchanged. By sheer analogy with the vindicated parts of quantum electrodynamics, neutral bosons were postulated as the carriers in weak interactions.

The most famous discovery of recent high energy physics is the failure of the conservation of parity. Contrary to the expectations of many physicists and philosophers, including Kant,[10] nature makes an absolute distinction between right-handedness and left-handedness. Apparently this happens only in weak interactions.

What we mean by right- or left-handed in nature has an element of convention. I remarked that electrons have spin. Imagine your right hand wrapped around a spinning particle with the fingers pointing in the direction of spin. Then your thumb is said to point in the direction of the spin vector. If such particles are traveling in a beam, consider the relation between the spin vector and the beam. If all the particles have their spin vector in the same direction as the beam, they have right-handed (linear) polarization, while if the spin vector is opposite to the beam direction, they have left-handed (linear) polarization.

The original discovery of parity violation showed that one kind of product of a particle decay, a so-called *muon neutrino*, exists only in left-handed polarization and never in right-handed polarization.

Parity violations have been found for weak *charged* interactions. What about weak *neutral* currents? The remarkable Weinberg-Salam model for the four kinds of force was proposed independently by Stephen Weinberg in 1967 and A. Salam in 1968. It implies a minute violation of parity in weak neutral interactions. Given that the model is sheer speculation, its success has been amazing, even awe inspiring. So it seemed worthwhile to try out the predicted failure of parity for weak neutral interactions. That would teach us more about those weak forces that act over so minute a distance.

The prediction is: Slightly more left-handed polarized electrons hitting certain targets will scatter, than right-handed electrons. Slightly more! The difference in relative frequency of the two kinds of scattering is one part in 10,000, comparable to a difference in probability between 0.50005 and 0.49995. Suppose one used the standard equipment available at the Stanford Linear Accelerator in the early 1970s, generating 120 pulses per second, each pulse providing one electron event. Then you would have to run the entire SLAC beam for 27 years in order to detect so small a difference in relative frequency. Considering that one uses the same beam for lots of experiments simultaneously, by letting different experiments use different pulses, and considering that no equipment remains stable for even a month, let alone 27 years, such an experiment is impossible. You need enormously more electrons coming off in each pulse. We need between 1000 and 10,000 more electrons per pulse than was once possible. The first attempt used an instrument now called PEGGY I. It had, in essence, a high-class version of J. J. Thomson's hot cathode. Some lithium was heated and electrons were boiled off. PEGGY II uses quite different principles.

PEGGY II

The basic idea began when C. Y. Prescott noticed, (by "chance"!) an article in an optics magazine about a crystalline substance called Gallium Arsenide. GaAs has a number of curious properties that make it important in laser technology. One of its quirks is that when it is struck by circularly polarized light of the right frequencies, it emits a lot of linearly polarized electrons. There is a good rough and ready quantum understanding of why this happens, and why half the emitted electrons will be polarized, ¾ polarized in one direction and ¼ polarized in the other.

PEGGY II uses this fact, plus the fact that GaAs emits lots of electrons due to features of its crystal structure. Then comes some engineering. It takes work to liberate an electron from a surface. We know that painting a surface with the right substance helps. In this case, a thin layer of cesium and oxygen is applied to the crystal. Moreover the less air pressure around the crystal, the more electrons will escape for a given amount of work. So the bombardment takes place in a good vacuum at the temperature of liquid nitrogen.

We need the right source of light. A laser with bursts of red light (7100 Ångstroms) is trained on the crystal. The light first goes through an ordinary polarizer, a very old-fashioned prism of calcite, or Iceland spar.[11] This gives longitudinally polarized light. We want circularly polarized light to hit the crystal. The polarized laser beam now goes through a cunning modern device, called a Pockel's cell. It electrically turns linearly polarized photons into circularly polarized ones. Being electric, it acts as a very fast switch. The direction of circular polarization depends on the direction of current in the cell. Hence the direction of polarization can be varied randomly. This is important, for we are trying to detect a minute asymmetry between right and left handed polarization. Randomizing helps us guard against any systematic "drift" in the equipment.[12] The randomization is generated by a radioactive decay device, and a computer records the direction of polarization for each pulse.

A circularly polarized pulse hits the GaAs crystal, resulting in a pulse of linearly polarized electrons. A beam of such pulses is maneuvered by magnets into the accelerator for the next bit of the experiment. It passes through a device that checks on a proportion of polarization along the way. The remainder of the experiment requires other devices and detectors of comparable ingenuity, but let us stop at PEGGY II.

Bugs

Short descriptions make it all sound too easy, so let us pause to reflect on debugging. Many of the bugs are never understood. They are eliminated by trial and error. Let us illustrate three different kinds: (1) The essential technical limitations that in the end have to be factored into the analysis of error. (2) Simpler mechanical defects you never think of until they are forced on you. (3) Hunches about what might go wrong.

1. Laser beams are not as constant as science fiction teaches, and there is always an irremediable amount of "jitter" in the beam over any stretch of time.

2. At a more humdrum level the electrons from the GaAs crystal are back-scattered and go back along the same channel as the laser beam used to hit the crystal. Most of them are then deflected magnetically. But some get reflected from the laser apparatus and get back into the system. So you have to eliminate these new ambient electrons. This is done by crude mechanical means, making them focus just off the crystal and so wander away.

3. Good experimenters guard against the absurd. Suppose that dust particles on an experimental surface lie down flat when a polarized pulse hits it, and then stand on their heads when hit by a pulse polarized in the opposite direction? Might that have a systematic effect, given that we are detecting a minute asymmetry? One of the team thought of this in the middle of the night and came down next morning frantically using antidust spray. They kept that up for a month, just in case.[13]

Results

Some 10^{11} events were needed to obtain a result that could be recognized above systematic and statistical error. Although the idea of systematic error presents interesting conceptual problems, it seems to be unknown to philosophers. There were systematic uncertainties in the detection of right-and left-handed polarization, there was some jitter, and there were other problems about the parameters of the two kinds of beam. These errors were analyzed and linearly added to the statistical error. To a student of statistical inference this is real seat-of-the-pants analysis with no rationale whatsoever. Be that as it may, thanks to PEGGY II the number of events was big enough to give a result that convinced the entire physics community.[14] Left-handed polarized electrons were scattered from deuterium slightly more frequently than right-handed electrons. This was the first convincing example of parity-violation in a weak neutral current interaction.

Comment

The making of PEGGY II was fairly non-theoretical. Nobody worked out in advance the polarizing properties of GaAs – that was found by a chance encounter with an unrelated experimental investigation. Although elementary quantum theory of crystals explains the polarization effect, it does not explain the properties of the actual crystal used. No one has been able to get a real crystal to polarize more than 37 percent of the electrons, although in principle 50 percent should be polarized.

Likewise although we have a general picture of why layers of cesium and oxygen will "produce negative electron affinity," i.e., make it easier for electrons to escape, we have no quantitative understanding of why this increases efficiency to a score of 37 percent.

Nor was there any guarantee that the bits and pieces would fit together. To give an even more current illustration, future experimental work, briefly described later in this paper, makes us want even more electrons per pulse than PEGGY II could give. When the parity experiment was reported in *The New York Times*, a group at Bell Laboratories read the newspaper and saw what was going on. They had been constructing a crystal lattice for totally unrelated purposes. It uses layers of GaAs and a related aluminum compound. The structure of this lattice leads one to expect that virtually all the electrons emitted would be polarized. So we might be able to double the efficiency of PEGGY II. But at present (July 1981) that nice idea has problems. The new lattice should also be coated in work-reducing paint. But the cesium oxygen stuff is applied at high temperature. Then the aluminum tends to ooze into the neighboring layer of GaAs, and the pretty artificial lattice becomes a bit uneven, limiting its fine polarized-electron-emitting properties. So perhaps this will never work.[15] The group are simultaneously reviving a souped up new thermionic cathode to try to get more electrons. Maybe PEGGY II would have shared the same fate, never working, and thermionic devices would have stolen the show.

Note, incidentally, that the Bell people did not need to know a lot of weak neutral current theory to send along their sample lattice. They just read *The New York Times*.

Moral

Once upon a time it made good sense to doubt that there are electrons. Even after Millikan had measured the charge on the electron, doubt made sense. Perhaps Millikan was engaging in "inference to the best explanation." The charges on his carefully selected oil drops were all small integral multiples of a least charge. He inferred that this is the real least charge in nature, and

hence it is the charge on the electron, and hence there are electrons, particles of least charge. In Millikan's day most (but not all) physicists did become increasingly convinced by one or more theories about the electron. However it is always admissible, at least for philosophers, to treat inferences to the best explanation in a purely instrumental way, without any commitment to the existence of entities used in the explanation.[16] But it is now seventy years after Millikan, and we no longer have to infer from explanatory success. Prescott *et al.* don't explain phenomena with electrons. They know a great deal about how to use them.

The group of experimenters do not know what electrons are, exactly. Inevitably they think in terms of particles. There is also a cloud picture of an electron which helps us think of complex wavefunctions of electrons in a bound state. The angular momentum and spin vector of a cloud make little sense outside a mathematical formalism. A beam of polarized clouds is fantasy so no experimenter uses that model — not because of doubting its truth, but because other models help more with the calculations. Nobody thinks that electrons "really" are just little spinning orbs about which you could, with a small enough hand, wrap the fingers and find the direction of spin along the thumb. There is instead a family of causal properties in terms of which gifted experimenters describe and deploy electrons in order to investigate something else, e.g., weak neutral currents and neutral bosons. We know an enormous amount about the behavior of electrons. We also know what does not matter to electrons. Thus we know that bending a polarized electron beam in magnetic coils does not affect polarization in any significant way. We have hunches, too strong to ignore although too trivial to test independently: e.g., dust might dance under changes of directions of polarization. Those hunches are based on a hard-won sense of the kinds of things electrons are. It does not matter at all to this hunch whether electrons are clouds or particles.

The experimentalist does not believe in electrons because, in the words retrieved from mediaeval science by Duhem, they "save the phenomena." On the contrary, we believe in them because we use them to *create* new phenomena, such as the phenomenon of parity violation in weak neutral current interactions.

When hypothetical entities become real

Note the complete contrast between electrons and neutral bosons. Nobody can yet manipulate a bunch of neutral bosons, if there are any. Even weak neutral currents are only just emerging from the mists of hypothesis. By 1980 a sufficient range of convincing experiments had made them the object of investigation. When might they lose their hypothetical status and become commonplace reality like electrons? When we use them to investigate something else.

I mentioned the desire to make a better gun than PEGGY II. Why? Because we now "know" that parity is violated in weak neutral interactions. Perhaps by an even more grotesque statistical analysis than that involved in the parity experiment, we can isolate just the weak interactions. That is, we have a lot of interactions, including say electromagnetic ones. We can censor these in various ways, but we can also statistically pick out a class of weak interactions as precisely those where parity is not conserved. This would possibly give us a road to quite deep investigations of matter and anti-matter. To do the statistics one needs even more electrons per pulse than PEGGY II could hope to generate. If such a project were to succeed, we should be beginning to use weak neutral currents as a manipulable tool for looking at something else. The next step towards a realism about such currents would have been made.

The message is general and could be extracted from almost any branch of physics. Dudley Shapere has recently used "observation" of the sun's hot core to illustrate how physicists employ the concept of observation. They collect neutrinos from the sun in an enormous disused underground mine that has been filled with old cleaning fluid (i.e., carbon tetrachloride). We would know a lot about the inside of the sun if we knew how many solar neutrinos arrive on the earth. So these are captured in the cleaning fluid; a few will form a new radioactive nucleus. The number that do this can be counted. Although the extent of neutrino manipulation is much less than electron manipulation in the PEGGY II experiment, here we are plainly using neutrinos to investigate something else. Yet not many years ago, neutrinos were about as hypothetical as an entity could get. After 1946 it was realized that when mesons disintegrate, giving off, among other things, highly energized electrons, one needed an extra nonionizing particle to conserve momentum and energy. At that time this postulated "neutrino" was thoroughly hypothetical, but now it is routinely used to examine other things.

Changing times

Although realisms and anti-realisms are part of the philosophy of science well back into Greek prehistory, our present versions mostly descend from debates about atomism at the end of the nineteenth century. Anti-realism about atoms was partly a matter of physics: the energeticists thought energy was at the bottom of everything, not tiny bits of matter. It also was connected with the positivism of Comte, Mach, Pearson and even J. S. Mill. Mill's young associate Alexander Bain states the point in a characteristic way, apt for 1870:

Some hypotheses consist of assumptions as to the minute structure and operations of bodies. From the nature of the case these assumptions can never be proved by direct means. Their merit is their suitability to express phenomena. They are Representative Fictions.[17]

"All assertions as to the ultimate structure of the particles of matter," continues Bain, "are and ever must be hypothetical. . . ." The kinetic theory of heat, he says, "serves an important intellectual function." But we cannot hold it to be a true description of the world. It is a Representative Fiction.

Bain was surely right a century ago. Assumptions about the minute structure of matter could not be proved then. The only proof could be indirect, namely that hypotheses seemed to provide some explanation and helped make good predictions. Such inferences need never produce conviction in the philosopher inclined to instrumentalism or some other brand of idealism.

Indeed the situation is quite similar to seventeenth-century epistemology. At that time knowledge was thought of as correct representation. But then one could never get outside the representations to be sure that they corresponded to the world. Every test of a representation is just another representation. "Nothing is so much like an idea as an idea," as Bishop Berkeley had it. To attempt to argue for scientific realism at the level of theory, testing, explanation, predictive success, convergence of theories and so forth is to be locked into a world of representations. No wonder that scientific anti-realism is so permanently in the race. It is a variant on "The Spectator Theory of Knowledge."

Scientists, as opposed to philosophers, did in general become realists about atoms by 1910. Michael Gardner, in one of the finest studies of real-life scientific realism, details many of the factors that went into that change in climate of opinion.[18] Despite the changing climate, some

variety of instrumentalism or fictionalism remained a strong philosophical alternative in 1910 and in 1930. That is what the history of philosophy teaches us. Its most recent lesson is Bas van Fraassen's *The Scientific Image*, whose "constructive empiricism" is another theory-oriented anti-realism. The lesson is: think about practice, not theory.

Anti-realism about atoms was very sensible when Bain wrote a century ago. Anti-realism about *any* sub-microscopic entities was a sound doctrine in those days. Things are different now. The "direct" proof of electrons and the like is our ability to manipulate them using well understood low-level causal properties. I do not of course claim that "reality" is constituted by human manipulability. We can, however, call something real, in the sense in which it matters to scientific realism, only when we understand quite well what its causal properties are. The best evidence for this kind of understanding is that we can set out, from scratch, to build machines that will work fairly reliably, taking advantage of this or that causal nexus. Hence, engineering, not theorizing, is the proof of scientific realism about entities.[19]

Notes

1   C.W.F. Everitt and Ian Hacking, "Which Comes First, Theory or Experiment?"

2   Ian Hacking, "Spekulation, Berechnung und die Erschaffung der Phänomenen," in *Versuchungen: Aufsätze zur Philosophie Paul Feyerabends*, (P. Duerr, ed.), Frankfurt, 1981, Bd 2, 126–58.

3   Nancy Cartwright makes a similar distinction in a sequence of papers, including "When Explanation Leads to Inference," *Philosophical Topics* **13**, (1982), 111–122. She approaches realism from the top, distinguishing theoretical laws (which do not state the facts) from phenomenological laws (which do). She believes in some "theoretical" entities and rejects much theory on the basis of a subtle analysis of modeling in physics. I proceed

in the opposite direction, from experimental practice. Both approaches share an interest in real-life physics as opposed to philosophical fantasy science. My own approach owes an enormous amount to Cartwright's parallel developments, which have often preceded my own. My use of the two kinds of realism is a case in point.

4 Hilary Putnam, "How Not To Talk About Meaning," "The Meaning of 'Meaning'," and other papers in *Mind, Language and Reality, Philosophical Papers*, Vol. 2. Cambridge, 1975.

5 These terms occur in e.g., Hilary Putnam, *Meaning and the Moral Sciences*, London, 1978, 123–30.

6 Francis Bacon, *The Great Instauration*, in *The Philosophical Works of Francis Bacon* (J.M. Robertson, ed.; Ellis and Spedding, Trans.), London, 1905, p. 252.

7 Ian Hacking, "Do We See Through a Microscope?" *Pacific Philosophical Quarterly*, winter 1981.

8 I thank Melissa Franklin, of the Stanford Linear Accelerator, for introducing me to PEGGY II and telling me how it works. She also arranged discussions with members of the PEGGY II group, some of which are mentioned below. The report of experiment E-122 described here is "Parity Non-conservation in Inelastic Electron Scattering," C.Y. Prescott *et al.*, *Physics Letters*. I have relied heavily on the in-house journal, the *SLAC Beam Line*, Report No. 8, October, 1978, "Parity Violation in Polarized Electron Scattering." This was prepared by the in-house science writer Bill Kirk, who is the clearest, most readable popularizer of difficult new experimental physics that I have come across.

9 The odd-sounding bosons are named after the Indian physicist S. N. Bose (1894–1974), also remembered in the name "Bose-Einstein statistics" (which bosons satisfy).

10 But excluding Leibniz, who "knew" there had to be some real, natural difference between right- and left-handedness.

11 Iceland spar is an elegant example of how experimental phenomena persist even while theories about them undergo revolutions. Mariners brought calcite from Iceland to Scandinavia. Erasmus Batholinus experimented with it and wrote about it in 1609. When you look through these beautiful crystals you see double, thanks to the so-called ordinary and extraordinary rays. Calcite is a natural polarizer. It was our entry to polarized light which for 300 years was the chief route to improved theoretical and experimental understanding of light and then electromagnetism. The use of calcite in PEGGY II is a happy reminder of a great tradition.

12 It also turns GaAs, a ¾–¼ left/right hand polarizer, into a 50–50 polarizer.

13 I owe these examples to conversation with Roger Miller of SLAC.

14 The concept of a "convincing experiment" is fundamental. Peter Gallison has done important work on this idea, studying European and American experiments on weak neutral currents conducted during the 1970s.

15 I owe this information to Charles Sinclair of SLAC.

16 My attitude to "inference to the best explanation" is one of many learned from Cartwright. See, for example, her paper on this topic cited in n. 11.

17 Alexander Bain, *Logic, Deductive and Inductive*, London and New York, 1870, p. 362.

18 Michael Gardner, "Realism and Instrumentalism in 19th-century Atomism," *Philosophy of Science* **46**, (1979), 1–34.

19 (Added in proof, February, 1983). As indicated in the text, this is a paper of July, 1981, and hence is out of date. For example, neutral bosons are described as purely hypothetical. Their status has changed since CERN announced on Jan. 23, 1983, that a group there had found W, the weak intermediary boson, in proton-antiproton decay at 540 GeV. These experimental issues are further discussed in my forthcoming book, *Representing and Intervening* (Cambridge, 1983).

Chapter 39

John Worrall

STRUCTURAL REALISM: THE BEST OF BOTH WORLDS?

Presently accepted physical theories postulate a curved space-time structure, fundamental particles and forces of various sorts. What we can know for sure on the basis of observation, at most, are only facts about the motions of macroscopic bodies, the tracks that appear in cloud chambers in certain circumstances and so on. Most of the content of the basic theories in physics goes "beyond" the "directly observational" – no matter how liberal a conception of the "directly observational" is adopted. What is the status of the genuinely theoretical, observation-transcendent content of our presently accepted theories? Most of us unreflectingly take it that the statements in this observation-transcendent part of the theory are attempted descriptions of a reality lying "behind" the observable phenomena: that those theories really do straightforwardly assert that spacetime is curved in the presence of matter, that electrons, neutrinos and the rest exist and do various funny things. Furthermore, most of us unreflectingly take it that the enormous empirical success of these theories legitimises the assumption that these descriptions of an underlying reality are accurate or at any rate "essentially" or "approximately" accurate. The main problem of scientific realism, as I understand it, is that of whether or not there are, after reflection, good reasons for holding this view that most of us unreflectingly adopt.

There are, of course, several anti-realist alternatives on offer. The most widely canvassed is some version of the pragmatic or instrumentalist view that the observation-transcendent content of our theories is not in fact, and despite its apparent logical form, descriptive at all, but instead simply "scaffolding" for the experimental laws. Theories are codification-schemes, theoretical terms like 'electron' or 'weak force' or whatever should not be taken as even intended to refer to real entities but instead as fictional names introduced simply to order our experimental laws into a system.[1] A more recent anti-realist position – that of van Fraassen – holds that theoretical terms do, at any rate purportedly, refer to real entities (and are not, for example, simply shorthand for complex observational terms) but that there is no reason to assume that even our best theories are true nor even "approximately" true, nor even that the aim of science is to produce true theories; instead acceptance of a theory should be taken to involve only the claim that the theory is 'empirically adequate', that it 'saves the phenomena'.[2]

I can find no essentially new arguments in the recent discussions.[3] What seem to me the two most persuasive arguments are very old – both are certainly to be found in Poincaré and in Duhem. The main interest in the problem of scientific realism lies, I think, in the fact that these two persuasive arguments appear to pull in opposite directions; one seems to speak for realism and the other against it; yet a really satisfactory position would need to have both arguments on its side. The concern of the present paper is to investigate this tension between the two arguments and to *suggest* (no more) that an old and hitherto mostly neglected position may offer the best hope of reconciling the two.

The main argument (perhaps 'consideration' would be more accurate) likely to incline someone towards realism I shall call the 'no miracles' argument (although a version of it is nowadays sometimes called the 'ultimate argument' for realism[4]). Very roughly, this argument goes as follows. It would be a miracle, a coincidence on a near cosmic scale, if a theory made as many correct empirical predictions as, say, the general theory of relativity or the photon theory of light *without* what that theory says about the fundamental structure of the universe being correct or "essentially" or "basically" correct. But we shouldn't accept miracles, not at any rate if there is a non-miraculous alternative. If what these theories say is going on "behind" the phenomena is indeed true or "approximately true" then it is no wonder that they get the phenomena right. So it is plausible to conclude that presently accepted theories are indeed "essentially" correct. After all, quantum theory gets certain phenomena, like the 'Lamb shift' correct to, whatever it is, 6 or 7 decimal places; in the view of some scientists, only a philosopher, overly impressed by merely logical possibilities, could believe that this is compatible with the quantum theory's failing to be a fundamentally correct description of reality.

Notice, by the way, that the argument requires the empirical success of a theory to be understood in a particular way. Not every empirical consequence that a theory has and which happens to be correct will give intuitive support for the idea that the theory must somehow or other have latched onto the "universal blueprint". Specifically, any empirical consequence which was *written into* the theory *post hoc* must be excluded. Clearly it is no miracle if a theory gets right a fact which was already known to hold and which the theory had been engineered to yield. If the fact concerned was used in the construction of the theory – for example, to fix the value of some initially free parameter – then the theory was *bound* to get that fact right. (On the other hand, if the experimental result concerned was *not* written into the theory, then the support it lends to the idea that the theory is "essentially correct" is surely independent of whether or not the result was already known when the theory was formulated.[5])

This intuitive 'no miracles' argument can be made more precise in various ways – all of them problematic and some of them more problematic than others. It is for instance often run as a form of an 'inference to the best explanation' or Peircian 'abduction'.[6] But, as Laudan and Fine have both pointed out,[7] since the anti-realist is precisely in the business of denying the validity of inference to the best explanation in science, he is hardly likely to allow it in philosophy as a means of arguing for realism. Perhaps more importantly, and despite the attempts of some philosophers to claim scientific status for realism itself on the basis of its explanatory power,[8] there is surely a crucial, pragmatic difference between a good scientific explanation, and the "explanation" afforded by the thesis of realism for the success of our present theories. A requirement for a convincing *scientific* explanation is *independent* testability – Newton's explanation of the planetary orbits is such a good one because the theory yields so much else that is testable besides the orbits: the oblateness of the earth, return of Halley's comet and so on. Yet in the case of realism's "explanation" of the success of our current theories there can of course be no question of any independent tests. Scientific realism can surely not be *inferred* in any interesting sense from science's success. The 'no miracles' argument cannot *establish* scientific realism; the claim is only that, other things being equal, a theory's predictive success supplies a *prima facie* plausibility argument in favour of its somehow or other having latched onto the truth.

Certainly the psychological force of the argument was sharply felt even by the philosophers who are usually (though as we shall see mistakenly) regarded as the great champions of anti-realism or instrumentalism: Pierre Duhem and Henri Poincaré. Here, for example, is Duhem:

> The highest test, therefore of [a theory] is to ask it to indicate in advance things which the future alone will reveal. And when the experiment is made and confirms the predictions obtained from our theory, we feel strengthened in our conviction that the relations established by our reason among abstract notions only correspond to the relations among things.[9]

And here Poincaré:

> Have we any right, for instance, to enunciate Newton's law? No doubt numerous observations are in agreement with it, but is not that a simple fact of chance? And how do we know besides, that this law which has been true for so many generations, will not be untrue in the next? To this objection the only answer you can give is: It is very improbable.[10]

So the 'no miracles' argument is likely, I think, to incline a commonsensical sort of person towards some sort of scientific realist view. But he is likely to feel those realist sentiments evaporating if he takes a close look at the *history* of science

and particularly at the phenomenon of *scientific revolutions*.

Newton's theory of gravitation had a stunning range of predictive success: the perturbations of the planetary orbits away from strict Keplerian ellipses, the variation of gravity over the earth's surface, the return of Halley's comet, precession of the equinoxes, and so on. Newtonians even turned empirical difficulties (like the initially anomalous motion of Uranus) into major successes (in this case the prediction of a hitherto unknown trans-Uranian planet subsequently christened Neptune). Physicists were wont to bemoan their fate at having been born after Newton – there was only one truth to be discovered about the 'system of the world' and Newton had discovered it. Certainly an apparently hugely convincing no-miracles argument could be – and was – constructed on behalf of Newton's theory. It would be a miracle if Newton's theory got the planetary motions so precisely right, that it should be right about Neptune and about Halley's comet, that the motion of incredibly distant objects like some binary stars should be in accordance with the theory – it would be a miracle if this were true but the theory is not. However, as we all know, Newton's theory was rejected in favour of Einstein's in the early twentieth century.

This would pose no problem if Einstein's theory were simply an extension of Newton's, that is, if it simply incorporated Newton's theory as a special case and then went on to say more. In general if the development of science were cumulative then scientific change would pose no problem either for the realist or for his 'no miracles' argument. The reason why Newton's theory got so many of the phenomena correct could still be that it was true, just not the whole truth.

Unfortunately Einstein's theory is not simply an extension of Newton's. The two theories are logically inconsistent: if Einstein's theory is true, then Newton's has to be false.[11] This is of course accepted by all presentday realists. The recognition that scientific progress, even in the "successful", "mature" sciences, is not strictly cumulative at the theoretical level but instead involves at least an element of modification and revision is the reason why no presentday realist would claim that we have grounds for holding that presently accepted theories are *true*. Instead the claim is only that we have grounds for holding that those theories are "approximately" or "essentially" true. This last claim might be called 'modified realism'. I shall, for convenience, drop the 'modified' in what follows, but it should be understood that my realists claim only that we have grounds for holding that our present theories in *mature* science are *approximately* true.

This realist claim involves two terms which are notoriously difficult to clarify. I shall propose my own rough characterisation of the "mature" sciences shortly. As for "approximately true", well known and major difficulties stand in the way of any attempt at precise analysis. Indeed various attempted characterisations (such as Popper's in terms of 'increasing verisimilitude') have turned out to be formally deeply flawed.[12] Although we do often operate quite happily at the intuitive level with the notion of approximate truth, it is surely not the sort of notion which can happily be left as a primitive. For one thing: if the notion is going to do the work that realists need it to do it is going to have to be transitive. Realists need to claim that although some presently accepted theory may subsequently be modified and replaced, it will still look "approximately true" in the light, not just of the next theory which supersedes it, but also in the light of the theory (if any) which supersedes the theory which supersedes it, *etc*. But is transitivity a property that the notion of approximate truth possesses even intuitively?

But there is anyway an important *prior* question here: that of whether or not, talking intuitively, in advance of formal analysis, the history of science (or some selected part of it) speaks in favour of successive scientific theories being increasingly good "approximations to the truth". This clearly depends on just *how radical* theory-change has standardly been in science. Again of course we are dealing in unfortunately vague terms. But surely the realist claim – that we have grounds for holding that our present theories are approximately true – is plausible only to the extent that it seems reasonable to say that Newton's theory, for example, "approximates" Einstein's, and that, in general, the development of science (at any rate the development of successful, "mature" science) has been "essentially" cumulative, that the deposed theories themselves, *and not just their successful empirical consequences*, have generally lived on, albeit in "modified form", after the "revolution". If, on the contrary, theory change in science has often involved "radical" shifts – something like the complete rejection of the genuinely theoretical assumptions (though combined of course with retention of the successful empirical content) – then realism is in dire straits. Before going further let's be clear on the dependence of realism on the claim that theory change has been "essentially cumulative".

Assume first that the realist has convinced us that the development of theoretical science has indeed been "essentially cumulative". He could then argue for his realism roughly as follows. The development of the "mature" sciences has so far been "essentially" cumulative at all levels – theoretical as well as observational. It seems reasonable therefore to infer inductively that that development will continue to be "essentially cumulative" in the future. This presumably means that, even should our present theories be replaced, they will continue to appear "approximately" correct in the light of the successor theories. Such a development is, of course, logically compatible with the genuinely theoretical assumptions, both of presently accepted theories and of those destined to be accepted in the future, being entirely untrue. However this is highly implausible since it would make the empirical success of all these theories entirely mysterious; while, on the other hand, the assumption that our present theories are approximately true is enough to explain the empirical success as non-miraculous.

No one, I take it (reiterating the point made earlier), would claim that this argument is completely watertight. The inductive "inference" from "essential cumulativity" in the past to "essential cumulativity" in the future could of course be questioned. Moreover, there is still the problem of what exactly is involved in approximate truth; and indeed the problem of whether or not the assumption of the approximate truth of our present theories really would explain their empirical success. It might seem plausible, intuitively speaking, to suppose that if a theory is "approximately" or "essentially" true then it is likely that most of its consequences will themselves be "essentially" correct. To take a straightforwardly empirical example, say that I make a slight arithmetical error in totting up my bank balance and come to the strictly mistaken view that my total worldly fortune is £100 when the truth is that it is £103. Will it seem 'miraculous' if this strictly false theory nonetheless supplies a quite reliable guide to life? After all, it might be claimed, most of the consequences that I am likely to be interested in – for example that I can't afford a month's holiday in Switzerland – will in fact be consequences both of the false theory, that I hold, and of the truth. Nonetheless plausible or not, there are formidable formal difficulties here.[13] Every false theory, of course, has infinitely many false consequences (as well as infinitely many true ones) and there are things that my "nearly true" theory gets

totally wrong. For example, the truth is that my total fortune expressed in pounds sterling is a prime number, whereas the "nearly true" theory I hold says – entirely incorrectly – that it's composite. Moreover, the argument seems committed to the claim that if theory T "approximates" theory T', which in turn "approximates" T", then T "approximates" T". (The theories which eventually supersede our presently accepted ones, might themselves – presumably *will* – eventually be superseded by still further theories. The realist needs to be assured that any presently accepted theory will continue to look approximately correct, even in the light of the further theories in the sequence, not just in the light of its immediate successor.) But is this transitivity assumption correct? After all, if we took a series of photographs at one-second intervals say of a developing tadpole, each photograph in the sequence would presumably "approximate" its predecessor and yet we start with a tadpole and finish with a frog. Does a frog "approximate" a tadpole? I propose, however, that, for present purposes, we put all these difficulties into abeyance. If he can sustain the claim that the development of the "mature" sciences has been "essentially cumulative", then the realist has at least some sort of argument for his claim.

If, on the contrary, the realist is forced to concede that there has been *radical* change at the theoretical level in the history of even the mature sciences then he surely is in deep trouble. Suppose that there are cases of mature theories which were once accepted, were predictively successful, and whose underlying theoretical assumptions nonetheless now seem unequivocally entirely false. The realist would have encouraged the earlier theorist to regard his theory's empirical success as giving him grounds for regarding the theory itself as approximately true. He now encourages scientists to regard their newer theory's empirical success as giving them grounds for regarding that newer theory as approximately true. The older and newer theories are radically at odds with one another at the theoretical level. Presumably if we have good grounds for thinking a theory T approximately true we equally have good grounds for thinking that any theory T′ radically at odds with T is false (plain false, not "approximately true"). So the realist would be in the unenviable position of telling us that we now have good grounds to regard as false a theory which he earlier would have told us we had good grounds to believe approximately true. Why should not his proposed judgment about presently accepted theories turn out to be similarly mistaken?

Assuming, then, that the realist is not talking about 'good grounds' in some defeasible, conjectural sense,[14] realism is not compatible with the existence of radical theoretical changes in science (or at any rate in mature science). The chief argument against realism − the argument from scientific revolutions − is based precisely on the claim that revolutionary changes have occurred in accepted scientific theories, changes in which the old theory could be said to "approximate" the new only by stretching the admittedly vague and therefore elastic notion of 'approximation' beyond breaking point.

At first glance this claim appears to be correct. Consider, for example, the history of optics. Even if we restrict this history to the modern era, there have been fundamental shifts in our theory about the basic constitution of light. The theory that a beam of light consists of a shower of tiny material particles was widely held in the eighteenth century. Some of its empirical consequences − such as those about simple reflection, refraction and prismatic dispersion − were correct. The theory was, however, rejected in favour of the idea that light consists, not of matter, but of certain vibratory motions set up by luminous bodies and carried by an all-pervading medium, the 'luminiferous aether'. It

would clearly be difficult to argue that the theory that light is a wave in a mechanical medium is an "extension" or even an "extension with slight modifications" of the idea that light consists of material particles: waves in a mechanical medium and particles travelling through empty space seem more like chalk and cheese than do chalk and cheese themselves. Nor was that all: Fresnel's wave theory itself was soon replaced by Maxwell's electromagnetic theory. Maxwell, as is well known, strove manfully to give an account of the electromagnetic field in terms of some underlying mechanical medium; but his attempts and those of others failed and it came to be accepted that the electromagnetic field is a primitive. So again a fundamental change in the accepted account of the basic structure of light seems to have occurred − instead of vibrations carried through an elastic medium, it becomes a series of wave-like changes in a disembodied electromagnetic field. A mechanical vibration and an electric ('displacement') current are surely radically different sorts of thing. Finally, the acceptance of the photon theory had light consisting again of discrete entities but ones which obey an entirely new mechanics.

In the meanwhile, as *theories* were changing light from chalk to cheese and then to super-chalk, there was a steady basically cumulative development in the captured and systematised empirical content of optics.[15] The material particle theory dealt satisfactorily with simple reflection and refraction and little else, the classical wave theory added interference and diffraction and eventually polarisation effects too, the electromagnetic theory added various results connecting light with electrical and magnetic effects, the photon theory added the photoelectric effect and much else besides. The process at the empirical level (properly construed) was essentially cumulative. There were *temporary* problems (for example over whether or not the classical wave theory could deal with the phenomena which had previously been taken to support the ("essentially") rectilinear propagation of light) but these were invariably settled quickly and positively.[16]

Or take the Newton-Einstein case again. At the *empirical* level it does seem intuitively reasonable to say that Einstein's theory is a sort of "extension with modifications" of Newton's. It is true that, even at this level, if we take the maximally precise consequences about the motion of a given body yielded by the two theories they will always strictly speaking contradict one another. But for a whole range of cases (those cases, of course, in which the velocities involved are fairly small compared to the velocity of light) the predictions of the two theories will be strictly different but observationally indistinguishable. It is also true, of course, that Newton's equations are limiting cases of corresponding relativistic equations. However, there is much more to Newton's theory than the laws of motion and the principle of universal gravitation considered simply as mathematical equations. These equations were interpreted within a set of very general theoretical assumptions which involved amongst other things the assumption that space is infinite, that time is absolute so that two events simultaneous for one observer are simultaneous for all, and that the inertial mass of a body is constant. Einstein's theory entails on the contrary that space is finite (though unbounded), that time is not absolute in the Newtonian sense and that the mass of a body increases with its velocity. All these are surely out and out contradictions.

The picture of the development of science certainly seems, then, to be one of essential cumulativity at the empirical level, accompanied by sharp changes of an entirely non-cumulative kind at the top theoretical levels.[17] This picture of theory-change in the past would seem to supply good inductive grounds for holding that those theories presently accepted in science will,

within a reasonably brief period, themselves be replaced by theories which retain (and extend) the empirical success of present theories, but do so on the basis of underlying theoretical assumptions entirely at odds with those presently accepted. This is, of course, the so-called *pessimistic induction* – usually regarded as a recent methodological discovery but in fact already stated clearly by Poincaré.[18] How can there be good grounds for holding our present theories to be "approximately" or "essentially" true, and at the same time seemingly strong historical-inductive grounds for regarding those theories as (probably) ontologically false?

Unless this picture of theory-change is shown to be inaccurate, then realism is surely untenable and basically only two (very different) possibilities open. The first can be motivated as follows. Science is the field in which rationality reigns. There can be no rational acceptance of claims of a kind which history gives us grounds to think are likely later to be rejected. The successful empirical content of a once-accepted theory *is* in general carried over to the new theory, but its basic theoretical claims are not. Theories, then, are best construed as making no real claims beyond their directly empirical consequences; or, if they *are* so construed, acceptance of these theoretical claims as true or approximately true is no part of the rational procedures of science. We are thus led into some sort of either pragmatic or 'constructive' anti-realism.

Such a position restores a pleasing, cumulative (or quasi-cumulative) development to science (that is, to the "real part" of science); but it does so at the expense of sacrificing the 'no miracles' argument entirely. After all, the theoretical science which the pragmatist alleges to be insubstantial and to play a purely codificatory role has, as a matter of fact, often proved fruitful. That is, interpreted literally and therefore treated as claims about the structure of the world, theories have yielded testable consequences over and above those they were introduced to codify and those consequences have turned out to be correct when checked empirically. Why? The pragmatist asserts that there is no answer.

The other alternative for someone who accepts the empirically cumulative, theoretically non-cumulative picture of scientific change, but who wishes to avoid pragmatism is pure, Popperian *conjectural realism*. This is Popper's view stripped of all the verisimilitude ideas, which always sat rather uncomfortably with the main theses. On this conjectural realist view the genuinely theoretical, observation-transcendent parts of scientific theories are not just codificatory schemes, they are *attempted* descriptions of the reality hidden behind the phenomena. And our present best theories are our present best shots at the truth. We certainly have reason to think that our presently best theories are our present best shots at the truth (they stand up to the present evidence better than any known rival), but we have no real reason to think that those present theories are true or even closer to the truth that their rejected predecessors. Indeed it can be accepted that the history of science makes it very unlikely that our present theories are even "approximately" true. They do, of course, standardly capture more empirical results than any of their predecessors, but this is no indication at all that they are any closer to capturing "God's blueprint of the universe". The fully methodologically-aware theoretical scientist nobly pursues his unended quest for the truth knowing that he will almost certainly fail and that even if he succeeds he will never know, nor even have any real indication, that he has succeeded.

Conjectural realism is certainly a modest unassuming position. It can be formulated as a version of realism in the senses we have so far discussed – as saying in fact that we do have the best possible grounds for holding our present best theories to be true (they are best confirmed or best 'corroborated' by the present evidence); we should not even ask for better grounds than these; but since the best corroborated theory tomorrow may fundamentally contradict the best corroborated theory of today, the grounds that we have for thinking the theories true are inevitably conjectural and (practically, not just in principle) defeasible. I defended this conjectural realist view myself in an earlier paper: presentations of the view frequently (almost invariably) met with the response that there is little, if any, difference of substance between it and anti-realism.[19] The main problem, I take it, is again that conjectural realism makes no concessions to the 'no miracles' argument. On the conjectural realist view, Newton's theory does assert that space and time are absolute, that there are action-at-a-distance forces of gravity, and that inertial mass is constant; all this was entirely wrong and *yet* the theory based on these assumptions was highly empirically adequate. This just has to be recorded as a fact. And if you happen to find it a rather surprising fact, then that's your own business – perhaps due to failure to internalise the elementary logical fact that all false theories have true consequences (in fact infinitely many of them).

Both the pragmatist and the conjectural realist can point out that we can't, on pain of infinite regress, account for everything and one of the things we can't account for is why this stuff that allegedly does no more than streamline the machinery of scientific proof or that turns out to be radically false should have turned out to be fruitful. There obviously can be no question of any 'knockdown refutation' of either view. Nonetheless if a position could be developed which accommodated some of the intuitions underlying the 'no miracles' argument and yet which, at the same time, cohered with the historical facts about theory-change in science, then it would arguably be more plausible than either pragmatism or conjectural realism.

Is it possible to have the best of both worlds, to account (no matter how tentatively) for the empirical success of theoretical science without running foul of the historical facts about theory-change? Richard Boyd and occasionally Hilary Putnam have claimed that realism is itself already the best of both worlds. They have claimed, more or less explicitly, that the picture of scientific change that I have painted is inaccurate, and so the argument from scientific revolutions is based on a false premise: the history of science is *not* in fact marked by radical theoretical revolutions (at any rate not the history of "mature" science). On the contrary, claims Boyd

> The historical progress of the mature sciences is largely a matter of successively more accurate approximations to the truth about both observable and unobservable phenomena. Later theories typically build upon the (observational and theoretical) knowledge embodied in previous theories.[20]

Elsewhere he asserts that scientists generally adopt the (realist) principle that 'new theories should ... resemble current theories with respect to their accounts of causal relations among theoretical entities'.[21] Similarly Putnam once claimed that many historical cases of theory-change show that 'what scientists try to do' is to preserve 'as often as possible' the 'mechanisms of the earlier theory' or 'to show that they are "limiting cases" of new mechanisms'.[22] I want first to explain why I think that these claims are wrong as they stand. I shall then argue that valid intuitions underlie the claims, but these intuitions are better captured in a rather different position which might be called *structural* or *syntactic realism*.

Larry Laudan has objected to Boyd and Putnam's claims by citing a whole list of theoretical entities, like phlogiston, caloric and a range of ethers which, he insists, once figured in successful theories but have now been *totally* rejected.[23] How, Laudan wants to know, can newer theories resemble older theories 'with respect to their accounts of causal relations among theoretical entities' if the newer theories entirely reject the theoretical entities of the old? How can relativistic physics be said to preserve 'the mechanisms' of, say, Fresnel's account of the transmission of light, when according to Fresnel's account transmission occurs via periodic disturbances in an all-pervading elastic medium, while according to relativity theory no such medium exists at all? How can later scientists be said to have applied to Fresnel's theory the principle that 'new theories should ... resemble current theories with respect to their accounts of causal relations among theoretical entities' when these later theories entirely deny the existence of the core theoretical entity in Fresnel's theory? Boyd alleges that the mechanisms of classical physics reappear as limiting cases of mechanisms in relativistic physics. Laudan replies that, although it is of course true that *some* classical laws are limiting cases of relativistic ones,

> there are other laws and general assertions made by the classical theory (e.g., claims about the density and fine structure of the ether, general laws about the character of the interaction between ether and matter, models and mechanisms detailing the compressibility of the ether) which could not conceivably be limiting cases of modern mechanics. The reason is a simple one: a theory cannot assign values to a variable that does not occur in that theory's language ... Classical ether physics contained a number of postulated mechanisms for dealing inter alia with the transmission of light through the ether. Such mechanisms could not possibly appear in a successor theory like the special theory of relativity which denies the very existence of

an ethereal medium and which accomplishes the explanatory tasks performed by the ether via very different mechanisms.[24]

Does the realist have any legitimate comeback to Laudan's criticisms? Certainly *some* of Laudan's examples can be dealt with fairly straightforwardly. Boyd and Putnam have been careful to restrict their claim of "essential" cumulativity to "mature" science only. Pre-Lavoisierian chemistry is their chief example of an immature science, so they would be happy to concede that phlogiston has been entirely rejected by later science.[25] Presumably some of the other items on Laudan's list of once scientifically accepted but now nonexistent entities would receive similar treatment.

The cogency of this reply clearly depends to a large extent on whether or not some reasonably precise account can be given of what it takes for a science to achieve "maturity". Neither Boyd nor Putnam has anything very precise to say on this score, and this has naturally engendered the suspicion that the realist has supplied himself with a very useful *ad hoc* device: whenever it seems clear that the basic claims of some previously accepted theory have now been totally rejected, the science to which that theory belonged is automatically counted as "immature" at the time that theory was accepted.

What is needed is a reasonably precise and *independent* criterion of maturity. And this can, it seems to me, in fact be 'read off' the chief sustaining argument for realism – the 'no miracles' argument. This argument, as I indicated before, applies only to theories which have enjoyed genuine predictive success. This must mean more than simply having correct empirical consequences – for these could have been forced into the framework of the theory concerned after the effects they describe had already been observed to occur. The undoubted fact that various chemical experimental results

could be incorporated into the phlogiston theory does not on its own found any argument, even of the intuitive kind we are considering, to the likely truth of the phlogiston theory. Similarly the fact that creationist biology can be made empirically adequate with respect to, say, the fossil record clearly founds no argument for the likely truth of the *Genesis* account of creation. Such empirical adequacy can of course easily be achieved – for example by simply making Gosse's assumption that God created the rocks with the "fossils" there already, just as they are found to be. (Perhaps God's purpose in doing this was to test our faith). But the fact that this elaborated version of creationism is then bound to imply the empirical details of the fossil record is, of course, neither a miracle nor an indication that the theory "is on the right track". The explanation for this predictive "success" is, of course, just that it is often easy to incorporate already known results *ad hoc* into a given framework. Nor is the success of a theory in predicting *particular* events of an *already known kind* enough on its own to sustain a 'no miracles' argument in favour of a theory. Even the most *ad hoc*, "cobbled up" theory will standardly be predictive in the sense that it will entail that the various results it has been made to absorb will continue to hold in the future. (For example, the heavily epicyclic corpuscular theory of light developed in the early nineteenth century by Biot having had various parameters fixed on the basis of certain results in crystal optics, implied, of course, that the "natural" generalisations of those results would continue to hold in the future). Theories will standardly exhibit this *weak* predictiveness because, Popper or no, scientists do instinctively inductively generalise on the results of well-controlled experiments which have so far always yielded the same results. But the success of such inductive manoeuvres, though no doubt miraculous enough in itself, does not speak in favour of the likely truth-likeness of any particular explanatory theory. The sort of predictive success which seems to elicit the intuitions underlying the 'no miracles' argument is a much stronger, more striking form of predictive success. In the stronger case, not just a new instance of an old empirical generalisation, but an entirely new empirical generalisation follows from some theory, and turns out to be experimentally confirmed. Instances of this are the prediction of the existence and orbit of a hitherto unknown planet by Newton's theory; and the prediction of the white spot at the centre of the shadow of an opaque disc and of the hitherto entirely unsuspected phenomenon of conical refraction by Fresnel's wave theory of light. So my suggestion is that, instead of leaving the notion of maturity as conveniently undefined, a realist should take it that a science counts as mature once it has theories within it which are predictive in this latter demanding sense — predictive of general *types* of phenomena, without these phenomena having been "written into" the theory.

With this somewhat more precise characterisation of maturity, Laudan's list of difficult cases for the modified realist can indeed be pared down considerably further. Laudan must be operating with some much weaker notion of empirical success than the idea of *predictive* success just explained when he cites the gravitational ether theories of Hartley and LeSage as examples of 'once successful' theories.[26] Presumably he means simply that these theories were able successfully to accommodate various already known observational results. But if we require *predictive* success of the strong kind indicated above, then surely neither Hartley's nor LeSage's speculative hypothesis scored any such success.

However there is no doubt that, no matter how hard-headed one is about predictive success, some of Laudan's examples remain to challenge the realist. Let's concentrate on what

seems to me (and to others)[27] the sharpest such challenge: the ether of classical physics. Indeed we can make the challenge still sharper by concentrating on the elastic solid ether involved in the classical wave theory of light proposed by Fresnel.

Fresnel's theory was based on the assumption that light consists in periodic disturbances originating in a source and transmitted by an all-pervading, mechanical medium. There can be no doubt that Fresnel himself believed in the 'real existence' of this medium — a highly attenuated and rare medium alright, but essentially an ordinary mechanical medium which generates elastic restoring forces on any of its 'parts' that are disturbed from their positions of equilibrium.[28] There is equally no doubt that Fresnel's theory enjoyed genuine predictive success — not least of course with the famous prediction of the white spot at the centre of the shadow of an opaque disc held in light diverging from a single slit. If Fresnel's theory does not count as "mature" science then it is difficult to see what does.[29]

Was Fresnel's elastic solid ether retained or "approximately retained" in later physical theories? Of course, as I have repeatedly said and as realists would admit, the notion of one theoretical entity approximating another, or of one causal mechanism being a limiting case of another is extremely vague and therefore enormously elastic. But if the notion is stretched too far, then the realist position surely becomes empty. If black "approximates" white, if a particle "approximates" a wave, if a spacetime curvature "approximates" an action-at-a-distance force, then no doubt the realist is right that we can be confident that future theories will be approximately like the ones we presently hold. This won't however be telling us very much. It does seem to me that the only clear-sighted judgment is that Fresnel's elastic solid ether was entirely overthrown in the course of later science. Indeed this occurred, long before the advent of relativity theory, when Maxwell's theory was accepted in its stead. It is true that Maxwell himself continued to hold out the hope that his electromagnetic field would one day be 'reduced' to an underlying mechanical substratum – essentially the ether as Fresnel had conceived it. But in view of the failure of a whole series of attempts at such a 'reduction', the field was eventually accepted as a primitive entity. Light became viewed as a periodic disturbance, *not* in an elastic medium, but in the 'disembodied' electromagnetic field. One would be hard pressed to cite two things more different than a displacement current, which is what this electromagnetic view makes light, and an elastic vibration through a medium, which is what Fresnel's theory had made it.

Hardin and Rosenberg, replying to Laudan, suggest that, rather than trying to claim that Fresnel's elastic solid ether was 'approximately preserved' in Maxwell's theory, the realist can 'reasonably' regard Fresnel as having been talking about the electromagnetic field all along.[30] This is certainly a striking suggestion! As someone influenced by Lakatos, I certainly would not want entirely to deny a role to rational reconstruction of history. Indeed it does seem reasonable for a historian to reserve the option of holding that a scientist did not fully understand his own theory; but to allow that he may have totally misunderstood it and indeed that it could not really be understood until some 50 years after his death, to hold that Fresnel was "really" talking about something of which we know he had not the slightest inkling, all this is surely taking 'rational reconstruction' too far. Even 'charity' can be overdone.[31] Fresnel was *obviously* claiming that the light-carrying 'luminiferous aether' is an elastic solid, obeying, in essence, the ordinary laws of the mechanics of such bodies: the ether has 'parts', restoring elastic forces are brought into play when a part is disturbed out of its

equilibrium position. He was obviously claiming this, and it turned out that, if later science is right, Fresnel was wrong. Hardin and Rosenberg's claim has a definite air of desperation about it.

Nonetheless there is *something* right about what they, and Boyd, say. There *was* an important element of continuity in the shift from Fresnel to Maxwell – and this was much more than a simple question of carrying over the successful *empirical* content into the new theory. At the same time it was rather less than a carrying over of the full theoretical content or full theoretical mechanisms (even in "approximate" form). And what was carried over can be captured without making the very far-fetched assumption of Hardin and Rosenberg that Fresnel's theory was "really" about the electromagnetic field all along. There was continuity or accumulation in the shift, but the continuity is one of *form* or *structure*, not of content. In fact this claim was already made and defended by Poincaré. And Poincaré used the example of the switch from Fresnel to Maxwell to argue for a general sort of *syntactic* or *structural realism* quite different from the anti-realist instrumentalism which is often attributed to him.[32] This largely forgotten thesis of Poincaré's seems to me to offer the only hopeful way of *both* underwriting the 'no miracles' argument *and* accepting an accurate account of the extent of theory change in science. Roughly speaking, it seems right to say that Fresnel completely misidentified the *nature* of light, but nonetheless it is no miracle that his theory enjoyed the empirical predictive success that it did; it is no miracle because Fresnel's theory, as science later saw it, attributed to light the right *structure*.

Poincaré's view is summarised in the following passage from *Science and Hypothesis*[33] which begins by clearly anticipating the currently fashionable 'pessimistic induction':

The ephemeral nature of scientific theories takes by surprise the man of the world. Their brief period of prosperity ended, he sees them abandoned one after the other; he sees ruins piled upon ruins; he predicts that the theories in fashion today will in a short time succumb in their turn, and he concludes that they are absolutely in vain. This is what he calls the *bankruptcy of science*.

But this passage continues:

His scepticism is superficial; he does not take into account the object of scientific theories and the part they play, or he would understand that the ruins may still be good for something. No theory seemed established on firmer ground than Fresnel's, which attributed light to the movements of the ether. Then if Maxwell's theory is preferred today, does it mean that Fresnel's work was in vain? No; for Fresnel's object was not to know whether there really is an ether, if it is or is not formed of atoms, if these atoms really move in this way or that; his object was to predict optical phenomena.[34]

This Fresnel's theory enables us to do today as well as it did before Maxwell's time. The differential equations are always true, they may be always integrated by the same methods, and the results of this integration still preserve their value.

So far, of course, this might seem a perfect statement of positivistic instrumentalism: Fresnel's theory is really just its empirical content and this is preserved in later theories. However Poincaré goes on to make it quite explicit that this is *not* his position.

It cannot be said that this is reducing physical theories to simple practical recipes; these equations express relations, and if the equations remain true, it is because the relations preserve their reality. They teach us now, as

they did then, that there is such and such a relation between this thing and that; only the something which we then called *motion*, we now call *electric current*. But these are merely names of the images we substituted for the real objects which Nature will hide for ever from our eyes. The true relations between these real objects are the only reality we can attain . . .

Poincaré is claiming that, although from the point of view of Maxwell's theory, Fresnel entirely misidentified the *nature* of light, his theory accurately described not just light's observable effects but its *structure*. There is no elastic solid ether. There is, however, from the later point of view, a (disembodied) electromagnetic field. The field in no clear sense approximates the ether, but disturbances in it do obey *formally* similar laws to those obeyed by elastic disturbances in a mechanical medium. Although Fresnel was quite wrong about *what* oscillates, he was, from this later point of view, right, not just about the optical phenomena, but right also that these phenomena depend on the oscillations of something or other at right angles to the light.

Thus if we restrict ourselves to the level of mathematical equations – *not* notice the phenomenal level – there is in fact complete continuity between Fresnel's and Maxwell's theories. Fresnel developed a famous set of equations for the relative intensities of the reflected and refracted light beams in various circumstances. Ordinary unpolarised light can be analysed into two components: one polarised in the plane of incidence, the other polarised at right angles to it. Let I^2, R^2, and X^2 be the intensities of the components polarised in the plane of incidence of the incident, reflected and refracted beams respectively; while I'^2, R'^2 and X'^2 are the components polarised at right angles to the plane of incidence. Finally let i and r be the angles made by the incident and refracted beams with the normal to a plane reflecting surface. Fresnel's equations then state

$$R/I = \tan(i - r)/\tan(i + r)$$

$$R'/I' = \sin(i - r)/\sin(i + r)$$

$$X/I = (2 \sin r. \cos i)/(\sin(i + r) \cos(i - r))$$

$$X'/I' = 2 \sin r.\cos i/\sin(i + r)$$

Fresnel developed these equations on the basis of the following picture of light. Light consists of vibrations transmitted through a mechanical medium. These vibrations occur at right angles to the direction of the transmission of light through the medium. In an unpolarised beam vibrations occur in all planes at right angles to the direction of transmission – but the overall beam can be described by regarding it as the composition of two vibrations: one occurring in the plane of incidence and one occurring in the plane at right angles to it. The bigger the vibrations, that is, the larger the maximum distance the particles are forced from their equilibrium positions by the vibration, the more intense the light. I, R, X etc. in fact measure the amplitudes of these vibrations and the intensities of the light are given by the squares of these amplitudes.

From the vantage point of Maxwell's theory as eventually accepted this account, to repeat, is entirely wrong. How could it be anything else when there is no elastic ether to do any vibrating? Nonetheless from this vantage point, Fresnel's theory has exactly the right structure – it's "just" that what vibrates according to Maxwell's theory, are the electric and magnetic field strengths. And in fact if we interpret I, R, X etc. as the amplitudes of the "vibration" of the relevant electric vectors, then Fresnel's equations are directly and fully entailed by Maxwell's theory. It wasn't, then, just that Fresnel's theory *happened* to make certain correct predictions, it made them because it had

accurately identified certain relations between optical phenomena. From the standpoint of this superseding theory, Fresnel was quite wrong about the nature of light, the theoretical mechanisms he postulated are *not* approximations to, or limiting cases of, the theoretical mechanisms of the newer theory. Nonetheless Fresnel was quite right not just about a whole range of optical phenomena but right that these phenomena depend on something or other that undergoes periodic change at right angles to the light.

But then, Poincaré argued, his contemporaries had no more justification for regarding Maxwell as having definitively discovered the nature of light, as having discovered that it *really* consists in vibrations of the electromagnetic field, than Fresnel's contemporaries had had for regarding Fresnel as having discovered the nature of light. At any rate this attitude towards Maxwell would be mistaken if it meant any more than that Maxwell built on the relations revealed by Fresnel and showed that further relations existed between phenomena hitherto regarded as purely optical on the one hand and electric and magnetic phenomena on the other.

This example of an important theory-change in science certainly appears, then, to exhibit cumulative growth at the structural level combined with radical replacement of the previous ontological ideas. It speaks, then, in favour of a *structural* realism. Is this simply a feature of this particular example or is preservation of structure a general feature of theory change in mature (that is, successfully predictive) science?

This particular example is in fact unrepresentative in at least one important respect: Fresnel's equations are taken over completely intact into the superseding theory – reappearing there newly interpreted but, as mathematical equations, entirely unchanged. The much more common pattern is that the old equations reappear as *limiting cases* of the new – that is, the old and new equations are strictly inconsistent, but the new tend to the old as some quantity tends to some limit.

The rule in the history of physics seems to be that, whenever a theory replaces a predecessor, which has however itself enjoyed genuine predictive success, the 'correspondence principle' applies. This requires the *mathematical equations* of the old theory to reemerge as limiting cases of the mathematical equations of the new. As is being increasingly realized,[35] the principle operates, not just as an after-the-event requirement on a new theory if it is to count as better than the current theory, but often also as a heuristic tool in the actual development of the new theory. Boyd in fact cites the general applicability of the correspondence principle as evidence for his realism.[36] But the principle applies *purely* at the mathematical level and hence is quite compatible with the new theory's basic theoretical assumptions (which *interpret* the terms in the equations) being entirely at odds with those of the old. I can see no clear sense in which an action-at-a-distance force of gravity is a "limiting case" of, or "approximates" a spacetime curvature. Or in which the 'theoretical mechanisms' of action-at-a-distance gravitational theory are "carried over" into general relativity theory. Yet Einstein's equations undeniably go over to Newton's in certain limiting special cases. In this sense, there is "approximate continuity" of *structure* in this case. As Boyd points out, a new theory could capture its predecessor's successful empirical content in ways other than yielding the equations of that predecessor as special cases of its own equations.[37] But the general applicability of the correspondence principle certainly is not evidence for full-blown realism – but instead only for structural realism.

Much clarificatory work needs to be done on this position, especially concerning the notion of one theory's structure approximating that of

another. But I hope that what I have said is enough to show that Poincaré's is the only available account of the status of scientific theories which holds out realistic promise of delivering the best of both worlds: of underwriting the 'no miracles' argument, while accepting the full impact of the historical facts about theory-change in science. It captures what is right about Boyd's realism (there is "essential accumulation" in "mature" science at levels higher than the purely empirical) and at the same time what is right about Laudan's criticism of realism (the accumulation does not extend to the fully interpreted top theoretical levels).

As one step towards clarifying the position further, let me end by suggesting that one criticism which, rightly or wrongly, has been levelled at scientific realism does not affect the structural version. Arthur Fine has strikingly claimed that

> Realism is dead . . . Its death was hastened by the debates over the interpretation of quantum theory where Bohr's nonrealist philosophy was seen to win out over Einstein's passionate realism.[38]

But realism has been pronounced dead before. Some eighteenth century scientists believed (implicitly, of course, they would not have expressed it in this way) that realism's death had been hastened by debates over the foundations of the theory of universal gravitation. But it is now surely clear that in this case realism was 'killed' by first saddling it with an extra claim which then proved a convenient target for the assassin's bullet. This extra claim was that a scientific theory could not invoke "unintelligible" notions, such as that of action-at-a-distance, as primitives. A realist interpretation required intelligibility and intelligibility required interpretation of the basic theoretical notions in terms of some antecedently accepted (and *allegedly* antecedently "understood") metaphysical framework (in the Newtonian case of course this was the framework of Cartesian action-by-contact mechanics). Without claiming to be an expert in the foundations of quantum mechanics (and with all due respect for the peculiarities of that theory) it does seem to me that, by identifying the realist position on quantum mechanics with Einstein's position, Fine is similarly saddling realism with a claim it in fact has no need to make. The realist is forced to claim that quantum-mechanical states cannot be taken as primitive, but must somehow be understood or reduced to or defined in classical terms.

But the structural realist at least is committed to no such claim – indeed he explicitly disowns it. He insists that it is a mistake to think that we can ever "understand" the *nature* of the basic furniture of the universe. He applauds what eventually happened in the Newtonian case. There the theory proved so persistently successful empirically and so persistently resistant to 'mechanistic reduction' that gravity (understood as a genuine action-at-a-distance force) became accepted as a primitive irreducible notion. (And action-at-a-distance forces became perfectly acceptable, and realistically interpreted, components of other scientific theories such as electrostatics). On the structural realist view what Newton really discovered are the relationships between phenomena expressed in the mathematical equations of his theory, the theoretical terms of which should be understood as genuine primitives.[39]

Is there any reason why a similar structural realist attitude cannot be adopted towards quantum mechanics? This view would be explicitly divorced from the 'classical' metaphysical prejudices of Einstein: that dynamical variables must always have sharp values and that all physical events are fully determined by antecedent conditions. Instead the view would simply be that quantum mechanics does seem to have

latched onto the real structure of the universe, that all sorts of phenomena exhibited by microsystems really do depend on the system's quantum state, which really does evolve and change in the way quantum mechanics describes. It is, of course, true that this state changes discontinuously in a way which the theory does not further explain when the system interacts with a 'macroscopic system' – but then Newton's theory does not *explain* gravitational interaction but simply postulates that it occurs. (Indeed no theory, of course, can explain everything on pain of infinite regress.) If such discontinuous changes of state seem to cry out for explanation this is because of the deeply ingrained nature of certain classical metaphysical assumptions (just as the idea that action-at-a-distance "cried out" for explanation was a reflection of a deeply ingrained prejudice for Cartesian style mechanics).

The structural realist simply asserts, in other words, that, in view of the theory's enormous empirical success, the structure of the universe is (probably) something like quantum mechanical. It is a mistake to think that we need to understand the nature of the quantum state at all; and *a fortiori* a mistake to think that we need to understand it in classical terms. (Of course this is not to assert that hidden variables programmes were obvious non-starters, that working on them was somehow obviously mistaken. No more than the structural realist needed to assert that the attempts at a Cartesian reduction of gravity were doomed from the start. The only claim is that ultimately evidence leads the way: if, despite all efforts, no scientific theory can be constructed which incorporates our favourite metaphysical assumptions, then no matter how firmly entrenched those principles might be, and no matter how fruitful they may have proved in the past, they must ultimately be given up.)

It seems to me then that, so long as we are talking about *structural* realism, the reports of realism's death at the hands of quantum mechanics are greatly exaggerated.[40]

Notes

I wish to thank John Watkins for some suggested improvements to an earlier draft; Elie Zahar for numerous enlightening discussions on the topic of this paper; and Howard Stein for his comments on the version delivered at Neuchâtel.

1 According to a famous remark of Quine's, for instance, the theoretical entities involved in current science (like electrons) are epistemologically on a par with the Greek gods – both are convenient fictions introduced in the attempt to order (empirical) reality. (Quine [1953], p. 44)

2 Van Fraassen [1980]. Van Fraassen calls his position 'constructive empiricism' (for criticisms see my [1983] review of his book).

3 Worrall [1982].

4 See Musgrave [1988].

5 I have argued for this notion of empirical support and against the idea that temporal novelty is epistemically important in my [1985], and especially in my [1989], which includes a detailed historical analysis of the famous 'white spot' episode involving Fresnel and Poisson, and often taken to provide support for the 'novel facts count more' thesis.

6 This form of the argument is strongly criticised by Larry Laudan in his [1981]. Strong and cogent reservations about the alleged explanation that realism supplies of science's success were also expressed in Howard Stein's paper delivered to the Neuchâtel conference.

7 Laudan, op. cit.; Fine [1984].

8 This position seems to have been held by Boyd, Niiniluoto and others. It is disowned by Putnam in his [1978]: '. . . I think that realism is like an empirical hypothesis in that it could be false, and that facts are relevant to its support (or to criticizing it); but that doesn't mean that realism is scientific (in any standard sense of 'scientific'), or that realism is a hypothesis.'

9 Duhem [1906], p. 28

10 Poincaré [1905], p. 186.

11 Professor Agazzi in his paper at Neuchâtel took the view that Newtonian physics remains true of objects in its intended domain and that quantum and relativistic physics are true of objects in quite different domains. But this position is surely untenable. Newton's theory was not about (its 'intended referent' was not) macroscopic objects moving with velocities small compared with that of light. It was about *all* material objects moving with *any* velocity you like. And that theory is *wrong* (or so we now think), gloriously wrong, of course, but wrong. Moreover, it isn't even, strictly speaking, right about certain bodies and certain motions and 'only' wrong when we are dealing with microscopic objects or bodies moving at very high velocities. If relativity and quantum theory are correct then Newton's theory's predictions about the motion of *any* body, even the most macroscopic and slowest moving, are *strictly false*. It's just that their falsity lies well within experimental error. That is, what is true is that Newton's theory is an *empirically* faultless approximation for a whole range of cases. It's also true, as Agazzi claimed, that scientists and engineers still often see themselves as applying classical physics in a whole range of areas. But the only clear-sighted account of what they are doing is, I think, that they are *in fact* applying the best supported theories available to them – namely, quantum mechanics and relativity theory. It's just that they know that these theories themselves entail the meta-result that, for their purposes (of sending rockets to the moon or whatever), it will make no practical difference to act *as if* they were applying classical physics, and indeed that it would be from the empirical point of view a waste of effort to apply the mathematically more demanding newer theories only for that sophistication to become entirely irrelevant when it comes to *empirical* application.

12 See Tichy [1974] and Miller [1974].

13 Two recent attempts to overcome these difficulties are Oddie [1986] and Niiniluoto [1987] – though both attempts involve substantive, non-logical and therefore challengeable, assumptions.

14 See *below*, p. 772 paras. 2–4.

15 Genuine examples of 'Kuhn loss' of captured *empirical* content are remarkably thin on the ground – *provided*, that is, that *empirical* content is properly understood. Feyerabend and Kuhn both use examples of "lost" content which are either clearly highly theoretical (Feyerabend even uses 'The Brownian particle is a perpetual motion machine of the second kind' as an example of an empirical statement!) or highly vague (Kuhn claims, for example, that while phlogiston theory could explain why metals are "similar" to one another, the superseding oxygen theory could not). For a criticism of Feyerabend on facts see my [1991], for a criticism of Kuhn see my [1989a].

16 The case of rectilinear propagation of light provides an illustrative example both of the essential empirical continuity of 'mature' science and of what it is about this process that leads Feyerabend and Kuhn to misrepresent it. Certain *theories* become so firmly entrenched at certain stages of the development of science, so much parts of 'background knowledge' that they, or at any rate particular experimental situations *interpreted in their light*, are readily talked of as 'facts'. This was certainly true of the 'fact' that light, if left to itself, is rectilinearly propagated. Here then is surely a 'fact' which was 'lost' in the wave revolution, since Fresnel's theory entails that light is *always* diffracted – it's just that in most circumstances the difference between the diffraction pattern and the predictions of geometrical optics is well below the observational level. But this last remark gives the game away. The idea that light is (*rigidly*) rectilinearly propagated was never an empirical result (not a 'crude fact' in Poincaré's terminology). The real empirical results – certain 'ray tracings', inability to see round corners or through bent opaque tubes *etc.* – were not 'lost' but simply reexplained as a result of the shift to the wave theory.

17 That this is the intuitive picture was fully emphasised by Poincaré and Duhem, rather lost sight of by the logical positivists and reemphasised by Popper and those influenced by him (such as John Watkins and Paul Feyerabend).

18 See Putnam [1978], pp. 25; and Poincaré [1905], p. 160 [quoted *below* p. 776–7].

19 For my defence of conjectural realism see my [1982]. The response of 'no real difference' between conjectural and anti-realism was made many times in seminars and private discussions (by van Fraassen amongst others). See also, for example, Newton Smith [1981], where realism is *defined* as including an 'epistemological ingredient' foreign to this conjecturalist approach. I should add that I am of course giving up the conjectural realist position in the present paper only in the sense that I am now inclined to think that a *stronger* position can be defended.

20 Boyd [1984], pp. 41–2. In discussion Richard Boyd acknowledged that he made no claim of approximate continuity for the 'metaphysical' components of accepted scientific theories. But I had thought that was what the debate is all about: does the empirical success of theories give us grounds to think that their basic ('metaphysical', observation-transcendent) description of the reality underlying the phenomena is at any rate approximately correct? Several of Richard Boyd's comments suggested to me at least that he defends not a full-blown realism, but something like the structural realism that I try to formulate below.

21 Boyd [1973], p. 8.

22 Putnam [1978], p. 20.

23 Laudan [1982], p. 231.

24 Laudan op. cit., pp. 237–8.

25 '. . . we do not carry [the principle of the benefit of the doubt] so far as to say that 'phlogiston' referred.' Putnam [1978], p. 25.

26 I have criticised Laudan on this point in my [1988b].

27 See, for example, Hardin and Rosenberg [1982] which tackles this challenge on behalf of the realist (see *below*, p. 776 para. 2).

28 This is not to deny, of course, that Fresnel was also guided by what was already known empirically about light. It is also true that at the time of Fresnel's work much remained to be discovered about the dynamical properties of elastic solids. As a result, Fresnel's theory was dynamically deficient in certain respects (especially when viewed in hindsight). But the fact that he failed to construct a fully dynamically adequate theory of light as a

disturbance in an elastic solid medium (or better: the fact that his theory ran into certain fundamental dynamical problems) does *not* mean that Fresnel did not even aim at such a theory, nor that he did not intend the theory he produced to be interpreted in this way. He clearly thought of light as a disturbance in an elastic medium and dynamical and mechanical considerations (often of an abstract, mathematical sort) certainly guided his research, along with the empirical data on light.

There is no doubt that, as Whittaker pointed out in his [1951], p. 116, some aspects of Fresnel's theory — in particular the discontinuity of the normal component of the displacement across the interface between two media — cohere rather better with Maxwell's notion of a displacement current than they do with the idea of an ordinary dynamical displacement. But, *contra* Hardin and Rosenberg (who cite Whittaker), this doesn't mean that Fresnel was talking about displacement currents all along; instead he was talking — in a flawed and problematic way — about elastic displacements.

29 Compare Laudan [1982], p. 225: 'If that [Fresnel's prediction of the 'white spot'] does not count as empirical success, nothing does!'

30 Hardin and Rosenberg [1982].

31 Putnam has a well-known (and notoriously vague) 'principle of charity' (or 'benefit of the doubt') which says that 'when speakers specify a referent for a term they use by a *description* and, because of mistaken factual beliefs that those speakers have, that description fails to refer, we should assume that they would accept reasonable reformulations of their descriptions . . .' ([1978], pp. 23–4).

32 One critic who explicitly does not classify Poincaré as an instrumentalist is Zahar (see his [1983a]). The term 'structural realism' was also used by Grover Maxwell for a position which he derived from Russell's later philosophy (see Maxwell [1970a] and [1970b]). Maxwell's position grows out of different (more 'philosophical') concerns, though it is clearly related to that of Poincaré (one of the points for further research is to clarify this relationship).

33 [1905], pp. 160–2.

34 Poincaré is quite wrong about Fresnel's 'object' (see *above* note 28). However, the normative

philosophical question of how a theory *ought* to be interpreted is, of course, logically independent of the historical, psychological question of what its creator in fact believed.

35 See, for example, Zahar [1983b] and Worrall [1985]; as well as Boyd [1984].

36 See his [1984].

37 Putnam gives this account of Boyd's position in his [1978], adding that applying the correspondence principle 'is often the *hardest* way to get a theory that keeps the old observational predictions'. I find this last remark very difficult to understand. How exactly could it be done otherwise? (I am assuming that what comes out is required to be a theory in some recognisable sense rather than simply any old collection of empirical statements.) Zahar has shown (see note 35) how the correspondence principle can be used as a definite heuristic principle supplying the scientist with real guidance. But suppose a scientist set out to obtain a theory which shares the successful empirical consequences of its predecessor in some other way than by yielding it predecessor's equations as limiting cases — surely he would be operating completely in the dark without any clear idea of how to go about the task. (I am assuming that various logical 'tricks' are excluded on the grounds that they would fail to produce anything that anyone (including the anti-realist) would regard as a theory.)

38 Fine [1984], p. 83.

39 See, in particular, Poincaré's discussion of the notion of force in his [1905], pp. 89–139.

40 It is not in fact clear to me that Fine's NOA (the natural ontological attitude) is substantially different from structural realism. Structural realism perhaps supplies a banner under which *both* those who regard themselves as realists *and* those who regard themselves as anti-realists of various sorts can unite.

Similar remarks about the "anti-realist" consequences of quantum mechanics are made — though without reference to Fine — by McMullin ([1984], p. 13). In allegedly defending realism, McMullin *also* seems to me in fact to defend structural realism. See my [1988a] review of the Leplin volume in which McMullin's article appears.

These last remarks on quantum mechanics were modified and elaborated in an attempt to meet the interesting objections raised in discussion at Neuchâtel by Professor d'Espagnat.

References

Boyd R., [1973]: 'Realism, Underdetermination and a Causal Theory of Evidence', Noûs, March 1973. **1**, pp. 1–12.

Boyd R., [1984]: 'The Current Status of Scientific Realism' in Leplin (ed.) [1984], pp. 41–82.

Duhem P., [1906]: The Aim and Structure of Physical Theory. (Page references to the Atheneum Press reprint).

Fine A., [1984]: 'The Natural Ontological Attitude' in Leplin (ed.) [1984], pp. 83–107.

Hardin C. and Rosenberg A., [1982]: 'In Defence of Convergent Realism', Philosophy of Science, **49**, 604–615.

Laudan L., [1981]: 'A Confutation of Convergent Realism', Philosophy of Science, 48 (page references to the reprint in Leplin (ed.) [1984]).

Leplin J. (ed.), [1984]: Scientific Realism. University of California.

Maxwell G., [1970a]: 'Structural Realism and the Meaning of Theoretical Terms' in S. Winokur and M. Radner (eds): Minnesota Studies in the Philosophy of Science IV. University of Minnesota Press.

Maxwell G., [1970b]: 'Theories, Perception and Structural Realism' in R.G. Colodny (ed.): The Nature and Function of Scientific Theories, 3–34. University of Pittsburgh.

McMullin E., [1984]: 'A Case for Scientific Realism' in Leplin (ed.) [1984].

Miller D., [1974]: 'Popper's Qualitative Theory of Verisimilitude', British Journal for the Philosophy of Science, **25**, 166–77.

Musgrave A., [1988]: 'The Ultimate Argument for Scientific Realism', forthcoming.

Newton Smith W., [1981]: The Rationality of Science. London, Routledge & Kegan Paul.

Niiniluoto I., [1987]: Truthlikeness. Dordrecht, Reidel.

Oddie G., [1986]: Likeness to Truth. Dordrecht, Reidel.

Poincaré H., [1905]: Science and Hypothesis. (Page references to the Dover reprint of the first English translation of 1905).

Putnam H., [1978]: Meaning and the Moral Sciences. London, Routledge & Kegan Paul.

Quine W.V.O., [1953]: 'Two Dogmas of Empiricism' in From a Logical Point of View. Harvard University Press.

Tichy P., [1974]: 'On Popper's Definition of Verisimilitude', British Journal for the Philosophy of Science, **25**, 155–60.

van Fraassen B., [1980]: The Scientific Image. Oxford, Clarendon Press.

Whittaker E.T., [1951]: History of Theories of Aether and Electricity. The Classical Theories. London, Thomas Nelson.

Worrall J., [1979]: 'Is the Empirical Content of a Theory Dependent on its Rivals?' in I. Niiniluoto and R. Tuomela (eds): The Logic and Epistemology of Scientific Change. Acta Philosophica Fennica. Amsterdam, North-Holland.

Worrall J., [1982]: 'Scientific Realism and Scientific Change', Philosophical Quarterly, **32**, 201–231.

Worrall J., [1983]: 'An Unreal Image', British Journal for the Philosophy of Science, **34**, 65–80.

Worrall J., [1985]: 'Scientific Discovery and Theory Confirmation' in J. Pitt (ed.): Change and Progress in Modern Science. Dordrecht, Reidel.

Worrall J., [1988a]: Review of Leplin [1984], Philosophical Quarterly, **38**.

Worrall J., [1988b]: 'The Value of a Fixed Methodology', British Journal for the Philosophy of Science, **39**, 263–275.

Worrall J., [1989]: 'Fresnel, Poisson and the White Spot: the Role of Successful Prediction in Theory-Acceptance' in D. Gooding, S. Shaffer and T. Pinch (eds): The Uses of Experiment – Studies of Experimentation in Natural Science. Cambridge, Cambridge University Press.

Worrall J., [1989a]: 'Scientific Revolutions and Scientific Rationality: the Case of the "Elderly Hold-Out" ' in C. Wade Savage (ed.): The Justification, Discovery and Evolution of Scientific Theories. Minnesota. Forthcoming.

Worrall J., [1991]: 'Feyerabend and the Facts', in G. Munévar (ed.) Beyond Reason, 329–353. Dordrecht, Kluwer Academic Press.

Zahar E.G., [1983a]: 'Poincaré's Independent Discovery of the Relativity Principle', Fundamenta Scientiae, **4**, 147–175.

Zahar E.G., [1983b]: 'Logic of Discovery or Psychology of Invention?', British Journal for the Philosophy of Science, **34**, 243–61.

Index

Numbers in **bold** indicate figures, tables and boxes